Pearson International Edition

Computer Systems
A Programmer's Perspective
Randal E. Bryant   David R. O'Hallaron
Second Edition

PEARSON

Pearson Education Limited
Edinburgh Gate
Harlow
Essex CM20 2JE
England and Associated Companies throughout the world

Visit us on the World Wide Web at: www.pearsoned.co.uk

© Pearson Education Limited 2014

 ISBN 10: 1-292-02584-0
ISBN 13: 978-1-292-02584-1

British Library Cataloguing-in-Publication Data
A catalogue record for this book is available from the British Library

Printed in the United States of America

Table of Contents

A Tour of Computer Systems

A *computer system* consists of hardware and systems software that work together to run application programs. Specific implementations of systems change over time, but the underlying concepts do not. All computer systems have similar hardware and software components that perform similar functions. Programmers get better at their craft by understanding how these components work and how they affect the correctness and performance of their programs.

You are poised for an exciting journey. If you dedicate yourself to learning the concepts in this text, then you will be on your way to becoming a rare "power programmer," enlightened by an understanding of the underlying computer system and its impact on your application programs.

You may learn practical skills such as how to avoid strange numerical errors caused by the way that computers represent numbers. You may learn how to optimize your C code by using clever tricks that exploit the designs of modern processors and memory systems. You may learn how the compiler implements procedure calls and how to use this knowledge to avoid the security holes from buffer overflow vulnerabilities that plague network and Internet software. You may learn how to recognize and avoid the nasty errors during linking that confound the average programmer. You may learn how to write your own Unix shell, your own dynamic storage allocation package, and even your own Web server. You may learn the promises and pitfalls of concurrency, a topic of increasing importance as multiple processor cores are integrated onto single chips.

In their classic text on the C programming language [58], Kernighan and Ritchie introduce readers to C using the `hello` program shown in Figure 1. Although `hello` is a very simple program, every major part of the system must work in concert in order for it to run to completion. In a sense, our goal is to help you understand what happens and why, when you run `hello` on your system.

We begin our study of systems by tracing the lifetime of the `hello` program, from the time it is created by a programmer, until it runs on a system, prints its simple message, and terminates. As we follow the lifetime of the program, we will briefly introduce the key concepts, terminology, and components that come into play.

---

code/intro/hello.c

```
1   #include <stdio.h>
2
3   int main()
4   {
5       printf("hello, world\n");
6   }
```

code/intro/hello.c

Figure 1 **The `hello` program.**

1   Information Is Bits + Context

Our `hello` program begins life as a *source program* (or *source file*) that the programmer creates with an editor and saves in a text file called `hello.c`. The source program is a sequence of bits, each with a value of 0 or 1, organized in 8-bit chunks called *bytes*. Each byte represents some text character in the program.

Most modern systems represent text characters using the ASCII standard that represents each character with a unique byte-sized integer value. For example, Figure 2 shows the ASCII representation of the `hello.c` program.

The `hello.c` program is stored in a file as a sequence of bytes. Each byte has an integer value that corresponds to some character. For example, the first byte has the integer value 35, which corresponds to the character '#'. The second byte has the integer value 105, which corresponds to the character 'i', and so on. Notice that each text line is terminated by the invisible *newline* character '\n', which is represented by the integer value 10. Files such as `hello.c` that consist exclusively of ASCII characters are known as *text files*. All other files are known as *binary files*.

The representation of `hello.c` illustrates a fundamental idea: All information in a system — including disk files, programs stored in memory, user data stored in memory, and data transferred across a network — is represented as a bunch of bits. The only thing that distinguishes different data objects is the context in which we view them. For example, in different contexts, the same sequence of bytes might represent an integer, floating-point number, character string, or machine instruction.

As programmers, we need to understand machine representations of numbers because they are not the same as integers and real numbers. They are finite approximations that can behave in unexpected ways.

#	i	n	c	l	u	d	e	\<sp\>	<	s	t	d	i	o	.
35	105	110	99	108	117	100	101	32	60	115	116	100	105	111	46

h	>	\n	\n	i	n	t	\<sp\>	m	a	i	n	(	)	\n	{
104	62	10	10	105	110	116	32	109	97	105	110	40	41	10	123

\n	\<sp\>	\<sp\>	\<sp\>	\<sp\>	p	r	i	n	t	f	(	"	h	e	l
10	32	32	32	32	112	114	105	110	116	102	40	34	104	101	108

l	o	,	\<sp\>	w	o	r	l	d	\	n	"	)	;	\n	}
108	111	44	32	119	111	114	108	100	92	110	34	41	59	10	125

Figure 2   **The ASCII text representation of** `hello.c`.

> **Aside** Origins of the C programming language
>
> C was developed from 1969 to 1973 by Dennis Ritchie of Bell Laboratories. The American National Standards Institute (ANSI) ratified the ANSI C standard in 1989, and this standardization later became the responsibility of the International Standards Organization (ISO). The standards define the C language and a set of library functions known as the *C standard library*. Kernighan and Ritchie describe ANSI C in their classic book, which is known affectionately as "K&R" [58]. In Ritchie's words [88], C is "quirky, flawed, and an enormous success." So why the success?
>
> - *C was closely tied with the Unix operating system.* C was developed from the beginning as the system programming language for Unix. Most of the Unix kernel, and all of its supporting tools and libraries, were written in C. As Unix became popular in universities in the late 1970s and early 1980s, many people were exposed to C and found that they liked it. Since Unix was written almost entirely in C, it could be easily ported to new machines, which created an even wider audience for both C and Unix.
> - *C is a small, simple language.* The design was controlled by a single person, rather than a committee, and the result was a clean, consistent design with little baggage. The K&R book describes the complete language and standard library, with numerous examples and exercises, in only 261 pages. The simplicity of C made it relatively easy to learn and to port to different computers.
> - *C was designed for a practical purpose.* C was designed to implement the Unix operating system. Later, other people found that they could write the programs they wanted, without the language getting in the way.
>
> C is the language of choice for system-level programming, and there is a huge installed base of application-level programs as well. However, it is not perfect for all programmers and all situations. C pointers are a common source of confusion and programming errors. C also lacks explicit support for useful abstractions such as classes, objects, and exceptions. Newer languages such as C++ and Java address these issues for application-level programs.

2 Programs Are Translated by Other Programs into Different Forms

The `hello` program begins life as a high-level C program because it can be read and understood by human beings in that form. However, in order to run `hello.c` on the system, the individual C statements must be translated by other programs into a sequence of low-level *machine-language* instructions. These instructions are then packaged in a form called an *executable object program* and stored as a binary disk file. Object programs are also referred to as *executable object files*.

On a Unix system, the translation from source file to object file is performed by a *compiler driver*:

```
unix> gcc -o hello hello.c
```

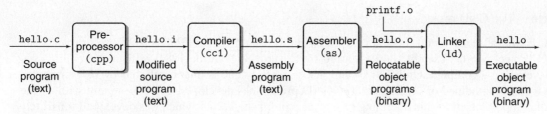

Figure 3 **The compilation system.**

Here, the GCC compiler driver reads the source file `hello.c` and translates it into an executable object file `hello`. The translation is performed in the sequence of four phases shown in Figure 3. The programs that perform the four phases (*preprocessor*, *compiler*, *assembler*, and *linker*) are known collectively as the *compilation system*.

- *Preprocessing phase*. The preprocessor (`cpp`) modifies the original C program according to directives that begin with the # character. For example, the `#include <stdio.h>` command in line 1 of `hello.c` tells the preprocessor to read the contents of the system header file `stdio.h` and insert it directly into the program text. The result is another C program, typically with the `.i` suffix.

- *Compilation phase*. The compiler (`cc1`) translates the text file `hello.i` into the text file `hello.s`, which contains an *assembly-language program*. Each statement in an assembly-language program exactly describes one low-level machine-language instruction in a standard text form. Assembly language is useful because it provides a common output language for different compilers for different high-level languages. For example, C compilers and Fortran compilers both generate output files in the same assembly language.

- *Assembly phase*. Next, the assembler (`as`) translates `hello.s` into machine-language instructions, packages them in a form known as a *relocatable object program*, and stores the result in the object file `hello.o`. The `hello.o` file is a binary file whose bytes encode machine language instructions rather than characters. If we were to view `hello.o` with a text editor, it would appear to be gibberish.

- *Linking phase*. Notice that our `hello` program calls the `printf` function, which is part of the *standard C library* provided by every C compiler. The `printf` function resides in a separate precompiled object file called `printf.o`, which must somehow be merged with our `hello.o` program. The linker (`ld`) handles this merging. The result is the `hello` file, which is an *executable object file* (or simply *executable*) that is ready to be loaded into memory and executed by the system.

Aside   The GNU project

GCC is one of many useful tools developed by the GNU (short for GNU's Not Unix) project. The GNU project is a tax-exempt charity started by Richard Stallman in 1984, with the ambitious goal of developing a complete Unix-like system whose source code is unencumbered by restrictions on how it can be modified or distributed. The GNU project has developed an environment with all the major components of a Unix operating system, except for the kernel, which was developed separately by the Linux project. The GNU environment includes the EMACS editor, GCC compiler, GDB debugger, assembler, linker, utilities for manipulating binaries, and other components. The GCC compiler has grown to support many different languages, with the ability to generate code for many different machines. Supported languages include C, C++, Fortran, Java, Pascal, Objective-C, and Ada.

The GNU project is a remarkable achievement, and yet it is often overlooked. The modern open-source movement (commonly associated with Linux) owes its intellectual origins to the GNU project's notion of *free software* ("free" as in "free speech," not "free beer"). Further, Linux owes much of its popularity to the GNU tools, which provide the environment for the Linux kernel.

3   It Pays to Understand How Compilation Systems Work

For simple programs such as `hello.c`, we can rely on the compilation system to produce correct and efficient machine code. However, there are some important reasons why programmers need to understand how compilation systems work:

* *Optimizing program performance.* Modern compilers are sophisticated tools that usually produce good code. As programmers, we do not need to know the inner workings of the compiler in order to write efficient code. However, in order to make good coding decisions in our C programs, we do need a basic understanding of machine-level code and how the compiler translates different C statements into machine code. For example, is a `switch` statement always more efficient than a sequence of `if-else` statements? How much overhead is incurred by a function call? Is a `while` loop more efficient than a `for` loop? Are pointer references more efficient than array indexes? Why does our loop run so much faster if we sum into a local variable instead of an argument that is passed by reference? How can a function run faster when we simply rearrange the parentheses in an arithmetic expression?

  Later in your studies, you may encounter two related machine languages: IA32, the 32-bit code that has become ubiquitous on machines running Linux, Windows, and more recently the Macintosh operating systems, and x86-64, a 64-bit extension found in more recent microprocessors. Compilers translate different C constructs into these languages. You may also learn how to tune the performance of your C programs by making simple transformations to the C code that help the compiler do its job better. Additionally, you may learn about the hierarchical nature of the memory system, how C compilers store data arrays in memory, and how your C programs can exploit this knowledge to run more efficiently.

- *Understanding link-time errors.* In our experience, some of the most perplexing programming errors are related to the operation of the linker, especially when you are trying to build large software systems. For example, what does it mean when the linker reports that it cannot resolve a reference? What is the difference between a static variable and a global variable? What happens if you define two global variables in different C files with the same name? What is the difference between a static library and a dynamic library? Why does it matter what order we list libraries on the command line? And scariest of all, why do some linker-related errors not appear until run time?

- *Avoiding security holes.* For many years, *buffer overflow vulnerabilities* have accounted for the majority of security holes in network and Internet servers. These vulnerabilities exist because too few programmers understand the need to carefully restrict the quantity and forms of data they accept from untrusted sources. A first step in learning secure programming is to understand the consequences of the way data and control information are stored on the program stack.

4 Processors Read and Interpret Instructions Stored in Memory

At this point, our `hello.c` source program has been translated by the compilation system into an executable object file called `hello` that is stored on disk. To run the executable file on a Unix system, we type its name to an application program known as a *shell*:

```
unix> ./hello
hello, world
unix>
```

The shell is a command-line interpreter that prints a prompt, waits for you to type a command line, and then performs the command. If the first word of the command line does not correspond to a built-in shell command, then the shell assumes that it is the name of an executable file that it should load and run. So in this case, the shell loads and runs the `hello` program and then waits for it to terminate. The `hello` program prints its message to the screen and then terminates. The shell then prints a prompt and waits for the next input command line.

4.1 Hardware Organization of a System

To understand what happens to our `hello` program when we run it, we need to understand the hardware organization of a typical system, which is shown in Figure 4. This particular picture is modeled after the family of Intel Pentium

Figure 4
Hardware organization of a typical system. CPU: Central Processing Unit, ALU: Arithmetic/Logic Unit, PC: Program counter, USB: Universal Serial Bus.

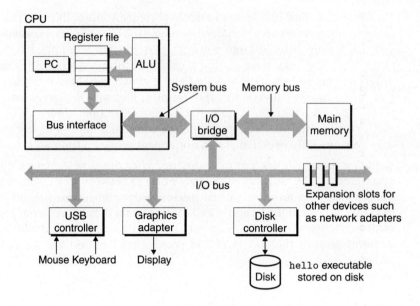

systems, but all systems have a similar look and feel. Don't worry about the complexity of this figure just now.

Buses

Running throughout the system is a collection of electrical conduits called *buses* that carry bytes of information back and forth between the components. Buses are typically designed to transfer fixed-sized chunks of bytes known as *words*. The number of bytes in a word (the *word size*) is a fundamental system parameter that varies across systems. Most machines today have word sizes of either 4 bytes (32 bits) or 8 bytes (64 bits). For the sake of our discussion here, we will assume a word size of 4 bytes, and we will assume that buses transfer only one word at a time.

I/O Devices

Input/output (I/O) devices are the system's connection to the external world. Our example system has four I/O devices: a keyboard and mouse for user input, a display for user output, and a disk drive (or simply disk) for long-term storage of data and programs. Initially, the executable hello program resides on the disk.

Each I/O device is connected to the I/O bus by either a *controller* or an *adapter*. The distinction between the two is mainly one of packaging. Controllers are chip sets in the device itself or on the system's main printed circuit board (often called the *motherboard*). An adapter is a card that plugs into a slot on the motherboard. Regardless, the purpose of each is to transfer information back and forth between the I/O bus and an I/O device.

Main Memory

The *main memory* is a temporary storage device that holds both a program and the data it manipulates while the processor is executing the program. Physically, main memory consists of a collection of *dynamic random access memory* (DRAM) chips. Logically, memory is organized as a linear array of bytes, each with its own unique address (array index) starting at zero. In general, each of the machine instructions that constitute a program can consist of a variable number of bytes. The sizes of data items that correspond to C program variables vary according to type. For example, on an IA32 machine running Linux, data of type short requires two bytes, types int, float, and long four bytes, and type double eight bytes.

Processor

The *central processing unit* (CPU), or simply *processor*, is the engine that interprets (or *executes*) instructions stored in main memory. At its core is a word-sized storage device (or *register*) called the *program counter* (PC). At any point in time, the PC points at (contains the address of) some machine-language instruction in main memory.[1]

From the time that power is applied to the system, until the time that the power is shut off, a processor repeatedly executes the instruction pointed at by the program counter and updates the program counter to point to the next instruction. A processor *appears to* operate according to a very simple instruction execution model, defined by its *instruction set architecture*. In this model, instructions execute in strict sequence, and executing a single instruction involves performing a series of steps. The processor reads the instruction from memory pointed at by the program counter (PC), interprets the bits in the instruction, performs some simple operation dictated by the instruction, and then updates the PC to point to the next instruction, which may or may not be contiguous in memory to the instruction that was just executed.

There are only a few of these simple operations, and they revolve around main memory, the *register file*, and the *arithmetic/logic unit* (ALU). The register file is a small storage device that consists of a collection of word-sized registers, each with its own unique name. The ALU computes new data and address values. Here are some examples of the simple operations that the CPU might carry out at the request of an instruction:

---

1. PC is also a commonly used acronym for "personal computer." However, the distinction between the two should be clear from the context.

- *Load:* Copy a byte or a word from main memory into a register, overwriting the previous contents of the register.
- *Store:* Copy a byte or a word from a register to a location in main memory, overwriting the previous contents of that location.
- *Operate:* Copy the contents of two registers to the ALU, perform an arithmetic operation on the two words, and store the result in a register, overwriting the previous contents of that register.
- *Jump:* Extract a word from the instruction itself and copy that word into the program counter (PC), overwriting the previous value of the PC.

We say that a processor appears to be a simple implementation of its instruction set architecture, but in fact modern processors use far more complex mechanisms to speed up program execution. Thus, we can distinguish the processor's instruction set architecture, describing the effect of each machine-code instruction, from its *microarchitecture*, describing how the processor is actually implemented.

4.2   Running the `hello` Program

Given this simple view of a system's hardware organization and operation, we can begin to understand what happens when we run our example program. We must omit a lot of details here that will be filled in later, but for now we will be content with the big picture.

Initially, the shell program is executing its instructions, waiting for us to type a command. As we type the characters ". /hello" at the keyboard, the shell program reads each one into a register, and then stores it in memory, as shown in Figure 5.

When we hit the enter key on the keyboard, the shell knows that we have finished typing the command. The shell then loads the executable `hello` file by executing a sequence of instructions that copies the code and data in the `hello` object file from disk to main memory. The data include the string of characters "hello, world\n" that will eventually be printed out.

Using a technique known as *direct memory access* (DMA), the data travels directly from disk to main memory, without passing through the processor. This step is shown in Figure 6.

Once the code and data in the `hello` object file are loaded into memory, the processor begins executing the machine-language instructions in the `hello` program's main routine. These instructions copy the bytes in the "hello, world\n" string from memory to the register file, and from there to the display device, where they are displayed on the screen. This step is shown in Figure 7.

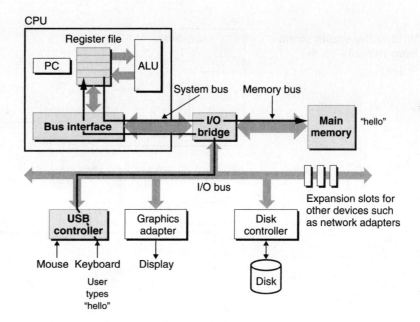

Figure 5
Reading the `hello` command from the keyboard.

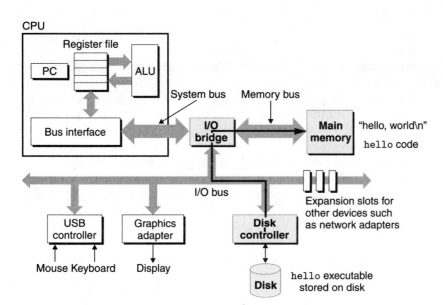

Figure 6 **Loading the executable from disk into main memory.**

Figure 7
Writing the output string from memory to the display.

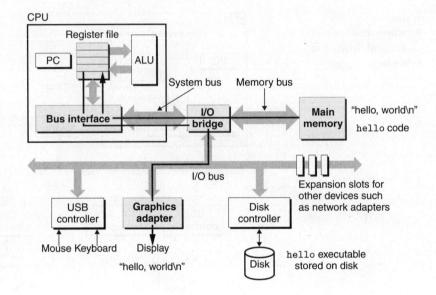

5 Caches Matter

An important lesson from this simple example is that a system spends a lot of time moving information from one place to another. The machine instructions in the `hello` program are originally stored on disk. When the program is loaded, they are copied to main memory. As the processor runs the program, instructions are copied from main memory into the processor. Similarly, the data string "`hello,world\n`", originally on disk, is copied to main memory, and then copied from main memory to the display device. From a programmer's perspective, much of this copying is overhead that slows down the "real work" of the program. Thus, a major goal for system designers is to make these copy operations run as fast as possible.

Because of physical laws, larger storage devices are slower than smaller storage devices. And faster devices are more expensive to build than their slower counterparts. For example, the disk drive on a typical system might be 1000 times larger than the main memory, but it might take the processor 10,000,000 times longer to read a word from disk than from memory.

Similarly, a typical register file stores only a few hundred bytes of information, as opposed to billions of bytes in the main memory. However, the processor can read data from the register file almost 100 times faster than from memory. Even more troublesome, as semiconductor technology progresses over the years, this *processor-memory gap* continues to increase. It is easier and cheaper to make processors run faster than it is to make main memory run faster.

To deal with the processor-memory gap, system designers include smaller faster storage devices called *cache memories* (or simply caches) that serve as temporary staging areas for information that the processor is likely to need in the near future. Figure 8 shows the cache memories in a typical system. An *L1*

Figure 8
Cache memories.

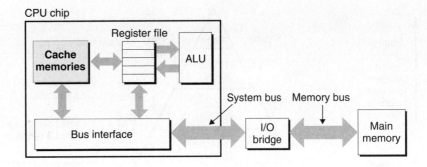

cache on the processor chip holds tens of thousands of bytes and can be accessed nearly as fast as the register file. A larger *L2 cache* with hundreds of thousands to millions of bytes is connected to the processor by a special bus. It might take 5 times longer for the process to access the L2 cache than the L1 cache, but this is still 5 to 10 times faster than accessing the main memory. The L1 and L2 caches are implemented with a hardware technology known as *static random access memory* (SRAM). Newer and more powerful systems even have three levels of cache: L1, L2, and L3. The idea behind caching is that a system can get the effect of both a very large memory and a very fast one by exploiting *locality*, the tendency for programs to access data and code in localized regions. By setting up caches to hold data that is likely to be accessed often, we can perform most memory operations using the fast caches.

One of the most important lessons in this book is that application programmers who are aware of cache memories can exploit them to improve the performance of their programs by an order of magnitude.

6 Storage Devices Form a Hierarchy

This notion of inserting a smaller, faster storage device (e.g., cache memory) between the processor and a larger slower device (e.g., main memory) turns out to be a general idea. In fact, the storage devices in every computer system are organized as a *memory hierarchy* similar to Figure 9. As we move from the top of the hierarchy to the bottom, the devices become slower, larger, and less costly per byte. The register file occupies the top level in the hierarchy, which is known as level 0, or L0. We show three levels of caching L1 to L3, occupying memory hierarchy levels 1 to 3. Main memory occupies level 4, and so on.

The main idea of a memory hierarchy is that storage at one level serves as a cache for storage at the next lower level. Thus, the register file is a cache for the L1 cache. Caches L1 and L2 are caches for L2 and L3, respectively. The L3 cache is a cache for the main memory, which is a cache for the disk. On some networked systems with distributed file systems, the local disk serves as a cache for data stored on the disks of other systems.

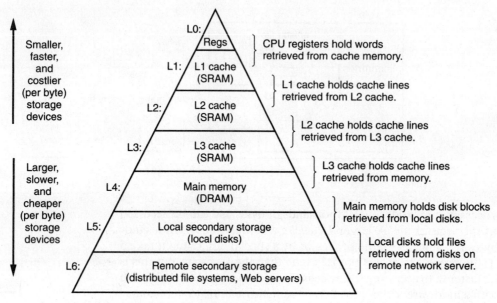

Figure 9 **An example of a memory hierarchy.**

Just as programmers can exploit knowledge of the different caches to improve performance, programmers can exploit their understanding of the entire memory hierarchy.

7 The Operating System Manages the Hardware

Back to our hello example. When the shell loaded and ran the hello program, and when the hello program printed its message, neither program accessed the keyboard, display, disk, or main memory directly. Rather, they relied on the services provided by the *operating system*. We can think of the operating system as a layer of software interposed between the application program and the hardware, as shown in Figure 10. All attempts by an application program to manipulate the hardware must go through the operating system.

The operating system has two primary purposes: (1) to protect the hardware from misuse by runaway applications, and (2) to provide applications with simple and uniform mechanisms for manipulating complicated and often wildly different low-level hardware devices. The operating system achieves both goals via the

Figure 10
Layered view of a computer system.

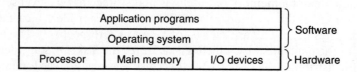

14

Figure 11
Abstractions provided by an operating system.

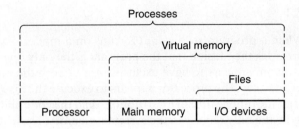

fundamental abstractions shown in Figure 11: *processes*, *virtual memory*, and *files*. As this figure suggests, files are abstractions for I/O devices, virtual memory is an abstraction for both the main memory and disk I/O devices, and processes are abstractions for the processor, main memory, and I/O devices. We will discuss each in turn.

Aside Unix and Posix

The 1960s was an era of huge, complex operating systems, such as IBM's OS/360 and Honeywell's Multics systems. While OS/360 was one of the most successful software projects in history, Multics dragged on for years and never achieved wide-scale use. Bell Laboratories was an original partner in the Multics project, but dropped out in 1969 because of concern over the complexity of the project and the lack of progress. In reaction to their unpleasant Multics experience, a group of Bell Labs researchers—Ken Thompson, Dennis Ritchie, Doug McIlroy, and Joe Ossanna—began work in 1969 on a simpler operating system for a DEC PDP-7 computer, written entirely in machine language. Many of the ideas in the new system, such as the hierarchical file system and the notion of a shell as a user-level process, were borrowed from Multics but implemented in a smaller, simpler package. In 1970, Brian Kernighan dubbed the new system "Unix" as a pun on the complexity of "Multics." The kernel was rewritten in C in 1973, and Unix was announced to the outside world in 1974 [89].

Because Bell Labs made the source code available to schools with generous terms, Unix developed a large following at universities. The most influential work was done at the University of California at Berkeley in the late 1970s and early 1980s, with Berkeley researchers adding virtual memory and the Internet protocols in a series of releases called Unix 4.xBSD (Berkeley Software Distribution). Concurrently, Bell Labs was releasing their own versions, which became known as System V Unix. Versions from other vendors, such as the Sun Microsystems Solaris system, were derived from these original BSD and System V versions.

Trouble arose in the mid 1980s as Unix vendors tried to differentiate themselves by adding new and often incompatible features. To combat this trend, IEEE (Institute for Electrical and Electronics Engineers) sponsored an effort to standardize Unix, later dubbed "Posix" by Richard Stallman. The result was a family of standards, known as the Posix standards, that cover such issues as the C language interface for Unix system calls, shell programs and utilities, threads, and network programming. As more systems comply more fully with the Posix standards, the differences between Unix versions are gradually disappearing.

7.1 Processes

When a program such as `hello` runs on a modern system, the operating system provides the illusion that the program is the only one running on the system. The program appears to have exclusive use of both the processor, main memory, and I/O devices. The processor appears to execute the instructions in the program, one after the other, without interruption. And the code and data of the program appear to be the only objects in the system's memory. These illusions are provided by the notion of a process, one of the most important and successful ideas in computer science.

A *process* is the operating system's abstraction for a running program. Multiple processes can run concurrently on the same system, and each process appears to have exclusive use of the hardware. By *concurrently*, we mean that the instructions of one process are interleaved with the instructions of another process. In most systems, there are more processes to run than there are CPUs to run them. Traditional systems could only execute one program at a time, while newer *multicore* processors can execute several programs simultaneously. In either case, a single CPU can appear to execute multiple processes concurrently by having the processor switch among them. The operating system performs this interleaving with a mechanism known as *context switching*. To simplify the rest of this discussion, we consider only a *uniprocessor system* containing a single CPU. We will return to the discussion of *multiprocessor* systems in Section 9.1.

The operating system keeps track of all the state information that the process needs in order to run. This state, which is known as the *context*, includes information such as the current values of the PC, the register file, and the contents of main memory. At any point in time, a uniprocessor system can only execute the code for a single process. When the operating system decides to transfer control from the current process to some new process, it performs a *context switch* by saving the context of the current process, restoring the context of the new process, and then passing control to the new process. The new process picks up exactly where it left off. Figure 12 shows the basic idea for our example `hello` scenario.

There are two concurrent processes in our example scenario: the shell process and the `hello` process. Initially, the shell process is running alone, waiting for input on the command line. When we ask it to run the `hello` program, the shell carries

Figure 12

Process context switching.

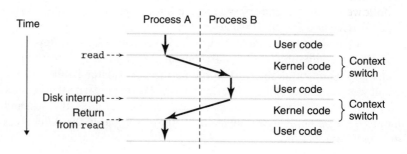

out our request by invoking a special function known as a *system call* that passes control to the operating system. The operating system saves the shell's context, creates a new `hello` process and its context, and then passes control to the new `hello` process. After `hello` terminates, the operating system restores the context of the shell process and passes control back to it, where it waits for the next command line input.

Implementing the process abstraction requires close cooperation between both the low-level hardware and the operating system software.

7.2 Threads

Although we normally think of a process as having a single control flow, in modern systems a process can actually consist of multiple execution units, called *threads*, each running in the context of the process and sharing the same code and global data. Threads are an increasingly important programming model because of the requirement for concurrency in network servers, because it is easier to share data between multiple threads than between multiple processes, and because threads are typically more efficient than processes. Multi-threading is also one way to make programs run faster when multiple processors are available, as we will discuss in Section 9.1.

7.3 Virtual Memory

Virtual memory is an abstraction that provides each process with the illusion that it has exclusive use of the main memory. Each process has the same uniform view of memory, which is known as its *virtual address space*. The virtual address space for Linux processes is shown in Figure 13. (Other Unix systems use a similar layout.) In Linux, the topmost region of the address space is reserved for code and data in the operating system that is common to all processes. The lower region of the address space holds the code and data defined by the user's process. Note that addresses in the figure increase from the bottom to the top.

The virtual address space seen by each process consists of a number of well-defined areas, each with a specific purpose. It will be helpful to look briefly at each, starting with the lowest addresses and working our way up:

- *Program code and data.* Code begins at the same fixed address for all processes, followed by data locations that correspond to global C variables. The code and data areas are initialized directly from the contents of an executable object file, in our case the `hello` executable.

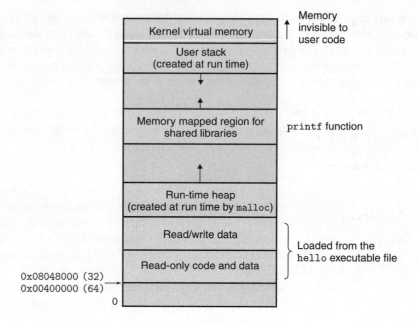

Figure 13

Process virtual address space.

For virtual memory to work, a sophisticated interaction is required between the hardware and the operating system software, including a hardware translation of every address generated by the processor. The basic idea is to store the contents

- *Heap.* The code and data areas are followed immediately by the run-time *heap.* Unlike the code and data areas, which are fixed in size once the process begins running, the heap expands and contracts dynamically at run time as a result of calls to C standard library routines such as malloc and free.

- *Shared libraries.* Near the middle of the address space is an area that holds the code and data for *shared libraries* such as the C standard library and the math library. The notion of a shared library is a powerful but somewhat difficult concept.

- *Stack.* At the top of the user's virtual address space is the *user stack* that the compiler uses to implement function calls. Like the heap, the user stack expands and contracts dynamically during the execution of the program. In particular, each time we call a function, the stack grows. Each time we return from a function, it contracts.

- *Kernel virtual memory.* The *kernel* is the part of the operating system that is always resident in memory. The top region of the address space is reserved for the kernel. Application programs are not allowed to read or write the contents of this area or to directly call functions defined in the kernel code.

For virtual memory to work, a sophisticated interaction is required between the hardware and the operating system software, including a hardware translation of every address generated by the processor. The basic idea is to store the contents of a process's virtual memory on disk, and then use the main memory as a cache for the disk.

7.4 Files

A *file* is a sequence of bytes, nothing more and nothing less. Every I/O device, including disks, keyboards, displays, and even networks, is modeled as a file. All input and output in the system is performed by reading and writing files, using a small set of system calls known as *Unix I/O*.

This simple and elegant notion of a file is nonetheless very powerful because it provides applications with a uniform view of all of the varied I/O devices that might be contained in the system. For example, application programmers who manipulate the contents of a disk file are blissfully unaware of the specific disk technology. Further, the same program will run on different systems that use different disk technologies.

Aside The Linux project

In August 1991, a Finnish graduate student named Linus Torvalds modestly announced a new Unix-like operating system kernel:

```
From: torvalds@klaava.Helsinki.FI (Linus Benedict Torvalds)
Newsgroups: comp.os.minix
Subject: What would you like to see most in minix?
Summary: small poll for my new operating system
Date: 25 Aug 91 20:57:08 GMT

Hello everybody out there using minix -
I'm doing a (free) operating system (just a hobby, won't be big and
professional like gnu) for 386(486) AT clones. This has been brewing
since April, and is starting to get ready. I'd like any feedback on
things people like/dislike in minix, as my OS resembles it somewhat
(same physical layout of the file-system (due to practical reasons)
among other things).

I've currently ported bash(1.08) and gcc(1.40), and things seem to work.
This implies that I'll get something practical within a few months, and
I'd like to know what features most people would want. Any suggestions
are welcome, but I won't promise I'll implement them :-)

Linus (torvalds@kruuna.helsinki.fi)
```

The rest, as they say, is history. Linux has evolved into a technical and cultural phenomenon. By combining forces with the GNU project, the Linux project has developed a complete, Posix-compliant version of the Unix operating system, including the kernel and all of the supporting infrastructure. Linux is available on a wide array of computers, from hand-held devices to mainframe computers. A group at IBM has even ported Linux to a wristwatch!

8 Systems Communicate with Other Systems Using Networks

Up to this point in our tour of systems, we have treated a system as an isolated collection of hardware and software. In practice, modern systems are often linked to other systems by networks. From the point of view of an individual system, the network can be viewed as just another I/O device, as shown in Figure 14. When the system copies a sequence of bytes from main memory to the network adapter, the data flows across the network to another machine, instead of, say, to a local disk drive. Similarly, the system can read data sent from other machines and copy this data to its main memory.

With the advent of global networks such as the Internet, copying information from one machine to another has become one of the most important uses of computer systems. For example, applications such as email, instant messaging, the World Wide Web, FTP, and telnet are all based on the ability to copy information over a network.

Returning to our `hello` example, we could use the familiar telnet application to run `hello` on a remote machine. Suppose we use a telnet *client* running on our

Figure 14

A network is another I/O device.

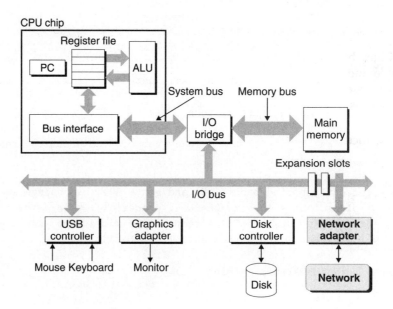

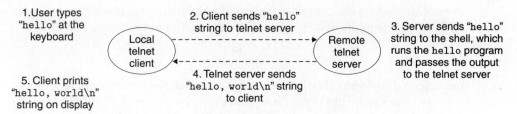

1.User types
"hello" at the
keyboard

2. Client sends "hello"
string to telnet server

3. Server sends "hello"
string to the shell, which
runs the hello program
and passes the output
to the telnet server

Local
telnet
client

Remote
telnet
server

5. Client prints
"hello, world\n"
string on display

4. Telnet server sends
"hello, world\n" string
to client

Figure 15 **Using telnet to run hello remotely over a network.**

local machine to connect to a telnet *server* on a remote machine. After we log in
to the remote machine and run a shell, the remote shell is waiting to receive an
input command. From this point, running the hello program remotely involves
the five basic steps shown in Figure 15.

After we type the "hello" string to the telnet client and hit the enter key,
the client sends the string to the telnet server. After the telnet server receives the
string from the network, it passes it along to the remote shell program. Next, the
remote shell runs the hello program, and passes the output line back to the telnet
server. Finally, the telnet server forwards the output string across the network to
the telnet client, which prints the output string on our local terminal.

This type of exchange between clients and servers is typical of all network
applications.

9   Important Themes

This concludes our initial whirlwind tour of systems. An important idea to take
away from this discussion is that a system is more than just hardware. It is a
collection of intertwined hardware and systems software that must cooperate in
order to achieve the ultimate goal of running application programs. The rest of
this book will fill in some details about the hardware and the software, and it will
show how, by knowing these details, you can write programs that are faster, more
reliable, and more secure.

To close out this chapter, we highlight several important concepts that cut
across all aspects of computer systems.

9.1   Concurrency and Parallelism

Throughout the history of digital computers, two demands have been constant
forces driving improvements: we want them to do more, and we want them to
run faster. Both of these factors improve when the processor does more things at
once. We use the term *concurrency* to refer to the general concept of a system with
multiple, simultaneous activities, and the term *parallelism* to refer to the use of
concurrency to make a system run faster. Parallelism can be exploited at multiple

levels of abstraction in a computer system. We highlight three levels here, working from the highest to the lowest level in the system hierarchy.

Thread-Level Concurrency

Building on the process abstraction, we are able to devise systems where multiple programs execute at the same time, leading to *concurrency*. With threads, we can even have multiple control flows executing within a single process. Support for concurrent execution has been found in computer systems since the advent of time-sharing in the early 1960s. Traditionally, this concurrent execution was only *simulated*, by having a single computer rapidly switch among its executing processes, much as a juggler keeps multiple balls flying through the air. This form of concurrency allows multiple users to interact with a system at the same time, such as when many people want to get pages from a single Web server. It also allows a single user to engage in multiple tasks concurrently, such as having a Web browser in one window, a word processor in another, and streaming music playing at the same time. Until recently, most actual computing was done by a single processor, even if that processor had to switch among multiple tasks. This configuration is known as a *uniprocessor system*.

When we construct a system consisting of multiple processors all under the control of a single operating system kernel, we have a *multiprocessor system*. Such systems have been available for large-scale computing since the 1980s, but they have more recently become commonplace with the advent of *multi-core* processors and *hyperthreading*. Figure 16 shows a taxonomy of these different processor types.

Multi-core processors have several CPUs (referred to as "cores") integrated onto a single integrated-circuit chip. Figure 17 illustrates the organization of an Intel Core i7 processor, where the microprocessor chip has four CPU cores, each with its own L1 and L2 caches but sharing the higher levels of cache as well as the interface to main memory. Industry experts predict that they will be able to have dozens, and ultimately hundreds, of cores on a single chip.

Hyperthreading, sometimes called *simultaneous multi-threading*, is a technique that allows a single CPU to execute multiple flows of control. It involves having multiple copies of some of the CPU hardware, such as program counters and register files, while having only single copies of other parts of the hardware, such as the units that perform floating-point arithmetic. Whereas a conventional

Figure 16

Categorizing different processor configurations. Multiprocessors are becoming prevalent with the advent of multi-core processors and hyperthreading.

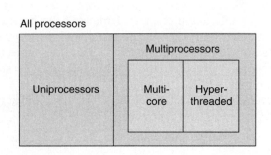

Figure 17

Intel Core i7 organization. Four processor cores are integrated onto a single chip.

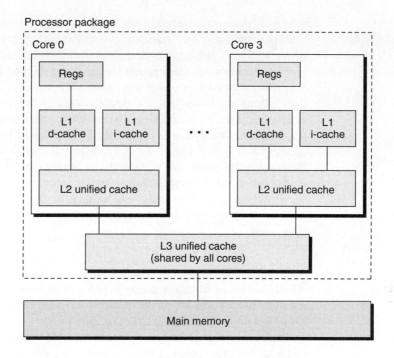

processor requires around 20,000 clock cycles to shift between different threads, a hyperthreaded processor decides which of its threads to execute on a cycle-by-cycle basis. It enables the CPU to make better advantage of its processing resources. For example, if one thread must wait for some data to be loaded into a cache, the CPU can proceed with the execution of a different thread. As an example, the Intel Core i7 processor can have each core executing two threads, and so a four-core system can actually execute eight threads in parallel.

The use of multiprocessing can improve system performance in two ways. First, it reduces the need to simulate concurrency when performing multiple tasks. As mentioned, even a personal computer being used by a single person is expected to perform many activities concurrently. Second, it can run a single application program faster, but only if that program is expressed in terms of multiple threads that can effectively execute in parallel. Thus, although the principles of concurrency have been formulated and studied for over 50 years, the advent of multi-core and hyperthreaded systems has greatly increased the desire to find ways to write application programs that can exploit the thread-level parallelism available with the hardware.

Instruction-Level Parallelism

At a much lower level of abstraction, modern processors can execute multiple instructions at one time, a property known as *instruction-level parallelism*. For example, early microprocessors, such as the 1978-vintage Intel 8086 required multiple (typically, 3–10) clock cycles to execute a single instruction. More recent processors can sustain execution rates of 2–4 instructions per clock cycle. Any given instruction requires much longer from start to finish, perhaps 20 cycles or more, but the processor uses a number of clever tricks to process as many as 100 instructions at a time. In your studies, you may encounter *pipelining*, where the actions required to execute an instruction are partitioned into different steps and the processor hardware is organized as a series of stages, each performing one of these steps. The stages can operate in parallel, working on different parts of different instructions. We will see that a fairly simple hardware design can sustain an execution rate close to one instruction per clock cycle.

Processors that can sustain execution rates faster than one instruction per cycle are known as *superscalar* processors. Most modern processors support superscalar operation.

Single-Instruction, Multiple-Data (SIMD) Parallelism

At the lowest level, many modern processors have special hardware that allows a single instruction to cause multiple operations to be performed in parallel, a mode known as *single-instruction, multiple-data*, or "SIMD" parallelism. For example, recent generations of Intel and AMD processors have instructions that can add four pairs of single-precision floating-point numbers (C data type `float`) in parallel.

These SIMD instructions are provided mostly to speed up applications that process image, sound, and video data. Although some compilers attempt to automatically extract SIMD parallelism from C programs, a more reliable method is to write programs using special *vector* data types supported in compilers such as GCC.

9.2 The Importance of Abstractions in Computer Systems

The use of *abstractions* is one of the most important concepts in computer science. For example, one aspect of good programming practice is to formulate a simple application-program interface (API) for a set of functions that allow programmers to use the code without having to delve into its inner workings. Different programming languages provide different forms and levels of support for abstraction, such as Java class declarations and C function prototypes.

We have already been introduced to several of the abstractions seen in computer systems, as indicated in Figure 18. On the processor side, the *instruction set architecture* provides an abstraction of the actual processor hardware. With this abstraction, a machine-code program behaves as if it were executed on a processor

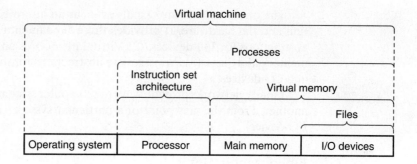

Virtual machine

Figure 18

Some abstractions provided by a computer system. A major theme in computer systems is to provide abstract representations at different levels to hide the complexity of the actual implementations.

that performs just one instruction at a time. The underlying hardware is far more elaborate, executing multiple instructions in parallel, but always in a way that is consistent with the simple, sequential model. By keeping the same execution model, different processor implementations can execute the same machine code, while offering a range of cost and performance.

On the operating system side, we have introduced three abstractions: *files* as an abstraction of I/O, *virtual memory* as an abstraction of program memory, and *processes* as an abstraction of a running program. To these abstractions we add a new one: the *virtual machine*, providing an abstraction of the entire computer, including the operating system, the processor, and the programs. The idea of a virtual machine was introduced by IBM in the 1960s, but it has become more prominent recently as a way to manage computers that must be able to run programs designed for multiple operating systems (such as Microsoft Windows, MacOS, and Linux) or different versions of the same operating system.

10 Summary

A computer system consists of hardware and systems software that cooperate to run application programs. Information inside the computer is represented as groups of bits that are interpreted in different ways, depending on the context. Programs are translated by other programs into different forms, beginning as ASCII text and then translated by compilers and linkers into binary executable files.

Processors read and interpret binary instructions that are stored in main memory. Since computers spend most of their time copying data between memory, I/O devices, and the CPU registers, the storage devices in a system are arranged in a hierarchy, with the CPU registers at the top, followed by multiple levels of hardware cache memories, DRAM main memory, and disk storage. Storage devices that are higher in the hierarchy are faster and more costly per bit than those lower in the hierarchy. Storage devices that are higher in the hierarchy serve as caches for devices that are lower in the hierarchy. Programmers can optimize the performance of their C programs by understanding and exploiting the memory hierarchy.

The operating system kernel serves as an intermediary between the application and the hardware. It provides three fundamental abstractions: (1) Files are abstractions for I/O devices. (2) Virtual memory is an abstraction for both main memory and disks. (3) Processes are abstractions for the processor, main memory, and I/O devices.

Finally, networks provide ways for computer systems to communicate with one another. From the viewpoint of a particular system, the network is just another I/O device.

Bibliographic Notes

Ritchie has written interesting first hand accounts of the early days of C and Unix [87, 88]. Ritchie and Thompson presented the first published account of Unix [89]. Silberschatz, Gavin, and Gagne [98] provide a comprehensive history of the different flavors of Unix. The GNU (www.gnu.org) and Linux (www.linux.org) Web pages have loads of current and historical information. The Posix standards are available online at (www.unix.org).

Representing and Manipulating Information

Modern computers store and process information represented as 2-valued signals. These lowly binary digits, or *bits*, form the basis of the digital revolution. The familiar decimal, or base-10, representation has been in use for over 1000 years, having been developed in India, improved by Arab mathematicians in the 12th century, and brought to the West in the 13th century by the Italian mathematician Leonardo Pisano (c. 1170 – c. 1250), better known as Fibonacci. Using decimal notation is natural for ten-fingered humans, but binary values work better when building machines that store and process information. Two-valued signals can readily be represented, stored, and transmitted, for example, as the presence or absence of a hole in a punched card, as a high or low voltage on a wire, or as a magnetic domain oriented clockwise or counterclockwise. The electronic circuitry for storing and performing computations on 2-valued signals is very simple and reliable, enabling manufacturers to integrate millions, or even billions, of such circuits on a single silicon chip.

In isolation, a single bit is not very useful. When we group bits together and apply some *interpretation* that gives meaning to the different possible bit patterns, however, we can represent the elements of any finite set. For example, using a binary number system, we can use groups of bits to encode nonnegative numbers. By using a standard character code, we can encode the letters and symbols in a document. We cover both of these encodings in this chapter, as well as encodings to represent negative numbers and to approximate real numbers.

We consider the three most important representations of numbers. *Unsigned* encodings are based on traditional binary notation, representing numbers greater than or equal to 0. *Two's-complement* encodings are the most common way to represent *signed* integers, that is, numbers that may be either positive or negative. *Floating-point* encodings are a base-two version of scientific notation for representing real numbers. Computers implement arithmetic operations, such as addition and multiplication, with these different representations, similar to the corresponding operations on integers and real numbers.

Computer representations use a limited number of bits to encode a number, and hence some operations can *overflow* when the results are too large to be represented. This can lead to some surprising results. For example, on most of today's computers (those using a 32-bit representation of data type int), computing the expression

```
200 * 300 * 400 * 500
```

yields −884,901,888. This runs counter to the properties of integer arithmetic—computing the product of a set of positive numbers has yielded a negative result.

On the other hand, integer computer arithmetic satisfies many of the familiar properties of true integer arithmetic. For example, multiplication is associative and commutative, so that computing any of the following C expressions yields −884,901,888:

```
(500  *  400) * (300 * 200)
((500 *  400) * 300) * 200
((200 *  500) * 300) * 400
400   * (200 * (300 * 500))
```

The computer might not generate the expected result, but at least it is consistent!

Floating-point arithmetic has altogether different mathematical properties. The product of a set of positive numbers will always be positive, although overflow will yield the special value $+\infty$. Floating-point arithmetic is not associative, due to the finite precision of the representation. For example, the C expression (3.14+1e20)-1e20 will evaluate to 0.0 on most machines, while 3.14+(1e20-1e20) will evaluate to 3.14. The different mathematical properties of integer vs. floating-point arithmetic stem from the difference in how they handle the finiteness of their representations—integer representations can encode a comparatively small range of values, but do so precisely, while floating-point representations can encode a wide range of values, but only approximately.

By studying the actual number representations, we can understand the ranges of values that can be represented and the properties of the different arithmetic operations. This understanding is critical to writing programs that work correctly over the full range of numeric values and that are portable across different combinations of machine, operating system, and compiler. As we will describe, a number of computer security vulnerabilities have arisen due to some of the subtleties of computer arithmetic. Whereas in an earlier era program bugs would only inconvenience people when they happened to be triggered, there are now legions of hackers who try to exploit any bug they can find to obtain unauthorized access to other people's systems. This puts a higher level of obligation on programmers to understand how their programs work and how they can be made to behave in undesirable ways.

Computers use several different binary representations to encode numeric values. You will need to be familiar with these representations as you progress into machine-level programming. We describe these encodings in this chapter and show you how to reason about number representations.

We derive several ways to perform arithmetic operations by directly manipulating the bit-level representations of numbers. Understanding these techniques will be important for understanding the machine-level code generated by compilers in their attempt to optimize the performance of arithmetic expression evaluation.

Our treatment of this material is based on a core set of mathematical principles. We start with the basic definitions of the encodings and then derive such properties as the range of representable numbers, their bit-level representations, and the properties of the arithmetic operations. We believe it is important for you to examine the material from this abstract viewpoint, because programmers need to have a clear understanding of how computer arithmetic relates to the more familiar integer and real arithmetic.

Aside   How to read this chapter

If you find equations and formulas daunting, do not let that stop you from getting the most out of this chapter! We provide full derivations of mathematical ideas for completeness, but the best way to read this material is often to skip over the derivation on your initial reading. Instead, study the examples provided, and be sure to work *all* of the practice problems. The examples will give you an intuition behind the ideas, and the practice problems engage you in *active learning*, helping you put thoughts into action. With these as background, you will find it much easier to go back and follow the derivations. Be assured, as well, that the mathematical skills required to understand this material are within reach of someone with good grasp of high school algebra.

The C++ programming language is built upon C, using the exact same numeric representations and operations. Everything said in this chapter about C also holds for C++. The Java language definition, on the other hand, created a new set of standards for numeric representations and operations. Whereas the C standards are designed to allow a wide range of implementations, the Java standard is quite specific on the formats and encodings of data. We highlight the representations and operations supported by Java at several places in the chapter.

Aside The evolution of the C programming language

The C programming language was first developed by Dennis Ritchie of Bell Laboratories for use with the Unix operating system (also developed at Bell Labs). At the time, most system programs, such as operating systems, had to be written largely in assembly code, in order to have access to the low-level representations of different data types. For example, it was not feasible to write a memory allocator, such as is provided by the `malloc` library function, in other high-level languages of that era.

The original Bell Labs version of C was documented in the first edition of the book by Brian Kernighan and Dennis Ritchie [57]. Over time, C has evolved through the efforts of several standardization groups. The first major revision of the original Bell Labs C led to the ANSI C standard in 1989, by a group working under the auspices of the American National Standards Institute. ANSI C was a major departure from Bell Labs C, especially in the way functions are declared. ANSI C is described in the second edition of Kernighan and Ritchie's book [58], which is still considered one of the best references on C.

The International Standards Organization took over responsibility for standardizing the C language, adopting a version that was substantially the same as ANSI C in 1990 and hence is referred to as "ISO C90." This same organization sponsored an updating of the language in 1999, yielding "ISO C99." Among other things, this version introduced some new data types and provided support for text strings requiring characters not found in the English language.

The GNU Compiler Collection (GCC) can compile programs according to the conventions of several different versions of the C language, based on different command line options, as shown in Figure 1. For example, to compile program `prog.c` according to ISO C99, we could give the command line

```
unix> gcc -std=c99 prog.c
```

The options `-ansi` and `-std=c89` have the same effect—the code is compiled according to the ANSI or ISO C90 standard. (C90 is sometimes referred to as "C89," since its standardization effort began in 1989.) The option `-std=c99` causes the compiler to follow the ISO C99 convention.

C version	GCC command line option
GNU 89	*none*, -std=gnu89
ANSI, ISO C90	-ansi, -std=c89
ISO C99	-std=c99
GNU 99	-std=gnu99

Figure 1 **Specifying different versions of C to GCC.**

As of the writing of this book, when no option is specified, the program will be compiled according to a version of C based on ISO C90, but including some features of C99, some of C++, and others specific to GCC. This version can be specified explicitly using the option -std=gnu89. The GNU project is developing a version that combines ISO C99, plus other features, that can be specified with command line option -std=gnu99. (Currently, this implementation is incomplete.) This will become the default version.

1 Information Storage

Rather than accessing individual bits in memory, most computers use blocks of eight bits, or *bytes*, as the smallest addressable unit of memory. A machine-level program views memory as a very large array of bytes, referred to as *virtual memory*. Every byte of memory is identified by a unique number, known as its *address*, and the set of all possible addresses is known as the *virtual address space*. As indicated by its name, this virtual address space is just a conceptual image presented to the machine-level program. The actual implementation uses a combination of random-access memory (RAM), disk storage, special hardware, and operating system software to provide the program with what appears to be a monolithic byte array.

In subsequent chapters, we will cover how the compiler and run-time system partitions this memory space into more manageable units to store the different *program objects*, that is, program data, instructions, and control information. Various mechanisms are used to allocate and manage the storage for different parts of the program. This management is all performed within the virtual address space. For example, the value of a pointer in C—whether it points to an integer, a structure, or some other program object—is the virtual address of the first byte of some block of storage. The C compiler also associates *type* information with each pointer, so that it can generate different machine-level code to access the value stored at the location designated by the pointer depending on the type of that value. Although the C compiler maintains this type information, the actual machine-level program it generates has no information about data types. It simply treats each program object as a block of bytes, and the program itself as a sequence of bytes.

New to C?   The role of pointers in C

Pointers are a central feature of C. They provide the mechanism for referencing elements of data structures, including arrays. Just like a variable, a pointer has two aspects: its *value* and its *type*. The value indicates the location of some object, while its type indicates what kind of object (e.g., integer or floating-point number) is stored at that location.

1.1 Hexadecimal Notation

A single byte consists of 8 bits. In binary notation, its value ranges from 00000000_2 to 11111111_2. When viewed as a decimal integer, its value ranges from 0_{10} to 255_{10}. Neither notation is very convenient for describing bit patterns. Binary notation is too verbose, while with decimal notation, it is tedious to convert to and from bit patterns. Instead, we write bit patterns as base-16, or *hexadecimal* numbers. Hexadecimal (or simply "hex") uses digits '0' through '9' along with characters 'A' through 'F' to represent 16 possible values. Figure 2 shows the decimal and binary values associated with the 16 hexadecimal digits. Written in hexadecimal, the value of a single byte can range from 00_{16} to FF_{16}.

In C, numeric constants starting with 0x or 0X are interpreted as being in hexadecimal. The characters 'A' through 'F' may be written in either upper or lower case. For example, we could write the number $FA1D37B_{16}$ as 0xFA1D37B, as 0xfa1d37b, or even mixing upper and lower case, e.g., 0xFa1D37b. We will use the C notation for representing hexadecimal values.

A common task in working with machine-level programs is to manually convert between decimal, binary, and hexadecimal representations of bit patterns. Converting between binary and hexadecimal is straightforward, since it can be performed one hexadecimal digit at a time. Digits can be converted by referring to a chart such as that shown in Figure 2. One simple trick for doing the conversion in your head is to memorize the decimal equivalents of hex digits A, C, and F. The hex values B, D, and E can be translated to decimal by computing their values relative to the first three.

For example, suppose you are given the number 0x173A4C. You can convert this to binary format by expanding each hexadecimal digit, as follows:

Hex digit	0	1	2	3	4	5	6	7
Decimal value	0	1	2	3	4	5	6	7
Binary value	0000	0001	0010	0011	0100	0101	0110	0111
Hex digit	8	9	A	B	C	D	E	F
Decimal value	8	9	10	11	12	13	14	15
Binary value	1000	1001	1010	1011	1100	1101	1110	1111

Figure 2   **Hexadecimal notation.** Each Hex digit encodes one of 16 values.

Hexadecimal	1	7	3	A	4	C
Binary	0001	0111	0011	1010	0100	1100

This gives the binary representation 000101110011101001001100.

Conversely, given a binary number 1111001010110110110011, you convert it to hexadecimal by first splitting it into groups of 4 bits each. Note, however, that if the total number of bits is not a multiple of 4, you should make the *leftmost* group be the one with fewer than 4 bits, effectively padding the number with leading zeros. Then you translate each group of 4 bits into the corresponding hexadecimal digit:

Binary	11	1100	1010	1101	1011	0011
Hexadecimal	3	C	A	D	B	3

Practice Problem 1

Perform the following number conversions:

A. 0x39A7F8 to binary

B. Binary 1100100101111011 to hexadecimal

C. 0xD5E4C to binary

D. Binary 1001101110011110110101 to hexadecimal

When a value x is a power of two, that is, $x = 2^n$ for some nonnegative integer n, we can readily write x in hexadecimal form by remembering that the binary representation of x is simply 1 followed by n zeros. The hexadecimal digit 0 represents four binary zeros. So, for n written in the form $i + 4j$, where $0 \le i \le 3$, we can write x with a leading hex digit of 1 ($i = 0$), 2 ($i = 1$), 4 ($i = 2$), or 8 ($i = 3$), followed by j hexadecimal 0s. As an example, for $x = 2048 = 2^{11}$, we have $n = 11 = 3 + 4 \cdot 2$, giving hexadecimal representation 0x800.

Practice Problem 2

Fill in the blank entries in the following table, giving the decimal and hexadecimal representations of different powers of 2:

n	2^n (Decimal)	2^n (Hexadecimal)
9	512	0x200
19		
	16,384	
		0x10000
17		
	32	
		0x80

Converting between decimal and hexadecimal representations requires using multiplication or division to handle the general case. To convert a decimal number x to hexadecimal, we can repeatedly divide x by 16, giving a quotient q and a remainder r, such that $x = q \cdot 16 + r$. We then use the hexadecimal digit representing r as the least significant digit and generate the remaining digits by repeating the process on q. As an example, consider the conversion of decimal 314156:

$$
\begin{aligned}
314156 &= 19634 \cdot 16 + 12 \quad &\text{(C)} \\
19634 &= 1227 \cdot 16 + 2 \quad &\text{(2)} \\
1227 &= 76 \cdot 16 + 11 \quad &\text{(B)} \\
76 &= 4 \cdot 16 + 12 \quad &\text{(C)} \\
4 &= 0 \cdot 16 + 4 \quad &\text{(4)}
\end{aligned}
$$

From this we can read off the hexadecimal representation as 0x4CB2C.

Conversely, to convert a hexadecimal number to decimal, we can multiply each of the hexadecimal digits by the appropriate power of 16. For example, given the number 0x7AF, we compute its decimal equivalent as $7 \cdot 16^2 + 10 \cdot 16 + 15 = 7 \cdot 256 + 10 \cdot 16 + 15 = 1792 + 160 + 15 = 1967$.

Practice Problem 3

A single byte can be represented by two hexadecimal digits. Fill in the missing entries in the following table, giving the decimal, binary, and hexadecimal values of different byte patterns:

Decimal	Binary	Hexadecimal
0	0000 0000	0x00
167		
62		
188		
	0011 0111	
	1000 1000	
	1111 0011	
		0x52
		0xAC
		0xE7

Aside Converting between decimal and hexadecimal

For converting larger values between decimal and hexadecimal, it is best to let a computer or calculator do the work. For example, the following script in the Perl language converts a list of numbers (given on the command line) from decimal to hexadecimal:

---
bin/d2h

```perl
1   #!/usr/local/bin/perl
2   # Convert list of decimal numbers into hex
3
4   for ($i = 0; $i < @ARGV; $i++) {
5       printf("%d\t= 0x%x\n", $ARGV[$i], $ARGV[$i]);
6   }
```
---
bin/d2h

Once this file has been set to be executable, the command

```
unix> ./d2h 100 500 751
```

yields output

```
100 = 0x64
500 = 0x1f4
751 = 0x2ef
```

Similarly, the following script converts from hexadecimal to decimal:

---
bin/h2d

```perl
1   #!/usr/local/bin/perl
2   # Convert list of hex numbers into decimal
3
4   for ($i = 0; $i < @ARGV; $i++) {
5     $val = hex($ARGV[$i]);
6     printf("0x%x = %d\n", $val, $val);
7   }
```
---
bin/h2d

Practice Problem 4

Without converting the numbers to decimal or binary, try to solve the following arithmetic problems, giving the answers in hexadecimal. **Hint:** Just modify the methods you use for performing decimal addition and subtraction to use base 16.

A. 0x503c + 0x8 = _____

B. 0x503c − 0x40 = _____

C. 0x503c + 64 = _____

D. 0x50ea − 0x503c = _____

1.2 Words

Every computer has a *word size*, indicating the nominal size of integer and pointer data. Since a virtual address is encoded by such a word, the most important system parameter determined by the word size is the maximum size of the virtual address space. That is, for a machine with a w-bit word size, the virtual addresses can range from 0 to $2^w - 1$, giving the program access to at most 2^w bytes.

Most personal computers today have a 32-bit word size. This limits the virtual address space to 4 gigabytes (written 4 GB), that is, just over 4×10^9 bytes. Although this is ample space for most applications, we have reached the point where many large-scale scientific and database applications require larger amounts of storage. Consequently, high-end machines with 64-bit word sizes are becoming increasingly common as storage costs decrease. As hardware costs drop over time, even desktop and laptop machines will switch to 64-bit word sizes, and so we will consider the general case of a w-bit word size, as well as the special cases of $w = 32$ and $w = 64$.

1.3 Data Sizes

Computers and compilers support multiple data formats using different ways to encode data, such as integers and floating point, as well as different lengths. For example, many machines have instructions for manipulating single bytes, as well as integers represented as 2-, 4-, and 8-byte quantities. They also support floating-point numbers represented as 4- and 8-byte quantities.

The C language supports multiple data formats for both integer and floating-point data. The C data type char represents a single byte. Although the name "char" derives from the fact that it is used to store a single character in a text string, it can also be used to store integer values. The C data type int can also be prefixed by the qualifiers short, long, and recently long long, providing integer representations of various sizes. Figure 3 shows the number of bytes allocated

C declaration	32-bit	64-bit
char	1	1
short int	2	2
int	4	4
long int	4	8
long long int	8	8
char *	4	8
float	4	4
double	8	8

Figure 3 **Sizes (in bytes) of C numeric data types.** The number of bytes allocated varies with machine and compiler. This chart shows the values typical of 32-bit and 64-bit machines.

for different C data types. The exact number depends on both the machine and the compiler. We show typical sizes for 32-bit and 64-bit machines. Observe that "short" integers have 2-byte allocations, while an unqualified `int` is 4 bytes. A "long" integer uses the full word size of the machine. The "long long" integer data type, introduced in ISO C99, allows the full range of 64-bit integers. For 32-bit machines, the compiler must compile operations for this data type by generating code that performs sequences of 32-bit operations.

Figure 3 also shows that a pointer (e.g., a variable declared as being of type "`char *`") uses the full word size of the machine. Most machines also support two different floating-point formats: single precision, declared in C as `float`, and double precision, declared in C as `double`. These formats use 4 and 8 bytes, respectively.

New to C?   Declaring pointers

For any data type T, the declaration

`T *p;`

indicates that p is a pointer variable, pointing to an object of type T. For example,

`char *p;`

is the declaration of a pointer to an object of type `char`.

Programmers should strive to make their programs portable across different machines and compilers. One aspect of portability is to make the program insensitive to the exact sizes of the different data types. The C standards set lower bounds on the numeric ranges of the different data types, as will be covered later, but there are no upper bounds. Since 32-bit machines have been the standard since around 1980, many programs have been written assuming the allocations listed for this word size in Figure 3. Given the increasing availability of 64-bit machines, many hidden word size dependencies will show up as bugs in migrating these programs to new machines. For example, many programmers assume that a program object declared as type `int` can be used to store a pointer. This works fine for most 32-bit machines but leads to problems on a 64-bit machine.

1.4   Addressing and Byte Ordering

For program objects that span multiple bytes, we must establish two conventions: what the address of the object will be, and how we will order the bytes in memory. In virtually all machines, a multi-byte object is stored as a contiguous sequence of bytes, with the address of the object given by the smallest address of the bytes used. For example, suppose a variable x of type `int` has address 0x100, that is, the value of the address expression &x is 0x100. Then the 4 bytes of x would be stored in memory locations 0x100, 0x101, 0x102, and 0x103.

For ordering the bytes representing an object, there are two common conventions. Consider a w-bit integer having a bit representation $[x_{w-1}, x_{w-2}, \ldots, x_1, x_0]$, where x_{w-1} is the most significant bit and x_0 is the least. Assuming w is a multiple of 8, these bits can be grouped as bytes, with the most significant byte having bits $[x_{w-1}, x_{w-2}, \ldots, x_{w-8}]$, the least significant byte having bits $[x_7, x_6, \ldots, x_0]$, and the other bytes having bits from the middle. Some machines choose to store the object in memory ordered from least significant byte to most, while other machines store them from most to least. The former convention—where the least significant byte comes first—is referred to as *little endian*. This convention is followed by most Intel-compatible machines. The latter convention—where the most significant byte comes first—is referred to as *big endian*. This convention is followed by most machines from IBM and Sun Microsystems. Note that we said "most." The conventions do not split precisely along corporate boundaries. For example, both IBM and Sun manufacture machines that use Intel-compatible processors and hence are little endian. Many recent microprocessors are *bi-endian*, meaning that they can be configured to operate as either little- or big-endian machines.

Continuing our earlier example, suppose the variable x of type int and at address 0x100 has a hexadecimal value of 0x01234567. The ordering of the bytes within the address range 0x100 through 0x103 depends on the type of machine:

Big endian

	0x100	0x101	0x102	0x103	
· · ·	01	23	45	67	· · ·

Little endian

	0x100	0x101	0x102	0x103	
· · ·	67	45	23	01	· · ·

Note that in the word 0x01234567 the high-order byte has hexadecimal value 0x01, while the low-order byte has value 0x67.

People get surprisingly emotional about which byte ordering is the proper one. In fact, the terms "little endian" and "big endian" come from the book *Gulliver's Travels* by Jonathan Swift, where two warring factions could not agree as to how a soft-boiled egg should be opened—by the little end or by the big. Just like the egg issue, there is no technological reason to choose one byte ordering convention over the other, and hence the arguments degenerate into bickering about socio-political issues. As long as one of the conventions is selected and adhered to consistently, the choice is arbitrary.

Aside   Origin of "endian"

Here is how Jonathan Swift, writing in 1726, described the history of the controversy between big and little endians:

> . . . Lilliput and Blefuscu . . . have, as I was going to tell you, been engaged in a most obstinate war for six-and-thirty moons past. It began upon the following occasion. It is allowed on all hands, that the primitive way of breaking eggs, before we eat them, was upon the larger end; but his present majesty's grandfather, while he was a boy, going to eat an egg, and breaking it according to the ancient practice, happened to cut one of his fingers. Whereupon the emperor his father published an edict, commanding all his subjects, upon great penalties, to break the smaller end of their eggs. The people so highly resented this law, that our histories tell us, there have been six rebellions raised on that account; wherein one emperor lost his life, and another his crown. These civil commotions were constantly fomented by the monarchs of Blefuscu; and when they were quelled, the exiles always fled for refuge to that empire. It is computed that eleven thousand persons have at several times suffered death, rather than submit to break their eggs at the smaller end. Many hundred large volumes have been published upon this controversy: but the books of the Big-endians have been long forbidden, and the whole party rendered incapable by law of holding employments.

> In his day, Swift was satirizing the continued conflicts between England (Lilliput) and France (Blefuscu). Danny Cohen, an early pioneer in networking protocols, first applied these terms to refer to byte ordering [25], and the terminology has been widely adopted.

For most application programmers, the byte orderings used by their machines are totally invisible; programs compiled for either class of machine give identical results. At times, however, byte ordering becomes an issue. The first is when binary data are communicated over a network between different machines. A common problem is for data produced by a little-endian machine to be sent to a big-endian machine, or vice versa, leading to the bytes within the words being in reverse order for the receiving program. To avoid such problems, code written for networking applications must follow established conventions for byte ordering to make sure the sending machine converts its internal representation to the network standard, while the receiving machine converts the network standard to its internal representation.

A second case where byte ordering becomes important is when looking at the byte sequences representing integer data. This occurs often when inspecting machine-level programs. As an example, the following line occurs in a file that gives a text representation of the machine-level code for an Intel IA32 processor:

```
 80483bd:   01 05 64 94 04 08       add    %eax,0x8049464
```

This line was generated by a *disassembler*, a tool that determines the instruction sequence represented by an executable program file. Note that this line states that the hexadecimal byte sequence 01 05 64 94 04 08 is the byte-level representation of an instruction that adds a word of data to the value stored at address 0x8049464. If we take the final 4 bytes of the sequence, 64 94 04 08, and write them in reverse order, we have 08 04 94 64. Dropping the leading 0, we have the value 0x8049464, the numeric value written on the right. Having bytes appear in reverse order is a common occurrence when reading machine-level program representations generated for little-endian machines such

```
1   #include <stdio.h>
2
3   typedef unsigned char *byte_pointer;
4
5   void show_bytes(byte_pointer start, int len) {
6       int i;
7       for (i = 0; i < len; i++)
8           printf(" %.2x", start[i]);
9       printf("\n");
10  }
11
12  void show_int(int x) {
13      show_bytes((byte_pointer) &x, sizeof(int));
14  }
15
16  void show_float(float x) {
17      show_bytes((byte_pointer) &x, sizeof(float));
18  }
19
20  void show_pointer(void *x) {
21      show_bytes((byte_pointer) &x, sizeof(void *));
22  }
```

Figure 4 **Code to print the byte representation of program objects.** This code uses casting to circumvent the type system. Similar functions are easily defined for other data types.

as this one. The natural way to write a byte sequence is to have the lowest-numbered byte on the left and the highest on the right, but this is contrary to the normal way of writing numbers with the most significant digit on the left and the least on the right.

A third case where byte ordering becomes visible is when programs are written that circumvent the normal type system. In the C language, this can be done using a *cast* to allow an object to be referenced according to a different data type from which it was created. Such coding tricks are strongly discouraged for most application programming, but they can be quite useful and even necessary for system-level programming.

Figure 4 shows C code that uses casting to access and print the byte representations of different program objects. We use typedef to define data type byte_pointer as a pointer to an object of type "unsigned char." Such a byte pointer references a sequence of bytes where each byte is considered to be a non-negative integer. The first routine show_bytes is given the address of a sequence of bytes, indicated by a byte pointer, and a byte count. It prints the individual bytes in hexadecimal. The C formatting directive "%.2x" indicates that an integer should be printed in hexadecimal with at least two digits.

New to C?   Naming data types with `typedef`

The `typedef` declaration in C provides a way of giving a name to a data type. This can be a great help in improving code readability, since deeply nested type declarations can be difficult to decipher.

   The syntax for `typedef` is exactly like that of declaring a variable, except that it uses a type name rather than a variable name. Thus, the declaration of `byte_pointer` in Figure 4 has the same form as the declaration of a variable of type "`unsigned char *`."

   For example, the declaration

```
typedef int *int_pointer;
int_pointer ip;
```

defines type "`int_pointer`" to be a pointer to an `int`, and declares a variable `ip` of this type. Alternatively, we could declare this variable directly as

```
int *ip;
```

New to C?   Formatted printing with `printf`

The `printf` function (along with its cousins `fprintf` and `sprintf`) provides a way to print information with considerable control over the formatting details. The first argument is a *format string*, while any remaining arguments are values to be printed. Within the format string, each character sequence starting with '%' indicates how to format the next argument. Typical examples include '%d' to print a decimal integer, '%f' to print a floating-point number, and '%c' to print a character having the character code given by the argument.

New to C?   Pointers and arrays

In function `show_bytes` (Figure 4), we see the close connection between pointers and arrays. We see that this function has an argument `start` of type `byte_pointer` (which has been defined to be a pointer to `unsigned char`), but we see the array reference `start[i]` on line 8. In C, we can dereference a pointer with array notation, and we can reference array elements with pointer notation. In this example, the reference `start[i]` indicates that we want to read the byte that is i positions beyond the location pointed to by `start`.

   Procedures `show_int`, `show_float`, and `show_pointer` demonstrate how to use procedure `show_bytes` to print the byte representations of C program objects of type `int`, `float`, and `void *`, respectively. Observe that they simply pass `show_bytes` a pointer `&x` to their argument `x`, casting the pointer to be of type "`unsigned char *`." This cast indicates to the compiler that the program should consider the pointer to be to a sequence of bytes rather than to an object of the original data type. This pointer will then be to the lowest byte address occupied by the object.

New to C? Pointer creation and dereferencing

In lines 13, 17, and 21 of Figure 4, we see uses of two operations that give C (and therefore C++) its distinctive character. The C "address of" operator & creates a pointer. On all three lines, the expression &x creates a pointer to the location holding the object indicated by variable x. The type of this pointer depends on the type of x, and hence these three pointers are of type int *, float *, and void **, respectively. (Data type void * is a special kind of pointer with no associated type information.)

The cast operator converts from one data type to another. Thus, the cast (byte_pointer) &x indicates that whatever type the pointer &x had before, the program will now reference a pointer to data of type unsigned char. The casts shown here do not change the actual pointer; they simply direct the compiler to refer to the data being pointed to according to the new data type.

These procedures use the C sizeof operator to determine the number of bytes used by the object. In general, the expression sizeof(T) returns the number of bytes required to store an object of type T. Using sizeof rather than a fixed value is one step toward writing code that is portable across different machine types.

We ran the code shown in Figure 5 on several different machines, giving the results shown in Figure 6. The following machines were used:

Linux 32:   Intel IA32 processor running Linux
Windows: Intel IA32 processor running Windows
Sun: Sun Microsystems SPARC processor running Solaris
Linux 64:   Intel x86-64 processor running Linux

Our argument 12,345 has hexadecimal representation 0x00003039. For the int data, we get identical results for all machines, except for the byte ordering. In particular, we can see that the least significant byte value of 0x39 is printed first for Linux 32, Windows, and Linux 64, indicating little-endian machines, and last for Sun, indicating a big-endian machine. Similarly, the bytes of the float data are identical, except for the byte ordering. On the other hand, the pointer values are completely different. The different machine/operating system configurations use

code/data/show-bytes.c

```
1   void test_show_bytes(int val) {
2       int ival = val;
3       float fval = (float) ival;
4       int *pval = &ival;
5       show_int(ival);
6       show_float(fval);
7       show_pointer(pval);
8   }
```

code/data/show-bytes.c

Figure 5 **Byte representation examples.** This code prints the byte representations of sample data objects.

Machine	Value	Type	Bytes (hex)
Linux 32	12,345	int	39 30 00 00
Windows	12,345	int	39 30 00 00
Sun	12,345	int	00 00 30 39
Linux 64	12,345	int	39 30 00 00
Linux 32	12,345.0	float	00 e4 40 46
Windows	12,345.0	float	00 e4 40 46
Sun	12,345.0	float	46 40 e4 00
Linux 64	12,345.0	float	00 e4 40 46
Linux 32	&ival	int *	e4 f9 ff bf
Windows	&ival	int *	b4 cc 22 00
Sun	&ival	int *	ef ff fa 0c
Linux 64	&ival	int *	b8 11 e5 ff ff 7f 00 00

Figure 6 **Byte representations of different data values.** Results for int and float are identical, except for byte ordering. Pointer values are machine dependent.

different conventions for storage allocation. One feature to note is that the Linux 32, Windows, and Sun machines use 4-byte addresses, while the Linux 64 machine uses 8-byte addresses.

Observe that although the floating-point and the integer data both encode the numeric value 12,345, they have very different byte patterns: 0x00003039 for the integer, and 0x4640E400 for floating point. In general, these two formats use different encoding schemes. If we expand these hexadecimal patterns into binary form and shift them appropriately, we find a sequence of 13 matching bits, indicated by a sequence of asterisks, as follows:

```
   0   0   0   0   3   0   3   9
000000000000000000011000000111001
                 *************
             4   6   4   0   E   4   0   0
         01000110010000001110010000000000
```

This is not coincidental. We will return to this example when we study floating-point formats.

Practice Problem 5

Consider the following three calls to show_bytes:

```
int val = 0x87654321;
byte_pointer valp = (byte_pointer) &val;
show_bytes(valp, 1); /* A. */
show_bytes(valp, 2); /* B. */
show_bytes(valp, 3); /* C. */
```

Indicate which of the following values will be printed by each call on a little-endian machine and on a big-endian machine:

A. Little endian: _____   Big endian: _____

B. Little endian: _____   Big endian: _____

C. Little endian: _____   Big endian: _____

Practice Problem 6

Using show_int and show_float, we determine that the integer 3510593 has hexadecimal representation 0x00359141, while the floating-point number 3510593.0 has hexadecimal representation 0x4A564504.

A. Write the binary representations of these two hexadecimal values.

B. Shift these two strings relative to one another to maximize the number of matching bits. How many bits match?

C. What parts of the strings do not match?

1.5 Representing Strings

A string in C is encoded by an array of characters terminated by the null (having value 0) character. Each character is represented by some standard encoding, with the most common being the ASCII character code. Thus, if we run our routine show_bytes with arguments "12345" and 6 (to include the terminating character), we get the result 31 32 33 34 35 00. Observe that the ASCII code for decimal digit x happens to be 0x3x, and that the terminating byte has the hex representation 0x00. This same result would be obtained on any system using ASCII as its character code, independent of the byte ordering and word size conventions. As a consequence, text data is more platform-independent than binary data.

Aside Generating an ASCII table

You can display a table showing the ASCII character code by executing the command man ascii.

Practice Problem 7

What would be printed as a result of the following call to show_bytes?

```
const char *s = "abcdef";
show_bytes((byte_pointer) s, strlen(s));
```

Note that letters 'a' through 'z' have ASCII codes 0x61 through 0x7A.

Aside   The Unicode standard for text encoding

The ASCII character set is suitable for encoding English-language documents, but it does not have much in the way of special characters, such as the French 'ç.' It is wholly unsuited for encoding documents in languages such as Greek, Russian, and Chinese. Over the years, a variety of methods have been developed to encode text for different languages. The Unicode Consortium has devised the most comprehensive and widely accepted standard for encoding text. The current Unicode standard (version 5.0) has a repertoire of nearly 100,000 characters supporting languages ranging from Albanian to Xamtanga (a language spoken by the Xamir people of Ethiopia).

The base encoding, known as the "Universal Character Set" of Unicode, uses a 32-bit representation of characters. This would seem to require every string of text to consist of 4 bytes per character. However, alternative codings are possible where common characters require just 1 or 2 bytes, while less common ones require more. In particular, the UTF-8 representation encodes each character as a sequence of bytes, such that the standard ASCII characters use the same single-byte encodings as they have in ASCII, implying that all ASCII byte sequences have the same meaning in UTF-8 as they do in ASCII.

The Java programming language uses Unicode in its representations of strings. Program libraries are also available for C to support Unicode.

1.6   Representing Code

Consider the following C function:

```
1    int sum(int x, int y) {
2        return x + y;
3    }
```

When compiled on our sample machines, we generate machine code having the following byte representations:

Linux 32:	55 89 e5 8b 45 0c 03 45 08 c9 c3
Windows:	55 89 e5 8b 45 0c 03 45 08 5d c3
Sun:	81 c3 e0 08 90 02 00 09
Linux 64:	55 48 89 e5 89 7d fc 89 75 f8 03 45 fc c9 c3

Here we find that the instruction codings are different. Different machine types use different and incompatible instructions and encodings. Even identical processors running different operating systems have differences in their coding conventions and hence are not binary compatible. Binary code is seldom portable across different combinations of machine and operating system.

A fundamental concept of computer systems is that a program, from the perspective of the machine, is simply a sequence of bytes. The machine has no information about the original source program, except perhaps some auxiliary tables maintained to aid in debugging.

~		&	0	1	|	0	1	^	0	1
0	1	0	0	0	0	0	1	0	0	1
1	0	1	0	1	1	1	1	1	1	0

Figure 7 **Operations of Boolean algebra.** Binary values 1 and 0 encode logic values TRUE and FALSE, while operations ~, &, |, and ^ encode logical operations NOT, AND, OR, and EXCLUSIVE-OR, respectively.

1.7 Introduction to Boolean Algebra

Since binary values are at the core of how computers encode, store, and manipulate information, a rich body of mathematical knowledge has evolved around the study of the values 0 and 1. This started with the work of George Boole (1815–1864) around 1850 and thus is known as *Boolean algebra*. Boole observed that by encoding logic values TRUE and FALSE as binary values 1 and 0, he could formulate an algebra that captures the basic principles of logical reasoning.

The simplest Boolean algebra is defined over the 2-element set $\{0, 1\}$. Figure 7 defines several operations in this algebra. Our symbols for representing these operations are chosen to match those used by the C bit-level operations, as will be discussed later. The Boolean operation ~ corresponds to the logical operation NOT, denoted by the symbol $\neg$. That is, we say that $\neg P$ is true when P is not true, and vice versa. Correspondingly, $\sim p$ equals 1 when p equals 0, and vice versa. Boolean operation & corresponds to the logical operation AND, denoted by the symbol $\wedge$. We say that $P \wedge Q$ holds when both P is true and Q is true. Correspondingly, p & q equals 1 only when $p = 1$ and $q = 1$. Boolean operation | corresponds to the logical operation OR, denoted by the symbol $\vee$. We say that $P \vee Q$ holds when either P is true or Q is true. Correspondingly, p | q equals 1 when either $p = 1$ or $q = 1$. Boolean operation ^ corresponds to the logical operation EXCLUSIVE-OR, denoted by the symbol $\oplus$. We say that $P \oplus Q$ holds when either P is true or Q is true, but not both. Correspondingly, p ^ q equals 1 when either $p = 1$ and $q = 0$, or $p = 0$ and $q = 1$.

Claude Shannon (1916–2001), who later founded the field of information theory, first made the connection between Boolean algebra and digital logic. In his 1937 master's thesis, he showed that Boolean algebra could be applied to the design and analysis of networks of electromechanical relays. Although computer technology has advanced considerably since, Boolean algebra still plays a central role in the design and analysis of digital systems.

We can extend the four Boolean operations to also operate on *bit vectors*, strings of zeros and ones of some fixed length w. We define the operations over bit vectors according their applications to the matching elements of the arguments. Let a and b denote the bit vectors $[a_{w-1}, a_{w-2}, \ldots, a_0]$ and $[b_{w-1}, b_{w-2}, \ldots, b_0]$, respectively. We define a & b to also be a bit vector of length w, where the ith element equals a_i & b_i, for $0 \leq i < w$. The operations |, ^, and ~ are extended to bit vectors in a similar fashion.

As examples, consider the case where $w = 4$, and with arguments $a = [0110]$ and $b = [1100]$. Then the four operations $a \ \& \ b, a \ | \ b, a \ \hat{} \ b$, and $\sim b$ yield

0110	0110	0110	
& 1100	| 1100	^ 1100	~ 1100
0100	1110	1010	0011

Practice Problem 8

Fill in the following table showing the results of evaluating Boolean operations on bit vectors.

Operation	Result
a	[01101001]
b	[01010101]
$\sim a$	_____
$\sim b$	_____
$a \ \& \ b$	_____
$a \ | \ b$	_____
$a \ \hat{} \ b$	_____

Web Aside DATA:BOOL   More on Boolean algebra and Boolean rings

The Boolean operations $|$, $\&$, and $\sim$ operating on bit vectors of length w form a *Boolean algebra*, for any integer $w > 0$. The simplest is the case where $w = 1$, and there are just two elements, but for the more general case there are 2^w bit vectors of length w. Boolean algebra has many of the same properties as arithmetic over integers. For example, just as multiplication distributes over addition, written $a \cdot (b + c) = (a \cdot b) + (a \cdot c)$, Boolean operation $\&$ distributes over $|$, written $a \ \& \ (b \ | \ c) = (a \ \& \ b) \ | \ (a \ \& \ c)$. In addition, however, Boolean operation $|$ distributes over $\&$, and so we can write $a \ | \ (b \ \& \ c) = (a \ | \ b) \ \& \ (a \ | \ c)$, whereas we cannot say that $a + (b \cdot c) = (a + b) \cdot (a + c)$ holds for all integers.

When we consider operations $\hat{}$, $\&$, and $\sim$ operating on bit vectors of length w, we get a different mathematical form, known as a *Boolean ring*. Boolean rings have many properties in common with integer arithmetic. For example, one property of integer arithmetic is that every value x has an *additive inverse* $-x$, such that $x + -x = 0$. A similar property holds for Boolean rings, where $\hat{}$ is the "addition" operation, but in this case each element is its own additive inverse. That is, $a \ \hat{} \ a = 0$ for any value a, where we use 0 here to represent a bit vector of all zeros. We can see this holds for single bits, since $0 \ \hat{} \ 0 = 1 \ \hat{} \ 1 = 0$, and it extends to bit vectors as well. This property holds even when we rearrange terms and combine them in a different order, and so $(a \ \hat{} \ b) \ \hat{} \ a = b$. This property leads to some interesting results and clever tricks, as we will explore in Problem 10.

One useful application of bit vectors is to represent finite sets. We can encode any subset $A \subseteq \{0, 1, \ldots, w - 1\}$ with a bit vector $[a_{w-1}, \ldots, a_1, a_0]$, where $a_i = 1$ if and only if $i \in A$. For example, recalling that we write a_{w-1} on the left and a_0 on the right, bit vector $a \doteq$ [01101001] encodes the set $A = \{0, 3, 5, 6\}$, while bit vector $b \doteq$ [01010101] encodes the set $B = \{0, 2, 4, 6\}$. With this way of encoding sets, Boolean operations | and & correspond to set union and intersection, respectively, and ~ corresponds to set complement. Continuing our earlier example, the operation a & b yields bit vector [01000001], while $A \cap B = \{0, 6\}$.

One sees the encoding of sets by bit vectors in a number of practical applications. For example, there are a number of different *signals* that can interrupt the execution of a program. We can selectively enable or disable different signals by specifying a bit-vector mask, where a 1 in bit position i indicates that signal i is enabled, and a 0 indicates that it is disabled. Thus, the mask represents the set of enabled signals.

Practice Problem 9

Computers generate color pictures on a video screen or liquid crystal display by mixing three different colors of light: red, green, and blue. Imagine a simple scheme, with three different lights, each of which can be turned on or off, projecting onto a glass screen:

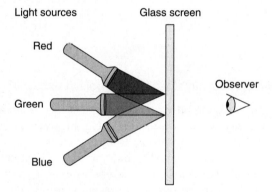

We can then create eight different colors based on the absence (0) or presence (1) of light sources R, G, and B:

R	G	B	Color
0	0	0	Black
0	0	1	Blue
0	1	0	Green
0	1	1	Cyan
1	0	0	Red
1	0	1	Magenta
1	1	0	Yellow
1	1	1	White

Each of these colors can be represented as a bit vector of length 3, and we can apply Boolean operations to them.

A. The complement of a color is formed by turning off the lights that are on and turning on the lights that are off. What would be the complement of each of the eight colors listed above?

B. Describe the effect of applying Boolean operations on the following colors:

Blue	|	Green	=	_____
Yellow	&	Cyan	=	_____
Red	^	Magenta	=	_____

1.8 Bit-Level Operations in C

One useful feature of C is that it supports bit-wise Boolean operations. In fact, the symbols we have used for the Boolean operations are exactly those used by C: | for OR, & for AND, ~ for NOT, and ^ for EXCLUSIVE-OR. These can be applied to any "integral" data type, that is, one declared as type char or int, with or without qualifiers such as short, long, long long, or unsigned. Here are some examples of expression evaluation for data type char:

C expression	Binary expression	Binary result	Hexadecimal result
~0x41	~[0100 0001]	[1011 1110]	0xBE
~0x00	~[0000 0000]	[1111 1111]	0xFF
0x69 & 0x55	[0110 1001] & [0101 0101]	[0100 0001]	0x41
0x69 | 0x55	[0110 1001] | [0101 0101]	[0111 1101]	0x7D

As our examples show, the best way to determine the effect of a bit-level expression is to expand the hexadecimal arguments to their binary representations, perform the operations in binary, and then convert back to hexadecimal.

Practice Problem 10

As an application of the property that $a \text{ ^ } a = 0$ for any bit vector a, consider the following program:

```
1   void inplace_swap(int *x, int *y) {
2       *y = *x ^ *y;   /* Step 1 */
3       *x = *x ^ *y;   /* Step 2 */
4       *y = *x ^ *y;   /* Step 3 */
5   }
```

As the name implies, we claim that the effect of this procedure is to swap the values stored at the locations denoted by pointer variables x and y. Note that unlike the usual technique for swapping two values, we do not need a third location to temporarily store one value while we are moving the other. There is no performance advantage to this way of swapping; it is merely an intellectual amusement.

Starting with values a and b in the locations pointed to by x and y, respectively, fill in the table that follows, giving the values stored at the two locations after each step of the procedure. Use the properties of ^ to show that the desired effect is achieved. Recall that every element is its own additive inverse (that is, $a \char`\^ a = 0$).

Step	*x	*y
Initially	a	b
Step 1		
Step 2		
Step 3		

Practice Problem 11

Armed with the function `inplace_swap` from Problem 10, you decide to write code that will reverse the elements of an array by swapping elements from opposite ends of the array, working toward the middle.

You arrive at the following function:

```
1    void reverse_array(int a[], int cnt) {
2        int first, last;
3        for (first = 0, last = cnt-1;
4             first <= last;
5             first++,last--)
6             inplace_swap(&a[first], &a[last]);
7    }
```

When you apply your function to an array containing elements 1, 2, 3, and 4, you find the array now has, as expected, elements 4, 3, 2, and 1. When you try it on an array with elements 1, 2, 3, 4, and 5, however, you are surprised to see that the array now has elements 5, 4, 0, 2, and 1. In fact, you discover that the code always works correctly on arrays of even length, but it sets the middle element to 0 whenever the array has odd length.

A. For an array of odd length $cnt = 2k + 1$, what are the values of variables `first` and `last` in the final iteration of function `reverse_array`?

B. Why does this call to function `xor_swap` set the array element to 0?

C. What simple modification to the code for `reverse_array` would eliminate this problem?

One common use of bit-level operations is to implement *masking* operations, where a mask is a bit pattern that indicates a selected set of bits within a word. As an example, the mask 0xFF (having ones for the least significant 8 bits) indicates the low-order byte of a word. The bit-level operation x & 0xFF yields a value consisting of the least significant byte of x, but with all other bytes set to 0. For example, with x = 0x89ABCDEF, the expression would yield 0x000000EF. The expression ~0 will yield a mask of all ones, regardless of the word size of the machine. Although the same mask can be written 0xFFFFFFFF for a 32-bit machine, such code is not as portable.

Practice Problem 12

Write C expressions, in terms of variable x, for the following values. Your code should work for any word size $w \geq 8$. For reference, we show the result of evaluating the expressions for x = 0x87654321, with $w = 32$.

A. The least significant byte of x, with all other bits set to 0. [0x00000021].

B. All but the least significant byte of x complemented, with the least significant byte left unchanged. [0x789ABC21].

C. The least significant byte set to all 1s, and all other bytes of x left unchanged. [0x876543FF].

Practice Problem 13

The Digital Equipment VAX computer was a very popular machine from the late 1970s until the late 1980s. Rather than instructions for Boolean operations AND and OR, it had instructions bis (bit set) and bic (bit clear). Both instructions take a data word x and a mask word m. They generate a result z consisting of the bits of x modified according to the bits of m. With bis, the modification involves setting z to 1 at each bit position where m is 1. With bic, the modification involves setting z to 0 at each bit position where m is 1.

To see how these operations relate to the C bit-level operations, assume we have functions bis and bic implementing the bit set and bit clear operations, and that we want to use these to implement functions computing bit-wise operations | and ^, without using any other C operations. Fill in the missing code below. **Hint:** Write C expressions for the operations bis and bic.

```
/* Declarations of functions implementing operations bis and bic */
int bis(int x, int m);
int bic(int x, int m);

/* Compute x|y using only calls to functions bis and bic */
int bool_or(int x, int y) {
  int result = _____ ;
  return result;
}

/* Compute x^y using only calls to functions bis and bic */
int bool_xor(int x, int y) {
  int result = _____ ;
  return result;
}
```

1.9 Logical Operations in C

C also provides a set of *logical* operators ||, &&, and !, which correspond to the OR, AND, and NOT operations of logic. These can easily be confused with the bit-level operations, but their function is quite different. The logical operations treat any nonzero argument as representing TRUE and argument 0 as representing FALSE. They return either 1 or 0, indicating a result of either TRUE or FALSE, respectively. Here are some examples of expression evaluation:

Expression	Result
!0x41	0x00
!0x00	0x01
!!0x41	0x01
0x69 && 0x55	0x01
0x69 || 0x55	0x01

Observe that a bit-wise operation will have behavior matching that of its logical counterpart only in the special case in which the arguments are restricted to 0 or 1.

A second important distinction between the logical operators && and || versus their bit-level counterparts & and | is that the logical operators do not evaluate their second argument if the result of the expression can be determined by evaluating the first argument. Thus, for example, the expression a && 5/a will never cause a division by zero, and the expression p && *p++ will never cause the dereferencing of a null pointer.

Practice Problem 14

Suppose that x and y have byte values 0x66 and 0x39, respectively. Fill in the following table indicating the byte values of the different C expressions:

Expression	Value	Expression	Value
x & y	_____	x && y	_____
x | y	_____	x || y	_____
~x | ~y	_____	!x || !y	_____
x & !y	_____	x && ~y	_____

Practice Problem 15

Using only bit-level and logical operations, write a C expression that is equivalent to x == y. In other words, it will return 1 when x and y are equal, and 0 otherwise.

1.10 Shift Operations in C

C also provides a set of *shift* operations for shifting bit patterns to the left and to the right. For an operand x having bit representation $[x_{n-1}, x_{n-2}, \ldots, x_0]$, the C expression x << k yields a value with bit representation $[x_{n-k-1}, x_{n-k-2},$

$\ldots, x_0, 0, \ldots 0]$. That is, x is shifted k bits to the left, dropping off the k most significant bits and filling the right end with k zeros. The shift amount should be a value between 0 and $n - 1$. Shift operations associate from left to right, so x << j << k is equivalent to (x << j) << k.

There is a corresponding right shift operation x >> k, but it has a slightly subtle behavior. Generally, machines support two forms of right shift: *logical* and *arithmetic*. A logical right shift fills the left end with k zeros, giving a result $[0, \ldots, 0, x_{n-1}, x_{n-2}, \ldots x_k]$. An arithmetic right shift fills the left end with k repetitions of the most significant bit, giving a result $[x_{n-1}, \ldots, x_{n-1}, x_{n-1}, x_{n-2}, \ldots x_k]$. This convention might seem peculiar, but as we will see it is useful for operating on signed integer data.

As examples, the following table shows the effect of applying the different shift operations to some sample 8-bit data:

Operation	Values	
Argument x	[01100011]	[10010101]
x << 4	[0011*0000*]	[0101*0000*]
x >> 4 (logical)	[*00000*110]	[*0000*1001]
x >> 4 (arithmetic)	[*00000*110]	[*1111*1001]

The italicized digits indicate the values that fill the right (left shift) or left (right shift) ends. Observe that all but one entry involves filling with zeros. The exception is the case of shifting [10010101] right arithmetically. Since its most significant bit is 1, this will be used as the fill value.

The C standards do not precisely define which type of right shift should be used. For unsigned data (i.e., integral objects declared with the qualifier unsigned), right shifts must be logical. For signed data (the default), either arithmetic or logical shifts may be used. This unfortunately means that any code assuming one form or the other will potentially encounter portability problems. In practice, however, almost all compiler/machine combinations use arithmetic right shifts for signed data, and many programmers assume this to be the case.

Java, on the other hand, has a precise definition of how right shifts should be performed. The expression x >> k shifts x arithmetically by k positions, while x >>> k shifts it logically.

Aside Shifting by k, for large values of k

For a data type consisting of w bits, what should be the effect of shifting by some value $k \geq w$? For example, what should be the effect of computing the following expressions on a 32-bit machine:

```
int      lval = 0xFEDCBA98  << 32;
int      aval = 0xFEDCBA98  >> 36;
unsigned uval = 0xFEDCBA98u >> 40;
```

The C standards carefully avoid stating what should be done in such a case. On many machines, the shift instructions consider only the lower $\log_2 w$ bits of the shift amount when shifting a w-bit value, and so the shift amount is effectively computed as $k \bmod w$. For example, on a 32-bit machine following this convention, the above three shifts are computed as if they were by amounts 0, 4, and 8, respectively, giving results

```
lval    0xFEDCBA98
aval    0xFFEDCBA9
uval    0x00FEDCBA
```

This behavior is not guaranteed for C programs, however, and so shift amounts should be kept less than the word size.

Java, on the other hand, specifically requires that shift amounts should be computed in the modular fashion we have shown.

Aside   Operator precedence issues with shift operations

It might be tempting to write the expression 1<<2 + 3<<4, intending it to mean (1<<2) + (3<<4). But, in C, the former expression is equivalent to 1 << (2+3) << 4, since addition (and subtraction) have higher precedence than shifts. The left-to-right associativity rule then causes this to be parenthesized as (1 << (2+3)) << 4, giving value 512, rather than the intended 52.

Getting the precedence wrong in C expressions is a common source of program errors, and often these are difficult to spot by inspection. When in doubt, put in parentheses!

Practice Problem 16

Fill in the table below showing the effects of the different shift operations on single-byte quantities. The best way to think about shift operations is to work with binary representations. Convert the initial values to binary, perform the shifts, and then convert back to hexadecimal. Each of the answers should be 8 binary digits or 2 hexadecimal digits.

x		x << 3		(Logical) x >> 2		(Arithmetic) x >> 2	
Hex	Binary	Binary	Hex	Binary	Hex	Binary	Hex
0xC3							
0x75							
0x87							
0x66							

2   Integer Representations

In this section, we describe two different ways bits can be used to encode integers — one that can only represent nonnegative numbers, and one that can represent

C data type	Minimum	Maximum
char	−128	127
unsigned char	0	255
short [int]	−32,768	32,767
unsigned short [int]	0	65,535
int	−2,147,483,648	2,147,483,647
unsigned [int]	0	4,294,967,295
long [int]	−2,147,483,648	2,147,483,647
unsigned long [int]	0	4,294,967,295
long long [int]	−9,223,372,036,854,775,808	9,223,372,036,854,775,807
unsigned long long [int]	0	18,446,744,073,709,551,615

Figure 8 **Typical ranges for C integral data types on a 32-bit machine.** Text in square brackets is optional.

C data type	Minimum	Maximum
char	−128	127
unsigned char	0	255
short [int]	−32,768	32,767
unsigned short [int]	0	65,535
int	−2,147,483,648	2,147,483,647
unsigned [int]	0	4,294,967,295
long [int]	−9,223,372,036,854,775,808	9,223,372,036,854,775,807
unsigned long [int]	0	18,446,744,073,709,551,615
long long [int]	−9,223,372,036,854,775,808	9,223,372,036,854,775,807
unsigned long long [int]	0	18,446,744,073,709,551,615

Figure 9 **Typical ranges for C integral data types on a 64-bit machine.** Text in square brackets is optional.

negative, zero, and positive numbers. We will see later that they are strongly related both in their mathematical properties and their machine-level implementations. We also investigate the effect of expanding or shrinking an encoded integer to fit a representation with a different length.

2.1 Integral Data Types

C supports a variety of *integral* data types—ones that represent finite ranges of integers. These are shown in Figures 8 and 9, along with the ranges of values they can have for "typical" 32- and 64-bit machines. Each type can specify a size with keyword char, short, long, or long long, as well as an indication of whether the represented numbers are all nonnegative (declared as unsigned), or possibly

C data type	Minimum	Maximum
char	−127	127
unsigned char	0	255
short [int]	−32,767	32,767
unsigned short [int]	0	65,535
int	−32,767	32,767
unsigned [int]	0	65,535
long [int]	−2,147,483,647	2,147,483,647
unsigned long [int]	0	4,294,967,295
long long [int]	−9,223,372,036,854,775,807	9,223,372,036,854,775,807
unsigned long long [int]	0	18,446,744,073,709,551,615

Figure 10 **Guaranteed ranges for C integral data types.** Text in square brackets is optional. The C standards require that the data types have at least these ranges of values.

negative (the default). As we saw in Figure 3, the number of bytes allocated for the different sizes vary according to machine's word size and the compiler. Based on the byte allocations, the different sizes allow different ranges of values to be represented. The only machine-dependent range indicated is for size designator long. Most 64-bit machines use an 8-byte representation, giving a much wider range of values than the 4-byte representation used on 32-bit machines.

One important feature to note in Figures 8 and 9 is that the ranges are not symmetric—the range of negative numbers extends one further than the range of positive numbers. We will see why this happens when we consider how negative numbers are represented.

The C standards define minimum ranges of values that each data type must be able to represent. As shown in Figure 10, their ranges are the same or smaller than the typical implementations shown in Figures 8 and 9. In particular, we see that they require only a symmetric range of positive and negative numbers. We also see that data type int could be implemented with 2-byte numbers, although this is mostly a throwback to the days of 16-bit machines. We also see that size long could be implemented with 4-byte numbers, as is often the case. Data type long long was introduced with ISO C99, and it requires at least an 8-byte representation.

New to C? Signed and unsigned numbers in C, C++, and Java

Both C and C++ support signed (the default) and unsigned numbers. Java supports only signed numbers.

2.2 Unsigned Encodings

Assume we have an integer data type of w bits. We write a bit vector as either $\vec{x}$, to denote the entire vector, or as $[x_{w-1}, x_{w-2}, \ldots, x_0]$, to denote the individual bits within the vector. Treating $\vec{x}$ as a number written in binary notation, we obtain the

Figure 11

Unsigned number examples for $w = 4$. When bit i in the binary representation has value 1, it contributes 2^i to the value.

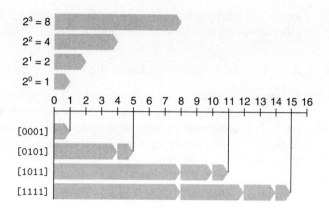

unsigned interpretation of $\vec{x}$. We express this interpretation as a function $B2U_w$ (for "binary to unsigned," length w):

$$B2U_w(\vec{x}) \doteq \sum_{i=0}^{w-1} x_i 2^i \tag{1}$$

In this equation, the notation "$\doteq$" means that the left-hand side is defined to be equal to the right-hand side. The function $B2U_w$ maps strings of zeros and ones of length w to nonnegative integers. As examples, Figure 11 shows the mapping, given by $B2U$, from bit vectors to integers for the following cases:

$$
\begin{aligned}
B2U_4([0001]) &= 0 \cdot 2^3 + 0 \cdot 2^2 + 0 \cdot 2^1 + 1 \cdot 2^0 &= 0 + 0 + 0 + 1 &= 1 \\
B2U_4([0101]) &= 0 \cdot 2^3 + 1 \cdot 2^2 + 0 \cdot 2^1 + 1 \cdot 2^0 &= 0 + 4 + 0 + 1 &= 5 \\
B2U_4([1011]) &= 1 \cdot 2^3 + 0 \cdot 2^2 + 1 \cdot 2^1 + 1 \cdot 2^0 &= 8 + 0 + 2 + 1 &= 11 \\
B2U_4([1111]) &= 1 \cdot 2^3 + 1 \cdot 2^2 + 1 \cdot 2^1 + 1 \cdot 2^0 &= 8 + 4 + 2 + 1 &= 15
\end{aligned}
\tag{2}
$$

In the figure, we represent each bit position i by a rightward-pointing blue bar of length 2^i. The numeric value associated with a bit vector then equals the combined length of the bars for which the corresponding bit values are 1.

Let us consider the range of values that can be represented using w bits. The least value is given by bit vector $[00 \cdots 0]$ having integer value 0, and the greatest value is given by bit vector $[11 \cdots 1]$ having integer value $UMax_w \doteq \sum_{i=0}^{w-1} 2^i = 2^w - 1$. Using the 4-bit case as an example, we have $UMax_4 = B2U_4([1111]) = 2^4 - 1 = 15$. Thus, the function $B2U_w$ can be defined as a mapping $B2U_w: \{0, 1\}^w \rightarrow \{0, \ldots, 2^w - 1\}$.

The unsigned binary representation has the important property that every number between 0 and $2^w - 1$ has a unique encoding as a w-bit value. For example, there is only one representation of decimal value 11 as an unsigned, 4-bit number, namely $[1011]$. This property is captured in mathematical terms by stating that function $B2U_w$ is a *bijection*—it associates a unique value to each bit vector of length w; conversely, each integer between 0 and $2^w - 1$ has a unique binary representation as a bit vector of length w.

2.3 Two's-Complement Encodings

For many applications, we wish to represent negative values as well. The most common computer representation of signed numbers is known as *two's-complement* form. This is defined by interpreting the most significant bit of the word to have negative weight. We express this interpretation as a function $B2T_w$ (for "binary to two's-complement" length w):

$$B2T_w(\vec{x}) \doteq -x_{w-1}2^{w-1} + \sum_{i=0}^{w-2} x_i 2^i \tag{3}$$

The most significant bit x_{w-1} is also called the *sign bit*. Its "weight" is -2^{w-1}, the negation of its weight in an unsigned representation. When the sign bit is set to 1, the represented value is negative, and when set to 0 the value is nonnegative. As examples, Figure 12 shows the mapping, given by $B2T$, from bit vectors to integers for the following cases:

$$
\begin{aligned}
B2T_4([0001]) &= -0 \cdot 2^3 + 0 \cdot 2^2 + 0 \cdot 2^1 + 1 \cdot 2^0 &= 0+0+0+1 &= 1 \\
B2T_4([0101]) &= -0 \cdot 2^3 + 1 \cdot 2^2 + 0 \cdot 2^1 + 1 \cdot 2^0 &= 0+4+0+1 &= 5 \\
B2T_4([1011]) &= -1 \cdot 2^3 + 0 \cdot 2^2 + 1 \cdot 2^1 + 1 \cdot 2^0 &= -8+0+2+1 &= -5 \\
B2T_4([1111]) &= -1 \cdot 2^3 + 1 \cdot 2^2 + 1 \cdot 2^1 + 1 \cdot 2^0 &= -8+4+2+1 &= -1
\end{aligned}
\tag{4}
$$

In the figure, we indicate that the sign bit has negative weight by showing it as a leftward-pointing gray bar. The numeric value associated with a bit vector is then given by the combination of the possible leftward-pointing gray bar and the rightward-pointing blue bars.

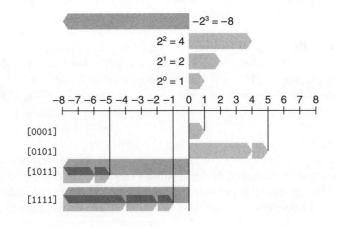

Figure 12

Two's-complement number examples for $w = 4$. Bit 3 serves as a sign bit, and so, when set to 1, it contributes $-2^3 = -8$ to the value. This weighting is shown as a leftward-pointing gray bar.

We see that the bit patterns are identical for Figures 11 and 12 (as well as for Equations 2 and 4), but the values differ when the most significant bit is 1, since in one case it has weight $+8$, and in the other case it has weight -8.

Let us consider the range of values that can be represented as a w-bit two's-complement number. The least representable value is given by bit vector $[10 \cdots 0]$ (set the bit with negative weight, but clear all others), having integer value $TMin_w \doteq -2^{w-1}$. The greatest value is given by bit vector $[01 \cdots 1]$ (clear the bit with negative weight, but set all others), having integer value $TMax_w \doteq \sum_{i=0}^{w-2} 2^i = 2^{w-1} - 1$. Using the 4-bit case as an example, we have $TMin_4 = B2T_4([1000]) = -2^3 = -8$, and $TMax_4 = B2T_4([0111]) = 2^2 + 2^1 + 2^0 = 4 + 2 + 1 = 7$.

We can see that $B2T_w$ is a mapping of bit patterns of length w to numbers between $TMin_w$ and $TMax_w$, written as $B2T_w: \{0, 1\}^w \to \{-2^{w-1}, \ldots, 2^{w-1} - 1\}$. As we saw with the unsigned representation, every number within the representable range has a unique encoding as a w-bit two's-complement number. In mathematical terms, we say that the function $B2T_w$ is a *bijection*—it associates a unique value to each bit vector of length w; conversely, each integer between -2^{w-1} and $2^{w-1} - 1$ has a unique binary representation as a bit vector of length w.

Practice Problem 17

Assuming $w = 4$, we can assign a numeric value to each possible hexadecimal digit, assuming either an unsigned or a two's-complement interpretation. Fill in the following table according to these interpretations by writing out the nonzero powers of two in the summations shown in Equations 1 and 3:

$\vec{x}$			
Hexadecimal	Binary	$B2U_4(\vec{x})$	$B2T_4(\vec{x})$
0xE	[1110]	$2^3 + 2^2 + 2^1 = 14$	$-2^3 + 2^2 + 2^1 = -2$
0x0			
0x5			
0x8			
0xD			
0xF			

Figure 13 shows the bit patterns and numeric values for several important numbers for different word sizes. The first three give the ranges of representable integers in terms of the values of $UMax_w$, $TMin_w$, and $TMax_w$. We will refer to these three special values often in the ensuing discussion. We will drop the subscript w and refer to the values $UMax$, $TMin$, and $TMax$ when w can be inferred from context or is not central to the discussion.

A few points are worth highlighting about these numbers. First, as observed in Figures 8 and 9, the two's-complement range is asymmetric: $|TMin| = |TMax| + 1$, that is, there is no positive counterpart to $TMin$. As we shall see, this leads to some peculiar properties of two's-complement arithmetic and can be

Value	Word size w			
	8	16	32	64
$UMax_w$	0xFF	0xFFFF	0xFFFFFFFF	0xFFFFFFFFFFFFFFFF
	255	65,535	4,294,967,295	18,446,744,073,709,551,615
$TMin_w$	0x80	0x8000	0x80000000	0x8000000000000000
	−128	−32,768	−2,147,483,648	−9,223,372,036,854,775,808
$TMax_w$	0x7F	0x7FFF	0x7FFFFFFF	0x7FFFFFFFFFFFFFFF
	127	32,767	2,147,483,647	9,223,372,036,854,775,807
−1	0xFF	0xFFFF	0xFFFFFFFF	0xFFFFFFFFFFFFFFFF
0	0x00	0x0000	0x00000000	0x0000000000000000

Figure 13 **Important numbers.** Both numeric values and hexadecimal representations are shown.

the source of subtle program bugs. This asymmetry arises, because half the bit patterns (those with the sign bit set to 1) represent negative numbers, while half (those with the sign bit set to 0) represent nonnegative numbers. Since 0 is nonnegative, this means that it can represent one less positive number than negative. Second, the maximum unsigned value is just over twice the maximum two's-complement value: $UMax = 2TMax + 1$. All of the bit patterns that denote negative numbers in two's-complement notation become positive values in an unsigned representation. Figure 13 also shows the representations of constants −1 and 0. Note that −1 has the same bit representation as $UMax$—a string of all ones. Numeric value 0 is represented as a string of all zeros in both representations.

The C standards do not require signed integers to be represented in two's-complement form, but nearly all machines do so. Programmers who are concerned with maximizing portability across all possible machines should not assume any particular range of representable values, beyond the ranges indicated in Figure 10, nor should they assume any particular representation of signed numbers. On the other hand, many programs are written assuming a two's-complement representation of signed numbers, and the "typical" ranges shown in Figures 8 and 9, and these programs are portable across a broad range of machines and compilers. The file <limits.h> in the C library defines a set of constants delimiting the ranges of the different integer data types for the particular machine on which the compiler is running. For example, it defines constants INT_MAX, INT_MIN, and UINT_MAX describing the ranges of signed and unsigned integers. For a two's-complement machine in which data type int has w bits, these constants correspond to the values of $TMax_w$, $TMin_w$, and $UMax_w$.

Aside   Exact-size integer types

For some programs, it is essential that data types be encoded using representations with specific sizes. For example, when writing programs to enable a machine to communicate over the Internet according to a standard protocol, it is important to have data types compatible with those specified by the protocol.

We have seen that some C data types, especially `long`, have different ranges on different machines, and in fact the C standards only specify the minimum ranges for any data type, and not the exact ranges. Although we can choose data types that will be compatible with standard representations on most machines, there is not guarantee of portability.

The ISO C99 standard introduces another class of integer types in the file `stdint.h`. This file defines a set of data types with declarations of the form `int`N`_t` and `uint`N`_t`, specifying N-bit signed and unsigned integers, for different values of N. The exact values of N are implementation dependent, but most compilers allow values of 8, 16, 32, and 64. Thus, we can unambiguously declare an unsigned, 16-bit variable by giving it type `uint16_t`, and a signed variable of 32 bits as `int32_t`.

Along with these data types are a set of macros defining the minimum and maximum values for each value of N. These have names of the form `INT`N`_MIN`, `INT`N`_MAX`, and `UINT`N`_MAX`.

The Java standard is quite specific about integer data type ranges and representations. It requires a two's-complement representation with the exact ranges shown for the 64-bit case (Figure 9). In Java, the single-byte data type is called `byte` instead of `char`, and there is no `long long` data type. These detailed requirements are intended to enable Java programs to behave identically regardless of the machines running them.

Aside Alternative representations of signed numbers

There are two other standard representations for signed numbers:

Ones' Complement: This is the same as two's complement, except that the most significant bit has weight $-(2^{w-1} - 1)$ rather than -2^{w-1}:

$$B2O_w(\vec{x}) \doteq -x_{w-1}(2^{w-1} - 1) + \sum_{i=0}^{w-2} x_i 2^i$$

Sign-Magnitude: The most significant bit is a sign bit that determines whether the remaining bits should be given negative or positive weight:

$$B2S_w(\vec{x}) \doteq (-1)^{x_{w-1}} \cdot \left(\sum_{i=0}^{w-2} x_i 2^i \right)$$

Both of these representations have the curious property that there are two different encodings of the number 0. For both representations, $[00 \cdots 0]$ is interpreted as $+0$. The value -0 can be represented in sign-magnitude form as $[10 \cdots 0]$ and in ones'-complement as $[11 \cdots 1]$. Although machines based on ones'-complement representations were built in the past, almost all modern machines use two's complement. We will see that sign-magnitude encoding is used with floating-point numbers.

Note the different position of apostrophes: *Two's* complement versus *Ones'* complement. The term "two's complement" arises from the fact that for nonnegative x we compute a w-bit representation of $-x$ as $2^w - x$ (a single two). The term "ones' complement" comes from the property that we can compute $-x$ in this notation as $[111 \cdots 1] - x$ (multiple ones).

Weight	12,345		-12,345		53,191	
	Bit	Value	Bit	Value	Bit	Value
1	1	1	1	1	1	1
2	0	0	1	2	1	2
4	0	0	1	4	1	4
8	1	8	0	0	0	0
16	1	16	0	0	0	0
32	1	32	0	0	0	0
64	0	0	1	64	1	64
128	0	0	1	128	1	128
256	0	0	1	256	1	256
512	0	0	1	512	1	512
1,024	0	0	1	1,024	1	1,024
2,048	0	0	1	2,048	1	2,048
4,096	1	4,096	0	0	0	0
8,192	1	8,192	0	0	0	0
16,384	0	0	1	16,384	1	16,384
±32,768	0	0	1	-32,768	1	32,768
Total		12,345		-12,345		53,191

Figure 14 **Two's-complement representations of 12,345 and -12,345, and unsigned representation of 53,191.** Note that the latter two have identical bit representations.

As an example, consider the following code:

```
1    short x = 12345;
2    short mx = -x;
3
4    show_bytes((byte_pointer) &x, sizeof(short));
5    show_bytes((byte_pointer) &mx, sizeof(short));
```

When run on a big-endian machine, this code prints 30 39 and cf c7, indicating that x has hexadecimal representation 0x3039, while mx has hexadecimal representation 0xCFC7. Expanding these into binary, we get bit patterns [0011000000111001] for x and [1100111111000111] for mx. As Figure 14 shows, Equation 3 yields values 12,345 and -12,345 for these two bit patterns.

Practice Problem 18

A *disassembler* is a program that converts an executable program file back to a more readable ASCII form. These files contain many hexadecimal numbers, typically representing values in two's-complement form. Being able to recognize these numbers and understand their significance (for example, whether they are negative or positive) is an important skill.

For the lines labeled A–J (on the right) in the following listing, convert the hexadecimal values (in 32-bit two's-complement form) shown to the right of the instruction names (sub, mov, and add) into their decimal equivalents:

```
8048337:    81 ec b8 01 00 00      sub    $0x1b8,%esp             A.
804833d:    8b 55 08               mov    0x8(%ebp),%edx
8048340:    83 c2 14               add    $0x14,%edx              B.
8048343:    8b 85 58 fe ff ff      mov    0xfffffe58(%ebp),%eax   C.
8048349:    03 02                  add    (%edx),%eax
804834b:    89 85 74 fe ff ff      mov    %eax,0xfffffe74(%ebp)   D.
8048351:    8b 55 08               mov    0x8(%ebp),%edx
8048354:    83 c2 44               add    $0x44,%edx              E.
8048357:    8b 85 c8 fe ff ff      mov    0xfffffec8(%ebp),%eax   F.
804835d:    89 02                  mov    %eax,(%edx)
804835f:    8b 45 10               mov    0x10(%ebp),%eax         G.
8048362:    03 45 0c               add    0xc(%ebp),%eax          H.
8048365:    89 85 ec fe ff ff      mov    %eax,0xfffffeec(%ebp)   I.
804836b:    8b 45 08               mov    0x8(%ebp),%eax
804836e:    83 c0 20               add    $0x20,%eax              J.
8048371:    8b 00                  mov    (%eax),%eax
```

2.4 Conversions Between Signed and Unsigned

C allows casting between different numeric data types. For example, suppose variable x is declared as int and u as unsigned. The expression (unsigned) x converts the value of x to an unsigned value, and (int) u converts the value of u to a signed integer. What should be the effect of casting signed value to unsigned, or vice versa? From a mathematical perspective, one can imagine several different conventions. Clearly, we want to preserve any value that can be represented in both forms. On the other hand, converting a negative value to unsigned might yield zero. Converting an unsigned value that is too large to be represented in two's-complement form might yield *TMax*. For most implementations of C, however, the answer to this question is based on a bit-level perspective, rather than on a numeric one.

For example, consider the following code:

```
1    short    int    v   = -12345;
2    unsigned short uv = (unsigned short) v;
3    printf("v = %d, uv = %u\n", v, uv);
```

When run on a two's-complement machine, it generates the following output:

```
v = -12345, uv = 53191
```

What we see here is that the effect of casting is to keep the bit values identical but change how these bits are interpreted. We saw in Figure 14 that the 16-bit two's-complement representation of $-12{,}345$ is identical to the 16-bit unsigned representation of $53{,}191$. Casting from short int to unsigned short changed the numeric value, but not the bit representation.

Similarly, consider the following code:

```
1       unsigned u = 4294967295u;    /* UMax_32 */
2       int     tu = (int) u;
3       printf("u = %u, tu = %d\n", u, tu);
```

When run on a two's-complement machine, it generates the following output:

```
u = 4294967295, tu = -1
```

We can see from Figure 13 that, for a 32-bit word size, the bit patterns representing $4{,}294{,}967{,}295$ ($UMax_{32}$) in unsigned form and -1 in two's-complement form are identical. In casting from unsigned int to int, the underlying bit representation stays the same.

This is a general rule for how most C implementations handle conversions between signed and unsigned numbers with the same word size—the numeric values might change, but the bit patterns do not. Let us capture this principle in a more mathematical form. Since both $B2U_w$ and $B2T_w$ are bijections, they have well-defined inverses. Define $U2B_w$ to be $B2U_w^{-1}$, and $T2B_w$ to be $B2T_w^{-1}$. These functions give the unsigned or two's-complement bit patterns for a numeric value. That is, given an integer x in the range $0 \leq x < 2^w$, the function $U2B_w(x)$ gives the unique w-bit unsigned representation of x. Similarly, when x is in the range $-2^{w-1} \leq x < 2^{w-1}$, the function $T2B_w(x)$ gives the unique w-bit two's-complement representation of x. Observe that for values in the range $0 \leq x < 2^{w-1}$, both of these functions will yield the same bit representation—the most significant bit will be 0, and hence it does not matter whether this bit has positive or negative weight.

Now define the function $U2T_w$ as $U2T_w(x) \doteq B2T_w(U2B_w(x))$. This function takes a number between 0 and $2^w - 1$ and yields a number between -2^{w-1} and $2^{w-1} - 1$, where the two numbers have identical bit representations, except that the argument is unsigned, while the result has a two's-complement representation. Similarly, for x between -2^{w-1} and $2^{w-1} - 1$, the function $T2U_w$, defined as $T2U_w(x) \doteq B2U_w(T2B_w(x))$, yields the number having the same unsigned representation as the two's-complement representation of x.

Pursuing our earlier examples, we see from Figure 14 that $T2U_{16}(-12{,}345) = 53{,}191$, and $U2T_{16}(53{,}191) = -12{,}345$. That is, the 16-bit pattern written in hexadecimal as 0xCFC7 is both the two's-complement representation of $-12{,}345$ and the unsigned representation of $53{,}191$. Similarly, from Figure 13, we see that $T2U_{32}(-1) = 4{,}294{,}967{,}295$, and $U2T_{32}(4{,}294{,}967{,}295) = -1$. That is, $UMax$ has the same bit representation in unsigned form as does -1 in two's-complement form.

We see, then, that function $U2T$ describes the conversion of an unsigned number to its 2-complement counterpart, while $T2U$ converts in the opposite direction. These describe the effect of casting between these data types in most C implementations.

Practice Problem 19

Using the table you filled in when solving Problem 17, fill in the following table describing the function $T2U_4$:

x	$T2U_4(x)$
-8	
-3	
-2	
-1	
0	
5	

To get a better understanding of the relation between a signed number x and its unsigned counterpart $T2U_w(x)$, we can use the fact that they have identical bit representations to derive a numerical relationship. Comparing Equations 1 and 3, we can see that for bit pattern $\vec{x}$, if we compute the difference $B2U_w(\vec{x}) - B2T_w(\vec{x})$, the weighted sums for bits from 0 to $w - 2$ will cancel each other, leaving a value: $B2U_w(\vec{x}) - B2T_w(\vec{x}) = x_{w-1}(2^{w-1} - -2^{w-1}) = x_{w-1}2^w$. This gives a relationship $B2U_w(\vec{x}) = x_{w-1}2^w + B2T_w(\vec{x})$. If we let $\vec{x} = T2B_w(x)$, we then have

$$B2U_w(T2B_w(x)) = T2U_w(x) = x_{w-1}2^w + x \tag{5}$$

This relationship is useful for proving relationships between unsigned and two's-complement arithmetic. In the two's-complement representation of x, bit x_{w-1} determines whether or not x is negative, giving

$$T2U_w(x) = \begin{cases} x + 2^w, & x < 0 \\ x, & x \geq 0 \end{cases} \tag{6}$$

As examples, Figure 15 compares how functions $B2U$ and $B2T$ assign values to bit patterns for $w = 4$. For the two's-complement case, the most significant bit serves as the sign bit, which we diagram as a gray, leftward-pointing bar. For the unsigned case, this bit has positive weight, which we show as a black, rightward-pointing bar. In going from two's complement to unsigned, the most significant bit changes its weight from -8 to $+8$. As a consequence, the values that are negative in a two's-complement representation increase by $2^4 = 16$ with an unsigned representation. Thus, -5 becomes $+11$, and -1 becomes $+15$.

Figure 16 illustrates the general behavior of function $T2U$. As it shows, when mapping a signed number to its unsigned counterpart, negative numbers are converted to large positive numbers, while nonnegative numbers remain unchanged.

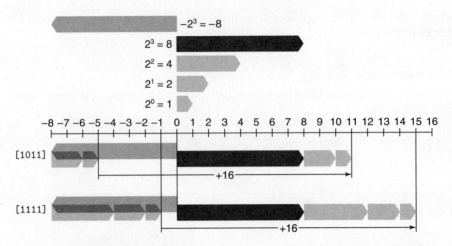

Figure 15 **Comparing unsigned and two's-complement representations for** $w = 4$. The weight of the most significant bit is -8 for two's complement, and $+8$ for unsigned, yielding a net difference of 16.

Figure 16

Conversion from two's complement to unsigned. Function *T2U* converts negative numbers to large positive numbers.

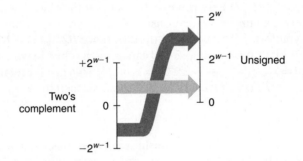

Practice Problem 20

Explain how Equation 6 applies to the entries in the table you generated when solving Problem 19.

Going in the other direction, we wish to derive the relationship between an unsigned number u and its signed counterpart $U2T_w(u)$, both having bit representations $\vec{u} = U2B_w(u)$. We have

$$B2T_w(U2B_w(u)) = U2T_w(u) = -u_{w-1}2^w + u \tag{7}$$

In the unsigned representation of u, bit u_{w-1} determines whether or not u is greater than or equal to 2^{w-1}, giving

$$U2T_w(u) = \begin{cases} u, & u < 2^{w-1} \\ u - 2^w, & u \geq 2^{w-1} \end{cases} \tag{8}$$

Figure 17

Conversion from unsigned to two's complement. Function *U2T* converts numbers greater than $2^{w-1} - 1$ to negative values.

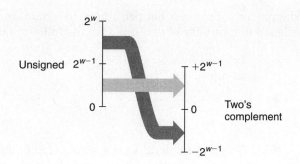

This behavior is illustrated in Figure 17. For small ($< 2^{w-1}$) numbers, the conversion from unsigned to signed preserves the numeric value. Large ($\geq 2^{w-1}$) numbers are converted to negative values.

To summarize, we considered the effects of converting in both directions between unsigned and two's-complement representations. For values x in the range $0 \leq x < 2^{w-1}$, we have $T2U_w(x) = x$ and $U2T_w(x) = x$. That is, numbers in this range have identical unsigned and two's-complement representations. For values outside of this range, the conversions either add or subtract 2^w. For example, we have $T2U_w(-1) = -1 + 2^w = UMax_w$—the negative number closest to 0 maps to the largest unsigned number. At the other extreme, one can see that $T2U_w(TMin_w) = -2^{w-1} + 2^w = 2^{w-1} = TMax_w + 1$—the most negative number maps to an unsigned number just outside the range of positive, two's-complement numbers. Using the example of Figure 14, we can see that $T2U_{16}(-12,345) = 65,536 + -12,345 = 53,191$.

2.5 Signed vs. Unsigned in C

As indicated in Figures 8 and 9, C supports both signed and unsigned arithmetic for all of its integer data types. Although the C standard does not specify a particular representation of signed numbers, almost all machines use two's complement. Generally, most numbers are signed by default. For example, when declaring a constant such as 12345 or 0x1A2B, the value is considered signed. Adding character 'U' or 'u' as a suffix creates an unsigned constant, e.g., 12345U or 0x1A2Bu.

C allows conversion between unsigned and signed. The rule is that the underlying bit representation is not changed. Thus, on a two's-complement machine, the effect is to apply the function $U2T_w$ when converting from unsigned to signed, and $T2U_w$ when converting from signed to unsigned, where w is the number of bits for the data type.

Conversions can happen due to explicit casting, such as in the following code:

```
1    int tx, ty;
2    unsigned ux, uy;
3
4    tx = (int) ux;
5    uy = (unsigned) ty;
```

Alternatively, they can happen implicitly when an expression of one type is assigned to a variable of another, as in the following code:

```
1       int tx, ty;
2       unsigned ux, uy;
3
4       tx = ux; /* Cast to signed */
5       uy = ty; /* Cast to unsigned */
```

When printing numeric values with printf, the directives %d, %u, and %x are used to print a number as a signed decimal, an unsigned decimal, and in hexadecimal format, respectively. Note that printf does not make use of any type information, and so it is possible to print a value of type int with directive %u and a value of type unsigned with directive %d. For example, consider the following code:

```
1       int x = -1;
2       unsigned u = 2147483648; /* 2 to the 31st */
3
4       printf("x = %u = %d\n", x, x);
5       printf("u = %u = %d\n", u, u);
```

When run on a 32-bit machine, it prints the following:

```
x = 4294967295 = -1
u = 2147483648 = -2147483648
```

In both cases, printf prints the word first as if it represented an unsigned number, and second as if it represented a signed number. We can see the conversion routines in action: $T2U_{32}(-1) = UMax_{32} = 2^{32} - 1$ and $U2T_{32}(2^{31}) = 2^{31} - 2^{32} = -2^{31} = TMin_{32}$.

Some of the peculiar behavior arises due to C's handling of expressions containing combinations of signed and unsigned quantities. When an operation is performed where one operand is signed and the other is unsigned, C implicitly casts the signed argument to unsigned and performs the operations assuming the numbers are nonnegative. As we will see, this convention makes little difference for standard arithmetic operations, but it leads to nonintuitive results for relational operators such as < and >. Figure 18 shows some sample relational expressions and their resulting evaluations, assuming a 32-bit machine using two's-complement representation. Consider the comparison -1 < 0U. Since the second operand is unsigned, the first one is implicitly cast to unsigned, and hence the expression is equivalent to the comparison 4294967295U < 0U (recall that $T2U_w(-1) = UMax_w$), which of course is false. The other cases can be understood by similar analyses.

Practice Problem 21

Assuming the expressions are evaluated on a 32-bit machine that uses two's-complement arithmetic, fill in the following table describing the effect of casting and relational operations, in the style of Figure 18:

Expression	Type	Evaluation
-2147483647-1 == 2147483648U		
-2147483647-1 < 2147483647		
-2147483647-1U < 2147483647		
-2147483647-1 < -2147483647		
-2147483647-1U < -2147483647		

Web Aside DATA:TMIN   Writing *TMin* in C

In Figure 18 and in Problem 21, we carefully wrote the value of $TMin_{32}$ as -2147483647-1. Why not simply write it as either -2147483648 or 0x80000000? Looking at the C header file `limits.h`, we see that they use a similar method as we have to write $TMin_{32}$ and $TMax_{32}$:

```
/* Minimum and maximum values a 'signed int' can hold. */
#define INT_MAX    2147483647
#define INT_MIN    (-INT_MAX - 1)
```

Unfortunately, a curious interaction between the asymmetry of the two's-complement representation and the conversion rules of C force us to write $TMin_{32}$ in this unusual way. Although understanding this issue requires us to delve into one of the murkier corners of the C language standards, it will help us appreciate some of the subtleties of integer data types and representations.

2.6   Expanding the Bit Representation of a Number

One common operation is to convert between integers having different word sizes while retaining the same numeric value. Of course, this may not be possible when the destination data type is too small to represent the desired value. Converting from a smaller to a larger data type, however, should always be possible. To convert

Expression	Type	Evaluation
0 == 0U	unsigned	1
-1 < 0	signed	1
-1 < 0U	unsigned	0 *
2147483647 > -2147483647-1	signed	1
2147483647U > -2147483647-1	unsigned	0 *
2147483647 > (int) 2147483648U	signed	1 *
-1 > -2	signed	1
(unsigned) -1 > -2	unsigned	1

Figure 18   **Effects of C promotion rules.** Nonintuitive cases marked by '*'. When either operand of a comparison is unsigned, the other operand is implicitly cast to unsigned. See Web Aside DATA:TMIN for why we write $TMin_{32}$ as -2147483647-1.

an unsigned number to a larger data type, we can simply add leading zeros to the representation; this operation is known as *zero extension*. For converting a two's-complement number to a larger data type, the rule is to perform a *sign extension*, adding copies of the most significant bit to the representation. Thus, if our original value has bit representation $[x_{w-1}, x_{w-2}, \ldots, x_0]$, the expanded representation is $[x_{w-1}, \ldots, x_{w-1}, x_{w-1}, x_{w-2}, \ldots, x_0]$. (We show the sign bit x_{w-1} in blue to highlight its role in sign extension.)

As an example, consider the following code:

```
1    short sx = -12345;          /* -12345 */
2    unsigned short usx = sx;     /*  53191 */
3    int   x = sx;                /* -12345 */
4    unsigned  ux = usx;          /*  53191 */
5
6    printf("sx  = %d:\t", sx);
7    show_bytes((byte_pointer) &sx, sizeof(short));
8    printf("usx = %u:\t", usx);
9    show_bytes((byte_pointer) &usx, sizeof(unsigned short));
10   printf("x   = %d:\t", x);
11   show_bytes((byte_pointer) &x, sizeof(int));
12   printf("ux  = %u:\t", ux);
13   show_bytes((byte_pointer) &ux, sizeof(unsigned));
```

When run on a 32-bit big-endian machine using a two's-complement representation, this code prints the output

```
sx  = -12345:  cf c7
usx = 53191:   cf c7
x   = -12345:  ff ff cf c7
ux  = 53191:   00 00 cf c7
```

We see that although the two's-complement representation of −12,345 and the unsigned representation of 53,191 are identical for a 16-bit word size, they differ for a 32-bit word size. In particular, −12,345 has hexadecimal representation 0xFFFFCFC7, while 53,191 has hexadecimal representation 0x0000CFC7. The former has been sign extended—16 copies of the most significant bit 1, having hexadecimal representation 0xFFFF, have been added as leading bits. The latter has been extended with 16 leading zeros, having hexadecimal representation 0x0000.

As an illustration, Figure 19 shows the result of applying expanding from word size $w = 3$ to $w = 4$ by sign extension. Bit vector [101] represents the value $-4 + 1 = -3$. Applying sign extension gives bit vector [1101] representing the value $-8 + 4 + 1 = -3$. We can see that, for $w = 4$, the combined value of the two most significant bits is $-8 + 4 = -4$, matching the value of the sign bit for $w = 3$. Similarly, bit vectors [111] and [1111] both represent the value −1.

Can we justify that sign extension works? What we want to prove is that

$$B2T_{w+k}([\underbrace{x_{w-1}, \ldots, x_{w-1}}_{k \text{ times}}, x_{w-1}, x_{w-2}, \ldots, x_0]) = B2T_w([x_{w-1}, x_{w-2}, \ldots, x_0])$$

Figure 19

Examples of sign extension from $w = 3$ to $w = 4$. For $w = 4$, the combined weight of the upper 2 bits is $-8 + 4 = -4$, matching that of the sign bit for $w = 3$.

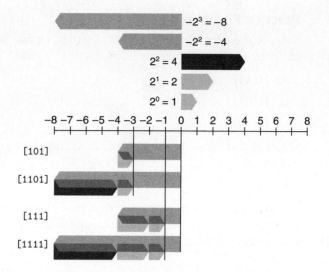

where, in the expression on the left-hand side, we have made k additional copies of bit x_{w-1}. The proof follows by induction on k. That is, if we can prove that sign extending by 1 bit preserves the numeric value, then this property will hold when sign extending by an arbitrary number of bits. Thus, the task reduces to proving that

$$B2T_{w+1}([x_{w-1}, x_{w-1}, x_{w-2}, \ldots, x_0]) = B2T_w([x_{w-1}, x_{w-2}, \ldots, x_0])$$

Expanding the left-hand expression with Equation 3 gives the following:

$$B2T_{w+1}([x_{w-1}, x_{w-1}, x_{w-2}, \ldots, x_0]) = -x_{w-1}2^w + \sum_{i=0}^{w-1} x_i 2^i$$

$$= -x_{w-1}2^w + x_{w-1}2^{w-1} + \sum_{i=0}^{w-2} x_i 2^i$$

$$= -x_{w-1}\left(2^w - 2^{w-1}\right) + \sum_{i=0}^{w-2} x_i 2^i$$

$$= -x_{w-1}2^{w-1} + \sum_{i=0}^{w-2} x_i 2^i$$

$$= B2T_w([x_{w-1}, x_{w-2}, \ldots, x_0])$$

The key property we exploit is that $2^w - 2^{w-1} = 2^{w-1}$. Thus, the combined effect of adding a bit of weight -2^w and of converting the bit having weight -2^{w-1} to be one with weight 2^{w-1} is to preserve the original numeric value.

Practice Problem 22

Show that each of the following bit vectors is a two's-complement representation of -5 by applying Equation 3:

A. [1011]

B. [11011]

C. [111011]

Observe that the second and third bit vectors can be derived from the first by sign extension.

One point worth making is that the relative order of conversion from one data size to another and between unsigned and signed can affect the behavior of a program. Consider the following code:

```
1    short sx = -12345;        /* -12345  */
2    unsigned  uy = sx;        /* Mystery! */
3
4    printf("uy  = %u:\t", uy);
5    show_bytes((byte_pointer) &uy, sizeof(unsigned));
```

When run on a big-endian machine, this code causes the following output to be printed:

```
uy = 4294954951:  ff ff cf c7
```

This shows that when converting from short to unsigned, we first change the size and then from signed to unsigned. That is, (unsigned) sx is equivalent to (unsigned) (int) sx, evaluating to 4,294,954,951, not (unsigned) (unsigned short) sx, which evaluates to 53,191. Indeed this convention is required by the C standards.

Practice Problem 23

Consider the following C functions:

```
int fun1(unsigned word) {
    return (int) ((word << 24) >> 24);
}

int fun2(unsigned word) {
    return ((int) word << 24) >> 24;
}
```

Assume these are executed on a machine with a 32-bit word size that uses two's-complement arithmetic. Assume also that right shifts of signed values are performed arithmetically, while right shifts of unsigned values are performed logically.

A. Fill in the following table showing the effect of these functions for several
   example arguments. You will find it more convenient to work with a hexadecimal representation. Just remember that hex digits 8 through F have their
   most significant bits equal to 1.

w	fun1(w)	fun2(w)
0x00000076		
0x87654321		
0x000000C9		
0xEDCBA987		

B. Describe in words the useful computation each of these functions performs.

2.7 Truncating Numbers

Suppose that, rather than extending a value with extra bits, we reduce the number
of bits representing a number. This occurs, for example, in the code:

```
1        int    x = 53191;
2        short sx = (short) x;   /* -12345 */
3        int    y = sx;          /* -12345 */
```

On a typical 32-bit machine, when we cast x to be short, we truncate the 32-
bit int to be a 16-bit short int. As we saw before, this 16-bit pattern is the
two's-complement representation of $-12,345$. When we cast this back to int,
sign extension will set the high-order 16 bits to ones, yielding the 32-bit two's-
complement representation of $-12,345$.

When truncating a w-bit number $\vec{x} = [x_{w-1}, x_{w-2}, \ldots, x_0]$ to a k-bit number,
we drop the high-order $w - k$ bits, giving a bit vector $\vec{x}' = [x_{k-1}, x_{k-2}, \ldots, x_0]$.
Truncating a number can alter its value—a form of overflow. We now investigate
what numeric value will result. For an unsigned number x, the result of truncating
it to k bits is equivalent to computing $x \bmod 2^k$. This can be seen by applying the
modulus operation to Equation 1:

$$B2U_w([x_{w-1}, x_{w-2}, \ldots, x_0]) \bmod 2^k = \left[\sum_{i=0}^{w-1} x_i 2^i \right] \bmod 2^k$$

$$= \left[\sum_{i=0}^{k-1} x_i 2^i \right] \bmod 2^k$$

$$= \sum_{i=0}^{k-1} x_i 2^i$$

$$= B2U_k([x_{k-1}, x_{k-2}, \ldots, x_0])$$

In this derivation, we make use of the property that $2^i \bmod 2^k = 0$ for any $i \geq k$, and that $\sum_{i=0}^{k-1} x_i 2^i \leq \sum_{i=0}^{k-1} 2^i = 2^k - 1 < 2^k$.

For a two's-complement number x, a similar argument shows that $B2T_w([x_{w-1}, x_{w-2}, \ldots, x_0]) \bmod 2^k = B2U_k([x_{k-1}, x_{k-2}, \ldots, x_0])$. That is, $x \bmod 2^k$ can be represented by an unsigned number having bit-level representation $[x_{k-1}, x_{k-2}, \ldots, x_0]$. In general, however, we treat the truncated number as being signed. This will have numeric value $U2T_k(x \bmod 2^k)$.

Summarizing, the effect of truncation for unsigned numbers is

$$B2U_k([x_{k-1}, x_{k-2}, \ldots, x_0]) = B2U_w([x_{w-1}, x_{w-2}, \ldots, x_0]) \bmod 2^k, \qquad (9)$$

while the effect for two's-complement numbers is

$$B2T_k([x_{k-1}, x_{k-2}, \ldots, x_0]) = U2T_k(B2U_w([x_{w-1}, x_{w-2}, \ldots, x_0]) \bmod 2^k) \qquad (10)$$

Practice Problem 24

Suppose we truncate a 4-bit value (represented by hex digits 0 through F) to a 3-bit value (represented as hex digits 0 through 7). Fill in the table below showing the effect of this truncation for some cases, in terms of the unsigned and two's-complement interpretations of those bit patterns.

Hex		Unsigned		Two's complement	
Original	Truncated	Original	Truncated	Original	Truncated
0	0	0		0	
2	2	2		2	
9	1	9		−7	
B	3	11		−5	
F	7	15		−1	

Explain how Equations 9 and 10 apply to these cases.

2.8 Advice on Signed vs. Unsigned

As we have seen, the implicit casting of signed to unsigned leads to some non-intuitive behavior. Nonintuitive features often lead to program bugs, and ones involving the nuances of implicit casting can be especially difficult to see. Since the casting takes place without any clear indication in the code, programmers often overlook its effects.

The following two practice problems illustrate some of the subtle errors that can arise due to implicit casting and the unsigned data type.

Practice Problem 25

Consider the following code that attempts to sum the elements of an array a, where the number of elements is given by parameter length:

```
1    /* WARNING: This is buggy code */
2    float sum_elements(float a[], unsigned length) {
3        int i;
4        float result = 0;
5
6        for (i = 0; i <= length-1; i++)
7            result += a[i];
8        return result;
9    }
```

When run with argument length equal to 0, this code should return 0.0. Instead it encounters a memory error. Explain why this happens. Show how this code can be corrected.

Practice Problem 26

You are given the assignment of writing a function that determines whether one string is longer than another. You decide to make use of the string library function strlen having the following declaration:

```
/* Prototype for library function strlen */
size_t strlen(const char *s);
```

Here is your first attempt at the function:

```
/* Determine whether string s is longer than string t */
/* WARNING: This function is buggy */
int strlonger(char *s, char *t) {
    return strlen(s) - strlen(t) > 0;
}
```

When you test this on some sample data, things do not seem to work quite right. You investigate further and determine that data type size_t is defined (via typedef) in header file stdio.h to be unsigned int.

A. For what cases will this function produce an incorrect result?

B. Explain how this incorrect result comes about.

C. Show how to fix the code so that it will work reliably.

Aside Security vulnerability in `getpeername`

In 2002, programmers involved in the FreeBSD open source operating systems project realized that their implementation of the `getpeername` library function had a security vulnerability. A simplified version of their code went something like this:

```
1   /*
2    * Illustration of code vulnerability similar to that found in
3    * FreeBSD's implementation of getpeername()
4    */
5
6   /* Declaration of library function memcpy */
7   void *memcpy(void *dest, void *src, size_t n);
8
9   /* Kernel memory region holding user-accessible data */
10  #define KSIZE 1024
11  char kbuf[KSIZE];
12
13  /* Copy at most maxlen bytes from kernel region to user buffer */
14  int copy_from_kernel(void *user_dest, int maxlen) {
15      /* Byte count len is minimum of buffer size and maxlen */
16      int len = KSIZE < maxlen ? KSIZE : maxlen;
17      memcpy(user_dest, kbuf, len);
18      return len;
19  }
```

In this code, we show the prototype for library function `memcpy` on line 7, which is designed to copy a specified number of bytes n from one region of memory to another.

The function `copy_from_kernel`, starting at line 14, is designed to copy some of the data maintained by the operating system kernel to a designated region of memory accessible to the user. Most of the data structures maintained by the kernel should not be readable by a user, since they may contain sensitive information about other users and about other jobs running on the system, but the region shown as kbuf was intended to be one that the user could read. The parameter maxlen is intended to be the length of the buffer allocated by the user and indicated by argument user_dest. The computation at line 16 then makes sure that no more bytes are copied than are available in either the source or the destination buffer.

Suppose, however, that some malicious programmer writes code that calls `copy_from_kernel` with a negative value of maxlen. Then the minimum computation on line 16 will compute this value for len, which will then be passed as the parameter n to memcpy. Note, however, that parameter n is declared as having data type size_t. This data type is declared (via typedef) in the library file stdio.h. Typically it is defined to be unsigned int on 32-bit machines. Since argument n is unsigned, memcpy will treat it as a very large, positive number and attempt to copy that many bytes from the kernel region to the user's buffer. Copying that many bytes (at least 2^{31}) will not actually work, because the program will encounter invalid addresses in the process, but the program could read regions of the kernel memory for which it is not authorized.

We can see that this problem arises due to the mismatch between data types: in one place the length parameter is signed; in another place it is unsigned. Such mismatches can be a source of bugs and, as this example shows, can even lead to security vulnerabilities. Fortunately, there were no reported cases where a programmer had exploited the vulnerability in FreeBSD. They issued a security advisory, "FreeBSD-SA-02:38.signed-error," advising system administrators on how to apply a patch that would remove the vulnerability. The bug can be fixed by declaring parameter `maxlen` to `copy_from_kernel` to be of type `size_t`, to be consistent with parameter n of `memcpy`. We should also declare local variable `len` and the return value to be of type `size_t`.

We have seen multiple ways in which the subtle features of unsigned arithmetic, and especially the implicit conversion of signed to unsigned, can lead to errors or vulnerabilities. One way to avoid such bugs is to never use unsigned numbers. In fact, few languages other than C support unsigned integers. Apparently these other language designers viewed them as more trouble than they are worth. For example, Java supports only signed integers, and it requires that they be implemented with two's-complement arithmetic. The normal right shift operator >> is guaranteed to perform an arithmetic shift. The special operator >>> is defined to perform a logical right shift.

Unsigned values are very useful when we want to think of words as just collections of bits with no numeric interpretation. This occurs, for example, when packing a word with *flags* describing various Boolean conditions. Addresses are naturally unsigned, so systems programmers find unsigned types to be helpful. Unsigned values are also useful when implementing mathematical packages for modular arithmetic and for multiprecision arithmetic, in which numbers are represented by arrays of words.

3   Integer Arithmetic

Many beginning programmers are surprised to find that adding two positive numbers can yield a negative result, and that the comparison x < y can yield a different result than the comparison x-y < 0. These properties are artifacts of the finite nature of computer arithmetic. Understanding the nuances of computer arithmetic can help programmers write more reliable code.

3.1   Unsigned Addition

Consider two nonnegative integers x and y, such that $0 \leq x, y \leq 2^w - 1$. Each of these numbers can be represented by w-bit unsigned numbers. If we compute their sum, however, we have a possible range $0 \leq x + y \leq 2^{w+1} - 2$. Representing this sum could require $w + 1$ bits. For example, Figure 20 shows a plot of the function $x + y$ when x and y have 4-bit representations. The arguments (shown on the horizontal axes) range from 0 to 15, but the sum ranges from 0 to 30. The shape of the function is a sloping plane (the function is linear in both dimensions). If we were

Integer addition

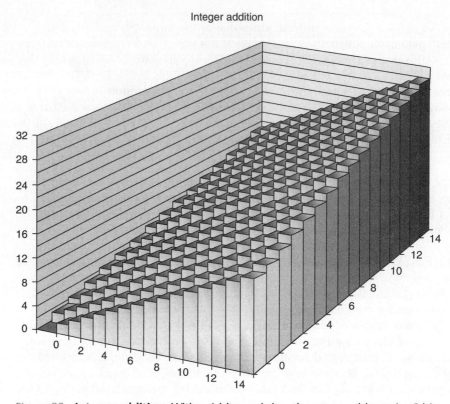

Figure 20   **Integer addition.** With a 4-bit word size, the sum could require 5 bits.

to maintain the sum as a $w+1$-bit number and add it to another value, we may require $w + 2$ bits, and so on. This continued "word size inflation" means we cannot place any bound on the word size required to fully represent the results of arithmetic operations. Some programming languages, such as Lisp, actually support *infinite precision* arithmetic to allow arbitrary (within the memory limits of the machine, of course) integer arithmetic. More commonly, programming languages support fixed-precision arithmetic, and hence operations such as "addition" and "multiplication" differ from their counterpart operations over integers.

Unsigned arithmetic can be viewed as a form of modular arithmetic. Unsigned addition is equivalent to computing the sum modulo 2^w. This value can be computed by simply discarding the high-order bit in the $w+1$-bit representation of $x + y$. For example, consider a 4-bit number representation with $x = 9$ and $y = 12$, having bit representations [1001] and [1100], respectively. Their sum is 21, having a 5-bit representation [10101]. But if we discard the high-order bit, we get [0101], that is, decimal value 5. This matches the value 21 mod 16 = 5.

In general, we can see that if $x + y < 2^w$, the leading bit in the $w+1$-bit representation of the sum will equal 0, and hence discarding it will not change the numeric value. On the other hand, if $2^w \leq x + y < 2^{w+1}$, the leading bit in

Figure 21

Relation between integer addition and unsigned addition. When $x + y$ is greater than $2^w - 1$, the sum overflows.

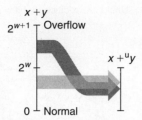

the $w+1$-bit representation of the sum will equal 1, and hence discarding it is equivalent to subtracting 2^w from the sum. These two cases are illustrated in Figure 21. This will give us a value in the range $0 \le x + y - 2^w < 2^{w+1} - 2^w = 2^w$, which is precisely the modulo 2^w sum of x and y. Let us define the operation $+_w^u$ for arguments x and y such that $0 \le x, y < 2^w$ as

$$x +_w^u y = \begin{cases} x + y, & x + y < 2^w \\ x + y - 2^w, & 2^w \le x + y < 2^{w+1} \end{cases} \tag{11}$$

This is precisely the result we get in C when performing addition on two w-bit unsigned values.

An arithmetic operation is said to *overflow* when the full integer result cannot fit within the word size limits of the data type. As Equation 11 indicates, overflow occurs when the two operands sum to 2^w or more. Figure 22 shows a plot of the unsigned addition function for word size $w = 4$. The sum is computed modulo $2^4 = 16$. When $x + y < 16$, there is no overflow, and $x +_4^u y$ is simply $x + y$. This is shown as the region forming a sloping plane labeled "Normal." When $x + y \ge 16$, the addition overflows, having the effect of decrementing the sum by 16. This is shown as the region forming a sloping plane labeled "Overflow."

When executing C programs, overflows are not signaled as errors. At times, however, we might wish to determine whether overflow has occurred. For example, suppose we compute $s \doteq x +_w^u y$, and we wish to determine whether s equals $x + y$. We claim that overflow has occurred if and only if $s < x$ (or equivalently, $s < y$). To see this, observe that $x + y \ge x$, and hence if s did not overflow, we will surely have $s \ge x$. On the other hand, if s did overflow, we have $s = x + y - 2^w$. Given that $y < 2^w$, we have $y - 2^w < 0$, and hence $s = x + (y - 2^w) < x$. In our earlier example, we saw that $9 +_4^u 12 = 5$. We can see that overflow occurred, since $5 < 9$.

Practice Problem 27

Write a function with the following prototype:

```
/* Determine whether arguments can be added without overflow */
int uadd_ok(unsigned x, unsigned y);
```

This function should return 1 if arguments x and y can be added without causing overflow.

Unsigned addition (4–bit word)

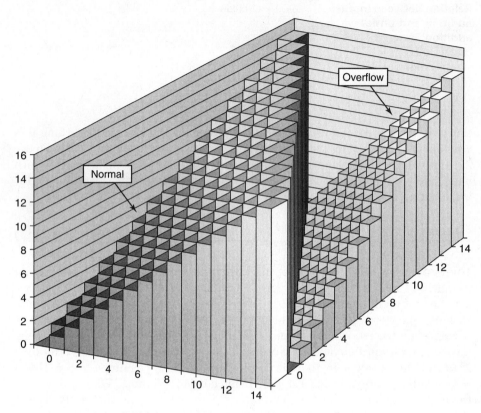

Figure 22 **Unsigned addition.** With a 4-bit word size, addition is performed modulo 16.

Modular addition forms a mathematical structure known as an *abelian group*, named after the Danish mathematician Niels Henrik Abel (1802–1829). That is, it is commutative (that's where the "abelian" part comes in) and associative; it has an identity element 0, and every element has an additive inverse. Let us consider the set of w-bit unsigned numbers with addition operation $+_w^u$. For every value x, there must be some value $-_w^u x$ such that $-_w^u x +_w^u x = 0$. When $x = 0$, the additive inverse is clearly 0. For $x > 0$, consider the value $2^w - x$. Observe that this number is in the range $0 < 2^w - x < 2^w$, and $(x + 2^w - x) \bmod 2^w = 2^w \bmod 2^w = 0$. Hence, it is the inverse of x under $+_w^u$. These two cases lead to the following equation for $0 \le x < 2^w$:

$$-_w^u x = \begin{cases} x, & x = 0 \\ 2^w - x, & x > 0 \end{cases} \tag{12}$$

Practice Problem 28

We can represent a bit pattern of length $w = 4$ with a single hex digit. For an unsigned interpretation of these digits, use Equation 12 to fill in the following table giving the values and the bit representations (in hex) of the unsigned additive inverses of the digits shown.

	x		$-_4^u x$	
Hex	Decimal		Decimal	Hex
0				
5				
8				
D				
F				

3.2 Two's-Complement Addition

With two's-complement addition, we must decide what to do when the result is either too large (positive) or too small (negative) to represent. Given integer values x and y in the range $-2^{w-1} \leq x, y \leq 2^{w-1} - 1$, their sum is in the range $-2^w \leq x + y \leq 2^w - 2$, potentially requiring $w + 1$ bits to represent exactly. As before, we avoid ever-expanding data sizes by truncating the representation to w bits. The result is not as familiar mathematically as modular addition, however.

The w-bit two's-complement sum of two numbers has the exact same bit-level representation as the unsigned sum. In fact, most computers use the same machine instruction to perform either unsigned or signed addition. Thus, we can define two's-complement addition for word size w, denoted as $+_w^t$, on operands x and y such that $-2^{w-1} \leq x, y < 2^{w-1}$ as

$$x +_w^t y \doteq U2T_w(T2U_w(x) +_w^u T2U_w(y)) \tag{13}$$

By Equation 5, we can write $T2U_w(x)$ as $x_{w-1}2^w + x$, and $T2U_w(y)$ as $y_{w-1}2^w + y$. Using the property that $+_w^u$ is simply addition modulo 2^w, along with the properties of modular addition, we then have

$$
\begin{aligned}
x +_w^t y &= U2T_w(T2U_w(x) +_w^u T2U_w(y)) \\
&= U2T_w[(x_{w-1}2^w + x + y_{w-1}2^w + y) \bmod 2^w] \\
&= U2T_w[(x + y) \bmod 2^w]
\end{aligned}
$$

The terms $x_{w-1}2^w$ and $y_{w-1}2^w$ drop out since they equal 0 modulo 2^w.

To better understand this quantity, let us define z as the integer sum $z \doteq x + y$, z' as $z' \doteq z \bmod 2^w$, and z'' as $z'' \doteq U2T_w(z')$. The value z'' is equal to $x +_w^t y$. We can divide the analysis into four cases, as illustrated in Figure 23:

1. $-2^w \leq z < -2^{w-1}$. Then we will have $z' = z + 2^w$. This gives $0 \leq z' < -2^{w-1} + 2^w = 2^{w-1}$. Examining Equation 8, we see that z' is in the range such that $z'' = z'$. This case is referred to as *negative overflow*. We have added two negative numbers x and y (that's the only way we can have $z < -2^{w-1}$) and obtained a nonnegative result $z'' = x + y + 2^w$.

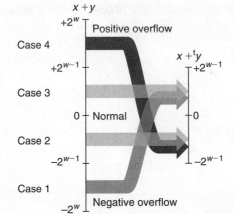

Figure 23

Relation between integer and two's-complement addition. When $x + y$ is less than -2^{w-1}, there is a negative overflow. When it is greater than $2^{w-1} + 1$, there is a positive overflow.

2. $-2^{w-1} \leq z < 0$. Then we will again have $z' = z + 2^w$, giving $-2^{w-1} + 2^w = 2^{w-1} \leq z' < 2^w$. Examining Equation 8, we see that z' is in such a range that $z'' = z' - 2^w$, and therefore $z'' = z' - 2^w = z + 2^w - 2^w = z$. That is, our two's-complement sum z'' equals the integer sum $x + y$.

3. $0 \leq z < 2^{w-1}$. Then we will have $z' = z$, giving $0 \leq z' < 2^{w-1}$, and hence $z'' = z' = z$. Again, the two's-complement sum z'' equals the integer sum $x + y$.

4. $2^{w-1} \leq z < 2^w$. We will again have $z' = z$, giving $2^{w-1} \leq z' < 2^w$. But in this range we have $z'' = z' - 2^w$, giving $z'' = x + y - 2^w$. This case is referred to as *positive overflow*. We have added two positive numbers x and y (that's the only way we can have $z \geq 2^{w-1}$) and obtained a negative result $z'' = x + y - 2^w$.

By the preceding analysis, we have shown that when operation $+_w^t$ is applied to values x and y in the range $-2^{w-1} \leq x, y \leq 2^{w-1} - 1$, we have

$$
x +_w^t y = \begin{cases} x + y - 2^w, & 2^{w-1} \leq x + y & \text{Positive overflow} \\ x + y, & -2^{w-1} \leq x + y < 2^{w-1} & \text{Normal} \\ x + y + 2^w, & x + y < -2^{w-1} & \text{Negative overflow} \end{cases} \quad (14)
$$

As an illustration, Figure 24 shows some examples of 4-bit two's-complement addition. Each example is labeled by the case to which it corresponds in the derivation of Equation 14. Note that $2^4 = 16$, and hence negative overflow yields a result 16 more than the integer sum, and positive overflow yields a result 16 less. We include bit-level representations of the operands and the result. Observe that the result can be obtained by performing binary addition of the operands and truncating the result to four bits.

Figure 25 illustrates two's-complement addition for word size $w = 4$. The operands range between -8 and 7. When $x + y < -8$, two's-complement addition has a negative underflow, causing the sum to be incremented by 16. When $-8 \leq x + y < 8$, the addition yields $x + y$. When $x + y \geq 8$, the addition has a negative overflow, causing the sum to be decremented by 16. Each of these three ranges forms a sloping plane in the figure.

x	y	$x + y$	$x +^{t}_{4} y$	Case
−8	−5	−13	3	1
[1000]	[1011]	[10011]	[0011]	
−8	−8	−16	0	1
[1000]	[1000]	[10000]	[0000]	
−8	5	−3	−3	2
[1000]	[0101]	[11101]	[1101]	
2	5	7	7	3
[0010]	[0101]	[00111]	[0111]	
5	5	10	−6	4
[0101]	[0101]	[01010]	[1010]	

Figure 24 **Two's-complement addition examples.** The bit-level representation of the 4-bit two's-complement sum can be obtained by performing binary addition of the operands and truncating the result to 4 bits.

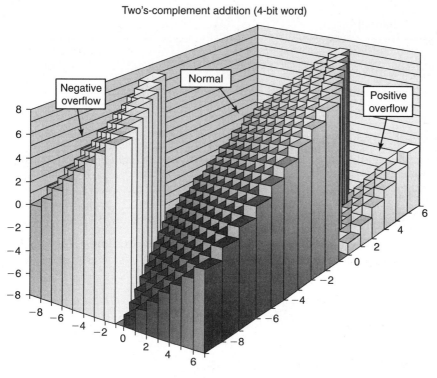

Figure 25 **Two's-complement addition.** With a 4-bit word size, addition can have a negative overflow when $x + y < -8$ and a positive overflow when $x + y \geq 8$.

Equation 14 also lets us identify the cases where overflow has occurred. When both x and y are negative but $x +_w^t y \geq 0$, we have negative overflow. When both x and y are positive but $x +_w^t y < 0$, we have positive overflow.

Practice Problem 29

Fill in the following table in the style of Figure 24. Give the integer values of the 5-bit arguments, the values of both their integer and two's-complement sums, the bit-level representation of the two's-complement sum, and the case from the derivation of Equation 14.

x	y	$x + y$	$x +_5^t y$	Case
[10100]	[10001]			
[11000]	[11000]			
[10111]	[01000]			
[00010]	[00101]			
[01100]	[00100]			

Practice Problem 30

Write a function with the following prototype:

```
/* Determine whether arguments can be added without overflow */
int tadd_ok(int x, int y);
```

This function should return 1 if arguments x and y can be added without causing overflow.

Practice Problem 31

Your coworker gets impatient with your analysis of the overflow conditions for two's-complement addition and presents you with the following implementation of tadd_ok:

```
/* Determine whether arguments can be added without overflow */
/* WARNING: This code is buggy. */
int tadd_ok(int x, int y) {
    int sum = x+y;
    return (sum-x == y) && (sum-y == x);
}
```

You look at the code and laugh. Explain why.

Practice Problem 32

You are assigned the task of writing code for a function `tsub_ok`, with arguments `x` and `y`, that will return 1 if computing `x-y` does not cause overflow. Having just written the code for Problem 30, you write the following:

```
/* Determine whether arguments can be subtracted without overflow */
/* WARNING: This code is buggy. */
int tsub_ok(int x, int y) {
    return tadd_ok(x, -y);
}
```

For what values of `x` and `y` will this function give incorrect results? Writing a correct version of this function is left as an exercise (Problem 74).

3.3 Two's-Complement Negation

We can see that every number x in the range $-2^{w-1} \leq x < 2^{w-1}$ has an additive inverse under $+_w^t$ as follows. First, for $x \neq -2^{w-1}$, we can see that its additive inverse is simply $-x$. That is, we have $-2^{w-1} < -x < 2^{w-1}$ and $-x +_w^t x = -x + x = 0$. For $x = -2^{w-1} = TMin_w$, on the other hand, $-x = 2^{w-1}$ cannot be represented as a w-bit number. We claim that this special value has itself as the additive inverse under $+_w^t$. The value of $-2^{w-1} +_w^t -2^{w-1}$ is given by the third case of Equation 14, since $-2^{w-1} + -2^{w-1} = -2^w$. This gives $-2^{w-1} +_w^t -2^{w-1} = -2^w + 2^w = 0$. From this analysis, we can define the two's-complement negation operation $-_w^t$ for x in the range $-2^{w-1} \leq x < 2^{w-1}$ as

$$-_w^t x = \begin{cases} -2^{w-1}, & x = -2^{w-1} \\ -x, & x > -2^{w-1} \end{cases} \qquad (15)$$

Practice Problem 33

We can represent a bit pattern of length $w = 4$ with a single hex digit. For a two's-complement interpretation of these digits, fill in the following table to determine the additive inverses of the digits shown:

x		$-_4^t x$	
Hex	Decimal	Decimal	Hex
0			
5			
8			
D			
F			

What do you observe about the bit patterns generated by two's-complement and unsigned (Problem 28) negation?

Web Aside DATA:TNEG   Bit-level representation of two's-complement negation

There are several clever ways to determine the two's-complement negation of a value represented at the bit level. These techniques are both useful, such as when one encounters the value 0xfffffffa when debugging a program, and they lend insight into the nature of the two's-complement representation.

One technique for performing two's-complement negation at the bit level is to complement the bits and then increment the result. In C, we can state that for any integer value x, computing the expressions -x and ~x + 1 will give identical results.

Here are some examples with a 4-bit word size:

$\vec{x}$		$\sim\vec{x}$		$incr(\sim\vec{x})$	
[0101]	5	[1010]	−6	[1011]	−5
[0111]	7	[1000]	−8	[1001]	−7
[1100]	−4	[0011]	3	[0100]	4
[0000]	0	[1111]	−1	[0000]	0
[1000]	−8	[0111]	7	[1000]	−8

For our earlier example, we know that the complement of 0xf is 0x0, and the complement of 0xa is 0x5, and so 0xfffffffa is the two's-complement representation of −6.

A second way to perform two's-complement negation of a number x is based on splitting the bit vector into two parts. Let k be the position of the rightmost 1, so the bit-level representation of x has the form $[x_{w-1}, x_{w-2}, \ldots, x_{k+1}, 1, 0, \ldots 0]$. (This is possible as long as $x \neq 0$.) The negation is then written in binary form as $[\sim x_{w-1}, \sim x_{w-2}, \ldots \sim x_{k+1}, 1, 0, \ldots, 0]$. That is, we complement each bit to the left of bit position k.

We illustrate this idea with some 4-bit numbers, where we highlight the rightmost pattern $1, 0, \ldots, 0$ in italics:

x		$-x$	
[1*100*]	−4	[0*100*]	4
[*1000*]	−8	[*1000*]	−8
[010*1*]	5	[101*1*]	−5
[011*1*]	7	[100*1*]	−7

3.4 Unsigned Multiplication

Integers x and y in the range $0 \leq x, y \leq 2^w - 1$ can be represented as w-bit unsigned numbers, but their product $x \cdot y$ can range between 0 and $(2^w - 1)^2 = 2^{2w} - 2^{w+1} + 1$. This could require as many as $2w$ bits to represent. Instead, unsigned multiplication in C is defined to yield the w-bit value given by the low-order w bits of the $2w$-bit integer product. By Equation 9, this can be seen to be equivalent to computing the product modulo 2^w. Thus, the effect of the w-bit unsigned multiplication operation $*^u_w$ is

$$x *^u_w y = (x \cdot y) \bmod 2^w \tag{16}$$

3.5 Two's-Complement Multiplication

Integers x and y in the range $-2^{w-1} \le x, y \le 2^{w-1} - 1$ can be represented as w-bit two's-complement numbers, but their product $x \cdot y$ can range between $-2^{w-1} \cdot (2^{w-1} - 1) = -2^{2w-2} + 2^{w-1}$ and $-2^{w-1} \cdot -2^{w-1} = 2^{2w-2}$. This could require as many as $2w$ bits to represent in two's-complement form—most cases would fit into $2w - 1$ bits, but the special case of 2^{2w-2} requires the full $2w$ bits (to include a sign bit of 0). Instead, signed multiplication in C generally is performed by truncating the $2w$-bit product to w bits. By Equation 10, the effect of the w-bit two's-complement multiplication operation $*_w^t$ is

$$x *_w^t y = U2T_w((x \cdot y) \bmod 2^w) \tag{17}$$

We claim that the bit-level representation of the product operation is identical for both unsigned and two's-complement multiplication. That is, given bit vectors $\vec{x}$ and $\vec{y}$ of length w, the bit-level representation of the unsigned product $B2U_w(\vec{x}) *_w^u B2U_w(\vec{y})$ is identical to the bit-level representation of the two's-complement product $B2T_w(\vec{x}) *_w^t B2T_w(\vec{y})$. This implies that the machine can use a single type of multiply instruction to multiply both signed and unsigned integers.

As illustrations, Figure 26 shows the results of multiplying different 3-bit numbers. For each pair of bit-level operands, we perform both unsigned and two's-complement multiplication, yielding 6-bit products, and then truncate these to 3 bits. The unsigned truncated product always equals $x \cdot y \bmod 8$. The bit-level representations of both truncated products are identical for both unsigned and two's-complement multiplication, even though the full 6-bit representations differ.

To show that the low-order bits of the two products (unsigned and two's complement) are identical, let $x = B2T_w(\vec{x})$ and $y = B2T_w(\vec{y})$ be the two's-complement values denoted by these bit patterns, and let $x' = B2U_w(\vec{x})$ and $y' = B2U_w(\vec{y})$ be the unsigned values. From Equation 5, we have $x' = x + x_{w-1}2^w$,

Mode	x		y		$x \cdot y$		Truncated $x \cdot y$	
Unsigned	5	[101]	3	[011]	15	[001111]	7	[111]
Two's comp.	−3	[101]	3	[011]	−9	[110111]	−1	[111]
Unsigned	4	[100]	7	[111]	28	[011100]	4	[100]
Two's comp.	−4	[100]	−1	[111]	4	[000100]	−4	[100]
Unsigned	3	[011]	3	[011]	9	[001001]	1	[001]
Two's comp.	3	[011]	3	[011]	9	[001001]	1	[001]

Figure 26 **Three-bit unsigned and two's-complement multiplication examples.** Although the bit-level representations of the full products may differ, those of the truncated products are identical.

and $y' = y + y_{w-1}2^w$. Computing the product of these values modulo 2^w gives the following:

$$(x' \cdot y') \bmod 2^w = [(x + x_{w-1}2^w) \cdot (y + y_{w-1}2^w)] \bmod 2^w \tag{18}$$

$$= [x \cdot y + (x_{w-1}y + y_{w-1}x)2^w + x_{w-1}y_{w-1}2^{2w}] \bmod 2^w$$

$$= (x \cdot y) \bmod 2^w$$

All of the terms with weight 2^w drop out due to the modulus operator, and so we have shown that the low-order w bits of $x \cdot y$ and $x' \cdot y'$ are identical.

Practice Problem 34

Fill in the following table showing the results of multiplying different 3-bit numbers, in the style of Figure 26:

Mode	x		y		$x \cdot y$	Truncated $x \cdot y$
Unsigned		[100]		[101]		
Two's comp.		[100]		[101]		
Unsigned		[010]		[111]		
Two's comp.		[010]		[111]		
Unsigned		[110]		[110]		
Two's comp.		[110]		[110]		

We can see that unsigned arithmetic and two's-complement arithmetic over w-bit numbers are isomorphic—the operations $+^u_w$, $-^u_w$, and $*^u_w$ have the exact same effect at the bit level as do $+^t_w$, $-^t_w$, and $*^t_w$.

Practice Problem 35

You are given the assignment to develop code for a function `tmult_ok` that will determine whether two arguments can be multiplied without causing overflow. Here is your solution:

```
/* Determine whether arguments can be multiplied without overflow */
int tmult_ok(int x, int y) {
    int p = x*y;
    /* Either x is zero, or dividing p by x gives y */
    return !x || p/x == y;
}
```

You test this code for a number of values of x and y, and it seems to work properly. Your coworker challenges you, saying, "If I can't use subtraction to test whether addition has overflowed (see Problem 31), then how can you use division to test whether multiplication has overflowed?"

Devise a mathematical justification of your approach, along the following lines. First, argue that the case $x = 0$ is handled correctly. Otherwise, consider

w-bit numbers x $(x \neq 0)$, y, p, and q, where p is the result of performing two's-complement multiplication on x and y, and q is the result of dividing p by x.

1. Show that $x \cdot y$, the integer product of x and y, can be written in the form $x \cdot y = p + t2^w$, where $t \neq 0$ if and only if the computation of p overflows.
2. Show that p can be written in the form $p = x \cdot q + r$, where $|r| < |x|$.
3. Show that $q = y$ if and only if $r = t = 0$.

Practice Problem 36

For the case where data type `int` has 32 bits, devise a version of `tmult_ok` (Problem 35) that uses the 64-bit precision of data type `long long`, without using division.

Aside   Security vulnerability in the XDR library

In 2002, it was discovered that code supplied by Sun Microsystems to implement the XDR library, a widely used facility for sharing data structures between programs, had a security vulnerability arising from the fact that multiplication can overflow without any notice being given to the program.

Code similar to that containing the vulnerability is shown below:

```
1   /*
2    * Illustration of code vulnerability similar to that found in
3    * Sun's XDR library.
4    */
5   void* copy_elements(void *ele_src[], int ele_cnt, size_t ele_size) {
6       /*
7        * Allocate buffer for ele_cnt objects, each of ele_size bytes
8        * and copy from locations designated by ele_src
9        */
10      void *result = malloc(ele_cnt * ele_size);
11      if (result == NULL)
12          /* malloc failed */
13          return NULL;
14      void *next = result;
15      int i;
16      for (i = 0; i < ele_cnt; i++) {
17          /* Copy object i to destination */
18          memcpy(next, ele_src[i], ele_size);
19          /* Move pointer to next memory region */
20          next += ele_size;
21      }
22      return result;
23  }
```

The function `copy_elements` is designed to copy `ele_cnt` data structures, each consisting of `ele_size` bytes into a buffer allocated by the function on line 10. The number of bytes required is computed as `ele_cnt * ele_size`.

Imagine, however, that a malicious programmer calls this function with `ele_cnt` being 1,048,577 ($2^{20} + 1$) and `ele_size` being 4,096 (2^{12}). Then the multiplication on line 10 will overflow, causing only 4096 bytes to be allocated, rather than the 4,294,971,392 bytes required to hold that much data. The loop starting at line 16 will attempt to copy all of those bytes, overrunning the end of the allocated buffer, and therefore corrupting other data structures. This could cause the program to crash or otherwise misbehave.

The Sun code was used by almost every operating system, and in such widely used programs as Internet Explorer and the Kerberos authentication system. The Computer Emergency Response Team (CERT), an organization run by the Carnegie Mellon Software Engineering Institute to track security vulnerabilities and breaches, issued advisory "CA-2002-25," and many companies rushed to patch their code. Fortunately, there were no reported security breaches caused by this vulnerability.

A similar vulnerability existed in many implementations of the library function `calloc`. These have since been patched.

Practice Problem 37

You are given the task of patching the vulnerability in the XDR code shown above. You decide to eliminate the possibility of the multiplication overflowing (on a 32-bit machine, at least) by computing the number of bytes to allocate using data type `long long unsigned`. You replace the original call to `malloc` (line 10) as follows:

```
long long unsigned asize =
    ele_cnt * (long long unsigned) ele_size;
void *result = malloc(asize);
```

A. Does your code provide any improvement over the original?

B. How would you change the code to eliminate the vulnerability, assuming data type `size_t` is the same as `unsigned int`, and these are 32 bits long?

3.6 Multiplying by Constants

On most machines, the integer multiply instruction is fairly slow, requiring 10 or more clock cycles, whereas other integer operations—such as addition, subtraction, bit-level operations, and shifting—require only 1 clock cycle. As a consequence, one important optimization used by compilers is to attempt to replace multiplications by constant factors with combinations of shift and addition operations. We will first consider the case of multiplying by a power of 2, and then generalize this to arbitrary constants.

Let x be the unsigned integer represented by bit pattern $[x_{w-1}, x_{w-2}, \ldots, x_0]$. Then for any $k \geq 0$, we claim the bit-level representation of $x2^k$ is given by

$[x_{w-1}, x_{w-2}, \ldots, x_0, 0, \ldots, 0]$, where k zeros have been added to the right. This property can be derived using Equation 1:

$$B2U_{w+k}([x_{w-1}, x_{w-2}, \ldots, x_0, 0, \ldots, 0]) = \sum_{i=0}^{w-1} x_i 2^{i+k}$$

$$= \left[\sum_{i=0}^{w-1} x_i 2^i\right] \cdot 2^k$$

$$= x 2^k$$

For $k < w$, we can truncate the shifted bit vector to be of length w, giving $[x_{w-k-1}, x_{w-k-2}, \ldots, x_0, 0, \ldots, 0]$. By Equation 9, this bit vector has numeric value $x 2^k \bmod 2^w = x *_w^u 2^k$. Thus, for unsigned variable x, the C expression x << k is equivalent to x * pwr2k, where pwr2k equals 2^k. In particular, we can compute pwr2k as 1U << k.

By similar reasoning, we can show that for a two's-complement number x having bit pattern $[x_{w-1}, x_{w-2}, \ldots, x_0]$, and any k in the range $0 \le k < w$, bit pattern $[x_{w-k-1}, \ldots, x_0, 0, \ldots, 0]$ will be the two's-complement representation of $x *_w^t 2^k$. Therefore, for signed variable x , the C expression x << k is equivalent to x * pwr2k, where pwr2k equals 2^k.

Note that multiplying by a power of 2 can cause overflow with either unsigned or two's-complement arithmetic. Our result shows that even then we will get the same effect by shifting.

Given that integer multiplication is much more costly than shifting and adding, many C compilers try to remove many cases where an integer is being multiplied by a constant with combinations of shifting, adding, and subtracting. For example, suppose a program contains the expression x*14. Recognizing that $14 = 2^3 + 2^2 + 2^1$, the compiler can rewrite the multiplication as (x<<3) + (x<<2) + (x<<1), replacing one multiplication with three shifts and two additions. The two computations will yield the same result, regardless of whether x is unsigned or two's complement, and even if the multiplication would cause an overflow. (This can be shown from the properties of integer arithmetic.) Even better, the compiler can also use the property $14 = 2^4 - 2^1$ to rewrite the multiplication as (x<<4) – (x<<1), requiring only two shifts and a subtraction.

Practice Problem 38

The LEA instruction can perform computations of the form (a<<k) + b, where k is either 0, 1, 2, or 3, and b is either 0 or some program value. The compiler often uses this instruction to perform multiplications by constant factors. For example, we can compute 3*a as (a<<1) + a.

Considering cases where b is either 0 or equal to a, and all possible values of k, what multiples of a can be computed with a single LEA instruction?

Generalizing from our example, consider the task of generating code for the expression x * K, for some constant K. The compiler can express the binary representation of K as an alternating sequence of zeros and ones:

$$[(0 \ldots 0)(1 \ldots 1)(0 \ldots 0) \cdots (1 \ldots 1)].$$

For example, 14 can be written as $[(0 \ldots 0)(111)(0)]$. Consider a run of ones from bit position n down to bit position m ($n \geq m$). (For the case of 14, we have $n = 3$ and $m = 1$.) We can compute the effect of these bits on the product using either of two different forms:

Form A: (x<<n) + (x<<$n-1$) + $\cdots$ + (x<<m)
Form B: (x<<$n+1$) - (x<<m)

By adding together the results for each run, we are able to compute x * K without any multiplications. Of course, the trade-off between using combinations of shifting, adding, and subtracting versus a single multiplication instruction depends on the relative speeds of these instructions, and these can be highly machine dependent. Most compilers only perform this optimization when a small number of shifts, adds, and subtractions suffice.

Practice Problem 39

How could we modify the expression for form B for the case where bit position n is the most significant bit?

Practice Problem 40

For each of the following values of K, find ways to express x * K using only the specified number of operations, where we consider both additions and subtractions to have comparable cost. You may need to use some tricks beyond the simple form A and B rules we have considered so far.

K	Shifts	Add/Subs	Expression
6	2	1	
31	1	1	
−6	2	1	
55	2	2	

Practice Problem 41

For a run of 1s starting at bit position n down to bit position m ($n \geq m$), we saw that we can generate two forms of code, A and B. How should the compiler decide which form to use?

k	>> k (Binary)	Decimal	$12340/2^k$
0	0011000000110100	12340	12340.0
1	*0*001100000011010	6170	6170.0
4	*0000*001100000011	771	771.25
8	*00000000*00110000	48	48.203125

Figure 27 **Dividing unsigned numbers by powers of 2.** The examples illustrate how performing a logical right shift by k has the same effect as dividing by 2^k and then rounding toward zero.

3.7 Dividing by Powers of Two

Integer division on most machines is even slower than integer multiplication— requiring 30 or more clock cycles. Dividing by a power of 2 can also be performed using shift operations, but we use a right shift rather than a left shift. The two different shifts—logical and arithmetic—serve this purpose for unsigned and two's-complement numbers, respectively.

Integer division always rounds toward zero. For $x \geq 0$ and $y > 0$, the result should be $\lfloor x/y \rfloor$, where for any real number a, $\lfloor a \rfloor$ is defined to be the unique integer a' such that $a' \leq a < a' + 1$. As examples, $\lfloor 3.14 \rfloor = 3$, $\lfloor -3.14 \rfloor = -4$, and $\lfloor 3 \rfloor = 3$.

Consider the effect of applying a logical right shift by k to an unsigned number. We claim this gives the same result as dividing by 2^k. As examples, Figure 27 shows the effects of performing logical right shifts on a 16-bit representation of 12,340 to perform division by 1, 2, 16, and 256. The zeros shifted in from the left are shown in italics. We also show the result we would obtain if we did these divisions with real arithmetic. These examples show that the result of shifting consistently rounds toward zero, as is the convention for integer division.

To show this relation between logical right shifting and dividing by a power of 2, let x be the unsigned integer represented by bit pattern $[x_{w-1}, x_{w-2}, \ldots, x_0]$, and k be in the range $0 \leq k < w$. Let x' be the unsigned number with $w-k$-bit representation $[x_{w-1}, x_{w-2}, \ldots, x_k]$, and x'' be the unsigned number with k-bit representation $[x_{k-1}, \ldots, x_0]$. We claim that $x' = \lfloor x/2^k \rfloor$. To see this, by Equation 1, we have $x = \sum_{i=0}^{w-1} x_i 2^i$, $x' = \sum_{i=k}^{w-1} x_i 2^{i-k}$, and $x'' = \sum_{i=0}^{k-1} x_i 2^i$. We can therefore write x as $x = 2^k x' + x''$. Observe that $0 \leq x'' \leq \sum_{i=0}^{k-1} 2^i = 2^k - 1$, and hence $0 \leq x'' < 2^k$, implying that $\lfloor x''/2^k \rfloor = 0$. Therefore, $\lfloor x/2^k \rfloor = \lfloor x' + x''/2^k \rfloor = x' + \lfloor x''/2^k \rfloor = x'$.

Performing a logical right shift of bit vector $[x_{w-1}, x_{w-2}, \ldots, x_0]$ by k yields the bit vector

$$[0, \ldots, 0, x_{w-1}, x_{w-2}, \ldots, x_k]$$

This bit vector has numeric value x'. Therefore, for unsigned variable x, the C expression x >> k is equivalent to x / pwr2k, where pwr2k equals 2^k.

k	>> k (Binary)	Decimal	$-12340/2^k$
0	1100111111001100	−12340	−12340.0
1	1110011111100110	−6170	−6170.0
4	1111110011111100	−772	−771.25
8	1111111111001111	−49	−48.203125

Figure 28 **Applying arithmetic right shift.** The examples illustrate that arithmetic right shift is similar to division by a power of 2, except that it rounds down rather than toward zero.

Now consider the effect of performing an *arithmetic* right shift on a two's-complement number. For a positive number, we have 0 as the most significant bit, and so the effect is the same as for a logical right shift. Thus, an arithmetic right shift by k is the same as division by 2^k for a nonnegative number. As an example of a negative number, Figure 28 shows the effect of applying arithmetic right shift to a 16-bit representation of −12,340 for different shift amounts. As we can see, the result is *almost* the same as dividing by a power of 2. For the case when no rounding is required ($k = 1$), the result is correct. But when rounding is required, shifting causes the result to be rounded downward rather than toward zero, as should be the convention. For example, the expression -7/2 should yield -3 rather than -4.

Let us better understand the effect of arithmetic right shifting and how we can use it to perform division by a power of 2. Let x be the two's-complement integer represented by bit pattern $[x_{w-1}, x_{w-2}, \ldots, x_0]$, and k be in the range $0 \le k < w$. Let x' be the two's-complement number represented by the $w - k$ bits $[x_{w-1}, x_{w-2}, \ldots, x_k]$, and x'' be the *unsigned* number represented by the low-order k bits $[x_{k-1}, \ldots, x_0]$. By a similar analysis as the unsigned case, we have $x = 2^k x' + x''$, and $0 \le x'' < 2^k$, giving $x' = \lfloor x/2^k \rfloor$. Furthermore, observe that shifting bit vector $[x_{w-1}, x_{w-2}, \ldots, x_0]$ right *arithmetically* by k yields the bit vector

$$[x_{w-1}, \ldots, x_{w-1}, x_{w-1}, x_{w-2}, \ldots, x_k]$$

which is the sign extension from $w - k$ bits to w bits of $[x_{w-1}, x_{w-2}, \ldots, x_k]$. Thus, this shifted bit vector is the two's-complement representation of $\lfloor x/2^k \rfloor$. This analysis confirms our findings from the examples of Figure 28.

For $x \ge 0$, or when no rounding is required ($x'' = 0$), our analysis shows that this shifted result is the desired value. For $x < 0$ and $y > 0$, however, the result of integer division should be $\lceil x/y \rceil$, where for any real number a, $\lceil a \rceil$ is defined to be the unique integer a' such that $a' - 1 < a \le a'$. That is, integer division should round negative results upward toward zero. Thus, right shifting a negative number by k is not equivalent to dividing it by 2^k when rounding occurs. This analysis also confirms our findings from the example of Figure 28.

We can correct for this improper rounding by "biasing" the value before shifting. This technique exploits the property that $\lceil x/y \rceil = \lfloor (x + y - 1)/y \rfloor$ for integers x and y such that $y > 0$. As examples, when $x = -30$ and $y = 4$, we have $x + y - 1 = -27$, and $\lceil -30/4 \rceil = -7 = \lfloor -27/4 \rfloor$. When $x = -32$ and $y = 4$,

k	Bias	−12,340 + Bias (Binary)	>> k (Binary)	Decimal	−12340/2^k
0	0	1100111111001100	1100111111001100	−12340	−12340.0
1	1	11001111110011*01*	*1*110011111100110	−6170	−6170.0
4	15	110011111101*1011*	*1111*110011111101	−771	−771.25
8	255	11010000*11001011*	*11111111*11010000	−48	−48.203125

Figure 29 **Dividing two's-complement numbers by powers of 2.** By adding a bias before the right shift, the result is rounded toward zero.

we have $x + y - 1 = -29$, and $\lceil -32/4 \rceil = -8 = \lfloor -29/4 \rfloor$. To see that this relation holds in general, suppose that $x = ky + r$, where $0 \le r < y$, giving $(x + y - 1)/y = k + (r + y - 1)/y$, and so $\lfloor (x + y - 1)/y \rfloor = k + \lfloor (r + y - 1)/y \rfloor$. The latter term will equal 0 when $r = 0$, and 1 when $r > 0$. That is, by adding a bias of $y - 1$ to x and then rounding the division downward, we will get k when y divides x and $k + 1$ otherwise. Thus, for $x < 0$, if we first add $2^k - 1$ to x before right shifting, we will get a correctly rounded result.

This analysis shows that for a two's-complement machine using arithmetic right shifts, the C expression

```
(x<0 ? x+(1<<k)-1 : x) >> k
```

is equivalent to x/pwr2k, where pwr2k equals 2^k.

Figure 29 demonstrates how adding the appropriate bias before performing the arithmetic right shift causes the result to be correctly rounded. In the third column, we show the result of adding the bias value to −12,340, with the lower k bits (those that will be shifted off to the right) shown in italics. We can see that the bits to the left of these may or may not be incremented. For the case where no rounding is required ($k = 1$), adding the bias only affects bits that are shifted off. For the cases where rounding is required, adding the bias causes the upper bits to be incremented, so that the result will be rounded toward zero.

Practice Problem 42

Write a function `div16` that returns the value x/16 for integer argument x. Your function should not use division, modulus, multiplication, any conditionals (`if` or `?:`), any comparison operators (e.g., `<`, `>`, or `==`), or any loops. You may assume that data type `int` is 32 bits long and uses a two's-complement representation, and that right shifts are performed arithmetically.

We now see that division by a power of 2 can be implemented using logical or arithmetic right shifts. This is precisely the reason the two types of right shifts are available on most machines. Unfortunately, this approach does not generalize to division by arbitrary constants. Unlike multiplication, we cannot express division by arbitrary constants K in terms of division by powers of 2.

Practice Problem 43

In the following code, we have omitted the definitions of constants M and N:

```
#define M        /* Mystery number 1 */
#define N        /* Mystery number 2 */
int arith(int x, int y) {
    int result = 0;
    result = x*M + y/N; /* M and N are mystery numbers. */
    return result;
}
```

We compiled this code for particular values of M and N. The compiler optimized the multiplication and division using the methods we have discussed. The following is a translation of the generated machine code back into C:

```
/* Translation of assembly code for arith */
int optarith(int x, int y) {
    int t = x;
    x <<= 5;
    x -= t;
    if (y < 0) y += 7;
    y >>= 3;  /* Arithmetic shift */
    return x+y;
}
```

What are the values of M and N?

3.8 Final Thoughts on Integer Arithmetic

As we have seen, the "integer" arithmetic performed by computers is really a form of modular arithmetic. The finite word size used to represent numbers limits the range of possible values, and the resulting operations can overflow. We have also seen that the two's-complement representation provides a clever way to be able to represent both negative and positive values, while using the same bit-level implementations as are used to perform unsigned arithmetic—operations such as addition, subtraction, multiplication, and even division have either identical or very similar bit-level behaviors whether the operands are in unsigned or two's-complement form.

We have seen that some of the conventions in the C language can yield some surprising results, and these can be sources of bugs that are hard to recognize or understand. We have especially seen that the unsigned data type, while conceptually straightforward, can lead to behaviors that even experienced programmers do not expect. We have also seen that this data type can arise in unexpected ways, for example, when writing integer constants and when invoking library routines.

Practice Problem 44

Assume we are running code on a 32-bit machine using two's-complement arithmetic for signed values. Right shifts are performed arithmetically for signed values and logically for unsigned values. The variables are declared and initialized as follows:

```
int x = foo();   /* Arbitrary value */
int y = bar();   /* Arbitrary value */

unsigned ux = x;
unsigned uy = y;
```

For each of the following C expressions, either (1) argue that it is true (evaluates to 1) for all values of x and y, or (2) give values of x and y for which it is false (evaluates to 0):

A. (x > 0) || (x-1 < 0)

B. (x & 7) != 7 || (x<<29 < 0)

C. (x * x) >= 0

D. x < 0 || -x <= 0

E. x > 0 || -x >= 0

F. x+y == uy+ux

G. x*~y + uy*ux == -x

4 Floating Point

A floating-point representation encodes rational numbers of the form $V = x \times 2^y$. It is useful for performing computations involving very large numbers ($|V| \gg 0$), numbers very close to 0 ($|V| \ll 1$), and more generally as an approximation to real arithmetic.

Up until the 1980s, every computer manufacturer devised its own conventions for how floating-point numbers were represented and the details of the operations performed on them. In addition, they often did not worry too much about the accuracy of the operations, viewing speed and ease of implementation as being more critical than numerical precision.

All of this changed around 1985 with the advent of IEEE Standard 754, a carefully crafted standard for representing floating-point numbers and the operations performed on them. This effort started in 1976 under Intel's sponsorship with the design of the 8087, a chip that provided floating-point support for the 8086 processor. They hired William Kahan, a professor at the University of California, Berkeley, as a consultant to help design a floating-point standard for its future processors. They allowed Kahan to join forces with a committee generating an industry-wide standard under the auspices of the Institute of Electrical and Electronics Engineers (IEEE). The committee ultimately adopted a standard close to the one Kahan had devised for Intel. Nowadays, virtually all computers support what has become known as *IEEE floating point*. This has greatly improved the portability of scientific application programs across different machines.

Aside   The IEEE

The Institute of Electrical and Electronic Engineers (IEEE—pronounced "Eye-Triple-Eee") is a professional society that encompasses all of electronic and computer technology. It publishes journals, sponsors conferences, and sets up committees to define standards on topics ranging from power transmission to software engineering.

In this section, we will see how numbers are represented in the IEEE floating-point format. We will also explore issues of *rounding*, when a number cannot be represented exactly in the format and hence must be adjusted upward or downward. We will then explore the mathematical properties of addition, multiplication, and relational operators. Many programmers consider floating point to be at best uninteresting and at worst arcane and incomprehensible. We will see that since the IEEE format is based on a small and consistent set of principles, it is really quite elegant and understandable.

4.1   Fractional Binary Numbers

A first step in understanding floating-point numbers is to consider binary numbers having fractional values. Let us first examine the more familiar decimal notation. Decimal notation uses a representation of the form $d_m d_{m-1} \cdots d_1 d_0 . d_{-1} d_{-2} \cdots d_{-n}$, where each decimal digit d_i ranges between 0 and 9. This notation represents a value d defined as

$$d = \sum_{i=-n}^{m} 10^i \times d_i$$

The weighting of the digits is defined relative to the decimal point symbol ('.'), meaning that digits to the left are weighted by positive powers of 10, giving integral values, while digits to the right are weighted by negative powers of 10, giving fractional values. For example, 12.34_{10} represents the number $1 \times 10^1 + 2 \times 10^0 + 3 \times 10^{-1} + 4 \times 10^{-2} = 12\frac{34}{100}$.

By analogy, consider a notation of the form $b_m b_{m-1} \cdots b_1 b_0 . b_{-1} b_{-2} \cdots b_{-n-1} b_{-n}$, where each binary digit, or bit, b_i ranges between 0 and 1, as is illustrated in Figure 30. This notation represents a number b defined as

$$b = \sum_{i=-n}^{m} 2^i \times b_i \qquad (19)$$

The symbol '.' now becomes a *binary point*, with bits on the left being weighted by positive powers of 2, and those on the right being weighted by negative powers of 2. For example, 101.11_2 represents the number $1 \times 2^2 + 0 \times 2^1 + 1 \times 2^0 + 1 \times 2^{-1} + 1 \times 2^{-2} = 4 + 0 + 1 + \frac{1}{2} + \frac{1}{4} = 5\frac{3}{4}$.

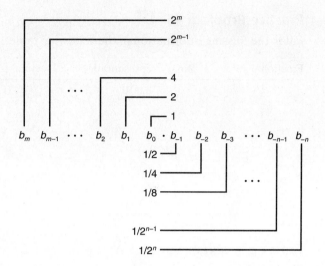

One can readily see from Equation 19 that shifting the binary point one position to the left has the effect of dividing the number by 2. For example, while 101.11_2 represents the number $5\frac{3}{4}$, 10.111_2 represents the number $2 + 0 + \frac{1}{2} + \frac{1}{4} + \frac{1}{8} = 2\frac{7}{8}$. Similarly, shifting the binary point one position to the right has the effect of multiplying the number by 2. For example, 1011.1_2 represents the number $8 + 0 + 2 + 1 + \frac{1}{2} = 11\frac{1}{2}$.

Note that numbers of the form $0.11\cdots1_2$ represent numbers just below 1. For example, 0.111111_2 represents $\frac{63}{64}$. We will use the shorthand notation $1.0 - \epsilon$ to represent such values.

Assuming we consider only finite-length encodings, decimal notation cannot represent numbers such as $\frac{1}{3}$ and $\frac{5}{7}$ exactly. Similarly, fractional binary notation can only represent numbers that can be written $x \times 2^y$. Other values can only be approximated. For example, the number $\frac{1}{5}$ can be represented exactly as the fractional decimal number 0.20. As a fractional binary number, however, we cannot represent it exactly and instead must approximate it with increasing accuracy by lengthening the binary representation:

Representation	Value	Decimal
0.0_2	$\frac{0}{2}$	0.0_{10}
0.01_2	$\frac{1}{4}$	0.25_{10}
0.010_2	$\frac{2}{8}$	0.25_{10}
0.0011_2	$\frac{3}{16}$	0.1875_{10}
0.00110_2	$\frac{6}{32}$	0.1875_{10}
0.001101_2	$\frac{13}{64}$	0.203125_{10}
0.0011010_2	$\frac{26}{128}$	0.203125_{10}
0.00110011_2	$\frac{51}{256}$	0.19921875_{10}

Practice Problem 45

Fill in the missing information in the following table:

Fractional value	Binary representation	Decimal representation
$\frac{1}{8}$	0.001	0.125
$\frac{3}{4}$		
$\frac{25}{16}$		
	10.1011	
	1.001	
		5.875
		3.1875

Practice Problem 46

The imprecision of floating-point arithmetic can have disastrous effects. On February 25, 1991, during the first Gulf War, an American Patriot Missile battery in Dharan, Saudi Arabia, failed to intercept an incoming Iraqi Scud missile. The Scud struck an American Army barracks and killed 28 soldiers. The U.S. General Accounting Office (GAO) conducted a detailed analysis of the failure [72] and determined that the underlying cause was an imprecision in a numeric calculation. In this exercise, you will reproduce part of the GAO's analysis.

The Patriot system contains an internal clock, implemented as a counter that is incremented every 0.1 seconds. To determine the time in seconds, the program would multiply the value of this counter by a 24-bit quantity that was a fractional binary approximation to $\frac{1}{10}$. In particular, the binary representation of $\frac{1}{10}$ is the nonterminating sequence $0.000110011[0011]\cdots_2$, where the portion in brackets is repeated indefinitely. The program approximated 0.1, as a value x, by considering just the first 23 bits of the sequence to the right of the binary point: $x = 0.00011001100110011001100$. (See Problem 51 for a discussion of how they could have approximated 0.1 more precisely.)

A. What is the binary representation of $0.1 - x$?

B. What is the approximate decimal value of $0.1 - x$?

C. The clock starts at 0 when the system is first powered up and keeps counting up from there. In this case, the system had been running for around 100 hours. What was the difference between the actual time and the time computed by the software?

D. The system predicts where an incoming missile will appear based on its velocity and the time of the last radar detection. Given that a Scud travels at around 2000 meters per second, how far off was its prediction?

Normally, a slight error in the absolute time reported by a clock reading would not affect a tracking computation. Instead, it should depend on the relative time between two successive readings. The problem was that the Patriot software had been upgraded to use a more accurate function for reading time, but not all of the function calls had been replaced by the new code. As a result, the tracking software used the accurate time for one reading and the inaccurate time for the other [100].

4.2 IEEE Floating-Point Representation

Positional notation such as considered in the previous section would not be efficient for representing very large numbers. For example, the representation of 5×2^{100} would consist of the bit pattern 101 followed by 100 zeros. Instead, we would like to represent numbers in a form $x \times 2^y$ by giving the values of x and y.

The IEEE floating-point standard represents a number in a form $V = (-1)^s \times M \times 2^E$:

- The *sign* s determines whether the number is negative ($s = 1$) or positive ($s = 0$), where the interpretation of the sign bit for numeric value 0 is handled as a special case.
- The *significand* M is a fractional binary number that ranges either between 1 and $2 - \epsilon$ or between 0 and $1 - \epsilon$.
- The *exponent* E weights the value by a (possibly negative) power of 2.

The bit representation of a floating-point number is divided into three fields to encode these values:

- The single sign bit s directly encodes the sign s.
- The k-bit exponent field $\mathtt{exp} = e_{k-1} \cdots e_1 e_0$ encodes the exponent E.
- The n-bit fraction field $\mathtt{frac} = f_{n-1} \cdots f_1 f_0$ encodes the significand M, but the value encoded also depends on whether or not the exponent field equals 0.

Figure 31 shows the packing of these three fields into words for the two most common formats. In the single-precision floating-point format (a `float` in C), fields s, exp, and frac are 1, $k = 8$, and $n = 23$ bits each, yielding a 32-bit representation. In the double-precision floating-point format (a `double` in C), fields s, exp, and frac are 1, $k = 11$, and $n = 52$ bits each, yielding a 64-bit representation.

The value encoded by a given bit representation can be divided into three different cases (the latter having two variants), depending on the value of exp. These are illustrated in Figure 32 for the single-precision format.

Case 1: Normalized Values

This is the most common case. It occurs when the bit pattern of exp is neither all zeros (numeric value 0) nor all ones (numeric value 255 for single precision, 2047 for double). In this case, the exponent field is interpreted as representing a signed integer in *biased* form. That is, the exponent value is $E = e - Bias$ where e is the unsigned number having bit representation $e_{k-1} \cdots e_1 e_0$, and *Bias* is a bias

Single precision

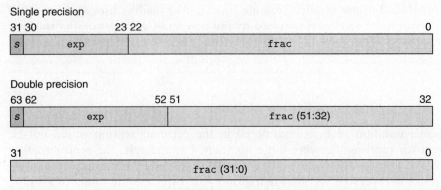

Double precision

Figure 31 **Standard floating-point formats.** Floating-point numbers are represented by three fields. For the two most common formats, these are packed in 32-bit (single precision) or 64-bit (double precision) words.

1. Normalized

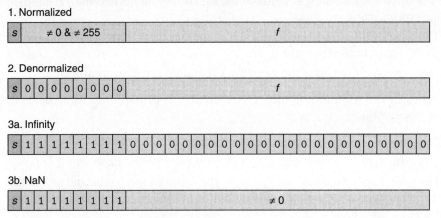

2. Denormalized

3a. Infinity

3b. NaN

Figure 32 **Categories of single-precision, floating-point values.** The value of the exponent determines whether the number is (1) normalized, (2) denormalized, or a (3) special value.

value equal to $2^{k-1} - 1$ (127 for single precision and 1023 for double). This yields exponent ranges from -126 to $+127$ for single precision and -1022 to $+1023$ for double precision.

The fraction field `frac` is interpreted as representing the fractional value f, where $0 \le f < 1$, having binary representation $0.f_{n-1} \cdots f_1 f_0$, that is, with the binary point to the left of the most significant bit. The significand is defined to be $M = 1 + f$. This is sometimes called an *implied leading 1* representation, because we can view M to be the number with binary representation $1.f_{n-1}f_{n-2} \cdots f_0$. This representation is a trick for getting an additional bit of precision for free, since we can always adjust the exponent E so that significand M is in the range $1 \le M < 2$ (assuming there is no overflow). We therefore do not need to explicitly represent the leading bit, since it always equals 1.

Case 2: Denormalized Values

When the exponent field is all zeros, the represented number is in *denormalized* form. In this case, the exponent value is $E = 1 - Bias$, and the significand value is $M = f$, that is, the value of the fraction field without an implied leading 1.

> **Aside** Why set the bias this way for denormalized values?
>
> Having the exponent value be $1 - Bias$ rather than simply $-Bias$ might seem counterintuitive. We will see shortly that it provides for smooth transition from denormalized to normalized values.

Denormalized numbers serve two purposes. First, they provide a way to represent numeric value 0, since with a normalized number we must always have $M \geq 1$, and hence we cannot represent 0. In fact the floating-point representation of $+0.0$ has a bit pattern of all zeros: the sign bit is 0, the exponent field is all zeros (indicating a denormalized value), and the fraction field is all zeros, giving $M = f = 0$. Curiously, when the sign bit is 1, but the other fields are all zeros, we get the value -0.0. With IEEE floating-point format, the values -0.0 and $+0.0$ are considered different in some ways and the same in others.

A second function of denormalized numbers is to represent numbers that are very close to 0.0. They provide a property known as *gradual underflow* in which possible numeric values are spaced evenly near 0.0.

Case 3: Special Values

A final category of values occurs when the exponent field is all ones. When the fraction field is all zeros, the resulting values represent infinity, either $+\infty$ when $s = 0$, or $-\infty$ when $s = 1$. Infinity can represent results that *overflow*, as when we multiply two very large numbers, or when we divide by zero. When the fraction field is nonzero, the resulting value is called a "*NaN*," short for "Not a Number." Such values are returned as the result of an operation where the result cannot be given as a real number or as infinity, as when computing $\sqrt{-1}$ or $\infty - \infty$. They can also be useful in some applications for representing uninitialized data.

4.3 Example Numbers

Figure 33 shows the set of values that can be represented in a hypothetical 6-bit format having $k = 3$ exponent bits and $n = 2$ fraction bits. The bias is $2^{3-1} - 1 = 3$. Part A of the figure shows all representable values (other than *NaN*). The two infinities are at the extreme ends. The normalized numbers with maximum magnitude are ± 14. The denormalized numbers are clustered around 0. These can be seen more clearly in part B of the figure, where we show just the numbers between -1.0 and $+1.0$. The two zeros are special cases of denormalized numbers. Observe that the representable numbers are not uniformly distributed—they are denser nearer the origin.

Figure 34 shows some examples for a hypothetical 8-bit floating-point format having $k = 4$ exponent bits and $n = 3$ fraction bits. The bias is $2^{4-1} - 1 = 7$. The

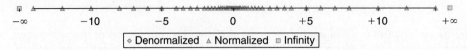

(a) Complete range

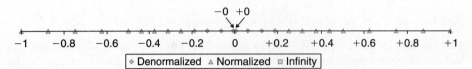

(b) Values between −1.0 and +1.0

Figure 33 **Representable values for 6-bit floating-point format.** There are $k = 3$ exponent bits and $n = 2$ fraction bits. The bias is 3.

Description	Bit representation	Exponent			Fraction		Value		
		e	E	2^E	f	M	$2^E \times M$	V	Decimal
Zero	0 0000 000	0	−6	$\frac{1}{64}$	$\frac{0}{8}$	$\frac{0}{8}$	$\frac{0}{512}$	0	0.0
Smallest pos.	0 0000 001	0	−6	$\frac{1}{64}$	$\frac{1}{8}$	$\frac{1}{8}$	$\frac{1}{512}$	$\frac{1}{512}$	0.001953
	0 0000 010	0	−6	$\frac{1}{64}$	$\frac{2}{8}$	$\frac{2}{8}$	$\frac{2}{512}$	$\frac{1}{256}$	0.003906
	0 0000 011	0	−6	$\frac{1}{64}$	$\frac{3}{8}$	$\frac{3}{8}$	$\frac{3}{512}$	$\frac{3}{512}$	0.005859
	$\vdots$								
Largest denorm.	0 0000 111	0	−6	$\frac{1}{64}$	$\frac{7}{8}$	$\frac{7}{8}$	$\frac{7}{512}$	$\frac{7}{512}$	0.013672
Smallest norm.	0 0001 000	1	−6	$\frac{1}{64}$	$\frac{0}{8}$	$\frac{8}{8}$	$\frac{8}{512}$	$\frac{1}{64}$	0.015625
	0 0001 001	1	−6	$\frac{1}{64}$	$\frac{1}{8}$	$\frac{9}{8}$	$\frac{9}{512}$	$\frac{9}{512}$	0.017578
	$\vdots$								
	0 0110 110	6	−1	$\frac{1}{2}$	$\frac{6}{8}$	$\frac{14}{8}$	$\frac{14}{16}$	$\frac{7}{8}$	0.875
	0 0110 111	6	−1	$\frac{1}{2}$	$\frac{7}{8}$	$\frac{15}{8}$	$\frac{15}{16}$	$\frac{15}{16}$	0.9375
One	0 0111 000	7	0	1	$\frac{0}{8}$	$\frac{8}{8}$	$\frac{8}{8}$	1	1.0
	0 0111 001	7	0	1	$\frac{1}{8}$	$\frac{9}{8}$	$\frac{9}{8}$	$\frac{9}{8}$	1.125
	0 0111 010	7	0	1	$\frac{2}{8}$	$\frac{10}{8}$	$\frac{10}{8}$	$\frac{5}{4}$	1.25
	$\vdots$								
	0 1110 110	14	7	128	$\frac{6}{8}$	$\frac{14}{8}$	$\frac{1792}{8}$	224	224.0
Largest norm.	0 1110 111	14	7	128	$\frac{7}{8}$	$\frac{15}{8}$	$\frac{1920}{8}$	240	240.0
Infinity	0 1111 000	—	—	—	—	—	—	∞	—

Figure 34 **Example nonnegative values for 8-bit floating-point format.** There are $k = 4$ exponent bits and $n = 3$ fraction bits. The bias is 7.

figure is divided into three regions representing the three classes of numbers. The different columns show how the exponent field encodes the exponent E, while the fraction field encodes the significand M, and together they form the represented value $V = 2^E \times M$. Closest to 0 are the denormalized numbers, starting with 0 itself. Denormalized numbers in this format have $E = 1 - 7 = -6$, giving a weight $2^E = \frac{1}{64}$. The fractions f and significands M range over the values $0, \frac{1}{8}, \ldots, \frac{7}{8}$, giving numbers V in the range 0 to $\frac{1}{64} \times \frac{7}{8} = \frac{7}{512}$.

The smallest normalized numbers in this format also have $E = 1 - 7 = -6$, and the fractions also range over the values $0, \frac{1}{8}, \ldots \frac{7}{8}$. However, the significands then range from $1 + 0 = 1$ to $1 + \frac{7}{8} = \frac{15}{8}$, giving numbers V in the range $\frac{8}{512} = \frac{1}{64}$ to $\frac{15}{512}$.

Observe the smooth transition between the largest denormalized number $\frac{7}{512}$ and the smallest normalized number $\frac{8}{512}$. This smoothness is due to our definition of E for denormalized values. By making it $1 - Bias$ rather than $-Bias$, we compensate for the fact that the significand of a denormalized number does not have an implied leading 1.

As we increase the exponent, we get successively larger normalized values, passing through 1.0 and then to the largest normalized number. This number has exponent $E = 7$, giving a weight $2^E = 128$. The fraction equals $\frac{7}{8}$, giving a significand $M = \frac{15}{8}$. Thus, the numeric value is $V = 240$. Going beyond this overflows to $+\infty$.

One interesting property of this representation is that if we interpret the bit representations of the values in Figure 34 as unsigned integers, they occur in ascending order, as do the values they represent as floating-point numbers. This is no accident—the IEEE format was designed so that floating-point numbers could be sorted using an integer sorting routine. A minor difficulty occurs when dealing with negative numbers, since they have a leading 1, and they occur in descending order, but this can be overcome without requiring floating-point operations to perform comparisons (see Problem 83).

Practice Problem 47

Consider a 5-bit floating-point representation based on the IEEE floating-point format, with one sign bit, two exponent bits ($k = 2$), and two fraction bits ($n = 2$). The exponent bias is $2^{2-1} - 1 = 1$.

The table that follows enumerates the entire nonnegative range for this 5-bit floating-point representation. Fill in the blank table entries using the following directions:

> e: The value represented by considering the exponent field to be an unsigned integer
>
> E: The value of the exponent after biasing
>
> 2^E: The numeric weight of the exponent
>
> f: The value of the fraction

M: The value of the significand

$2^E \times M$: The (unreduced) fractional value of the number

V: The reduced fractional value of the number

Decimal: The decimal representation of the number

Express the values of $2^E, f, M, 2^E \times M$, and V either as integers (when possible) or as fractions of the form $\frac{x}{y}$, where y is a power of 2. You need not fill in entries marked "$-$".

Bits	e	E	2^E	f	M	$2^E \times M$	V	Decimal
0 00 00								
0 00 01								
0 00 10								
0 00 11								
0 01 00								
0 01 01	1	0	1	$\frac{1}{4}$	$\frac{5}{4}$	$\frac{5}{4}$	$\frac{5}{4}$	1.25
0 01 10								
0 01 11								
0 10 00								
0 10 01								
0 10 10								
0 10 11								
0 11 00	$-$	$-$	$-$	$-$	$-$	$-$		$-$
0 11 01	$-$	$-$	$-$	$-$	$-$	$-$		$-$
0 11 10	$-$	$-$	$-$	$-$	$-$	$-$		$-$
0 11 11	$-$	$-$	$-$	$-$	$-$	$-$		$-$

Figure 35 shows the representations and numeric values of some important single- and double-precision floating-point numbers. As with the 8-bit format shown in Figure 34, we can see some general properties for a floating-point representation with a k-bit exponent and an n-bit fraction:

- The value $+0.0$ always has a bit representation of all zeros.
- The smallest positive denormalized value has a bit representation consisting of a 1 in the least significant bit position and otherwise all zeros. It has a fraction (and significand) value $M = f = 2^{-n}$ and an exponent value $E = -2^{k-1} + 2$. The numeric value is therefore $V = 2^{-n-2^{k-1}+2}$.
- The largest denormalized value has a bit representation consisting of an exponent field of all zeros and a fraction field of all ones. It has a fraction (and significand) value $M = f = 1 - 2^{-n}$ (which we have written $1 - \epsilon$) and an exponent value $E = -2^{k-1} + 2$. The numeric value is therefore $V = (1 - 2^{-n}) \times 2^{-2^{k-1}+2}$, which is just slightly smaller than the smallest normalized value.

Description	exp	frac	Single precision		Double precision	
			Value	Decimal	Value	Decimal
Zero	$00\cdots00$	$0\cdots00$	0	0.0	0	0.0
Smallest denorm.	$00\cdots00$	$0\cdots01$	$2^{-23}\times2^{-126}$	1.4×10^{-45}	$2^{-52}\times2^{-1022}$	4.9×10^{-324}
Largest denorm.	$00\cdots00$	$1\cdots11$	$(1-\epsilon)\times2^{-126}$	1.2×10^{-38}	$(1-\epsilon)\times2^{-1022}$	2.2×10^{-308}
Smallest norm.	$00\cdots01$	$0\cdots00$	1×2^{-126}	1.2×10^{-38}	1×2^{-1022}	2.2×10^{-308}
One	$01\cdots11$	$0\cdots00$	1×2^{0}	1.0	1×2^{0}	1.0
Largest norm.	$11\cdots10$	$1\cdots11$	$(2-\epsilon)\times2^{127}$	3.4×10^{38}	$(2-\epsilon)\times2^{1023}$	1.8×10^{308}

Figure 35 **Examples of nonnegative floating-point numbers.**

- The smallest positive normalized value has a bit representation with a 1 in the least significant bit of the exponent field and otherwise all zeros. It has a significand value $M = 1$ and an exponent value $E = -2^{k-1} + 2$. The numeric value is therefore $V = 2^{-2^{k-1}+2}$.

- The value 1.0 has a bit representation with all but the most significant bit of the exponent field equal to 1 and all other bits equal to 0. Its significand value is $M = 1$ and its exponent value is $E = 0$.

- The largest normalized value has a bit representation with a sign bit of 0, the least significant bit of the exponent equal to 0, and all other bits equal to 1. It has a fraction value of $f = 1 - 2^{-n}$, giving a significand $M = 2 - 2^{-n}$ (which we have written $2 - \epsilon$). It has an exponent value $E = 2^{k-1} - 1$, giving a numeric value $V = (2 - 2^{-n}) \times 2^{2^{k-1}-1} = (1 - 2^{-n-1}) \times 2^{2^{k-1}}$.

One useful exercise for understanding floating-point representations is to convert sample integer values into floating-point form. For example, we saw in Figure 14 that 12,345 has binary representation [11000000111001]. We create a normalized representation of this by shifting 13 positions to the right of a binary point, giving $12345 = 1.1000000111001_2 \times 2^{13}$. To encode this in IEEE single-precision format, we construct the fraction field by dropping the leading 1 and adding 10 zeros to the end, giving binary representation [10000001110010000000000]. To construct the exponent field, we add bias 127 to 13, giving 140, which has binary representation [10001100]. We combine this with a sign bit of 0 to get the floating-point representation in binary of [01000110010000001110010000000000]. Recall from Section 1.4 that we observed the following correlation in the bit-level representations of the integer value 12345 (0x3039) and the single-precision floating-point value 12345.0 (0x4640E400):

```
 0   0   0   0   3   0   3   9
00000000000000000011000000111001
              *************
        4   6   4   0   E   4   0   0
        01000110010000001110010000000000
```

We can now see that the region of correlation corresponds to the low-order bits of the integer, stopping just before the most significant bit equal to 1 (this bit forms the implied leading 1), matching the high-order bits in the fraction part of the floating-point representation.

Practice Problem 48

As mentioned in Problem 6, the integer 3,510,593 has hexadecimal representation 0x00359141, while the single-precision, floating-point number 3510593.0 has hexadecimal representation 0x4A564504. Derive this floating-point representation and explain the correlation between the bits of the integer and floating-point representations.

Practice Problem 49

A. For a floating-point format with an n-bit fraction, give a formula for the smallest positive integer that cannot be represented exactly (because it would require an $n+1$-bit fraction to be exact). Assume the exponent field size k is large enough that the range of representable exponents does not provide a limitation for this problem.

B. What is the numeric value of this integer for single-precision format $(n = 23)$?

4.4 Rounding

Floating-point arithmetic can only approximate real arithmetic, since the representation has limited range and precision. Thus, for a value x, we generally want a systematic method of finding the "closest" matching value x' that can be represented in the desired floating-point format. This is the task of the *rounding* operation. One key problem is to define the direction to round a value that is halfway between two possibilities. For example, if I have $1.50 and want to round it to the nearest dollar, should the result be $1 or $2? An alternative approach is to maintain a lower and an upper bound on the actual number. For example, we could determine representable values x^- and x^+ such that the value x is guaranteed to lie between them: $x^- \leq x \leq x^+$. The IEEE floating-point format defines four different *rounding modes*. The default method finds a closest match, while the other three can be used for computing upper and lower bounds.

Figure 36 illustrates the four rounding modes applied to the problem of rounding a monetary amount to the nearest whole dollar. Round-to-even (also called round-to-nearest) is the default mode. It attempts to find a closest match. Thus, it rounds $1.40 to $1 and $1.60 to $2, since these are the closest whole dollar values. The only design decision is to determine the effect of rounding values that are halfway between two possible results. Round-to-even mode adopts the

Mode	$1.40	$1.60	$1.50	$2.50	$−1.50
Round-to-even	$1	$2	$2	$2	$−2
Round-toward-zero	$1	$1	$1	$2	$−1
Round-down	$1	$1	$1	$2	$−2
Round-up	$2	$2	$2	$3	$−1

Figure 36 **Illustration of rounding modes for dollar rounding.** The first rounds to a nearest value, while the other three bound the result above or below.

convention that it rounds the number either upward or downward such that the least significant digit of the result is even. Thus, it rounds both $1.50 and $2.50 to $2.

The other three modes produce guaranteed bounds on the actual value. These can be useful in some numerical applications. Round-toward-zero mode rounds positive numbers downward and negative numbers upward, giving a value $\hat{x}$ such that $|\hat{x}| \leq |x|$. Round-down mode rounds both positive and negative numbers downward, giving a value x^- such that $x^- \leq x$. Round-up mode rounds both positive and negative numbers upward, giving a value x^+ such that $x \leq x^+$.

Round-to-even at first seems like it has a rather arbitrary goal—why is there any reason to prefer even numbers? Why not consistently round values halfway between two representable values upward? The problem with such a convention is that one can easily imagine scenarios in which rounding a set of data values would then introduce a statistical bias into the computation of an average of the values. The average of a set of numbers that we rounded by this means would be slightly higher than the average of the numbers themselves. Conversely, if we always rounded numbers halfway between downward, the average of a set of rounded numbers would be slightly lower than the average of the numbers themselves. Rounding toward even numbers avoids this statistical bias in most real-life situations. It will round upward about 50% of the time and round downward about 50% of the time.

Round-to-even rounding can be applied even when we are not rounding to a whole number. We simply consider whether the least significant digit is even or odd. For example, suppose we want to round decimal numbers to the nearest hundredth. We would round 1.2349999 to 1.23 and 1.2350001 to 1.24, regardless of rounding mode, since they are not halfway between 1.23 and 1.24. On the other hand, we would round both 1.2350000 and 1.2450000 to 1.24, since 4 is even.

Similarly, round-to-even rounding can be applied to binary fractional numbers. We consider least significant bit value 0 to be even and 1 to be odd. In general, the rounding mode is only significant when we have a bit pattern of the form $XX \cdots X.YY \cdots Y100 \cdots$, where X and Y denote arbitrary bit values with the rightmost Y being the position to which we wish to round. Only bit patterns of this form denote values that are halfway between two possible results. As examples, consider the problem of rounding values to the nearest quarter (i.e., 2 bits to the right of the binary point). We would round 10.000011_2 $(2\frac{3}{32})$ down to 10.00_2 (2), and 10.00110_2 ($2\frac{3}{16}$) up to 10.01_2 ($2\frac{1}{4}$), because these values are not halfway between two possible values. We would round 10.11100_2 ($2\frac{7}{8}$) up to 11.00_2 (3) and 10.10100_2 ($2\frac{5}{8}$) down to 10.10_2 ($2\frac{1}{2}$), since these values are halfway between two possible results, and we prefer to have the least significant bit equal to zero.

Practice Problem 50

Show how the following binary fractional values would be rounded to the nearest half (1 bit to the right of the binary point), according to the round-to-even rule. In each case, show the numeric values, both before and after rounding.

A. 10.010_2

B. 10.011_2

C. 10.110_2

D. 11.001_2

Practice Problem 51

We saw in Problem 46 that the Patriot missile software approximated 0.1 as $x = 0.00011001100110011001100_2$. Suppose instead that they had used IEEE round-to-even mode to determine an approximation x' to 0.1 with 23 bits to the right of the binary point.

A. What is the binary representation of x'?

B. What is the approximate decimal value of $x' - 0.1$?

C. How far off would the computed clock have been after 100 hours of operation?

D. How far off would the program's prediction of the position of the Scud missile have been?

Practice Problem 52

Consider the following two 7-bit floating-point representations based on the IEEE floating point format. Neither has a sign bit—they can only represent nonnegative numbers.

1. Format A
   - There are $k = 3$ exponent bits. The exponent bias is 3.
   - There are $n = 4$ fraction bits.

2. Format B
   - There are $k = 4$ exponent bits. The exponent bias is 7.
   - There are $n = 3$ fraction bits.

Below, you are given some bit patterns in Format A, and your task is to convert them to the closest value in Format B. If necessary, you should apply the round-to-even rounding rule. In addition, give the values of numbers given by the Format A and Format B bit patterns. Give these as whole numbers (e.g., 17) or as fractions (e.g., 17/64).

Format A		Format B	
Bits	Value	Bits	Value
011 0000	1	0111 000	1
101 1110			
010 1001			
110 1111			
000 0001			

4.5 Floating-Point Operations

The IEEE standard specifies a simple rule for determining the result of an arithmetic operation such as addition or multiplication. Viewing floating-point values x and y as real numbers, and some operation $\odot$ defined over real numbers, the computation should yield $Round(x \odot y)$, the result of applying rounding to the exact result of the real operation. In practice, there are clever tricks floating-point unit designers use to avoid performing this exact computation, since the computation need only be sufficiently precise to guarantee a correctly rounded result. When one of the arguments is a special value such as -0, ∞, or NaN, the standard specifies conventions that attempt to be reasonable. For example, $1/-0$ is defined to yield $-\infty$, while $1/+0$ is defined to yield $+\infty$.

One strength of the IEEE standard's method of specifying the behavior of floating-point operations is that it is independent of any particular hardware or software realization. Thus, we can examine its abstract mathematical properties without considering how it is actually implemented.

We saw earlier that integer addition, both unsigned and two's complement, forms an abelian group. Addition over real numbers also forms an abelian group, but we must consider what effect rounding has on these properties. Let us define $x +^f y$ to be $Round(x + y)$. This operation is defined for all values of x and y, although it may yield infinity even when both x and y are real numbers due to overflow. The operation is commutative, with $x +^f y = y +^f x$ for all values of x and y. On the other hand, the operation is not associative. For example, with single-precision floating point the expression (3.14+1e10)-1e10 evaluates to 0.0—the value 3.14 is lost due to rounding. On the other hand, the expression 3.14+(1e10-1e10) evaluates to 3.14. As with an abelian group, most values have inverses under floating-point addition, that is, $x +^f -x = 0$. The exceptions are infinities (since $+\infty - \infty = NaN$), and NaN's, since $NaN +^f x = NaN$ for any x.

The lack of associativity in floating-point addition is the most important group property that is lacking. It has important implications for scientific programmers and compiler writers. For example, suppose a compiler is given the following code fragment:

```
x = a + b + c;
y = b + c + d;
```

The compiler might be tempted to save one floating-point addition by generating the following code:

```
t = b + c;
x = a + t;
y = t + d;
```

However, this computation might yield a different value for x than would the original, since it uses a different association of the addition operations. In most applications, the difference would be so small as to be inconsequential. Unfortunately, compilers have no way of knowing what trade-offs the user is willing to make between efficiency and faithfulness to the exact behavior of the original program. As a result, they tend to be very conservative, avoiding any optimizations that could have even the slightest effect on functionality.

On the other hand, floating-point addition satisfies the following monotonicity property: if $a \geq b$ then $x + a \geq x + b$ for any values of a, b, and x other than NaN. This property of real (and integer) addition is not obeyed by unsigned or two's-complement addition.

Floating-point multiplication also obeys many of the properties one normally associates with multiplication. Let us define $x *^f y$ to be $Round(x \times y)$. This operation is closed under multiplication (although possibly yielding infinity or NaN), it is commutative, and it has 1.0 as a multiplicative identity. On the other hand, it is not associative, due to the possibility of overflow or the loss of precision due to rounding. For example, with single-precision floating point, the expression (1e20*1e20)*1e-20 evaluates to $+\infty$, while 1e20*(1e20*1e-20) evaluates to 1e20. In addition, floating-point multiplication does not distribute over addition. For example, with single-precision floating point, the expression 1e20*(1e20-1e20) evaluates to 0.0, while 1e20*1e20-1e20*1e20 evaluates to NaN.

On the other hand, floating-point multiplication satisfies the following monotonicity properties for any values of a, b, and c other than NaN:

$$a \geq b \text{ and } c \geq 0 \Rightarrow a *^f c \geq b *^f c$$

$$a \geq b \text{ and } c \leq 0 \Rightarrow a *^f c \leq b *^f c$$

In addition, we are also guaranteed that $a *^f a \geq 0$, as long as $a \neq NaN$. As we saw earlier, none of these monotonicity properties hold for unsigned or two's-complement multiplication.

This lack of associativity and distributivity is of serious concern to scientific programmers and to compiler writers. Even such a seemingly simple task as writing code to determine whether two lines intersect in 3-dimensional space can be a major challenge.

4.6 Floating Point in C

All versions of C provide two different floating-point data types: float and double. On machines that support IEEE floating point, these data types correspond to single- and double-precision floating point. In addition, the machines use the round-to-even rounding mode. Unfortunately, since the C standards do not require the machine to use IEEE floating point, there are no standard methods to change the rounding mode or to get special values such as $-0, +\infty, -\infty$, or *NaN*. Most systems provide a combination of include ('.h') files and procedure libraries to provide access to these features, but the details vary from one system to another. For example, the GNU compiler GCC defines program constants INFINITY (for $+\infty$) and NAN (for *NaN*) when the following sequence occurs in the program file:

```
#define _GNU_SOURCE 1
#include <math.h>
```

More recent versions of C, including ISO C99, include a third floating-point data type, long double. For many machines and compilers, this data type is equivalent to the double data type. For Intel-compatible machines, however, GCC implements this data type using an 80-bit "extended precision" format, providing a much larger range and precision than does the standard 64-bit format. The properties of this format are investigated in Problem 85.

Practice Problem 53

Fill in the following macro definitions to generate the double-precision values $+\infty, -\infty$, and 0:

```
#define POS_INFINITY
#define NEG_INFINITY
#define NEG_ZERO
```

You cannot use any include files (such as math.h), but you can make use of the fact that the largest finite number that can be represented with double precision is around 1.8×10^{308}.

When casting values between int, float, and double formats, the program changes the numeric values and the bit representations as follows (assuming a 32-bit int):

- From int to float, the number cannot overflow, but it may be rounded.
- From int or float to double, the exact numeric value can be preserved because double has both greater range (i.e., the range of representable values), as well as greater precision (i.e., the number of significant bits).
- From double to float, the value can overflow to $+\infty$ or $-\infty$, since the range is smaller. Otherwise, it may be rounded, because the precision is smaller.
- From float or double to int the value will be rounded toward zero. For example, 1.999 will be converted to 1, while -1.999 will be converted to -1. Furthermore, the value may overflow. The C standards do not specify a fixed result for this case. Intel-compatible microprocessors designate the

bit pattern $[10\cdots00]$ ($TMin_w$ for word size w) as an *integer indefinite* value. Any conversion from floating point to integer that cannot assign a reasonable integer approximation yields this value. Thus, the expression (int) +1e10 yields −21483648, generating a negative value from a positive one.

Web Aside DATA:IA32-FP   Intel IA32 floating-point arithmetic

In the next chapter, we will begin an in-depth study of Intel IA32 processors, the processor found in many of today's personal computers. Here we highlight an idiosyncrasy of these machines that can seriously affect the behavior of programs operating on floating-point numbers when compiled with GCC.

IA32 processors, like most other processors, have special memory elements called *registers* for holding floating-point values as they are being computed and used. The unusual feature of IA32 is that the floating-point registers use a special 80-bit *extended-precision* format to provide a greater range and precision than the normal 32-bit single-precision and 64-bit double-precision formats used for values held in memory. (See Problem 85.) All single- and double-precision numbers are converted to this format as they are loaded from memory into floating-point registers. The arithmetic is always performed in extended precision. Numbers are converted from extended precision to single- or double-precision format as they are stored in memory.

This extension to 80 bits for all register data and then contraction to a smaller format for memory data has some undesirable consequences for programmers. It means that storing a number from a register to memory and then retrieving it back into the register can cause it to change, due to rounding, underflow, or overflow. This storing and retrieving is not always visible to the C programmer, leading to some very peculiar results.

More recent versions of Intel processors, including both IA32 and newer 64-bit machines, provide direct hardware support for single- and double-precision floating-point operations. The peculiarities of the historic IA32 approach will diminish in importance with new hardware and with compilers that generate code based on the newer floating-point instructions.

Aside   Ariane 5: the high cost of floating-point overflow

Converting large floating-point numbers to integers is a common source of programming errors. Such an error had disastrous consequences for the maiden voyage of the Ariane 5 rocket, on June 4, 1996. Just 37 seconds after liftoff, the rocket veered off its flight path, broke up, and exploded. Communication satellites valued at $500 million were on board the rocket.

A later investigation [69, 39] showed that the computer controlling the inertial navigation system had sent invalid data to the computer controlling the engine nozzles. Instead of sending flight control information, it had sent a diagnostic bit pattern indicating that an overflow had occurred during the conversion of a 64-bit floating-point number to a 16-bit signed integer.

The value that overflowed measured the horizontal velocity of the rocket, which could be more than 5 times higher than that achieved by the earlier Ariane 4 rocket. In the design of the Ariane 4 software, they had carefully analyzed the numeric values and determined that the horizontal velocity would never overflow a 16-bit number. Unfortunately, they simply reused this part of the software in the Ariane 5 without checking the assumptions on which it had been based.

© Fourmy/REA/SABA/Corbis

Practice Problem 54

Assume variables x, f, and d are of type int, float, and double, respectively. Their values are arbitrary, except that neither f nor d equals $+\infty$, $-\infty$, or *NaN*. For each of the following C expressions, either argue that it will always be true (i.e., evaluate to 1) or give a value for the variables such that it is not true (i.e., evaluates to 0).

A. x == (int)(double) x

B. x == (int)(float) x

C. d == (double)(float) d

D. f == (float)(double) f

E. f == -(-f)

F. 1.0/2 == 1/2.0

G. d*d >= 0.0

H. (f+d)-f == d

5 Summary

Computers encode information as bits, generally organized as sequences of bytes. Different encodings are used for representing integers, real numbers, and character strings. Different models of computers use different conventions for encoding numbers and for ordering the bytes within multi-byte data.

The C language is designed to accommodate a wide range of different implementations in terms of word sizes and numeric encodings. Most current machines have 32-bit word sizes, although high-end machines increasingly have 64-bit words. Most machines use two's-complement encoding of integers and IEEE encoding of floating point. Understanding these encodings at the bit level, as well as understanding the mathematical characteristics of the arithmetic operations, is important for writing programs that operate correctly over the full range of numeric values.

When casting between signed and unsigned integers of the same size, most C implementations follow the convention that the underlying bit pattern does not change. On a two's-complement machine, this behavior is characterized by functions $T2U_w$ and $U2T_w$, for a w-bit value. The implicit casting of C gives results that many programmers do not anticipate, often leading to program bugs.

Due to the finite lengths of the encodings, computer arithmetic has properties quite different from conventional integer and real arithmetic. The finite length can cause numbers to overflow, when they exceed the range of the representation. Floating-point values can also underflow, when they are so close to 0.0 that they are changed to zero.

The finite integer arithmetic implemented by C, as well as most other programming languages, has some peculiar properties compared to true integer arithmetic. For example, the expression x*x can evaluate to a negative number due to overflow. Nonetheless, both unsigned and two's-complement arithmetic satisfy many of the other properties of integer arithmetic, including associativity, commutativity, and distributivity. This allows compilers to do many optimizations. For example, in replacing the expression 7*x by (x<<3)-x, we make use of the associative, commutative, and distributive properties, along with the relationship between shifting and multiplying by powers of 2.

We have seen several clever ways to exploit combinations of bit-level operations and arithmetic operations. For example, we saw that with two's-complement arithmetic ~x+1 is equivalent to -x. As another example, suppose we want a bit pattern of the form $[0, \ldots, 0, 1, \ldots, 1]$, consisting of $w - k$ zeros followed by k ones. Such bit patterns are useful for masking operations. This pattern can be generated by the C expression (1<<k)-1, exploiting the property that the desired bit pattern has numeric value $2^k - 1$. For example, the expression (1<<8)-1 will generate the bit pattern 0xFF.

Floating-point representations approximate real numbers by encoding numbers of the form $x \times 2^y$. The most common floating-point representation is defined by IEEE Standard 754. It provides for several different precisions, with the most common being single (32 bits) and double (64 bits). IEEE floating point also has representations for special values representing plus and minus infinity, as well as not-a-number.

Floating-point arithmetic must be used very carefully, because it has only limited range and precision, and because it does not obey common mathematical properties such as associativity.

Bibliographic Notes

Reference books on C [48, 58] discuss properties of the different data types and operations. (Of these two, only Steele and Harbison [48] cover the newer features found in ISO C99.) The C standards do not specify details such as precise word sizes or numeric encodings. Such details are intentionally omitted to make it possible to implement C on a wide range of different machines. Several books have been written giving advice to C programmers [59, 70] that warn about problems with overflow, implicit casting to unsigned, and some of the other pitfalls we have covered in this chapter. These books also provide helpful advice on variable naming, coding styles, and code testing. Seacord's book on security issues in C and C++ programs [94], combines information about C programs, how they are compiled and executed, and how vulnerabilities may arise. Books on Java (we recommend the one coauthored by James Gosling, the creator of the language [4]) describe the data formats and arithmetic operations supported by Java.

Most books on logic design [56, 115] have a section on encodings and arithmetic operations. Such books describe different ways of implementing arithmetic circuits. Overton's book on IEEE floating point [78] provides a detailed description of the format as well as the properties from the perspective of a numerical applications programmer.

Homework Problems

55 ◆

Compile and run the sample code that uses `show_bytes` (file `show-bytes.c`) on different machines to which you have access. Determine the byte orderings used by these machines.

56 ◆

Try running the code for `show_bytes` for different sample values.

57 ◆

Write procedures `show_short`, `show_long`, and `show_double` that print the byte representations of C objects of types `short int`, `long int`, and `double`, respectively. Try these out on several machines.

58 ◆◆

Write a procedure `is_little_endian` that will return 1 when compiled and run on a little-endian machine, and will return 0 when compiled and run on a big-endian machine. This program should run on any machine, regardless of its word size.

59 ◆◆

Write a C expression that will yield a word consisting of the least significant byte of x and the remaining bytes of y. For operands x = 0x89ABCDEF and y = 0x76543210, this would give 0x765432EF.

60 ◆◆

Suppose we number the bytes in a w-bit word from 0 (least significant) to $w/8 - 1$ (most significant). Write code for the following C function, which will return an unsigned value in which byte i of argument x has been replaced by byte b:

```
unsigned replace_byte (unsigned x, int i, unsigned char b);
```

Here are some examples showing how the function should work:

```
replace_byte(0x12345678, 2, 0xAB) --> 0x12AB5678
replace_byte(0x12345678, 0, 0xAB) --> 0x123456AB
```

Bit-level integer coding rules

In several of the following problems, we will artificially restrict what programming constructs you can use to help you gain a better understanding of the bit-level, logic, and arithmetic operations of C. In answering these problems, your code must follow these rules:

- Assumptions
    - Integers are represented in two's-complement form.
    - Right shifts of signed data are performed arithmetically.
    - Data type int is w bits long. For some of the problems, you will be given a specific value for w, but otherwise your code should work as long as w is a multiple of 8. You can use the expression sizeof(int)<<3 to compute w.

- Forbidden
    - Conditionals (if or ?:), loops, switch statements, function calls, and macro invocations.
    - Division, modulus, and multiplication.
    - Relative comparison operators (<, >, <=, and >=).
    - Casting, either explicit or implicit.

- Allowed operations
    - All bit-level and logic operations.
    - Left and right shifts, but only with shift amounts between 0 and $w - 1$.
    - Addition and subtraction.
    - Equality (==) and inequality (!=) tests. (Some of the problems do not allow these.)
    - Integer constants INT_MIN and INT_MAX.

Even with these rules, you should try to make your code readable by choosing descriptive variable names and using comments to describe the logic behind your solutions. As an example, the following code extracts the most significant byte from integer argument x:

```
/* Get most significant byte from x */
int get_msb(int x) {
    /* Shift by w-8 */
    int shift_val = (sizeof(int)-1)<<3;
    /* Arithmetic shift */
    int xright = x >> shift_val;
    /* Zero all but LSB */
    return xright & 0xFF;
}
```

61 ◆◆

Write C expressions that evaluate to 1 when the following conditions are true, and to 0 when they are false. Assume x is of type `int`.

A. Any bit of x equals 1.

B. Any bit of x equals 0.

C. Any bit in the least significant byte of x equals 1.

D. Any bit in the most significant byte of x equals 0.

Your code should follow the bit-level integer coding rules, with the additional restriction that you may not use equality (`==`) or inequality (`!=`) tests.

62 ◆◆◆

Write a function `int_shifts_are_arithmetic()` that yields 1 when run on a machine that uses arithmetic right shifts for `int`s, and 0 otherwise. Your code should work on a machine with any word size. Test your code on several machines.

63 ◆◆◆

Fill in code for the following C functions. Function `srl` performs a logical right shift using an arithmetic right shift (given by value `xsra`), followed by other operations not including right shifts or division. Function `sra` performs an arithmetic right shift using a logical right shift (given by value `xsrl`), followed by other operations not including right shifts or division. You may use the computation `8*sizeof(int)` to determine w, the number of bits in data type `int`. The shift amount k can range from 0 to $w - 1$.

```
unsigned srl(unsigned x, int k) {
    /* Perform shift arithmetically */
    unsigned xsra = (int) x >> k;
    .
    .
    .
}
```

```
int sra(int x, int k) {
    /* Perform shift logically */
    int xsrl = (unsigned) x >> k;
      .
      .
      .
}
```

64 ◆

Write code to implement the following function:

```
/* Return 1 when any odd bit of x equals 1; 0 otherwise.
   Assume w=32. */
int any_odd_one(unsigned x);
```

Your function should follow the bit-level integer coding rules, except that you may assume that data type int has $w = 32$ bits.

65 ◆◆◆◆

Write code to implement the following function:

```
/* Return 1 when x contains an odd number of 1s; 0 otherwise.
   Assume w=32. */
int odd_ones(unsigned x);
```

Your function should follow the bit-level integer coding rules, except that you may assume that data type int has $w = 32$ bits.

   Your code should contain a total of at most 12 arithmetic, bit-wise, and logical operations.

66 ◆◆◆

Write code to implement the following function:

```
/*
 * Generate mask indicating leftmost 1 in x.  Assume w=32.
 * For example 0xFF00 -> 0x8000, and 0x6600 --> 0x4000.
 * If x = 0, then return 0.
 */
int leftmost_one(unsigned x);
```

Your function should follow the bit-level integer coding rules, except that you may assume that data type int has $w = 32$ bits.

   Your code should contain a total of at most 15 arithmetic, bit-wise, and logical operations.

   Hint: First transform x into a bit vector of the form $[0 \cdots 011 \cdots 1]$.

67 ◆◆

You are given the task of writing a procedure int_size_is_32() that yields 1 when run on a machine for which an int is 32 bits, and yields 0 otherwise. You are not allowed to use the sizeof operator. Here is a first attempt:

```
1   /* The following code does not run properly on some machines */
2   int bad_int_size_is_32() {
3       /* Set most significant bit (msb) of 32-bit machine */
4       int set_msb = 1 << 31;
5       /* Shift past msb of 32-bit word */
6       int beyond_msb = 1 << 32;
7
8       /* set_msb is nonzero when word size >= 32
9          beyond_msb is zero when word size <= 32 */
10      return set_msb && !beyond_msb;
11  }
```

When compiled and run on a 32-bit SUN SPARC, however, this procedure returns 0. The following compiler message gives us an indication of the problem:

```
warning: left shift count >= width of type
```

A. In what way does our code fail to comply with the C standard?

B. Modify the code to run properly on any machine for which data type int is at least 32 bits.

C. Modify the code to run properly on any machine for which data type int is at least 16 bits.

68 ◆◆

Write code for a function with the following prototype:

```
/*
 * Mask with least signficant n bits set to 1
 * Examples: n = 6 --> 0x2F, n = 17 --> 0x1FFFF
 * Assume 1 <= n <= w
 */
int lower_one_mask(int n);
```

Your function should follow the bit-level integer coding rules. Be careful of the case $n = w$.

69 ◆◆◆

Write code for a function with the following prototype:

```
/*
 * Do rotating left shift.  Assume 0 <= n < w
 * Examples when x = 0x12345678 and w = 32:
 *    n=4 -> 0x23456781, n=20 -> 0x67812345
 */
unsigned rotate_left(unsigned x, int n);
```

Your function should follow the bit-level integer coding rules. Be careful of the case $n = 0$.

70 ◆◆

Write code for the function with the following prototype:

```
/*
 * Return 1 when x can be represented as an n-bit, 2's complement
 * number; 0 otherwise
 * Assume 1 <= n <= w
 */
int fits_bits(int x, int n);
```

Your function should follow the bit-level integer coding rules.

71 ◆

You just started working for a company that is implementing a set of procedures to operate on a data structure where 4 signed bytes are packed into a 32-bit unsigned. Bytes within the word are numbered from 0 (least significant) to 3 (most significant). You have been assigned the task of implementing a function for a machine using two's-complement arithmetic and arithmetic right shifts with the following prototype:

```
/* Declaration of data type where 4 bytes are packed
   into an unsigned */
typedef unsigned packed_t;
```

```
/* Extract byte from word.  Return as signed integer */
int xbyte(packed_t word, int bytenum);
```

That is, the function will extract the designated byte and sign extend it to be a 32-bit int.

Your predecessor (who was fired for incompetence) wrote the following code:

```
/* Failed attempt at xbyte */
int xbyte(packed_t word, int bytenum)
{
    return (word >> (bytenum << 3)) & 0xFF;
}
```

A. What is wrong with this code?

B. Give a correct implementation of the function that uses only left and right shifts, along with one subtraction.

72 ◆◆

You are given the task of writing a function that will copy an integer val into a buffer buf, but it should do so only if enough space is available in the buffer.

Here is the code you write:

```
/* Copy integer into buffer if space is available */
/* WARNING: The following code is buggy */
```

```
void copy_int(int val, void *buf, int maxbytes) {
    if (maxbytes-sizeof(val) >= 0)
            memcpy(buf, (void *) &val, sizeof(val));
}
```

This code makes use of the library function memcpy. Although its use is a bit artificial here, where we simply want to copy an int, it illustrates an approach commonly used to copy larger data structures.

You carefully test the code and discover that it *always* copies the value to the buffer, even when maxbytes is too small.

A. Explain why the conditional test in the code always succeeds. **Hint:** The sizeof operator returns a value of type size_t.

B. Show how you can rewrite the conditional test to make it work properly.

73 ◆◆
Write code for a function with the following prototype:

```
/* Addition that saturates to TMin or TMax */
int saturating_add(int x, int y);
```

Instead of overflowing the way normal two's-complement addition does, saturating addition returns *TMax* when there would be positive overflow, and *TMin* when there would be negative overflow. Saturating arithmetic is commonly used in programs that perform digital signal processing.

Your function should follow the bit-level integer coding rules.

74 ◆◆
Write a function with the following prototype:

```
/* Determine whether arguments can be subtracted without overflow */
int tsub_ok(int x, int y);
```

This function should return 1 if the computation x − y does not overflow.

75 ◆◆◆
Suppose we want to compute the complete $2w$-bit representation of $x \cdot y$, where both x and y are unsigned, on a machine for which data type unsigned is w bits. The low-order w bits of the product can be computed with the expression x*y, so we only require a procedure with prototype

```
unsigned int unsigned_high_prod(unsigned x, unsigned y);
```

that computes the high-order w bits of $x \cdot y$ for unsigned variables.

We have access to a library function with prototype

```
int signed_high_prod(int x, int y);
```

that computes the high-order w bits of $x \cdot y$ for the case where x and y are in two's-complement form. Write code calling this procedure to implement the function for unsigned arguments. Justify the correctness of your solution.

Hint: Look at the relationship between the signed product $x \cdot y$ and the unsigned product $x' \cdot y'$ in the derivation of Equation 18.

76 ◆◆

Suppose we are given the task of generating code to multiply integer variable x by various different constant factors K. To be efficient, we want to use only the operations +, -, and <<. For the following values of K, write C expressions to perform the multiplication using at most three operations per expression.

A. $K = 17$:
B. $K = -7$:
C. $K = 60$:
D. $K = -112$:

77 ◆◆

Write code for a function with the following prototype:

```
/* Divide by power of two.  Assume 0 <= k < w-1 */
int divide_power2(int x, int k);
```

The function should compute $x/2^k$ with correct rounding, and it should follow the bit-level integer coding rules.

78 ◆◆

Write code for a function `mul3div4` that, for integer argument x, computes `3*x/4`, but following the bit-level integer coding rules. Your code should replicate the fact that the computation `3*x` can cause overflow.

79 ◆◆◆

Write code for a function `threefourths` which, for integer argument x, computes the value of $\frac{3}{4}x$, rounded toward zero. It should not overflow. Your function should follow the bit-level integer coding rules.

80 ◆◆

Write C expressions to generate the bit patterns that follow, where a^k represents k repetitions of symbol a. Assume a w-bit data type. Your code may contain references to parameters j and k, representing the values of j and k, but not a parameter representing w.

A. $1^{w-k}0^k$
B. $0^{w-k-j}1^k0^j$

81 ◆

We are running programs on a machine where values of type int are 32 bits. They are represented in two's complement, and they are right shifted arithmetically. Values of type unsigned are also 32 bits.

We generate arbitrary values x and y, and convert them to unsigned values as follows:

```
/* Create some arbitrary values */
int x = random();
int y = random();
/* Convert to unsigned */
unsigned ux = (unsigned) x;
unsigned uy = (unsigned) y;
```

For each of the following C expressions, you are to indicate whether or not the expression *always* yields 1. If it always yields 1, describe the underlying mathematical principles. Otherwise, give an example of arguments that make it yield 0.

A. `(x<y) == (-x>-y)`

B. `((x+y)<<4) + y-x == 17*y+15*x`

C. `~x+~y+1 == ~(x+y)`

D. `(ux-uy) == -(unsigned)(y-x)`

E. `((x >> 2) << 2) <= x`

82 ◆◆

Consider numbers having a binary representation consisting of an infinite string of the form $0.y\,y\,y\,y\,y\,y \cdots$, where y is a k-bit sequence. For example, the binary representation of $\frac{1}{3}$ is $0.01010101 \cdots$ ($y = 01$), while the representation of $\frac{1}{5}$ is $0.001100110011 \cdots$ ($y = 0011$).

A. Let $Y = B2U_k(y)$, that is, the number having binary representation y. Give a formula in terms of Y and k for the value represented by the infinite string. **Hint:** Consider the effect of shifting the binary point k positions to the right.

B. What is the numeric value of the string for the following values of y?

(a) 101

(b) 0110

(c) 010011

83 ◆

Fill in the return value for the following procedure, which tests whether its first argument is less than or equal to its second. Assume the function f2u returns an unsigned 32-bit number having the same bit representation as its floating-point argument. You can assume that neither argument is *NaN*. The two flavors of zero, $+0$ and -0, are considered equal.

```
int float_le(float x, float y) {
    unsigned ux = f2u(x);
    unsigned uy = f2u(y);
```

```
/* Get the sign bits */
unsigned sx = ux >> 31;
unsigned sy = uy >> 31;

/* Give an expression using only ux, uy, sx, and sy */
return _____ ;
}
```

84 ◆

Given a floating-point format with a k-bit exponent and an n-bit fraction, write formulas for the exponent E, significand M, the fraction f, and the value V for the quantities that follow. In addition, describe the bit representation.

A. The number 7.0

B. The largest odd integer that can be represented exactly

C. The reciprocal of the smallest positive normalized value

85 ◆

Intel-compatible processors also support an "extended precision" floating-point format with an 80-bit word divided into a sign bit, $k = 15$ exponent bits, a single *integer* bit, and $n = 63$ fraction bits. The integer bit is an explicit copy of the implied bit in the IEEE floating-point representation. That is, it equals 1 for normalized values and 0 for denormalized values. Fill in the following table giving the approximate values of some "interesting" numbers in this format:

Description	Extended precision	
	Value	Decimal
Smallest positive denormalized		
Smallest positive normalized		
Largest normalized		

86 ◆

Consider a 16-bit floating-point representation based on the IEEE floating-point format, with one sign bit, seven exponent bits ($k = 7$), and eight fraction bits ($n = 8$). The exponent bias is $2^{7-1} - 1 = 63$.

Fill in the table that follows for each of the numbers given, with the following instructions for each column:

Hex: The four hexadecimal digits describing the encoded form.

M: The value of the significand. This should be a number of the form x or $\frac{x}{y}$, where x is an integer, and y is an integral power of 2. Examples include 0, $\frac{67}{64}$, and $\frac{1}{256}$.

E: The integer value of the exponent.

V: The numeric value represented. Use the notation x or $x \times 2^z$, where x and z are integers.

As an example, to represent the number $\frac{7}{8}$, we would have $s = 0$, $M = \frac{7}{4}$, and $E = -1$. Our number would therefore have an exponent field of 0x3E (decimal value $63 - 1 = 62$) and a significand field 0xC0 (binary 11000000_2), giving a hex representation 3EC0.

You need not fill in entries marked "—".

Description	Hex	M	E	V
-0				—
Smallest value > 2				
512				—
Largest denormalized				
$-\infty$		—	—	—
Number with hex representation 3BB0	—			

87 ◆◆

Consider the following two 9-bit floating-point representations based on the IEEE floating-point format.

1. Format A
   - There is one sign bit.
   - There are $k = 5$ exponent bits. The exponent bias is 15.
   - There are $n = 3$ fraction bits.

2. Format B
   - There is one sign bit.
   - There are $k = 4$ exponent bits. The exponent bias is 7.
   - There are $n = 4$ fraction bits.

Below, you are given some bit patterns in Format A, and your task is to convert them to the closest value in Format B. If rounding is necessary, you should *round toward* $+\infty$. In addition, give the values of numbers given by the Format A and Format B bit patterns. Give these as whole numbers (e.g., 17) or as fractions (e.g., $17/64$ or $17/2^6$).

Format A		Format B	
Bits	Value	Bits	Value
1 01111 001	$\frac{-9}{8}$	1 0111 0010	$\frac{-9}{8}$
0 10110 011			
1 00111 010			
0 00000 111			
1 11100 000			
0 10111 100			

88 ◆

We are running programs on a machine where values of type int have a 32-bit two's-complement representation. Values of type float use the 32-bit IEEE format, and values of type double use the 64-bit IEEE format.

We generate arbitrary integer values x, y, and z, and convert them to values of type double as follows:

```
/* Create some arbitrary values */
int x = random();
int y = random();
int z = random();
/* Convert to double */
double    dx = (double) x;
double    dy = (double) y;
double    dz = (double) z;
```

For each of the following C expressions, you are to indicate whether or not the expression *always* yields 1. If it always yields 1, describe the underlying mathematical principles. Otherwise, give an example of arguments that make it yield 0. Note that you cannot use an IA32 machine running GCC to test your answers, since it would use the 80-bit extended-precision representation for both float and double.

A. (float) x == (float) dx

B. dx – dy == (double) (x-y)

C. (dx + dy) + dz == dx + (dy + dz)

D. (dx * dy) * dz == dx * (dy * dz)

E. dx / dx == dz / dz

89 ◆

You have been assigned the task of writing a C function to compute a floating-point representation of 2^x. You decide that the best way to do this is to directly construct the IEEE single-precision representation of the result. When x is too small, your routine will return 0.0. When x is too large, it will return $+\infty$. Fill in the blank portions of the code that follows to compute the correct result. Assume the function u2f returns a floating-point value having an identical bit representation as its unsigned argument.

```
float fpwr2(int x)
{

    /* Result exponent and fraction */
    unsigned exp, frac;
    unsigned u;

    if (x < _____) {
        /* Too small.  Return 0.0 */
        exp = _____;
        frac = _____;
    } else if (x < _____) {
```

```
            /* Denormalized result */
            exp = _____;
            frac = _____;
    } else if (x < _____) {
            /* Normalized result. */
            exp = _____;
            frac = _____;
    } else {
            /* Too big.  Return +oo */
            exp = _____;
            frac = _____;
    }

    /* Pack exp and frac into 32 bits */
    u = exp << 23 | frac;
    /* Return as float */
    return u2f(u);
}
```

90 ◆

Around 250 B.C., the Greek mathematician Archimedes proved that $\frac{223}{71} < \pi < \frac{22}{7}$. Had he had access to a computer and the standard library `<math.h>`, he would have been able to determine that the single-precision floating-point approximation of π has the hexadecimal representation 0x40490FDB. Of course, all of these are just approximations, since π is not rational.

  A. What is the fractional binary number denoted by this floating-point value?

  B. What is the fractional binary representation of $\frac{22}{7}$? **Hint:** See Problem 82.

  C. At what bit position (relative to the binary point) do these two approximations to π diverge?

Bit-level floating-point coding rules

In the following problems, you will write code to implement floating-point functions, operating directly on bit-level representations of floating-point numbers. Your code should exactly replicate the conventions for IEEE floating-point operations, including using round-to-even mode when rounding is required.

Toward this end, we define data type `float_bits` to be equivalent to unsigned:

```
/* Access bit-level representation floating-point number */
typedef unsigned float_bits;
```

Rather than using data type `float` in your code, you will use `float_bits`. You may use both `int` and `unsigned` data types, including unsigned and integer constants and operations. You may not use any unions, structs, or arrays. Most significantly, you may not use any floating-point data types, operations, or constants. Instead, your code should perform the bit manipulations that implement the specified floating-point operations.

The following function illustrates the use of these coding rules. For argument f, it returns ± 0 if f is denormalized (preserving the sign of f) and returns f otherwise.

```
/* If f is denorm, return 0.  Otherwise, return f */
float_bits float_denorm_zero(float_bits f) {
    /* Decompose bit representation into parts */
    unsigned sign = f>>31;
    unsigned exp =  f>>23 & 0xFF;
    unsigned frac = f     & 0x7FFFFF;
    if (exp == 0) {
        /* Denormalized.  Set fraction to 0 */
        frac = 0;
    }
    /* Reassemble bits */
    return (sign << 31) | (exp << 23) | frac;
}
```

91 ◆◆

Following the bit-level floating-point coding rules, implement the function with the following prototype:

```
/* Compute -f.  If f is NaN, then return f. */
float_bits float_negate(float_bits f);
```

For floating-point number f, this function computes $-f$. If f is NaN, your function should simply return f.

Test your function by evaluating it for all 2^{32} values of argument f and comparing the result to what would be obtained using your machine's floating-point operations.

92 ◆◆

Following the bit-level floating-point coding rules, implement the function with the following prototype:

```
/* Compute |f|.  If f is NaN, then return f. */
float_bits float_absval(float_bits f);
```

For floating-point number f, this function computes $|f|$. If f is NaN, your function should simply return f.

Test your function by evaluating it for all 2^{32} values of argument f and comparing the result to what would be obtained using your machine's floating-point operations.

93 ◆◆◆

Following the bit-level floating-point coding rules, implement the function with the following prototype:

```
/* Compute 2*f.  If f is NaN, then return f. */
float_bits float_twice(float_bits f);
```

For floating-point number f, this function computes $2.0 \cdot f$. If f is *NaN*, your function should simply return f.

Test your function by evaluating it for all 2^{32} values of argument f and comparing the result to what would be obtained using your machine's floating-point operations.

94 ◆◆◆

Following the bit-level floating-point coding rules, implement the function with the following prototype:

```
/* Compute 0.5*f.  If f is NaN, then return f. */
float_bits float_half(float_bits f);
```

For floating-point number f, this function computes $0.5 \cdot f$. If f is *NaN*, your function should simply return f.

Test your function by evaluating it for all 2^{32} values of argument f and comparing the result to what would be obtained using your machine's floating-point operations.

95 ◆◆◆◆

Following the bit-level floating-point coding rules, implement the function with the following prototype:

```
/*
 * Compute (int) f.
 * If conversion causes overflow or f is NaN, return 0x80000000
 */
int float_f2i(float_bits f);
```

For floating-point number f, this function computes (int) f. Your function should round toward zero. If f cannot be represented as an integer (e.g., it is out of range, or it is *NaN*), then the function should return 0x80000000.

Test your function by evaluating it for all 2^{32} values of argument f and comparing the result to what would be obtained using your machine's floating-point operations.

96 ◆◆◆◆

Following the bit-level floating-point coding rules, implement the function with the following prototype:

```
/* Compute (float) i */
float_bits float_i2f(int i);
```

For argument `i`, this function computes the bit-level representation of `(float) i`.

Test your function by evaluating it for all 2^{32} values of argument `f` and comparing the result to what would be obtained using your machine's floating-point operations.

Solutions to Practice Problems

Solution to Problem 1

Understanding the relation between hexadecimal and binary formats will be important once we start looking at machine-level programs. The method for doing these conversions is in the text, but it takes a little practice to become familiar.

A. 0x39A7F8 to binary:

Hexadecimal	3	9	A	7	F	8
Binary	0011	1001	1010	0111	1111	1000

B. Binary 1100100101111011 to hexadecimal:

Binary	1100	1001	0111	1011
Hexadecimal	C	9	7	B

C. 0xD5E4C to binary:

Hexadecimal	D	5	E	4	C
Binary	1101	0101	1110	0100	1100

D. Binary 1001101110011110110101 to hexadecimal:

Binary	10	0110	1110	0111	1011	0101
Hexadecimal	2	6	E	7	B	5

Solution to Problem 2

This problem gives you a chance to think about powers of 2 and their hexadecimal representations.

n	2^n (Decimal)	2^n (Hexadecimal)
9	512	0x200
19	524,288	0x80000
14	16,384	0x4000
16	65,536	0x10000
17	131,072	0x20000
5	32	0x20
7	128	0x80

Solution to Problem 3

This problem gives you a chance to try out conversions between hexadecimal and decimal representations for some smaller numbers. For larger ones, it becomes much more convenient and reliable to use a calculator or conversion program.

Decimal	Binary	Hexadecimal
0	0000 0000	0x00
$167 = 10 \cdot 16 + 7$	1010 0111	0xA7
$62 = 3 \cdot 16 + 14$	0011 1110	0x3E
$188 = 11 \cdot 16 + 12$	1011 1100	0xBC
$3 \cdot 16 + 7 = 55$	0011 0111	0x37
$8 \cdot 16 + 8 = 136$	1000 1000	0x88
$15 \cdot 16 + 3 = 243$	1111 0011	0xF3
$5 \cdot 16 + 2 = 82$	0101 0010	0x52
$10 \cdot 16 + 12 = 172$	1010 1100	0xAC
$14 \cdot 16 + 7 = 231$	1110 0111	0xE7

Solution to Problem 4

When you begin debugging machine-level programs, you will find many cases where some simple hexadecimal arithmetic would be useful. You can always convert numbers to decimal, perform the arithmetic, and convert them back, but being able to work directly in hexadecimal is more efficient and informative.

A. 0x503c + 0x8 = 0x5044. Adding 8 to hex c gives 4 with a carry of 1.

B. 0x503c − 0x40 = 0x4ffc. Subtracting 4 from 3 in the second digit position requires a borrow from the third. Since this digit is 0, we must also borrow from the fourth position.

C. 0x503c + 64 = 0x507c. Decimal 64 (2^6) equals hexadecimal 0x40.

D. 0x50ea − 0x503c = 0xae. To subtract hex c (decimal 12) from hex a (decimal 10), we borrow 16 from the second digit, giving hex e (decimal 14). In the second digit, we now subtract 3 from hex d (decimal 13), giving hex a (decimal 10).

Solution to Problem 5

This problem tests your understanding of the byte representation of data and the two different byte orderings.

Little endian: 21   Big endian: 87
Little endian: 21 43   Big endian: 87 65
Little endian: 21 43 65   Big endian: 87 65 43

Recall that show_bytes enumerates a series of bytes starting from the one with lowest address and working toward the one with highest address. On a little-endian machine, it will list the bytes from least significant to most. On a big-endian machine, it will list bytes from the most significant byte to the least.

Solution to Problem 6

This problem is another chance to practice hexadecimal to binary conversion. It also gets you thinking about integer and floating-point representations. We will explore these representations in more detail later in this chapter.

A. Using the notation of the example in the text, we write the two strings as follows:

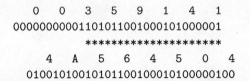

```
    0   0   3   5   9   1   4   1
000000000011010110010001010000001
    *******************
    4   A   5   6   4   5   0   4
0100101001010110010001010000100
```

B. With the second word shifted two positions to the right relative to the first, we find a sequence with 21 matching bits.

C. We find all bits of the integer embedded in the floating-point number, except for the most significant bit having value 1. Such is the case for the example in the text as well. In addition, the floating-point number has some nonzero high-order bits that do not match those of the integer.

Solution to Problem 7

It prints 61 62 63 64 65 66. Recall also that the library routine `strlen` does not count the terminating null character, and so `show_bytes` printed only through the character 'f'.

Solution to Problem 8

This problem is a drill to help you become more familiar with Boolean operations.

Operation	Result
a	[01101001]
b	[01010101]
~a	[10010110]
~b	[10101010]
a & b	[01000001]
a | b	[01111101]
a ^ b	[00111100]

Solution to Problem 9

This problem illustrates how Boolean algebra can be used to describe and reason about real-world systems. We can see that this color algebra is identical to the Boolean algebra over bit vectors of length 3.

A. Colors are complemented by complementing the values of R, G, and B. From this, we can see that White is the complement of Black, Yellow is the complement of Blue, Magenta is the complement of Green, and Cyan is the complement of Red.

B. We perform Boolean operations based on a bit-vector representation of the colors. From this we get the following:

Blue (001)	|	Green (010)	=	Cyan (011)
Yellow (110)	&	Cyan (011)	=	Green (010)
Red (100)	^	Magenta (101)	=	Blue (001)

Solution to Problem 10

This procedure relies on the fact that EXCLUSIVE-OR is commutative and associative, and that $a \char94 a = 0$ for any a.

Step	*x	*y
Initially	a	b
Step 1	a	$a \char94 b$
Step 2	$a \char94 (a \char94 b) = (a \char94 a) \char94 b = b$	$a \char94 b$
Step 3	b	$b \char94 (a \char94 b) = (b \char94 b) \char94 a = a$

See Problem 11 for a case where this function will fail.

Solution to Problem 11

This problem illustrates a subtle and interesting feature of our inplace swap routine.

A. Both `first` and `last` have value k, so we are attempting to swap the middle element with itself.

B. In this case, arguments x and y to `inplace_swap` both point to the same location. When we compute *x ^ *y, we get 0. We then store 0 as the middle element of the array, and the subsequent steps keep setting this element to 0. We can see that our reasoning in Problem 10 implicitly assumed that x and y denote different locations.

C. Simply replace the test in line 4 of `reverse_array` to be `first < last`, since there is no need to swap the middle element with itself.

Solution to Problem 12

Here are the expressions:

A. `x & 0xFF`

B. `x ^ ~0xFF`

C. `x | 0xFF`

These expressions are typical of the kind commonly found in performing low-level bit operations. The expression `~0xFF` creates a mask where the 8 least-significant bits equal 0 and the rest equal 1. Observe that such a mask will be generated regardless of the word size. By contrast, the expression `0xFFFFFF00` would only work on a 32-bit machine.

Solution to Problem 13

These problems help you think about the relation between Boolean operations and typical ways that programmers apply masking operations. Here is the code:

```
/* Declarations of functions implementing operations bis and bic */
int bis(int x, int m);
int bic(int x, int m);

/* Compute x|y using only calls to functions bis and bic */
int bool_or(int x, int y) {
  int result = bis(x,y);
  return result;
}

/* Compute x^y using only calls to functions bis and bic */
int bool_xor(int x, int y) {
  int result = bis(bic(x,y), bic(y,x));
  return result;
}
```

The bis operation is equivalent to Boolean OR—a bit is set in z if either this bit is set in x or it is set in m. On the other hand, bic(x, m) is equivalent to x&~m; we want the result to equal 1 only when the corresponding bit of x is 1 and of m is 0.

Given that, we can implement | with a single call to bis. To implement ^, we take advantage of the property

$$x \text{ \textasciicircum } y = (x \text{ \& } \sim y) \mid (\sim x \text{ \& } y).$$

Solution to Problem 14

This problem highlights the relation between bit-level Boolean operations and logic operations in C. A common programming error is to use a bit-level operation when a logic one is intended, or vice versa.

Expression	Value	Expression	Value
x & y	0x20	x && y	0x01
x | y	0x7F	x || y	0x01
~x | ~y	0xDF	!x || !y	0x00
x & !y	0x00	x && ~y	0x01

Solution to Problem 15

The expression is !(x ^ y).

That is, x^y will be zero if and only if every bit of x matches the corresponding bit of y. We then exploit the ability of ! to determine whether a word contains any nonzero bit.

There is no real reason to use this expression rather than simply writing x == y, but it demonstrates some of the nuances of bit-level and logical operations.

Solution to Problem 16

This problem is a drill to help you understand the different shift operations.

x		x << 3		(Logical) x >> 2		(Arithmetic) x >> 2	
Hex	Binary	Binary	Hex	Binary	Hex	Binary	Hex
0xC3	[11000011]	[00011000]	0x18	[00110000]	0x30	[11110000]	0xF0
0x75	[01110101]	[10101000]	0xA8	[00011101]	0x1D	[00011101]	0x1D
0x87	[10000111]	[00111000]	0x38	[00100001]	0x21	[11100001]	0xE1
0x66	[01100110]	[00110000]	0x30	[00011001]	0x19	[00011001]	0x19

Solution to Problem 17

In general, working through examples for very small word sizes is a very good way to understand computer arithmetic.

The unsigned values correspond to those in Figure 2. For the two's-complement values, hex digits 0 through 7 have a most significant bit of 0, yielding nonnegative values, while hex digits 8 through F have a most significant bit of 1, yielding a negative value.

$\vec{x}$			
Hexadecimal	Binary	$B2U_4(\vec{x})$	$B2T_4(\vec{x})$
0xE	[1110]	$2^3 + 2^2 + 2^1 = 14$	$-2^3 + 2^2 + 2^1 = -2$
0x0	[0000]	0	0
0x5	[0101]	$2^2 + 2^0 = 5$	$2^2 + 2^0 = 5$
0x8	[1000]	$2^3 = 8$	$-2^3 = -8$
0xD	[1101]	$2^3 + 2^2 + 2^0 = 13$	$-2^3 + 2^2 + 2^0 = -3$
0xF	[1111]	$2^3 + 2^2 + 2^1 + 2^0 = 15$	$-2^3 + 2^2 + 2^1 + 2^0 = -1$

Solution to Problem 18

For a 32-bit machine, any value consisting of eight hexadecimal digits beginning with one of the digits 8 through f represents a negative number. It is quite common to see numbers beginning with a string of f's, since the leading bits of a negative number are all ones. You must look carefully, though. For example, the number 0x8048337 has only seven digits. Filling this out with a leading zero gives 0x08048337, a positive number.

```
8048337:   81 ec b8 01 00 00   sub   $0x1b8,%esp          A.    440
804833d:   8b 55 08            mov   0x8(%ebp),%edx
8048340:   83 c2 14            add   $0x14,%edx           B.     20
8048343:   8b 85 58 fe ff ff   mov   0xfffffe58(%ebp),%eax  C.  -424
8048349:   03 02               add   (%edx),%eax
804834b:   89 85 74 fe ff ff   mov   %eax,0xfffffe74(%ebp)  D.  -396
8048351:   8b 55 08            mov   0x8(%ebp),%edx
8048354:   83 c2 44            add   $0x44,%edx           E.     68
8048357:   8b 85 c8 fe ff ff   mov   0xfffffec8(%ebp),%eax  F.  -312
```

```
804835d:   89 02                  mov      %eax,(%edx)
804835f:   8b 45 10               mov      0x10(%ebp),%eax      G.    16
8048362:   03 45 0c               add      0xc(%ebp),%eax       H.    12
8048365:   89 85 ec fe ff ff      mov      %eax,0xfffffeec(%ebp)  I.  -276
804836b:   8b 45 08               mov      0x8(%ebp),%eax
804836e:   83 c0 20               add      $0x20,%eax           J.    32
8048371:   8b 00                  mov      (%eax),%eax
```

Solution to Problem 19

The functions *T2U* and *U2T* are very peculiar from a mathematical perspective. It is important to understand how they behave.

We solve this problem by reordering the rows in the solution of Problem 17 according to the two's-complement value and then listing the unsigned value as the result of the function application. We show the hexadecimal values to make this process more concrete.

$\vec{x}$ (hex)	x	$T2U_4(x)$
0x8	−8	8
0xD	−3	13
0xE	−2	14
0xF	−1	15
0x0	0	0
0x5	5	5

Solution to Problem 20

This exercise tests your understanding of Equation 6.

For the first four entries, the values of x are negative and $T2U_4(x) = x + 2^4$. For the remaining two entries, the values of x are nonnegative and $T2U_4(x) = x$.

Solution to Problem 21

This problem reinforces your understanding of the relation between two's-complement and unsigned representations, and the effects of the C promotion rules. Recall that $TMin_{32}$ is $-2{,}147{,}483{,}648$, and that when cast to unsigned it becomes $2{,}147{,}483{,}648$. In addition, if either operand is unsigned, then the other operand will be cast to unsigned before comparing.

Expression	Type	Evaluation
-2147483647-1 == 2147483648U	unsigned	1
-2147483647-1 < 2147483647	signed	1
-2147483647-1U < 2147483647	unsigned	0
-2147483647-1 < -2147483647	signed	1
-2147483647-1U < -2147483647	unsigned	1

Solution to Problem 22

This exercise provides a concrete demonstration of how sign extension preserves the numeric value of a two's-complement representation.

A.	[1011]:	$-2^3 + 2^1 + 2^0$	$=$	$-8 + 2 + 1$	$=$	-5
B.	[11011]:	$-2^4 + 2^3 + 2^1 + 2^0$	$=$	$-16 + 8 + 2 + 1$	$=$	-5
C.	[111011]:	$-2^5 + 2^4 + 2^3 + 2^1 + 2^0$	$=$	$-32 + 16 + 8 + 2 + 1$	$=$	-5

Solution to Problem 23

The expressions in these functions are common program "idioms" for extracting values from a word in which multiple bit fields have been packed. They exploit the zero-filling and sign-extending properties of the different shift operations. Note carefully the ordering of the cast and shift operations. In fun1, the shifts are performed on unsigned variable word, and hence are logical. In fun2, shifts are performed after casting word to int, and hence are arithmetic.

A.

w	fun1(w)	fun2(w)
0x00000076	0x00000076	0x00000076
0x87654321	0x00000021	0x00000021
0x000000C9	0x000000C9	0xFFFFFFC9
0xEDCBA987	0x00000087	0xFFFFFF87

B. Function fun1 extracts a value from the low-order 8 bits of the argument, giving an integer ranging between 0 and 255. Function fun2 extracts a value from the low-order 8 bits of the argument, but it also performs sign extension. The result will be a number between -128 and 127.

Solution to Problem 24

The effect of truncation is fairly intuitive for unsigned numbers, but not for two's-complement numbers. This exercise lets you explore its properties using very small word sizes.

Hex		Unsigned		Two's complement	
Original	Truncated	Original	Truncated	Original	Truncated
0	0	0	0	0	0
2	2	2	2	2	2
9	1	9	1	-7	1
B	3	11	3	-5	3
F	7	15	7	-1	-1

As Equation 9 states, the effect of this truncation on unsigned values is to simply find their residue, modulo 8. The effect of the truncation on signed values is a bit more complex. According to Equation 10, we first compute the modulo 8 residue of the argument. This will give values 0 through 7 for arguments 0 through 7, and also for arguments -8 through -1. Then we apply function $U2T_3$ to these residues, giving two repetitions of the sequences 0 through 3 and -4 through -1.

Solution to Problem 25

This problem is designed to demonstrate how easily bugs can arise due to the implicit casting from signed to unsigned. It seems quite natural to pass parameter `length` as an unsigned, since one would never want to use a negative length. The stopping criterion `i <= length-1` also seems quite natural. But combining these two yields an unexpected outcome!

Since parameter `length` is unsigned, the computation $0 - 1$ is performed using unsigned arithmetic, which is equivalent to modular addition. The result is then *UMax*. The $\leq$ comparison is also performed using an unsigned comparison, and since any number is less than or equal to *UMax*, the comparison always holds! Thus, the code attempts to access invalid elements of array a.

The code can be fixed either by declaring `length` to be an `int`, or by changing the test of the `for` loop to be `i < length`.

Solution to Problem 26

This example demonstrates a subtle feature of unsigned arithmetic, and also the property that we sometimes perform unsigned arithmetic without realizing it. This can lead to very tricky bugs.

A. *For what cases will this function produce an incorrect result?* The function will incorrectly return 1 when s is shorter than t.

B. *Explain how this incorrect result comes about.* Since `strlen` is defined to yield an unsigned result, the difference and the comparison are both computed using unsigned arithmetic. When s is shorter than t, the difference `strlen(s) - strlen(t)` should be negative, but instead becomes a large, unsigned number, which is greater than 0.

C. *Show how to fix the code so that it will work reliably.* Replace the test with the following:

```
return strlen(s) > strlen(t);
```

Solution to Problem 27

This function is a direct implementation of the rules given to determine whether or not an unsigned addition overflows.

```
/* Determine whether arguments can be added without overflow */
int uadd_ok(unsigned x, unsigned y) {
    unsigned sum = x+y;
    return sum >= x;
}
```

Solution to Problem 28

This problem is a simple demonstration of arithmetic modulo 16. The easiest way to solve it is to convert the hex pattern into its unsigned decimal value. For nonzero values of x, we must have $(-_4^u x) + x = 16$. Then we convert the complemented value back to hex.

x		$-^u_4 x$	
Hex	Decimal	Decimal	Hex
0	0	0	0
5	5	11	B
8	8	8	8
D	13	3	3
F	15	1	1

Solution to Problem 29

This problem is an exercise to make sure you understand two's-complement addition.

x	y	$x + y$	$x +^t_5 y$	Case
−12	−15	−27	5	1
[10100]	[10001]	[100101]	[00101]	
−8	−8	−16	−16	2
[11000]	[11000]	[110000]	[10000]	
−9	8	−1	−1	2
[10111]	[01000]	[111111]	[11111]	
2	5	7	7	3
[00010]	[00101]	[000111]	[00111]	
12	4	16	−16	4
[01100]	[00100]	[010000]	[10000]	

Solution to Problem 30

This function is a direct implementation of the rules given to determine whether or not a two's-complement addition overflows.

```
/* Determine whether arguments can be added without overflow */
int tadd_ok(int x, int y) {
    int sum = x+y;
    int neg_over = x <  0 && y <  0 && sum >= 0;
    int pos_over = x >= 0 && y >= 0 && sum <  0;
    return !neg_over && !pos_over;
}
```

Solution to Problem 31

Your coworker could have learned, by studying Section 3.2, that two's-complement addition forms an abelian group, and so the expression (x+y)-x will evaluate to y regardless of whether or not the addition overflows, and that (x+y)-y will always evaluate to x.

Solution to Problem 32

This function will give correct values, except when y is *TMin*. In this case, we will have -y also equal to *TMin*, and so function tadd_ok will consider there to be negative overflow any time x is negative. In fact, x-y does not overflow for these cases.

One lesson to be learned from this exercise is that *TMin* should be included as one of the cases in any test procedure for a function.

Solution to Problem 33

This problem helps you understand two's-complement negation using a very small word size.

For $w = 4$, we have $TMin_4 = -8$. So -8 is its own additive inverse, while other values are negated by integer negation.

x		$-^t_4 x$	
Hex	Decimal	Decimal	Hex
0	0	0	0
5	5	−5	B
8	−8	−8	8
D	−3	3	3
F	−1	1	1

The bit patterns are the same as for unsigned negation.

Solution to Problem 34

This problem is an exercise to make sure you understand two's-complement multiplication.

Mode	x		y		$x \cdot y$		Truncated $x \cdot y$	
Unsigned	4	[100]	5	[101]	20	[010100]	4	[100]
Two's comp.	−4	[100]	−3	[101]	12	[001100]	−4	[100]
Unsigned	2	[010]	7	[111]	14	[001110]	6	[110]
Two's comp.	2	[010]	−1	[111]	−2	[111110]	−2	[110]
Unsigned	6	[110]	6	[110]	36	[100100]	4	[100]
Two's comp.	−2	[110]	−2	[110]	4	[000100]	−4	[100]

Solution to Problem 35

It's not realistic to test this function for all possible values of x and y. Even if you could run 10 billion tests per second, it would require over 58 years to test all combinations when data type int is 32 bits. On the other hand, it is feasible to test your code by writing the function with data type short or char and then testing it exhaustively.

Here's a more principled approach, following the proposed set of arguments:

1. We know that $x \cdot y$ can be written as a $2w$-bit two's-complement number. Let u denote the unsigned number represented by the lower w bits, and v denote the two's-complement number represented by the upper w bits. Then, based on Equation 3, we can see that $x \cdot y = v2^w + u$.

We also know that $u = T2U_w(p)$, since they are unsigned and two's-complement numbers arising from the same bit pattern, and so by Equation 5, we can write $u = p + p_{w-1}2^w$, where p_{w-1} is the most significant bit of p. Letting $t = v + p_{w-1}$, we have $x \cdot y = p + t2^w$.

When $t = 0$, we have $x \cdot y = p$; the multiplication does not overflow. When $t \neq 0$, we have $x \cdot y \neq p$; the multiplication does overflow.

2. By definition of integer division, dividing p by nonzero x gives a quotient q and a remainder r such that $p = x \cdot q + r$, and $|r| < |x|$. (We use absolute values here, because the signs of x and r may differ. For example, dividing -7 by 2 gives quotient -3 and remainder -1.)

3. Suppose $q = y$. Then we have $x \cdot y = x \cdot y + r + t2^w$. From this, we can see that $r + t2^w = 0$. But $|r| < |x| \leq 2^w$, and so this identity can hold only if $t = 0$, in which case $r = 0$.

   Suppose $r = t = 0$. Then we will have $x \cdot y = x \cdot q$, implying that $y = q$.

When x equals 0, multiplication does not overflow, and so we see that our code provides a reliable way to test whether or not two's-complement multiplication causes overflow.

Solution to Problem 36

With 64 bits, we can perform the multiplication without overflowing. We then test whether casting the product to 32 bits changes the value:

```
1    /* Determine whether arguments can be multiplied without overflow */
2    int tmult_ok(int x, int y) {
3        /* Compute product without overflow */
4        long long pll = (long long) x*y;
5        /* See if casting to int preserves value */
6        return pll == (int) pll;
7    }
```

Note that the casting on the right-hand side of line 4 is critical. If we instead wrote the line as

```
long long pll = x*y;
```

the product would be computed as a 32-bit value (possibly overflowing) and then sign extended to 64 bits.

Solution to Problem 37

A. This change does not help at all. Even though the computation of asize will be accurate, the call to malloc will cause this value to be converted to a 32-bit unsigned number, and so the same overflow conditions will occur.

B. With malloc having a 32-bit unsigned number as its argument, it cannot possibly allocate a block of more than 2^{32} bytes, and so there is no point attempting to allocate or copy this much memory. Instead, the function should abort and return NULL, as illustrated by the following replacement to the original call to malloc (line 10):

```
long long unsigned required_size =
    ele_cnt * (long long unsigned) ele_size;
size_t request_size = (size_t) required_size;
if (required_size != request_size)
    /* Overflow must have occurred.  Abort operation */
    return NULL;
void *result = malloc(request_size);
if (result == NULL)
    /* malloc failed */
    return NULL;
```

Solution to Problem 38

The LEA instruction is provided to support pointer arithmetic, but the C compiler often uses it as a way to perform multiplication by small constants.

For each value of k, we can compute two multiples: 2^k (when b is 0) and $2^k + 1$ (when b is a). Thus, we can compute multiples 1, 2, 3, 4, 5, 8, and 9.

Solution to Problem 39

The expression simply becomes $-(x<<m)$. To see this, let the word size be w so that $n = w-1$. Form B states that we should compute $(x<<w) - (x<<m)$, but shifting x to the left by w will yield the value 0.

Solution to Problem 40

This problem requires you to try out the optimizations already described and also to supply a bit of your own ingenuity.

K	Shifts	Add/Subs	Expression
6	2	1	(x<<2) + (x<<1)
31	1	1	(x<<5) - x
-6	2	1	(x<<1) - (x<<3)
55	2	2	(x<<6) - (x<<3) - x

Observe that the fourth case uses a modified version of form B. We can view the bit pattern [110111] as having a run of 6 ones with a zero in the middle, and so we apply the rule for form B, but then we subtract the term corresponding to the middle zero bit.

Solution to Problem 41

Assuming that addition and subtraction have the same performance, the rule is to choose form A when $n = m$, either form when $n = m + 1$, and form B when $n > m + 1$.

The justification for this rule is as follows. Assume first that $m > 1$. When $n = m$, form A requires only a single shift, while form B requires two shifts and a subtraction. When $n = m + 1$, both forms require two shifts and either an addition or a subtraction. When $n > m + 1$, form B requires only two shifts and one subtraction, while form A requires $n - m + 1 > 2$ shifts and $n - m > 1$ additions. For the case of $m = 1$, we get one fewer shift for both forms A and B, and so the same rules apply for choosing between the two.

Solution to Problem 42

The only challenge here is to compute the bias without any testing or conditional operations. We use the trick that the expression x >> 31 generates a word with all ones if x is negative, and all zeros otherwise. By masking off the appropriate bits, we get the desired bias value.

```
int div16(int x) {
    /* Compute bias to be either 0 (x >= 0) or 15 (x < 0) */
    int bias = (x >> 31) & 0xF;
    return (x + bias) >> 4;
}
```

Solution to Problem 43

We have found that people have difficulty with this exercise when working directly with assembly code. It becomes more clear when put in the form shown in optarith.

We can see that M is 31; x*M is computed as (x<<5)-x.

We can see that N is 8; a bias value of 7 is added when y is negative, and the right shift is by 3.

Solution to Problem 44

These "C puzzle" problems provide a clear demonstration that programmers must understand the properties of computer arithmetic:

A. (x > 0) || (x-1 < 0)
   False. Let x be $-2{,}147{,}483{,}648$ ($TMin_{32}$). We will then have x-1 equal to 2147483647 ($TMax_{32}$).

B. (x & 7) != 7 || (x<<29 < 0)
   True. If (x & 7) != 7 evaluates to 0, then we must have bit x_2 equal to 1. When shifted left by 29, this will become the sign bit.

C. (x * x) >= 0
   False. When x is 65,535 (0xFFFF), x*x is $-131{,}071$ (0xFFFE0001).

D. x < 0 || -x <= 0
   True. If x is nonnegative, then -x is nonpositive.

E. x > 0 || -x >= 0
   False. Let x be $-2{,}147{,}483{,}648$ ($TMin_{32}$). Then both x and -x are negative.

F. `x+y == uy+ux`
True. Two's-complement and unsigned addition have the same bit-level behavior, and they are commutative.

G. `x*~y + uy*ux == -x`
True. `~y` equals `-y-1`. `uy*ux` equals `x*y`. Thus, the left hand side is equivalent to `x*-y-x+x*y`.

Solution to Problem 45

Understanding fractional binary representations is an important step to understanding floating-point encodings. This exercise lets you try out some simple examples.

Fractional value	Binary representation	Decimal representation
$\frac{1}{8}$	0.001	0.125
$\frac{3}{4}$	0.11	0.75
$\frac{25}{16}$	1.1001	1.5625
$\frac{43}{16}$	10.1011	2.6875
$\frac{9}{8}$	1.001	1.125
$\frac{47}{8}$	101.111	5.875
$\frac{51}{16}$	11.0011	3.1875

One simple way to think about fractional binary representations is to represent a number as a fraction of the form $\frac{x}{2^k}$. We can write this in binary using the binary representation of x, with the binary point inserted k positions from the right. As an example, for $\frac{25}{16}$, we have $25_{10} = 11001_2$. We then put the binary point four positions from the right to get 1.1001_2.

Solution to Problem 46

In most cases, the limited precision of floating-point numbers is not a major problem, because the *relative* error of the computation is still fairly low. In this example, however, the system was sensitive to the *absolute* error.

A. We can see that $0.1 - x$ has binary representation

$$0.00000000000000000000000001100[1100]\cdots_2$$

B. Comparing this to the binary representation of $\frac{1}{10}$, we can see that it is simply $2^{-20} \times \frac{1}{10}$, which is around 9.54×10^{-8}.

C. $9.54 \times 10^{-8} \times 100 \times 60 \times 60 \times 10 \approx 0.343$ seconds.

D. $0.343 \times 2000 \approx 687$ meters.

Solution to Problem 47

Working through floating-point representations for very small word sizes helps clarify how IEEE floating point works. Note especially the transition between denormalized and normalized values.

Bits	e	E	2^E	f	M	$2^E \times M$	V	Decimal
0 00 00	0	0	1	$\frac{0}{4}$	$\frac{0}{4}$	$\frac{0}{4}$	0	0.0
0 00 01	0	0	1	$\frac{1}{4}$	$\frac{1}{4}$	$\frac{1}{4}$	$\frac{1}{4}$	0.25
0 00 10	0	0	1	$\frac{2}{4}$	$\frac{2}{4}$	$\frac{2}{4}$	$\frac{1}{2}$	0.5
0 00 11	0	0	1	$\frac{3}{4}$	$\frac{3}{4}$	$\frac{3}{4}$	$\frac{3}{4}$	0.75
0 01 00	1	0	1	$\frac{0}{4}$	$\frac{4}{4}$	$\frac{4}{4}$	1	1.0
0 01 01	1	0	1	$\frac{1}{4}$	$\frac{5}{4}$	$\frac{5}{4}$	$\frac{5}{4}$	1.25
0 01 10	1	0	1	$\frac{2}{4}$	$\frac{6}{4}$	$\frac{6}{4}$	$\frac{3}{2}$	1.5
0 01 11	1	0	1	$\frac{3}{4}$	$\frac{7}{4}$	$\frac{7}{4}$	$\frac{7}{4}$	1.75
0 10 00	2	1	2	$\frac{0}{4}$	$\frac{4}{4}$	$\frac{8}{4}$	2	2.0
0 10 01	2	1	2	$\frac{1}{4}$	$\frac{5}{4}$	$\frac{10}{4}$	$\frac{5}{2}$	2.5
0 10 10	2	1	2	$\frac{2}{4}$	$\frac{6}{4}$	$\frac{12}{4}$	3	3.0
0 10 11	2	1	2	$\frac{3}{4}$	$\frac{7}{4}$	$\frac{14}{4}$	$\frac{7}{2}$	3.5
0 11 00	—	—	—	—	—	—	∞	—
0 11 01	—	—	—	—	—	—	NaN	—
0 11 10	—	—	—	—	—	—	NaN	—
0 11 11	—	—	—	—	—	—	NaN	—

Solution to Problem 48

Hexadecimal value 0x359141 is equivalent to binary [11010110010000101000001]. Shifting this right 21 places gives $1.1010110010000101000001_2 \times 2^{21}$. We form the fraction field by dropping the leading 1 and adding two 0s, giving [10101100100010100000100]. The exponent is formed by adding bias 127 to 21, giving 148 (binary [10010100]). We combine this with a sign field of 0 to give a binary representation

$$[01001010010101100100010100000100].$$

We see that the matching bits in the two representations correspond to the low-order bits of the integer, up to the most significant bit equal to 1 matching the high-order 21 bits of the fraction:

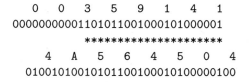

```
   0    0    3    5    9    1    4    1
000000000011010110010000101000001
         ********************
    4    A    5    6    4    5    0    4
   01001010010101100100010100000100
```

Solution to Problem 49

This exercise helps you think about what numbers cannot be represented exactly in floating point.

A. The number has binary representation 1, followed by n 0s, followed by 1, giving value $2^{n+1} + 1$.

B. When $n = 23$, the value is $2^{24} + 1 = 16,777,217$.

Solution to Problem 50

Performing rounding by hand helps reinforce the idea of round-to-even with binary numbers.

Original		Rounded	
10.010_2	$2\frac{1}{4}$	10.0	2
10.011_2	$2\frac{3}{8}$	10.1	$2\frac{1}{2}$
10.110_2	$2\frac{3}{4}$	11.0	3
11.001_2	$3\frac{1}{8}$	11.0	3

Solution to Problem 51

A. Looking at the nonterminating sequence for 1/10, we can see that the 2 bits to the right of the rounding position are 1, and so a better approximation to 1/10 would be obtained by incrementing x to get $x' = 0.00011001100110011001101_2$, which is larger than 0.1.

B. We can see that $x' - 0.1$ has binary representation:

$$0.00000000000000000000000[1100].$$

Comparing this to the binary representation of $\frac{1}{10}$, we can see that it is $2^{-22} \times \frac{1}{10}$, which is around 2.38×10^{-8}.

C. $2.38 \times 10^{-8} \times 100 \times 60 \times 60 \times 10 \approx 0.086$ seconds, a factor of 4 less than the error in the Patriot system.

D. $0.343 \times 2000 \approx 171$ meters.

Solution to Problem 52

This problem tests a lot of concepts about floating-point representations, including the encoding of normalized and denormalized values, as well as rounding.

Format A		Format B		
Bits	Value	Bits	Value	Comments
011 0000	1	0111 000	1	
101 1110	$\frac{15}{2}$	1001 111	$\frac{15}{2}$	
010 1001	$\frac{25}{32}$	0110 100	$\frac{3}{4}$	Round down
110 1111	$\frac{31}{2}$	1011 000	16	Round up
000 0001	$\frac{1}{64}$	0001 000	$\frac{1}{64}$	Denorm $\rightarrow$ norm

Solution to Problem 53

In general, it is better to use a library macro rather than inventing your own code. This code seems to work on a variety of machines, however.

We assume that the value 1e400 overflows to infinity.

```
#define POS_INFINITY 1e400
#define NEG_INFINITY (-POS_INFINITY)
#define NEG_ZERO (-1.0/POS_INFINITY)
```

Solution to Problem 54

Exercises such as this one help you develop your ability to reason about floating-point operations from a programmer's perspective. Make sure you understand each of the answers.

A. `x == (int)(double) x`
Yes, since `double` has greater precision and range than `int`.

B. `x == (int)(float) x`
No. For example, when x is *TMax*.

C. `d == (double)(float) d`
No. For example, when d is 1e40, we will get $+\infty$ on the right.

D. `f == (float)(double) f`
Yes, since `double` has greater precision and range than `float`.

E. `f == -(-f)`
Yes, since a floating-point number is negated by simply inverting its sign bit.

F. `1.0/2 == 1/2.0`
Yes, the numerators and denominators will both be converted to floating-point representations before the division is performed.

G. `d*d >= 0.0`
Yes, although it may overflow to $+\infty$.

H. `(f+d)-f == d`
No, for example when f is `1.0e20` and d is `1.0`, the expression `f+d` will be rounded to `1.0e20`, and so the expression on the left-hand side will evaluate to 0.0, while the right-hand side will be 1.0.

Machine-Level Representation of Programs

Machine-Level Representation of Programs

Computers execute *machine code*, sequences of bytes encoding the low-level operations that manipulate data, manage memory, read and write data on storage devices, and communicate over networks. A compiler generates machine code through a series of stages, based on the rules of the programming language, the instruction set of the target machine, and the conventions followed by the operating system. The GCC C compiler generates its output in the form of *assembly code*, a textual representation of the machine code giving the individual instructions in the program. GCC then invokes both an *assembler* and a *linker* to generate the executable machine code from the assembly code. In this chapter, we will take a close look at machine code and its human-readable representation as assembly code.

When programming in a high-level language such as C, and even more so in Java, we are shielded from the detailed, machine-level implementation of our program. In contrast, when writing programs in assembly code (as was done in the early days of computing) a programmer must specify the low-level instructions the program uses to carry out a computation. Most of the time, it is much more productive and reliable to work at the higher level of abstraction provided by a high-level language. The type checking provided by a compiler helps detect many program errors and makes sure we reference and manipulate data in consistent ways. With modern, optimizing compilers, the generated code is usually at least as efficient as what a skilled, assembly-language programmer would write by hand. Best of all, a program written in a high-level language can be compiled and executed on a number of different machines, whereas assembly code is highly machine specific.

So why should we spend our time learning machine code? Even though compilers do most of the work in generating assembly code, being able to read and understand it is an important skill for serious programmers. By invoking the compiler with appropriate command-line parameters, the compiler will generate a file showing its output in assembly-code form. By reading this code, we can understand the optimization capabilities of the compiler and analyze the underlying inefficiencies in the code. Programmers seeking to maximize the performance of a critical section of code often try different variations of the source code, each time compiling and examining the generated assembly code to get a sense of how efficiently the program will run. Furthermore, there are times when the layer of abstraction provided by a high-level language hides information about the run-time behavior of a program that we need to understand. For example, when writing concurrent programs using a thread package, it is important to know what region of memory is used to hold the different program variables. This information is visible at the assembly-code level. As another example, many of the ways programs can be attacked, allowing worms and viruses to infest a system, involve nuances of the way programs store their run-time control information. Many attacks involve exploiting weaknesses in system programs to overwrite information and thereby take control of the system. Understanding how these vulnerabilities arise and how to guard against them requires a knowledge of the machine-level representation of programs. The need for programmers to learn assembly code has shifted over the years from one of being able to write programs directly in assembly to one of being able to read and understand the code generated by compilers.

In this chapter, we will learn the details of two particular assembly languages and see how C programs get compiled into these forms of machine code. Reading the assembly code generated by a compiler involves a different set of skills than writing assembly code by hand. We must understand the transformations typical compilers make in converting the constructs of C into machine code. Relative to the computations expressed in the C code, optimizing compilers can rearrange execution order, eliminate unneeded computations, replace slow operations with faster ones, and even change recursive computations into iterative ones. Understanding the relation between source code and the generated assembly can often be a challenge—it's much like putting together a puzzle having a slightly different design than the picture on the box. It is a form of *reverse engineering*— trying to understand the process by which a system was created by studying the system and working backward. In this case, the system is a machine-generated assembly-language program, rather than something designed by a human. This simplifies the task of reverse engineering, because the generated code follows fairly regular patterns, and we can run experiments, having the compiler generate code for many different programs. In our presentation, we give many examples and provide a number of exercises illustrating different aspects of assembly language and compilers. This is a subject where mastering the details is a prerequisite to understanding the deeper and more fundamental concepts. Those who say "I understand the general principles, I don't want to bother learning the details" are deluding themselves. It is critical for you to spend time studying the examples, working through the exercises, and checking your solutions with those provided.

Our presentation is based on two related machine languages: Intel IA32, the dominant language of most computers today, and x86-64, its extension to run on 64-bit machines. Our focus starts with IA32. Intel processors have grown from primitive 16-bit processors in 1978 to the mainstream machines for today's desktop, laptop, and server computers. The architecture has grown correspondingly, with new features added and with the 16-bit architecture transformed to become IA32, supporting 32-bit data and addresses. The result is a rather peculiar design with features that make sense only when viewed from a historical perspective. It is also laden with features providing backward compatibility that are not used by modern compilers and operating systems. We will focus on the subset of the features used by GCC and Linux. This allows us to avoid much of the complexity and arcane features of IA32.

Our technical presentation starts with a quick tour to show the relation between C, assembly code, and machine code. We then proceed to the details of IA32, starting with the representation and manipulation of data and the implementation of control. We see how control constructs in C, such as if, while, and switch statements, are implemented. We then cover the implementation of procedures, including how the program maintains a run-time stack to support the passing of data and control between procedures, as well as storage for local variables. Next, we consider how data structures such as arrays, structures, and unions are implemented at the machine level. With this background in machine-level programming, we can examine the problems of out of bounds memory references and the vulnerability of systems to buffer overflow attacks. We finish this part of the presentation with some tips on using the GDB debugger for examining the run-time behavior of a machine-level program.

As we will discuss, the extension of IA32 to 64 bits, termed x86-64, was originally developed by Advanced Micro Devices (AMD), Intel's biggest competitor. Whereas a 32-bit machine can only make use of around 4 gigabytes (2^{32} bytes) of random-access memory, current 64-bit machines can use up to 256 terabytes (2^{48} bytes). The computer industry is currently in the midst of a transition from 32-bit to 64-bit machines. Most of the microprocessors in recent server and desktop machines, as well as in many laptops, support either 32-bit or 64-bit operation. However, most of the operating systems running on these machines support only 32-bit applications, and so the capabilities of the hardware are not fully utilized. As memory prices drop, and the desire to perform computations involving very large data sets increases, 64-bit machines and applications will become commonplace. It is therefore appropriate to take a close look at x86-64. We will see that in making the transition from 32 to 64 bits, the engineers at AMD also incorporated features that make the machines better targets for optimizing compilers and that improve system performance.

We provide Web Asides to cover material intended for dedicated machine-language enthusiasts. In one, we examine the code generated when code is compiled using higher degrees of optimization. Each successive version of the GCC compiler implements more sophisticated optimization algorithms, and these can radically transform a program to the point where it is difficult to understand the relation between the original source code and the generated machine-level program. Another Web Aside gives a brief presentation of ways to incorporate assembly code into C programs. For some applications, the programmer must drop down to assembly code to access low-level features of the machine. One approach is to write entire functions in assembly code and combine them with C functions during the linking stage. A second is to use GCC's support for embedding assembly code directly within C programs. We provide separate Web Asides for two different machine languages for floating-point code. The "x87" floating-point instructions have been available since the early days of Intel processors. This implementation of floating point is particularly arcane, and so we advise that only people determined to work with floating-point code on older machines attempt to study this section. The more recent "SSE" instructions were developed to support *multimedia applications*, but in their more recent versions (version 2 and later), and with more recent versions of GCC, SSE has become the preferred method for mapping floating point onto both IA32 and x86-64 machines.

1 A Historical Perspective

The Intel processor line, colloquially referred to as *x86*, has followed a long, evolutionary development. It started with one of the first single-chip, 16-bit microprocessors, where many compromises had to be made due to the limited capabilities of integrated circuit technology at the time. Since then, it has grown to take advantage of technology improvements as well as to satisfy the demands for higher performance and for supporting more advanced operating systems.

The list that follows shows some models of Intel processors and some of their key features, especially those affecting machine-level programming. We use the number of transistors required to implement the processors as an indication of how they have evolved in complexity (K denotes 1000, and M denotes 1,000,000).

8086: (1978, 29 K transistors). One of the first single-chip, 16-bit microprocessors. The 8088, a variant of the 8086 with an 8-bit external bus, formed the heart of the original IBM personal computers. IBM contracted with then-tiny Microsoft to develop the MS-DOS operating system. The original models came with 32,768 bytes of memory and two floppy drives (no hard drive). Architecturally, the machines were limited to a 655,360-byte address space—addresses were only 20 bits long (1,048,576 bytes addressable), and the operating system reserved 393,216 bytes for its own use. In 1980, Intel introduced the 8087 floating-point coprocessor (45 K transistors) to operate alongside an 8086 or 8088 processor, executing the floating-point instructions. The 8087 established the floating-point model for the x86 line, often referred to as "x87."

80286: (1982, 134 K transistors). Added more (and now obsolete) addressing modes. Formed the basis of the IBM PC-AT personal computer, the original platform for MS Windows.

i386: (1985, 275 K transistors). Expanded the architecture to 32 bits. Added the flat addressing model used by Linux and recent versions of the Windows family of operating system. This was the first machine in the series that could support a Unix operating system.

i486: (1989, 1.2 M transistors). Improved performance and integrated the floating-point unit onto the processor chip but did not significantly change the instruction set.

Pentium: (1993, 3.1 M transistors). Improved performance, but only added minor extensions to the instruction set.

PentiumPro: (1995, 5.5 M transistors). Introduced a radically new processor design, internally known as the *P6* microarchitecture. Added a class of "conditional move" instructions to the instruction set.

Pentium II: (1997, 7 M transistors). Continuation of the P6 microarchitecture.

Pentium III: (1999, 8.2 M transistors). Introduced SSE, a class of instructions for manipulating vectors of integer or floating-point data. Each datum can be 1, 2, or 4 bytes, packed into vectors of 128 bits. Later versions of this chip went up to 24 M transistors, due to the incorporation of the level-2 cache on chip.

Pentium 4: (2000, 42 M transistors). Extended SSE to SSE2, adding new data types (including double-precision floating point), along with 144 new instructions for these formats. With these extensions, compilers can use SSE instructions, rather than x87 instructions, to compile floating-point code. Introduced the *NetBurst* microarchitecture, which could operate at very high clock speeds, but at the cost of high power consumption.

Pentium 4E: (2004, 125 M transistors). Added *hyperthreading*, a method to run two programs simultaneously on a single processor, as well as EM64T, Intel's implementation of a 64-bit extension to IA32 developed by Advanced Micro Devices (AMD), which we refer to as x86-64.

Core 2: (2006, 291 M transistors). Returned back to a microarchitecture similar to P6. First *multi-core* Intel microprocessor, where multiple processors are implemented on a single chip. Did not support hyperthreading.

Core i7: (2008, 781 M transistors). Incorporated both hyperthreading and multi-core, with the initial version supporting two executing programs on each core and up to four cores on each chip.

Each successive processor has been designed to be backward compatible—able to run code compiled for any earlier version. As we will see, there are many strange artifacts in the instruction set due to this evolutionary heritage. Intel has had several names for their processor line, including *IA32*, for "Intel Architecture 32-bit," and most recently *Intel64*, the 64-bit extension to IA32, which we will refer to as *x86-64*. We will refer to the overall line by the commonly used colloquial name "x86," reflecting the processor naming conventions up through the i486.

Aside Moore's law

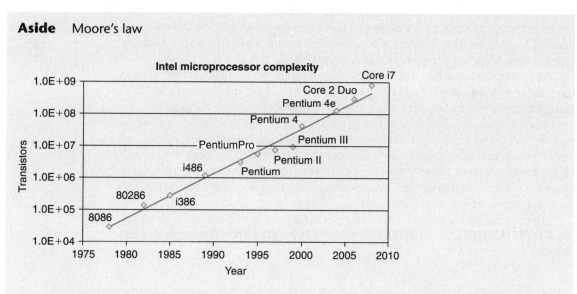

If we plot the number of transistors in the different Intel processors versus the year of introduction, and use a logarithmic scale for the *y*-axis, we can see that the growth has been phenomenal. Fitting a line through the data, we see that the number of transistors increases at an annual rate of approximately 38%, meaning that the number of transistors doubles about every 26 months. This growth has been sustained over the multiple-decade history of x86 microprocessors.

In 1965, Gordon Moore, a founder of Intel Corporation, extrapolated from the chip technology of the day, in which they could fabricate circuits with around 64 transistors on a single chip, to predict that the number of transistors per chip would double every year for the next 10 years. This predication became known as *Moore's law*. As it turns out, his prediction was just a little bit optimistic, but also too short-sighted. Over more than 45 years, the semiconductor industry has been able to double transistor counts on average every 18 months.

Similar exponential growth rates have occurred for other aspects of computer technology—disk capacities, memory-chip capacities, and processor performance. These remarkable growth rates have been the major driving forces of the computer revolution.

Over the years, several companies have produced processors that are compatible with Intel processors, capable of running the exact same machine-level programs. Chief among these is Advanced Micro Devices (AMD). For years, AMD lagged just behind Intel in technology, forcing a marketing strategy where they produced processors that were less expensive although somewhat lower in performance. They became more competitive around 2002, being the first to break the 1-gigahertz clock-speed barrier for a commercially available microprocessor, and introducing x86-64, the widely adopted 64-bit extension to IA32. Although we will talk about Intel processors, our presentation holds just as well for the compatible processors produced by Intel's rivals.

Much of the complexity of x86 is not of concern to those interested in programs for the Linux operating system as generated by the GCC compiler. The memory model provided in the original 8086 and its extensions in the 80286 are obsolete. Instead, Linux uses what is referred to as *flat* addressing, where the entire memory space is viewed by the programmer as a large array of bytes.

As we can see in the list of developments, a number of formats and instructions have been added to x86 for manipulating vectors of small integers and floating-point numbers. These features were added to allow improved performance on multimedia applications, such as image processing, audio and video encoding and decoding, and three-dimensional computer graphics. In its default invocation for 32-bit execution, GCC assumes it is generating code for an i386, even though there are very few of these 1985-era microprocessors running any longer. Only by giving specific command-line options, or by compiling for 64-bit operation, will the compiler make use of the more recent extensions.

For the next part of our presentation, we will focus only on the IA32 instruction set. We will then look at the extension to 64 bits via x86-64 toward the end of the chapter.

2 Program Encodings

Suppose we write a C program as two files p1.c and p2.c. We can then compile this code on an IA32 machine using a Unix command line:

```
unix> gcc -O1 -o p p1.c p2.c
```

The command gcc indicates the GCC C compiler. Since this is the default compiler on Linux, we could also invoke it as simply cc. The command-line option -O1 instructs the compiler to apply level-one optimizations. In general, increasing the level of optimization makes the final program run faster, but at a risk of increased compilation time and difficulties running debugging tools on the code. As we will also see, invoking higher levels of optimization can generate code that is so heavily transformed that the relationship between the generated machine code and the original source code is difficult to understand. We will therefore use level-one optimization as a learning tool and then see what happens as we increase the level of optimization. In practice, level-two optimization (specified with the option -O2) is considered a better choice in terms of the resulting program performance.

The gcc command actually invokes a sequence of programs to turn the source code into executable code. First, the C *preprocessor* expands the source code to include any files specified with #include commands and to expand any macros, specified with #define declarations. Second, the *compiler* generates assembly-code versions of the two source files having names p1.s and p2.s. Next, the *assembler* converts the assembly code into binary *object-code* files p1.o and p2.o. Object code is one form of machine code—it contains binary representations of all of the instructions, but the addresses of global values are not yet filled in. Finally, the *linker* merges these two object-code files along with code implementing library functions (e.g., printf) and generates the final executable code file p. Executable code is the second form of machine code we will consider—it is the exact form of code that is executed by the processor.

2.1 Machine-Level Code

Computer systems employ several different forms of abstraction, hiding details of an implementation through the use of a simpler, abstract model. Two of these are especially important for machine-level programming. First, the format and behavior of a machine-level program is defined by the *instruction set architecture*, or "ISA," defining the processor state, the format of the instructions, and the effect each of these instructions will have on the state. Most ISAs, including IA32 and x86-64, describe the behavior of a program as if each instruction is executed in sequence, with one instruction completing before the next one begins. The processor hardware is far more elaborate, executing many instructions concurrently, but they employ safeguards to ensure that the overall behavior matches the sequential operation dictated by the ISA. Second, the memory addresses used by a machine-level program are virtual addresses, providing a memory model that appears to be a very large byte array. The actual implementation of the memory system involves a combination of multiple hardware memories and operating system software.

The compiler does most of the work in the overall compilation sequence, transforming programs expressed in the relatively abstract execution model provided by C into the very elementary instructions that the processor executes. The assembly-code representation is very close to machine code. Its main feature is that it is in a more readable textual format, as compared to the binary format of machine code. Being able to understand assembly code and how it relates to the original C code is a key step in understanding how computers execute programs.

IA32 machine code differs greatly from the original C code. Parts of the processor state are visible that normally are hidden from the C programmer:

- The *program counter* (commonly referred to as the "PC," and called %eip in IA32) indicates the address in memory of the next instruction to be executed.

- The integer *register file* contains eight named locations storing 32-bit values. These registers can hold addresses (corresponding to C pointers) or integer data. Some registers are used to keep track of critical parts of the program state, while others are used to hold temporary data, such as the local variables of a procedure, and the value to be returned by a function.

- The condition code registers hold status information about the most recently executed arithmetic or logical instruction. These are used to implement conditional changes in the control or data flow, such as is required to implement if and while statements.

- A set of floating-point registers store floating-point data.

Whereas C provides a model in which objects of different data types can be declared and allocated in memory, machine code views the memory as simply a large, byte-addressable array. Aggregate data types in C such as arrays and structures are represented in machine code as contiguous collections of bytes. Even for scalar data types, assembly code makes no distinctions between signed or unsigned integers, between different types of pointers, or even between pointers and integers.

The program memory contains the executable machine code for the program, some information required by the operating system, a run-time stack for managing procedure calls and returns, and blocks of memory allocated by the user (for example, by using the malloc library function). As mentioned earlier, the program memory is addressed using virtual addresses. At any given time, only limited subranges of virtual addresses are considered valid. For example, although the 32-bit addresses of IA32 potentially span a 4-gigabyte range of address values, a typical program will only have access to a few megabytes. The operating system manages this virtual address space, translating virtual addresses into the physical addresses of values in the actual processor memory.

A single machine instruction performs only a very elementary operation. For example, it might add two numbers stored in registers, transfer data between memory and a register, or conditionally branch to a new instruction address. The compiler must generate sequences of such instructions to implement program constructs such as arithmetic expression evaluation, loops, or procedure calls and returns.

Aside The ever-changing forms of generated code

In our presentation, we will show the code generated by a particular version of GCC with particular settings of the command-line options. If you compile code on your own machine, chances are you will be using a different compiler or a different version of GCC and hence will generate different code. The open-source community supporting GCC keeps changing the code generator, attempting to generate more efficient code according to changing code guidelines provided by the microprocessor manufacturers.

Our goal in studying the examples shown in our presentation is to demonstrate how to examine assembly code and map it back to the constructs found in high-level programming languages. You will need to adapt these techniques to the style of code generated by your particular compiler.

2.2 Code Examples

Suppose we write a C code file code.c containing the following procedure definition:

```
1    int accum = 0;
2
3    int sum(int x, int y)
4    {
5        int t = x + y;
6        accum += t;
7        return t;
8    }
```

To see the assembly code generated by the C compiler, we can use the "-S" option on the command line:

```
unix> gcc -O1 -S code.c
```

This will cause GCC to run the compiler, generating an assembly file code.s, and go no further. (Normally it would then invoke the assembler to generate an object-code file.)

The assembly-code file contains various declarations including the set of lines:

```
sum:
    pushl   %ebp
    movl    %esp, %ebp
    movl    12(%ebp), %eax
    addl    8(%ebp), %eax
    addl    %eax, accum
    popl    %ebp
    ret
```

Each indented line in the above code corresponds to a single machine instruction. For example, the pushl instruction indicates that the contents of register %ebp should be pushed onto the program stack. All information about local variable names or data types has been stripped away. We still see a reference to the global variable `accum`, since the compiler has not yet determined where in memory this variable will be stored.

If we use the '-c' command-line option, GCC will both compile and assemble the code:

```
unix> gcc -O1 -c code.c
```

This will generate an object-code file `code.o` that is in binary format and hence cannot be viewed directly. Embedded within the 800 bytes of the file `code.o` is a 17-byte sequence having hexadecimal representation

```
55 89 e5 8b 45 0c 03 45 08 01 05 00 00 00 00 5d c3
```

This is the object-code corresponding to the assembly instructions listed above. A key lesson to learn from this is that the program actually executed by the machine is simply a sequence of bytes encoding a series of instructions. The machine has very little information about the source code from which these instructions were generated.

> **Aside** How do I find the byte representation of a program?
>
> To generate these bytes, we used a *disassembler* (to be described shortly) to determine that the code for `sum` is 17 bytes long. Then we ran the GNU debugging tool GDB on file `code.o` and gave it the command
>
> ```
> (gdb) x/17xb sum
> ```
>
> telling it to examine (abbreviated 'x') 17 hex-formatted (also abbreviated 'x') bytes (abbreviated 'b'). You will find that GDB has many useful features for analyzing machine-level programs, as will be discussed in Section 11.

To inspect the contents of machine-code files, a class of programs known as *disassemblers* can be invaluable. These programs generate a format similar to assembly code from the machine code. With Linux systems, the program OBJDUMP (for "object dump") can serve this role given the '-d' command-line flag:

```
unix> objdump -d code.o
```

The result is (where we have added line numbers on the left and annotations in italicized text) as follows:

```
    Disassembly of function sum in binary file code.o
1   00000000 <sum>:
    Offset  Bytes                    Equivalent assembly language
2      0:   55                        push   %ebp
3      1:   89 e5                     mov    %esp,%ebp
4      3:   8b 45 0c                  mov    0xc(%ebp),%eax
5      6:   03 45 08                  add    0x8(%ebp),%eax
6      9:   01 05 00 00 00 00         add    %eax,0x0
7      f:   5d                        pop    %ebp
8     10:   c3                        ret
```

On the left, we see the 17 hexadecimal byte values listed in the byte sequence earlier, partitioned into groups of 1 to 6 bytes each. Each of these groups is a single instruction, with the assembly-language equivalent shown on the right.

Several features about machine code and its disassembled representation are worth noting:

- IA32 instructions can range in length from 1 to 15 bytes. The instruction encoding is designed so that commonly used instructions and those with fewer operands require a smaller number of bytes than do less common ones or ones with more operands.

- The instruction format is designed in such a way that from a given starting position, there is a unique decoding of the bytes into machine instructions. For example, only the instruction pushl %ebp can start with byte value 55.

- The disassembler determines the assembly code based purely on the byte sequences in the machine-code file. It does not require access to the source or assembly-code versions of the program.

- The disassembler uses a slightly different naming convention for the instructions than does the assembly code generated by GCC. In our example, it has omitted the suffix 'l' from many of the instructions. These suffixes are size designators and can be omitted in most cases.

Generating the actual executable code requires running a linker on the set of object-code files, one of which must contain a function main. Suppose in file main.c we had the following function:

```
1    int main()
2    {
3        return sum(1, 3);
4    }
```

Then, we could generate an executable program prog as follows:

unix> *gcc -O1 -o prog code.o main.c*

The file prog has grown to 9,123 bytes, since it contains not just the code for our two procedures but also information used to start and terminate the program as well as to interact with the operating system. We can also disassemble the file prog:

unix> *objdump -d prog*

The disassembler will extract various code sequences, including the following:

```
     Disassembly of function sum in executable file prog
1    08048394 <sum>:
     Offset      Bytes              Equivalent assembly language
2    8048394:    55                 push    %ebp
3    8048395:    89 e5              mov     %esp,%ebp
4    8048397:    8b 45 0c           mov     0xc(%ebp),%eax
```

```
5    804839a:   03 45 08                        add      0x8(%ebp),%eax
6    804839d:   01 05 18 a0 04 08               add      %eax,0x804a018
7    80483a3:   5d                              pop      %ebp
8    80483a4:   c3                              ret
```

This code is almost identical to that generated by the disassembly of code.c. One important difference is that the addresses listed along the left are different—the linker has shifted the location of this code to a different range of addresses. A second difference is that the linker has determined the location for storing global variable accum. On line 6 of the disassembly for code.o, the address of accum was listed as 0. In the disassembly of prog, the address has been set to 0x804a018. This is shown in the assembly-code rendition of the instruction. It can also be seen in the last 4 bytes of the instruction, listed from least-significant to most as 18 a0 04 08.

2.3 Notes on Formatting

The assembly code generated by GCC is difficult for a human to read. On one hand, it contains information with which we need not be concerned, while on the other hand, it does not provide any description of the program or how it works. For example, suppose the file simple.c contains the following code:

```
1    int simple(int *xp, int y)
2    {
3        int t = *xp + y;
4        *xp = t;
5        return t;
6    }
```

When GCC is run with flags '-S' and '-O1', it generates the following file for simple.s:

```
.file    "simple.c"
.text
.globl simple
.type    simple, @function
simple:
pushl    %ebp
movl     %esp, %ebp
movl     8(%ebp), %edx
movl     12(%ebp), %eax
addl     (%edx), %eax
movl     %eax, (%edx)
popl     %ebp
ret
.size    simple, .-simple
.ident   "GCC: (Ubuntu 4.3.2-1ubuntu11) 4.3.2"
.section         .note.GNU-stack,"",@progbits
```

All of the lines beginning with '.' are directives to guide the assembler and linker. We can generally ignore these. On the other hand, there are no explanatory remarks about what the instructions do or how they relate to the source code.

To provide a clearer presentation of assembly code, we will show it in a form that omits most of the directives, while including line numbers and explanatory annotations. For our example, an annotated version would appear as follows:

```
1   simple:
2       pushl   %ebp                Save frame pointer
3       movl    %esp, %ebp          Create new frame pointer
4       movl    8(%ebp), %edx       Retrieve xp
5       movl    12(%ebp), %eax      Retrieve y
6       addl    (%edx), %eax        Add *xp to get t
7       movl    %eax, (%edx)        Store t at xp
8       popl    %ebp                Restore frame pointer
9       ret                         Return
```

We typically show only the lines of code relevant to the point being discussed. Each line is numbered on the left for reference and annotated on the right by a brief description of the effect of the instruction and how it relates to the computations of the original C code. This is a stylized version of the way assembly-language programmers format their code.

Aside   ATT versus Intel assembly-code formats

In our presentation, we show assembly code in ATT (named after "AT&T," the company that operated Bell Laboratories for many years) format, the default format for GCC, OBJDUMP, and the other tools we will consider. Other programming tools, including those from Microsoft as well as the documentation from Intel, show assembly code in *Intel* format. The two formats differ in a number of ways. As an example, GCC can generate code in Intel format for the sum function using the following command line:

```
unix> gcc -O1 -S -masm=intel code.c
```

This gives the following assembly code:

```
    Assembly code for simple in Intel format
1   simple:
2       push    ebp
3       mov     ebp, esp
4       mov     edx, DWORD PTR [ebp+8]
5       mov     eax, DWORD PTR [ebp+12]
6       add     eax, DWORD PTR [edx]
7       mov     DWORD PTR [edx], eax
8       pop     ebp
9       ret
```

We see that the Intel and ATT formats differ in the following ways:

- The Intel code omits the size designation suffixes. We see instruction mov instead of movl.
- The Intel code omits the '%' character in front of register names, using esp instead of %esp.
- The Intel code has a different way of describing locations in memory, for example 'DWORD PTR [ebp+8]' rather than '8(%ebp)'.
- Instructions with multiple operands list them in the reverse order. This can be very confusing when switching between the two formats.

Although we will not be using Intel format in our presentation, you will encounter it in IA32 documentation from Intel and Windows documentation from Microsoft.

3 Data Formats

Due to its origins as a 16-bit architecture that expanded into a 32-bit one, Intel uses the term "word" to refer to a 16-bit data type. Based on this, they refer to 32-bit quantities as "double words." They refer to 64-bit quantities as "quad words." Most instructions we will encounter operate on bytes or double words.

Figure 1 shows the IA32 representations used for the primitive data types of C. Most of the common data types are stored as double words. This includes both regular and long int's, whether or not they are signed. In addition, all pointers (shown here as char *) are stored as 4-byte double words. Bytes are commonly used when manipulating string data. More recent extensions of the C language include the data type long long, which is represented using 8 bytes. IA32 does not support this data type in hardware. Instead, the compiler must generate sequences of instructions that operate on these data 32 bits at a time. Floating-point numbers

C declaration	Intel data type	Assembly code suffix	Size (bytes)
char	Byte	b	1
short	Word	w	2
int	Double word	l	4
long int	Double word	l	4
long long int	—	—	4
char *	Double word	l	4
float	Single precision	s	4
double	Double precision	l	8
long double	Extended precision	t	10/12

Figure 1 **Sizes of C data types in IA32.** IA32 does not provide hardware support for 64-bit integer arithmetic. Compiling code with long long data requires generating sequences of operations to perform the arithmetic in 32-bit chunks.

come in three different forms: single-precision (4-byte) values, corresponding to C data type `float`; double-precision (8-byte) values, corresponding to C data type `double`; and extended-precision (10-byte) values. GCC uses the data type `long double` to refer to extended-precision floating-point values. It also stores them as 12-byte quantities to improve memory system performance, as will be discussed later. Using the `long double` data type (introduced in ISO C99) gives us access to the extended-precision capability of x86. For most other machines, this data type will be represented using the same 8-byte format of the ordinary `double` data type.

As the table indicates, most assembly-code instructions generated by GCC have a single-character suffix denoting the size of the operand. For example, the data movement instruction has three variants: `movb` (move byte), `movw` (move word), and `movl` (move double word). The suffix '1' is used for double words, since 32-bit quantities are considered to be "long words," a holdover from an era when 16-bit word sizes were standard. Note that the assembly code uses the suffix '1' to denote both a 4-byte integer as well as an 8-byte double-precision floating-point number. This causes no ambiguity, since floating point involves an entirely different set of instructions and registers.

4 Accessing Information

An IA32 central processing unit (CPU) contains a set of eight *registers* storing 32-bit values. These registers are used to store integer data as well as pointers. Figure 2 diagrams the eight registers. Their names all begin with %e, but otherwise, they have peculiar names. With the original 8086, the registers were 16 bits and each had a specific purpose. The names were chosen to reflect these different purposes. With flat addressing, the need for specialized registers is greatly reduced. For the most part, the first six registers can be considered general-purpose registers

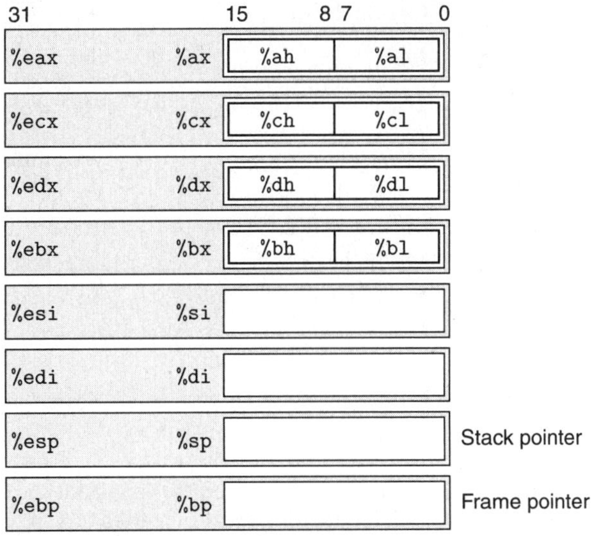

Figure 2

IA32 integer registers. All eight registers can be accessed as either 16 bits (word) or 32 bits (double word). The 2 low-order bytes of the first four registers can be accessed independently.

with no restrictions placed on their use. We said "for the most part," because some instructions use fixed registers as sources and/or destinations. In addition, within procedures there are different conventions for saving and restoring the first three registers (%eax, %ecx, and %edx) than for the next three (%ebx, %edi, and %esi). This will be discussed in Section 7. The final two registers (%ebp and %esp) contain pointers to important places in the program stack. They should only be altered according to the set of standard conventions for stack management.

As indicated in Figure 2, the low-order 2 bytes of the first four registers can be independently read or written by the byte operation instructions. This feature was provided in the 8086 to allow backward compatibility to the 8008 and 8080—two 8-bit microprocessors that date back to 1974. When a byte instruction updates one of these single-byte "register elements," the remaining 3 bytes of the register do not change. Similarly, the low-order 16 bits of each register can be read or written by word operation instructions. This feature stems from IA32's evolutionary heritage as a 16-bit microprocessor and is also used when operating on integers with size designator short.

4.1 Operand Specifiers

Most instructions have one or more *operands*, specifying the source values to reference in performing an operation and the destination location into which to place the result. IA32 supports a number of operand forms (see Figure 3). Source values can be given as constants or read from registers or memory. Results can be stored in either registers or memory. Thus, the different operand possibilities can be classified into three types. The first type, *immediate*, is for constant values. In ATT-format assembly code, these are written with a '$' followed by an integer using standard C notation, for example, $-577 or $0x1F. Any value that fits into a 32-bit word can be used, although the assembler will use 1- or 2-byte encodings

Type	Form	Operand value	Name
Immediate	$*Imm*	*Imm*	Immediate
Register	E_a	$R[E_a]$	Register
Memory	*Imm*	$M[Imm]$	Absolute
Memory	(E_a)	$M[R[E_a]]$	Indirect
Memory	$Imm(E_b)$	$M[Imm + R[E_b]]$	Base + displacement
Memory	(E_b, E_i)	$M[R[E_b] + R[E_i]]$	Indexed
Memory	$Imm(E_b, E_i)$	$M[Imm + R[E_b] + R[E_i]]$	Indexed
Memory	$(, E_i, s)$	$M[R[E_i] \cdot s]$	Scaled indexed
Memory	$Imm(, E_i, s)$	$M[Imm + R[E_i] \cdot s]$	Scaled indexed
Memory	(E_b, E_i, s)	$M[R[E_b] + R[E_i] \cdot s]$	Scaled indexed
Memory	$Imm(E_b, E_i, s)$	$M[Imm + R[E_b] + R[E_i] \cdot s]$	Scaled indexed

Figure 3 **Operand forms.** Operands can denote immediate (constant) values, register values, or values from memory. The scaling factor s must be either 1, 2, 4, or 8.

when possible. The second type, *register*, denotes the contents of one of the registers, either one of the eight 32-bit registers (e.g., %eax) for a double-word operation, one of the eight 16-bit registers (e.g., %ax) for a word operation, or one of the eight single-byte register elements (e.g., %al) for a byte operation. In Figure 3, we use the notation E_a to denote an arbitrary register a, and indicate its value with the reference $R[E_a]$, viewing the set of registers as an array R indexed by register identifiers.

The third type of operand is a *memory* reference, in which we access some memory location according to a computed address, often called the *effective address*. Since we view the memory as a large array of bytes, we use the notation $M_b[Addr]$ to denote a reference to the b-byte value stored in memory starting at address $Addr$. To simplify things, we will generally drop the subscript b.

As Figure 3 shows, there are many different *addressing modes* allowing different forms of memory references. The most general form is shown at the bottom of the table with syntax $Imm(E_b, E_i, s)$. Such a reference has four components: an immediate offset Imm, a base register E_b, an index register E_i, and a scale factor s, where s must be 1, 2, 4, or 8. The effective address is then computed as $Imm + R[E_b] + R[E_i] \cdot s$. This general form is often seen when referencing elements of arrays. The other forms are simply special cases of this general form where some of the components are omitted. As we will see, the more complex addressing modes are useful when referencing array and structure elements.

Practice Problem 1

Assume the following values are stored at the indicated memory addresses and registers:

Address	Value		Register	Value
0x100	0xFF		%eax	0x100
0x104	0xAB		%ecx	0x1
0x108	0x13		%edx	0x3
0x10C	0x11			

Fill in the following table showing the values for the indicated operands:

Operand	Value
%eax	
0x104	
$0x108	
(%eax)	
4(%eax)	
9(%eax,%edx)	
260(%ecx,%edx)	
0xFC(,%ecx,4)	
(%eax,%edx,4)	

Instruction		Effect	Description
MOV	S, D	$D \leftarrow S$	Move
movb		Move byte	
movw		Move word	
movl		Move double word	
MOVS	S, D	$D \leftarrow \text{SignExtend}(S)$	Move with sign extension
movsbw		Move sign-extended byte to word	
movsbl		Move sign-extended byte to double word	
movswl		Move sign-extended word to double word	
MOVZ	S, D	$D \leftarrow \text{ZeroExtend}(S)$	Move with zero extension
movzbw		Move zero-extended byte to word	
movzbl		Move zero-extended byte to double word	
movzwl		Move zero-extended word to double word	
pushl	S	$R[\%esp] \leftarrow R[\%esp] - 4;$ $M[R[\%esp]] \leftarrow S$	Push double word
popl	D	$D \leftarrow M[R[\%esp]];$ $R[\%esp] \leftarrow R[\%esp] + 4$	Pop double word

Figure 4 **Data movement instructions.**

4.2 Data Movement Instructions

Among the most heavily used instructions are those that copy data from one location to another. The generality of the operand notation allows a simple data movement instruction to perform what in many machines would require a number of instructions. Figure 4 lists the important data movement instructions. As can be seen, we group the many different instructions into *instruction classes*, where the instructions in a class perform the same operation, but with different operand sizes. For example, the MOV class consists of three instructions: movb, movw, and movl. All three of these instructions perform the same operation; they differ only in that they operate on data of size 1, 2, and 4 bytes, respectively.

The instructions in the MOV class copy their source values to their destinations. The source operand designates a value that is immediate, stored in a register, or stored in memory. The destination operand designates a location that is either a register or a memory address. IA32 imposes the restriction that a move instruction cannot have both operands refer to memory locations. Copying a value from one memory location to another requires two instructions—the first to load the source value into a register, and the second to write this register value to the destination. Referring to Figure 2, the register operands for these instructions can be any of the eight 32-bit registers (%eax–%ebp) for movl, any of the eight 16-bit registers (%ax–%bp) for movw, and any of the single-byte register elements (%ah–%bh, %al–%bl) for movb. The following MOV instruction examples show the five possible combinations of source and destination types. Recall that the source operand comes first and the destination second:

```
1    movl $0x4050,%eax         Immediate--Register,  4 bytes
2    movw  %bp,%sp             Register--Register,   2 bytes
3    movb (%edi,%ecx),%ah      Memory--Register,     1 byte
4    movb $-17,(%esp)          Immediate--Memory,    1 byte
5    movl %eax,-12(%ebp)       Register--Memory,     4 bytes
```

Both the MOVS and the MOVZ instruction classes serve to copy a smaller amount of source data to a larger data location, filling in the upper bits by either sign expansion (MOVS) or by zero expansion (MOVZ). With sign expansion, the upper bits of the destination are filled in with copies of the most significant bit of the source value. With zero expansion, the upper bits are filled with zeros. As can be seen, there are three instructions in each of these classes, covering all cases of 1- and 2-byte source sizes and 2- and 4-byte destination sizes (omitting the redundant combinations movsww and movzww, of course).

Aside   Comparing byte movement instructions

Observe that the three byte-movement instructions movb, movsbl, and movzbl differ from each other in subtle ways. Here is an example:

```
     Assume initially that %dh = CD, %eax = 98765432
1    movb %dh,%al            %eax = 987654CD
2    movsbl %dh,%eax         %eax = FFFFFFCD
3    movzbl %dh,%eax         %eax = 000000CD
```

In these examples, all set the low-order byte of register %eax to the second byte of %edx. The movb instruction does not change the other 3 bytes. The movsbl instruction sets the other 3 bytes to either all ones or all zeros, depending on the high-order bit of the source byte. The movzbl instruction sets the other 3 bytes to all zeros in any case.

The final two data movement operations are used to push data onto and pop data from the program stack. As we will see, the stack plays a vital role in the handling of procedure calls. By way of background, a stack is a data structure where values can be added or deleted, but only according to a "last-in, first-out" discipline. We add data to a stack via a *push* operation and remove it via a *pop* operation, with the property that the value popped will always be the value that was most recently pushed and is still on the stack. A stack can be implemented as an array, where we always insert and remove elements from one end of the array. This end is called the *top* of the stack. With IA32, the program stack is stored in some region of memory. As illustrated in Figure 5, the stack grows downward such that the top element of the stack has the lowest address of all stack elements. (By convention, we draw stacks upside down, with the stack "top" shown at the bottom of the figure). The stack pointer %esp holds the address of the top stack element.

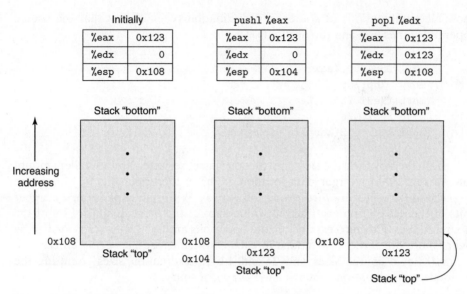

Figure 5 Illustration of stack operation. By convention, we draw stacks upside down, so that the "top" of the stack is shown at the bottom. IA32 stacks grow toward lower addresses, so pushing involves decrementing the stack pointer (register %esp) and storing to memory, while popping involves reading from memory and incrementing the stack pointer.

The pushl instruction provides the ability to push data onto the stack, while the popl instruction pops it. Each of these instructions takes a single operand—the data source for pushing and the data destination for popping.

Pushing a double-word value onto the stack involves first decrementing the stack pointer by 4 and then writing the value at the new top of stack address. Therefore, the behavior of the instruction pushl %ebp is equivalent to that of the pair of instructions

```
subl $4,%esp        Decrement stack pointer
movl %ebp,(%esp)    Store %ebp on stack
```

except that the pushl instruction is encoded in the machine code as a single byte, whereas the pair of instructions shown above requires a total of 6 bytes. The first two columns in Figure 5 illustrate the effect of executing the instruction pushl %eax when %esp is 0x108 and %eax is 0x123. First %esp is decremented by 4, giving 0x104, and then 0x123 is stored at memory address 0x104.

Popping a double word involves reading from the top of stack location and then incrementing the stack pointer by 4. Therefore, the instruction popl %eax is equivalent to the following pair of instructions:

```
movl (%esp),%eax    Read %eax from stack
addl $4,%esp        Increment stack pointer
```

The third column of Figure 5 illustrates the effect of executing the instruction popl %edx immediately after executing the pushl. Value 0x123 is read from memory and written to register %edx. Register %esp is incremented back to 0x108. As shown in the figure, the value 0x123 remains at memory location 0x104 until it is overwritten (e.g., by another push operation). However, the stack top is always considered to be the address indicated by %esp. Any value stored beyond the stack top is considered invalid.

Since the stack is contained in the same memory as the program code and other forms of program data, programs can access arbitrary positions within the stack using the standard memory addressing methods. For example, assuming the topmost element of the stack is a double word, the instruction movl 4(%esp),%edx will copy the second double word from the stack to register %edx.

Practice Problem 2

For each of the following lines of assembly language, determine the appropriate instruction suffix based on the operands. (For example, mov can be rewritten as movb, movw, or movl.)

```
1    mov    %eax, (%esp)
2    mov    (%eax), %dx
3    mov    $0xFF, %bl
4    mov    (%esp,%edx,4), %dh
5    push   $0xFF
6    mov    %dx, (%eax)
7    pop    %edi
```

Practice Problem 3

Each of the following lines of code generates an error message when we invoke the assembler. Explain what is wrong with each line.

```
1    movb $0xF, (%bl)
2    movl %ax, (%esp)
3    movw (%eax),4(%esp)
4    movb %ah,%sh
5    movl %eax,$0x123
6    movl %eax,%dx
7    movb %si, 8(%ebp)
```

4.3 Data Movement Example

As an example of code that uses data movement instructions, consider the data exchange routine shown in Figure 6, both as C code and as assembly code generated by GCC. We omit the portion of the assembly code that allocates space on the run-time stack on procedure entry and deallocates it prior to return. The details of this set-up and completion code will be covered when we discuss procedure linkage. The code we are left with is called the "body."

New to C?   Some examples of pointers

Function exchange (Figure 6) provides a good illustration of the use of pointers in C. Argument xp is a pointer to an integer, while y is an integer itself. The statement

```
int x = *xp;
```

indicates that we should read the value stored in the location designated by xp and store it as a local variable named x. This read operation is known as pointer *dereferencing*. The C operator * performs pointer dereferencing.

    The statement

```
*xp = y;
```

does the reverse—it writes the value of parameter y at the location designated by xp. This is also a form of pointer dereferencing (and hence the operator *), but it indicates a write operation since it is on the left-hand side of the assignment.

    The following is an example of exchange in action:

```
int a = 4;
int b = exchange(&a, 3);
printf("a = %d, b = %d\n", a, b);
```

This code will print

```
a = 3, b = 4
```

The C operator & (called the "address of" operator) *creates* a pointer, in this case to the location holding local variable a. Function exchange then overwrote the value stored in a with 3 but returned 4 as the function value. Observe how by passing a pointer to exchange, it could modify data held at some remote location.

    When the body of the procedure starts execution, procedure parameters xp and y are stored at offsets 8 and 12 relative to the address in register %ebp. Instructions 1 and 2 read parameter xp from memory and store it in register

(a) C code

```
1    int exchange(int *xp, int y)
2    {
3        int x = *xp;
4
5        *xp = y;
6        return x;
7    }
```

(b) Assembly code

```
     xp at %ebp+8, y at %ebp+12
1        movl    8(%ebp), %edx      Get xp
         By copying to %eax below, x becomes the return value
2        movl    (%edx), %eax       Get x at xp
3        movl    12(%ebp), %ecx     Get y
4        movl    %ecx, (%edx)       Store y at xp
```

Figure 6 **C and assembly code for exchange routine body.** The stack set-up and completion portions have been omitted.

%edx. Instruction 2 uses register %edx and reads x into register %eax, a direct implementation of the operation x = *xp in the C program. Later, register %eax will be used to return a value from this function, and so the return value will be x. Instruction 3 loads parameter y into register %ecx. Instruction 4 then writes this value to the memory location designated by xp in register %edx, a direct implementation of the operation *xp = y. This example illustrates how the MOV instructions can be used to read from memory to a register (instructions 1 to 3), and to write from a register to memory (instruction 4).

Two features about this assembly code are worth noting. First, we see that what we call "pointers" in C are simply addresses. Dereferencing a pointer involves copying that pointer into a register, and then using this register in a memory reference. Second, local variables such as x are often kept in registers rather than stored in memory locations. Register access is much faster than memory access.

Practice Problem 4

Assume variables v and p declared with types

```
src_t v;
dest_t *p;
```

where src_t and dest_t are data types declared with typedef. We wish to use the appropriate data movement instruction to implement the operation

```
*p = (dest_t) v;
```

where v is stored in the appropriately named portion of register %eax (i.e., %eax, %ax, or %al), while pointer p is stored in register %edx.

For the following combinations of src_t and dest_t, write a line of assembly code that does the appropriate transfer. Recall that when performing a cast that involves both a size change and a change of "signedness" in C, the operation should change the signedness first.

src_t	dest_t	Instruction
int	int	movl %eax, (%edx)
char	int	
char	unsigned	
unsigned char	int	
int	char	
unsigned	unsigned char	
unsigned	int	

Practice Problem 5

You are given the following information. A function with prototype

```
void decode1(int *xp, int *yp, int *zp);
```

is compiled into assembly code. The body of the code is as follows:

```
xp at %ebp+8, yp at %ebp+12, zp at %ebp+16
1   movl    8(%ebp), %edi
2   movl    12(%ebp), %edx
3   movl    16(%ebp), %ecx
4   movl    (%edx), %ebx
5   movl    (%ecx), %esi
6   movl    (%edi), %eax
7   movl    %eax, (%edx)
8   movl    %ebx, (%ecx)
9   movl    %esi, (%edi)
```

Parameters xp, yp, and zp are stored at memory locations with offsets 8, 12, and 16, respectively, relative to the address in register %ebp.

Write C code for decode1 that will have an effect equivalent to the assembly code above.

5   Arithmetic and Logical Operations

Figure 7 lists some of the integer and logic operations. Most of the operations are given as instruction classes, as they can have different variants with different operand sizes. (Only leal has no other size variants.) For example, the instruction class ADD consists of three addition instructions: addb, addw, and addl, adding bytes, words, and double words, respectively. Indeed, each of the instruction classes shown has instructions for operating on byte, word, and double-word data. The operations are divided into four groups: load effective address, unary, binary, and shifts. *Binary* operations have two operands, while *unary* operations have one operand. These operands are specified using the same notation as described in Section 4.

5.1   Load Effective Address

The *load effective address* instruction leal is actually a variant of the movl instruction. It has the form of an instruction that reads from memory to a register, but it does not reference memory at all. Its first operand appears to be a memory reference, but instead of reading from the designated location, the instruction copies the effective address to the destination. We indicate this computation in Figure 7 using the C address operator &S. This instruction can be used to generate pointers for later memory references. In addition, it can be used to compactly describe common arithmetic operations. For example, if register %edx contains value x, then the instruction leal 7(%edx,%edx,4), %eax will set register %eax to $5x + 7$. Compilers often find clever uses of leal that have nothing to do with effective address computations. The destination operand must be a register.

Instruction		Effect	Description
leal	S, D	$D \leftarrow \&S$	Load effective address
INC	D	$D \leftarrow D + 1$	Increment
DEC	D	$D \leftarrow D - 1$	Decrement
NEG	D	$D \leftarrow -D$	Negate
NOT	D	$D \leftarrow \sim D$	Complement
ADD	S, D	$D \leftarrow D + S$	Add
SUB	S, D	$D \leftarrow D - S$	Subtract
IMUL	S, D	$D \leftarrow D * S$	Multiply
XOR	S, D	$D \leftarrow D \char`^ S$	Exclusive-or
OR	S, D	$D \leftarrow D \mid S$	Or
AND	S, D	$D \leftarrow D \mathbin{\&} S$	And
SAL	k, D	$D \leftarrow D << k$	Left shift
SHL	k, D	$D \leftarrow D << k$	Left shift (same as SAL)
SAR	k, D	$D \leftarrow D >>_A k$	Arithmetic right shift
SHR	k, D	$D \leftarrow D >>_L k$	Logical right shift

Figure 7 **Integer arithmetic operations.** The load effective address (`leal`) instruction is commonly used to perform simple arithmetic. The remaining ones are more standard unary or binary operations. We use the notation $>>_A$ and $>>_L$ to denote arithmetic and logical right shift, respectively. Note the nonintuitive ordering of the operands with ATT-format assembly code.

Practice Problem 6

Suppose register %eax holds value x and %ecx holds value y. Fill in the table below with formulas indicating the value that will be stored in register %edx for each of the given assembly code instructions:

Instruction	Result
`leal 6(%eax), %edx`	
`leal (%eax,%ecx), %edx`	
`leal (%eax,%ecx,4), %edx`	
`leal 7(%eax,%eax,8), %edx`	
`leal 0xA(,%ecx,4), %edx`	
`leal 9(%eax,%ecx,2), %edx`	

5.2 Unary and Binary Operations

Operations in the second group are unary operations, with the single operand serving as both source and destination. This operand can be either a register or a memory location. For example, the instruction incl (%esp) causes the 4-byte element on the top of the stack to be incremented. This syntax is reminiscent of the C increment (++) and decrement (−−) operators.

The third group consists of binary operations, where the second operand is used as both a source and a destination. This syntax is reminiscent of the C assignment operators, such as x += y. Observe, however, that the source operand is given first and the destination second. This looks peculiar for noncommutative operations. For example, the instruction subl %eax,%edx decrements register %edx by the value in %eax. (It helps to read the instruction as "Subtract %eax from %edx.") The first operand can be either an immediate value, a register, or a memory location. The second can be either a register or a memory location. As with the movl instruction, however, the two operands cannot both be memory locations.

Practice Problem 7

Assume the following values are stored at the indicated memory addresses and registers:

Address	Value
0x100	0xFF
0x104	0xAB
0x108	0x13
0x10C	0x11

Register	Value
%eax	0x100
%ecx	0x1
%edx	0x3

Fill in the following table showing the effects of the following instructions, both in terms of the register or memory location that will be updated and the resulting value:

Instruction	Destination	Value
addl %ecx,(%eax)		
subl %edx,4(%eax)		
imull $16,(%eax,%edx,4)		
incl 8(%eax)		
decl %ecx		
subl %edx,%eax		

5.3 Shift Operations

The final group consists of shift operations, where the shift amount is given first, and the value to shift is given second. Both arithmetic and logical right shifts are possible. The shift amount is encoded as a single byte, since only shift amounts between 0 and 31 are possible (only the low-order 5 bits of the shift amount are considered). The shift amount is given either as an immediate or in the single-byte register element %cl. (These instructions are unusual in only allowing this specific register as operand.) As Figure 7 indicates, there are two names for the left shift instruction: SAL and SHL. Both have the same effect, filling from the right with zeros. The right shift instructions differ in that SAR performs an arithmetic shift (fill with copies of the sign bit), whereas SHR performs a logical shift (fill with zeros). The destination operand of a shift operation can be either a register or a memory location. We denote the two different right shift operations in Figure 7 as $>>_A$ (arithmetic) and $>>_L$ (logical).

Practice Problem 8

Suppose we want to generate assembly code for the following C function:

```
int shift_left2_rightn(int x, int n)
{
    x <<= 2;
    x >>= n;
    return x;
}
```

The code that follows is a portion of the assembly code that performs the actual shifts and leaves the final value in register %eax. Two key instructions have been omitted. Parameters x and n are stored at memory locations with offsets 8 and 12, respectively, relative to the address in register %ebp.

```
1    movl    8(%ebp), %eax    Get x
2    _____          x <<= 2
3    movl    12(%ebp), %ecx   Get n
4    _____          x >>= n
```

Fill in the missing instructions, following the annotations on the right. The right shift should be performed arithmetically.

5.4 Discussion

We see that most of the instructions shown in Figure 7 can be used for either unsigned or two's-complement arithmetic. Only right shifting requires instructions that differentiate between signed versus unsigned data. This is one of the features that makes two's-complement arithmetic the preferred way to implement signed integer arithmetic.

Figure 8 shows an example of a function that performs arithmetic operations and its translation into assembly code. As before, we have omitted the stack set-up and completion portions. Function arguments x, y, and z are stored in memory at offsets 8, 12, and 16 relative to the address in register %ebp, respectively.

The assembly code instructions occur in a different order than in the C source code. Instructions 2 and 3 compute the expression z*48 by a combination of leal and shift instructions. Line 5 computes the value of x+y. Line 6 computes the AND of t1 and 0xFFFF. The final multiply is computed by line 7. Since the destination of the multiply is register %eax, this will be the value returned by the function.

(a) C code

```
1    int arith(int x,
2              int y,
3              int z)
4    {
5        int t1 = x+y;
6        int t2 = z*48;
7        int t3 = t1 & 0xFFFF;
8        int t4 = t2 * t3;
9        return t4;
10   }
```

(b) Assembly code

```
     x at %ebp+8, y at %ebp+12, z at %ebp+16
1    movl    16(%ebp), %eax      z
2    leal    (%eax,%eax,2), %eax z*3
3    sall    $4, %eax            t2 = z*48
4    movl    12(%ebp), %edx      y
5    addl    8(%ebp), %edx       t1 = x+y
6    andl    $65535, %edx        t3 = t1&0xFFFF
7    imull   %edx, %eax          Return t4 = t2*t3
```

Figure 8 **C and assembly code for arithmetic routine body.** The stack set-up and completion portions have been omitted.

In the assembly code of Figure 8, the sequence of values in register %eax corresponds to program values z, 3*z, z*48, and t4 (as the return value). In general, compilers generate code that uses individual registers for multiple program values and moves program values among the registers.

Practice Problem 9

In the following variant of the function of Figure 8(a), the expressions have been replaced by blanks:

```
1    int arith(int x,
2              int y,
3              int z)
4    {
5        int t1 = _____;
6        int t2 = _____;
7        int t3 = _____;
8        int t4 = _____;
9        return t4;
10   }
```

The portion of the generated assembly code implementing these expressions is as follows:

```
     x at %ebp+8, y at %ebp+12, z at %ebp+16
1    movl    12(%ebp), %eax
2    xorl    8(%ebp), %eax
3    sarl    $3, %eax
4    notl    %eax
5    subl    16(%ebp), %eax
```

Based on this assembly code, fill in the missing portions of the C code.

Practice Problem 10

It is common to find assembly code lines of the form

```
xorl %edx,%edx
```

in code that was generated from C where no EXCLUSIVE-OR operations were present.

A. Explain the effect of this particular EXCLUSIVE-OR instruction and what useful operation it implements.

B. What would be the more straightforward way to express this operation in assembly code?

C. Compare the number of bytes to encode these two different implementations of the same operation.

5.5 Special Arithmetic Operations

Figure 9 describes instructions that support generating the full 64-bit product of two 32-bit numbers, as well as integer division.

The imull instruction, a member of the IMUL instruction class listed in Figure 7, is known as a "two-operand" multiply instruction. It generates a 32-bit product from two 32-bit operands, implementing the operations $*_{32}^u$ and $*_{32}^t$. Recall that when truncating the product to 32 bits, both unsigned multiply and two's-complement multiply have the same bit-level behavior. IA32 also provides two different "one-operand" multiply instructions to compute the full 64-bit product of two 32-bit values—one for unsigned (mull), and one for two's-complement (imull) multiplication. For both of these, one argument must be in register %eax, and the other is given as the instruction source operand. The product is then stored

Instruction		Effect	Description
imull	S	$R[\%edx]{:}R[\%eax] \leftarrow S \times R[\%eax]$	Signed full multiply
mull	S	$R[\%edx]{:}R[\%eax] \leftarrow S \times R[\%eax]$	Unsigned full multiply
cltd		$R[\%edx]{:}R[\%eax] \leftarrow \text{SignExtend}(R[\%eax])$	Convert to quad word
idivl	S	$R[\%edx] \leftarrow R[\%edx]{:}R[\%eax] \bmod S;$ $R[\%eax] \leftarrow R[\%edx]{:}R[\%eax] \div S$	Signed divide
divl	S	$R[\%edx] \leftarrow R[\%edx]{:}R[\%eax] \bmod S;$ $R[\%eax] \leftarrow R[\%edx]{:}R[\%eax] \div S$	Unsigned divide

Figure 9 **Special arithmetic operations.** These operations provide full 64-bit multiplication and division, for both signed and unsigned numbers. The pair of registers %edx and %eax are viewed as forming a single 64-bit quad word.

in registers %edx (high-order 32 bits) and %eax (low-order 32 bits). Although the name imull is used for two distinct multiplication operations, the assembler can tell which one is intended by counting the number of operands.

As an example, suppose we have signed numbers x and y stored at positions 8 and 12 relative to %ebp, and we want to store their full 64-bit product as 8 bytes on top of the stack. The code would proceed as follows:

```
      x at %ebp+8, y at %ebp+12
1        movl    12(%ebp), %eax    Put y in %eax
2        imull   8(%ebp)           Multiply by x
3        movl    %eax, (%esp)      Store low-order 32 bits
4        movl    %edx, 4(%esp)     Store high-order 32 bits
```

Observe that the locations in which we store the two registers are correct for a little-endian machine—the high-order bits in register %edx are stored at offset 4 relative to the low-order bits in %eax. With the stack growing toward lower addresses, that means that the low-order bits are at the top of the stack.

Our earlier table of arithmetic operations (Figure 7) does not list any division or modulus operations. These operations are provided by the single-operand divide instructions similar to the single-operand multiply instructions. The signed division instruction idivl takes as dividend the 64-bit quantity in registers %edx (high-order 32 bits) and %eax (low-order 32 bits). The divisor is given as the instruction operand. The instruction stores the quotient in register %eax and the remainder in register %edx.

As an example, suppose we have signed numbers x and y stored at positions 8 and 12 relative to %ebp, and we want to store values x/y and x mod y on the stack. GCC generates the following code:

```
      x at %ebp+8, y at %ebp+12
1        movl    8(%ebp), %edx     Put x in %edx
2        movl    %edx, %eax        Copy x to %eax
3        sarl    $31, %edx         Sign extend x in %edx
4        idivl   12(%ebp)          Divide by y
5        movl    %eax, 4(%esp)     Store x / y
6        movl    %edx, (%esp)      Store x % y
```

The move instruction on line 1 and the arithmetic shift on line 3 have the combined effect of setting register %edx to either all zeros or all ones depending on the sign of x, while the move instruction on line 2 copies x into %eax. Thus, we have the combined registers %edx and %eax storing a 64-bit, sign-extended version of x. Following the idivl instruction, the quotient and remainder are copied to the top two stack locations (instructions 5 and 6).

A more conventional method of setting up the divisor makes use of the `cltd`[1] instruction. This instruction sign extends %eax into %edx. With this instruction, the code sequence shown above becomes

```
    x at %ebp+8, y at %ebp+12
1       movl    8(%ebp),%eax        Load x into %eax
2       cltd                        Sign extend into %edx
3       idivl   12(%ebp)            Divide by y
4       movl    %eax, 4(%esp)       Store x / y
5       movl    %edx, (%esp)        Store x % y
```

We can see that the first two instructions have the same overall effect as the first three instructions in our earlier code sequence. Different versions of GCC generate these two different ways of setting up the dividend for integer division.

Unsigned division makes use of the `divl` instruction. Typically register %edx is set to 0 beforehand.

Practice Problem 11

Modify the assembly code shown for signed division so that it computes the unsigned quotient and remainder of numbers x and y and stores the results on the stack.

Practice Problem 12

Consider the following C function prototype, where `num_t` is a data type declared using `typedef`:

```
void store_prod(num_t *dest, unsigned x, num_t y) {
    *dest = x*y;
}
```

GCC generates the following assembly code implementing the body of the computation:

```
    dest at %ebp+8, x at %ebp+12, y at %ebp+16
1       movl    12(%ebp), %eax
2       movl    20(%ebp), %ecx
3       imull   %eax, %ecx
4       mull    16(%ebp)
5       leal    (%ecx,%edx), %edx
6       movl    8(%ebp), %ecx
7       movl    %eax, (%ecx)
8       movl    %edx, 4(%ecx)
```

---

1. This instruction is called `cdq` in the Intel documentation, one of the few cases where the ATT-format name for an instruction bears no relation to the Intel name.

Observe that this code requires two memory reads to fetch argument y (lines 2 and 4), two multiplies (lines 3 and 4), and two memory writes to store the result (lines 7 and 8).

A. What data type is num_t?

B. Describe the algorithm used to compute the product and argue that it is correct.

---

6 Control

So far, we have only considered the behavior of *straight-line* code, where instructions follow one another in sequence. Some constructs in C, such as conditionals, loops, and switches, require conditional execution, where the sequence of operations that gets performed depends on the outcomes of tests applied to the data. Machine code provides two basic low-level mechanisms for implementing conditional behavior: it tests data values and then either alters the control flow or the data flow based on the result of these tests.

Data-dependent control flow is the more general and more common approach for implementing conditional behavior, and so we will examine this first. Normally, both statements in C and instructions in machine code are executed *sequentially*, in the order they appear in the program. The execution order of a set of machine-code instructions can be altered with a *jump* instruction, indicating that control should pass to some other part of the program, possibly contingent on the result of some test. The compiler must generate instruction sequences that build upon this low-level mechanism to implement the control constructs of C.

In our presentation, we first cover the machine-level mechanisms and then show how the different control constructs of C are implemented with them. We then return to the use of conditional data transfer to implement data-dependent behavior.

6.1 Condition Codes

In addition to the integer registers, the CPU maintains a set of single-bit *condition code* registers describing attributes of the most recent arithmetic or logical operation. These registers can then be tested to perform conditional branches. The most useful condition codes are:

CF: Carry Flag. The most recent operation generated a carry out of the most significant bit. Used to detect overflow for unsigned operations.

ZF: Zero Flag. The most recent operation yielded zero.

SF: Sign Flag. The most recent operation yielded a negative value.

OF: Overflow Flag. The most recent operation caused a two's-complement overflow—either negative or positive.

Instruction		Based on	Description
CMP	S_2, S_1	$S_1 - S_2$	Compare
cmpb		Compare byte	
cmpw		Compare word	
cmpl		Compare double word	
TEST	S_2, S_1	S_1 & S_2	Test
testb		Test byte	
testw		Test word	
testl		Test double word	

Figure 10 **Comparison and test instructions.** These instructions set the condition codes without updating any other registers.

For example, suppose we used one of the ADD instructions to perform the equivalent of the C assignment t=a+b, where variables a, b, and t are integers. Then the condition codes would be set according to the following C expressions:

CF:	(unsigned) t < (unsigned) a	Unsigned overflow
ZF:	(t == 0)	Zero
SF:	(t < 0)	Negative
OF:	(a < 0 == b < 0) && (t < 0 != a < 0)	Signed overflow

The leal instruction does not alter any condition codes, since it is intended to be used in address computations. Otherwise, all of the instructions listed in Figure 7 cause the condition codes to be set. For the logical operations, such as XOR, the carry and overflow flags are set to 0. For the shift operations, the carry flag is set to the last bit shifted out, while the overflow flag is set to 0. For reasons that we will not delve into, the INC and DEC instructions set the overflow and zero flags, but they leave the carry flag unchanged.

In addition to the setting of condition codes by the instructions of Figure 7, there are two instruction classes (having 8, 16, and 32-bit forms) that set condition codes without altering any other registers; these are listed in Figure 10. The CMP instructions set the condition codes according to the differences of their two operands. They behave in the same way as the SUB instructions, except that they set the condition codes without updating their destinations. With ATT format, the operands are listed in reverse order, making the code difficult to read. These instructions set the zero flag if the two operands are equal. The other flags can be used to determine ordering relations between the two operands. The TEST instructions behave in the same manner as the AND instructions, except that they set the condition codes without altering their destinations. Typically, the same operand is repeated (e.g., testl %eax,%eax to see whether %eax is negative, zero, or positive), or one of the operands is a mask indicating which bits should be tested.

Instruction		Synonym	Effect	Set condition
sete	*D*	setz	*D* ← ZF	Equal / zero
setne	*D*	setnz	*D* ← ~ZF	Not equal / not zero
sets	*D*		*D* ← SF	Negative
setns	*D*		*D* ← ~SF	Nonnegative
setg	*D*	setnle	*D* ← ~(SF ^ OF) & ~ZF	Greater (signed >)
setge	*D*	setnl	*D* ← ~(SF ^ OF)	Greater or equal (signed >=)
setl	*D*	setnge	*D* ← SF ^ OF	Less (signed <)
setle	*D*	setng	*D* ← (SF ^ OF) | ZF	Less or equal (signed <=)
seta	*D*	setnbe	*D* ← ~CF & ~ZF	Above (unsigned >)
setae	*D*	setnb	*D* ← ~CF	Above or equal (unsigned >=)
setb	*D*	setnae	*D* ← CF	Below (unsigned <)
setbe	*D*	setna	*D* ← CF | ZF	Below or equal (unsigned <=)

Figure 11 **The set instructions.** Each instruction sets a single byte to 0 or 1 based on some combination of the condition codes. Some instructions have "synonyms," i.e., alternate names for the same machine instruction.

6.2 Accessing the Condition Codes

Rather than reading the condition codes directly, there are three common ways of using the condition codes: (1) we can set a single byte to 0 or 1 depending on some combination of the condition codes, (2) we can conditionally jump to some other part of the program, or (3) we can conditionally transfer data. For the first case, the instructions described in Figure 11 set a single byte to 0 or to 1 depending on some combination of the condition codes. We refer to this entire class of instructions as the SET instructions; they differ from one another based on which combinations of condition codes they consider, as indicated by the different suffixes for the instruction names. It is important to recognize that the suffixes for these instructions denote different conditions and not different operand sizes. For example, instructions setl and setb denote "set less" and "set below," not "set long word" or "set byte."

A SET instruction has either one of the eight single-byte register elements (Figure 2) or a single-byte memory location as its destination, setting this byte to either 0 or 1. To generate a 32-bit result, we must also clear the high-order 24 bits. A typical instruction sequence to compute the C expression a < b, where a and b are both of type int, proceeds as follows:

```
     a is in %edx, b is in %eax
1    cmpl    %eax, %edx     Compare a:b
2    setl    %al            Set low order byte of %eax to 0 or 1
3    movzbl  %al, %eax      Set remaining bytes of %eax to 0
```

The movzbl instruction clears the high-order 3 bytes of %eax.

For some of the underlying machine instructions, there are multiple possible names, which we list as "synonyms." For example, both setg (for "set greater") and setnle (for "set not less or equal") refer to the same machine instruction. Compilers and disassemblers make arbitrary choices of which names to use.

Although all arithmetic and logical operations set the condition codes, the descriptions of the different SET instructions apply to the case where a comparison instruction has been executed, setting the condition codes according to the computation t = a-b. More specifically, let a, b, and t be the integers represented in two's-complement form by variables a, b, and t, respectively, and so $t = a -_w^t b$, where w depends on the sizes associated with a and b.

Consider the sete, or "set when equal" instruction. When $a = b$, we will have $t = 0$, and hence the zero flag indicates equality. Similarly, consider testing for signed comparison with the setl, or "set when less," instruction. When no overflow occurs (indicated by having OF set to 0), we will have $a < b$ when $a -_w^t b < 0$, indicated by having SF set to 1, and $a \geq b$ when $a -_w^t b \geq 0$, indicated by having SF set to 0. On the other hand, when overflow occurs, we will have $a < b$ when $a -_w^t b > 0$ (positive overflow) and $a > b$ when $a -_w^t b < 0$ (negative overflow). We cannot have overflow when $a = b$. Thus, when OF is set to 1, we will have $a < b$ if and only if SF is set to 0. Combining these cases, the EXCLUSIVE-OR of the overflow and sign bits provides a test for whether $a < b$. The other signed comparison tests are based on other combinations of SF ^ OF and ZF.

For the testing of unsigned comparisons, we now let a and b be the integers represented in unsigned form by variables a and b. In performing the computation t = a-b, the carry flag will be set by the CMP instruction when the $a - b < 0$, and so the unsigned comparisons use combinations of the carry and zero flags.

It is important to note how machine code distinguishes between signed and unsigned values. Unlike in C, it does not associate a data type with each program value. Instead, it mostly uses the same instructions for the two cases, because many arithmetic operations have the same bit-level behavior for unsigned and two's-complement arithmetic. Some circumstances require different instructions to handle signed and unsigned operations, such as using different versions of right shifts, division and multiplication instructions, and different combinations of condition codes.

Practice Problem 13

The following C code

```
int comp(data_t a, data_t b) {
    return a COMP b;
}
```

shows a general comparison between arguments a and b, where we can set the data type of the arguments by declaring data_t with a typedef declaration, and we can set the comparison by defining COMP with a #define declaration.

Suppose a is in %edx and b is in %eax. For each of the following instruction sequences, determine which data types data_t and which comparisons COMP could cause the compiler to generate this code. (There can be multiple correct answers; you should list them all.)

A. cmpl   %eax, %edx
    setl   %al

B. cmpw   %ax, %dx
    setge   %al

C. cmpb   %al, %dl
    setb   %al

D. cmpl   %eax, %edx
    setne   %al

Practice Problem 14

The following C code

```
int test(data_t a) {
    return a TEST 0;
}
```

shows a general comparison between argument a and 0, where we can set the data type of the argument by declaring data_t with a typedef, and the nature of the comparison by declaring TEST with a #define declaration. For each of the following instruction sequences, determine which data types data_t and which comparisons TEST could cause the compiler to generate this code. (There can be multiple correct answers; list all correct ones.)

A. testl   %eax, %eax
    setne   %al

B. testw   %ax, %ax
    sete   %al

C. testb   %al, %al
    setg   %al

D. testw   %ax, %ax
    seta   %al

6.3   Jump Instructions and Their Encodings

Under normal execution, instructions follow each other in the order they are listed. A *jump* instruction can cause the execution to switch to a completely new position in the program. These jump destinations are generally indicated in

Instruction		Synonym	Jump condition	Description
jmp	*Label*		1	Direct jump
jmp	*Operand*		1	Indirect jump
je	*Label*	jz	ZF	Equal / zero
jne	*Label*	jnz	~ZF	Not equal / not zero
js	*Label*		SF	Negative
jns	*Label*		~SF	Nonnegative
jg	*Label*	jnle	~(SF ^ OF) & ~ZF	Greater (signed >)
jge	*Label*	jnl	~(SF ^ OF)	Greater or equal (signed >=)
jl	*Label*	jnge	SF ^ OF	Less (signed <)
jle	*Label*	jng	(SF ^ OF) | ZF	Less or equal (signed <=)
ja	*Label*	jnbe	~CF & ~ZF	Above (unsigned >)
jae	*Label*	jnb	~CF	Above or equal (unsigned >=)
jb	*Label*	jnae	CF	Below (unsigned <)
jbe	*Label*	jna	CF | ZF	Below or equal (unsigned <=)

Figure 12 **The jump instructions.** These instructions jump to a labeled destination when the jump condition holds. Some instructions have "synonyms," alternate names for the same machine instruction.

assembly code by a *label*. Consider the following (very contrived) assembly-code sequence:

```
1      movl $0,%eax          Set %eax to 0
2      jmp .L1               Goto .L1
3      movl (%eax),%edx      Null pointer dereference
4    .L1:
5      popl %edx
```

The instruction `jmp .L1` will cause the program to skip over the `movl` instruction and instead resume execution with the `popl` instruction. In generating the object-code file, the assembler determines the addresses of all labeled instructions and encodes the *jump targets* (the addresses of the destination instructions) as part of the jump instructions.

Figure 12 shows the different jump instructions. The `jmp` instruction jumps unconditionally. It can be either a *direct* jump, where the jump target is encoded as part of the instruction, or an *indirect* jump, where the jump target is read from a register or a memory location. Direct jumps are written in assembly by giving a label as the jump target, e.g., the label ".L1" in the code shown. Indirect jumps are written using '*' followed by an operand specifier using one of the formats described in Section 4.1. As examples, the instruction

`jmp *%eax`

uses the value in register `%eax` as the jump target, and the instruction

`jmp *(%eax)`

reads the jump target from memory, using the value in %eax as the read address.

The remaining jump instructions in the table are *conditional*—they either jump or continue executing at the next instruction in the code sequence, depending on some combination of the condition codes. The names of these instructions and the conditions under which they jump match those of the SET instructions (see Figure 11). As with the SET instructions, some of the underlying machine instructions have multiple names. Conditional jumps can only be direct.

Although we will not concern ourselves with the detailed format of machine code, understanding how the targets of jump instructions are encoded will become important when we study linking. In addition, it helps when interpreting the output of a disassembler. In assembly code, jump targets are written using symbolic labels. The assembler, and later the linker, generate the proper encodings of the jump targets. There are several different encodings for jumps, but some of the most commonly used ones are *PC relative*. That is, they encode the difference between the address of the target instruction and the address of the instruction immediately following the jump. These offsets can be encoded using 1, 2, or 4 bytes. A second encoding method is to give an "absolute" address, using 4 bytes to directly specify the target. The assembler and linker select the appropriate encodings of the jump destinations.

As an example of PC-relative addressing, the following fragment of assembly code was generated by compiling a file silly.c. It contains two jumps: the jle instruction on line 1 jumps forward to a higher address, while the jg instruction on line 8 jumps back to a lower one.

```
1       jle     .L2             if <=, goto dest2
2   .L5:                        dest1:
3       movl    %edx, %eax
4       sarl    %eax
5       subl    %eax, %edx
6       leal    (%edx,%edx,2), %edx
7       testl   %edx, %edx
8       jg      .L5             if >, goto dest1
9   .L2:                        dest2:
10      movl    %edx, %eax
```

The disassembled version of the ".o" format generated by the assembler is as follows:

```
1    8:   7e 0d           jle   17 <silly+0x17>   Target = dest2
2    a:   89 d0           mov   %edx,%eax         dest1:
3    c:   d1 f8           sar   %eax
4    e:   29 c2           sub   %eax,%edx
5   10:   8d 14 52        lea   (%edx,%edx,2),%edx
6   13:   85 d2           test  %edx,%edx
7   15:   7f f3           jg    a <silly+0xa>     Target = dest1
8   17:   89 d0           mov   %edx,%eax         dest2:
```

In the annotations generated by the disassembler on the right, the jump targets are indicated as 0x17 for the jump instruction on line 1 and 0xa for the jump instruction on line 7. Looking at the byte encodings of the instructions, however, we see that the target of the first jump instruction is encoded (in the second byte) as 0xd (decimal 13). Adding this to 0xa (decimal 10), the address of the following instruction, we get jump target address 0x17 (decimal 23), the address of the instruction on line 8.

Similarly, the target of the second jump instruction is encoded as 0xf3 (decimal −13) using a single-byte, two's-complement representation. Adding this to 0x17 (decimal 23), the address of the instruction on line 8, we get 0xa (decimal 10), the address of the instruction on line 2.

As these examples illustrate, the value of the program counter when performing PC-relative addressing is the address of the instruction following the jump, not that of the jump itself. This convention dates back to early implementations, when the processor would update the program counter as its first step in executing an instruction.

The following shows the disassembled version of the program after linking:

```
1    804839c:   7e 0d              jle     80483ab <silly+0x17>
2    804839e:   89 d0              mov     %edx,%eax
3    80483a0:   d1 f8              sar     %eax
4    80483a2:   29 c2              sub     %eax,%edx
5    80483a4:   8d 14 52           lea     (%edx,%edx,2),%edx
6    80483a7:   85 d2              test    %edx,%edx
7    80483a9:   7f f3              jg      804839e <silly+0xa>
8    80483ab:   89 d0              mov     %edx,%eax
```

The instructions have been relocated to different addresses, but the encodings of the jump targets in lines 1 and 7 remain unchanged. By using a PC-relative encoding of the jump targets, the instructions can be compactly encoded (requiring just 2 bytes), and the object code can be shifted to different positions in memory without alteration.

Practice Problem 15

In the following excerpts from a disassembled binary, some of the information has been replaced by Xs. Answer the following questions about these instructions.

A. What is the target of the je instruction below? (You don't need to know anything about the call instruction here.)

```
804828f:        74 05                    je      XXXXXXX
8048291:        e8 1e 00 00 00           call    80482b4
```

B. What is the target of the jb instruction below?

```
8048357:        72 e7                    jb      XXXXXXX
8048359:        c6 05 10 a0 04 08 01     movb    $0x1,0x804a010
```

C. What is the address of the `mov` instruction?

```
XXXXXXX:     74 12                  je     8048391
XXXXXXX:     b8 00 00 00 00         mov    $0x0,%eax
```

D. In the code that follows, the jump target is encoded in PC-relative form as a 4-byte, two's-complement number. The bytes are listed from least significant to most, reflecting the little-endian byte ordering of IA32. What is the address of the jump target?

```
80482bf:     e9 e0 ff ff ff         jmp    XXXXXXX
80482c4:     90                      nop
```

E. Explain the relation between the annotation on the right and the byte coding on the left.

```
80482aa:     ff 25 fc 9f 04 08      jmp    *0x8049ffc
```

To implement the control constructs of C via conditional control transfer, the compiler must use the different types of jump instructions we have just seen. We will go through the most common constructs, starting from simple conditional branches, and then consider loops and `switch` statements.

6.4 Translating Conditional Branches

The most general way to translate conditional expressions and statements from C into machine code is to use combinations of conditional and unconditional jumps. (As an alternative, we will see in Section 6.6 that some conditionals can be implemented by conditional transfers of data rather than control.) For example, Figure 13(a) shows the C code for a function that computes the absolute value of the difference of two numbers.[2] GCC generates the assembly code shown as Figure 13(c). We have created a version in C, called `gotodiff` (Figure 13(b)), that more closely follows the control flow of this assembly code. It uses the `goto` statement in C, which is similar to the unconditional jump of assembly code. The statement `goto x_ge_y` on line 4 causes a jump to the label `x_ge_y` (since it occurs when $x \geq y$) on line 7, skipping the computation of `y-x` on line 5. If the test fails, the program computes the result as `y-x` and then transfers unconditionally to the end of the code. Using `goto` statements is generally considered a bad programming style, since their use can make code very difficult to read and debug. We use them in our presentation as a way to construct C programs that describe the control flow of assembly-code programs. We call this style of programming "goto code."

The assembly-code implementation first compares the two operands (line 3), setting the condition codes. If the comparison result indicates that x is greater

---

2. Actually, it can return a negative value if one of the subtractions overflows. Our interest here is to demonstrate machine code, not to implement robust code.

(a) Original C code

```
1    int absdiff(int x, int y) {
2        if (x < y)
3            return y - x;
4        else
5            return x - y;
6    }
```

(b) Equivalent goto version

```
1    int gotodiff(int x, int y) {
2        int result;
3        if (x >= y)
4            goto x_ge_y;
5        result = y - x;
6        goto done;
7    x_ge_y:
8        result = x - y;
9    done:
10       return result;
11   }
```

(c) Generated assembly code

```
     x at %ebp+8, y at %ebp+12
1        movl    8(%ebp), %edx       Get x
2        movl    12(%ebp), %eax      Get y
3        cmpl    %eax, %edx          Compare x:y
4        jge     .L2                 if >= goto x_ge_y
5        subl    %edx, %eax          Compute result = y-x
6        jmp     .L3                 Goto done
7    .L2:                            x_ge_y:
8        subl    %eax, %edx          Compute result = x-y
9        movl    %edx, %eax          Set result as return value
10   .L3:                            done: Begin completion code
```

Figure 13 Compilation of conditional statements. C procedure `absdiff` (part (a)) contains an if-else statement. The generated assembly code is shown (part (c)), along with a C procedure `gotodiff` (part (b)) that mimics the control flow of the assembly code. The stack set-up and completion portions of the assembly code have been omitted.

than or equal to y, it then jumps to a block of code that computes x–y (line 8). Otherwise, it continues with the execution of code that computes y–x (line 5). In both cases, the computed result is stored in register %eax, and the program reaches line 10, at which point it executes the stack completion code (not shown).

The general form of an if-else statement in C is given by the template

```
if (test-expr)
    then-statement
else
    else-statement
```

where *test-expr* is an integer expression that evaluates either to 0 (interpreted as meaning "false") or to a nonzero value (interpreted as meaning "true"). Only one of the two branch statements (*then-statement* or *else-statement*) is executed.

For this general form, the assembly implementation typically adheres to the following form, where we use C syntax to describe the control flow:

```
    t = test-expr;
    if (!t)
        goto false;
    then-statement
    goto done;
false:
    else-statement
done:
```

That is, the compiler generates separate blocks of code for *then-statement* and *else-statement*. It inserts conditional and unconditional branches to make sure the correct block is executed.

Practice Problem 16

When given the C code

```
1    void cond(int a, int *p)
2    {
3        if (p && a > 0)
4            *p += a;
5    }
```

GCC generates the following assembly code for the body of the function:

```
     a %ebp+8, p at %ebp+12
1      movl    8(%ebp), %edx
2      movl    12(%ebp), %eax
3      testl   %eax, %eax
4      je      .L3
5      testl   %edx, %edx
6      jle     .L3
7      addl    %edx, (%eax)
8    .L3:
```

A. Write a goto version in C that performs the same computation and mimics the control flow of the assembly code, in the style shown in Figure 13(b). You might find it helpful to first annotate the assembly code as we have done in our examples.

B. Explain why the assembly code contains two conditional branches, even though the C code has only one if statement.

Practice Problem 17

An alternate rule for translating if statements into goto code is as follows:

```
    t = test-expr;
    if (t)
        goto true;
    else-statement
    goto done;
true:
    then-statement
done:
```

A. Rewrite the goto version of absdiff based on this alternate rule.

B. Can you think of any reasons for choosing one rule over the other?

Practice Problem 18

Starting with C code of the form

```
1    int test(int x, int y) {
2        int val = _____ ;
3        if ( _____ ) {
4            if ( _____ )
5                val = _____ ;
6            else
7                val = _____ ;
8        } else if ( _____ )
9            val = _____ ;
10       return val;
11   }
```

GCC generates the following assembly code:

```
     x at %ebp+8, y at %ebp+12
1        movl    8(%ebp), %eax
2        movl    12(%ebp), %edx
3        cmpl    $-3, %eax
4        jge     .L2
5        cmpl    %edx, %eax
6        jle     .L3
7        imull   %edx, %eax
8        jmp     .L4
9    .L3:
10       leal    (%edx,%eax), %eax
11       jmp     .L4
12   .L2:
```

```
13      cmpl    $2, %eax
14      jg      .L5
15      xorl    %edx, %eax
16      jmp     .L4
17  .L5:
18      subl    %edx, %eax
19  .L4:
```

Fill in the missing expressions in the C code. To make the code fit into the C code template, you will need to undo some of the reordering of computations done by GCC.

6.5 Loops

C provides several looping constructs—namely, do-while, while, and for. No corresponding instructions exist in machine code. Instead, combinations of conditional tests and jumps are used to implement the effect of loops. Most compilers generate loop code based on the do-while form of a loop, even though this form is relatively uncommon in actual programs. Other loops are transformed into do-while form and then compiled into machine code. We will study the translation of loops as a progression, starting with do-while and then working toward ones with more complex implementations.

Do-While Loops

The general form of a do-while statement is as follows:

```
do
    body-statement
    while (test-expr);
```

The effect of the loop is to repeatedly execute *body-statement*, evaluate *test-expr*, and continue the loop if the evaluation result is nonzero. Observe that *body-statement* is executed at least once.

This general form can be translated into conditionals and goto statements as follows:

```
loop:
    body-statement
    t = test-expr;
    if (t)
        goto loop;
```

That is, on each iteration the program evaluates the body statement and then the test expression. If the test succeeds, we go back for another iteration.

As an example, Figure 14(a) shows an implementation of a routine to compute the factorial of its argument, written $n!$, with a do-while loop. This function only computes the proper value for $n > 0$.

(a) C code

```
1    int fact_do(int n)
2    {
3        int result = 1;
4        do {
5            result *= n;
6            n = n-1;
7        } while (n > 1);
8        return result;
9    }
```

(b) Register usage

Register	Variable	Initially
%eax	result	1
%edx	n	*n*

(c) Corresponding assembly-language code

```
     Argument: n at %ebp+8
     Registers: n in %edx, result in %eax
1        movl    8(%ebp), %edx      Get n
2        movl    $1, %eax           Set result = 1
3    .L2:                           loop:
4        imull   %edx, %eax         Compute result *= n
5        subl    $1, %edx           Decrement n
6        cmpl    $1, %edx           Compare n:1
7        jg      .L2                If >, goto loop
     Return result
```

Figure 14 **Code for do-while version of factorial program.** The C code, the generated assembly code, and a table of register usage is shown.

Practice Problem 19

A. What is the maximum value of *n* for which we can represent *n*! with a 32-bit int?

B. What about for a 64-bit long long int?

The assembly code shown in Figure 14(c) shows a standard implementation of a do-while loop. Following the initialization of register %edx to hold *n* and %eax to hold result, the program begins looping. It first executes the *body* of the loop, consisting here of the updates to variables result and *n* (lines 4–5). It then tests whether $n > 1$, and, if so, it jumps back to the beginning of the loop. We see here that the conditional jump (line 7) is the key instruction in implementing a loop. It determines whether to continue iterating or to exit the loop.

Determining which registers are used for which program values can be challenging, especially with loop code. We have shown such a mapping in Figure 14. In this case, the mapping is fairly simple to determine: we can see *n* getting loaded into register %edx on line 1, getting decremented on line 5, and being tested on line 6. We therefore conclude that this register holds *n*.

We can see register %eax getting initialized to 1 (line 2), and being updated by multiplication on line 4. Furthermore, since %eax is used to return the function value, it is often chosen to hold program values that are returned. We therefore conclude that %eax corresponds to program value result.

Aside   Reverse engineering loops

A key to understanding how the generated assembly code relates to the original source code is to find a mapping between program values and registers. This task was simple enough for the loop of Figure 14, but it can be much more challenging for more complex programs. The C compiler will often rearrange the computations, so that some variables in the C code have no counterpart in the machine code, and new values are introduced into the machine code that do not exist in the source code. Moreover, it will often try to minimize register usage by mapping multiple program values onto a single register.

The process we described for `fact_do` works as a general strategy for reverse engineering loops. Look at how registers are initialized before the loop, updated and tested within the loop, and used after the loop. Each of these provides a clue that can be combined to solve a puzzle. Be prepared for surprising transformations, some of which are clearly cases where the compiler was able to optimize the code, and others where it is hard to explain why the compiler chose that particular strategy. In our experience, GCC often makes transformations that provide no performance benefit and can even decrease code performance.

Practice Problem 20

For the C code

```
1    int dw_loop(int x, int y, int n) {
2        do {
3            x += n;
4            y *= n;
5            n--;
6        } while ((n > 0) && (y < n));
7        return x;
8    }
```

GCC generates the following assembly code:

```
     x at %ebp+8, y at %ebp+12, n at %ebp+16
1        movl    8(%ebp), %eax
2        movl    12(%ebp), %ecx
3        movl    16(%ebp), %edx
4    .L2:
5        addl    %edx, %eax
6        imull   %edx, %ecx
7        subl    $1, %edx
8        testl   %edx, %edx
9        jle     .L5
10       cmpl    %edx, %ecx
11       jl      .L2
12   .L5:
```

A. Make a table of register usage, similar to the one shown in Figure 14(b).

B. Identify *test-expr* and *body-statement* in the C code, and the corresponding lines in the assembly code.

C. Add annotations to the assembly code describing the operation of the program, similar to those shown in Figure 14(b).

While Loops

The general form of a `while` statement is as follows:

```
while (test-expr)
    body-statement
```

It differs from `do-while` in that *test-expr* is evaluated and the loop is potentially terminated before the first execution of *body-statement*. There are a number of ways to translate a `while` loop into machine code. One common approach, also used by GCC, is to transform the code into a `do-while` loop by using a conditional branch to skip the first execution of the body if needed:

```
if (!test-expr)
    goto done;
do
    body-statement
    while (test-expr);
done:
```

This, in turn, can be transformed into goto code as

```
    t = test-expr;
    if (!t)
        goto done;
loop:
    body-statement
    t = test-expr;
    if (t)
        goto loop;
done:
```

Using this implementation strategy, the compiler can often optimize the initial test, for example determining that the test condition will always hold.

As an example, Figure 15 shows an implementation of the factorial function using a `while` loop (Figure 15(a)). This function correctly computes $0! = 1$. The adjacent function `fact_while_goto` (Figure 15(b)) is a C rendition of the assembly code generated by GCC. Comparing the code generated for `fact_while` (Figure 15) to that for `fact_do` (Figure 14), we see that they are nearly identical. The only difference is the initial test (line 3) and the jump around the loop (line 4). The compiler closely followed our template for converting a `while` loop to a `do-while` loop, and for translating this loop to goto code.

(a) C code

```
1   int fact_while(int n)
2   {
3       int result = 1;
4       while (n > 1) {
5           result *= n;
6           n = n-1;
7       }
8       return result;
9   }
```

(b) Equivalent goto version

```
1   int fact_while_goto(int n)
2   {
3       int result = 1;
4       if (n <= 1)
5           goto done;
6   loop:
7       result *= n;
8       n = n-1;
9       if (n > 1)
10          goto loop;
11  done:
12      return result;
13  }
```

(c) Corresponding assembly-language code

```
    Argument: n at %ebp+8
    Registers: n in %edx, result in %eax
1       movl    8(%ebp), %edx       Get n
2       movl    $1, %eax            Set result = 1
3       cmpl    $1, %edx            Compare n:1
4       jle     .L7                 If <=, goto done
5   .L10:                           loop:
6       imull   %edx, %eax          Compute result *= n
7       subl    $1, %edx            Decrement n
8       cmpl    $1, %edx            Compare n:1
9       jg      .L10                If >, goto loop
10  .L7:                            done:
    Return result
```

Figure 15 **C and assembly code for while version of factorial.** The fact_while_goto function illustrates the operation of the assembly code version.

Practice Problem 21

For the C code

```
1   int loop_while(int a, int b)
2   {
3       int result = 1;
4       while (a < b) {
5           result *= (a+b);
6           a++;
7       }
8       return result;
9   }
```

GCC generates the following assembly code:

```
    a at %ebp+8, b at %ebp+12
1       movl    8(%ebp), %ecx
2       movl    12(%ebp), %ebx
3       movl    $1, %eax
4       cmpl    %ebx, %ecx
5       jge     .L11
6       leal    (%ebx,%ecx), %edx
7       movl    $1, %eax
8   .L12:
9       imull   %edx, %eax
10      addl    $1, %ecx
11      addl    $1, %edx
12      cmpl    %ecx, %ebx
13      jg      .L12
14  .L11:
```

In generating this code, GCC makes an interesting transformation that, in effect, introduces a new program variable.

A. Register %edx is initialized on line 6 and updated within the loop on line 11. Consider this to be a new program variable. Describe how it relates to the variables in the C code.

B. Create a table of register usage for this function.

C. Annotate the assembly code to describe how it operates.

D. Write a goto version of the function (in C) that mimics how the assembly code program operates.

Practice Problem 22

A function, fun_a, has the following overall structure:

```
int fun_a(unsigned x) {
    int val = 0;
    while (          ) {
              ;
    }
    return        ;
}
```

The GCC C compiler generates the following assembly code:

```
    x at %ebp+8
1       movl    8(%ebp), %edx
2       movl    $0, %eax
3       testl   %edx, %edx
```

```
4       je      .L7
5   .L10:
6       xorl    %edx, %eax
7       shrl    %edx              Shift right by 1
8       jne     .L10
9   .L7:
10      andl    $1, %eax
```

Reverse engineer the operation of this code and then do the following:

A. Use the assembly-code version to fill in the missing parts of the C code.

B. Describe in English what this function computes.

For Loops

The general form of a for loop is as follows:

```
for (init-expr; test-expr; update-expr)
    body-statement
```

The C language standard states (with one exception, highlighted in Problem 24) that the behavior of such a loop is identical to the following code, which uses a while loop:

```
init-expr;
while (test-expr) {
    body-statement
    update-expr;
}
```

The program first evaluates the initialization expression *init-expr*. It enters a loop where it first evaluates the test condition *test-expr*, exiting if the test fails, then executes the body of the loop *body-statement*, and finally evaluates the update expression *update-expr*.

The compiled form of this code is based on the transformation from while to do-while described previously, first giving a do-while form:

```
init-expr;
if (!test-expr)
    goto done;
do {
    body-statement
    update-expr;
} while (test-expr);
done:
```

This, in turn, can be transformed into goto code as

```
    init-expr;
    t = test-expr;
    if (!t)
        goto done;
loop:
    body-statement
    update-expr;
    t = test-expr;
    if (t)
        goto loop;
done:
```

As an example, consider a factorial function written with a for loop:

```
1   int fact_for(int n)
2   {
3       int i;
4       int result = 1;
5       for (i = 2; i <= n; i++)
6           result *= i;
7       return result;
8   }
```

As shown, the natural way of writing a factorial function with a for loop is to multiply factors from 2 up to n, and so this function is quite different from the code we showed using either a while or a do-while loop.

We can identify the different components of the for loop in this code as follows:

init-expr	i = 2
test-expr	i <= n
update-expr	i++
body-statement	result *= i;

Substituting these components into the template we have shown yields the following version in goto code:

```
1   int fact_for_goto(int n)
2   {
3       int i = 2;
4       int result = 1;
5       if (!(i <= n))
6           goto done;
7   loop:
8       result *= i;
9       i++;
```

```
10          if (i <= n)
11              goto loop;
12      done:
13          return result;
14      }
```

Indeed, a close examination of the assembly code produced by GCC closely follows this template:

```
    Argument: n at %ebp+8
    Registers: n in %ecx, i in %edx, result in %eax
1       movl    8(%ebp), %ecx       Get n
2       movl    $2, %edx            Set i to 2          (init)
3       movl    $1, %eax            Set result to 1
4       cmpl    $1, %ecx            Compare n:1         (!test)
5       jle     .L14                If <=, goto done
6   .L17:                       loop:
7       imull   %edx, %eax          Compute result *= i (body)
8       addl    $1, %edx            Increment i         (update)
9       cmpl    %edx, %ecx          Compare n:i         (test)
10      jge     .L17                If >=, goto loop
11  .L14:                       done:
```

We see from this presentation that all three forms of loops in C—do-while, while, and for—can be translated by a single strategy, generating code that contains one or more conditional branches. Conditional transfer of control provides the basic mechanism for translating loops into machine code.

Practice Problem 23

A function fun_b has the following overall structure:

```
int fun_b(unsigned x) {
    int val = 0;
    int i;
    for ( _____ ; _____ ; _____ ) {

    }
    return val;
}
```

The GCC C compiler generates the following assembly code:

```
    x at %ebp+8
1       movl    8(%ebp), %ebx
2       movl    $0, %eax
3       movl    $0, %ecx
4   .L13:
```

```
5     leal    (%eax,%eax), %edx
6     movl    %ebx, %eax
7     andl    $1, %eax
8     orl     %edx, %eax
9     shrl    %ebx              Shift right by 1
10    addl    $1, %ecx
11    cmpl    $32, %ecx
12    jne     .L13
```

Reverse engineer the operation of this code and then do the following:

A. Use the assembly-code version to fill in the missing parts of the C code.

B. Describe in English what this function computes.

Practice Problem 24

Executing a continue statement in C causes the program to jump to the end of the current loop iteration. The stated rule for translating a for loop into a while loop needs some refinement when dealing with continue statements. For example, consider the following code:

```
/* Example of for loop using a continue statement */
/* Sum even numbers between 0 and 9 */
int sum = 0;
int i;
for (i = 0; i < 10; i++) {
    if (i & 1)
        continue;
    sum += i;
}
```

A. What would we get if we naively applied our rule for translating the for loop into a while loop? What would be wrong with this code?

B. How could you replace the continue statement with a goto statement to ensure that the while loop correctly duplicates the behavior of the for loop?

6.6 Conditional Move Instructions

The conventional way to implement conditional operations is through a conditional transfer of *control*, where the program follows one execution path when a condition holds and another when it does not. This mechanism is simple and general, but it can be very inefficient on modern processors.

An alternate strategy is through a conditional transfer of *data*. This approach computes both outcomes of a conditional operation, and then selects one based on whether or not the condition holds. This strategy makes sense only in restricted cases, but it can then be implemented by a simple *conditional move* instruction that is better matched to the performance characteristics of modern processors.

We will examine this strategy and its implementation with more recent versions of IA32 processors.

Starting with the PentiumPro in 1995, recent generations of IA32 processors have had conditional move instructions that either do nothing or copy a value to a register, depending on the values of the condition codes. For years, these instructions have been largely unused. With its default settings, GCC did not generate code that used them, because that would prevent backward compatibility, even though almost all x86 processors manufactured by Intel and its competitors since 1997 have supported these instructions. More recently, for systems running on processors that are certain to support conditional moves, such as Intel-based Apple Macintosh computers (introduced in 2006) and the 64-bit versions of Linux and Windows, GCC will generate code using conditional moves. By giving special command-line parameters on other machines, we can indicate to GCC that the target machine supports conditional move instructions.

As an example, Figure 16(a) shows a variant form of the function absdiff we used in Figure 13 to illustrate conditional branching. This version uses a conditional *expression* rather than a conditional *statement* to illustrate the concepts behind conditional data transfers more clearly, but in fact GCC

(a) Original C code

```
1    int absdiff(int x, int y) {
2        return x < y ? y-x : x-y;
3    }
```

(b) Implementation using conditional assignment

```
1    int cmovdiff(int x, int y) {
2        int tval = y-x;
3        int rval = x-y;
4        int test = x < y;
5        /* Line below requires
6           single instruction: */
7        if (test) rval = tval;
8        return rval;
9    }
```

(c) Generated assembly code

```
     x at %ebp+8, y at %ebp+12
1        movl     8(%ebp), %ecx       Get x
2        movl     12(%ebp), %edx      Get y
3        movl     %edx, %ebx          Copy y
4        subl     %ecx, %ebx          Compute y-x
5        movl     %ecx, %eax          Copy x
6        subl     %edx, %eax          Compute x-y and set as return value
7        cmpl     %edx, %ecx          Compare x:y
8        cmovl    %ebx, %eax          If <, replace return value with y-x
```

Figure 16 **Compilation of conditional statements using conditional assignment.** C function absdiff (a) contains a conditional expression. The generated assembly code is shown (c), along with a C function cmovdiff (b) that mimics the operation of the assembly code. The stack set-up and completion portions of the assembly code have been omitted.

generates identical code for this version as it does for the version of Figure 13. If we compile this giving GCC the command-line option '-march=i686',[3] we generate the assembly code shown in Figure 16(c), having an approximate form shown by the C function cmovdiff shown in Figure 16(b). Studying the C version, we can see that it computes both y-x and x-y, naming these tval and rval, respectively. It then tests whether x is less than y, and if so, copies tval to rval before returning rval. The assembly code in Figure 16(c) follows the same logic. The key is that the single cmovl instruction (line 8) of the assembly code implements the conditional assignment (line 7) of cmovdiff. This instruction has the same syntax as a MOV instruction, except that it only performs the data movement if the specified condition holds. (The suffix 'l' in cmovl stands for "less," not for "long.")

To understand why code based on conditional data transfers can outperform code based on conditional control transfers (as in Figure 13), we must understand something about how modern processors operate. Processors achieve high performance through *pipelining*, where an instruction is processed via a sequence of stages, each performing one small portion of the required operations (e.g., fetching the instruction from memory, determining the instruction type, reading from memory, performing an arithmetic operation, writing to memory, and updating the program counter.) This approach achieves high performance by overlapping the steps of the successive instructions, such as fetching one instruction while performing the arithmetic operations for a previous instruction. To do this requires being able to determine the sequence of instructions to be executed well ahead of time in order to keep the pipeline full of instructions to be executed. When the machine encounters a conditional jump (referred to as a "branch"), it often cannot determine yet whether or not the jump will be followed. Processors employ sophisticated *branch prediction logic* to try to guess whether or not each jump instruction will be followed. As long as it can guess reliably (modern microprocessor designs try to achieve success rates on the order of 90%), the instruction pipeline will be kept full of instructions. Mispredicting a jump, on the other hand, requires that the processor discard much of the work it has already done on future instructions and then begin filling the pipeline with instructions starting at the correct location. As we will see, such a misprediction can incur a serious penalty, say, 20–40 clock cycles of wasted effort, causing a serious degradation of program performance.

As an example, we ran timings of the absdiff function on an Intel Core i7 processor using both methods of implementing the conditional operation. In a typical application, the outcome of the test x < y is highly unpredictable, and so even the most sophisticated branch prediction hardware will guess correctly only around 50% of the time. In addition, the computations performed in each of the two code sequences require only a single clock cycle. As a consequence, the branch misprediction penalty dominates the performance of this function. For the IA32 code with conditional jumps, we found that the function requires around 13 clock

---

3. In GCC terminology, the Pentium should be considered model "586" and the PentiumPro should be considered model "686" of the x86 line.

cycles per call when the branching pattern is easily predictable, and around 35 clock cycles per call when the branching pattern is random. From this we can infer that the branch misprediction penalty is around 44 clock cycles. That means time required by the function ranges between around 13 and 57 cycles, depending on whether or not the branch is predicted correctly.

Aside How did you determine this penalty?

Assume the probability of misprediction is p, the time to execute the code without misprediction is T_{OK}, and the misprediction penalty is T_{MP}. Then the average time to execute the code as a function of p is $T_{avg}(p) = (1 - p)T_{OK} + p(T_{OK} + T_{MP}) = T_{OK} + pT_{MP}$. We are given T_{OK} and T_{ran}, the average time when $p = 0.5$, and we want to determine T_{MP}. Substituting into the equation, we get $T_{ran} = T_{avg}(0.5) = T_{OK} + 0.5T_{MP}$, and therefore $T_{MP} = 2(T_{ran} - T_{MP})$. So, for $T_{OK} = 13$ and $T_{ran} = 35$, we get $T_{MP} = 44$.

On the other hand, the code compiled using conditional moves requires around 14 clock cycles regardless of the data being tested. The flow of control does not depend on data, and this makes it easier for the processor to keep its pipeline full.

Practice Problem 25

Running on a Pentium 4, our code required around 16 cycles when the branching pattern was highly predictable, and around 31 cycles when the pattern was random.

A. What is the approximate miss penalty?

B. How many cycles would the function require when the branch is mispredicted?

Figure 17 illustrates some of the conditional move instructions added to the IA32 instruction set with the introduction of the PentiumPro microprocessor and supported by most IA32 processors manufactured by Intel and its competitors since 1997. Each of these instructions has two operands: a source register or memory location S, and a destination register R. As with the different SET (Section 6.2) and jump instructions (Section 6.3), the outcome of these instructions depends on the values of the condition codes. The source value is read from either memory or the source register, but it is copied to the destination only if the specified condition holds.

For IA32, the source and destination values can be 16 or 32 bits long. Single-byte conditional moves are not supported. Unlike the unconditional instructions, where the operand length is explicitly encoded in the instruction name (e.g., movw and movl), the assembler can infer the operand length of a conditional move instruction from the name of the destination register, and so the same instruction name can be used for all operand lengths.

Unlike conditional jumps, the processor can execute conditional move instructions without having to predict the outcome of the test. The processor simply

Instruction	Synonym	Move condition	Description
cmove *S, R*	cmovz	ZF	Equal / zero
cmovne *S, R*	cmovnz	~ZF	Not equal / not zero
cmovs *S, R*		SF	Negative
cmovns *S, R*		~SF	Nonnegative
cmovg *S, R*	cmovnle	~(SF ^ OF) & ~ZF	Greater (signed >)
cmovge *S, R*	cmovnl	~(SF ^ OF)	Greater or equal (signed >=)
cmovl *S, R*	cmovnge	SF ^ OF	Less (signed <)
cmovle *S, R*	cmovng	(SF ^ OF) | ZF	Less or equal (signed <=)
cmova *S, R*	cmovnbe	~CF & ~ZF	Above (unsigned >)
cmovae *S, R*	cmovnb	~CF	Above or equal (Unsigned >=)
cmovb *S, R*	cmovnae	CF	Below (unsigned <)
cmovbe *S, R*	cmovna	CF | ZF	below or equal (unsigned <=)

Figure 17 **The conditional move instructions.** These instructions copy the source value *S* to its destination *R* when the move condition holds. Some instructions have "synonyms," alternate names for the same machine instruction.

reads the source value (possibly from memory), checks the condition code, and then either updates the destination register or keeps it the same.

To understand how conditional operations can be implemented via conditional data transfers, consider the following general form of conditional expression and assignment:

v = *test-expr* ? *then-expr* : *else-expr*;

With traditional IA32, the compiler generates code having a form shown by the following abstract code:

```
    if (!test-expr)
        goto false;
    v = true-expr;
    goto done;
  false:
    v = else-expr;
done:
```

This code contains two code sequences—one evaluating *then-expr* and one evaluating *else-expr*. A combination of conditional and unconditional jumps is used to ensure that just one of the sequences is evaluated.

For the code based on conditional move, both the *then-expr* and the *else-expr* are evaluated, with the final value chosen based on the evaluation *test-expr*. This can be described by the following abstract code:

```
vt = then-expr;
v  = else-expr;
t  = test-expr;
if (t) v = vt;
```

The final statement in this sequence is implemented with a conditional move—value vt is copied to v only if test condition t holds.

Not all conditional expressions can be compiled using conditional moves. Most significantly, the abstract code we have shown evaluates both *then-expr* and *else-expr* regardless of the test outcome. If one of those two expressions could possibly generate an error condition or a side effect, this could lead to invalid behavior. As an illustration, consider the following C function:

```
int cread(int *xp) {
    return (xp ? *xp : 0);
}
```

At first, this seems like a good candidate to compile using a conditional move to read the value designated by pointer xp, as shown in the following assembly code:

```
      Invalid implementation of function cread
      xp in register %edx
1     movl    $0, %eax          Set 0 as return value
2     testl   %edx, %edx        Test xp
3     cmovne  (%edx), %eax      if !0, dereference xp to get return value
```

This implementation is invalid, however, since the dereferencing of xp by the cmovne instruction (line 3) occurs even when the test fails, causing a null pointer dereferencing error. Instead, this code must be compiled using branching code.

A similar case holds when either of the two branches causes a side effect, as illustrated by the following function:

```
1    /* Global variable */
2    int lcount = 0;
3    int absdiff_se(int x, int y) {
4        return x < y ? (lcount++, y-x) : x-y;
5    }
```

This function increments global variable lcount as part of *then-expr*. Thus, branching code must be used to ensure this side effect only occurs when the test condition holds.

Using conditional moves also does not always improve code efficiency. For example, if either the *then-expr* or the *else-expr* evaluation requires a significant computation, then this effort is wasted when the corresponding condition does not hold. Compilers must take into account the relative performance of wasted computation versus the potential for performance penalty due to branch misprediction. In truth, they do not really have enough information to make this decision reliably; for example, they do not know how well the branches will follow predictable patterns. Our experiments with GCC indicate that it only uses conditional moves when the two expressions can be computed very easily, for example, with single add instructions. In our experience, GCC uses conditional control transfers even in many cases where the cost of branch misprediction would exceed even more complex computations.

Overall, then, we see that conditional data transfers offer an alternative strategy to conditional control transfers for implementing conditional operations. They can only be used in restricted cases, but these cases are fairly common and provide a much better match to the operation of modern processors.

Practice Problem 26

In the following C function, we have left the definition of operation OP incomplete:

```
#define OP _____   /* Unknown operator */

int arith(int x) {
    return x OP 4;
}
```

When compiled, GCC generates the following assembly code:

```
    Register: x in %edx
1       leal    3(%edx), %eax
2       testl   %edx, %edx
3       cmovns  %edx, %eax
4       sarl    $2, %eax        Return value in %eax
```

A. What operation is OP?

B. Annotate the code to explain how it works.

Practice Problem 27

Starting with C code of the form

```
1   int test(int x, int y) {
2       int val = _____ ;
3       if (_____) {
4           if (_____)
5               val = _____ ;
6           else
7               val = _____ ;
```

```
8      } else if (_____)
9          val = _____;
10     return val;
11  }
```

GCC, with the command-line setting '-march=i686', generates the following assembly code:

```
      x at %ebp+8, y at %ebp+12
1      movl    8(%ebp), %ebx
2      movl    12(%ebp), %ecx
3      testl   %ecx, %ecx
4      jle     .L2
5      movl    %ebx, %edx
6      subl    %ecx, %edx
7      movl    %ecx, %eax
8      xorl    %ebx, %eax
9      cmpl    %ecx, %ebx
10     cmovl   %edx, %eax
11     jmp     .L4
12  .L2:
13     leal    0(,%ebx,4), %edx
14     leal    (%ecx,%ebx), %eax
15     cmpl    $-2, %ecx
16     cmovge  %edx, %eax
17  .L4:
```

Fill in the missing expressions in the C code.

6.7 Switch Statements

A switch statement provides a multi-way branching capability based on the value of an integer index. They are particularly useful when dealing with tests where there can be a large number of possible outcomes. Not only do they make the C code more readable, they also allow an efficient implementation using a data structure called a *jump table*. A jump table is an array where entry *i* is the address of a code segment implementing the action the program should take when the switch index equals *i*. The code performs an array reference into the jump table using the switch index to determine the target for a jump instruction. The advantage of using a jump table over a long sequence of if-else statements is that the time taken to perform the switch is independent of the number of switch cases. GCC selects the method of translating a switch statement based on the number of cases and the sparsity of the case values. Jump tables are used when there are a number of cases (e.g., four or more) and they span a small range of values.

Figure 18(a) shows an example of a C switch statement. This example has a number of interesting features, including case labels that do not span a contiguous

(a) Switch statement

```
1    int switch_eg(int x, int n) {
2        int result = x;
3
4        switch (n) {
5
6        case 100:
7            result *= 13;
8            break;
9
10       case 102:
11           result += 10;
12           /* Fall through */
13
14       case 103:
15           result += 11;
16           break;
17
18       case 104:
19       case 106:
20           result *= result;
21           break;
22
23       default:
24           result = 0;
25       }
26
27       return result;
28   }
```

(b) Translation into extended C

```
1    int switch_eg_impl(int x, int n) {
2        /* Table of code pointers */
3        static void *jt[7] = {
4            &&loc_A, &&loc_def, &&loc_B,
5            &&loc_C, &&loc_D, &&loc_def,
6            &&loc_D
7        };
8
9        unsigned index = n - 100;
10       int result;
11
12       if (index > 6)
13           goto loc_def;
14
15       /* Multiway branch */
16       goto *jt[index];
17
18   loc_def:  /* Default case*/
19       result = 0;
20       goto done;
21
22   loc_C:    /* Case 103 */
23       result = x;
24       goto rest;
25
26   loc_A:    /* Case 100 */
27       result = x * 13;
28       goto done;
29
30   loc_B:    /* Case 102 */
31       result = x + 10;
32       /* Fall through */
33
34   rest:     /* Finish case 103 */
35       result += 11;
36       goto done;
37
38   loc_D:    /* Cases 104, 106 */
39       result = x * x;
40       /* Fall through */
41
42   done:
43       return result;
44   }
```

Figure 18 **Switch statement example with translation into extended C.** The translation shows the structure of jump table jt and how it is accessed. Such tables are supported by GCC as an extension to the C language.

```
    x at %ebp+8, n at %ebp+12
1       movl    8(%ebp), %edx            Get x
2       movl    12(%ebp), %eax           Get n
    Set up jump table access
3       subl    $100, %eax               Compute index = n-100
4       cmpl    $6, %eax                 Compare index:6
5       ja      .L2                      If >, goto loc_def
6       jmp     *.L7(,%eax,4)            Goto *jt[index]
    Default case
7   .L2:                                 loc_def:
8       movl    $0, %eax                   result = 0;
9       jmp     .L8                        Goto done
    Case 103
10  .L5:                                 loc_C:
11      movl    %edx, %eax                 result = x;
12      jmp     .L9                        Goto rest
    Case 100
13  .L3:                                 loc_A:
14      leal    (%edx,%edx,2), %eax        result = x*3;
15      leal    (%edx,%eax,4), %eax        result = x+4*result
16      jmp     .L8                        Goto done
    Case 102
17  .L4:                                 loc_B:
18      leal    10(%edx), %eax             result = x+10
    Fall through
19  .L9:                                 rest:
20      addl    $11, %eax                  result += 11;
21      jmp     .L8                        Goto done
    Cases 104, 106
22  .L6:                                 loc_D
23      movl    %edx, %eax                 result = x
24      imull   %edx, %eax                 result *= x
    Fall through
25  .L8:                                 done:
    Return result
```

Figure 19 **Assembly code for** switch **statement example in Figure 18.**

range (there are no labels for cases 101 and 105), cases with multiple labels (cases 104 and 106), and cases that *fall through* to other cases (case 102) because the code for the case does not end with a break statement.

Figure 19 shows the assembly code generated when compiling switch_eg. The behavior of this code is shown in C as the procedure switch_eg_impl in Figure 18(b). This code makes use of support provided by GCC for jump tables, as an extension to the C language. The array jt contains seven entries, each of which is the address of a block of code. These locations are defined by labels in the code, and indicated in the entries in jt by code pointers, consisting of the labels prefixed by '&&.' (Recall that the operator & creates a pointer for a data value. In making this extension, the authors of GCC created a new operator && to create a pointer for a code location.) We recommend that you study the C procedure switch_eg_impl and how it relates assembly code version.

Our original C code has cases for values 100, 102–104, and 106, but the switch variable n can be an arbitrary int. The compiler first shifts the range to between 0 and 6 by subtracting 100 from n, creating a new program variable that we call index in our C version. It further simplifies the branching possibilities by treating index as an *unsigned* value, making use of the fact that negative numbers in a two's-complement representation map to large positive numbers in an unsigned representation. It can therefore test whether index is outside of the range 0–6 by testing whether it is greater than 6. In the C and assembly code, there are five distinct locations to jump to, based on the value of index. These are: loc_A (identified in the assembly code as .L3), loc_B (.L4), loc_C (.L5), loc_D (.L6), and loc_def (.L2), where the latter is the destination for the default case. Each of these labels identifies a block of code implementing one of the case branches. In both the C and the assembly code, the program compares index to 6 and jumps to the code for the default case if it is greater.

The key step in executing a switch statement is to access a code location through the jump table. This occurs in line 16 in the C code, with a goto statement that references the jump table jt. This *computed goto* is supported by GCC as an extension to the C language. In our assembly-code version, a similar operation occurs on line 6, where the jmp instruction's operand is prefixed with '*', indicating an indirect jump, and the operand specifies a memory location indexed by register %eax, which holds the value of index. (We will see in Section 8 how array references are translated into machine code.)

Our C code declares the jump table as an array of seven elements, each of which is a pointer to a code location. These elements span values 0–6 of index, corresponding to values 100–106 of n. Observe the jump table handles duplicate cases by simply having the same code label (loc_D) for entries 4 and 6, and it handles missing cases by using the label for the default case (loc_def) as entries 1 and 5.

In the assembly code, the jump table is indicated by the following declarations, to which we have added comments:

```
1    .section    .rodata
2    .align 4            Align address to multiple of 4
3    .L7:
4    .long   .L3         Case 100: loc_A
5    .long   .L2         Case 101: loc_def
6    .long   .L4         Case 102: loc_B
7    .long   .L5         Case 103: loc_C
```

```
8        .long    .L6      Case 104: loc_D
9        .long    .L2      Case 105: loc_def
10       .long    .L6      Case 106: loc_D
```

These declarations state that within the segment of the object-code file called ".rodata" (for "Read-Only Data"), there should be a sequence of seven "long" (4-byte) words, where the value of each word is given by the instruction address associated with the indicated assembly code labels (e.g., .L3). Label .L7 marks the start of this allocation. The address associated with this label serves as the base for the indirect jump (line 6).

The different code blocks (C labels loc_A through loc_D and loc_def) implement the different branches of the switch statement. Most of them simply compute a value for result and then go to the end of the function. Similarly, the assembly-code blocks compute a value for register %eax and jump to the position indicated by label .L8 at the end of the function. Only the code for case labels 102 and 103 do not follow this pattern, to account for the way that case 102 falls through to 103 in the original C code. This is handled in the assembly code and switch_eg_impl by having separate destinations for the two cases (loc_C and loc_B in C, .L5 and .L4 in assembly), where both of these blocks then converge on code that increments result by 11 (labeled rest in C and .L9 in assembly).

Examining all of this code requires careful study, but the key point is to see that the use of a jump table allows a very efficient way to implement a multiway branch. In our case, the program could branch to five distinct locations with a single jump table reference. Even if we had a switch statement with hundreds of cases, they could be handled by a single jump table access.

Practice Problem 28

In the C function that follows, we have omitted the body of the switch statement. In the C code, the case labels did not span a contiguous range, and some cases had multiple labels.

```
int switch2(int x) {
    int result = 0;
    switch (x) {
        /* Body of switch statement omitted */
    }
    return result;
}
```

In compiling the function, GCC generates the assembly code that follows for the initial part of the procedure and for the jump table. Variable x is initially at offset 8 relative to register %ebp.

```
        x at %ebp+8                              Jump table for switch2
1       movl     8(%ebp), %eax           1      .L8:
        Set up jump table access          2         .long    .L3
2       addl     $2, %eax                3         .long    .L2
3       cmpl     $6, %eax                4         .long    .L4
4       ja       .L2                     5         .long    .L5
5       jmp      *.L8(,%eax,4)           6         .long    .L6
                                         7         .long    .L6
                                         8         .long    .L7
```

Based on this information, answer the following questions:

A. What were the values of the case labels in the switch statement body?

B. What cases had multiple labels in the C code?

Practice Problem 29

For a C function switcher with the general structure

```
1    int switcher(int a, int b, int c)
2    {
3        int answer;
4        switch(a) {
5        case _____:          /* Case A */
6            c = _____;
7            /* Fall through */
8        case _____:          /* Case B */
9            answer = _____;
10           break;
11       case _____:          /* Case C */
12       case _____:          /* Case D */
13           answer = _____;
14           break;
15       case _____:          /* Case E */
16           answer = _____;
17           break;
18       default:
19           answer = _____;
20       }
21       return answer;
22   }
```

GCC generates the assembly code and jump table shown in Figure 20.

Fill in the missing parts of the C code. Except for the ordering of case labels C and D, there is only one way to fit the different cases into the template.

```
a at %ebp+8,  b at %ebp+12,  c at %ebp+16
1      movl    8(%ebp), %eax          1    .L7:
2      cmpl    $7, %eax               2      .long   .L3
3      ja      .L2                    3      .long   .L2
4      jmp     *.L7(,%eax,4)          4      .long   .L4
5    .L2:                             5      .long   .L2
6      movl    12(%ebp), %eax         6      .long   .L5
7      jmp     .L8                    7      .long   .L6
8    .L5:                             8      .long   .L2
9      movl    $4, %eax               9      .long   .L4
10     jmp     .L8
11   .L6:
12     movl    12(%ebp), %eax
13     xorl    $15, %eax
14     movl    %eax, 16(%ebp)
15   .L3:
16     movl    16(%ebp), %eax
17     addl    $112, %eax
18     jmp     .L8
19   .L4:
20     movl    16(%ebp), %eax
21     addl    12(%ebp), %eax
22     sall    $2, %eax
23   .L8:
```

Figure 20 **Assembly code and jump table for Problem 29.**

7 Procedures

A procedure call involves passing both data (in the form of procedure parameters and return values) and control from one part of a program to another. In addition, it must allocate space for the local variables of the procedure on entry and deallocate them on exit. Most machines, including IA32, provide only simple instructions for transferring control to and from procedures. The passing of data and the allocation and deallocation of local variables is handled by manipulating the program stack.

7.1 Stack Frame Structure

IA32 programs make use of the program stack to support procedure calls. The machine uses the stack to pass procedure arguments, to store return information, to save registers for later restoration, and for local storage. The portion of the stack allocated for a single procedure call is called a *stack frame*. Figure 21 diagrams the general structure of a stack frame. The topmost stack frame is delimited by two pointers, with register %ebp serving as the *frame pointer*, and register %esp serving

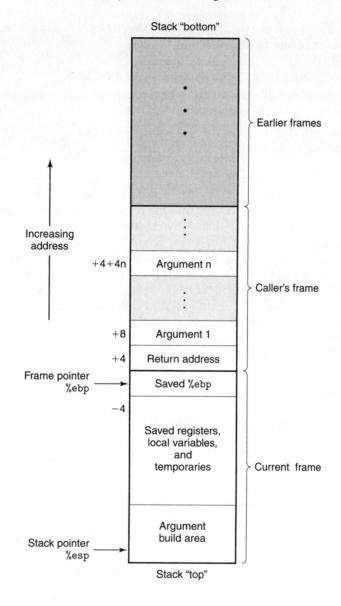

Figure 21
Stack frame structure. The stack is used for passing arguments, for storing return information, for saving registers, and for local storage.

as the *stack pointer*. The stack pointer can move while the procedure is executing, and hence most information is accessed relative to the frame pointer.

Suppose procedure P (the *caller*) calls procedure Q (the *callee*). The arguments to Q are contained within the stack frame for P. In addition, when P calls Q, the *return address* within P where the program should resume execution when it returns from Q is pushed onto the stack, forming the end of P's stack frame. The stack frame for Q starts with the saved value of the frame pointer (a copy of register %ebp), followed by copies of any other saved register values.

Procedure Q also uses the stack for any local variables that cannot be stored in registers. This can occur for the following reasons:

- There are not enough registers to hold all of the local data.
- Some of the local variables are arrays or structures and hence must be accessed by array or structure references.
- The address operator '&' is applied to a local variable, and hence we must be able to generate an address for it.

In addition, Q uses the stack frame for storing arguments to any procedures it calls. As illustrated in Figure 21, within the called procedure, the first argument is positioned at offset 8 relative to %ebp, and the remaining arguments (assuming their data types require no more than 4 bytes) are stored in successive 4-byte blocks, so that argument i is at offset $4 + 4i$ relative to %ebp. Larger arguments (such as structures and larger numeric formats) require larger regions on the stack.

As described earlier, the stack grows toward lower addresses and the stack pointer %esp points to the top element of the stack. Data can be stored on and retrieved from the stack using the pushl and popl instructions. Space for data with no specified initial value can be allocated on the stack by simply decrementing the stack pointer by an appropriate amount. Similarly, space can be deallocated by incrementing the stack pointer.

7.2 Transferring Control

The instructions supporting procedure calls and returns are shown in the following table:

Instruction		Description
call	*Label*	Procedure call
call	*Operand*	Procedure call
leave		Prepare stack for return
ret		Return from call

The call instruction has a target indicating the address of the instruction where the called procedure starts. Like jumps, a call can either be direct or indirect. In assembly code, the target of a direct call is given as a label, while the target of an indirect call is given by a * followed by an operand specifier using one of the formats described in Section 4.1.

The effect of a call instruction is to push a return address on the stack and jump to the start of the called procedure. The return address is the address of the instruction immediately following the call in the program, so that execution will resume at this location when the called procedure returns. The ret instruction pops an address off the stack and jumps to this location. The proper use of this instruction is to have prepared the stack so that the stack pointer points to the place where the preceding call instruction stored its return address.

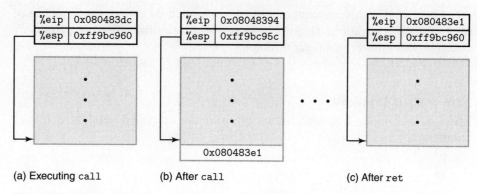

(a) Executing `call`   (b) After `call`   (c) After `ret`

Figure 22 **Illustration of `call` and `ret` functions.** The `call` instruction transfers control to the start of a function, while the `ret` instruction returns back to the instruction following the call.

Figure 22 illustrates the execution of the `call` and `ret` instructions for the `sum` and `main` functions introduced in Section 2.2. The following are excerpts of the disassembled code for the two functions:

```
    Beginning of function sum
1   08048394 <sum>:
2     8048394:  55                          push    %ebp
    . . .
    Return from function sum
3     80483a4:  c3                          ret
    . . .
    Call to sum from main
4     80483dc:  e8 b3 ff ff ff              call    8048394 <sum>
5     80483e1:  83 c4 14                    add     $0x14,%esp
```

In this code, we can see that the `call` instruction with address 0x080483dc in `main` calls function `sum`. This status is shown in Figure 22(a), with the indicated values for the stack pointer `%esp` and the program counter `%eip`. The effect of the `call` is to push the return address 0x080483e1 onto the stack and to jump to the first instruction in function `sum`, at address 0x08048394 (Figure 22(b)). The execution of function `sum` continues until it hits the `ret` instruction at address 0x080483a4. This instruction pops the value 0x080483e1 from the stack and jumps to this address, resuming the execution of `main` just after the `call` instruction in `sum` (Figure 22(c)).

The `leave` instruction can be used to prepare the stack for returning. It is equivalent to the following code sequence:

```
1   movl %ebp, %esp    Set stack pointer to beginning of frame
2   popl %ebp          Restore saved %ebp and set stack ptr to end of caller's frame
```

Alternatively, this preparation can be performed by an explicit sequence of move and pop operations. Register %eax is used for returning the value from any function that returns an integer or pointer.

Practice Problem 30

The following code fragment occurs often in the compiled version of library routines:

```
1     call next
2  next:
3     popl %eax
```

A. To what value does register %eax get set?

B. Explain why there is no matching ret instruction to this call.

C. What useful purpose does this code fragment serve?

7.3 Register Usage Conventions

The set of program registers acts as a single resource shared by all of the procedures. Although only one procedure can be active at a given time, we must make sure that when one procedure (the *caller*) calls another (the *callee*), the callee does not overwrite some register value that the caller planned to use later. For this reason, IA32 adopts a uniform set of conventions for register usage that must be respected by all procedures, including those in program libraries.

By convention, registers %eax, %edx, and %ecx are classified as *caller-save* registers. When procedure Q is called by P, it can overwrite these registers without destroying any data required by P. On the other hand, registers %ebx, %esi, and %edi are classified as *callee-save* registers. This means that Q must save the values of any of these registers on the stack before overwriting them, and restore them before returning, because P (or some higher-level procedure) may need these values for its future computations. In addition, registers %ebp and %esp must be maintained according to the conventions described here.

As an example, consider the following code:

```
1  int P(int x)
2  {
3      int y = x*x;
4      int z = Q(y);
5      return y + z;
6  }
```

Procedure P computes y before calling Q, but it must also ensure that the value of y is available after Q returns. It can do this by one of two means:

- It can store the value of y in its own stack frame before calling Q; when Q returns, procedure P can then retrieve the value of y from the stack. In other words, P, the *caller,* saves the value.

- It can store the value of y in a callee-save register. If Q, or any procedure called by Q, wants to use this register, it must save the register value in its stack frame and restore the value before it returns (in other words, the *callee* saves the value). When Q returns to P, the value of y will be in the callee-save register, either because the register was never altered or because it was saved and restored.

Either convention can be made to work, as long as there is agreement as to which function is responsible for saving which value. IA32 follows both approaches, partitioning the registers into one set that is caller-save, and another set that is callee-save.

Practice Problem 31

The following code sequence occurs right near the beginning of the assembly code generated by GCC for a C procedure:

```
1      subl    $12, %esp
2      movl    %ebx, (%esp)
3      movl    %esi, 4(%esp)
4      movl    %edi, 8(%esp)
5      movl    8(%ebp), %ebx
6      movl    12(%ebp), %edi
7      movl    (%ebx), %esi
8      movl    (%edi), %eax
9      movl    16(%ebp), %edx
10     movl    (%edx), %ecx
```

We see that just three registers (%ebx, %esi, and %edi) are saved on the stack (lines 2–4). The program modifies these and three other registers (%eax, %ecx, and %edx). At the end of the procedure, the values of registers %edi, %esi, and %ebx are restored (not shown), while the other three are left in their modified states.

Explain this apparent inconsistency in the saving and restoring of register states.

7.4 Procedure Example

As an example, consider the C functions defined in Figure 23, where function caller includes a call to function swap_add. Figure 24 shows the stack frame structure both just before caller calls function swap_add and while swap_add

```
1    int swap_add(int *xp, int *yp)
2    {
3        int x = *xp;
4        int y = *yp;
5
6        *xp = y;
7        *yp = x;
8        return x + y;
9    }
10
11   int caller()
12   {
13       int arg1 = 534;
14       int arg2 = 1057;
15       int sum = swap_add(&arg1, &arg2);
16       int diff = arg1 - arg2;
17
18       return sum * diff;
19   }
```

Figure 23 **Example of procedure definition and call.**

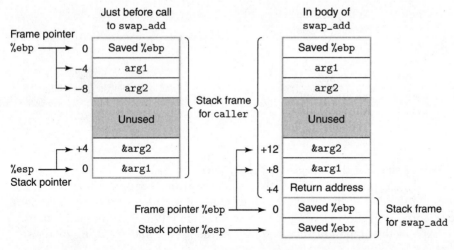

Figure 24 **Stack frames for `caller` and `swap_add`. Procedure `swap_add` retrieves its arguments from the stack frame for `caller`.**

is running. Some of the instructions access stack locations relative to the stack pointer %esp while others access locations relative to the base pointer %ebp. These offsets are identified by the lines shown relative to the two pointers.

New to C?   Passing parameters to a function

Some languages, such as Pascal, provide two different ways to pass parameters to procedures—by *value*, where the caller provides the actual parameter value, and by *reference*, where the caller provides a pointer to the value. In C, all parameters are passed by value, but we can mimic the effect of a reference parameter by explicitly generating a pointer to a value and passing this pointer to a procedure. We can see this with the call by `caller` to `swap_add` (Figure 23). By passing pointers to `arg1` and `arg2`, `caller` provides a way for `swap_add` to modify these values.

One of the ways in which C++ extends C is the inclusion of reference parameters.

The stack frame for `caller` includes storage for local variables `arg1` and `arg2`, at positions -4 and -8 relative to the frame pointer. These variables must be stored on the stack, since the code must associate an address with them. The following assembly code from the compiled version of `caller` shows how it calls `swap_add`:

```
1    caller:
2        pushl    %ebp                Save old %ebp
3        movl     %esp, %ebp          Set %ebp as frame pointer
4        subl     $24, %esp           Allocate 24 bytes on stack
5        movl     $534, -4(%ebp)      Set arg1 to 534
6        movl     $1057, -8(%ebp)     Set arg2 to 1057
7        leal     -8(%ebp), %eax      Compute &arg2
8        movl     %eax, 4(%esp)       Store on stack
9        leal     -4(%ebp), %eax      Compute &arg1
10       movl     %eax, (%esp)        Store on stack
11       call     swap_add            Call the swap_add function
```

This code saves a copy of %ebp and sets %ebp to the beginning of the stack frame (lines 2–3). It then allocates 24 bytes on the stack by decrementing the stack pointer (recall that the stack grows toward lower addresses). It initializes `arg1` and `arg2` to 534 and 1057, respectively (lines 5–6), and computes the values of `&arg2` and `&arg1` and stores these on the stack to form the arguments to `swap_add` (lines 7–10). It stores these arguments relative to the stack pointer, at offsets 0 and +4 for later access by `swap_add`. It then calls `swap_add`. Of the 24 bytes allocated for the stack frame, 8 are used for the local variables, 8 are used for passing parameters to `swap_add`, and 8 are not used for anything.

Aside   Why does GCC allocate space that never gets used?

We see that the code generated by GCC for `caller` allocates 24 bytes on the stack even though it only makes use of 16 of them. We will see many examples of this apparent wastefulness. GCC adheres to an x86 programming guideline that the total stack space used by the function should be a multiple of 16 bytes. Including the 4 bytes for the saved value of %ebp and the 4 bytes for the return address, `caller` uses a total of 32 bytes. The motivation for this convention is to ensure a proper *alignment* for accessing data. We will explain the reason for having alignment conventions and how they are implemented in Section 9.3.

The compiled code for `swap_add` has three parts: the "setup," where the stack frame is initialized; the "body," where the actual computation of the procedure is performed; and the "finish," where the stack state is restored and the procedure returns.

The following is the setup code for `swap_add`. Recall that before reaching this part of the code, the `call` instruction will have pushed the return address onto the stack.

```
1   swap_add:
2       pushl   %ebp            Save old %ebp
3       movl    %esp, %ebp      Set %ebp as frame pointer
4       pushl   %ebx            Save %ebx
```

Function `swap_add` requires register `%ebx` for temporary storage. Since this is a callee-save register, it pushes the old value onto the stack as part of the stack frame setup. At this point, the state of the stack is as shown on the right-hand side of Figure 24. Register `%ebp` has been shifted to serve as the frame pointer for `swap_add`.

The following is the body code for `swap_add`:

```
5       movl    8(%ebp), %edx    Get xp
6       movl    12(%ebp), %ecx   Get yp
7       movl    (%edx), %ebx     Get x
8       movl    (%ecx), %eax     Get y
9       movl    %eax, (%edx)     Store y at xp
10      movl    %ebx, (%ecx)     Store x at yp
11      addl    %ebx, %eax       Return value = x+y
```

This code retrieves its arguments from the stack frame for `caller`. Since the frame pointer has shifted, the locations of these arguments has shifted from positions +4 and 0 relative to the old value of `%esp` to positions +12 and +8 relative to new value of `%ebp`. The sum of variables x and y is stored in register `%eax` to be passed as the returned value.

The following is the finishing code for `swap_add`:

```
12      popl    %ebx             Restore %ebx
13      popl    %ebp             Restore %ebp
14      ret                      Return
```

This code restores the values of registers `%ebx` and `%ebp`, while also resetting the stack pointer so that it points to the stored return address, so that the `ret` instruction transfers control back to `caller`.

The following code in `caller` comes immediately after the instruction calling `swap_add`:

```
12      movl    -4(%ebp), %edx
13      subl    -8(%ebp), %edx
14      imull   %edx, %eax
15      leave
16      ret
```

This code retrieves the values of arg1 and arg2 from the stack in order to compute diff, and uses register %eax as the return value from swap_add. Observe the use of the leave instruction to reset both the stack and the frame pointer prior to return. We have seen in our code examples that the code generated by GCC sometimes uses a leave instruction to deallocate a stack frame, and sometimes it uses one or two popl instructions. Either approach is acceptable, and the guidelines from Intel and AMD as to which is preferable change over time.

We can see from this example that the compiler generates code to manage the stack structure according to a simple set of conventions. Arguments are passed to a function on the stack, where they can be retrieved using positive offsets (+8, +12, ...) relative to %ebp. Space can be allocated on the stack either by using push instructions or by subtracting offsets from the stack pointer. Before returning, a function must restore the stack to its original condition by restoring any callee-saved registers and %ebp, and by resetting %esp so that it points to the return address. It is important for all procedures to follow a consistent set of conventions for setting up and restoring the stack in order for the program to execute properly.

Practice Problem 32

A C function fun has the following code body:

```
*p = d;
return x-c;
```

The IA32 code implementing this body is as follows:

```
1    movsbl   12(%ebp),%edx
2    movl     16(%ebp), %eax
3    movl     %edx, (%eax)
4    movswl   8(%ebp),%eax
5    movl     20(%ebp), %edx
6    subl     %eax, %edx
7    movl     %edx, %eax
```

Write a prototype for function fun, showing the types and ordering of the arguments p, d, x, and c.

Practice Problem 33

Given the C function

```
1    int proc(void)
2    {
3        int x,y;
4        scanf("%x %x", &y, &x);
5        return x-y;
6    }
```

GCC generates the following assembly code:

```
1   proc:
2       pushl   %ebp
3       movl    %esp, %ebp
4       subl    $40, %esp
5       leal    -4(%ebp), %eax
6       movl    %eax, 8(%esp)
7       leal    -8(%ebp), %eax
8       movl    %eax, 4(%esp)
9       movl    $.LC0, (%esp)     Pointer to string "%x %x"
10      call    scanf
         Diagram stack frame at this point
11      movl    -4(%ebp), %eax
12      subl    -8(%ebp), %eax
13      leave
14      ret
```

Assume that procedure proc starts executing with the following register values:

Register	Value
%esp	0x800040
%ebp	0x800060

Suppose proc calls scanf (line 10), and that scanf reads values 0x46 and 0x53 from the standard input. Assume that the string "%x %x" is stored at memory location 0x300070.

A. What value does %ebp get set to on line 3?

B. What value does %esp get set to on line 4?

C. At what addresses are local variables x and y stored?

D. Draw a diagram of the stack frame for proc right after scanf returns. Include as much information as you can about the addresses and the contents of the stack frame elements.

E. Indicate the regions of the stack frame that are not used by proc.

7.5 Recursive Procedures

The stack and linkage conventions described in the previous section allow procedures to call themselves recursively. Since each call has its own private space on the stack, the local variables of the multiple outstanding calls do not interfere with one another. Furthermore, the stack discipline naturally provides the proper policy for allocating local storage when the procedure is called and deallocating it when it returns.

```
1   int rfact(int n)
2   {
3       int result;
4       if (n <= 1)
5           result = 1;
6       else
7           result = n * rfact(n-1);
8       return result;
9   }
```

Figure 25 **C code for recursive factorial program.**

Figure 25 shows the C code for a recursive factorial function. The assembly code generated by GCC is shown in Figure 26. Let us examine how the machine code will operate when called with argument n. The set-up code (lines 2–5) creates a stack frame containing the old version of %ebp, the saved value for callee-save register %ebx, and 4 bytes to hold the argument when it calls itself recursively, as illustrated in Figure 27. It uses register %ebx to save a copy of n (line 6). It sets the return value in register %eax to 1 (line 7) in anticipation of the case where $n \leq 1$, in which event it will jump to the completion code.

For the recursive case, it computes $n - 1$, stores it on the stack, and calls itself (lines 10–12). Upon completion of the code, we can assume (1) register %eax holds

```
        Argument: n at %ebp+8
        Registers: n in %ebx, result in %eax
1   rfact:
2       pushl   %ebp                Save old %ebp
3       movl    %esp, %ebp          Set %ebp as frame pointer
4       pushl   %ebx                Save callee save register %ebx
5       subl    $4, %esp            Allocate 4 bytes on stack
6       movl    8(%ebp), %ebx       Get n
7       movl    $1, %eax            result = 1
8       cmpl    $1, %ebx            Compare n:1
9       jle     .L53                If <=, goto done
10      leal    -1(%ebx), %eax      Compute n-1
11      movl    %eax, (%esp)        Store at top of stack
12      call    rfact               Call rfact(n-1)
13      imull   %ebx, %eax          Compute result = return value * n
14  .L53:                       done:
15      addl    $4, %esp            Deallocate 4 bytes from stack
16      popl    %ebx                Restore %ebx
17      popl    %ebp                Restore %ebp
18      ret                         Return result
```

Figure 26 **Assembly code for the recursive factorial program in Figure 25.**

Figure 27
Stack frame for recursive factorial function. The state of the frame is shown just before the recursive call.

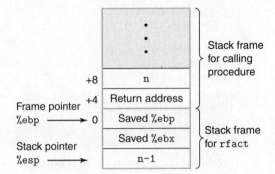

the value of $(n - 1)!$ and (2) callee-save register %ebx holds the parameter n. It therefore multiplies these two quantities (line 13) to generate the return value of the function.

For both cases—the terminal condition and the recursive call—the code proceeds to the completion section (lines 15–17) to restore the stack and callee-saved register, and then it returns.

We can see that calling a function recursively proceeds just like any other function call. Our stack discipline provides a mechanism where each invocation of a function has its own private storage for state information (saved values of the return location, frame pointer, and callee-save registers). If need be, it can also provide storage for local variables. The stack discipline of allocation and deallocation naturally matches the call-return ordering of functions. This method of implementing function calls and returns even works for more complex patterns, including mutual recursion (for example, when procedure P calls Q, which in turn calls P).

Practice Problem 34

For a C function having the general structure

```
int rfun(unsigned x) {
    if ( _____ )
        return _____;
    unsigned nx = _____;
    int rv = rfun(nx);
    return _____;
}
```

GCC generates the following assembly code (with the setup and completion code omitted):

```
1    movl    8(%ebp), %ebx
2    movl    $0, %eax
3    testl   %ebx, %ebx
4    je      .L3
```

```
5       movl    %ebx, %eax
6       shrl    %eax            Shift right by 1
7       movl    %eax, (%esp)
8       call    rfun
9       movl    %ebx, %edx
10      andl    $1, %edx
11      leal    (%edx,%eax), %eax
12   .L3:
```

A. What value does `rfun` store in the callee-save register `%ebx`?

B. Fill in the missing expressions in the C code shown above.

C. Describe in English what function this code computes.

8   Array Allocation and Access

Arrays in C are one means of aggregating scalar data into larger data types. C uses a particularly simple implementation of arrays, and hence the translation into machine code is fairly straightforward. One unusual feature of C is that we can generate pointers to elements within arrays and perform arithmetic with these pointers. These are translated into address computations in machine code.

Optimizing compilers are particularly good at simplifying the address computations used by array indexing. This can make the correspondence between the C code and its translation into machine code somewhat difficult to decipher.

8.1   Basic Principles

For data type T and integer constant N, the declaration

T `A[N];`

has two effects. First, it allocates a contiguous region of $L \cdot N$ bytes in memory, where L is the size (in bytes) of data type T. Let us denote the starting location as x_A. Second, it introduces an identifier `A` that can be used as a pointer to the beginning of the array. The value of this pointer will be x_A. The array elements can be accessed using an integer index ranging between 0 and $N-1$. Array element i will be stored at address $x_A + L \cdot i$.

As examples, consider the following declarations:

```
char    A[12];
char    *B[8];
double  C[6];
double  *D[5];
```

These declarations will generate arrays with the following parameters:

Array	Element size	Total size	Start address	Element i
A	1	12	x_A	$x_A + i$
B	4	32	x_B	$x_B + 4i$
C	8	48	x_C	$x_C + 8i$
D	4	20	x_D	$x_D + 4i$

Array A consists of 12 single-byte (char) elements. Array C consists of six double-precision floating-point values, each requiring 8 bytes. B and D are both arrays of pointers, and hence the array elements are 4 bytes each.

The memory referencing instructions of IA32 are designed to simplify array access. For example, suppose E is an array of int's, and we wish to evaluate E[i], where the address of E is stored in register %edx and i is stored in register %ecx. Then the instruction

```
movl (%edx,%ecx,4),%eax
```

will perform the address computation $x_E + 4i$, read that memory location, and copy the result to register %eax. The allowed scaling factors of 1, 2, 4, and 8 cover the sizes of the common primitive data types.

Practice Problem 35

Consider the following declarations:

```
short        S[7];
short       *T[3];
short      **U[6];
long double  V[8];
long double *W[4];
```

Fill in the following table describing the element size, the total size, and the address of element i for each of these arrays.

Array	Element size	Total size	Start address	Element i
S	_____	_____	x_S	_____
T	_____	_____	x_T	_____
U	_____	_____	x_U	_____
V	_____	_____	x_V	_____
W	_____	_____	x_W	_____

8.2 Pointer Arithmetic

C allows arithmetic on pointers, where the computed value is scaled according to the size of the data type referenced by the pointer. That is, if p is a pointer to data of type T, and the value of p is x_p, then the expression p+i has value $x_p + L \cdot i$, where L is the size of data type T.

The unary operators & and * allow the generation and dereferencing of pointers. That is, for an expression *Expr* denoting some object, &*Expr* is a pointer giving the address of the object. For an expression *AExpr* denoting an address, *$*AExpr$* gives the value at that address. The expressions *Expr* and $*\&Expr$ are therefore equivalent. The array subscripting operation can be applied to both arrays and pointers. The array reference A[i] is identical to the expression *(A+i). It computes the address of the ith array element and then accesses this memory location.

Expanding on our earlier example, suppose the starting address of integer array E and integer index i are stored in registers %edx and %ecx, respectively. The following are some expressions involving E. We also show an assembly-code implementation of each expression, with the result being stored in register %eax.

Expression	Type	Value	Assembly code
E	int *	x_E	movl %edx,%eax
E[0]	int	$M[x_E]$	movl (%edx),%eax
E[i]	int	$M[x_E + 4i]$	movl (%edx,%ecx,4),%eax
&E[2]	int *	$x_E + 8$	leal 8(%edx),%eax
E+i-1	int *	$x_E + 4i - 4$	leal -4(%edx,%ecx,4),%eax
*(E+i-3)	int *	$M[x_E + 4i - 12]$	movl -12(%edx,%ecx,4),%eax
&E[i]-E	int	i	movl %ecx,%eax

In these examples, the leal instruction is used to generate an address, while movl is used to reference memory (except in the first and last cases, where the former copies an address and the latter copies the index). The final example shows that one can compute the difference of two pointers within the same data structure, with the result divided by the size of the data type.

Practice Problem 36

Suppose the address of short integer array S and integer index i are stored in registers %edx and %ecx, respectively. For each of the following expressions, give its type, a formula for its value, and an assembly code implementation. The result should be stored in register %eax if it is a pointer and register element %ax if it is a short integer.

Expression	Type	Value	Assembly code
S+1			
S[3]			
&S[i]			
S[4*i+1]			
S+i-5			

231

8.3 Nested Arrays

The general principles of array allocation and referencing hold even when we create arrays of arrays. For example, the declaration

```
int A[5][3];
```

is equivalent to the declaration

```
typedef int row3_t[3];
row3_t A[5];
```

Data type `row3_t` is defined to be an array of three integers. Array `A` contains five such elements, each requiring 12 bytes to store the three integers. The total array size is then $4 \cdot 5 \cdot 3 = 60$ bytes.

Array `A` can also be viewed as a two-dimensional array with five rows and three columns, referenced as `A[0][0]` through `A[4][2]`. The array elements are ordered in memory in "row major" order, meaning all elements of row 0, which can be written `A[0]`, followed by all elements of row 1 (`A[1]`), and so on.

Row	Element	Address
A[0]	A[0][0]	x_A
	A[0][1]	$x_A + 4$
	A[0][2]	$x_A + 8$
A[1]	A[1][0]	$x_A + 12$
	A[1][1]	$x_A + 16$
	A[1][2]	$x_A + 20$
A[2]	A[2][0]	$x_A + 24$
	A[2][1]	$x_A + 28$
	A[2][2]	$x_A + 32$
A[3]	A[3][0]	$x_A + 36$
	A[3][1]	$x_A + 40$
	A[3][2]	$x_A + 44$
A[4]	A[4][0]	$x_A + 48$
	A[4][1]	$x_A + 52$
	A[4][2]	$x_A + 56$

This ordering is a consequence of our nested declaration. Viewing `A` as an array of five elements, each of which is an array of three `int`'s, we first have `A[0]`, followed by `A[1]`, and so on.

To access elements of multidimensional arrays, the compiler generates code to compute the offset of the desired element and then uses one of the mov instructions with the start of the array as the base address and the (possibly scaled) offset as an index. In general, for an array declared as

T D[R][C];

Machine-Level Representation of Programs

array element D[i][j] is at memory address

$$\&D[i][j] = x_D + L(C \cdot i + j), \tag{1}$$

where L is the size of data type T in bytes. As an example, consider the 5×3 integer array A defined earlier. Suppose $x_A, i,$ and j are at offsets 8, 12, and 16 relative to %ebp, respectively. Then array element A[i][j] can be copied to register %eax by the following code:

```
  A at %ebp+8, i at %ebp+12, j at %ebp+16
1     movl    12(%ebp), %eax        Get i
2     leal    (%eax,%eax,2), %eax   Compute 3*i
3     movl    16(%ebp), %edx        Get j
4     sall    $2, %edx              Compute j*4
5     addl    8(%ebp), %edx         Compute xA + 4j
6     movl    (%edx,%eax,4), %eax   Read from M[xA + 4j + 12i]
```

As can be seen, this code computes the element's address as $x_A + 4j + 12i = x_A + 4(3i + j)$ using a combination of shifting, adding, and scaling to avoid more costly multiplication instructions.

Practice Problem 37

Consider the following source code, where M and N are constants declared with #define:

```
1  int mat1[M][N];
2  int mat2[N][M];
3
4  int sum_element(int i, int j) {
5      return mat1[i][j] + mat2[j][i];
6  }
```

In compiling this program, GCC generates the following assembly code:

```
  i at %ebp+8, j at %ebp+12
1     movl    8(%ebp), %ecx
2     movl    12(%ebp), %edx
3     leal    0(,%ecx,8), %eax
4     subl    %ecx, %eax
5     addl    %edx, %eax
6     leal    (%edx,%edx,4), %edx
7     addl    %ecx, %edx
8     movl    mat1(,%eax,4), %eax
9     addl    mat2(,%edx,4), %eax
```

Use your reverse engineering skills to determine the values of M and N based on this assembly code.

8.4 Fixed-Size Arrays

The C compiler is able to make many optimizations for code operating on multi-dimensional arrays of fixed size. For example, suppose we declare data type `fix_matrix` to be 16×16 arrays of integers as follows:

```
1   #define N 16
2   typedef int fix_matrix[N][N];
```

(This example illustrates a good coding practice. Whenever a program uses some constant as an array dimension or buffer size, it is best to associate a name with it via a `#define` declaration, and then use this name consistently, rather than the numeric value. That way, if an occasion ever arises to change the value, it can be done by simply modifying the `#define` declaration.) The code in Figure 28(a) computes element i, k of the product of arrays A and B, according to the formula $\sum_{0 \leq j < N} a_{i,j} \cdot b_{j,k}$. The C compiler generates code that we then recoded into C, shown as function `fix_prod_ele_opt` in Figure 28(b). This code contains a number of clever optimizations. It recognizes that the loop will access just the elements of row i of array A, and so it creates a local pointer variable, which we have named `Arow`, to provide direct access to row i of the array. `Arow` is initialized to `&A[i][0]`, and so array element `A[i][j]` can be accessed as `Arow[j]`. It also recognizes that the loop will access the elements of array B as `B[0][k],B[1][k],...,B[15][k]` in sequence. These elements occupy positions in memory starting with the address of array element `B[0][k]` and spaced 64 bytes apart. The program can therefore use a pointer variable `Bptr` to access these successive locations. In C, this pointer is shown as being incremented by N (16), although in fact the actual address is incremented by $4 \cdot 16 = 64$.

The following is the actual assembly code for the loop. We see that four variables are maintained in registers within the loop: `Arow`, `Bptr`, `j`, and `result`.

```
    Registers: Arow in %esi, Bptr in %ecx, j in %edx, result in %ebx
1   .L6:                              loop:
2       movl    (%ecx), %eax          Get *Bptr
3       imull   (%esi,%edx,4), %eax   Multiply by Arow[j]
4       addl    %eax, %ebx            Add to result
5       addl    $1, %edx              Increment j
6       addl    $64, %ecx             Add 64 to Bptr
7       cmpl    $16, %edx             Compare j:16
8       jne     .L6                   If !=, goto loop
```

As can be seen, register `%ecx` is incremented by 64 within the loop (line 6). Machine code considers every pointer to be a byte address, and so in compiling pointer arithmetic, it must scale every increment by the size of the underlying data type.

Practice Problem 38

The following C code sets the diagonal elements of one of our fixed-size arrays to
val:

```
1   /* Set all diagonal elements to val */
2   void fix_set_diag(fix_matrix A, int val) {
3       int i;
4       for (i = 0; i < N; i++)
5           A[i][i] = val;
6   }
```

When compiled, GCC generates the following assembly code:

```
    A at %ebp+8, val at %ebp+12
1       movl    8(%ebp), %ecx
2       movl    12(%ebp), %edx
3       movl    $0, %eax
4   .L14:
5       movl    %edx, (%ecx,%eax)
6       addl    $68, %eax
7       cmpl    $1088, %eax
8       jne     .L14
```

Create a C-code program `fix_set_diag_opt` that uses optimizations similar
to those in the assembly code, in the same style as the code in Figure 28(b). Use
expressions involving the parameter N rather than integer constants, so that your
code will work correctly if N is redefined.

8.5 Variable-Size Arrays

Historically, C only supported multidimensional arrays where the sizes (with the
possible exception of the first dimension) could be determined at compile time.
Programmers requiring variable-sized arrays had to allocate storage for these
arrays using functions such as `malloc` or `calloc`, and had to explicitly encode the
mapping of multidimensional arrays into single-dimension ones via row-major
indexing, as expressed in Equation 1. ISO C99 introduced the capability to have
array dimensions be expressions that are computed as the array is being allocated,
and recent versions of GCC support most of the conventions for variable-sized
arrays in ISO C99.

In the C version of variable-size arrays, we can declare an array
int A[*expr1*][*expr2*], either as a local variable or as an argument to a function,
and then the dimensions of the array are determined by evaluating the expressions
expr1 and *expr2* at the time the declaration is encountered. So, for example, we
can write a function to access element i, j of an $n \times n$ array as follows:

```
1   int var_ele(int n, int A[n][n], int i, int j) {
2       return A[i][j];
3   }
```

(a) Original C code

```
1    /* Compute i,k of fixed matrix product */
2    int fix_prod_ele (fix_matrix A, fix_matrix B,  int i, int k) {
3        int j;
4        int result = 0;
5
6        for (j = 0; j < N; j++)
7            result += A[i][j] * B[j][k];
8
9        return result;
10   }
```

(b) Optimized C code

```
1    /* Compute i,k of fixed matrix product */
2    int fix_prod_ele_opt(fix_matrix A, fix_matrix B, int i, int k) {
3        int *Arow = &A[i][0];
4        int *Bptr = &B[0][k];
5        int result = 0;
6        int j;
7        for (j = 0; j != N; j++) {
8            result += Arow[j] * *Bptr;
9            Bptr += N;
10       }
11       return result;
12   }
```

Figure 28 **Original and optimized code to compute element** i, k **of matrix product for fixed-length arrays.** The compiler performs these optimizations automatically.

The parameter n must precede the parameter A[n][n], so that the function can compute the array dimensions as the parameter is encountered.

GCC generates code for this referencing function as

```
     n at %ebp+8, A at %ebp+12, i at %ebp+16, j at %ebp+20
1    movl    8(%ebp), %eax          Get n
2    sall    $2, %eax               Compute 4*n
3    movl    %eax, %edx             Copy 4*n
4    imull   16(%ebp), %edx         Compute 4*n*i
5    movl    20(%ebp), %eax         Get j
6    sall    $2, %eax               Compute 4*j
7    addl    12(%ebp), %eax         Compute x_A + 4*j
8    movl    (%eax,%edx), %eax      Read from x_A + 4*(n*i + j)
```

As the annotations show, this code computes the address of element i, j as $x_A + 4(n \cdot i + j)$. The address computation is similar to that of the fixed-size array, except that (1) the positions of the arguments on the stack are shifted due to the addition of parameter n, and (2) a multiply instruction is used (line 4) to

```
1    /* Compute i,k of variable matrix product */
2    int var_prod_ele(int n, int A[n][n], int B[n][n], int i, int k) {
3        int j;
4        int result = 0;
5
6        for (j = 0; j < n; j++)
7            result += A[i][j] * B[j][k];
8
9        return result;
10   }
```

Figure 29 **Code to compute element** i, k **of matrix product for variable-sized arrays.** The compiler performs optimizations similar to those for fixed-size arrays.

compute $n \cdot i$, rather than an leal instruction to compute $3i$. We see therefore that referencing variable-size arrays requires only a slight generalization over fixed-size ones. The dynamic version must use a multiplication instruction to scale i by n, rather than a series of shifts and adds. In some processors, this multiplication can incur a significant performance penalty, but it is unavoidable in this case.

When variable-sized arrays are referenced within a loop, the compiler can often optimize the index computations by exploiting the regularity of the access patterns. For example, Figure 29 shows C code to compute element i, k of the product of two $n \times n$ arrays A and B. The compiler generates code similar to what we saw for fixed-size arrays. In fact, the code bears close resemblance to that of Figure 28(b), except that it scales Bptr, the pointer to element B[j][k], by the variable value n rather than the fixed value N on each iteration.

The following is the assembly code for the loop of var_prod_ele:

```
     n stored at %ebp+8
     Registers: Arow in %esi, Bptr in %ecx, j in %edx,
         result in %ebx, %edi holds 4*n
1    .L30:                              loop:
2        movl    (%ecx), %eax           Get *Bptr
3        imull   (%esi,%edx,4), %eax    Multiply by Arow[j]
4        addl    %eax, %ebx             Add to result
5        addl    $1, %edx               Increment j
6        addl    %edi, %ecx             Add 4*n to Bptr
7        cmpl    %edx, 8(%ebp)          Compare n:j
8        jg      .L30                   If >, goto loop
```

We see that the program makes use of both a scaled value $4n$ (register %edi) for incrementing Bptr and the actual value of n stored at offset 8 from %ebp to check the loop bounds. The need for two values does not show up in the C code, due to the scaling of pointer arithmetic. The code retrieves the value of n from memory on each iteration to check for loop termination (line 7). This is an example of *register spilling*: there are not enough registers to hold all of the needed temporary data, and hence the compiler must keep some local variables in memory. In this case the compiler chose to spill n, because it is a "read-only" value—it does not change value within the loop. IA32 must often spill loop values to memory, since the processor has so few registers. In general, reading from memory can be done more readily than writing to memory, and so spilling read-only variables is preferable. See Problem 61 regarding how to improve this code to avoid register spilling.

9 Heterogeneous Data Structures

C provides two mechanisms for creating data types by combining objects of different types: *structures*, declared using the keyword struct, aggregate multiple objects into a single unit; *unions*, declared using the keyword union, allow an object to be referenced using several different types.

9.1 Structures

The C struct declaration creates a data type that groups objects of possibly different types into a single object. The different components of a structure are referenced by names. The implementation of structures is similar to that of arrays in that all of the components of a structure are stored in a contiguous region of memory, and a pointer to a structure is the address of its first byte. The compiler maintains information about each structure type indicating the byte offset of each field. It generates references to structure elements using these offsets as displacements in memory referencing instructions.

New to C? Representing an object as a struct

The struct data type constructor is the closest thing C provides to the objects of C++ and Java. It allows the programmer to keep information about some entity in a single data structure, and reference that information with names.

For example, a graphics program might represent a rectangle as a structure:

```
struct rect {
    int llx;    /* X coordinate of lower-left corner */
    int lly;    /* Y coordinate of lower-left corner */
    int color;  /* Coding of color                   */
    int width;  /* Width (in pixels)                 */
    int height; /* Height (in pixels)                */
};
```

We could declare a variable r of type struct rect and set its field values as follows:

```
struct rect r;
r.llx = r.lly = 0;
r.color = 0xFF00FF;
r.width = 10;
r.height = 20;
```

where the expression r.llx selects field llx of structure r.

Alternatively, we can both declare the variable and initialize its fields with a single statement:

```
struct rect r = { 0, 0, 0xFF00FF, 10, 20 };
```

It is common to pass pointers to structures from one place to another rather than copying them. For example, the following function computes the area of a rectangle, where a pointer to the rectangle struct is passed to the function:

```
int area(struct rect *rp)
{
    return (*rp).width * (*rp).height;
}
```

The expression `(*rp).width` dereferences the pointer and selects the `width` field of the resulting structure. Parentheses are required, because the compiler would interpret the expression `*rp.width` as `*(rp.width)`, which is not valid. This combination of dereferencing and field selection is so common that C provides an alternative notation using `->`. That is, `rp->width` is equivalent to the expression `(*rp).width`. For example, we could write a function that rotates a rectangle counterclockwise by 90 degrees as

```
void rotate_left(struct rect *rp)
{
    /* Exchange width and height */
    int t        = rp->height;
    rp->height = rp->width;
    rp->width  = t;
    /* Shift to new lower-left corner */
    rp->llx    -= t;
}
```

The objects of C++ and Java are more elaborate than structures in C, in that they also associate a set of *methods* with an object that can be invoked to perform computation. In C, we would simply write these as ordinary functions, such as the functions `area` and `rotate_left` shown above.

As an example, consider the following structure declaration:

```
struct rec {
    int i;
    int j;
    int a[3];
    int *p;
};
```

This structure contains four fields: two 4-byte `int`'s, an array consisting of three 4-byte `int`'s, and a 4-byte integer pointer, giving a total of 24 bytes:

Offset	0	4	8		20	24
Contents	i	j	a[0]	a[1]	a[2]	p

239

Observe that array a is embedded within the structure. The numbers along the top of the diagram give the byte offsets of the fields from the beginning of the structure.

To access the fields of a structure, the compiler generates code that adds the appropriate offset to the address of the structure. For example, suppose variable r of type struct rec * is in register %edx. Then the following code copies element r->i to element r->j:

```
1    movl    (%edx), %eax        Get r->i
2    movl    %eax, 4(%edx)       Store in r->j
```

Since the offset of field i is 0, the address of this field is simply the value of r. To store into field j, the code adds offset 4 to the address of r.

To generate a pointer to an object within a structure, we can simply add the field's offset to the structure address. For example, we can generate the pointer &(r->a[1]) by adding offset $8 + 4 \cdot 1 = 12$. For pointer r in register %eax and integer variable i in register %edx, we can generate the pointer value &(r->a[i]) with the single instruction

```
     Registers: r in %edx, i in %eax
1    leal    8(%edx,%eax,4), %eax      Set %eax to &r->a[i]
```

As a final example, the following code implements the statement

```
r->p = &r->a[r->i + r->j];
```

starting with r in register %edx:

```
1    movl    4(%edx), %eax              Get r->j
2    addl    (%edx), %eax               Add r->i
3    leal    8(%edx,%eax,4), %eax       Compute &r->a[r->i + r->j]
4    movl    %eax, 20(%edx)             Store in r->p
```

As these examples show, the selection of the different fields of a structure is handled completely at compile time. The machine code contains no information about the field declarations or the names of the fields.

Practice Problem 39

Consider the following structure declaration:

```
struct prob {
    int *p;
    struct {
        int x;
        int y;
    } s;
    struct prob *next;
};
```

This declaration illustrates that one structure can be embedded within another, just as arrays can be embedded within structures, and arrays can be embedded within arrays.

The following procedure (with some expressions omitted) operates on this structure:

```
void sp_init(struct prob *sp)
{
    sp->s.x  = _____ ;
    sp->p    = _____ ;
    sp->next = _____ ;
}
```

A. What are the offsets (in bytes) of the following fields?

p: _____

s.x: _____

s.y: _____

next: _____

B. How many total bytes does the structure require?

C. The compiler generates the following assembly code for the body of sp_init:

```
    sp at %ebp+8
1       movl    8(%ebp), %eax
2       movl    8(%eax), %edx
3       movl    %edx, 4(%eax)
4       leal    4(%eax), %edx
5       movl    %edx, (%eax)
6       movl    %eax, 12(%eax)
```

On the basis of this information, fill in the missing expressions in the code for sp_init.

9.2 Unions

Unions provide a way to circumvent the type system of C, allowing a single object to be referenced according to multiple types. The syntax of a union declaration is identical to that for structures, but its semantics are very different. Rather than having the different fields reference different blocks of memory, they all reference the same block.

Consider the following declarations:

```
struct S3 {
    char c;
    int i[2];
```

```
    double v;
};

union U3 {
    char c;
    int i[2];
    double v;
};
```

When compiled on an IA32 Linux machine, the offsets of the fields, as well as the total size of data types S3 and U3, are as shown in the following table:

Type	c	i	v	Size
S3	0	4	12	20
U3	0	0	0	8

(We will see shortly why i has offset 4 in S3 rather than 1, and we will discuss why the results would be different for a machine running Microsoft Windows.) For pointer p of type union U3 *, references p->c, p->i[0], and p->v would all reference the beginning of the data structure. Observe also that the overall size of a union equals the maximum size of any of its fields.

Unions can be useful in several contexts. However, they can also lead to nasty bugs, since they bypass the safety provided by the C type system. One application is when we know in advance that the use of two different fields in a data structure will be mutually exclusive. Then, declaring these two fields as part of a union rather than a structure will reduce the total space allocated.

For example, suppose we want to implement a binary tree data structure where each leaf node has a double data value, while each internal node has pointers to two children, but no data. If we declare this as

```
struct NODE_S {
    struct NODE_S *left;
    struct NODE_S *right;
    double data;
};
```

then every node requires 16 bytes, with half the bytes wasted for each type of node. On the other hand, if we declare a node as

```
union NODE_U {
    struct {
        union NODE_U *left;
        union NODE_U *right;
    } internal;
    double data;
};
```

then every node will require just 8 bytes. If n is a pointer to a node of type union
NODE *, we would reference the data of a leaf node as n->data, and the children
of an internal node as n->internal.left and n->internal.right.

With this encoding, however, there is no way to determine whether a given
node is a leaf or an internal node. A common method is to introduce an enumerated
type defining the different possible choices for the union, and then create a
structure containing a tag field and the union:

```
typedef enum { N_LEAF, N_INTERNAL } nodetype_t;

struct NODE_T {
    nodetype_t type;
    union {
        struct {
            struct NODE_T *left;
            struct NODE_T *right;
        } internal;
        double data;
    } info;
};
```

This structure requires a total of 12 bytes: 4 for type, and either 4 each for
info.internal.left and info.internal.right, or 8 for info.data. In this
case, the savings gain of using a union is small relative to the awkwardness of
the resulting code. For data structures with more fields, the savings can be more
compelling.

Unions can also be used to access the bit patterns of different data types.
For example, the following code returns the bit representation of a float as an
unsigned:

```
1   unsigned float2bit(float f)
2   {
3       union {
4           float f;
5           unsigned u;
6       } temp;
7       temp.f = f;
8       return temp.u;
9   };
```

In this code, we store the argument in the union using one data type, and access it
using another. Interestingly, the code generated for this procedure is identical to
that for the following procedure:

```
1   unsigned copy(unsigned u)
2   {
3       return u;
4   }
```

The body of both procedures is just a single instruction:

```
1    movl    8(%ebp), %eax
```

This demonstrates the lack of type information in machine code. The argument will be at offset 8 relative to %ebp regardless of whether it is a float or an unsigned. The procedure simply copies its argument as the return value without modifying any bits.

When using unions to combine data types of different sizes, byte-ordering issues can become important. For example, suppose we write a procedure that will create an 8-byte double using the bit patterns given by two 4-byte unsigned's:

```
1    double bit2double(unsigned word0, unsigned word1)
2    {
3        union {
4            double d;
5            unsigned u[2];
6        } temp;
7
8        temp.u[0] = word0;
9        temp.u[1] = word1;
10       return temp.d;
11   }
```

On a little-endian machine such as IA32, argument word0 will become the low-order 4 bytes of d, while word1 will become the high-order 4 bytes. On a big-endian machine, the role of the two arguments will be reversed.

Practice Problem 40

Suppose you are given the job of checking that a C compiler generates the proper code for structure and union access. You write the following structure declaration:

```
typedef union {
    struct {
        short  v;
        short  d;
        int    s;
    } t1;
    struct {
        int a[2];
        char *p;
    } t2;
} u_type;
```

You write a series of functions of the form

```
void get(u_type *up, TYPE *dest) {
    *dest =  EXPR;
}
```

with different access expressions EXPR, and with destination data type TYPE set according to type associated with EXPR. You then examine the code generated when compiling the functions to see if they match your expectations.

Suppose in these functions that up and dest are loaded into registers %eax and %edx, respectively. Fill in the following table with data type TYPE and sequences of 1–3 instructions to compute the expression and store the result at dest. Try to use just registers %eax and %edx, using register %ecx when these do not suffice.

EXPR	TYPE	Code
up->t1.s	int	movl 4(%eax), %eax
		movl %eax, (%edx)
up->t1.v		
&up->t1.d		
up->t2.a		
up->t2.a[up->t1.s]		
*up->t2.p		

9.3 Data Alignment

Many computer systems place restrictions on the allowable addresses for the primitive data types, requiring that the address for some type of object must be a multiple of some value K (typically 2, 4, or 8). Such *alignment restrictions* simplify the design of the hardware forming the interface between the processor and the memory system. For example, suppose a processor always fetches 8 bytes from memory with an address that must be a multiple of 8. If we can guarantee that any double will be aligned to have its address be a multiple of 8, then the value can be read or written with a single memory operation. Otherwise, we may need to perform two memory accesses, since the object might be split across two 8-byte memory blocks.

The IA32 hardware will work correctly regardless of the alignment of data. However, Intel recommends that data be aligned to improve memory system performance. Linux follows an alignment policy where 2-byte data types (e.g., short) must have an address that is a multiple of 2, while any larger data types (e.g., int, int *, float, and double) must have an address that is a multiple of 4. Note that this requirement means that the least significant bit of the address of an object of type short must equal zero. Similarly, any object of type int, or any pointer, must be at an address having the low-order 2 bits equal to zero.

Aside A case of mandatory alignment

For most IA32 instructions, keeping data aligned improves efficiency, but it does not affect program behavior. On the other hand, some of the SSE instructions for implementing multimedia operations will not work correctly with unaligned data. These instructions operate on 16-byte blocks of data, and the instructions that transfer data between the SSE unit and memory require the memory addresses to be multiples of 16. Any attempt to access memory with an address that does not satisfy this alignment will lead to an *exception*, with the default behavior for the program to terminate.

This is the motivation behind the IA32 convention of making sure that every stack frame is a multiple of 16 bytes long. The compiler can allocate storage within a stack frame in such a way that a block can be stored with a 16-byte alignment.

Aside Alignment with Microsoft Windows

Microsoft Windows imposes a stronger alignment requirement—any primitive object of K bytes, for $K = 2, 4,$ or 8, must have an address that is a multiple of K. In particular, it requires that the address of a double or a long long be a multiple of 8. This requirement enhances the memory performance at the expense of some wasted space. The Linux convention, where 8-byte values are aligned on 4-byte boundaries was probably good for the i386, back when memory was scarce and memory interfaces were only 4 bytes wide. With modern processors, Microsoft's alignment is a better design decision. Data type long double, for which GCC generates IA32 code allocating 12 bytes (even though the actual data type requires only 10 bytes) has a 4-byte alignment requirement with both Windows and Linux.

Alignment is enforced by making sure that every data type is organized and allocated in such a way that every object within the type satisfies its alignment restrictions. The compiler places directives in the assembly code indicating the desired alignment for global data. For example, the assembly-code declaration of the jump table prior to Practice Problem 28 contains the following directive on line 2:

```
.align 4
```

This ensures that the data following it (in this case the start of the jump table) will start with an address that is a multiple of 4. Since each table entry is 4 bytes long, the successive elements will obey the 4-byte alignment restriction.

Library routines that allocate memory, such as `malloc`, must be designed so that they return a pointer that satisfies the worst-case alignment restriction for the machine it is running on, typically 4 or 8. For code involving structures, the compiler may need to insert gaps in the field allocation to ensure that each structure element satisfies its alignment requirement. The structure then has some required alignment for its starting address.

For example, consider the following structure declaration:

```
struct S1 {
    int  i;
    char c;
    int  j;
};
```

Suppose the compiler used the minimal 9-byte allocation, diagrammed as follows:

Then it would be impossible to satisfy the 4-byte alignment requirement for both fields `i` (offset 0) and `j` (offset 5). Instead, the compiler inserts a 3-byte gap (shown here as shaded in blue) between fields `c` and `j`:

As a result, `j` has offset 8, and the overall structure size is 12 bytes. Furthermore, the compiler must ensure that any pointer p of type `struct S1*` satisfies a 4-byte alignment. Using our earlier notation, let pointer p have value x_p. Then x_p must be a multiple of 4. This guarantees that both p->i (address x_p) and p->j (address $x_p + 8$) will satisfy their 4-byte alignment requirements.

In addition, the compiler may need to add padding to the end of the structure so that each element in an array of structures will satisfy its alignment requirement. For example, consider the following structure declaration:

```
struct S2 {
    int  i;
    int  j;
    char c;
};
```

If we pack this structure into 9 bytes, we can still satisfy the alignment requirements for fields `i` and `j` by making sure that the starting address of the structure satisfies a 4-byte alignment requirement. Consider, however, the following declaration:

```
struct S2 d[4];
```

With the 9-byte allocation, it is not possible to satisfy the alignment requirement for each element of d, because these elements will have addresses x_d, $x_d + 9$, $x_d + 18$, and $x_d + 27$. Instead, the compiler allocates 12 bytes for structure S2, with the final 3 bytes being wasted space:

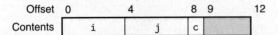

That way the elements of d will have addresses x_d, $x_d + 12$, $x_d + 24$, and $x_d + 36$. As long as x_d is a multiple of 4, all of the alignment restrictions will be satisfied.

Practice Problem 41

For each of the following structure declarations, determine the offset of each field, the total size of the structure, and its alignment requirement under Linux/IA32.

A. `struct P1 { int i; char c; int j; char d; };`

B. `struct P2 { int i; char c; char d; int j; };`

C. `struct P3 { short w[3]; char c[3] };`

D. `struct P4 { short w[3]; char *c[3] };`

E. `struct P3 { struct P1 a[2]; struct P2 *p };`

Practice Problem 42

For the structure declaration

```
struct {
    char      *a;
    short      b;
    double     c;
    char       d;
    float      e;
    char       f;
    long long  g;
    void      *h;
} foo;
```

suppose it was compiled on a Windows machine, where each primitive data type of K bytes must have an offset that is a multiple of K.

A. What are the byte offsets of all the fields in the structure?

B. What is the total size of the structure?

C. Rearrange the fields of the structure to minimize wasted space, and then show the byte offsets and total size for the rearranged structure.

10 Putting It Together: Understanding Pointers

Pointers are a central feature of the C programming language. They serve as a uniform way to generate references to elements within different data structures. Pointers are a source of confusion for novice programmers, but the underlying concepts are fairly simple. Here we highlight some key principles of pointers and their mapping into machine code.

- *Every pointer has an associated type.* This type indicates what kind of object the pointer points to. Using the following pointer declarations as illustrations,

  ```
  int *ip;
  char **cpp;
  ```

  variable `ip` is a pointer to an object of type `int`, while `cpp` is a pointer to an object that itself is a pointer to an object of type `char`. In general, if the object has type T, then the pointer has type $*T$. The special `void *` type represents a generic pointer. For example, the `malloc` function returns a generic pointer, which is converted to a typed pointer via either an explicit cast or by the implicit casting of the assignment operation. Pointer types are not part of machine code; they are an abstraction provided by C to help programmers avoid addressing errors.

- *Every pointer has a value.* This value is an address of some object of the designated type. The special NULL (0) value indicates that the pointer does not point anywhere.

- *Pointers are created with the & operator.* This operator can be applied to any C expression that is categorized as an *lvalue*, meaning an expression that can appear on the left side of an assignment. Examples include variables and the elements of structures, unions, and arrays. We have seen that the machine-code realization of the & operator often uses the `leal` instruction to compute the expression value, since this instruction is designed to compute the address of a memory reference.

- *Pointers are dereferenced with the * operator.* The result is a value having the type associated with the pointer. Dereferencing is implemented by a memory reference, either storing to or retrieving from the specified address.

- *Arrays and pointers are closely related.* The name of an array can be referenced (but not updated) as if it were a pointer variable. Array referencing (e.g., `a[3]`) has the exact same effect as pointer arithmetic and dereferencing (e.g., `*(a+3)`). Both array referencing and pointer arithmetic require scaling the offsets by the object size. When we write an expression `p+i` for pointer `p` with value p, the resulting address is computed as $p + L \cdot i$, where L is the size of the data type associated with `p`.

- *Casting from one type of pointer to another changes its type but not its value.* One effect of casting is to change any scaling of pointer arithmetic. So for example, if `p` is a pointer of type `char *` having value p, then the expression (int *) p+7 computes $p + 28$, while (int *) (p+7) computes $p + 7$. (Recall that casting has higher precedence than addition.)

- *Pointers can also point to functions.* This provides a powerful capability for storing and passing references to code, which can be invoked in some other part of the program. For example, if we have a function defined by the prototype

```
int fun(int x, int *p);
```

then we can declare and assign a pointer fp to this function by the following code sequence:

```
(int) (*fp)(int, int *);
fp  = fun;
```

We can then invoke the function using this pointer:

```
int y = 1;
int result = fp(3, &y);
```

The value of a function pointer is the address of the first instruction in the machine-code representation of the function.

New to C? Function pointers

The syntax for declaring function pointers is especially difficult for novice programmers to understand. For a declaration such as

```
int (*f)(int*);
```

it helps to read it starting from the inside (starting with "f") and working outward. Thus, we see that f is a pointer, as indicated by "(*f)." It is a pointer to a function that has a single int * as an argument, as indicated by "(*f)(int*)". Finally, we see that it is a pointer to a function that takes an int * as an argument and returns int.

The parentheses around *f are required, because otherwise the declaration

```
int *f(int*);
```

would be read as

```
(int *) f(int*);
```

That is, it would be interpreted as a function prototype, declaring a function f that has an int * as its argument and returns an int *.

Kernighan & Ritchie [58, Sect. 5.12] present a helpful tutorial on reading C declarations.

11 Life in the Real World: Using the GDB Debugger

The GNU debugger GDB provides a number of useful features to support the run-time evaluation and analysis of machine-level programs. With the examples and exercises in this book, we attempt to infer the behavior of a program by just looking at the code. Using GDB, it becomes possible to study the behavior by watching the program in action, while having considerable control over its execution.

Figure 30 shows examples of some GDB commands that help when working with machine-level, IA32 programs. It is very helpful to first run OBJDUMP to get a disassembled version of the program. Our examples are based on running GDB on the file prog. We start GDB with the following command line:

```
unix> gdb prog
```

The general scheme is to set breakpoints near points of interest in the program. These can be set to just after the entry of a function, or at a program address. When one of the breakpoints is hit during program execution, the program will halt and return control to the user. From a breakpoint, we can examine different registers and memory locations in various formats. We can also single-step the program, running just a few instructions at a time, or we can proceed to the next breakpoint.

As our examples suggest, GDB has an obscure command syntax, but the online help information (invoked within GDB with the help command) overcomes this shortcoming. Rather than using the command-line interface to GDB, many programmers prefer using DDD, an extension to GDB that provides a graphic user interface.

Web Aside ASM:OPT   Machine code generated with higher levels of optimization

In our presentation, we have looked at machine code generated with level-one optimization (specified with the command-line option '-O1'). In practice, most heavily used programs are compiled with higher levels of optimization. For example, all of the GNU libraries and packages are compiled with level-two optimization, specified with the command-line option '-O2'.

Recent versions of GCC employ an extensive set of optimizations at level two, making the mapping between the source code and the generated code more difficult to discern. Here are some examples of the optimizations that can be found at level two:

- The control structures become more entangled. Most procedures have multiple return points, and the stack management code to set up and complete a function is intermixed with the code implementing the operations of the procedure.

- Procedure calls are often *inlined*, replacing them by the instructions implementing the procedures. This eliminates much of the overhead involved in calling and returning from a function, and it enables optimizations that are specific to individual function calls. On the other hand, if we try to set a breakpoint for a function in a debugger, we might never encounter a call to this function.

Command	Effect
Starting and stopping	
quit	Exit GDB
run	Run your program (give command line arguments here)
kill	Stop your program
Breakpoints	
break sum	Set breakpoint at entry to function sum
break *0x8048394	Set breakpoint at address 0x8048394
delete 1	Delete breakpoint 1
delete	Delete all breakpoints
Execution	
stepi	Execute one instruction
stepi 4	Execute four instructions
nexti	Like stepi, but proceed through function calls
continue	Resume execution
finish	Run until current function returns
Examining code	
disas	Disassemble current function
disas sum	Disassemble function sum
disas 0x8048397	Disassemble function around address 0x8048397
disas 0x8048394 0x80483a4	Disassemble code within specified address range
print /x $eip	Print program counter in hex
Examining data	
print $eax	Print contents of %eax in decimal
print /x $eax	Print contents of %eax in hex
print /t $eax	Print contents of %eax in binary
print 0x100	Print decimal representation of 0x100
print /x 555	Print hex representation of 555
print /x ($ebp+8)	Print contents of %ebp plus 8 in hex
print *(int *) 0xfff076b0	Print integer at address 0xfff076b0
print *(int *) ($ebp+8)	Print integer at address %ebp + 8
x/2w 0xfff076b0	Examine two (4-byte) words starting at address 0xfff076b0
x/20b sum	Examine first 20 bytes of function sum
Useful information	
info frame	Information about current stack frame
info registers	Values of all the registers
help	Get information about GDB

Figure 30 **Example GDB commands.** These examples illustrate some of the ways GDB supports debugging of machine-level programs.

- Recursion is often replaced by iteration. For example, the recursive factorial function rfact (Figure 25) is compiled into code very similar to that generated for the while loop implementation (Figure 15). Again, this can lead to some surprises when we try to monitor program execution with a debugger.

These optimizations can significantly improve program performance, but they make the mapping between source and machine code much more difficult to discern. This can make the programs more difficult to debug. Nonetheless, these higher level optimizations have now become standard, and so those who study programs at the machine level must become familiar with the possible optimizations they may encounter.

12 Out-of-Bounds Memory References and Buffer Overflow

We have seen that C does not perform any bounds checking for array references, and that local variables are stored on the stack along with state information such as saved register values and return addresses. This combination can lead to serious program errors, where the state stored on the stack gets corrupted by a write to an out-of-bounds array element. When the program then tries to reload the register or execute a ret instruction with this corrupted state, things can go seriously wrong.

A particularly common source of state corruption is known as *buffer overflow*. Typically some character array is allocated on the stack to hold a string, but the size of the string exceeds the space allocated for the array. This is demonstrated by the following program example:

```
1   /* Sample implementation of library function gets() */
2   char *gets(char *s)
3   {
4       int c;
5       char *dest = s;
6       int gotchar = 0; /* Has at least one character been read? */
7       while ((c = getchar()) != '\n' && c != EOF) {
8           *dest++ = c; /* No bounds checking! */
9           gotchar = 1;
10      }
11      *dest++ = '\0';   /* Terminate string */
12      if (c == EOF && !gotchar)
13          return NULL; /* End of file or error */
14      return s;
15  }
16
```

```
17    /* Read input line and write it back */
18    void echo()
19    {
20        char buf[8];   /* Way too small! */
21        gets(buf);
22        puts(buf);
23    }
```

The preceding code shows an implementation of the library function gets to demonstrate a serious problem with this function. It reads a line from the standard input, stopping when either a terminating newline character or some error condition is encountered. It copies this string to the location designated by argument s, and terminates the string with a null character. We show the use of gets in the function echo, which simply reads a line from standard input and echoes it back to standard output.

The problem with gets is that it has no way to determine whether sufficient space has been allocated to hold the entire string. In our echo example, we have purposely made the buffer very small—just eight characters long. Any string longer than seven characters will cause an out-of-bounds write.

Examining the assembly code generated by GCC for echo shows how the stack is organized.

```
1     echo:
2         pushl   %ebp                    Save %ebp on stack
3         movl    %esp, %ebp
4         pushl   %ebx                    Save %ebx
5         subl    $20, %esp               Allocate 20 bytes on stack
6         leal    -12(%ebp), %ebx         Compute buf as %ebp-12
7         movl    %ebx, (%esp)            Store buf at top of stack
8         call    gets                    Call gets
9         movl    %ebx, (%esp)            Store buf at top of stack
10        call    puts                    Call puts
11        addl    $20, %esp               Deallocate stack space
12        popl    %ebx                    Restore %ebx
13        popl    %ebp                    Restore %ebp
14        ret                             Return
```

We can see in this example that the program stores the contents of registers %ebp and %ebx on the stack, and then allocates an additional 20 bytes by subtracting 20 from the stack pointer (line 5). The location of character array buf is computed as 12 bytes below %ebp (line 6), just below the stored value of %ebx, as illustrated in Figure 31. As long as the user types at most seven characters, the string returned by gets (including the terminating null) will fit within the space allocated for buf. A longer string, however, will cause gets to overwrite some of the information

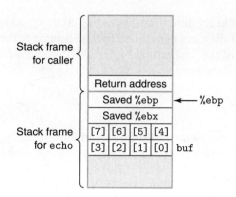

Figure 31

Stack organization for echo function. Character array buf is just below part of the saved state. An out-of-bounds write to buf can corrupt the program state.

stored on the stack. As the string gets longer, the following information will get corrupted:

Characters typed	Additional corrupted state
0–7	None
8–11	Saved value of %ebx
12–15	Saved value of %ebp
16–19	Return address
20+	Saved state in caller

As this table indicates, the corruption is cumulative—as the number of characters increases, more state gets corrupted. Depending on which portions of the state are affected, the program can misbehave in several different ways:

- If the stored value of %ebx is corrupted, then this register will not be restored properly in line 12, and so the caller will not be able to rely on the integrity of this register, even though it should be callee-saved.

- If the stored value of %ebp is corrupted, then this register will not be restored properly on line 13, and so the caller will not be able to reference its local variables or parameters properly.

- If the stored value of the return address is corrupted, then the ret instruction (line 14) will cause the program to jump to a totally unexpected location.

None of these behaviors would seem possible based on the C code. The impact of out-of-bounds writing to memory by functions such as gets can only be understood by studying the program at the machine-code level.

Our code for echo is simple but sloppy. A better version involves using the function fgets, which includes as an argument a count on the maximum number of bytes to read. Problem 68 asks you to write an echo function that can handle an input string of arbitrary length. In general, using gets or any function that can overflow storage is considered a bad programming practice. The C compiler even produces the following error message when compiling a file containing a call to gets: "The gets function is dangerous and should not be used." Unfortunately, a number of commonly used library functions, including `strcpy`, `strcat`, and `sprintf`, have the property that they can generate a byte sequence without being given any indication of the size of the destination buffer [94]. Such conditions can lead to vulnerabilities to buffer overflow.

Practice Problem 43

Figure 32 shows a (low-quality) implementation of a function that reads a line from standard input, copies the string to newly allocated storage, and returns a pointer to the result.

Consider the following scenario. Procedure `getline` is called with the return address equal to 0x8048643, register `%ebp` equal to 0xbffffc94, register `%ebx` equal to 0x1, register `%edi` is equal to 0x2, and register `%esi` is equal to 0x3. You type in the string " 012345678901234567890123". The program terminates with

(a) C code

```
1   /* This is very low-quality code.
2      It is intended to illustrate bad programming practices.
3      See Problem 43. */
4   char *getline()
5   {
6       char buf[8];
7       char *result;
8       gets(buf);
9       result = malloc(strlen(buf));
10      strcpy(result, buf);
11      return result;
12  }
```

(b) Disassembly up through call to gets

```
1    080485c0 <getline>:
2     80485c0:  55                 push   %ebp
3     80485c1:  89 e5              mov    %esp,%ebp
4     80485c3:  83 ec 28           sub    $0x28,%esp
5     80485c6:  89 5d f4           mov    %ebx,-0xc(%ebp)
6     80485c9:  89 75 f8           mov    %esi,-0x8(%ebp)
7     80485cc:  89 7d fc           mov    %edi,-0x4(%ebp)
      Diagram stack at this point
8     80485cf:  8d 75 ec           lea    -0x14(%ebp),%esi
9     80485d2:  89 34 24           mov    %esi,(%esp)
10    80485d5:  e8 a3 ff ff ff     call   804857d <gets>
      Modify diagram to show stack contents at this point
```

Figure 32 **C and disassembled code for Problem 43.**

a segmentation fault. You run GDB and determine that the error occurs during the execution of the ret instruction of getline.

A. Fill in the diagram that follows, indicating as much as you can about the stack just after executing the instruction at line 7 in the disassembly. Label the quantities stored on the stack (e.g., "Return address") on the right, and their hexadecimal values (if known) within the box. Each box represents 4 bytes. Indicate the position of %ebp.

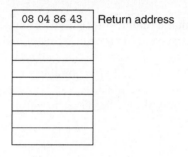

B. Modify your diagram to show the effect of the call to gets (line 10).

C. To what address does the program attempt to return?

D. What register(s) have corrupted value(s) when getline returns?

E. Besides the potential for buffer overflow, what two other things are wrong with the code for getline?

A more pernicious use of buffer overflow is to get a program to perform a function that it would otherwise be unwilling to do. This is one of the most common methods to attack the security of a system over a computer network. Typically, the program is fed with a string that contains the byte encoding of some executable code, called the *exploit code*, plus some extra bytes that overwrite the return address with a pointer to the exploit code. The effect of executing the ret instruction is then to jump to the exploit code.

In one form of attack, the exploit code then uses a system call to start up a shell program, providing the attacker with a range of operating system functions. In another form, the exploit code performs some otherwise unauthorized task, repairs the damage to the stack, and then executes ret a second time, causing an (apparently) normal return to the caller.

As an example, the famous Internet worm of November 1988 used four different ways to gain access to many of the computers across the Internet. One was a buffer overflow attack on the finger daemon fingerd, which serves requests by the FINGER command. By invoking FINGER with an appropriate string, the worm could make the daemon at a remote site have a buffer overflow and execute code that gave the worm access to the remote system. Once the worm gained access to a system, it would replicate itself and consume virtually all of the machine's computing resources. As a consequence, hundreds of machines were effectively paralyzed until security experts could determine how to eliminate the worm. The author of the worm was caught and prosecuted. He was sentenced to 3 years probation, 400 hours of community service, and a $10,500 fine. Even to this day, however, people continue to find security leaks in systems that leave them vulnerable to buffer overflow attacks. This highlights the need for careful programming. Any interface to the external environment should be made "bullet proof" so that no behavior by an external agent can cause the system to misbehave.

Aside Worms and viruses

Both worms and viruses are pieces of code that attempt to spread themselves among computers. As described by Spafford [102], a *worm* is a program that can run by itself and can propagate a fully working version of itself to other machines. A *virus* is a piece of code that adds itself to other programs, including operating systems. It cannot run independently. In the popular press, the term "virus" is used to refer to a variety of different strategies for spreading attacking code among systems, and so you will hear people saying "virus" for what more properly should be called a "worm."

12.1 Thwarting Buffer Overflow Attacks

Buffer overflow attacks have become so pervasive and have caused so many problems with computer systems that modern compilers and operating systems have implemented mechanisms to make it more difficult to mount these attacks and to limit the ways by which an intruder can seize control of a system via a buffer overflow attack. In this section, we will present ones that are provided by recent versions of GCC for Linux.

Stack Randomization

In order to insert exploit code into a system, the attacker needs to inject both the code as well as a pointer to this code as part of the attack string. Generating this pointer requires knowing the stack address where the string will be located. Historically, the stack addresses for a program were highly predictable. For all systems running the same combination of program and operating system version, the stack locations were fairly stable across many machines. So, for example, if an attacker could determine the stack addresses used by a common Web server, it could devise an attack that would work on many machines. Using infectious disease as an analogy, many systems were vulnerable to the exact same strain of a virus, a phenomenon often referred to as a *security monoculture* [93].

The idea of *stack randomization* is to make the position of the stack vary from one run of a program to another. Thus, even if many machines are running identical code, they would all be using different stack addresses. This is implemented by allocating a random amount of space between 0 and n bytes on the stack at the start of a program, for example, by using the allocation function `alloca`, which allocates space for a specified number of bytes on the stack. This allocated space is not used by the program, but it causes all subsequent stack locations to vary from one execution of a program to another. The allocation range n needs to be large enough to get sufficient variations in the stack addresses, yet small enough that it does not waste too much space in the program.

The following code shows a simple way to determine a "typical" stack address:

```
1    int main() {
2        int local;
3        printf("local at %p\n", &local);
4        return 0;
5    }
```

This code simply prints the address of a local variable in the main function. Running the code 10,000 times on a Linux machine in 32-bit mode, the addresses ranged from 0xff7fa7e0 to 0xffffd7e0, a range of around 2^{23}. By comparison, running on an older Linux system, the same address occurred every time. Running in 64-bit mode on the newer machine, the addresses ranged from 0x7fff00241914 to 0x7ffffff98664, a range of nearly 2^{32}.

Stack randomization has become standard practice in Linux systems. It is one of a larger class of techniques known as *address-space layout randomization*, or ASLR [95]. With ASLR, different parts of the program, including program code, library code, stack, global variables, and heap data, are loaded into different regions of memory each time a program is run. That means that a program running on one machine will have very different address mappings than the same program running on other machines. This can thwart some forms of attack.

Overall, however, a persistent attacker can overcome randomization by brute force, repeatedly attempting attacks with different addresses. A common trick is to include a long sequence of nop (pronounced "no op," short for "no operation") instructions before the actual exploit code. Executing this instruction has no effect, other than incrementing the program counter to the next instruction. As long as the attacker can guess an address somewhere within this sequence, the program will run through the sequence and then hit the exploit code. The common term for this sequence is a "nop sled" [94], expressing the idea that the program "slides" through the sequence. If we set up a 256-byte nop sled, then the randomization over $n = 2^{23}$ can be cracked by enumerating $2^{15} = 32,768$ starting addresses, which is entirely feasible for a determined attacker. For the 64-bit case, trying to enumerate $2^{24} = 16,777,216$ is a bit more daunting. We can see that stack randomization and other aspects of ASLR can increase the effort required to successfully attack a system, and therefore greatly reduce the rate at which a virus or worm can spread, but it cannot provide a complete safeguard.

Practice Problem 44

Running our stack-checking code 10,000 times on a system running Linux version 2.6.16, we obtained addresses ranging from a minimum of 0xffffb754 to a maximum of 0xffffd754.

A. What is the approximate range of addresses?

B. If we attempted a buffer overrun with a 128-byte nop sled, how many attempts would it take to exhaustively test all starting addresses?

Figure 33

Stack organization for echo function with stack protector enabled. A special "canary" value is positioned between array buf and the saved state. The code checks the canary value to determine whether or not the stack state has been corrupted.

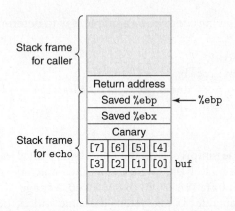

Stack Corruption Detection

A second line of defense is to be able to detect when a stack has been corrupted. We saw in the example of the echo function (Figure 31) that the corruption typically occurs when we overrun the bounds of a local buffer. In C, there is no reliable way to prevent writing beyond the bounds of an array. Instead, we can try to detect when such a write has occurred before any harmful effects can occur.

Recent versions of GCC incorporate a mechanism known as *stack protector* into the generated code to detect buffer overruns. The idea is to store a special *canary* value[4] in the stack frame between any local buffer and the rest of the stack state, as illustrated in Figure 33 [32, 94]. This canary value, also referred to as a *guard value*, is generated randomly each time the program is run, and so there is no easy way for an attacker to determine what it is. Before restoring the register state and returning from the function, the program checks if the canary has been altered by some operation of this function or one that it has called. If so, the program aborts with an error.

Recent versions of GCC try to determine whether a function is vulnerable to a stack overflow, and insert this type of overflow detection automatically. In fact, for our earlier demonstration of stack overflow, we had to give the command-line option "-fno-stack-protector" to prevent GCC from inserting this code. When we compile the function echo without this option, and hence with stack protector enabled, we get the following assembly code:

```
1   echo:
2       pushl   %ebp
3       movl    %esp, %ebp
4       pushl   %ebx
5       subl    $20, %esp
6       movl    %gs:20, %eax      Retrieve canary
7       movl    %eax, -8(%ebp)    Store on stack
```

---

4. The term "canary" refers to the historic use of these birds to detect the presence of dangerous gasses in coal mines.

```
8      xorl    %eax, %eax          Zero out register
9      leal    -16(%ebp), %ebx     Compute buf as %ebp-16
10     movl    %ebx, (%esp)        Store buf at top of stack
11     call    gets                Call gets
12     movl    %ebx, (%esp)        Store buf at top of stack
13     call    puts                Call puts
14     movl    -8(%ebp), %eax      Retrieve canary
15     xorl    %gs:20, %eax        Compare to stored value
16     je      .L19                If =, goto ok
17     call    __stack_chk_fail    Stack corrupted!
18   .L19:                         ok:
19     addl    $20, %esp           Normal return ...
20     popl    %ebx
21     popl    %ebp
22     ret
```

We see that this version of the function retrieves a value from memory (line 6) and stores it on the stack at offset -8 from %ebp. The instruction argument %gs:20 is an indication that the canary value is read from memory using *segmented addressing*, an addressing mechanism that dates back to the 80286 and is seldom found in programs running on modern systems. By storing the canary in a special segment, it can be marked as "read only," so that an attacker cannot overwrite the stored canary value. Before restoring the register state and returning, the function compares the value stored at the stack location with the canary value (via the xorl instruction on line 15.) If the two are identical, the xorl instruction will yield 0, and the function will complete in the normal fashion. A nonzero value indicates that the canary on the stack has been modified, and so the code will call an error routine.

Stack protection does a good job of preventing a buffer overflow attack from corrupting state stored on the program stack. It incurs only a small performance penalty, especially because GCC only inserts it when there is a local buffer of type char in the function. Of course, there are other ways to corrupt the state of an executing program, but reducing the vulnerability of the stack thwarts many common attack strategies.

Practice Problem 45

The function intlen, along with the functions len and iptoa, provides a very convoluted way of computing the number of decimal digits required to represent an integer. We will use this as a way to study some aspects of the GCC stack protector facility.

```
int len(char *s) {
    return strlen(s);
}

void iptoa(char *s, int *p)
```

```
{
    int val = *p;
    sprintf(s, "%d", val);
}

int intlen(int x) {
    int v;
    char buf[12];
    v = x;
    iptoa(buf, &v);
    return len(buf);
}
```

The following show portions of the code for `intlen`, compiled both with and without stack protector:

Without protector

```
1    subl    $36, %esp
2    movl    8(%ebp), %eax
3    movl    %eax, -8(%ebp)
4    leal    -8(%ebp), %eax
5    movl    %eax, 4(%esp)
6    leal    -20(%ebp), %ebx
7    movl    %ebx, (%esp)
8    call    iptoa
```

With protector

```
1     subl    $52, %esp
2     movl    %gs:20, %eax
3     movl    %eax, -8(%ebp)
4     xorl    %eax, %eax
5     movl    8(%ebp), %eax
6     movl    %eax, -24(%ebp)
7     leal    -24(%ebp), %eax
8     movl    %eax, 4(%esp)
9     leal    -20(%ebp), %ebx
10    movl    %ebx, (%esp)
11    call    iptoa
```

A. For both versions: What are the positions in the stack frame for buf, v, and (when present) the canary value?

B. How would the rearranged ordering of the local variables in the protected code provide greater security against a buffer overrun attack?

Limiting Executable Code Regions

A final step is to eliminate the ability of an attacker to insert executable code into a system. One method is to limit which memory regions hold executable code. In typical programs, only the portion of memory holding the code generated by the compiler need be executable. The other portions can be restricted to allow just reading and writing. The virtual memory space is logically divided into *pages*, typically with 2048 or 4096 bytes per page. The hardware supports different forms of *memory protection*, indicating the forms of access allowed by both user programs and by the operating system kernel. Many systems allow control over three forms of access: read (reading data from memory), write (storing data into memory), and execute (treating the memory contents as machine-level code). Historically, the x86 architecture merged the read and execute access controls into a single 1-bit flag, so that any page marked as readable was also executable. The stack had to be kept both readable and writable, and therefore the bytes on the stack were also executable. Various schemes were implemented to be able to limit some pages to being readable but not executable, but these generally introduced significant inefficiencies.

More recently, AMD introduced an "NX" (for "no-execute") bit into the memory protection for its 64-bit processors, separating the read and execute access modes, and Intel followed suit. With this feature, the stack can be marked as being readable and writable, but not executable, and the checking of whether a page is executable is performed in hardware, with no penalty in efficiency.

Some types of programs require the ability to dynamically generate and execute code. For example, "just-in-time" compilation techniques dynamically generate code for programs written in interpreted languages, such as Java, to improve execution performance. Whether or not we can restrict the executable code to just that part generated by the compiler in creating the original program depends on the language and the operating system.

The techniques we have outlined—randomization, stack protection, and limiting which portions of memory can hold executable code—are three of the most common mechanisms used to minimize the vulnerability of programs to buffer overflow attacks. They all have the properties that they require no special effort on the part of the programmer and incur very little or no performance penalty. Each separately reduces the level of vulnerability, and in combination they become even more effective. Unfortunately, there are still ways to attack computers [81, 94], and so worms and viruses continue to compromise the integrity of many machines.

Web Aside ASM:EASM   Combining assembly code with C programs

Although a C compiler does a good job of converting the computations we express in a program into machine code, there are some features of a machine that cannot be accessed by a C program. For example, IA32 machines have a condition code PF (for "parity flag") that is set to 1 when there is an even number of ones in the low-order 8 bits of the computed result. Computing this information in C requires at least seven shifting, masking, and exclusive-or operations. It is ironic that the hardware performs this computation as part of every arithmetic or logical operation, but there is no way for a C program to determine the value of the PF condition code.

There are two ways to incorporate assembly code into C programs. First, we can write an entire function as a separate assembly-code file and let the assembler and linker combine this with code we have written in C. Second, we can use the *inline assembly* feature of GCC, where brief sections of assembly code can be incorporated into a C program using the asm directive. This approach has the advantage that it minimizes the amount of machine-specific code.

Of course, including assembly code in a C program makes the code specific to a particular class of machines (such as IA32), and so it should only be used when the desired feature can only be accessed in this way.

13 x86-64: Extending IA32 to 64 Bits

Intel's IA32 instruction set architecture (ISA) has been the dominant instruction format for the world's computers for many years. IA32 has been the platform of choice for most Windows, Linux, and, since 2006, even Macintosh computers. The IA32 format used today was, for the most part, defined in 1985 with the introduction of the i386 microprocessor, extending the 16-bit instruction set defined by the original 8086 to 32 bits. Even though subsequent processor generations have introduced new instruction types and formats, many compilers, including GCC, have avoided using these features in the interest of maintaining backward compatibility. For example, we saw in Section 6.6 that the conditional move instructions, introduced by Intel in 1995, can yield significant efficiency improvements over more traditional conditional branches, yet in most configurations GCC will not generate these instructions.

A shift is underway to a 64-bit version of the Intel instruction set. Originally developed by Advanced Micro Devices (AMD) and named *x86-64*, it is now supported by most processors from AMD (who now call it *AMD64*) and by Intel, who refer to it as *Intel64*. Most people still refer to it as "x86-64," and we follow this convention. (Some vendors have shortened this to simply "x64".) Newer versions of Linux and Windows support this extension, although systems still run only 32-bit versions of these operating systems. In extending GCC to support x86-64, the developers saw an opportunity to also make use of some of the instruction-set features that had been added in more recent generations of IA32 processors.

This combination of new hardware and revised compiler makes x86-64 code substantially different in form and in performance than IA32 code. In creating the 64-bit extension, the AMD engineers adopted some of the features found in reduced instruction set computers (RISC) [49] that made them the favored targets for optimizing compilers. For example, there are now 16 general-purpose registers, rather than the performance-limiting 8 of the original 8086. The developers of GCC were able to exploit these features, as well as those of more recent generations of the IA32 architecture, to obtain substantial performance improvements. For example, procedure parameters are now passed via registers rather than on the stack, greatly reducing the number of memory read and write operations.

This section serves as a supplement to our description of IA32, describing the extensions in both the hardware and the software support to accommodate x86-64. We assume readers are already familiar with IA32. We start with a brief history of how AMD and Intel arrived at x86-64, followed by a summary of the main features that distinguish x86-64 code from IA32 code, and then work our way through the individual features.

13.1 History and Motivation for x86-64

Over the many years since introduction of the i386 in 1985, the capabilities of microprocessors have changed dramatically. In 1985, a fully configured high-end desktop computer, such as the Sun-3 workstation sold by Sun Microsystems, had at most 8 megabytes of random-access memory (RAM) and 100 megabytes of disk storage. It used a Motorola 68020 microprocessor (Intel microprocessors of that era did not have the necessary features and performance for high-end machines) with a 12.5-megahertz clock and ran around 4 million instructions per second. Nowadays, a typical high-end desktop system has 4 gigabytes of RAM ($512\times$ increase), 1 terabyte of disk storage ($10,000\times$ increase), and a nearly 4-gigahertz clock, running around 5 billion instructions per second ($1250\times$ increase). Microprocessor-based systems have become pervasive. Even today's supercomputers are based on harnessing the power of many microprocessors computing in parallel. Given these large quantitative improvements, it is remarkable that the world's computing base mostly runs code that is binary compatible with machines that existed back in 1985 (except that they did not have nearly enough memory to handle today's operating systems and applications).

The 32-bit word size of the IA32 has become a major limitation in growing the capacity of microprocessors. Most significantly, the word size of a machine defines the range of virtual addresses that programs can use, giving a 4-gigabyte virtual address space in the case of 32 bits. It is now feasible to buy more than this amount of RAM for a machine, but the system cannot make effective use of it. For applications that involve manipulating large data sets, such as scientific computing, databases, and data mining, the 32-bit word size makes life difficult for programmers. They must write code using *out-of-core* algorithms,[5] where the data reside on disk and are explicitly read into memory for processing.

Further progress in computing technology requires shifting to a larger word size. Following the tradition of growing word sizes by doubling, the next logical step is 64 bits. In fact, 64-bit machines have been available for some time. Digital Equipment Corporation introduced its Alpha processor in 1992, and it became a popular choice for high-end computing. Sun Microsystems introduced a 64-bit version of its SPARC architecture in 1995. At the time, however, Intel was not a serious contender for high-end computers, and so the company was under less pressure to switch to 64 bits.

---

5. The physical memory of a machine is often referred to as *core memory,* dating to an era when each bit of a random-access memory was implemented with a magnetized ferrite core.

Intel's first foray into 64-bit computers were the Itanium processors, based on a totally new instruction set, known as "IA64." Unlike Intel's historic strategy of maintaining backward compatibility as it introduced each new generation of microprocessor, IA64 is based on a radically new approach jointly developed with Hewlett-Packard. Its *Very Large Instruction Word* (VLIW) format packs multiple instructions into bundles, allowing higher degrees of parallel execution. Implementing IA64 proved to be very difficult, and so the first Itanium chips did not appear until 2001, and these did not achieve the expected level of performance on real applications. Although the performance of Itanium-based systems has improved, they have not captured a significant share of the computer market. Itanium machines can execute IA32 code in a compatibility mode, but not with very good performance. Most users have preferred to make do with less expensive, and often faster, IA32-based systems.

Meanwhile, Intel's archrival, Advanced Micro Devices (AMD), saw an opportunity to exploit Intel's misstep with IA64. For years, AMD had lagged just behind Intel in technology, and so they were relegated to competing with Intel on the basis of price. Typically, Intel would introduce a new microprocessor at a price premium. AMD would come along 6 to 12 months later and have to undercut Intel significantly to get any sales—a strategy that worked but yielded very low profits. In 2003, AMD introduced a 64-bit microprocessor based on its "x86-64" instruction set. As the name implies, x86-64 is an evolution of the Intel instruction set to 64 bits. It maintains full backward compatibility with IA32, but it adds new data formats, as well as other features that enable higher capacity and higher performance. With x86-64, AMD captured some of the high-end market that had historically belonged to Intel. AMD's recent generations of processors have indeed proved very successful as high-performance machines. Most recently, AMD has renamed this instruction set *AMD64*, but "x86-64" persists as a favored name.

Intel realized that its strategy of a complete shift from IA32 to IA64 was not working, and so began supporting their own variant of x86-64 in 2004 with processors in the Pentium 4 Xeon line. Since they had already used the name "IA64" to refer to Itanium, they then faced a difficulty in finding their own name for this 64-bit extension. In the end, they decided to describe x86-64 as an enhancement to IA32, and so they referred to it as *IA32-EM64T*, for "Enhanced Memory 64-bit Technology." In late 2006, they adopted the name *Intel64*.

On the compiler side, the developers of GCC steadfastly maintained binary compatibility with the i386, even as useful features were being added to the IA32 instruction set, including conditional moves and a more modern set of floating-point instructions. These features would only be used when code was compiled with special settings of command-line options. Switching to x86-64 as a target provided an opportunity for GCC to give up backward compatibility and instead exploit these newer features even with standard command-line options.

In this text, we use "IA32" to refer to the combination of hardware and GCC code found in traditional 32-bit versions of Linux running on Intel-based machines. We use "x86-64" to refer to the hardware and code combination running on the newer 64-bit machines from AMD and Intel. In the worlds of Linux and GCC, these two platforms are referred to as "i386" and "x86_64," respectively.

13.2 An Overview of x86-64

The combination of the new hardware supplied by Intel and AMD, and the new versions of GCC targeting these machines makes x86-64 code substantially different from that generated for IA32 machines. The main features include:

- Pointers and long integers are 64 bits long. Integer arithmetic operations support 8, 16, 32, and 64-bit data types.

- The set of general-purpose registers is expanded from 8 to 16.

- Much of the program state is held in registers rather than on the stack. Integer and pointer procedure arguments (up to 6) are passed via registers. Some procedures do not need to access the stack at all.

- Conditional operations are implemented using conditional move instructions when possible, yielding better performance than traditional branching code.

- Floating-point operations are implemented using the register-oriented instruction set introduced with SSE version 2, rather than the stack-based approach supported by IA32.

Data Types

Figure 34 shows the sizes of different C data types for x86-64, and compares them to the sizes for IA32 (rightmost column). We see that pointers (shown here as data type char *) require 8 bytes rather than 4. These are referred to as *quad words* by Intel, since they are 4 times longer than the nominal 16-bit "word." In principle, this gives programs the ability to access 2^{64} bytes, or 16 *exabytes*, of memory (around 18.4×10^{18} bytes). That seems like an astonishing amount of memory, but keep in mind that 4 gigabytes seemed like an extremely large amount of memory when the first 32-bit machines appeared in the late 1970s. In practice, most machines do not really support the full address range—the current generations of AMD and

C declaration	Intel data type	Assembly code suffix	x86-64 size (bytes)	IA32 Size
char	Byte	b	1	1
short	Word	w	2	2
int	Double word	l	4	4
long int	Quad word	q	8	4
long long int	Quad word	q	8	8
char *	Quad word	q	8	4
float	Single precision	s	4	4
double	Double precision	d	8	8
long double	Extended precision	t	10/16	10/12

Figure 34 **Sizes of standard data types with x86-64.** These are compared to the sizes for IA32. Both long integers and pointers require 8 bytes, as compared to 4 for IA32.

Intel x86-64 machines support 256 terabytes (2^{48} bytes) of virtual memory—but allocating a full 64 bits for pointers is a good idea for long-term compatibility.

We also see that the prefix "long" changes integers to 64 bits, allowing a considerably larger range of values. In fact, data type long becomes identical to long long. Moreover, the hardware provides registers that can hold 64-bit integers and instructions that can operate on these quad words.

As with IA32, the long prefix also changes a floating-point double to use the 80-bit format supported by IA32. These are stored in memory with an allocation of 16 bytes for x86-64, compared to 12 bytes for IA32. This improves the performance of memory read and write operations, which typically fetch 8 or 16 bytes at a time. Whether 12 or 16 bytes are allocated, only the low-order 10 bytes are actually used. Moreover, the long double data type is only supported by an older class of floating-point instructions that have some idiosyncratic properties (see Web Aside DATA:IA32-FP), while both the float and double data types are supported by the more recent SSE instructions. The long double data type should only be used by programs requiring the additional precision and range the extended-precision format provides over the double-precision format.

Practice Problem 46

The cost of DRAM, the memory technology used to implement the main memories of microprocessors, has dropped from around $8,000 per megabyte in 1980 to around $0.06 in 2010, roughly a factor of 1.48 every year, or around 51 every 10 years. Let us assume these trends will continue indefinitely (which may not be realistic), and that our budget for a machine's memory is around $1,000, so that we would have configured a machine with 128 kilobytes in 1980 and with 16.3 gigabytes in 2010.

A. Estimate when our $1,000 budget would pay for 256 terabytes of memory.

B. Estimate when our $1,000 budget would pay for 16 exabytes of memory.

C. How much earlier would these transition points occur if we raised our DRAM budget to $10,000?

Assembly-Code Example

In Section 2.3, we presented the IA32 assembly code generated by GCC for a function simple. Below is the C code for simple_l, similar to simple, except that it uses long integers:

```
long int simple_l(long int *xp, long int y)
{
    long int t = *xp + y;
    *xp = t;
    return t;
}
```

When GCC is run on an x86-64 Linux machine with the command line

```
unix> gcc -O1 -S -m32 code.c
```

it generates code that is compatible with any IA32 machine (we annotate the code to highlight which instructions read (R) data from memory and which instructions write (W) data to memory):

```
    IA32 implementation of function simple_1.
    xp at %ebp+8, y at %ebp+12
1   simple_1:
2     pushl   %ebp              Save frame pointer       (W)
3     movl    %esp, %ebp        Create new frame pointer
4     movl    8(%ebp), %edx     Retrieve xp              (R)
5     movl    12(%ebp), %eax    Retrieve yp              (R)
6     addl    (%edx), %eax      Add *xp to get t         (R)
7     movl    %eax, (%edx)      Store t at xp            (W)
8     popl    %ebp              Restore frame pointer    (R)
9     ret                       Return                   (R)
```

When we instruct GCC to generate x86-64 code

```
unix> gcc -O1 -S -m64 code.c
```

(on most machines, the flag -m64 is not required), we get very different code:

```
    x86-64 version of function simple_1.
    xp in %rdi, y in %rsi
1   simple_1:
2     movq    %rsi, %rax        Copy y
3     addq    (%rdi), %rax      Add *xp to get t   (R)
4     movq    %rax, (%rdi)      Store t at xp      (W)
5     ret                       Return             (R)
```

Some of the key differences include:

- Instead of movl and addl instructions, we see movq and addq. The pointers and variables declared as long integers are now 64 bits (quad words) rather than 32 bits (long words).

- We see the 64-bit versions of registers (e.g., %rsi and %rdi, rather than %esi and %edi). The procedure returns a value by storing it in register %rax.

- No stack frame gets generated in the x86-64 version. This eliminates the instructions that set up (lines 2–3) and remove (line 8) the stack frame in the IA32 code.

- Arguments xp and y are passed in registers (%rdi and %rsi, respectively) rather than on the stack. This eliminates the need to fetch the arguments from memory.

The net effect of these changes is that the IA32 code consists of eight instructions making seven memory references (five reads, two writes), while the x86-64 code consists of four instructions making three memory references (two reads, one write). The relative performance of the two versions depends greatly on the hardware on which they are executed. Running on an Intel Pentium 4E, one of the first Intel machines to support x86-64, we found that the IA32 version requires around 18 clock cycles per call to simple_l, while the x86-64 version requires only 12. This 50% performance improvement on the same machine with the same C code is quite striking. On a newer Intel Core i7 processor, we found that both versions required around 12 clock cycles, indicating no performance improvement. On other machines we have tried, the performance difference lies somewhere between these two extremes. In general, x86-64 code is more compact, requires fewer memory accesses, and runs more efficiently than the corresponding IA32 code.

13.3 Accessing Information

Figure 35 shows the set of general-purpose registers under x86-64. Compared to the registers for IA32 (Figure 2), we see a number of differences:

- The number of registers has been doubled to 16.
- All registers are 64 bits long. The 64-bit extensions of the IA32 registers are named %rax, %rcx, %rdx, %rbx, %rsi, %rdi, %rsp, and %rbp. The new registers are named %r8–%r15.
- The low-order 32 bits of each register can be accessed directly. This gives us the familiar registers from IA32: %eax, %ecx, %edx, %ebx, %esi, %edi, %esp, and %ebp, as well as eight new 32-bit registers: %r8d–%r15d.
- The low-order 16 bits of each register can be accessed directly, as is the case for IA32. The *word-size* versions of the new registers are named %r8w–%r15w.
- The low-order 8 bits of each register can be accessed directly. This is true in IA32 only for the first four registers (%al, %cl, %dl, %bl). The byte-size versions of the other IA32 registers are named %sil, %dil, %spl, and %bpl. The byte-size versions of the new registers are named %r8b–%r15b.
- For backward compatibility, the second byte of registers %rax, %rcx, %rdx, and %rbx can be directly accessed by instructions having single-byte operands.

As with IA32, most of the registers can be used interchangeably, but there are some special cases. Register %rsp has special status, in that it holds a pointer to the top stack element. Unlike in IA32, however, there is no frame pointer register; register %rbp is available for use as a general-purpose register. Particular conventions are used for passing procedure arguments via registers and for how registers are to be saved and restored during procedure calls, as is discussed in Section 13.4. In addition, some arithmetic instructions make special use of registers %rax and %rdx.

For the most part, the operand specifiers of x86-64 are just the same as those in IA32 (see Figure 3), except that the base and index register identifiers must

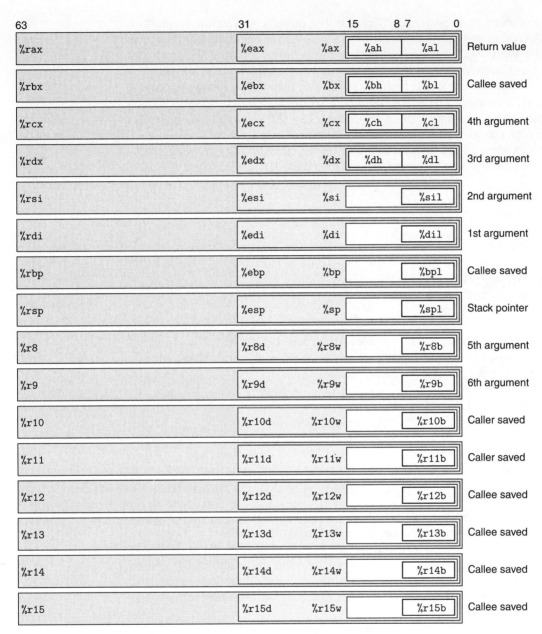

Figure 35 **Integer registers.** The existing eight registers are extended to 64-bit versions, and eight new registers are added. Each register can be accessed as either 8 bits (byte), 16 bits (word), 32 bits (double word), or 64 bits (quad word).

use the 'r' version of a register (e.g., %rax) rather than the 'e' version. In addition to the IA32 addressing forms, some forms of *PC-relative* operand addressing are supported. With IA32, this form of addressing is only supported for jump and other control transfer instructions (see Section 6.3). This mode is provided to compensate for the fact that the offsets (shown in Figure 3 as *Imm*) are only 32 bits long. By viewing this field as a 32-bit two's-complement number, instructions can access data within a window of around $\pm 2.15 \times 10^9$ relative to the program counter. With x86-64, the program counter is named %rip.

As an example of PC-relative data addressing, consider the following procedure, which calls the function simple_l examined earlier:

```
long int gval1 = 567;
long int gval2 = 763;

long int call_simple_l()
{
    long int z = simple_l(&gval1, 12L);
    return z + gval2;
}
```

This code references global variables gval1 and gval2. When this function is compiled, assembled, and linked, we get the following executable code (as generated by the disassembler objdump):

```
1   0000000000400541 <call_simple_l>:
2     400541:  be 0c 00 00 00         mov     $0xc,%esi           Load 12 as 2nd argument
3     400546:  bf 20 10 60 00         mov     $0x601020,%edi      Load &gval1 as 1st argument
4     40054b:  e8 c3 ff ff ff         callq   400513 <simple_l>   Call simple_l
5     400550:  48 03 05 d1 0a 20 00   add     0x200ad1(%rip),%rax Add gval2 to result
6     400557:  c3                     retq                        Return
```

The instruction on line 3 stores the address of global variable gval1 in register %rdi. It does this by copying the constant value 0x601020 into register %edi. The upper 32 bits of %rdi are automatically set to zero. The instruction on line 5 retrieves the value of gval2 and adds it to the value returned by the call to simple_l. Here we see PC-relative addressing—the immediate value 0x200ad1 is added to the address of the following instruction to get 0x200ad1 + 0x400557 = 0x601028.

Figure 36 documents some of the data movement instructions available with x86-64 beyond those found in IA32 (see Figure 4). Some instructions require the destination to be a register, indicated by *R*. Others can have either a register or a memory location as destination, indicated by *D*. Most of these instructions fall within a class of instructions seen with IA32. The movabsq instruction, on the other hand, has no counterpart in IA32. This instruction can copy a full 64-bit immediate value to its destination register. When the movq instruction has an immediate value as its source operand, it is limited to a 32-bit value, which is sign-extended to 64 bits.

Instruction		Effect	Description
movabsq	I, R	$R \leftarrow I$	Move absolute quad word
MOV	S, D	$D \leftarrow S$	Move
movq		Move quad word	
MOVS	S, D	$D \leftarrow \text{SignExtend}(S)$	Move with sign extension
movsbq		Move sign-extended byte to quad word	
movswq		Move sign-extended word to quad word	
movslq		Move sign-extended double word to quad word	
MOVZ	S, D	$D \leftarrow \text{ZeroExtend}(S)$	Move with zero extension
movzbq		Move zero-extended byte to quad word	
movzwq		Move zero-extended word to quad word	
pushq	S	$R[\%rsp] \leftarrow R[\%rsp] - 8;$ $M[R[\%rsp]] \leftarrow S$	Push quad word
popq	D	$D \leftarrow M[R[\%rsp]];$ $R[\%rsp] \leftarrow R[\%rsp] + 8$	Pop quad word

Figure 36 **Data movement instructions.** These supplement the movement instructions of IA32 (Figure 4). The movabsq instruction only allows immediate data (shown as I) as the source value. Others allow immediate data, a register, or memory (shown as S). Some instructions require the destination to be a register (shown as R), while others allow both register and memory destinations (shown as D).

Moving from a smaller data size to a larger one can involve either sign extension (MOVS) or zero extension (MOVZ). Perhaps unexpectedly, instructions that move or generate 32-bit register values also set the upper 32 bits of the register to zero. Consequently there is no need for an instruction movzlq. Similarly, the instruction movzbq has the exact same behavior as movzbl when the destination is a register—both set the upper 56 bits of the destination register to zero. This is in contrast to instructions that generate 8- or 16-bit values, such as movb; these instructions do not alter the other bits in the register. The new stack instructions pushq and popq allow pushing and popping of 64-bit values.

Practice Problem 47

The following C function converts an argument of type src_t to a return value of type dst_t, where these two types are defined using typedef:

```
dest_t cvt(src_t x)
{
    dest_t y = (dest_t) x;
    return y;
}
```

Machine-Level Representation of Programs

Assume argument x is in the appropriately named portion of register %rdi (i.e., %rdi, %edi, %di, or %dil), and that some form of data movement instruction is to be used to perform the type conversion and to copy the value to the appropriately named portion of register %rax. Fill in the following table indicating the instruction, the source register, and the destination register for the following combinations of source and destination type:

src_t	dest_t	Instruction	S	D
long	long	movq	%rdi	%rax
int	long			
char	long			
unsigned int	unsigned long			
unsigned char	unsigned long			
long	int			
unsigned long	unsigned			

Arithmetic Instructions

In Figure 7, we listed a number of arithmetic and logic instructions, using a class name, such as "ADD", to represent instructions for different operand sizes, such as addb (byte), addw (word), and addl (long word). To each of these classes we now add instructions that operate on quad words with the suffix 'q'. Examples of these quad-word instructions include leaq (load effective address), incq (increment), addq (add), and salq (shift left). These quad-word instructions have the same argument types as their shorter counterparts. As mentioned earlier, instructions that generate 32-bit register results, such as addl, also set the upper 32 bits of the register to zero. Instructions that generate 16-bit results, such as addw, only affect their 16-bit destination registers, and similarly for instructions that generate 8-bit results. As with the movq instruction, immediate operands are limited to 32-values, which are sign extended to 64 bits.

When mixing operands of different sizes, GCC must choose the right combinations of arithmetic instructions, sign extensions, and zero extensions. These depend on subtle aspects of type conversion and the behavior of the instructions for different operand sizes. This is illustrated by the following C function:

```
1    long int gfun(int x, int y)
2    {
3        long int t1 = (long) x + y;   /* 64-bit addition */
4        long int t2 = (long) (x + y); /* 32-bit addition */
5        return t1 | t2;
6    }
```

Given that integers are 32 bits and long integers are 64, the two additions in this function proceed as follows. Recall that casting has higher precedence than addition, and so line 3 calls for x to be converted to 64 bits, and by operand

274

promotion y is also converted. Value t1 is then computed using 64-bit addition. On the other hand, t2 is computed in line 4 by performing 32-bit addition and then extending this value to 64 bits.

The assembly code generated for this function is as follows:

```
1    gfun:
     x in %rdi, y in %rsi
2       leal    (%rsi,%rdi), %eax    Compute t2 as 32-bit sum of x and y
        cltq is equivalent to movslq %eax,%rax
3       cltq                         Sign extend to 64 bits
4       movslq  %esi,%rsi            Convert y to long
5       movslq  %edi,%rdi            Convert x to long
6       addq    %rdi, %rsi           Compute t1 (64-bit addition)
7       orq     %rsi, %rax           Set t1 | t2 as return value
8       ret                          Return
```

Local value t2 is computed with an leal instruction (line 2), which uses 32-bit arithmetic. It is then sign-extended to 64 bits using the cltq instruction, which we will see is a special instruction equivalent to executing the instruction movslq %eax,%rax. The movslq instructions on lines 4–5 take the lower 32 bits of the arguments and sign extend them to 64 bits in the same registers. The addq instruction on line 6 then performs 64-bit addition to get t1.

Practice Problem 48

A C function arithprob with arguments a, b, c, and d has the following body:

```
return a*b + c*d;
```

It compiles to the following x86-64 code:

```
1    arithprob:
2       movslq  %ecx,%rcx
3       imulq   %rdx, %rcx
4       movsbl  %sil,%esi
5       imull   %edi, %esi
6       movslq  %esi,%rsi
7       leaq    (%rcx,%rsi), %rax
8       ret
```

The arguments and return value are all signed integers of various lengths. Arguments a, b, c, and d are passed in the appropriate regions of registers %rdi, %rsi, %rdx, and %rcx, respectively. Based on this assembly code, write a function prototype describing the return and argument types for arithprob.

Figure 37 show instructions used to generate the full 128-bit product of two 64-bit words, as well as ones to support 64-bit division. They are similar to their 32-bit counterparts (Figure 9). Several of these instructions view the combination

Instruction		Effect	Description
imulq	S	R[%rdx]:R[%rax] ← S × R[%rax]	Signed full multiply
mulq	S	R[%rdx]:R[%rax] ← S × R[%rax]	Unsigned full multiply
cltq		R[%rax] ← SignExtend(R[%eax])	Convert %eax to quad word
cqto		R[%rdx]:R[%rax] ← SignExtend(R[%rax])	Convert to oct word
idivq	S	R[%rdx] ← R[%rdx]:R[%rax] mod S; R[%rax] ← R[%rdx]:R[%rax] ÷ S	Signed divide
divq	S	R[%rdx] ← R[%rdx]:R[%rax] mod S; R[%rax] ← R[%rdx]:R[%rax] ÷ S	Unsigned divide

Figure 37 **Special arithmetic operations.** These operations support full 64-bit multiplication and division, for both signed and unsigned numbers. The pair of registers %rdx and %rax are viewed as forming a single 128-bit oct word.

of registers %rdx and %rax as forming a 128-bit *oct word*. For example, the imulq and mulq instructions store the result of multiplying two 64-bit values—the first as given by the source operand and the second from register %rax.

The two divide instructions idivq and divq start with %rdx:%rax as the 128-bit dividend and the source operand as the 64-bit divisor. They then store the quotient in register %rax and the remainder in register %rdx. Preparing the dividend depends on whether unsigned (divq) or signed (idivq) division is to be performed. In the former case, register %rdx is simply set to zero. In the latter case, the instruction cqto is used to perform sign extension, copying the sign bit of %rax into every bit of %rdx.[6] Figure 37 also shows an instruction cltq to sign extend register %eax to %rax.[7] This instruction is just a shorthand for the instruction movslq %eax,%rax.

13.4 Control

The control instructions and methods of implementing control transfers in x86-64 are the same as those in IA32 (Section 6.) As shown in Figure 38, two new instructions, cmpq and testq, are added to compare and test quad words, augmenting those for byte, word, and double word sizes (Figure 10). GCC uses both conditional data transfer and conditional control transfer, since all x86-64 machines support conditional moves.

To illustrate the similarity between IA32 and x86-64 code, consider the assembly code generated by compiling an integer factorial function implemented with a while loop (Figure 15), as is shown in Figure 39. As can be seen, these two

---

6. ATT-format instruction cqto is called cqo in Intel and AMD documentation.

7. Instruction cltq is called cdqe in Intel and AMD documentation.

Instruction		Based on	Description
CMP	S_2, S_1	$S_1 - S_2$	Compare
cmpq		Compare quad word	
TEST	S_2, S_1	$S_1 \& S_2$	Test
testq		Test quad word	

Figure 38 **64-bit comparison and test instructions.** These instructions set the condition codes without updating any other registers.

(a) IA32 version

```
 1    fact_while:
        n at %ebp+8
 2        pushl    %ebp                Save frame pointer
 3        movl     %esp, %ebp          Create new frame pointer
 4        movl     8(%ebp), %edx       Get n
 5        movl     $1, %eax            Set result = 1
 6        cmpl     $1, %edx            Compare n:1
 7        jle      .L7                 If <=, goto done
 8    .L10:                            loop:
 9        imull    %edx, %eax          Compute result *= n
10        subl     $1, %edx            Decrement n
11        cmpl     $1, %edx            Compare n:1
12        jg       .L10                If >, goto loop
13    .L7:                             done:
14        popl     %ebp                Restore frame pointer
15        ret                          Return result
```

(b) x86-64 version

```
 1    fact_while:
        n in %rdi
 2        movl     $1, %eax            Set result = 1
 3        cmpl     $1, %edi            Compare n:1
 4        jle      .L7                 If <=, goto done
 5    .L10:                            loop:
 6        imull    %edi, %eax          Compute result *= n
 7        subl     $1, %edi            Decrement n
 8        cmpl     $1, %edi            Compare n:1
 9        jg       .L10                If >, goto loop
10    .L7:                             done:
11        rep                          (See explanation in aside)
12        ret                          Return result
```

Figure 39 **IA32 and x86-64 versions of factorial.** Both were compiled from the C code shown in Figure 15.

versions are very similar. They differ only in how arguments are passed (on the stack vs. in registers), and the absence of a stack frame or frame pointer in the x86-64 code.

Aside Why is there a rep instruction in this code?

On line 11 of the x86-64 code, we see the instruction rep precedes the return instruction ret. Looking at the Intel and AMD documentation for the rep instruction, we find that it is normally used to implement a repeating string operation [3, 29]. It seems completely inappropriate here. The answer to this puzzle can be seen in AMD's guidelines to compiler writers [1]. They recommend using the combination of rep followed by ret to avoid making the ret instruction be the destination of a conditional jump instruction. Without the rep instruction, the jg instruction would proceed to the ret instruction when the branch is not taken. According to AMD, their processors cannot properly predict the destination of a ret instruction when it is reached from a jump instruction. The rep instruction serves as a form of no-operation here, and so inserting it as the jump destination does not change behavior of the code, except to make it faster on AMD processors.

Practice Problem 49

A function fun_c has the following overall structure:

```c
long fun_c(unsigned long x) {
    long val = 0;
    int i;
    for ( _____ ; _____ ; _____ ) {
        _____ ;
    }
    _____ ;
    return _____ ;
}
```

The GCC C compiler generates the following assembly code:

```
1   fun_c:
    x in %rdi
2       movl      $0, %ecx
3       movl      $0, %edx
4       movabsq   $72340172838076673, %rsi
5   .L2:
6       movq      %rdi, %rax
7       andq      %rsi, %rax
8       addq      %rax, %rcx
9       shrq      %rdi            Shift right by 1
10      addl      $1, %edx
11      cmpl      $8, %edx
12      jne       .L2
```

```
13    movq    %rcx, %rax
14    sarq    $32, %rax
15    addq    %rcx, %rax
16    movq    %rax, %rdx
17    sarq    $16, %rdx
18    addq    %rax, %rdx
19    movq    %rdx, %rax
20    sarq    $8, %rax
21    addq    %rdx, %rax
22    andl    $255, %eax
23    ret
```

Reverse engineer the operation of this code. You will find it useful to convert the decimal constant on line 4 to hexadecimal.

    A. Use the assembly-code version to fill in the missing parts of the C code.

    B. Describe in English what this code computes.

Procedures

We have already seen in our code samples that the x86-64 implementation of procedure calls differs substantially from that of IA32. By doubling the register set, programs need not be so dependent on the stack for storing and retrieving procedure information. This can greatly reduce the overhead for procedure calls and returns.

Here are some of the highlights of how procedures are implemented with x86-64:

- Arguments (up to the first six) are passed to procedures via registers, rather than on the stack. This eliminates the overhead of storing and retrieving values on the stack.

- The callq instruction stores a 64-bit return address on the stack.

- Many functions do not require a stack frame. Only functions that cannot keep all local variables in registers need to allocate space on the stack.

- Functions can access storage on the stack up to 128 bytes beyond (i.e., at a lower address than) the current value of the stack pointer. This allows some functions to store information on the stack without altering the stack pointer.

- There is no frame pointer. Instead, references to stack locations are made relative to the stack pointer. Most functions allocate their total stack storage needs at the beginning of the call and keep the stack pointer at a fixed position.

- As with IA32, some registers are designated as callee-save registers. These must be saved and restored by any procedure that modifies them.

Operand	Argument Number					
size (bits)	1	2	3	4	5	6
64	%rdi	%rsi	%rdx	%rcx	%r8	%r9
32	%edi	%esi	%edx	%ecx	%r8d	%r9d
16	%di	%si	%dx	%cx	%r8w	%r9w
8	%dil	%sil	%dl	%cl	%r8b	%r9b

Figure 40 **Registers for passing function arguments.** The registers are used in a specified order and named according to the argument sizes.

Argument Passing

Up to six integral (i.e., integer and pointer) arguments can be passed via registers. The registers are used in a specified order, with the name used for a register depending on the size of the data type being passed. These are shown in Figure 40. Arguments are allocated to these registers according to their ordering in the argument list. Arguments smaller than 64 bits can be accessed using the appropriate subsection of the 64-bit register. For example, if the first argument is 32 bits, it can be accessed as %edi.

As an example of argument passing, consider the following C function having eight arguments:

```
void proc(long  a1, long  *a1p,
          int   a2, int   *a2p,
          short a3, short *a3p,
          char  a4, char  *a4p)
{
    *a1p += a1;
    *a2p += a2;
    *a3p += a3;
    *a4p += a4;
}
```

The arguments include a range of different-sized integers (64, 32, 16, and 8 bits) as well as different types of pointers, each of which is 64 bits.

This function is implemented in x86-64 as follows:

```
x86-64 implementation of function proc
Arguments passed as follows:
  a1  in %rdi      (64 bits)
  a1p in %rsi      (64 bits)
  a2  in %edx      (32 bits)
  a2p in %rcx      (64 bits)
  a3  in %r8w      (16 bits)
  a3p in %r9       (64 bits)
```

```
        a4  at %rsp+8      (8 bits)
        a4p at %rsp+16     (64 bits)
1   proc:
2       movq    16(%rsp), %r10      Fetch a4p  (64 bits)
3       addq    %rdi, (%rsi)        *a1p += a1 (64 bits)
4       addl    %edx, (%rcx)        *a2p += a2 (32 bits)
5       addw    %r8w, (%r9)         *a3p += a3 (16 bits)
6       movzbl  8(%rsp), %eax       Fetch a4   (8 bits)
7       addb    %al, (%r10)         *a4p += a4 (8 bits)
8       ret
```

The first six arguments are passed in registers, while the last two are at positions 8 and 16 relative to the stack pointer. Different versions of the ADD instruction are used according to the sizes of the operands: addq for a1 (long), addl for a2 (int), addw for a3 (short), and addb for a4 (char).

Practice Problem 50

A C function incrprob has arguments q, t, and x of different sizes, and each may be signed or unsigned. The function has the following body:

```
*t += x;
*q += *t;
```

It compiles to the following x86-64 code:

```
1   incrprob:
2       addl    (%rdx), %edi
3       movl    %edi, (%rdx)
4       movslq  %edi,%rdi
5       addq    %rdi, (%rsi)
6       ret
```

Determine all four valid function prototypes for incrprob by determining the ordering and possible types of the three parameters.

Stack Frames

We have already seen that many compiled functions do not require a stack frame. If all of the local variables can be held in registers, and the function does not call any other functions (sometimes referred to as a *leaf procedure*, in reference to the tree structure of procedure calls), then the only need for the stack is to save the return address.

On the other hand, there are several reasons a function may require a stack frame:

- There are too many local variables to hold in registers.
- Some local variables are arrays or structures.

- The function uses the address-of operator (&) to compute the address of a local variable.
- The function must pass some arguments on the stack to another function.
- The function needs to save the state of a callee-save register before modifying it.

When any of these conditions hold, we find the compiled code for the function creating a stack frame. Unlike the code for IA32, where the stack pointer fluctuates back and forth as values are pushed and popped, the stack frames for x86-64 procedures usually have a fixed size, set at the beginning of the procedure by decrementing the stack pointer (register %rsp). The stack pointer remains at a fixed position during the call, making it possible to access data using offsets relative to the stack pointer. As a consequence, the frame pointer (register %ebp) seen in IA32 code is no longer needed.

Whenever one function (the *caller*) calls another (the *callee*), the return address gets pushed onto the stack. By convention, we consider this part of the caller's stack frame, in that it encodes part of the caller's state. But this information gets popped from the stack as control returns to the caller, and so it does not affect the offsets used by the caller for accessing values within the stack frame.

The following function illustrates many aspects of the x86-64 stack discipline. Despite the length of this example, it is worth studying carefully.

```
long int call_proc()
{
    long  x1 = 1; int  x2 = 2;
    short x3 = 3; char x4 = 4;
    proc(x1, &x1, x2, &x2, x3, &x3, x4, &x4);
    return (x1+x2)*(x3-x4);
}
```

GCC generates the following x86-64 code.

```
    x86-64 implementation of call_proc
1   call_proc:
2     subq    $32, %rsp          Allocate 32-byte stack frame
3     movq    $1, 16(%rsp)       Store 1 in &x1
4     movl    $2, 24(%rsp)       Store 2 in &x2
5     movw    $3, 28(%rsp)       Store 3 in &x3
6     movb    $4, 31(%rsp)       Store 4 in &x4
7     leaq    24(%rsp), %rcx     Pass &x2 as argument 4
8     leaq    16(%rsp), %rsi     Pass &x1 as argument 2
9     leaq    31(%rsp), %rax     Compute &x4
10    movq    %rax, 8(%rsp)      Pass &x4 as argument 8
11    movl    $4, (%rsp)         Pass    4 as argument 7
12    leaq    28(%rsp), %r9      Pass &x3 as argument 6
13    movl    $3, %r8d           Pass    3 as argument 5
14    movl    $2, %edx           Pass    2 as argument 3
15    movl    $1, %edi           Pass    1 as argument 1
```

```
16    call    proc                    Call
17    movswl  28(%rsp),%eax           Get x3 and convert to int
18    movsbl  31(%rsp),%edx           Get x4 and convert to int
19    subl    %edx, %eax              Compute x3-x4
20    cltq                            Sign extend to long int
21    movslq  24(%rsp),%rdx           Get x2
22    addq    16(%rsp), %rdx          Compute x1+x2
23    imulq   %rdx, %rax              Compute (x1+x2)*(x3-x4)
24    addq    $32, %rsp               Deallocate stack frame
25    ret                             Return
```

Figure 41(a) illustrates the stack frame set up during the execution of call_proc. Function call_proc allocates 32 bytes on the stack by decrementing the stack pointer. It uses bytes 16–31 to hold local variables x1 (bytes 16–23), x2 (bytes 24–27), x3 (bytes 28–29), and x4 (byte 31). These allocations are sized according to the variable types. Byte 30 is unused. Bytes 0–7 and 8–15 of the stack frame are used to hold the seventh and eighth arguments to call_proc, since there are not enough argument registers. The parameters are allocated eight bytes each, even though parameter x4 requires only a single byte. In the code for call_proc, we can see instructions initializing the local variables and setting up the parameters (both in registers and on the stack) for the call to call_proc. After proc returns, the local variables are combined to compute the final expression, which is returned in register %rax. The stack space is deallocated by simply incrementing the stack pointer before the ret instruction.

Figure 41(b) illustrates the stack during the execution of proc. The call instruction pushed the return address onto the stack, and so the stack pointer is shifted down by 8 relative to its position during the execution of call_proc.

Figure 41

Stack frame structure for call_proc. The frame is required to hold local variables x1 through x4, as well as the seventh and eighth arguments to proc. During the execution of proc (b), the stack pointer is shifted down by 8.

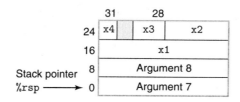

(a) Before call to proc

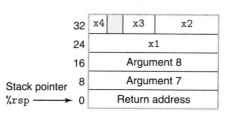

(b) During call to proc

Hence, within the code for proc, arguments 7 and 8 are accessed by offsets of 8 and 16 from the stack pointer.

Observe how call_proc changed the stack pointer only once during its execution. GCC determined that 32 bytes would suffice for holding all local variables and for holding the additional arguments to proc. Minimizing the amount of movement by the stack pointer simplifies the compiler's task of generating reference to stack elements using offsets from the stack pointer.

Register Saving Conventions

We saw in IA32 (Section 7.3) that some registers used for holding temporary values are designated as *caller-saved*, where a function is free to overwrite their values, while others are *callee-saved*, where a function must save their values on the stack before writing to them. With x86-64, the following registers are designated as being callee-saved: %rbx, %rbp, and %r12–%r15.

> **Aside**   Are there any caller-saved temporary registers?
>
> Of the 16 general-purpose registers, we've seen that 6 are designated for passing arguments, 6 are for callee-saved temporaries, 1 (%rax) holds the return value for a function, and 1 (%rsp) serves as the stack pointer. Only %r10 and %r11 are left as caller-saved temporary registers. Of course, an argument register can be used when there are fewer than six arguments or when the function is done using that argument, and %rax can be used multiple times before the final result is generated.

We illustrate the use of callee-saved registers with a somewhat unusual version of a recursive factorial function:

```c
/* Compute x! and store at resultp */
void sfact_helper(long int x, long int *resultp)
{
    if (x <= 1)
        *resultp = 1;
    else {
        long int nresult;
        sfact_helper(x-1, &nresult);
        *resultp = x * nresult;
    }
}
```

To compute the factorial of a value x, this function would be called at the top level as follows:

```c
long int sfact(long int x)
{
    long int result;
    sfact_helper(x, &result);
    return result;
}
```

The x86-64 code for sfact_helper is shown below.

```
      Arguments: x in %rdi, resultp in %rsi
1     sfact_helper:
2       movq    %rbx, -16(%rsp)      Save %rbx (callee save)
3       movq    %rbp, -8(%rsp)       Save %rbp (callee save)
4       subq    $40, %rsp            Allocate 40 bytes on stack
5       movq    %rdi, %rbx           Copy x to %rbx
6       movq    %rsi, %rbp           Copy resultp to %rbp
7       cmpq    $1, %rdi             Compare x:1
8       jg      .L14                 If >, goto recur
9       movq    $1, (%rsi)           Store 1 in *resultp
10      jmp     .L16                 Goto done
11    .L14:                          recur:
12      leaq    16(%rsp), %rsi       Compute &nresult as second argument
13      leaq    -1(%rdi), %rdi       Compute xm1 = x-1 as first argument
14      call    sfact_helper         Call sfact_helper(xm1, &nresult)
15      movq    %rbx, %rax           Copy x
16      imulq   16(%rsp), %rax       Compute x*nresult
17      movq    %rax, (%rbp)         Store at resultp
18    .L16:                          done:
19      movq    24(%rsp), %rbx       Restore %rbx
20      movq    32(%rsp), %rbp       Restore %rbp
21      addq    $40, %rsp            Deallocate stack
22      ret                          Return
```

Figure 42 illustrates how sfact_helper uses the stack to store the values of callee-saved registers and to hold the local variable nresult. This implementation

Figure 42

Stack frame for function sfact_helper. This function decrements the stack pointer *after* saving some of the state.

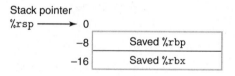

(a) Before decrementing the stack pointer

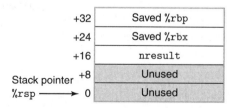

(b) After decrementing the stack pointer

has the interesting feature that the two callee-saved registers it uses (%rbx and %rbp) are saved on the stack (lines 2–3) *before* the stack pointer is decremented (line 4) to allocate the stack frame. As a consequence, the stack offset for %rbx shifts from −16 at the beginning to +24 at the end (line 19). Similarly, the offset for %rbp shifts from −8 to +32.

Being able to access memory beyond the stack pointer is an unusual feature of x86-64. It requires that the virtual memory management system allocate memory for that region. The x86-64 ABI [73] specifies that programs can use the 128 bytes beyond (i.e., at lower addresses than) the current stack pointer. The ABI refers to this area as the *red zone*. It must be kept available for reading and writing as the stack pointer moves.

Practice Problem 51

For the C program

```
long int local_array(int i)
{
    long int a[4] = {2L, 3L, 5L, 7L};
    int idx = i & 3;
    return a[idx];
}
```

GCC generates the following code:

```
    x86-64 implementation of local_array
    Argument: i in %edi
1   local_array:
2       movq    $2, -40(%rsp)
3       movq    $3, -32(%rsp)
4       movq    $5, -24(%rsp)
5       movq    $7, -16(%rsp)
6       andl    $3, %edi
7       movq    -40(%rsp,%rdi,8), %rax
8       ret
```

A. Draw a diagram indicating the stack locations used by this function and their offsets relative to the stack pointer.

B. Annotate the assembly code to describe the effect of each instruction.

C. What interesting feature does this example illustrate about the x86-64 stack discipline?

Practice Problem 52

For the recursive factorial program

```
long int rfact(long int x)
{
    if (x <= 0)
        return 1;
    else {
        long int xm1 = x-1;
        return x * rfact(xm1);
    }
}
```

GCC generates the following code:

```
     x86-64 implementation of recursive factorial function rfact
     Argument x in %rdi
1    rfact:
2        pushq    %rbx
3        movq     %rdi, %rbx
4        movl     $1, %eax
5        testq    %rdi, %rdi
6        jle      .L11
7        leaq     -1(%rdi), %rdi
8        call     rfact
9        imulq    %rbx, %rax
10   .L11:
11       popq     %rbx
12       ret
```

A. What value does the function store in %rbx?

B. What are the purposes of the pushq (line 2) and popq (line 11) instructions?

C. Annotate the assembly code to describe the effect of each instruction.

D. How does this function manage the stack frame differently from others we have seen?

13.5 Data Structures

Data structures follow the same principles in x86-64 as they do in IA32: arrays are allocated as sequences of identically sized blocks holding the array elements, structures are allocated as sequences of variably sized blocks holding the structure elements, and unions are allocated as a single block big enough to hold the largest union element.

One difference is that x86-64 follows a more stringent set of alignment requirements. For any scalar data type requiring K bytes, its starting address must be a multiple of K. Thus, data types `long` and `double` as well as pointers, must be aligned on 8-byte boundaries. In addition, data type `long double` uses a 16-byte alignment (and size allocation), even though the actual representation requires only 10 bytes. These alignment conditions are imposed to improve memory system performance—the memory interface is designed in most processors to read or write aligned blocks that are 8 or 16 bytes long.

Practice Problem 53

For each of the following structure declarations, determine the offset of each field, the total size of the structure, and its alignment requirement under x86-64.

A. `struct P1 { int i; char c; long j; char d; };`

B. `struct P2 { long i; char c; char d; int j; };`

C. `struct P3 { short w[3]; char c[3] };`

D. `struct P4 { short w[3]; char *c[3] };`

E. `struct P3 { struct P1 a[2]; struct P2 *p };`

13.6 Concluding Observations about x86-64

Both AMD and the authors of GCC deserve credit for moving x86 processors into a new era. The formulation of both the x86-64 hardware and the programming conventions changed the processor from one that relied heavily on the stack to hold program state to one where the most heavily used part of the state is held in the much faster and expanded register set. Finally, x86 has caught up to ideas developed for RISC processors in the early 1980s!

Processors capable of running either IA32 or x86-64 code are becoming commonplace. Many current desktop and laptop systems are still running 32-bit versions of their operating systems, and these machines are restricted to running only 32-bit applications, as well. Machines running 64-bit operating systems, and therefore capable of running both 32- and 64-bit applications, have become the widespread choice for high-end machines, such as for database servers and scientific computing. The biggest drawback in transforming applications from 32 bits to 64 bits is that the pointer variables double in size, and since many data structures contain pointers, this means that the overall memory requirement can nearly double. The transition from 32- to 64-bit applications has only occurred for ones having memory needs that exceed the 4-gigabyte address space limitation of IA32. History has shown that applications grow to use all available processing power and memory size, and so we can reliably predict that 64-bit processors running 64-bit operating systems and applications will become increasingly more commonplace.

14 Machine-Level Representations of Floating-Point Programs

Thus far, we have only considered programs that represent and operate on integer data types. In order to implement programs that make use of floating-point data, we must have some method of storing floating-point data and additional instructions to operate on floating-point values, to convert between floating-point and integer values, and to perform comparisons between floating-point values. We also require conventions on how to pass floating-point values as function arguments and to return them as function results. We call this combination of storage model, instructions, and conventions the *floating-point architecture* for a machine.

Due to its long evolutionary heritage, x86 processors provide multiple floating-point architectures, of which two are in current use. The first, referred to as "x87," dates back to the earliest days of Intel microprocessors and until recently was the standard implementation. The second, referred to as "SSE," is based on recent additions to x86 processors to support multimedia applications.

Web Aside ASM:X87   The x87 floating-point architecture

The historical x87 floating-point architecture is one of the least elegant features of the x87 architecture. In the original Intel machines, floating point was performed by a separate *coprocessor*, a unit with its own registers and processing capabilities that executes a subset of the instructions. This coprocessor was implemented as a separate chip, named the 8087, 80287, and i387, to accompany the processor chips 8086, 80286, and i386, respectively, and hence the colloquial name "x87." All x86 processors support the x87 architecture, and so this continues to be a possible target for compiling floating-point code.

x87 instructions operate on a shallow stack of floating-point registers. In a stack model, some instructions read values from memory and push them onto the stack; others pop operands from the stack, perform an operation, and then push the result; while others pop values from the stack and store them to memory. This approach has the advantage that there is a simple algorithm by which a compiler can map the evaluation of arithmetic expressions into stack code.

Modern compilers can make many optimizations that do not fit well within a stack model, for example, making use of a single computed result multiple times. Consequently, the x87 architecture implements an odd hybrid between a stack and a register model, where the different elements of the stack can be read and written explicitly, as well as shifted up and down by pushing and popping. In addition, the x87 stack is limited to a depth of eight values; when additional values are pushed, the ones at the bottom are simply discarded. Hence, the compiler must keep track of the stack depth. Furthermore, a compiler must treat all floating-point registers as being caller-save, since their values might disappear off the bottom if other procedures push more values onto the stack.

Web Aside ASM:SSE   The SSE floating-point architecture

Starting with the Pentium 4, the SSE2 instruction set, added to support multimedia applications, becomes a viable floating-point architecture for compiled C code. Unlike the stack-based architecture of x87, SSE-based floating point uses a straightforward register-based approach, a much better target for optimizing compilers. With SSE2, floating-point code is similar to integer code, except that it uses a different set of registers and instructions. When compiling for x86-64, GCC generates SSE code. On the other hand, its default is to generate x87 code for IA32, but it can be directed to generate SSE code by a suitable setting of the command-line parameters.

15 Summary

In this chapter, we have peered beneath the layer of abstraction provided by the C language to get a view of machine-level programming. By having the compiler generate an assembly-code representation of the machine-level program, we gain insights into both the compiler and its optimization capabilities, along with the machine, its data types, and its instruction set. Knowing the characteristics of a compiler can help when trying to write programs that have efficient mappings onto the machine. We have also gotten a more complete picture of how the program stores data in different memory regions. Application programmers often need to know whether a program variable is on the run-time stack, in some dynamically allocated data structure, or part of the global program data. Understanding how programs map onto machines makes it easier to understand the differences between these kinds of storage.

Machine-level programs, and their representation by assembly code, differ in many ways from C programs. There is minimal distinction between different data types. The program is expressed as a sequence of instructions, each of which performs a single operation. Parts of the program state, such as registers and the run-time stack, are directly visible to the programmer. Only low-level operations are provided to support data manipulation and program control. The compiler must use multiple instructions to generate and operate on different data structures and to implement control constructs such as conditionals, loops, and procedures. We have covered many different aspects of C and how it gets compiled. We have seen that the lack of bounds checking in C makes many programs prone to buffer overflows. This has made many systems vulnerable to attacks by malicious intruders, although recent safeguards provided by the run-time system and the compiler help make programs more secure.

We have only examined the mapping of C onto IA32 and x86-64, but much of what we have covered is handled in a similar way for other combinations of language and machine. For example, compiling C++ is very similar to compiling C. In fact, early implementations of C++ first performed a source-to-source conversion from C++ to C and generated object-code by running a C compiler on the result. C++ objects are represented by structures, similar to a C struct. Methods are represented by pointers to the code implementing the methods. By contrast, Java is implemented in an entirely different fashion. The object code of Java is a special binary representation known as *Java byte code*. This code can be viewed as a machine-level program for a *virtual machine*. As its name suggests, this machine is not implemented directly in hardware. Instead, software interpreters process the byte code, simulating the behavior of the virtual machine. Alternatively, an approach known as *just-in-time compilation* dynamically translates byte code sequences into machine instructions. This approach provides faster execution when code is executed multiple times, such as in loops. The advantage of using byte code as the low-level representation of a program is that the same code can be "executed" on many different machines, whereas the machine code we have considered runs only on x86 machines.

Bibliographic Notes

Both Intel and AMD provide extensive documentation on their processors. This includes general descriptions of an assembly-language programmer's view of the hardware [2, 27], as well as detailed references about the individual instructions [3, 28, 29]. Reading the instruction descriptions is complicated by the facts that (1) all documentation is based on the Intel assembly-code format, (2) there are many variations for each instruction due to the different addressing and execution modes, and (3) there are no illustrative examples. Still, these remain the authoritative references about the behavior of each instruction.

The organization amd64.org has been responsible for defining the *Application Binary Interface* (ABI) for x86-64 code running on Linux systems [73]. This interface describes details for procedure linkages, binary code files, and a number of other features that are required for machine-code programs to execute properly.

As we have discussed, the ATT format used by GCC is very different from the Intel format used in Intel documentation and by other compilers (including the Microsoft compilers). Blum's book [9] is one of the few references based on ATT format, and it provides an extensive description of how to embed assembly code into C programs using the asm directive.

Muchnick's book on compiler design [76] is considered the most comprehensive reference on code-optimization techniques. It covers many of the techniques we discuss here, such as register usage conventions and the advantages of generating code for loops based on their do-while form.

Much has been written about the use of buffer overflow to attack systems over the Internet. Detailed analyses of the 1988 Internet worm have been published by Spafford [102] as well as by members of the team at MIT who helped stop its spread [40]. Since then a number of papers and projects have generated ways both to create and to prevent buffer overflow attacks. Seacord's book [94] provides a wealth of information about buffer overflow and other attacks on code generated by C compilers.

Homework Problems

54 ◆
A function with prototype

```
int decode2(int x, int y, int z);
```

is compiled into IA32 assembly code. The body of the code is as follows:

```
      x at %ebp+8, y at %ebp+12, z at %ebp+16
1     movl    12(%ebp), %edx
2     subl    16(%ebp), %edx
3     movl    %edx, %eax
4     sall    $31, %eax
5     sarl    $31, %eax
6     imull   8(%ebp), %edx
7     xorl    %edx, %eax
```

Parameters x, y, and z are stored at memory locations with offsets 8, 12, and 16 relative to the address in register %ebp. The code stores the return value in register %eax.

Write C code for decode2 that will have an effect equivalent to our assembly code.

55 ♦

The following code computes the product of x and y and stores the result in memory. Data type ll_t is defined to be equivalent to long long.

```
typedef long long ll_t;

void store_prod(ll_t *dest, int x, ll_t y) {
    *dest = x*y;
}
```

GCC generates the following assembly code implementing the computation:

```
      dest at %ebp+8, x at %ebp+12, y at %ebp+16
1     movl    16(%ebp), %esi
2     movl    12(%ebp), %eax
3     movl    %eax, %edx
4     sarl    $31, %edx
5     movl    20(%ebp), %ecx
6     imull   %eax, %ecx
7     movl    %edx, %ebx
8     imull   %esi, %ebx
9     addl    %ebx, %ecx
10    mull    %esi
11    leal    (%ecx,%edx), %edx
12    movl    8(%ebp), %ecx
13    movl    %eax, (%ecx)
14    movl    %edx, 4(%ecx)
```

This code uses three multiplications to implement the multiprecision arithmetic required to implement 64-bit arithmetic on a 32-bit machine. Describe the algorithm used to compute the product, and annotate the assembly code to show how it realizes your algorithm. **Hint:** See Problem 12 and its solution.

56 ◆◆

Consider the following assembly code:

```
    x at %ebp+8, n at %ebp+12
1       movl    8(%ebp), %esi
2       movl    12(%ebp), %ebx
3       movl    $-1, %edi
4       movl    $1, %edx
5   .L2:
6       movl    %edx, %eax
7       andl    %esi, %eax
8       xorl    %eax, %edi
9       movl    %ebx, %ecx
10      sall    %cl, %edx
11      testl   %edx, %edx
12      jne     .L2
13      movl    %edi, %eax
```

The preceding code was generated by compiling C code that had the following overall form:

```
1   int loop(int x, int n)
2   {
3       int result = _____;
4       int mask;
5       for (mask = _____; mask _____; mask = _____) {
6           result = _____;
7       }
8       return result;
9   }
```

Your task is to fill in the missing parts of the C code to get a program equivalent to the generated assembly code. Recall that the result of the function is returned in register %eax. You will find it helpful to examine the assembly code before, during, and after the loop to form a consistent mapping between the registers and the program variables.

A. Which registers hold program values x, n, result, and mask?

B. What are the initial values of result and mask?

C. What is the test condition for mask?

D. How does mask get updated?

E. How does result get updated?

F. Fill in all the missing parts of the C code.

57 ◆◆

In Section 6.6, we examined the following code as a candidate for the use of conditional data transfer:

```
int cread(int *xp) {
    return (xp ? *xp : 0);
}
```

We showed a trial implementation using a conditional move instruction but argued that it was not valid, since it could attempt to read from a null address.

Write a C function `cread_alt` that has the same behavior as `cread`, except that it can be compiled to use conditional data transfer. When compiled with the command-line option '`-march=i686`', the generated code should use a conditional move instruction rather than one of the jump instructions.

58 ◆◆

The code that follows shows an example of branching on an enumerated type value in a switch statement. Recall that enumerated types in C are simply a way to introduce a set of names having associated integer values. By default, the values assigned to the names go from zero upward. In our code, the actions associated with the different case labels have been omitted.

```
/* Enumerated type creates set of constants numbered 0 and upward */
typedef enum {MODE_A, MODE_B, MODE_C, MODE_D, MODE_E} mode_t;

int switch3(int *p1, int *p2, mode_t action)
{
    int result = 0;
    switch(action) {
    case MODE_A:

    case MODE_B:

    case MODE_C:

    case MODE_D:

    case MODE_E:

    default:

    }
    return result;
}
```

The part of the generated assembly code implementing the different actions is shown in Figure 43. The annotations indicate the argument locations, the register values, and the case labels for the different jump destinations. Register `%edx` corresponds to program variable `result` and is initialized to −1.

Fill in the missing parts of the C code. Watch out for cases that fall through.

```
      Arguments: p1 at %ebp+8, p2 at %ebp+12, action at %ebp+16
      Registers: result in %edx (initialized to -1)
      The jump targets:
1     .L17:                          MODE_E
2        movl    $17, %edx
3        jmp     .L19
4     .L13:                          MODE_A
5        movl    8(%ebp), %eax
6        movl    (%eax), %edx
7        movl    12(%ebp), %ecx
8        movl    (%ecx), %eax
9        movl    8(%ebp), %ecx
10       movl    %eax, (%ecx)
11       jmp     .L19
12    .L14:                          MODE_B
13       movl    12(%ebp), %edx
14       movl    (%edx), %eax
15       movl    %eax, %edx
16       movl    8(%ebp), %ecx
17       addl    (%ecx), %edx
18       movl    12(%ebp), %eax
19       movl    %edx, (%eax)
20       jmp     .L19
21    .L15:                          MODE_C
22       movl    12(%ebp), %edx
23       movl    $15, (%edx)
24       movl    8(%ebp), %ecx
25       movl    (%ecx), %edx
26       jmp     .L19
27    .L16:                          MODE_D
28       movl    8(%ebp), %edx
29       movl    (%edx), %eax
30       movl    12(%ebp), %ecx
31       movl    %eax, (%ecx)
32       movl    $17, %edx
33    .L19:                          default
34       movl    %edx, %eax          Set return value
```

Figure 43 **Assembly code for Problem 58.** This code implements the different branches of a switch statement.

59 ◆◆

This problem will give you a chance to reverse engineer a switch statement from machine code. In the following procedure, the body of the switch statement has been removed:

```
1   int switch_prob(int x, int n)
2   {
3       int result = x;
4
5       switch(n) {
6
7           /* Fill in code here */
8       }
9
10      return result;
11  }
```

Figure 44 shows the disassembled machine code for the procedure. We can see in lines 4 and 5 that parameters x and n are loaded into registers %eax and %edx, respectively.

The jump table resides in a different area of memory. We can see from the indirect jump on line 9 that the jump table begins at address 0x80485d0. Using the GDB debugger, we can examine the six 4-byte words of memory comprising the jump table with the command x/6w 0x80485d0. GDB prints the following:

```
(gdb) x/6w 0x80485d0
0x80485d0:  0x08048438  0x08048448  0x08048438  0x0804843d
0x80485e0:  0x08048442  0x08048445
```

Fill in the body of the switch statement with C code that will have the same behavior as the machine code.

```
1   08048420 <switch_prob>:
2   8048420:    55                      push    %ebp
3   8048421:    89 e5                   mov     %esp,%ebp
4   8048423:    8b 45 08                mov     0x8(%ebp),%eax
5   8048426:    8b 55 0c                mov     0xc(%ebp),%edx
6   8048429:    83 ea 32                sub     $0x32,%edx
7   804842c:    83 fa 05                cmp     $0x5,%edx
8   804842f:    77 17                   ja      8048448 <switch_prob+0x28>
9   8048431:    ff 24 95 d0 85 04 08    jmp     *0x80485d0(,%edx,4)
10  8048438:    c1 e0 02                shl     $0x2,%eax
11  804843b:    eb 0e                   jmp     804844b <switch_prob+0x2b>
12  804843d:    c1 f8 02                sar     $0x2,%eax
13  8048440:    eb 09                   jmp     804844b <switch_prob+0x2b>
14  8048442:    8d 04 40                lea     (%eax,%eax,2),%eax
15  8048445:    0f af c0                imul    %eax,%eax
16  8048448:    83 c0 0a                add     $0xa,%eax
17  804844b:    5d                      pop     %ebp
18  804844c:    c3                      ret
```

Figure 44 **Disassembled code for Problem 59.**

60 ◆◆◆

Consider the following source code, where R, S, and T are constants declared with #define:

```
int A[R][S][T];

int store_ele(int i, int j, int k, int *dest)
{
    *dest = A[i][j][k];
    return sizeof(A);
}
```

In compiling this program, GCC generates the following assembly code:

```
i at %ebp+8, j at %ebp+12, k at %ebp+16, dest at %ebp+20
1    movl    12(%ebp), %edx
2    leal    (%edx,%edx,4), %eax
3    leal    (%edx,%eax,2), %eax
4    imull   $99, 8(%ebp), %edx
5    addl    %edx, %eax
6    addl    16(%ebp), %eax
7    movl    A(,%eax,4), %edx
8    movl    20(%ebp), %eax
9    movl    %edx, (%eax)
10   movl    $1980, %eax
```

A. Extend Equation 1 from two dimensions to three to provide a formula for the location of array element A[i][j][k].

B. Use your reverse engineering skills to determine the values of R, S, and T based on the assembly code.

61 ◆◆

The code generated by the C compiler for var_prod_ele (Figure 29) cannot fit all of the values it uses in the loop in registers, and so it must retrieve the value of n from memory on each iteration. Write C code for this function that incorporates optimizations similar to those performed by GCC, but such that the compiled code does not spill any loop values into memory.

Recall that the processor only has six registers available to hold temporary data, since registers %ebp and %esp cannot be used for this purpose. One of these registers must be used to hold the result of the multiply instruction. Hence, you must reduce the number of values in the loop from six (result, Arow, Bcol, j, n, and 4*n) to five.

You will need to find a strategy that works for your particular compiler. Keep trying different strategies until you find one that works.

62 ◆◆

The following code transposes the elements of an $M \times M$ array, where M is a constant defined by #define:

```
void transpose(int A[M][M]) {
    int i, j;
    for (i = 0; i < M; i++)
        for (j = 0; j < i; j++) {
            int t = A[i][j];
            A[i][j] = A[j][i];
            A[j][i] = t;
        }
}
```

When compiled with optimization level –O2, GCC generates the following code for the inner loop of the function:

```
1    .L3:
2        movl    (%ebx), %eax
3        movl    (%esi,%ecx,4), %edx
4        movl    %eax, (%esi,%ecx,4)
5        addl    $1, %ecx
6        movl    %edx, (%ebx)
7        addl    $52, %ebx
8        cmpl    %edi, %ecx
9        jl      .L3
```

A. What is the value of M?

B. What registers hold program values i and j?

C. Write a C code version of transpose that makes use of the optimizations that occur in this loop. Use the parameter M in your code rather than numeric constants.

63 ◆◆

Consider the following source code, where E1 and E2 are macro expressions declared with #define that compute the dimensions of array A in terms of parameter n. This code computes the sum of the elements of column j of the array.

```
1    int sum_col(int n, int A[E1(n)][E2(n)], int j) {
2        int i;
3        int result = 0;
4        for (i = 0; i < E1(n); i++)
5            result += A[i][j];
6        return result;
7    }
```

In compiling this program, GCC generates the following assembly code:

```
    n at %ebp+8, A at %ebp+12, j at %ebp+16
1    movl    8(%ebp), %eax
2    leal    (%eax,%eax), %edx
```

```
3       leal    (%edx,%eax), %ecx
4       movl    %edx, %ebx
5       leal    1(%edx), %eax
6       movl    $0, %edx
7       testl   %eax, %eax
8       jle     .L3
9       leal    0(,%ecx,4), %esi
10      movl    16(%ebp), %edx
11      movl    12(%ebp), %ecx
12      leal    (%ecx,%edx,4), %eax
13      movl    $0, %edx
14      movl    $1, %ecx
15      addl    $2, %ebx
16    .L4:
17      addl    (%eax), %edx
18      addl    $1, %ecx
19      addl    %esi, %eax
20      cmpl    %ebx, %ecx
21      jne     .L4
22    .L3:
23      movl    %edx, %eax
```

Use your reverse engineering skills to determine the definitions of E1 and E2.

64 ◆◆

For this exercise, we will examine the code generated by GCC for functions that have structures as arguments and return values, and from this see how these language features are typically implemented.

The following C code has a function word_sum having structures as argument and return values, and a function prod that calls word_sum:

```
typedef struct {
    int a;
    int *p;
} str1;

typedef struct {
    int sum;
    int diff;
} str2;

str2 word_sum(str1 s1) {
    str2 result;
    result.sum  = s1.a + *s1.p;
    result.diff = s1.a - *s1.p;
```

```
    return result;
}

int prod(int x, int y)
{
    str1 s1;
    str2 s2;
    s1.a = x;
    s1.p = &y;
    s2 = word_sum(s1);
    return s2.sum * s2.diff;
}
```

GCC generates the following code for these two functions:

```
1   word_sum:
2       pushl   %ebp
3       movl    %esp, %ebp
4       pushl   %ebx
5       movl    8(%ebp), %eax
6       movl    12(%ebp), %ebx
7       movl    16(%ebp), %edx
8       movl    (%edx), %edx
9       movl    %ebx, %ecx
10      subl    %edx, %ecx
11      movl    %ecx, 4(%eax)
12      addl    %ebx, %edx
13      movl    %edx, (%eax)
14      popl    %ebx
15      popl    %ebp
16      ret     $4
```

```
1   prod:
2       pushl   %ebp
3       movl    %esp, %ebp
4       subl    $20, %esp
5       leal    12(%ebp), %edx
6       leal    -8(%ebp), %ecx
7       movl    8(%ebp), %eax
8       movl    %eax, 4(%esp)
9       movl    %edx, 8(%esp)
10      movl    %ecx, (%esp)
11      call    word_sum
12      subl    $4, %esp
13      movl    -4(%ebp), %eax
14      imull   -8(%ebp), %eax
15      leave
16      ret
```

The instruction ret $4 is like a normal return instruction, but it increments the stack pointer by 8 (4 for the return address plus 4 additional), rather than 4.

A. We can see in lines 5–7 of the code for word_sum that it appears as if three values are being retrieved from the stack, even though the function has only a single argument. Describe what these three values are.

B. We can see in line 4 of the code for prod that 20 bytes are allocated in the stack frame. These get used as five fields of 4 bytes each. Describe how each of these fields gets used.

C. How would you describe the general strategy for passing structures as arguments to a function?

D. How would you describe the general strategy for handling a structure as a return value from a function?

65 ♦♦♦

In the following code, *A* and *B* are constants defined with #define:

```
typedef struct {
    short x[A][B]; /* Unknown constants A and B */
    int y;
} str1;

typedef struct {
    char array[B];
    int t;
    short s[B];
    int u;
} str2;

void setVal(str1 *p, str2 *q) {
    int v1 = q->t;
    int v2 = q->u;
    p->y = v1+v2;
}
```

GCC generates the following code for the body of setVal:

```
1    movl    12(%ebp), %eax
2    movl    36(%eax), %edx
3    addl    12(%eax), %edx
4    movl    8(%ebp), %eax
5    movl    %edx, 92(%eax)
```

What are the values of *A* and *B*? (The solution is unique.)

66 ♦♦♦

You are charged with maintaining a large C program, and you come across the following code:

```
1    typedef struct {
2        int left;
3        a_struct a[CNT];
4        int right;
5    } b_struct;
6
7    void test(int i, b_struct *bp)
8    {
9        int n = bp->left + bp->right;
10       a_struct *ap = &bp->a[i];
11       ap->x[ap->idx] = n;
12   }
```

```
 1      00000000 <test>:
 2          0:    55                        push    %ebp
 3          1:    89 e5                     mov     %esp,%ebp
 4          3:    8b 45 08                  mov     0x8(%ebp),%eax
 5          6:    8b 4d 0c                  mov     0xc(%ebp),%ecx
 6          9:    8d 04 80                  lea     (%eax,%eax,4),%eax
 7          c:    03 44 81 04               add     0x4(%ecx,%eax,4),%eax
 8         10:    8b 91 b8 00 00 00         mov     0xb8(%ecx),%edx
 9         16:    03 11                     add     (%ecx),%edx
10         18:    89 54 81 08               mov     %edx,0x8(%ecx,%eax,4)
11         1c:    5d                        pop     %ebp
12         1d:    c3                        ret
```

Figure 45   **Disassembled code for Problem 66.**

The declarations of the compile-time constant CNT and the structure a_struct are in a file for which you do not have the necessary access privilege. Fortunately, you have a copy of the '.o' version of code, which you are able to disassemble with the OBJDUMP program, yielding the disassembly shown in Figure 45.

Using your reverse engineering skills, deduce the following.

A. The value of CNT.

B. A complete declaration of structure a_struct. Assume that the only fields in this structure are idx and x.

67 ◆◆◆

Consider the following union declaration:

```
union ele {
    struct {
        int *p;
        int y;
    } e1;
    struct {
        int x;
        union ele *next;
    } e2;
};
```

This declaration illustrates that structures can be embedded within unions.

The following procedure (with some expressions omitted) operates on a linked list having these unions as list elements:

```
void proc (union ele *up)
{
    up->_____ = *(up->_____) - up->_____;
}
```

A. What would be the offsets (in bytes) of the following fields:

e1.p: _____

e1.y: _____

e2.x: _____

e2.next: _____

B. How many total bytes would the structure require?

C. The compiler generates the following assembly code for the body of proc:

```
        up at %ebp+8
1       movl    8(%ebp), %edx
2       movl    4(%edx), %ecx
3       movl    (%ecx), %eax
4       movl    (%eax), %eax
5       subl    (%edx), %eax
6       movl    %eax, 4(%ecx)
```

On the basis of this information, fill in the missing expressions in the code for proc. **Hint:** Some union references can have ambiguous interpretations. These ambiguities get resolved as you see where the references lead. There is only one answer that does not perform any casting and does not violate any type constraints.

68 ◆

Write a function good_echo that reads a line from standard input and writes it to standard output. Your implementation should work for an input line of arbitrary length. You may use the library function fgets, but you must make sure your function works correctly even when the input line requires more space than you have allocated for your buffer. Your code should also check for error conditions and return when one is encountered. Refer to the definitions of the standard I/O functions for documentation [48, 58].

69 ◆

The following declaration defines a class of structures for use in constructing binary trees:

```
1    typedef struct ELE *tree_ptr;
2
3    struct ELE {
4        long val;
5        tree_ptr left;
6        tree_ptr right;
7    };
```

For a function with the following prototype:

```
long trace(tree_ptr tp);
```

GCC generates the following x86-64 code:

```
1    trace:
     tp in %rdi
2        movl    $0, %eax
3        testq   %rdi, %rdi
4        je      .L3
5    .L5:
6        movq    (%rdi), %rax
7        movq    16(%rdi), %rdi
8        testq   %rdi, %rdi
9        jne     .L5
10   .L3:
11       rep
12       ret
```

A. Generate a C version of the function, using a while loop.

B. Explain in English what this function computes.

70 ◆◆

Using the same tree data structure we saw in Problem 69, and a function with the prototype

```
long traverse(tree_ptr tp);
```

GCC generates the following x86-64 code:

```
1    traverse:
     tp in %rdi
2        movq    %rbx, -24(%rsp)
3        movq    %rbp, -16(%rsp)
4        movq    %r12, -8(%rsp)
5        subq    $24, %rsp
6        movq    %rdi, %rbp
7        movabsq $-9223372036854775808, %rax
8        testq   %rdi, %rdi
9        je      .L9
10       movq    (%rdi), %rbx
11       movq    8(%rdi), %rdi
12       call    traverse
13       movq    %rax, %r12
14       movq    16(%rbp), %rdi
15       call    traverse
```

```
16        cmpq     %rax, %r12
17        cmovge   %r12, %rax
18        cmpq     %rbx, %rax
19        cmovl    %rbx, %rax
20     .L9:
21        movq     (%rsp), %rbx
22        movq     8(%rsp), %rbp
23        movq     16(%rsp), %r12
24        addq     $24, %rsp
25        ret
```

A. Generate a C version of the function.

B. Explain in English what this function computes.

Solutions to Practice Problems

Solution to Problem 1
This exercise gives you practice with the different operand forms.

Operand	Value	Comment
%eax	0x100	Register
0x104	0xAB	Absolute address
$0x108	0x108	Immediate
(%eax)	0xFF	Address 0x100
4(%eax)	0xAB	Address 0x104
9(%eax,%edx)	0x11	Address 0x10C
260(%ecx,%edx)	0x13	Address 0x108
0xFC(,%ecx,4)	0xFF	Address 0x100
(%eax,%edx,4)	0x11	Address 0x10C

Solution to Problem 2
As we have seen, the assembly code generated by GCC includes suffixes on the instructions, while the disassembler does not. Being able to switch between these two forms is an important skill to learn. One important feature is that memory references in IA32 are always given with double-word registers, such as %eax, even if the operand is a byte or single word.

Here is the code written with suffixes:

```
1        movl     %eax, (%esp)
2        movw     (%eax), %dx
3        movb     $0xFF, %bl
4        movb     (%esp,%edx,4), %dh
5        pushl    $0xFF
6        movw     %dx, (%eax)
7        popl     %edi
```

Machine-Level Representation of Programs

Solution to Problem 3

Since we will rely on GCC to generate most of our assembly code, being able to write correct assembly code is not a critical skill. Nonetheless, this exercise will help you become more familiar with the different instruction and operand types.

Here is the code with explanations of the errors:

```
1    movb $0xF, (%bl)        Cannot use %bl as address register
2    movl %ax, (%esp)        Mismatch between instruction suffix and register ID
3    movw (%eax),4(%esp)     Cannot have both source and destination be memory references
4    movb %ah,%sh            No register named %sh
5    movl %eax,$0x123        Cannot have immediate as destination
6    movl %eax,%dx           Destination operand incorrect size
7    movb %si, 8(%ebp)       Mismatch between instruction suffix and register ID
```

Solution to Problem 4

This exercise gives you more experience with the different data movement instructions and how they relate to the data types and conversion rules of C.

src_t	dest_t	Instruction
int	int	movl %eax, (%edx)
char	int	movsbl %al, (%edx)
char	unsigned	movsbl %al, (%edx)
unsigned char	int	movzbl %al, (%edx)
int	char	movb %al, (%edx)
unsigned	unsigned char	movb %al, (%edx)
unsigned	int	movl %eax, (%edx)

Solution to Problem 5

Reverse engineering is a good way to understand systems. In this case, we want to reverse the effect of the C compiler to determine what C code gave rise to this assembly code. The best way is to run a "simulation," starting with values x, y, and z at the locations designated by pointers xp, yp, and zp, respectively. We would then get the following behavior:

```
     xp at %ebp+8, yp at %ebp+12, zp at %ebp+16
1    movl    8(%ebp), %edi      Get xp
2    movl    12(%ebp), %edx     Get yp
3    movl    16(%ebp), %ecx     Get zp
4    movl    (%edx), %ebx       Get y
5    movl    (%ecx), %esi       Get z
6    movl    (%edi), %eax       Get x
7    movl    %eax, (%edx)       Store x at yp
8    movl    %ebx, (%ecx)       Store y at zp
9    movl    %esi, (%edi)       Store z at xp
```

From this, we can generate the following C code:

```
void decode1(int *xp, int *yp, int *zp)
{
    int tx = *xp;
    int ty = *yp;
    int tz = *zp;

    *yp = tx;
    *zp = ty;
    *xp = tz;
}
```

Solution to Problem 6

This exercise demonstrates the versatility of the `leal` instruction and gives you more practice in deciphering the different operand forms. Although the operand forms are classified as type "Memory" in Figure 3, no memory access occurs.

Instruction	Result
leal 6(%eax), %edx	$6 + x$
leal (%eax,%ecx), %edx	$x + y$
leal (%eax,%ecx,4), %edx	$x + 4y$
leal 7(%eax,%eax,8), %edx	$7 + 9x$
leal 0xA(,%ecx,4), %edx	$10 + 4y$
leal 9(%eax,%ecx,2), %edx	$9 + x + 2y$

Solution to Problem 7

This problem gives you a chance to test your understanding of operands and the arithmetic instructions. The instruction sequence is designed so that the result of each instruction does not affect the behavior of subsequent ones.

Instruction	Destination	Value
addl %ecx,(%eax)	0x100	0x100
subl %edx,4(%eax)	0x104	0xA8
imull $16,(%eax,%edx,4)	0x10C	0x110
incl 8(%eax)	0x108	0x14
decl %ecx	%ecx	0x0
subl %edx,%eax	%eax	0xFD

Solution to Problem 8

This exercise gives you a chance to generate a little bit of assembly code. The solution code was generated by GCC. By loading parameter n in register %ecx, it can then use byte register %cl to specify the shift amount for the `sarl` instruction:

```
1    movl    8(%ebp), %eax      Get x
2    sall    $2, %eax           x <<= 2
3    movl    12(%ebp), %ecx     Get n
4    sarl    %cl, %eax          x >>= n
```

Solution to Problem 9

This problem is fairly straightforward, since each of the expressions is implemented by a single instruction and there is no reordering of the expressions.

```
5        int t1 = xy;
6        int t2 = t1 >> 3;
7        int t3 = t2;
8        int t4 = t3-z;
```

Solution to Problem 10

A. This instruction is used to set register %edx to zero, exploiting the property that $x \wedge x = 0$ for any x. It corresponds to the C statement x = 0.

B. A more direct way of setting register %edx to zero is with the instruction movl $0,%edx.

C. Assembling and disassembling this code, however, we find that the version with xorl requires only 2 bytes, while the version with movl requires 5.

Solution to Problem 11

We can simply replace the cltd instruction with one that sets register %edx to 0, and use divl rather than idivl as our division instruction, yielding the following code:

```
x at %ebp+8, y at %ebp+12
    movl    8(%ebp),%eax       Load x into %eax
    movl    $0,%edx            Set high-order bits to 0
    divl    12(%ebp)           Unsigned divide by y
    movl    %eax, 4(%esp)      Store x / y
    movl    %edx, (%esp)       Store x % y
```

Solution to Problem 12

A. We can see that the program is performing multiprecision operations on 64-bit data. We can also see that the 64-bit multiply operation (line 4) uses unsigned arithmetic, and so we conclude that num_t is unsigned long long.

B. Let x denote the value of variable x, and let y denote the value of y, which we can write as $y = y_h \cdot 2^{32} + y_l$, where y_h and y_l are the values represented by the high- and low-order 32 bits, respectively. We can therefore compute $x \cdot y = x \cdot y_h \cdot 2^{32} + x \cdot y_l$. The full representation of the product would be 96 bits long, but we require only the low-order 64 bits. We can therefore let s be the low-order 32 bits of $x \cdot y_h$ and t be the full 64-bit product $x \cdot y_l$, which we can split into high- and low-order parts t_h and t_l. The final result has t_l as the low-order part, and $s + t_h$ as the high-order part.

Here is the annotated assembly code:

```
    dest at %ebp+8, x at %ebp+12, y at %ebp+16
1       movl    12(%ebp), %eax          Get x
2       movl    20(%ebp), %ecx          Get y_h
3       imull   %eax, %ecx              Compute s = x*y_h
4       mull    16(%ebp)                Compute t = x*y_l
5       leal    (%ecx,%edx), %edx       Add s to t_h
6       movl    8(%ebp), %ecx           Get dest
7       movl    %eax, (%ecx)            Store t_l
8       movl    %edx, 4(%ecx)           Store s+t_h
```

Solution to Problem 13

It is important to understand that assembly code does not keep track of the type of a program value. Instead, the different instructions determine the operand sizes and whether they are signed or unsigned. When mapping from instruction sequences back to C code, we must do a bit of detective work to infer the data types of the program values.

A. The suffix '1' and the register identifiers indicate 32-bit operands, while the comparison is for a two's complement '<'. We can infer that data_t must be int.

B. The suffix 'w' and the register identifiers indicate 16-bit operands, while the comparison is for a two's-complement '>='. We can infer that data_t must be short.

C. The suffix 'b' and the register identifiers indicate 8-bit operands, while the comparison is for an unsigned '<'. We can infer that data_t must be unsigned char.

D. The suffix '1' and the register identifiers indicate 32-bit operands, while the comparison is for '!=', which is the same whether the arguments are signed, unsigned, or pointers. We can infer that data_t could be either int, unsigned, or some form of pointer. For the first two cases, they could also have the long size designator.

Solution to Problem 14

This problem is similar to Problem 13, except that it involves TEST instructions rather than CMP instructions.

A. The suffix '1' and the register identifiers indicate 32-bit operands, while the comparison is for '!=', which is the same for signed or unsigned. We can infer that data_t must be either int, unsigned, or some type of pointer. For the first two cases, they could also have the long size designator.

B. The suffix 'w' and the register identifier indicate 16-bit operands, while the comparison is for '==', which is the same for signed or unsigned. We can infer that data_t must be either short or unsigned short.

C. The suffix 'b' and the register identifier indicate an 8-bit operand, while the comparison is for two's complement '>'. We can infer that data_t must be char.

D. The suffix 'w' and the register identifier indicate 16-bit operands, while the comparison is for unsigned '>'. We can infer that data_t must be unsigned short.

Solution to Problem 15

This exercise requires you to examine disassembled code in detail and reason about the encodings for jump targets. It also gives you practice in hexadecimal arithmetic.

A. The je instruction has as target 0x8048291 + 0x05. As the original disassembled code shows, this is 0x8048296:

```
804828f:        74 05                     je      8048296
8048291:        e8 1e 00 00 00            call    80482b4
```

B. The jb instruction has as target 0x8048359 − 25 (since 0xe7 is the 1-byte, two's-complement representation of −25). As the original disassembled code shows, this is 0x8048340:

```
8048357:        72 e7                     jb      8048340
8048359:        c6 05 10 a0 04 08 01      movb    $0x1,0x804a010
```

C. According to the annotation produced by the disassembler, the jump target is at absolute address 0x8048391. According to the byte encoding, this must be at an address 0x12 bytes beyond that of the mov instruction. Subtracting these gives address 0x804837f, as confirmed by the disassembled code:

```
804837d:        74 12                     je      8048391
804837f:        b8 00 00 00 00            mov     $0x0,%eax
```

D. Reading the bytes in reverse order, we see that the target offset is 0xffffffe0, or decimal −32. Adding this to 0x80482c4 (the address of the nop instruction) gives address 0x80482a4:

```
80482bf:        e9 e0 ff ff ff            jmp     80482a4
80482c4:        90                        nop
```

E. An indirect jump is denoted by instruction code ff 25. The address from which the jump target is to be read is encoded explicitly by the following 4 bytes. Since the machine is little endian, these are given in reverse order as fc 9f 04 08.

Solution to Problem 16

Annotating assembly code and writing C code that mimics its control flow are good first steps in understanding assembly-language programs. This problem gives you practice for an example with simple control flow. It also gives you a chance to examine the implementation of logical operations.

A. Here is the C code:

```
1   void goto_cond(int a, int *p) {
2       if (p == 0)
3           goto done;
4       if (a <= 0)
5           goto done;
6       *p += a;
7   done:
8       return;
9   }
```

B. The first conditional branch is part of the implementation of the && expression. If the test for p being non-null fails, the code will skip the test of a > 0.

Solution to Problem 17

This is an exercise to help you think about the idea of a general translation rule and how to apply it.

A. Converting to this alternate form involves only switching around a few lines of the code:

```
1   int gotodiff_alt(int x, int y) {
2       int result;
3       if (x < y)
4           goto true;
5       result = x - y;
6       goto done;
7   true:
8       result =  y - x;
9   done:
10      return result;
11  }
```

B. In most respects, the choice is arbitrary. But the original rule works better for the common case where there is no else statement. For this case, we can simply modify the translation rule to be as follows:

```
t = test-expr;
if (!t)
    goto done;
then-statement
done:
```

A translation based on the alternate rule is more cumbersome.

Solution to Problem 18

This problem requires that you work through a nested branch structure, where you will see how our rule for translating if statements has been applied. For the most part, the machine code is a straightforward translation of the C code. The only difference is that the initialization expression (line 2 in the C code) has been moved down (line 15 in the assembly code) so that it only gets computed when it is certain that this will be the returned value.

```
1    int test(int x, int y) {
2        int val = xy;
3        if (x < -3) {
4            if (y < x)
5                val = x*y;
6            else
7                val = x+y;
8        } else if (x > 2)
9            val = x-y;
10       return val;
11   }
```

Solution to Problem 19

A. If we build up a table of factorials computed with data type int, we get the following:

n	$n!$	OK?
1	1	Y
2	2	Y
3	6	Y
4	24	Y
5	120	Y
6	720	Y
7	5,040	Y
8	40,320	Y
9	362,880	Y
10	3,628,800	Y
11	39,916,800	Y
12	479,001,600	Y
13	1,932,053,504	Y
14	1,278,945,280	N

We can see that 14! has overflowed, since the numbers stopped growing. We can also test whether or not the computation of $n!$ has overflowed by computing $n!/n$ and seeing whether it equals $(n - 1)!$.

B. Doing the computation with data type long long lets us go up to 20!, yielding 2,432,902,008,176,640,000.

Solution to Problem 20

The code generated when compiling loops can be tricky to analyze, because the compiler can perform many different optimizations on loop code, and because it can be difficult to match program variables with registers. We start practicing this skill with a fairly simple loop.

A. The register usage can be determined by simply looking at how the arguments get fetched.

Register usage

Register	Variable	Initially
%eax	x	x
%ecx	y	y
%edx	n	n

B. The *body-statement* portion consists of lines 3 through 5 in the C code and lines 5 through 7 in the assembly code. The *test-expr* portion is on line 6 in the C code. In the assembly code, it is implemented by the instructions on lines 8 through 11.

C. The annotated code is as follows:

```
      x at %ebp+8, y at %ebp+12, n at %ebp+16
1        movl    8(%ebp), %eax      Get x
2        movl    12(%ebp), %ecx     Get y
3        movl    16(%ebp), %edx     Get n
4     .L2:                          loop:
5        addl    %edx, %eax         x += n
6        imull   %edx, %ecx         y *= n
7        subl    $1, %edx           n--
8        testl   %edx, %edx         Test n
9        jle     .L5                If <= 0, goto done
10       cmpl    %edx, %ecx         Compare y:n
11       jl      .L2                If <, goto loop
12    .L5:                          done:
```

As with the code of Problem 16, two conditional branches are required to implement the && operation.

Solution to Problem 21

This problem demonstrates how the transformations made by the compiler can make it difficult to decipher the generated assembly code.

A. We can see that the register is initialized to $a + b$ and then incremented on each iteration. Similarly, the value of a (held in register %ecx) is incremented on each iteration. We can therefore see that the value in register %edx will always equal $a + b$. Let us call this apb (for "a plus b").

B. Here is a table of register usage:

Register	Program value	Initial value
%ecx	a	a
%ebx	b	b
%eax	result	1
%edx	apb	$a + b$

C. The annotated code is as follows:

```
Arguments: a at %ebp+8, b at %ebp+12
Registers: a in %ecx, b in %ebx, result in %eax, %edx set to apb (a+b)
1    movl    8(%ebp), %ecx          Get a
2    movl    12(%ebp), %ebx         Get b
3    movl    $1, %eax               Set result = 1
4    cmpl    %ebx, %ecx             Compare a:b
5    jge     .L11                   If >=, goto done
6    leal    (%ebx,%ecx), %edx      Compute apb = a+b
7    movl    $1, %eax               Set result = 1
8  .L12:                            loop:
9    imull   %edx, %eax             Compute result *= apb
10   addl    $1, %ecx               Compute a++
11   addl    $1, %edx               Compute apb++
12   cmpl    %ecx, %ebx             Compare b:a
13   jg      .L12                   If >, goto loop
14 .L11:                            done:
     Return result
```

D. The equivalent goto code is as follows:

```
1    int loop_while_goto(int a, int b)
2    {
3        int result = 1;
4        if (a >= b)
5            goto done;
6        /* apb has same value as a+b in original code */
7        int apb = a+b;
8    loop:
9        result *= apb;
10       a++;
11       apb++;
12       if (b > a)
13           goto loop;
14   done:
15       return result;
16   }
```

Solution to Problem 22

Being able to work backward from assembly code to C code is a prime example of reverse engineering.

A. Here is the original C code:

```c
int fun_a(unsigned x) {
    int val = 0;
    while (x) {
        val = x;
        x >>= 1;
    }
    return val & 0x1;
}
```

B. This code computes the *parity* of argument x. That is, it returns 1 if there is an odd number of ones in x and 0 if there is an even number.

Solution to Problem 23

This problem is trickier than Problem 22, since the code within the loop is more complex and the overall operation is less familiar.

A. Here is the original C code:

```c
int fun_b(unsigned x) {
    int val = 0;
    int i;
    for (i = 0; i < 32; i++) {
        val = (val << 1) | (x & 0x1);
        x >>= 1;
    }
    return val;
}
```

B. This code reverses the bits in x, creating a mirror image. It does this by shifting the bits of x from left to right, and then filling these bits in as it shifts val from right to left.

Solution to Problem 24

Our stated rule for translating a `for` loop into a `while` loop is just a bit too simplistic—this is the only aspect that requires special consideration.

A. Applying our translation rule would yield the following code:

```c
/* Naive translation of for loop into while loop */
/* WARNING: This is buggy code */
int sum = 0;
int i = 0;
```

```
while (i < 10) {
    if (i & 1)
        /* This will cause an infinite loop */
        continue;
    sum += i;
    i++;
}
```

This code has an infinite loop, since the continue statement would prevent index variable i from being updated.

B. The general solution is to replace the continue statement with a goto statement that skips the rest of the loop body and goes directly to the update portion:

```
/* Correct translation of for loop into while loop */
int sum = 0;
int i = 0;
while (i < 10) {
    if (i & 1)
        goto update;
    sum += i;
update:
    i++;
}
```

Solution to Problem 25

This problem reinforces our method of computing the misprediction penalty.

A. We can apply our formula directly to get $T_{MP} = 2(31 - 16) = 30$.

B. When misprediction occurs, the function will require around $16 + 30 = 46$ cycles.

Solution to Problem 26

This problem provides a chance to study the use of conditional moves.

A. The operator is '/'. We see this is an example of dividing by a power of 2 by right shifting. Before shifting by $k = 2$, we must add a bias of $2^k - 1 = 3$ when the dividend is negative.

B. Here is an annotated version of the assembly code:

```
      Computation by function arith
      Register: x in %edx
1     leal    3(%edx), %eax    temp = x+3
2     testl   %edx, %edx       Test x
3     cmovns  %edx, %eax       If >= 0, temp = x
4     sarl    $2, %eax         Return temp >> 2 (= x/4)
```

The program creates a temporary value equal to $x + 3$, in anticipation of x being negative and therefore requiring biasing. The cmovns instruction conditionally changes this number to x when $x \geq 0$, and then it is shifted by 2 to generate $x/4$.

Solution to Problem 27

This problem is similar to Problem 18, except that some of the conditionals have been implemented by conditional data transfers. Although it might seem daunting to fit this code into the framework of the original C code, you will find that it follows the translation rules fairly closely.

```
1    int test(int x, int y) {
2        int val = 4*x;
3        if (y > 0) {
4            if (x < y)
5                val = x-y;
6            else
7                val = xy;
8        } else if (y < -2)
9            val = x+y;
10       return val;
11   }
```

Solution to Problem 28

This problem gives you a chance to reason about the control flow of a switch statement. Answering the questions requires you to combine information from several places in the assembly code.

1. Line 2 of the assembly code adds 2 to x to set the lower range of the cases to zero. That means that the minimum case label is -2.

2. Lines 3 and 4 cause the program to jump to the default case when the adjusted case value is greater than 6. This implies that the maximum case label is $-2 + 6 = 4$.

3. In the jump table, we see that the entry on line 3 (case value -1) has the same destination (.L2) as the jump instruction on line 4, indicating the default case behavior. Thus, case label -1 is missing in the switch statement body.

4. In the jump table, we see that the entries on lines 6 and 7 have the same destination. These correspond to case labels 2 and 3.

From this reasoning, we draw the following two conclusions:

A. The case labels in the switch statement body had values $-2, 0, 1, 2, 3$, and 4.

B. The case with destination .L6 had labels 2 and 3.

Solution to Problem 29

The key to reverse engineering compiled switch statements is to combine the information from the assembly code and the jump table to sort out the different cases. We can see from the ja instruction (line 3) that the code for the default case has label .L2. We can see that the only other repeated label in the jump table is .L4, and so this must be the code for the cases C and D. We can see that the code falls through at line 14, and so label .L6 must match case A and label .L3 must match case B. That leaves only label .L2 to match case E.

The original C code is as follows. Observe how the compiler optimized the case where a equals 4 by setting the return value to be 4, rather than a.

```
1    int switcher(int a, int b, int c)
2    {
3        int answer;
4        switch(a) {
6        case 5:
7            c = b  15;
8            /* Fall through */
9        case 0:
10           answer = c + 112;
11           break;
12       case 2:
13       case 7:
14           answer = (c + b) << 2;
15           break;
16       case 4:
17           answer = a;   /* equivalently, answer = 4 */
18           break;
19       default:
20           answer = b;
21       }
22       return answer;
23   }
```

Solution to Problem 30

This is another example of an assembly-code idiom. At first it seems quite peculiar—a call instruction with no matching ret. Then we realize that it is not really a procedure call after all.

  A. %eax is set to the address of the popl instruction.

  B. This is not a true procedure call, since the control follows the same ordering as the instructions and the return address is popped from the stack.

  C. This is the only way in IA32 to get the value of the program counter into an integer register.

Solution to Problem 31

This problem makes concrete the discussion of register usage conventions. Registers %edi, %esi, and %ebx are callee-save. The procedure must save them on the stack before altering their values and restore them before returning. The other three registers are caller-save. They can be altered without affecting the behavior of the caller.

Solution to Problem 32

One step in learning to read IA32 code is to become very familiar with the way arguments are passed on the stack. The key to solving this problem is to note that the storage of d at p is implemented by the instruction at line 3 of the assembly code, from which you work backward to determine the types and positions of arguments d and p. Similarly, the subtraction is performed at line 6, and from this you can work backward to determine the types and positions of arguments x and c.

The following is the function prototype:

```
int fun(short c, char d, int *p, int x);
```

As this example shows, reverse engineering is like solving a puzzle. It's important to identify the points where there is a unique choice, and then work around these points to fill in the rest of the details.

Solution to Problem 33

Being able to reason about how functions use the stack is a critical part of understanding compiler-generated code. As this example illustrates, the compiler may allocate a significant amount of space that never gets used.

A. We started with %esp having value 0x800040. The pushl instruction on line 2 decrements the stack pointer by 4, giving 0x80003C, and this becomes the new value of %ebp.

B. Line 4 decrements the stack pointer by 40 (hex 0x28), yielding 0x800014.

C. We can see how the two leal instructions (lines 5 and 7) compute the arguments to pass to scanf, while the two movl instructions (lines 6 and 8) store them on the stack. Since the function arguments appear on the stack at increasingly positive offsets from %esp, we can conclude that line 5 computes &x, while line 7 computes line &y. These have values 0x800038 and 0x800034, respectively.

D. The stack frame has the following structure and contents:

0x80003C	0x800060	← %ebp
0x800038	0x53	x
0x800034	0x46	y
0x800030		
0x80002C		
0x800028		
0x800024		
0x800020		
0x80001C	0x800038	
0x800018	0x800034	
0x800014	0x300070	← %esp

E. Byte addresses 0x800020 through 0x800033 are unused.

Solution to Problem 34

This problem provides a chance to examine the code for a recursive function. An important lesson to learn is that recursive code has the exact same structure as the other functions we have seen. The stack and register-saving disciplines suffice to make recursive functions operate correctly.

A. Register %ebx holds the value of parameter x, so that it can be used to compute the result expression.

B. The assembly code was generated from the following C code:

```
int rfun(unsigned x) {
    if (x == 0)
        return 0;
    unsigned nx = x>>1;
    int rv = rfun(nx);
    return (x & 0x1) + rv;
}
```

C. Like the code of Problem 49, this function computes the sum of the bits in argument x. It recursively computes the sum of all but the least significant bit, and then it adds the least significant bit to get the result.

Solution to Problem 35

This exercise tests your understanding of data sizes and array indexing. Observe that a pointer of any kind is 4 bytes long. For IA32, GCC allocates 12 bytes for data type long double, even though the actual format requires only 10 bytes.

Array	Element size	Total size	Start address	Element i
S	2	14	x_S	$x_S + 2i$
T	4	12	x_T	$x_T + 4i$
U	4	24	x_U	$x_U + 4i$
V	12	96	x_V	$x_V + 12i$
W	4	16	x_W	$x_W + 4i$

Solution to Problem 36

This problem is a variant of the one shown for integer array E. It is important to understand the difference between a pointer and the object being pointed to. Since data type short requires 2 bytes, all of the array indices are scaled by a factor of 2. Rather than using movl, as before, we now use movw.

Expression	Type	Value	Assembly
S+1	short *	$x_S + 2$	leal 2(%edx),%eax
S[3]	short	$M[x_S + 6]$	movw 6(%edx),%ax
&S[i]	short *	$x_S + 2i$	leal (%edx,%ecx,2),%eax
S[4*i+1]	short	$M[x_S + 8i + 2]$	movw 2(%edx,%ecx,8),%ax
S+i-5	short *	$x_S + 2i - 10$	leal -10(%edx,%ecx,2),%eax

Machine-Level Representation of Programs

Solution to Problem 37

This problem requires you to work through the scaling operations to determine the address computations, and to apply Equation 1 for row-major indexing. The first step is to annotate the assembly code to determine how the address references are computed:

```
1    movl    8(%ebp), %ecx              Get i
2    movl    12(%ebp), %edx             Get j
3    leal    0(,%ecx,8), %eax           8*i
4    subl    %ecx, %eax                 8*i-i = 7*i
5    addl    %edx, %eax                 7*i+j
6    leal    (%edx,%edx,4), %edx        5*j
7    addl    %ecx, %edx                 5*j+i
8    movl    mat1(,%eax,4), %eax        mat1[7*i+j]
9    addl    mat2(,%edx,4), %eax        mat2[5*j+i]
```

We can see that the reference to matrix mat1 is at byte offset $4(7i + j)$, while the reference to matrix mat2 is at byte offset $4(5j + i)$. From this, we can determine that mat1 has 7 columns, while mat2 has 5, giving $M = 5$ and $N = 7$.

Solution to Problem 38

This exercise requires that you be able to study compiler-generated assembly code to understand what optimizations have been performed. In this case, the compiler was clever in its optimizations.

Let us first study the following C code, and then see how it is derived from the assembly code generated for the original function.

```
1    /* Set all diagonal elements to val */
2    void fix_set_diag_opt(fix_matrix A, int val) {
3      int *Abase = &A[0][0];
4      int index = 0;
5      do {
6        Abase[index] = val;
7        index += (N+1);
8      } while (index != (N+1)*N);
9    }
```

This function introduces a variable Abase, of type int *, pointing to the start of array A. This pointer designates a sequence of 4-byte integers consisting of elements of A in row-major order. We introduce an integer variable index that steps through the diagonal elements of A, with the property that diagonal elements i and $i + 1$ are spaced $N + 1$ elements apart in the sequence, and that once we reach diagonal element N (index value $N(N + 1)$), we have gone beyond the end.

The actual assembly code follows this general form, but now the pointer increments must be scaled by a factor of 4. We label register %eax as holding a value `index4` equal to `index` in our C version, but scaled by a factor of 4. For $N = 16$, we can see that our stopping point for `index4` will be $4 \cdot 16(16 + 1) = 1088$.

```
    A at %ebp+8, val at %ebp+12
1       movl    8(%ebp), %ecx         Get Abase = &A[0][0]
2       movl    12(%ebp), %edx        Get val
3       movl    $0, %eax              Set index4 to 0
4   .L14:                             loop:
5       movl    %edx, (%ecx,%eax)     Set Abase[index4/4] to val
6       addl    $68, %eax             index4 += 4(N+1)
7       cmpl    $1088, %eax           Compare index4:4N(N+1)
8       jne     .L14                  If !=, goto loop
```

Solution to Problem 39

This problem gets you to think about structure layout and the code used to access structure fields. The structure declaration is a variant of the example shown in the text. It shows that nested structures are allocated by embedding the inner structures within the outer ones.

A. The layout of the structure is as follows:

Offset	0	4	8	12	16
Contents	p	s.x	s.y	next	

B. It uses 16 bytes.

C. As always, we start by annotating the assembly code:

```
    sp at %ebp+8
1       movl    8(%ebp), %eax        Get sp
2       movl    8(%eax), %edx        Get sp->s.y
3       movl    %edx, 4(%eax)        Store in sp->s.x
4       leal    4(%eax), %edx        Compute &(sp->s.x)
5       movl    %edx, (%eax)         Store in sp->p
6       movl    %eax, 12(%eax)       Store sp in sp->next
```

From this, we can generate C code as follows:

```
void sp_init(struct prob *sp)
{
    sp->s.x   = sp->s.y;
    sp->p     = &(sp->s.x);
    sp->next  = sp;
}
```

Solution to Problem 40

Structures and unions involve a simple set of concepts, but it takes practice to be comfortable with the different referencing patterns and their implementations.

EXPR	TYPE	Code
up->t1.s	int	movl 4(%eax), %eax movl %eax, (%edx)
up->t1.v	short	movw (%eax), %ax movw %ax, (%edx)
&up->t1.d	short *	leal 2(%eax), %eax movl %eax, (%edx)
up->t2.a	int *	movl %eax, (%edx)

up->t2.a[up->t1.s]	int	movl 4(%eax), %ecx
		movl (%eax,%ecx,4), %eax
		movl %eax, (%edx)

*up->t2.p	char	movl 8(%eax), %eax
		movb (%eax), %al
		movb %al, (%edx)

Solution to Problem 41

Understanding structure layout and alignment is very important for understanding how much storage different data structures require and for understanding the code generated by the compiler for accessing structures. This problem lets you work out the details of some example structures.

A. struct P1 { int i; char c; int j; char d; };

i	c	j	d	Total	Alignment
0	4	8	12	16	4

B. struct P2 { int i; char c; char d; int j; };

i	c	j	d	Total	Alignment
0	4	5	8	12	4

C. struct P3 { short w[3]; char c[3] };

w	c	Total	Alignment
0	6	10	2

D. struct P4 { short w[3]; char *c[3] };

w	c	Total	Alignment
0	8	20	4

E. struct P3 { struct P1 a[2]; struct P2 *p };

a	p	Total	Alignment
0	32	36	4

Solution to Problem 42

This is an exercise in understanding structure layout and alignment.

A. Here are the object sizes and byte offsets:

Field	a	b	c	d	e	f	g	h
Size	4	2	8	1	4	1	8	4
Offset	0	4	8	16	20	24	32	40

B. The structure is a total of 48 bytes long. The end of the structure must be padded by 4 bytes to satisfy the 8-byte alignment requirement.

C. One strategy that works, when all data elements have a length equal to a power of two, is to order the structure elements in descending order of size. This leads to a declaration,

```
struct {
    double     c;
    long long g;
    float      e;
    char      *a;
    void      *h;
    short      b;
    char       d;
    char       f;
} foo;
```

with the following offsets, for a total of 32 bytes:

Field	c	g	e	a	h	b	d	f
Size	8	8	4	4	4	2	1	1
Offset	0	8	16	20	24	28	30	31

Solution to Problem 43

This problem covers a wide range of topics, such as stack frames, string representations, ASCII code, and byte ordering. It demonstrates the dangers of out-of-bounds memory references and the basic ideas behind buffer overflow.

A. Stack after line 7:

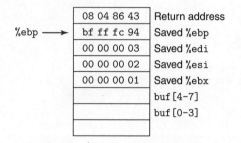

B. Stack after line 10:

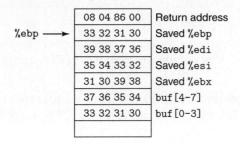

	08 04 86 00	Return address
%ebp →	33 32 31 30	Saved %ebp
	39 38 37 36	Saved %edi
	35 34 33 32	Saved %esi
	31 30 39 38	Saved %ebx
	37 36 35 34	buf [4-7]
	33 32 31 30	buf [0-3]

C. The program is attempting to return to address 0x08048600. The low-order byte was overwritten by the terminating null character.

D. The saved values of the following registers were altered:

Register	Value
%ebp	33323130
%edi	39383736
%esi	35343332
%ebx	31303938

These values will be loaded into the registers before getline returns.

E. The call to malloc should have had strlen(buf)+1 as its argument, and the code should also check that the returned value is not equal to NULL.

Solution to Problem 44

A. This corresponds to a range of around 2^{13} addresses.

B. A 128-byte nop sled would cover 2^7 addresses with each test, and so we would only require $2^6 = 64$ attempts.

This example clearly shows that the degree of randomization in this version of Linux would provide only minimal deterrence against an overflow attack.

Solution to Problem 45

This problem gives you another chance to see how IA32 code manages the stack, and to also better understand how to defend against buffer overflow attacks.

A. For the unprotected code, we can see that lines 4 and 6 compute the positions of v and buf to be at offsets −8 and −20 relative to %ebp. In the protected code, the canary is stored at offset −8 (line 3), while v and buf are at offsets −24 and −20 (lines 7 and 9).

B. In the protected code, local variable v is positioned closer to the top of the stack than buf, and so an overrun of buf will not corrupt the value of v.

In fact, buf is positioned so that any buffer overrun will corrupt the canary value.

Solution to Problem 46

Achieving a factor of 51 price improvement every 10 years over 3 decades has been truly remarkable, and it helps explain why computers have become so pervasive in our society.

A. Assuming the baseline of 16.3 gigabytes in 2010, 256 terabytes represents an increase by a factor of 1.608×10^4, which would take around 25 years, giving us 2035.

B. Sixteen exabytes is an increase of 1.054×10^9 over 16.3 gigabytes. This would take around 53 years, giving us 2063.

C. Increasing the budget by a factor of 10 cuts about 6 years off our schedule, making it possible to meet the two memory-size goals in years 2029 and 2057, respectively.

These numbers, of course, should not be taken too literally. It would require scaling memory technology well beyond what are believed to be fundamental physical limits of the current technology. Nonetheless, it indicates that, within the lifetimes of many readers of this book, there will be systems with exabyte-scale memory systems.

Solution to Problem 47

This problem illustrates some of the subtleties of type conversion and the different move instructions. In some cases, we make use of the property that the movl instruction will set the upper 32 bits of the destination register to zeros. Some of the problems have multiple solutions.

src_t	dest_t	Instruction	S	D	Explanation
long	long	movq	%rdi	%rax	No conversion
int	long	movslq	%edi	%rax	Sign extend
char	long	movsbq	%dil	%rax	Sign extend
unsigned int	unsigned long	movl	%edi	%eax	Zero extend to 64 bits
unsigned char	unsigned long	movzbq	%dil	%rax	Zero extend to 64
unsigned char	unsigned long	movzbl	%dil	%eax	Zero extend to 64 bits
long	int	movslq	%edi	%rax	Sign extend to 64 bits
long	int	movl	%edi	%eax	Zero extend to 64 bits
unsigned long	unsigned	movl	%edi	%eax	Zero extend to 64 bits

We show that the long to int conversion can use either movslq or movl, even though one will sign extend the upper 32 bits, while the other will zero extend it. This is because the values of the upper 32 bits are ignored for any subsequent instruction having %eax as an operand.

Solution to Problem 48

We can step through the code for `arithprob` and determine the following:

1. The first `movslq` instruction sign extends `d` to a long integer prior to its multiplication by `c`. This implies that `d` has type `int` and `c` has type `long`.

2. The `movsbl` instruction (line 4) sign extends `b` to an integer prior to its multiplication by `a`. This means that `b` has type `char` and `a` has type `int`.

3. The sum is computed using a `leaq` instruction, indicating that the return value has type `long`.

From this, we can determine that the unique prototype for `arithprob` is

```
long arithprob(int a, char b, long c, int d);
```

Solution to Problem 49

This problem demonstrates a clever way to count the number of 1 bits in a word. It uses several tricks that look fairly obscure at the assembly-code level.

A. Here is the original C code:

```
long fun_c(unsigned long x) {
    long val = 0;
    int i;
    for (i = 0; i < 8; i++) {
        val += x & 0x0101010101010101L;
        x >>= 1;
    }
    val += (val >> 32);
    val += (val >> 16);
    val += (val >> 8);
    return val & 0xFF;
}
```

B. This code sums the bits in `x` by computing 8 single-byte sums in parallel, using all 8 bytes of `val`. It then sums the two halves of `val`, then the two low-order 16 bits, and then the 2 low-order bytes of this sum to get the final amount in the low-order byte. It masks off the high-order bits to get the final result. This approach has the advantage that it requires only 8 iterations, rather than the more typical 64.

Solution to Problem 50

We can step through the code for `incrprob` and determine the following:

1. The `addl` instruction fetches a 32-bit integer from the location given by the third argument register and adds it to the 32-bit version of the first argument register. From this, we can infer that `t` is the third argument and `x` is the first argument. We can see that `t` must be a pointer to a signed or unsigned integer, but `x` could be either signed or unsigned, and it could either be 32 bits or 64 (since when adding it to `*t`, the code should truncate it to 32 bits).

2. The `movslq` instruction sign extends the sum (a copy of `*t`) to a long integer. From this, we can infer that `t` must be a pointer to a signed integer.

3. The `addq` instruction adds the sign-extended value of the previous sum to the location indicated by the second argument register. From this, we can infer that `q` is the second argument and that it is a pointer to a long integer.

There are four valid prototypes for `incrprob`, depending on whether or not `x` is long, and whether it is signed or unsigned. We show these as four different prototypes:

```
void incrprob_s(int x, long *q, int *t);
void incrprob_u(unsigned x, long *q, int *t);
void incrprob_sl(long x, long *q, int *t);
void incrprob_ul(unsigned long x, long *q, int *t);
```

Solution to Problem 51

This function is an example of a leaf function that requires local storage. It can use space beyond the stack pointer for its local storage, never altering the stack pointer.

A. Stack locations used:

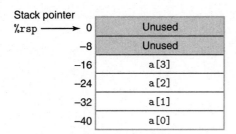

B.
```
        x86-64 implementation of local_array
        Argument i in %edi
1   local_array:
2       movq    $2, -40(%rsp)       Store 2 in a[0]
3       movq    $3, -32(%rsp)       Store 3 in a[1]
4       movq    $5, -24(%rsp)       Store 5 in a[2]
5       movq    $7, -16(%rsp)       Store 7 in a[3]
6       andl    $3, %edi            Compute idx = i & 3
7       movq    -40(%rsp,%rdi,8), %rax   Compute a[idx] as return value
8       ret                         Return
```

C. The function never changes the stack pointer. It stores all of its local values in the region beyond the stack pointer.

Solution to Problem 52

A. Register `%rbx` is used to hold the parameter x.

B. Since %rbx is callee-saved, it must be stored on the stack. Since this is the only use of the stack for this function, the code uses push and pop instructions to save and restore the register.

C.
```
      x86-64 implementation of recursive factorial function rfact
      Argument: x in %rdi
1   rfact:
2       pushq    %rbx                    Save %rbx (callee save)
3       movq     %rdi, %rbx              Copy x to %rbx
4       movl     $1, %eax                result = 1
5       testq    %rdi, %rdi              Test x
6       jle      .L11                    If <=0, goto done
7       leaq     -1(%rdi), %rdi          Compute xm1 = x-1
8       call     rfact                   Call rfact(xm1)
9       imulq    %rbx, %rax              Compute result = x*rfact(xm1)
10  .L11:                                done:
11      popq     %rbx                    Restore %rbx
12      ret                              Return
```

D. Instead of explicitly decrementing and incrementing the stack pointer, the code can use pushq and popq to both modify the stack pointer and to save and restore register state.

Solution to Problem 53

This problem is similar to Problem 41, but updated for x86-64.

A. struct P1 { int i; char c; long j; char d; };

i	c	j	d	Total	Alignment
0	4	8	16	24	8

B. struct P2 { long i; char c; char d; int j; };

i	c	d	j	Total	Alignment
0	8	9	12	16	8

C. struct P3 { short w[3]; char c[3] };

w	c	Total	Alignment
0	6	10	2

D. struct P4 { short w[3]; char *c[3] };

w	c	Total	Alignment
0	8	32	8

E. struct P3 { struct P1 a[2]; struct P2 *p };

a	p	Total	Alignment
0	48	56	8

Processor Architecture

From Chapter 4 of *Computer Systems: A Programmer's Perspective*, Second Edition. Randal E. Bryant and David R. O'Hallaron. Copyright © 2011 by Pearson Education, Inc. Published by Prentice Hall. All rights reserved.

Modern microprocessors are among the most complex systems ever created by humans. A single silicon chip, roughly the size of a fingernail, can contain a complete high-performance processor, large cache memories, and the logic required to interface it to external devices. In terms of performance, the processors implemented on a single chip today dwarf the room-sized supercomputers that cost over $10 million just 20 years ago. Even the embedded processors found in everyday appliances such as cell phones, personal digital assistants, and handheld game systems are far more powerful than the early developers of computers ever envisioned.

Thus far, we have only viewed computer systems down to the level of machine-language programs. We have seen that a processor must execute a sequence of instructions, where each instruction performs some primitive operation, such as adding two numbers. An instruction is encoded in binary form as a sequence of 1 or more bytes. The instructions supported by a particular processor and their byte-level encodings are known as its *instruction-set architecture* (ISA). Different "families" of processors, such as Intel IA32, IBM/Freescale PowerPC, and the ARM processor family have different ISAs. A program compiled for one type of machine will not run on another. On the other hand, there are many different models of processors within a single family. Each manufacturer produces processors of ever-growing performance and complexity, but the different models remain compatible at the ISA level. Popular families, such as IA32, have processors supplied by multiple manufacturers. Thus, the ISA provides a conceptual layer of abstraction between compiler writers, who need only know what instructions are permitted and how they are encoded, and processor designers, who must build machines that execute those instructions.

In this chapter, we take a brief look at the design of processor hardware. We study the way a hardware system can execute the instructions of a particular ISA. This view will give you a better understanding of how computers work and the technological challenges faced by computer manufacturers. One important concept is that the actual way a modern processor operates can be quite different from the model of computation implied by the ISA. The ISA model would seem to imply *sequential* instruction execution, where each instruction is fetched and executed to completion before the next one begins. By executing different parts of multiple instructions simultaneously, the processor can achieve higher performance than if it executed just one instruction at a time. Special mechanisms are used to make sure the processor computes the same results as it would with sequential execution. This idea of using clever tricks to improve performance while maintaining the functionality of a simpler and more abstract model is well known in computer science. Examples include the use of caching in Web browsers and information retrieval data structures such as balanced binary trees and hash tables.

Chances are you will never design your own processor. This is a task for experts working at fewer than 100 companies worldwide. Why, then, should you learn about processor design?

- *It is intellectually interesting and important.* There is an intrinsic value in learning how things work. It is especially interesting to learn the inner workings of a system that is such a part of the daily lives of computer scientists and engineers and yet remains a mystery to many. Processor design embodies many of the principles of good engineering practice. It requires creating a simple and regular structure to perform a complex task.

- *Understanding how the processor works aids in understanding how the overall computer system works.* Seeing the processor side of the processor-memory interface gives insight into the memory system and the techniques used to create an image of a very large memory with a very fast access time.

- *Although few people design processors, many design hardware systems that contain processors.* This has become commonplace as processors are embedded into real-world systems such as automobiles and appliances. Embedded-system designers must understand how processors work, because these systems are generally designed and programmed at a lower level of abstraction than is the case for desktop systems.

- *You just might work on a processor design.* Although the number of companies producing microprocessors is small, the design teams working on those processors are already large and growing. There can be over 1000 people involved in the different aspects of a major processor design.

In this chapter, we start by defining a simple instruction set that we use as a running example for our processor implementations. We call this the "Y86" instruction set, because it was inspired by the IA32 instruction set, which is colloquially referred to as "x86." Compared with IA32, the Y86 instruction set has fewer data types, instructions, and addressing modes. It also has a simpler byte-level encoding. Still, it is sufficiently complete to allow us to write simple programs manipulating integer data. Designing a processor to implement Y86 requires us to face many of the challenges faced by processor designers.

We then provide some background on digital hardware design. We describe the basic building blocks used in a processor and how they are connected together and operated. This presentation builds on the concepts of Boolean algebra and bit-level operations. We also introduce a simple language, HCL (for "Hardware Control Language"), to describe the control portions of hardware systems. We will later use this language to describe our processor designs. Even if you already have some background in logic design, read this section to understand our particular notation.

As a first step in designing a processor, we present a functionally correct, but somewhat impractical, Y86 processor based on *sequential* operation. This processor executes a complete Y86 instruction on every clock cycle. The clock must run slowly enough to allow an entire series of actions to complete within one cycle. Such a processor could be implemented, but its performance would be well below what could be achieved for this much hardware.

With the sequential design as a basis, we then apply a series of transformations to create a *pipelined* processor. This processor breaks the execution of each instruction into five steps, each of which is handled by a separate section or *stage* of the hardware. Instructions progress through the stages of the pipeline, with one instruction entering the pipeline on each clock cycle. As a result, the processor can be executing the different steps of up to five instructions simultaneously. Making this processor preserve the sequential behavior of the Y86 ISA requires handling a variety of *hazard* conditions, where the location or operands of one instruction depend on those of other instructions that are still in the pipeline.

We have devised a variety of tools for studying and experimenting with our processor designs. These include an assembler for Y86, a simulator for running Y86 programs on your machine, and simulators for two sequential and one pipelined processor design. The control logic for these designs is described by files in HCL notation. By editing these files and recompiling the simulator, you can alter and extend the simulator's behavior. A number of exercises are provided that involve implementing new instructions and modifying how the machine processes instructions. Testing code is provided to help you evaluate the correctness of your modifications. These exercises will greatly aid your understanding of the material and will give you an appreciation for the many different design alternatives faced by processor designers.

Web Aside ARCH:VLOG presents a representation of our pipelined Y86 processor in the Verilog hardware description language. This involves creating modules for the basic hardware building blocks and for the overall processor structure. We automatically translate the HCL description of the control logic into Verilog. By first debugging the HCL description with our simulators, we eliminate many of the tricky bugs that would otherwise show up in the hardware design. Given a Verilog description, there are commercial and open-source tools to support simulation and *logic synthesis*, generating actual circuit designs for the microprocessors. So, although much of the effort we expend here is to create pictorial and textual descriptions of a system, much as one would when writing software, the fact that these designs can be automatically synthesized demonstrates that we are indeed creating a system that can be realized as hardware.

1 The Y86 Instruction Set Architecture

Defining an instruction set architecture, such as Y86, includes defining the different state elements, the set of instructions and their encodings, a set of programming conventions, and the handling of exceptional events.

1.1 Programmer-Visible State

As Figure 1 illustrates, each instruction in a Y86 program can read and modify some part of the processor state. This is referred to as the *programmer-visible* state, where the "programmer" in this case is either someone writing programs in assembly code or a compiler generating machine-level code. We will see in our processor implementations that we do not need to represent and organize this state in exactly the manner implied by the ISA, as long as we can make sure that machine-level programs appear to have access to the programmer-visible state. The state for Y86 is similar to that for IA32. There are eight *program registers*:

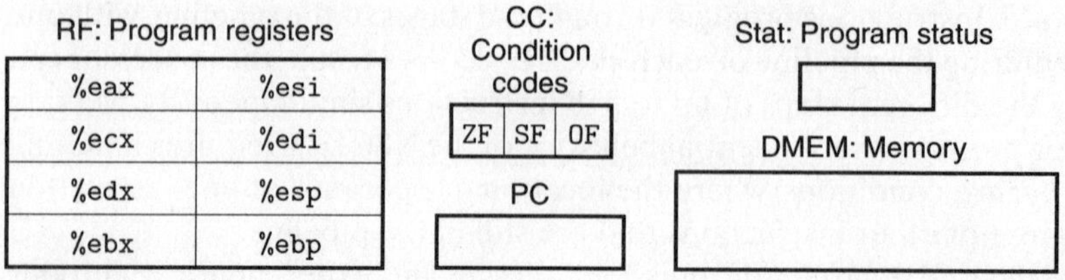

Figure 1 **Y86 programmer-visible state.** As with IA32, programs for Y86 access and modify the program registers, the condition code, the program counter (PC), and the memory. The status code indicates whether the program is running normally, or some special event has occurred.

%eax, %ecx, %edx, %ebx, %esi, %edi, %esp, and %ebp. Each of these stores a word. Register %esp is used as a stack pointer by the push, pop, call, and return instructions. Otherwise, the registers have no fixed meanings or values. There are three single-bit *condition codes*, ZF, SF, and OF, storing information about the effect of the most recent arithmetic or logical instruction. The program counter (PC) holds the address of the instruction currently being executed.

The *memory* is conceptually a large array of bytes, holding both program and data. Y86 programs reference memory locations using *virtual addresses*. A combination of hardware and operating system software translates these into the actual, or *physical*, addresses indicating where the values are actually stored in memory. We can think of the virtual memory system as providing Y86 programs with an image of a monolithic byte array.

A final part of the program state is a status code Stat, indicating the overall state of program execution. It will indicate either normal operation, or that some sort of *exception* has occurred, such as when an instruction attempts to read from an invalid memory address. The possible status codes and the handling of exceptions is described in Section 1.4.

1.2 Y86 Instructions

Figure 2 gives a concise description of the individual instructions in the Y86 ISA. We use this instruction set as a target for our processor implementations. The set of Y86 instructions is largely a subset of the IA32 instruction set. It includes only 4-byte integer operations, has fewer addressing modes, and includes a smaller set of operations. Since we only use 4-byte data, we can refer to these as "words" without any ambiguity. In this figure, we show the assembly-code representation of the instructions on the left and the byte encodings on the right. The assembly-code format is similar to the ATT format for IA32.

Here are some further details about the different Y86 instructions.

- The IA32 movl instruction is split into four different instructions: irmovl, rrmovl, mrmovl, and rmmovl, explicitly indicating the form of the source and destination. The source is either immediate (i), register (r), or memory (m).

Figure 2

Y86 instruction set.
Instruction encodings range between 1 and 6 bytes. An instruction consists of a 1-byte instruction specifier, possibly a 1-byte register specifier, and possibly a 4-byte constant word. Field fn specifies a particular integer operation (OP1), data movement condition (cmovXX), or branch condition (jXX). All numeric values are shown in hexadecimal.

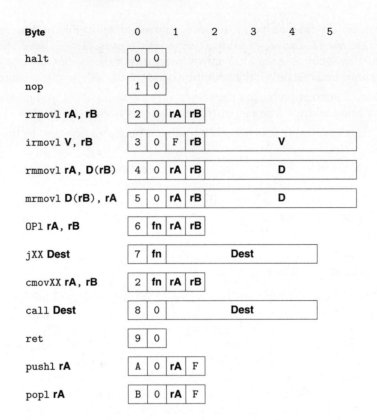

It is designated by the first character in the instruction name. The destination is either register (r) or memory (m). It is designated by the second character in the instruction name. Explicitly identifying the four types of data transfer will prove helpful when we decide how to implement them.

The memory references for the two memory movement instructions have a simple base and displacement format. We do not support the second index register or any scaling of a register's value in the address computation.

As with IA32, we do not allow direct transfers from one memory location to another. In addition, we do not allow a transfer of immediate data to memory.

- There are four integer operation instructions, shown in Figure 2 as OP1. These are addl, subl, andl, and xorl. They operate only on register data, whereas IA32 also allows operations on memory data. These instructions set the three condition codes ZF, SF, and OF (zero, sign, and overflow).

- The seven jump instructions (shown in Figure 2 as jXX) are jmp, jle, jl, je, jne, jge, and jg. Branches are taken according to the type of branch and the settings of the condition codes. The branch conditions are the same as with IA32.

- There are six conditional move instructions (shown in Figure 2 as cmovXX): cmovle, cmovl, cmove, cmovne, cmovge, and cmovg. These have the same format as the register-register move instruction rrmovl, but the destination register is updated only if the condition codes satisfy the required constraints.

- The call instruction pushes the return address on the stack and jumps to the destination address. The ret instruction returns from such a call.

- The pushl and popl instructions implement push and pop, just as they do in IA32.

- The halt instruction stops instruction execution. IA32 has a comparable instruction, called hlt. IA32 application programs are not permitted to use this instruction, since it causes the entire system to suspend operation. For Y86, executing the halt instruction causes the processor to stop, with the status code set to HLT. (See Section 1.4.)

1.3 Instruction Encoding

Figure 2 also shows the byte-level encoding of the instructions. Each instruction requires between 1 and 6 bytes, depending on which fields are required. Every instruction has an initial byte identifying the instruction type. This byte is split into two 4-bit parts: the high-order, or *code*, part, and the low-order, or *function*, part. As you can see in Figure 2, code values range from 0 to 0xB. The function values are significant only for the cases where a group of related instructions share a common code. These are given in Figure 3, showing the specific encodings of the integer operation, conditional move, and branch instructions. Observe that rrmovl has the same instruction code as the conditional moves. It can be viewed as an "unconditional move" just as the jmp instruction is an unconditional jump, both having function code 0.

As shown in Figure 4, each of the eight program registers has an associated *register identifier* (ID) ranging from 0 to 7. The numbering of registers in Y86 matches what is used in IA32. The program registers are stored within the CPU in a *register file*, a small random-access memory where the register IDs serve

Operations				Branches								Moves						
addl	6	0		jmp	7	0	jne	7	4			rrmovl	2	0		cmovne	2	4
subl	6	1		jle	7	1	jge	7	5			cmovle	2	1		cmovge	2	5
andl	6	2		jl	7	2	jg	7	6			cmovl	2	2		cmovg	2	6
xorl	6	3		je	7	3						cmove	2	3				

Figure 3 **Function codes for Y86 instruction set.** The code specifies a particular integer operation, branch condition, or data transfer condition. These instructions are shown as OP1, jXX, and cmovXX in Figure 2.

Number	Register name
0	%eax
1	%ecx
2	%edx
3	%ebx
4	%esp
5	%ebp
6	%esi
7	%edi
F	No register

Figure 4 **Y86 program register identifiers.** Each of the eight program registers has an associated identifier (ID) ranging from 0 to 7. ID 0xF in a register field of an instruction indicates the absence of a register operand.

as addresses. ID value 0xF is used in the instruction encodings and within our hardware designs when we need to indicate that no register should be accessed.

Some instructions are just 1 byte long, but those that require operands have longer encodings. First, there can be an additional *register specifier byte*, specifying either one or two registers. These register fields are called rA and rB in Figure 2. As the assembly-code versions of the instructions show, they can specify the registers used for data sources and destinations, as well as the base register used in an address computation, depending on the instruction type. Instructions that have no register operands, such as branches and call, do not have a register specifier byte. Those that require just one register operand (irmovl, pushl, and popl) have the other register specifier set to value 0xF. This convention will prove useful in our processor implementation.

Some instructions require an additional 4-byte *constant word*. This word can serve as the immediate data for irmovl, the displacement for rmmovl and mrmovl address specifiers, and the destination of branches and calls. Note that branch and call destinations are given as absolute addresses, rather than using the PC-relative addressing seen in IA32. Processors use PC-relative addressing to give more compact encodings of branch instructions and to allow code to be copied from one part of memory to another without the need to update all of the branch target addresses. Since we are more concerned with simplicity in our presentation, we use absolute addressing. As with IA32, all integers have a little-endian encoding. When the instruction is written in disassembled form, these bytes appear in reverse order.

As an example, let us generate the byte encoding of the instruction rmmovl %esp,0x12345(%edx) in hexadecimal. From Figure 2, we can see that rmmovl has initial byte 40. We can also see that source register %esp should be encoded in the rA field, and base register %edx should be encoded in the rB field. Using the register numbers in Figure 4, we get a register specifier byte of 42. Finally, the displacement is encoded in the 4-byte constant word. We first pad 0x12345 with leading zeros to fill out 4 bytes, giving a byte sequence of 00 01 23 45. We write this in byte-reversed order as 45 23 01 00. Combining these, we get an instruction encoding of 404245230100.

One important property of any instruction set is that the byte encodings must have a unique interpretation. An arbitrary sequence of bytes either encodes a unique instruction sequence or is not a legal byte sequence. This property holds for Y86, because every instruction has a unique combination of code and function in its initial byte, and given this byte, we can determine the length and meaning of any additional bytes. This property ensures that a processor can execute an object-code program without any ambiguity about the meaning of the code. Even if the code is embedded within other bytes in the program, we can readily determine the instruction sequence as long as we start from the first byte in the sequence. On the other hand, if we do not know the starting position of a code sequence, we cannot reliably determine how to split the sequence into individual instructions. This causes problems for disassemblers and other tools that attempt to extract machine-level programs directly from object-code byte sequences.

Practice Problem 1

Determine the byte encoding of the Y86 instruction sequence that follows. The line ".pos 0x100" indicates that the starting address of the object code should be 0x100.

```
.pos 0x100  # Start code at address 0x100
    irmovl $15,%ebx         #   Load 15 into %ebx
    rrmovl %ebx,%ecx        #   Copy 15 to %ecx
loop:                       # loop:
    rmmovl %ecx,-3(%ebx)    #   Save %ecx at address 15-3 = 12
    addl   %ebx,%ecx        #   Increment %ecx by 15
    jmp loop                #   Goto loop
```

Practice Problem 2

For each byte sequence listed, determine the Y86 instruction sequence it encodes. If there is some invalid byte in the sequence, show the instruction sequence up to that point and indicate where the invalid value occurs. For each sequence, we show the starting address, then a colon, and then the byte sequence.

A. 0x100:30f3fcffffff40630008000000

B. 0x200:a06f80080200000030f30a00000090

C. 0x300:50540700000010f0b01f

D. 0x400:6113730004000000

E. 0x500:6362a0f0

Aside Comparing IA32 to Y86 instruction encodings

Compared with the instruction encodings used in IA32, the encoding of Y86 is much simpler but also less compact. The register fields occur only in fixed positions in all Y86 instructions, whereas they are packed into various positions in the different IA32 instructions. We use a 4-bit encoding of registers, even though there are only eight possible registers. IA32 uses just 3 bits. Thus, IA32 can pack a push or pop instruction into just 1 byte, with a 5-bit field indicating the instruction type and the remaining 3 bits for the register specifier. IA32 can encode constant values in 1, 2, or 4 bytes, whereas Y86 always requires 4 bytes.

Aside RISC and CISC instruction sets

IA32 is sometimes labeled as a "complex instruction set computer" (CISC—pronounced "sisk"), and is deemed to be the opposite of ISAs that are classified as "reduced instruction set computers" (RISC—pronounced "risk"). Historically, CISC machines came first, having evolved from the earliest computers. By the early 1980s, instruction sets for mainframe and minicomputers had grown quite large, as machine designers incorporated new instructions to support high-level tasks, such as manipulating circular buffers, performing decimal arithmetic, and evaluating polynomials. The first microprocessors appeared in the early 1970s and had limited instruction sets, because the integrated-circuit technology then posed severe constraints on what could be implemented on a single chip. Microprocessors evolved quickly and, by the early 1980s, were following the path of increasing instruction-set complexity set by mainframes and minicomputers. The x86 family took this path, evolving into IA32, and more recently into x86-64. Even the x86 line continues to evolve as new classes of instructions are added based on the needs of emerging applications.

The RISC design philosophy developed in the early 1980s as an alternative to these trends. A group of hardware and compiler experts at IBM, strongly influenced by the ideas of IBM researcher John Cocke, recognized that they could generate efficient code for a much simpler form of instruction set. In fact, many of the high-level instructions that were being added to instruction sets were very difficult to generate with a compiler and were seldom used. A simpler instruction set could be implemented with much less hardware and could be organized in an efficient pipeline structure, similar to those described later in this chapter. IBM did not commercialize this idea until many years later, when it developed the Power and PowerPC ISAs.

The RISC concept was further developed by Professors David Patterson, of the University of California at Berkeley, and John Hennessy, of Stanford University. Patterson gave the name RISC to this new class of machines, and CISC to the existing class, since there had previously been no need to have a special designation for a nearly universal form of instruction set.

Comparing CISC with the original RISC instruction sets, we find the following general characteristics:

CISC	Early RISC
A large number of instructions. The Intel document describing the complete set of instructions [28, 29] is over 1200 pages long.	Many fewer instructions. Typically less than 100.
Some instructions with long execution times. These include instructions that copy an entire block from one part of memory to another and others that copy multiple registers to and from memory.	No instruction with a long execution time. Some early RISC machines did not even have an integer multiply instruction, requiring compilers to implement multiplication as a sequence of additions.

CISC	Early RISC
Variable-length encodings. IA32 instructions can range from 1 to 15 bytes.	Fixed-length encodings. Typically all instructions are encoded as 4 bytes.
Multiple formats for specifying operands. In IA32, a memory operand specifier can have many different combinations of displacement, base and index registers, and scale factors.	Simple addressing formats. Typically just base and displacement addressing.
Arithmetic and logical operations can be applied to both memory and register operands.	Arithmetic and logical operations only use register operands. Memory referencing is only allowed by *load* instructions, reading from memory into a register, and *store* instructions, writing from a register to memory. This convention is referred to as a *load/store architecture*.
Implementation artifacts hidden from machine-level programs. The ISA provides a clean abstraction between programs and how they get executed.	Implementation artifacts exposed to machine-level programs. Some RISC machines prohibit particular instruction sequences and have jumps that do not take effect until the following instruction is executed. The compiler is given the task of optimizing performance within these constraints.
Condition codes. Special flags are set as a side effect of instructions and then used for conditional branch testing.	No condition codes. Instead, explicit test instructions store the test results in normal registers for use in conditional evaluation.
Stack-intensive procedure linkage. The stack is used for procedure arguments and return addresses.	Register-intensive procedure linkage. Registers are used for procedure arguments and return addresses. Some procedures can thereby avoid any memory references. Typically, the processor has many more (up to 32) registers.

The Y86 instruction set includes attributes of both CISC and RISC instruction sets. On the CISC side, it has condition codes, variable-length instructions, and stack-intensive procedure linkages. On the RISC side, it uses a load-store architecture and a regular encoding. It can be viewed as taking a CISC instruction set (IA32) and simplifying it by applying some of the principles of RISC.

Aside The RISC versus CISC controversy

Through the 1980s, battles raged in the computer architecture community regarding the merits of RISC versus CISC instruction sets. Proponents of RISC claimed they could get more computing power for a given amount of hardware through a combination of streamlined instruction set design, advanced compiler technology, and pipelined processor implementation. CISC proponents countered that fewer CISC instructions were required to perform a given task, and so their machines could achieve higher overall performance.

Major companies introduced RISC processor lines, including Sun Microsystems (SPARC), IBM and Motorola (PowerPC), and Digital Equipment Corporation (Alpha). A British company, Acorn

Computers Ltd., developed its own architecture, ARM (originally an acronym for "Acorn RISC Machine"), which is widely used in embedded applications, such as cellphones.

In the early 1990s, the debate diminished as it became clear that neither RISC nor CISC in their purest forms were better than designs that incorporated the best ideas of both. RISC machines evolved and introduced more instructions, many of which take multiple cycles to execute. RISC machines today have hundreds of instructions in their repertoire, hardly fitting the name "reduced instruction set machine." The idea of exposing implementation artifacts to machine-level programs proved to be short-sighted. As new processor models were developed using more advanced hardware structures, many of these artifacts became irrelevant, but they still remained part of the instruction set. Still, the core of RISC design is an instruction set that is well-suited to execution on a pipelined machine.

More recent CISC machines also take advantage of high-performance pipeline structures. They fetch the CISC instructions and dynamically translate them into a sequence of simpler, RISC-like operations. For example, an instruction that adds a register to memory is translated into three operations: one to read the original memory value, one to perform the addition, and a third to write the sum to memory. Since the dynamic translation can generally be performed well in advance of the actual instruction execution, the processor can sustain a very high execution rate.

Marketing issues, apart from technological ones, have also played a major role in determining the success of different instruction sets. By maintaining compatibility with its existing processors, Intel with x86 made it easy to keep moving from one generation of processor to the next. As integrated-circuit technology improved, Intel and other x86 processor manufacturers could overcome the inefficiencies created by the original 8086 instruction set design, using RISC techniques to produce performance comparable to the best RISC machines. The evolution of IA32 into x86-64 provided an opportunity to incorporate several features of RISC into x86. In the areas of desktop and laptop computing, x86 has achieved total domination, and it is increasingly popular for high-end server machines.

RISC processors have done very well in the market for *embedded processors*, controlling such systems as cellular telephones, automobile brakes, and Internet appliances. In these applications, saving on cost and power is more important than maintaining backward compatibility. In terms of the number of processors sold, this is a very large and growing market.

1.4   Y86 Exceptions

The programmer-visible state for Y86 (Figure 1) includes a status code Stat describing the overall state of the executing program. The possible values for this code are shown in Figure 5. Code value 1, named AOK, indicates that the program is executing normally, while the other codes indicate that some type of *exception* has occurred. Code 2, named HLT, indicates that the processor has executed a halt instruction. Code 3, named ADR, indicates that the processor attempted to read from or write to an invalid memory address, either while fetching an instruction or while reading or writing data. We limit the maximum address (the exact limit varies by implementation), and any access to an address beyond this limit will trigger an ADR exception. Code 4, named INS, indicates that an invalid instruction code has been encountered.

Value	Name	Meaning
1	AOK	Normal operation
2	HLT	halt instruction encountered
3	ADR	Invalid address encountered
4	INS	Invalid instruction encountered

Figure 5 **Y86 status codes.** In our design, the processor halts for any code other than AOK.

For Y86, we will simply have the processor stop executing instructions when it encounters any of the exceptions listed. In a more complete design, the processor would typically invoke an *exception handler*, a procedure designated to handle the specific type of exception encountered. Exception handlers can be configured to have different effects, such as aborting the program or invoking a user-defined *signal handler*.

1.5 Y86 Programs

Figure 6 shows IA32 and Y86 assembly code for the following C function:

```
int Sum(int *Start, int Count)
{
    int sum = 0;
    while (Count) {
        sum += *Start;
        Start++;
        Count--;
    }
    return sum;
}
```

The IA32 code was generated by the GCC compiler. The Y86 code is essentially the same, except that Y86 sometimes requires two instructions to accomplish what can be done with a single IA32 instruction. If we had written the program using array indexing, however, the conversion to Y86 code would be more difficult, since Y86 does not have scaled addressing modes. This code follows many of the programming conventions we have seen for IA32, including the use of the stack and frame pointers. For simplicity, it does not follow the IA32 convention of having some registers designated as callee-save registers. This is just a programming convention that we can either adopt or ignore as we please.

Figure 7 shows an example of a complete program file written in Y86 assembly code. The program contains both data and instructions. Directives indicate where to place code or data and how to align it. The program specifies issues such

IA32 code

int Sum(int *Start, int Count)

```
1    Sum:
2      pushl %ebp
3      movl %esp,%ebp
4      movl 8(%ebp),%ecx      ecx = Start
5      movl 12(%ebp),%edx     edx = Count
6      xorl %eax,%eax         sum = 0
7      testl %edx,%edx
8      je .L34
9    .L35:
10     addl (%ecx),%eax       add *Start to sum
11     addl $4,%ecx           Start++
12     decl %edx              Count--
13     jnz .L35               Stop when 0
14   .L34:
15     movl %ebp,%esp
16     popl %ebp
17     ret
```

Y86 code

int Sum(int *Start, int Count)

```
1    Sum:
2      pushl %ebp
3      rrmovl %esp,%ebp
4      mrmovl 8(%ebp),%ecx      ecx = Start
5      mrmovl 12(%ebp),%edx     edx = Count
6      xorl %eax,%eax           sum = 0
7      andl    %edx,%edx        Set condition codes
8      je      End
9    Loop:
10     mrmovl (%ecx),%esi        get *Start
11     addl %esi,%eax            add to sum
12     irmovl $4,%ebx
13     addl %ebx,%ecx            Start++
14     irmovl $-1,%ebx
15     addl %ebx,%edx            Count--
16     jne     Loop              Stop when 0
17   End:
18     rrmovl %ebp,%esp
19     popl %ebp
20     ret
```

Figure 6 Comparison of Y86 and IA32 assembly programs. The Sum function computes the sum of an integer array. The Y86 code differs from the IA32 mainly in that it may require multiple instructions to perform what can be done with a single IA32 instruction.

as stack placement, data initialization, program initialization, and program termination.

In this program, words beginning with " . " are *assembler directives* telling the assembler to adjust the address at which it is generating code or to insert some words of data. The directive .pos 0 (line 2) indicates that the assembler should begin generating code starting at address 0. This is the starting address for all Y86 programs. The next two instructions (lines 3 and 4) initialize the stack and frame pointers. We can see that the label Stack is declared at the end of the program (line 47), to indicate address 0x100 using a .pos directive (line 46). Our stack will therefore start at this address and grow toward lower addresses. We must ensure that the stack does not grow so large that it overwrites the code or other program data.

Lines 9 to 13 of the program declare an array of four words, having values 0xd, 0xc0, 0xb00, and 0xa000. The label array denotes the start of this array, and is aligned on a 4-byte boundary (using the .align directive). Lines 17 to 6 show a "main" procedure that calls the function Sum on the four-word array and then halts.

```
 1    # Execution begins at address 0
 2          .pos 0
 3    init:  irmovl Stack, %esp      # Set up stack pointer
 4           irmovl Stack, %ebp      # Set up base pointer
 5           call Main               # Execute main program
 6           halt                    # Terminate program
 7
 8    # Array of 4 elements
 9           .align 4
10    array: .long 0xd
11           .long 0xc0
12           .long 0xb00
13           .long 0xa000
14
15    Main:  pushl %ebp
16           rrmovl %esp,%ebp
17           irmovl $4,%eax
18           pushl %eax              # Push 4
19           irmovl array,%edx
20           pushl %edx              # Push array
21           call Sum                # Sum(array, 4)
22           rrmovl %ebp,%esp
23           popl %ebp
24           ret
25
26           # int Sum(int *Start, int Count)
27    Sum:   pushl %ebp
28           rrmovl %esp,%ebp
29           mrmovl 8(%ebp),%ecx     # ecx = Start
30           mrmovl 12(%ebp),%edx    # edx = Count
31           xorl %eax,%eax          # sum = 0
32           andl  %edx,%edx         # Set condition codes
33           je    End
34    Loop:  mrmovl (%ecx),%esi      # get *Start
35           addl %esi,%eax          # add to sum
36           irmovl $4,%ebx          #
37           addl %ebx,%ecx          # Start++
38           irmovl $-1,%ebx         #
39           addl %ebx,%edx          # Count--
40           jne   Loop              # Stop when 0
41    End:   rrmovl %ebp,%esp
42           popl %ebp
43           ret
44
45    # The stack starts here and grows to lower addresses
46           .pos 0x100
47    Stack:
```

Figure 7 **Sample program written in Y86 assembly code.** The Sum function is called to compute the sum of a four-element array.

As this example shows, since our only tool for creating Y86 code is an assembler, the programmer must perform tasks we ordinarily delegate to the compiler, linker, and run-time system. Fortunately, we only do this for small programs, for which simple mechanisms suffice.

Figure 8 shows the result of assembling the code shown in Figure 7 by an assembler we call YAS. The assembler output is in ASCII format to make it more readable. On lines of the assembly file that contain instructions or data, the object code contains an address, followed by the values of between 1 and 6 bytes.

We have implemented an *instruction set simulator* we call YIS, the purpose of which is to model the execution of a Y86 machine-code program, without attempting to model the behavior of any specific processor implementation. This form of simulation is useful for debugging programs before actual hardware is available, and for checking the result of either simulating the hardware or running the program on the hardware itself. Running on our sample object code, YIS generates the following output:

```
Stopped in 52 steps at PC = 0x11.  Status 'HLT', CC Z=1 S=0 O=0
Changes to registers:
%eax:    0x00000000        0x0000abcd
%ecx:    0x00000000        0x00000024
%ebx:    0x00000000        0xffffffff
%esp:    0x00000000        0x00000100
%ebp:    0x00000000        0x00000100
%esi:    0x00000000        0x0000a000

Changes to memory:
0x00e8: 0x00000000         0x000000f8
0x00ec: 0x00000000         0x0000003d
0x00f0: 0x00000000         0x00000014
0x00f4: 0x00000000         0x00000004
0x00f8: 0x00000000         0x00000100
0x00fc: 0x00000000         0x00000011
```

The first line of the simulation output summarizes the execution and the resulting values of the PC and program status. In printing register and memory values, it only prints out words that change during simulation, either in registers or in memory. The original values (here they are all zero) are shown on the left, and the final values are shown on the right. We can see in this output that register %eax contains 0xabcd, the sum of the four-element array passed to subroutine Sum. In addition, we can see that the stack, which starts at address 0x100 and grows toward lower addresses, has been used, causing changes to words of memory at addresses 0xe8 through 0xfc. This is well away from 0x7c, the maximum address of the executable code.

```
                        | # Execution begins at address 0
0x000:                  |        .pos 0
0x000: 30f400010000     | init:  irmovl Stack, %esp      # Set up stack pointer
0x006: 30f500010000     |        irmovl Stack, %ebp      # Set up base pointer
0x00c: 8024000000       |        call Main               # Execute main program
0x011: 00               |        halt                    # Terminate program
                        |
                        | # Array of 4 elements
0x014:                  |        .align 4
0x014: 0d000000         | array: .long 0xd
0x018: c0000000         |        .long 0xc0
0x01c: 000b0000         |        .long 0xb00
0x020: 00a00000         |        .long 0xa000
                        |
0x024: a05f             | Main:  pushl %ebp
0x026: 2045             |        rrmovl %esp,%ebp
0x028: 30f004000000     |        irmovl $4,%eax
0x02e: a00f             |        pushl %eax              # Push 4
0x030: 30f214000000     |        irmovl array,%edx
0x036: a02f             |        pushl %edx              # Push array
0x038: 8042000000       |        call Sum                # Sum(array, 4)
0x03d: 2054             |        rrmovl %ebp,%esp
0x03f: b05f             |        popl %ebp
0x041: 90               |        ret
                        |
                        | # int Sum(int *Start, int Count)
0x042: a05f             | Sum:   pushl %ebp
0x044: 2045             |        rrmovl %esp,%ebp
0x046: 501508000000     |        mrmovl 8(%ebp),%ecx     # ecx = Start
0x04c: 50250c000000     |        mrmovl 12(%ebp),%edx    # edx = Count
0x052: 6300             |        xorl %eax,%eax          # sum = 0
0x054: 6222             |        andl   %edx,%edx        # Set condition codes
0x056: 7378000000       |        je     End
0x05b: 506100000000     | Loop:  mrmovl (%ecx),%esi      # get *Start
0x061: 6060             |        addl %esi,%eax          # add to sum
0x063: 30f304000000     |        irmovl $4,%ebx          #
0x069: 6031             |        addl %ebx,%ecx          # Start++
0x06b: 30f3ffffffff     |        irmovl $-1,%ebx         #
0x071: 6032             |        addl %ebx,%edx          # Count--
0x073: 745b000000       |        jne    Loop             # Stop when 0
0x078: 2054             | End:   rrmovl %ebp,%esp
0x07a: b05f             |        popl %ebp
0x07c: 90               |        ret
                        |
                        | # The stack starts here and grows to lower addresses
0x100:                  |        .pos 0x100
0x100:                  | Stack:
```

Figure 8 **Output of YAS assembler.** Each line includes a hexadecimal address and between 1 and 6 bytes of object code.

Practice Problem 3

Write Y86 code to implement a recursive sum function rSum, based on the following C code:

```
int rSum(int *Start, int Count)
{
    if (Count <= 0)
        return 0;
    return *Start + rSum(Start+1, Count-1);
}
```

You might find it helpful to compile the C code on an IA32 machine and then translate the instructions to Y86.

Practice Problem 4

Modify the Y86 code for the Sum function (Figure 6) to implement a function AbsSum that computes the sum of absolute values of an array. Use a *conditional jump* instruction within your inner loop.

Practice Problem 5

Modify the Y86 code for the Sum function (Figure 6) to implement a function AbsSum that computes the sum of absolute values of an array. Use a *conditional move* instruction within your inner loop.

1.6 Some Y86 Instruction Details

Most Y86 instructions transform the program state in a straightforward manner, and so defining the intended effect of each instruction is not difficult. Two unusual instruction combinations, however, require special attention.

The pushl instruction both decrements the stack pointer by 4 and writes a register value to memory. It is therefore not totally clear what the processor should do when executing the instruction pushl %esp, since the register being pushed is being changed by the same instruction. Two different conventions are possible: (1) push the original value of %esp, or (2) push the decremented value of %esp.

For the Y86 processor, let us adopt the same convention as is used with IA32, as determined in the following problem.

Practice Problem 6

Let us determine the behavior of the instruction pushl %esp for an IA32 processor. We could try reading the Intel documentation on this instruction, but a simpler approach is to conduct an experiment on an actual machine. The C compiler would not normally generate this instruction, so we must use hand-generated assembly code for this task. Here is a test function we have written (Web Aside ASM:EASM describes how to write programs that combine C code with hand-written assembly code):

```
1       .text
2     .globl pushtest
3     pushtest:
4       pushl   %ebp
5       movl    %esp, %ebp
6       movl    %esp, %eax      Copy stack pointer
7       pushl   %esp            Push stack pointer
8       popl    %edx            Pop it back
9       subl    %edx,%eax       Subtract new from old stack pointer
10      leave                   Restore stack & frame pointers
11      ret
```

In our experiments, we find that function pushtest always returns zero. What does this imply about the behavior of the instruction pushl %esp under IA32?

A similar ambiguity occurs for the instruction popl %esp. It could either set %esp to the value read from memory or to the incremented stack pointer. As with Problem 6, let us run an experiment to determine how an IA32 machine would handle this instruction, and then design our Y86 machine to follow the same convention.

Practice Problem 7

The following assembly-code function lets us determine the behavior of the instruction popl %esp for IA32:

```
1       .text
2     .globl poptest
3     poptest:
4       pushl   %ebp
5       movl    %esp, %ebp
6       pushl   $0xabcd         Push test value
7       popl    %esp            Pop to stack pointer
8       movl    %esp, %eax      Set popped value as return value
9       leave                   Restore stack and frame pointers
10      ret
```

We find this function always returns 0xabcd. What does this imply about the behavior of popl %esp? What other Y86 instruction would have the exact same behavior?

Aside   Getting the details right: Inconsistencies across x86 models

Problems 6 and 7 are designed to help us devise a consistent set of conventions for instructions that push or pop the stack pointer. There seems to be little reason why one would want to perform either of these operations, and so a natural question to ask is "Why worry about such picky details?"

Several useful lessons can be learned about the importance of consistency from the following excerpt from the Intel documentation of the POP instruction [29]:

For IA-32 processors from the Intel 286 on, the PUSH ESP instruction pushes the value of the ESP register as it existed before the instruction was executed. (This is also true for Intel 64 architecture, real-address and virtual-8086 modes of IA-32 architecture.) For the Intel® 8086 processor, the PUSH SP instruction pushes the new value of the SP register (that is the value after it has been decremented by 2).

What this note states is that different models of x86 processors do different things when instructed to push the stack pointer register. Some push the original value, while others push the decremented value. (Interestingly, there is no corresponding ambiguity about popping to the stack pointer register.) There are two drawbacks to this inconsistency:

- It decreases code portability. Programs may have different behavior depending on the processor model. Although the particular instruction is not at all common, even the potential for incompatibility can have serious consequences.

- It complicates the documentation. As we see here, a special note is required to try to clarify the differences. The documentation for x86 is already complex enough without special cases such as this one.

We conclude, therefore, that working out details in advance and striving for complete consistency can save a lot of trouble in the long run.

2   Logic Design and the Hardware Control Language HCL

In hardware design, electronic circuits are used to compute functions on bits and to store bits in different kinds of memory elements. Most contemporary circuit technology represents different bit values as high or low voltages on signal wires. In current technology, logic value 1 is represented by a high voltage of around 1.0 volt, while logic value 0 is represented by a low voltage of around 0.0 volts. Three major components are required to implement a digital system: combinational logic to compute functions on the bits, memory elements to store bits, and clock signals to regulate the updating of the memory elements.

In this section, we provide a brief description of these different components. We also introduce HCL (for "hardware control language"), the language that we use to describe the control logic of the different processor designs. We only describe HCL informally here. A complete reference for HCL can be found in Web Aside ARCH:HCL.

349

Aside Modern logic design

At one time, hardware designers created circuit designs by drawing schematic diagrams of logic circuits (first with paper and pencil, and later with computer graphics terminals). Nowadays, most designs are expressed in a *hardware description language* (HDL), a textual notation that looks similar to a programming language but that is used to describe hardware structures rather than program behaviors. The most commonly used languages are Verilog, having a syntax similar to C, and VHDL, having a syntax similar to the Ada programming language. These languages were originally designed for expressing simulation models of digital circuits. In the mid-1980s, researchers developed *logic synthesis* programs that could generate efficient circuit designs from HDL descriptions. There are now a number of commercial synthesis programs, and this has become the dominant technique for generating digital circuits. This shift from hand-designed circuits to synthesized ones can be likened to the shift from writing programs in assembly code to writing them in a high-level language and having a compiler generate the machine code.

Our HCL language expresses only the control portions of a hardware design, with only a limited set of operations and with no modularity. As we will see, however, the control logic is the most difficult part of designing a microprocessor. We have developed tools that can directly translate HCL into Verilog, and by combining this code with Verilog code for the basic hardware units, we can generate HDL descriptions from which actual working microprocessors can be synthesized. By carefully separating out, designing, and testing the control logic, we can create a working microprocessor with reasonable effort. Web Aside ARCH:VLOG describes how we can generate Verilog versions of a Y86 processor.

2.1 Logic Gates

Logic gates are the basic computing elements for digital circuits. They generate an output equal to some Boolean function of the bit values at their inputs. Figure 9 shows the standard symbols used for Boolean functions AND, OR, and NOT. HCL expressions are shown below the gates for the operators in C: && for AND, || for OR, and ! for NOT. We use these instead of the bit-level C operators &, |, and ~, because logic gates operate on single-bit quantities, not entire words. Although the figure illustrates only two-input versions of the AND and OR gates, it is common to see these being used as n-way operations for $n > 2$. We still write these in HCL using binary operators, though, so the operation of a three-input AND gate with inputs a, b, and c is described with the HCL expression a && b && c.

Logic gates are always active. If some input to a gate changes, then within some small amount of time, the output will change accordingly.

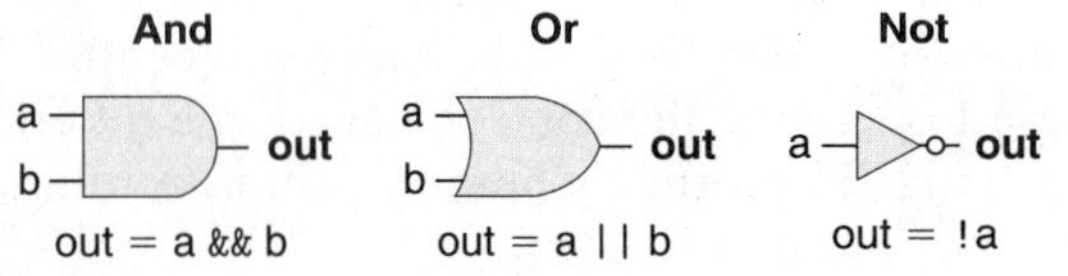

Figure 9

Logic gate types. Each gate generates output equal to some Boolean function of its inputs.

And
out = a && b

Or
out = a || b

Not
out = !a

Figure 10

Combinational circuit to test for bit equality. The output will equal 1 when both inputs are 0, or both are 1.

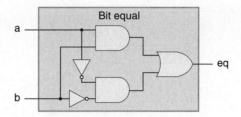

2.2 Combinational Circuits and HCL Boolean Expressions

By assembling a number of logic gates into a network, we can construct computational blocks known as *combinational circuits*. Two restrictions are placed on how the networks are constructed:

- The outputs of two or more logic gates cannot be connected together. Otherwise, the two could try to drive the wire in opposite directions, possibly causing an invalid voltage or a circuit malfunction.
- The network must be *acyclic*. That is, there cannot be a path through a series of gates that forms a loop in the network. Such loops can cause ambiguity in the function computed by the network.

Figure 10 shows an example of a simple combinational circuit that we will find useful. It has two inputs, a and b. It generates a single output eq, such that the output will equal 1 if either a and b are both 1 (detected by the upper AND gate) or are both 0 (detected by the lower AND gate). We write the function of this network in HCL as

```
bool eq = (a && b) || (!a && !b);
```

This code simply defines the bit-level (denoted by data type `bool`) signal eq as a function of inputs a and b. As this example shows, HCL uses C-style syntax, with '=' associating a signal name with an expression. Unlike C, however, we do not view this as performing a computation and assigning the result to some memory location. Instead, it is simply a way to give a name to an expression.

Practice Problem 8

Write an HCL expression for a signal xor, equal to the EXCLUSIVE-OR of inputs a and b. What is the relation between the signals xor and eq defined above?

Figure 11 shows another example of a simple but useful combinational circuit known as a *multiplexor* (commonly referred to as a "MUX"). A multiplexor selects a value from among a set of different data signals, depending on the value of a control input signal. In this single-bit multiplexor, the two data signals are the input bits a and b, while the control signal is the input bit s. The output will equal a when s is 1, and it will equal b when s is 0. In this circuit, we can see that the two AND gates determine whether to pass their respective data inputs to the OR gate.

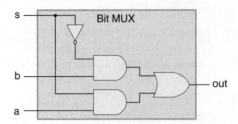

Figure 11

Single-bit multiplexor circuit. The output will equal input a if the control signal s is 1 and will equal input b when s is 0.

The upper AND gate passes signal b when s is 0 (since the other input to the gate is !s), while the lower AND gate passes signal a when s is 1. Again, we can write an HCL expression for the output signal, using the same operations as are present in the combinational circuit:

```
bool out = (s && a) || (!s && b);
```

Our HCL expressions demonstrate a clear parallel between combinational logic circuits and logical expressions in C. They both use Boolean operations to compute functions over their inputs. Several differences between these two ways of expressing computation are worth noting:

- Since a combinational circuit consists of a series of logic gates, it has the property that the outputs continually respond to changes in the inputs. If some input to the circuit changes, then after some delay, the outputs will change accordingly. In contrast, a C expression is only evaluated when it is encountered during the execution of a program.

- Logical expressions in C allow arguments to be arbitrary integers, interpreting 0 as FALSE and anything else as TRUE. In contrast, our logic gates only operate over the bit values 0 and 1.

- Logical expressions in C have the property that they might only be partially evaluated. If the outcome of an AND or OR operation can be determined by just evaluating the first argument, then the second argument will not be evaluated. For example, with the C expression

  ```
  (a && !a) && func(b,c)
  ```

  the function func will not be called, because the expression (a && !a) evaluates to 0. In contrast, combinational logic does not have any partial evaluation rules. The gates simply respond to changing inputs.

2.3 Word-Level Combinational Circuits and HCL Integer Expressions

By assembling large networks of logic gates, we can construct combinational circuits that compute much more complex functions. Typically, we design circuits that operate on data *words*. These are groups of bit-level signals that represent an integer or some control pattern. For example, our processor designs will contain numerous words, with word sizes ranging between 4 and 32 bits, representing integers, addresses, instruction codes, and register identifiers.

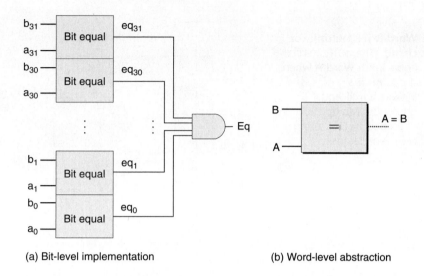

Figure 12

Word-level equality test circuit. The output will equal 1 when each bit from word A equals its counterpart from word B. Word-level equality is one of the operations in HCL.

(a) Bit-level implementation

(b) Word-level abstraction

Combinational circuits to perform word-level computations are constructed using logic gates to compute the individual bits of the output word, based on the individual bits of the input words. For example, Figure 12 shows a combinational circuit that tests whether two 32-bit words A and B are equal. That is, the output will equal 1 if and only if each bit of A equals the corresponding bit of B. This circuit is implemented using 32 of the single-bit equality circuits shown in Figure 10. The outputs of these single-bit circuits are combined with an AND gate to form the circuit output.

In HCL, we will declare any word-level signal as an int, without specifying the word size. This is done for simplicity. In a full-featured hardware description language, every word can be declared to have a specific number of bits. HCL allows words to be compared for equality, and so the functionality of the circuit shown in Figure 12 can be expressed at the word level as

```
bool Eq = (A == B);
```

where arguments A and B are of type int. Note that we use the same syntax conventions as in C, where '=' denotes assignment, while '==' denotes the equality operator.

As is shown on the right side of Figure 12, we will draw word-level circuits using medium-thickness lines to represent the set of wires carrying the individual bits of the word, and we will show the resulting Boolean signal as a dashed line.

Practice Problem 9

Suppose you want to implement a word-level equality circuit using the EXCLUSIVE-OR circuits from Problem 8 rather than from bit-level equality circuits. Design such a circuit for a 32-bit word consisting of 32 bit-level EXCLUSIVE-OR circuits and two additional logic gates.

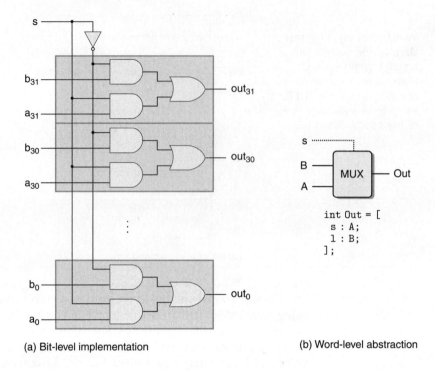

Figure 13

Word-level multiplexor circuit. The output will equal input word A when the control signal s is 1, and it will equal B otherwise. Multiplexors are described in HCL using case expressions.

(a) Bit-level implementation

(b) Word-level abstraction

```
int Out = [
    s : A;
    1 : B;
];
```

Figure 13 shows the circuit for a word-level multiplexor. This circuit generates a 32-bit word Out equal to one of the two input words, A or B, depending on the control input bit s. The circuit consists of 32 identical subcircuits, each having a structure similar to the bit-level multiplexor from Figure 11. Rather than simply replicating the bit-level multiplexor 32 times, the word-level version reduces the number of inverters by generating !s once and reusing it at each bit position.

We will use many forms of multiplexors in our processor designs. They allow us to select a word from a number of sources depending on some control condition. Multiplexing functions are described in HCL using *case expressions*. A case expression has the following general form:

```
[
    select_1    :    expr_1
    select_2    :    expr_2
              :
              :
    select_k    :    expr_k
]
```

The expression contains a series of cases, where each case *i* consists of a Boolean expression $select_i$, indicating when this case should be selected, and an integer expression $expr_i$, indicating the resulting value.

Unlike the switch statement of C, we do not require the different selection expressions to be mutually exclusive. Logically, the selection expressions are evaluated in sequence, and the case for the first one yielding 1 is selected. For example, the word-level multiplexor of Figure 13 can be described in HCL as

```
int Out = [
        s: A;
        1: B;
];
```

In this code, the second selection expression is simply 1, indicating that this case should be selected if no prior one has been. This is the way to specify a default case in HCL. Nearly all case expressions end in this manner.

Allowing nonexclusive selection expressions makes the HCL code more readable. An actual hardware multiplexor must have mutually exclusive signals controlling which input word should be passed to the output, such as the signals s and !s in Figure 13. To translate an HCL case expression into hardware, a logic synthesis program would need to analyze the set of selection expressions and resolve any possible conflicts by making sure that only the first matching case would be selected.

The selection expressions can be arbitrary Boolean expressions, and there can be an arbitrary number of cases. This allows case expressions to describe blocks where there are many choices of input signals with complex selection criteria. For example, consider the diagram of a four-way multiplexor shown in Figure 14. This circuit selects from among the four input words A, B, C, and D based on the control signals s1 and s0, treating the controls as a 2-bit binary number. We can express this in HCL using Boolean expressions to describe the different combinations of control bit patterns:

```
int Out4 = [
        !s1 && !s0 : A; # 00
        !s1        : B; # 01
        !s0        : C; # 10
        1          : D; # 11
];
```

The comments on the right (any text starting with # and running for the rest of the line is a comment) show which combination of s1 and s0 will cause the case to

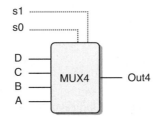

Figure 14

Four-way multiplexor.
The different combinations
of control signals s1 and
s0 determine which data
input is transmitted to the
output.

be selected. Observe that the selection expressions can sometimes be simplified, since only the first matching case is selected. For example, the second expression can be written !s1, rather than the more complete !s1 && s0, since the only other possibility having s1 equal to 0 was given as the first selection expression. Similarly, the third expression can be written as !s0, while the fourth can simply be written as 1.

As a final example, suppose we want to design a logic circuit that finds the minimum value among a set of words A, B, and C, diagrammed as follows:

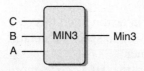

We can express this using an HCL case expression as

```
int Min3 = [
        A <= B && A <= C  :  A;
        B <= A && B <= C  :  B;
        1                 :  C;
];
```

Practice Problem 10

Write HCL code describing a circuit that for word inputs A, B, and C selects the *median* of the three values. That is, the output equals the word lying between the minimum and maximum of the three inputs.

Combinational logic circuits can be designed to perform many different types of operations on word-level data. The detailed design of these is beyond the scope of our presentation. One important combinational circuit, known as an *arithmetic/logic unit* (ALU), is diagrammed at an abstract level in Figure 15. This circuit has three inputs: two data inputs labeled A and B, and a control input. Depending on the setting of the control input, the circuit will perform different arithmetic or logical operations on the data inputs. Observe that the four

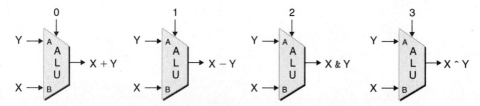

Figure 15 **Arithmetic/logic unit (ALU).** Depending on the setting of the function input, the circuit will perform one of four different arithmetic and logical operations.

operations diagrammed for this ALU correspond to the four different integer operations supported by the Y86 instruction set, and the control values match the function codes for these instructions (Figure 3). Note also the ordering of operands for subtraction, where the A input is subtracted from the B input. This ordering is chosen in anticipation of the ordering of arguments in the subl instruction.

2.4 Set Membership

In our processor designs, we will find many examples where we want to compare one signal against a number of possible matching signals, such as to test whether the code for some instruction being processed matches some category of instruction codes. As a simple example, suppose we want to generate the signals s1 and s0 for the four-way multiplexor of Figure 14 by selecting the high- and low-order bits from a 2-bit signal code, as follows:

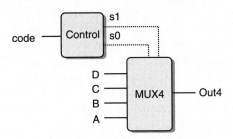

In this circuit, the 2-bit signal code would then control the selection among the four data words A, B, C, and D. We can express the generation of signals s1 and s0 using equality tests based on the possible values of code:

```
bool s1 = code == 2 || code == 3;

bool s0 = code == 1 || code == 3;
```

A more concise expression can be written that expresses the property that s1 is 1 when code is in the set {2, 3}, and s0 is 1 when code is in the set {1, 3}:

```
bool s1 = code in { 2, 3 };

bool s0 = code in { 1, 3 };
```

The general form of a set membership test is

$$iexpr \text{ in } \{iexpr_1, iexpr_2, \ldots, iexpr_k\}$$

where the value being tested, *iexpr*, and the candidate matches, *iexpr*$_1$ through *iexpr*$_k$, are all integer expressions.

2.5 Memory and Clocking

Combinational circuits, by their very nature, do not store any information. Instead, they simply react to the signals at their inputs, generating outputs equal to some function of the inputs. To create *sequential circuits*, that is, systems that have state and perform computations on that state, we must introduce devices that store information represented as bits. Our storage devices are all controlled by a single *clock*, a periodic signal that determines when new values are to be loaded into the devices. We consider two classes of memory devices:

Clocked registers (or simply *registers*) store individual bits or words. The clock signal controls the loading of the register with the value at its input.

Random-access memories (or simply *memories*) store multiple words, using an address to select which word should be read or written. Examples of random-access memories include (1) the virtual memory system of a processor, where a combination of hardware and operating system software make it appear to a processor that it can access any word within a large address space; and (2) the register file, where register identifiers serve as the addresses. In an IA32 or Y86 processor, the register file holds the eight program registers (%eax, %ecx, etc.).

As we can see, the word "register" means two slightly different things when speaking of hardware versus machine-language programming. In hardware, a register is directly connected to the rest of the circuit by its input and output wires. In machine-level programming, the registers represent a small collection of addressable words in the CPU, where the addresses consist of register IDs. These words are generally stored in the register file, although we will see that the hardware can sometimes pass a word directly from one instruction to another to avoid the delay of first writing and then reading the register file. When necessary to avoid ambiguity, we will call the two classes of registers "hardware registers" and "program registers," respectively.

Figure 16 gives a more detailed view of a hardware register and how it operates. For most of the time, the register remains in a fixed state (shown as x), generating an output equal to its current state. Signals propagate through the combinational logic preceding the register, creating a new value for the register input (shown as y), but the register output remains fixed as long as the clock is low. As the clock rises, the input signals are loaded into the register as its next state (y), and this becomes the new register output until the next rising clock edge. A key point is that the registers serve as barriers between the combinational logic in different parts of the circuit. Values only propagate from a register input to its output once every clock cycle at the rising clock edge. Our Y86 processors will

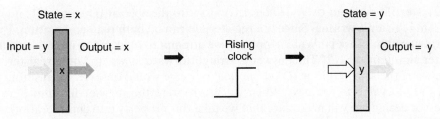

Figure 16 Register operation. The register outputs remain held at the current register state until the clock signal rises. When the clock rises, the values at the register inputs are captured to become the new register state.

use clocked registers to hold the program counter (PC), the condition codes (CC), and the program status (Stat).

The following diagram shows a typical register file:

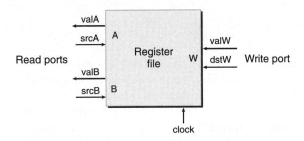

This register file has two *read ports*, named A and B, and one *write port*, named W. Such a *multiported* random-access memory allows multiple read and write operations to take place simultaneously. In the register file diagrammed, the circuit can read the values of two program registers and update the state of a third. Each port has an address input, indicating which program register should be selected, and a data output or input giving a value for that program register. The addresses are register identifiers, using the encoding shown in Figure 4. The two read ports have address inputs srcA and srcB (short for "source A" and "source B") and data outputs valA and valB (short for "value A" and "value B"). The write port has address input dstW (short for "destination W") and data input valW (short for "value W").

The register file is not a combinational circuit, since it has internal storage. In our implementation, however, data can be read from the register file as if it were a block of combinational logic having addresses as inputs and the data as outputs. When either srcA or srcB is set to some register ID, then, after some delay, the value stored in the corresponding program register will appear on either valA or valB. For example, setting srcA to 3 will cause the value of program register %ebx to be read, and this value will appear on output valA.

The writing of words to the register file is controlled by the clock signal in a manner similar to the loading of values into a clocked register. Every time the clock rises, the value on input valW is written to the program register indicated by the register ID on input dstW. When dstW is set to the special ID value 0xF, no program register is written. Since the register file can be both read and written, a natural question to ask is "What happens if we attempt to read and write the same register simultaneously?" The answer is straightforward: if we update a register while using the same register ID on the read port, we would observe a transition from the old value to the new. When we incorporate the register file into our processor design, we will make sure that we take this property into consideration.

Our processor has a random-access memory for storing program data, as illustrated below:

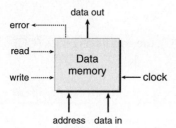

This memory has a single address input, a data input for writing, and a data output for reading. Like the register file, reading from our memory operates in a manner similar to combinational logic: If we provide an address on the address input and set the write control signal to 0, then after some delay, the value stored at that address will appear on data out. The error signal will be set to 1 if the address is out of range and to 0 otherwise. Writing to the memory is controlled by the clock: we set address to the desired address, data in to the desired value, and write to 1. When we then operate the clock, the specified location in the memory will be updated, as long as the address is valid. As with the read operation, the error signal will be set to 1 if the address is invalid. This signal is generated by combinational logic, since the required bounds checking is purely a function of the address input and does not involve saving any state.

Aside Real-life memory design

The memory system in a full-scale microprocessor is far more complex than the simple one we assume in our design. It consists of several forms of hardware memories, including several random-access memories plus magnetic disk, as well as a variety of hardware and software mechanisms for managing these devices.

Nonetheless, our simple memory design can be used for smaller systems, and it provides us with an abstraction of the interface between the processor and memory for more complex systems.

Our processor includes an additional read-only memory for reading instructions. In most actual systems, these memories are merged into a single memory with two ports: one for reading instructions and the other for reading or writing data.

3 Sequential Y86 Implementations

Now we have the components required to implement a Y86 processor. As a first step, we describe a processor called SEQ (for "sequential" processor). On each clock cycle, SEQ performs all the steps required to process a complete instruction. This would require a very long cycle time, however, and so the clock rate would be unacceptably low. Our purpose in developing SEQ is to provide a first step toward our ultimate goal of implementing an efficient, pipelined processor.

3.1 Organizing Processing into Stages

In general, processing an instruction involves a number of operations. We organize them in a particular sequence of stages, attempting to make all instructions follow a uniform sequence, even though the instructions differ greatly in their actions. The detailed processing at each step depends on the particular instruction being executed. Creating this framework will allow us to design a processor that makes best use of the hardware. The following is an informal description of the stages and the operations performed within them:

Fetch: The fetch stage reads the bytes of an instruction from memory, using the program counter (PC) as the memory address. From the instruction it extracts the two 4-bit portions of the instruction specifier byte, referred to as icode (the instruction code) and ifun (the instruction function). It possibly fetches a register specifier byte, giving one or both of the register operand specifiers rA and rB. It also possibly fetches a 4-byte constant word valC. It computes valP to be the address of the instruction following the current one in sequential order. That is, valP equals the value of the PC plus the length of the fetched instruction.

Decode: The decode stage reads up to two operands from the register file, giving values valA and/or valB. Typically, it reads the registers designated by instruction fields rA and rB, but for some instructions it reads register %esp.

Execute: In the execute stage, the arithmetic/logic unit (ALU) either performs the operation specified by the instruction (according to the value of ifun), computes the effective address of a memory reference, or increments or decrements the stack pointer. We refer to the resulting value as valE. The condition codes are possibly set. For a jump instruction, the stage tests the condition codes and branch condition (given by ifun) to see whether or not the branch should be taken.

Memory: The memory stage may write data to memory, or it may read data from memory. We refer to the value read as valM.

Write back: The write-back stage writes up to two results to the register file.

PC update: The PC is set to the address of the next instruction.

The processor loops indefinitely, performing these stages. In our simplified implementation, the processor will stop when any exception occurs: it executes a

```
 1    0x000: 30f209000000 |     irmovl $9,   %edx
 2    0x006: 30f315000000 |     irmovl $21,  %ebx
 3    0x00c: 6123         |     subl %edx, %ebx        # subtract
 4    0x00e: 30f480000000 |     irmovl $128,%esp       # Problem 11
 5    0x014: 404364000000 |     rmmovl %esp, 100(%ebx) # store
 6    0x01a: a02f         |     pushl %edx             # push
 7    0x01c: b00f         |     popl  %eax             # Problem 12
 8    0x01e: 7328000000   |     je done                # Not taken
 9    0x023: 8029000000   |     call proc              # Problem 16
10    0x028:              | done:
11    0x028: 00           |     halt
12    0x029:              | proc:
13    0x029: 90           |     ret                    # Return
```

Figure 17 **Sample Y86 instruction sequence.** We will trace the processing of these instructions through the different stages.

halt or invalid instruction, or it attempts to read or write an invalid address. In a more complete design, the processor would enter an exception-handling mode and begin executing special code determined by the type of exception.

As can be seen by the preceding description, there is a surprising amount of processing required to execute a single instruction. Not only must we perform the stated operation of the instruction, we must also compute addresses, update stack pointers, and determine the next instruction address. Fortunately, the overall flow can be similar for every instruction. Using a very simple and uniform structure is important when designing hardware, since we want to minimize the total amount of hardware, and we must ultimately map it onto the two-dimensional surface of an integrated-circuit chip. One way to minimize the complexity is to have the different instructions share as much of the hardware as possible. For example, each of our processor designs contains a single arithmetic/logic unit that is used in different ways depending on the type of instruction being executed. The cost of duplicating blocks of logic in hardware is much higher than the cost of having multiple copies of code in software. It is also more difficult to deal with many special cases and idiosyncrasies in a hardware system than with software.

Our challenge is to arrange the computing required for each of the different instructions to fit within this general framework. We will use the code shown in Figure 17 to illustrate the processing of different Y86 instructions. Figures 18 through 21 contain tables describing how the different Y86 instructions proceed through the stages. It is worth the effort to study these tables carefully. They are in a form that enables a straightforward mapping into the hardware. Each line in these tables describes an assignment to some signal or stored state (indicated by the assignment operation ←). These should be read as if they were evaluated in sequence from top to bottom. When we later map the computations to hardware, we will find that we do not need to perform these evaluations in strict sequential order.

Stage	OP1 rA, rB	rrmovl rA, rB	irmovl V, rB
Fetch	icode : ifun ← $M_1[PC]$ rA : rB ← $M_1[PC + 1]$ valP ← $PC + 2$	icode : ifun ← $M_1[PC]$ rA : rB ← $M_1[PC + 1]$ valP ← $PC + 2$	icode : ifun ← $M_1[PC]$ rA : rB ← $M_1[PC + 1]$ valC ← $M_4[PC + 2]$ valP ← $PC + 6$
Decode	valA ← R[rA] valB ← R[rB]	valA ← R[rA]	
Execute	valE ← valB OP valA Set CC	valE ← $0 +$ valA	valE ← $0 +$ valC
Memory			
Write back	R[rB] ← valE	R[rB] ← valE	R[rB] ← valE
PC update	PC ← valP	PC ← valP	PC ← valP

Figure 18 **Computations in sequential implementation of Y86 instructions** OP1, rrmovl, **and** irmovl. These instructions compute a value and store the result in a register. The notation icode : ifun indicates the two components of the instruction byte, while rA : rB indicates the two components of the register specifier byte. The notation $M_1[x]$ indicates accessing (either reading or writing) 1 byte at memory location x, while $M_4[x]$ indicates accessing 4 bytes.

Figure 18 shows the processing required for instruction types OP1 (integer and logical operations), rrmovl (register-register move), and irmovl (immediate-register move). Let us first consider the integer operations. Examining Figure 2, we can see that we have carefully chosen an encoding of instructions so that the four integer operations (addl, subl, andl, and xorl) all have the same value of icode. We can handle them all by an identical sequence of steps, except that the ALU computation must be set according to the particular instruction operation, encoded in ifun.

The processing of an integer-operation instruction follows the general pattern listed above. In the fetch stage, we do not require a constant word, and so valP is computed as $PC + 2$. During the decode stage, we read both operands. These are supplied to the ALU in the execute stage, along with the function specifier ifun, so that valE becomes the instruction result. This computation is shown as the expression valB OP valA, where OP indicates the operation specified by ifun. Note the ordering of the two arguments—this order is consistent with the conventions of Y86 (and IA32). For example, the instruction subl %eax,%edx is supposed to compute the value R[%edx] − R[%eax]. Nothing happens in the memory stage for these instructions, but valE is written to register rB in the write-back stage, and the PC is set to valP to complete the instruction execution.

Aside   Tracing the execution of a `subl` instruction

As an example, let us follow the processing of the `subl` instruction on line 3 of the object code shown in Figure 17. We can see that the previous two instructions initialize registers `%edx` and `%ebx` to 9 and 21, respectively. We can also see that the instruction is located at address 0x00c and consists of 2 bytes, having values 0x61 and 0x23. The stages would proceed as shown in the following table, which lists the generic rule for processing an `OP1` instruction (Figure 18) on the left, and the computations for this specific instruction on the right.

Stage	Generic OP1 rA, rB	Specific subl %edx, %ebx
Fetch	icode : ifun $\leftarrow M_1[PC]$ rA : rB $\leftarrow M_1[PC+1]$ valP $\leftarrow PC + 2$	icode : ifun $\leftarrow M_1[0x00c] = 6:1$ rA : rB $\leftarrow M_1[0x00d] = 2:3$ valP $\leftarrow 0x00c + 2 = 0x00e$
Decode	valA $\leftarrow R[rA]$ valB $\leftarrow R[rB]$	valA $\leftarrow R[\%edx] = 9$ valB $\leftarrow R[\%ebx] = 21$
Execute	valE $\leftarrow$ valB OP valA Set CC	valE $\leftarrow 21 - 9 = 12$ ZF $\leftarrow 0$, SF $\leftarrow 0$, OF $\leftarrow 0$
Memory		
Write back	R[rB] $\leftarrow$ valE	R[%ebx] $\leftarrow$ valE $= 12$
PC update	PC $\leftarrow$ valP	PC $\leftarrow$ valP $= 0x00e$

As this trace shows, we achieve the desired effect of setting register `%ebx` to 12, setting all three condition codes to zero, and incrementing the PC by 2.

Executing an `rrmovl` instruction proceeds much like an arithmetic operation. We do not need to fetch the second register operand, however. Instead, we set the second ALU input to zero and add this to the first, giving valE = valA, which is then written to the register file. Similar processing occurs for `irmovl`, except that we use constant value valC for the first ALU input. In addition, we must increment the program counter by 6 for `irmovl` due to the long instruction format. Neither of these instructions changes the condition codes.

Practice Problem 11

Fill in the right-hand column of the following table to describe the processing of the `irmovl` instruction on line 4 of the object code in Figure 17:

Stage	Generic irmovl V, rB	Specific irmovl $128, %esp
Fetch	icode:ifun ← $M_1[PC]$ rA:rB ← $M_1[PC+1]$ valC ← $M_4[PC+2]$ valP ← PC + 6	_____ _____ _____ _____
Decode		
Execute	valE ← 0 + valC	_____
Memory		
Write back	R[rB] ← valE	_____
PC update	PC ← valP	_____

How does this instruction execution modify the registers and the PC?

Figure 19 shows the processing required for the memory write and read instructions rmmovl and mrmovl. We see the same basic flow as before, but using the ALU to add valC to valB, giving the effective address (the sum of the displacement

Stage	rmmovl rA, D(rB)	mrmovl D(rB), rA
Fetch	icode:ifun ← $M_1[PC]$ rA:rB ← $M_1[PC+1]$ valC ← $M_4[PC+2]$ valP ← PC + 6	icode:ifun ← $M_1[PC]$ rA:rB ← $M_1[PC+1]$ valC ← $M_4[PC+2]$ valP ← PC + 6
Decode	valA ← R[rA] valB ← R[rB]	valB ← R[rB]
Execute	valE ← valB + valC	valE ← valB + valC
Memory	M_4[valE] ← valA	valM ← M_4[valE]
Write back		R[rA] ← valM
PC update	PC ← valP	PC ← valP

Figure 19 **Computations in sequential implementation of Y86 instructions rmmovl and mrmovl. These instructions read or write memory.**

and the base register value) for the memory operation. In the memory stage we either write the register value valA to memory, or we read valM from memory.

Aside   Tracing the execution of an `rmmovl` instruction

Let us trace the processing of the `rmmovl` instruction on line 5 of the object code shown in Figure 17. We can see that the previous instruction initialized register `%esp` to 128, while `%ebx` still holds 12, as computed by the `subl` instruction (line 3). We can also see that the instruction is located at address 0x014 and consists of 6 bytes. The first 2 have values 0x40 and 0x43, while the final 4 are a byte-reversed version of the number 0x00000064 (decimal 100). The stages would proceed as follows:

Stage	Generic rmmovl rA, D(rB)	Specific rmmovl %esp, 100(%ebx)
Fetch	$\text{icode} : \text{ifun} \leftarrow M_1[PC]$ $rA : rB \leftarrow M_1[PC + 1]$ $\text{valC} \leftarrow M_4[PC + 2]$ $\text{valP} \leftarrow PC + 6$	$\text{icode} : \text{ifun} \leftarrow M_1[0x014] = 4 : 0$ $rA : rB \leftarrow M_1[0x015] = 4 : 3$ $\text{valC} \leftarrow M_4[0x016] = 100$ $\text{valP} \leftarrow 0x014 + 6 = 0x01a$
Decode	$\text{valA} \leftarrow R[rA]$ $\text{valB} \leftarrow R[rB]$	$\text{valA} \leftarrow R[\%esp] = 128$ $\text{valB} \leftarrow R[\%ebx] = 12$
Execute	$\text{valE} \leftarrow \text{valB} + \text{valC}$	$\text{valE} \leftarrow 12 + 100 = 112$
Memory	$M_4[\text{valE}] \leftarrow \text{valA}$	$M_4[112] \leftarrow 128$
Write back		
PC update	$PC \leftarrow \text{valP}$	$PC \leftarrow 0x01a$

As this trace shows, the instruction has the effect of writing 128 to memory address 112 and incrementing the PC by 6.

Figure 20 shows the steps required to process `pushl` and `popl` instructions. These are among the most difficult Y86 instructions to implement, because they involve both accessing memory and incrementing or decrementing the stack pointer. Although the two instructions have similar flows, they have important differences.

The `pushl` instruction starts much like our previous instructions, but in the decode stage we use `%esp` as the identifier for the second register operand, giving the stack pointer as value valB. In the execute stage, we use the ALU to decrement the stack pointer by 4. This decremented value is used for the memory write address and is also stored back to `%esp` in the write-back stage. By using valE as the address for the write operation, we adhere to the Y86 (and IA32) convention that `pushl` should decrement the stack pointer before writing, even though the actual updating of the stack pointer does not occur until after the memory operation has completed.

Stage	pushl rA	popl rA
Fetch	icode:ifun ← $M_1[PC]$ rA:rB ← $M_1[PC + 1]$ valP ← PC + 2	icode:ifun ← $M_1[PC]$ rA:rB ← $M_1[PC + 1]$ valP ← PC + 2
Decode	valA ← R[rA] valB ← R[%esp]	valA ← R[%esp] valB ← R[%esp]
Execute	valE ← valB + (−4)	valE ← valB + 4
Memory	$M_4[valE]$ ← valA	valM ← $M_4[valA]$
Write back	R[%esp] ← valE	R[%esp] ← valE R[rA] ← valM
PC update	PC ← valP	PC ← valP

Figure 20 **Computations in sequential implementation of Y86 instructions** pushl **and** popl. These instructions push and pop the stack.

Aside   Tracing the execution of a pushl instruction

Let us trace the processing of the pushl instruction on line 6 of the object code shown in Figure 17. At this point, we have 9 in register %edx and 128 in register %esp. We can also see that the instruction is located at address 0x01a and consists of 2 bytes having values 0xa0 and 0x28. The stages would proceed as follows:

Stage	Generic pushl rA	Specific pushl %edx
Fetch	icode:ifun ← $M_1[PC]$ rA:rB ← $M_1[PC + 1]$ valP ← PC + 2	icode:ifun ← $M_1[0x01a] = a : 0$ rA:rB ← $M_1[0x01b] = 2 : 8$ valP ← 0x01a + 2 = 0x01c
Decode	valA ← R[rA] valB ← R[%esp]	valA ← R[%edx] = 9 valB ← R[%esp] = 128
Execute	valE ← valB + (−4)	valE ← 128 + (−4) = 124
Memory	$M_4[valE]$ ← valA	$M_4[124]$ ← 9
Write back	R[%esp] ← valE	R[%esp] ← 124
PC update	PC ← valP	PC ← 0x01c

As this trace shows, the instruction has the effect of setting %esp to 124, writing 9 to address 124, and incrementing the PC by 2.

The popl instruction proceeds much like pushl, except that we read two copies of the stack pointer in the decode stage. This is clearly redundant, but we will see that having the stack pointer as both valA and valB makes the subsequent flow more similar to that of other instructions, enhancing the overall uniformity of the design. We use the ALU to increment the stack pointer by 4 in the execute stage, but use the unincremented value as the address for the memory operation. In the write-back stage, we update both the stack pointer register with the incremented stack pointer, and register rA with the value read from memory. Using the unincremented stack pointer as the memory read address preserves the Y86 (and IA32) convention that popl should first read memory and then increment the stack pointer.

Practice Problem 12

Fill in the right-hand column of the following table to describe the processing of the popl instruction on line 7 of the object code in Figure 17:

Stage	Generic popl rA	Specific popl %eax
Fetch	icode:ifun ← $M_1[PC]$	
	rA:rB ← $M_1[PC+1]$	
	valP ← PC + 2	
Decode	valA ← R[%esp]	
	valB ← R[%esp]	
Execute	valE ← valB + 4	
Memory	valM ← $M_4[valA]$	
Write back	R[%esp] ← valE	
	R[rA] ← valM	
PC update	PC ← valP	

What effect does this instruction execution have on the registers and the PC?

Practice Problem 13

What would be the effect of the instruction pushl %esp according to the steps listed in Figure 20? Does this conform to the desired behavior for Y86, as determined in Problem 6?

Practice Problem 14

Assume the two register writes in the write-back stage for popl occur in the order listed in Figure 20. What would be the effect of executing popl %esp? Does this conform to the desired behavior for Y86, as determined in Problem 7?

Figure 21 indicates the processing of our three control transfer instructions: the different jumps, call, and ret. We see that we can implement these instructions with the same overall flow as the preceding ones.

As with integer operations, we can process all of the jumps in a uniform manner, since they differ only when determining whether or not to take the branch. A jump instruction proceeds through fetch and decode much like the previous instructions, except that it does not require a register specifier byte. In the execute stage, we check the condition codes and the jump condition to determine whether or not to take the branch, yielding a 1-bit signal Cnd. During the PC update stage, we test this flag, and set the PC to valC (the jump target) if the flag is 1, and to valP (the address of the following instruction) if the flag is 0. Our notation $x \, ? \, a : b$ is similar to the conditional expression in C—it yields a when x is nonzero and b when x is zero.

Stage	jXX Dest	call Dest	ret
Fetch	$\text{icode:ifun} \leftarrow M_1[PC]$	$\text{icode:ifun} \leftarrow M_1[PC]$	$\text{icode:ifun} \leftarrow M_1[PC]$
	$\text{valC} \leftarrow M_4[PC+1]$	$\text{valC} \leftarrow M_4[PC+1]$	
	$\text{valP} \leftarrow PC + 5$	$\text{valP} \leftarrow PC + 5$	$\text{valP} \leftarrow PC + 1$
Decode			$\text{valA} \leftarrow R[\%esp]$
		$\text{valB} \leftarrow R[\%esp]$	$\text{valB} \leftarrow R[\%esp]$
Execute		$\text{valE} \leftarrow \text{valB} + (-4)$	$\text{valE} \leftarrow \text{valB} + 4$
	$\text{Cnd} \leftarrow \text{Cond(CC, ifun)}$		
Memory		$M_4[\text{valE}] \leftarrow \text{valP}$	$\text{valM} \leftarrow M_4[\text{valA}]$
Write back		$R[\%esp] \leftarrow \text{valE}$	$R[\%esp] \leftarrow \text{valE}$
PC update	$PC \leftarrow \text{Cnd ? valC : valP}$	$PC \leftarrow \text{valC}$	$PC \leftarrow \text{valM}$

Figure 21 **Computations in sequential implementation of Y86 instructions** jXX, call, **and** ret. These instructions cause control transfers.

Aside Tracing the execution of a `je` instruction

Let us trace the processing of the `je` instruction on line 8 of the object code shown in Figure 17. The condition codes were all set to zero by the `subl` instruction (line 3), and so the branch will not be taken. The instruction is located at address 0x01e and consists of 5 bytes. The first has value 0x73, while the remaining 4 are a byte-reversed version of the number 0x00000028, the jump target. The stages would proceed as follows:

Stage	Generic jXX Dest	Specific je 0x028
Fetch	icode:ifun ← $M_1[\text{PC}]$	icode:ifun ← $M_1[\text{0x01e}] = 7:3$
	valC ← $M_4[\text{PC}+1]$	valC ← $M_4[\text{0x01f}] = \text{0x028}$
	valP ← $\text{PC}+5$	valP ← $\text{0x01e}+5 = \text{0x023}$
Decode		
Execute		
	Cnd ← Cond(CC, ifun)	Cnd ← Cond($\langle 0,0,0\rangle$, 3) = 0
Memory		
Write back		
PC update	PC ← Cnd ? valC : valP	PC ← 0 ? 0x028 : 0x023 = 0x023

As this trace shows, the instruction has the effect of incrementing the PC by 5.

Practice Problem 15

We can see by the instruction encodings (Figures 2 and 3) that the `rmmovl` instruction is the unconditional version of a more general class of instructions that include the conditional moves. Show how you would modify the steps for the `rrmovl` instruction below to also handle the six conditional move instructions. You may find it useful to see how the implementation of the `jXX` instructions (Figure 21) handles conditional behavior.

Stage	cmovXX rA, rB
Fetch	icode:ifun ← $M_1[\text{PC}]$
	rA:rB ← $M_1[\text{PC}+1]$
	valP ← $\text{PC}+2$
Decode	valA ← R[rA]
Execute	valE ← $0 + \text{valA}$
Memory	
Write back	
	R[rB] ← valE
PC update	PC ← valP

Instructions call and ret bear some similarity to instructions pushl and popl, except that we push and pop program counter values. With instruction call, we push valP, the address of the instruction that follows the call instruction. During the PC update stage, we set the PC to valC, the call destination. With instruction ret, we assign valM, the value popped from the stack, to the PC in the PC update stage.

Practice Problem 16

Fill in the right-hand column of the following table to describe the processing of the call instruction on line 9 of the object code in Figure 17:

Stage	Generic call Dest	Specific call 0x029
Fetch	icode : ifun ← M_1[PC]	
	valC ← M_4[PC + 1]	
	valP ← PC + 5	
Decode		
	valB ← R[%esp]	
Execute	valE ← valB + (−4)	
Memory	M_4[valE] ← valP	
Write back	R[%esp] ← valE	
PC update	PC ← valC	

What effect would this instruction execution have on the registers, the PC, and the memory?

We have created a uniform framework that handles all of the different types of Y86 instructions. Even though the instructions have widely varying behavior, we can organize the processing into six stages. Our task now is to create a hardware design that implements the stages and connects them together.

Aside   Tracing the execution of a ret instruction

Let us trace the processing of the ret instruction on line 13 of the object code shown in Figure 17. The instruction address is 0x029 and is encoded by a single byte 0x90. The previous call instruction set %esp to 124 and stored the return address 0x028 at memory address 124. The stages would proceed as follows:

Stage	Generic ret	Specific ret
Fetch	icode : ifun ← $M_1[PC]$	icode : ifun ← $M_1[0x029] = 9 : 0$
	valP ← PC + 1	valP ← 0x029 + 1 = 0x02a
Decode	valA ← R[%esp]	valA ← R[%esp] = 124
	valB ← R[%esp]	valB ← R[%esp] = 124
Execute	valE ← valB + 4	valE ← 124 + 4 = 128
Memory	valM ← $M_4[valA]$	valM ← $M_4[124]$ = 0x028
Write back	R[%esp] ← valE	R[%esp] ← 128
PC update	PC ← valM	PC ← 0x028

As this trace shows, the instruction has the effect of setting the PC to 0x028, the address of the halt instruction. It also sets %esp to 128.

3.2 SEQ Hardware Structure

The computations required to implement all of the Y86 instructions can be organized as a series of six basic stages: fetch, decode, execute, memory, write back, and PC update. Figure 22 shows an abstract view of a hardware structure that can perform these computations. The program counter is stored in a register, shown in the lower left-hand corner (labeled "PC"). Information then flows along wires (shown grouped together as a heavy black line), first upward and then around to the right. Processing is performed by *hardware units* associated with the different stages. The feedback paths coming back down on the right-hand side contain the updated values to write to the register file and the updated program counter. In SEQ, all of the processing by the hardware units occurs within a single clock cycle, as is discussed in Section 3.3. This diagram omits some small blocks of combinational logic as well as all of the control logic needed to operate the different hardware units and to route the appropriate values to the units. We will add this detail later. Our method of drawing processors with the flow going from bottom to top is unconventional. We will explain the reason for our convention when we start designing pipelined processors.

The hardware units are associated with the different processing stages:

Fetch: Using the program counter register as an address, the instruction memory reads the bytes of an instruction. The PC incrementer computes valP, the incremented program counter.

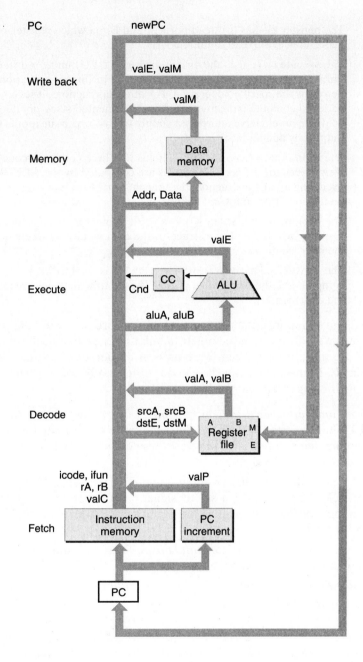

Figure 22
Abstract view of SEQ, a sequential implementation. The information processed during execution of an instruction follows a clockwise flow starting with an instruction fetch using the program counter (PC), shown in the lower left-hand corner of the figure.

Decode: The register file has two read ports, A and B, via which register values valA and valB are read simultaneously.

Execute: The execute stage uses the arithmetic/logic (ALU) unit for different purposes according to the instruction type. For integer operations, it performs the specified operation. For other instructions, it serves as an adder to compute an incremented or decremented stack pointer, to compute an effective address, or simply to pass one of its inputs to its outputs by adding zero.

The condition code register (CC) holds the three condition-code bits. New values for the condition codes are computed by the ALU. When executing a jump instruction, the branch signal Cnd is computed based on the condition codes and the jump type.

Memory: The data memory reads or writes a word of memory when executing a memory instruction. The instruction and data memories access the same memory locations, but for different purposes.

Write back: The register file has two write ports. Port E is used to write values computed by the ALU, while port M is used to write values read from the data memory.

Figure 23 gives a more detailed view of the hardware required to implement SEQ (although we will not see the complete details until we examine the individual stages). We see the same set of hardware units as earlier, but now the wires are shown explicitly. In this figure, as well as in our other hardware diagrams, we use the following drawing conventions:

* *Hardware units are shown as light blue boxes.* These include the memories, the ALU, and so forth. We will use the same basic set of units for all of our processor implementations. We will treat these units as "black boxes" and not go into their detailed designs.

* *Control logic blocks are drawn as gray rounded rectangles.* These blocks serve to select from among a set of signal sources, or to compute some Boolean function. We will examine these blocks in complete detail, including developing HCL descriptions.

* *Wire names are indicated in white round boxes.* These are simply labels on the wires, not any kind of hardware element.

* *Word-wide data connections are shown as medium lines.* Each of these lines actually represents a bundle of 32 wires, connected in parallel, for transferring a word from one part of the hardware to another.

* *Byte and narrower data connections are shown as thin lines.* Each of these lines actually represents a bundle of four or eight wires, depending on what type of values must be carried on the wires.

* *Single-bit connections are shown as dotted lines.* These represent control values passed between the units and blocks on the chip.

All of the computations we have shown in Figures 18 through 21 have the property that each line represents either the computation of a specific value, such

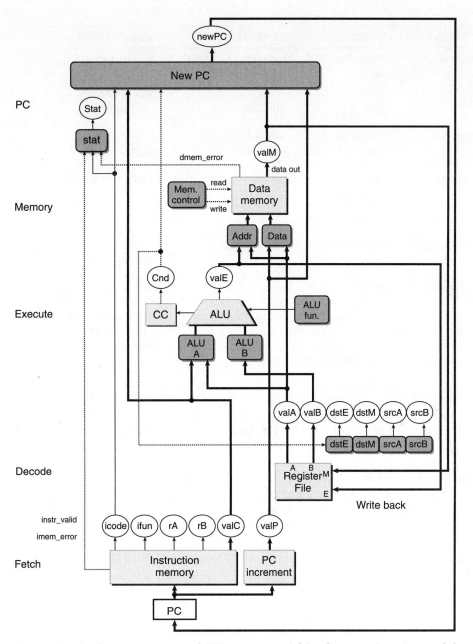

Figure 23 **Hardware structure of SEQ, a sequential implementation.** Some of the control signals, as well as the register and control word connections, are not shown.

Stage	Computation	OP1 rA, rB	mrmovl D(rB), rA
Fetch	icode, ifun	icode : ifun ← $M_1[PC]$	icode : ifun ← $M_1[PC]$
	rA, rB	rA : rB ← $M_1[PC + 1]$	rA : rB ← $M_1[PC + 1]$
	valC		valC ← $M_4[PC + 2]$
	valP	valP ← $PC + 2$	valP ← $PC + 6$
Decode	valA, srcA	valA ← R[rA]	
	valB, srcB	valB ← R[rB]	valB ← R[rB]
Execute	valE	valE ← valB OP valA	valE ← valB + valC
	Cond. codes	Set CC	
Memory	read/write		valM ← M_4[valE]
Write back	E port, dstE	R[rB] ← valE	
	M port, dstM		R[rA] ← valM
PC update	PC	PC ← valP	PC ← valP

Figure 24   **Identifying the different computation steps in the sequential implementation.** The second column identifies the value being computed or the operation being performed in the stages of SEQ. The computations for instructions OP1 and mrmovl are shown as examples of the computations.

as valP, or the activation of some hardware unit, such as the memory. These computations and actions are listed in the second column of Figure 24. In addition to the signals we have already described, this list includes four register ID signals: srcA, the source of valA; srcB, the source of valB; dstE, the register to which valE gets written; and dstM, the register to which valM gets written.

The two right-hand columns of this figure show the computations for the OP1 and mrmovl instructions to illustrate the values being computed. To map the computations into hardware, we want to implement control logic that will transfer the data between the different hardware units and operate these units in such a way that the specified operations are performed for each of the different instruction types. That is the purpose of the control logic blocks, shown as gray rounded boxes in Figure 23. Our task is to proceed through the individual stages and create detailed designs for these blocks.

3.3   SEQ Timing

In introducing the tables of Figures 18 through 21, we stated that they should be read as if they were written in a programming notation, with the assignments performed in sequence from top to bottom. On the other hand, the hardware structure of Figure 23 operates in a fundamentally different way, with a single clock transition triggering a flow through combinational logic to execute an entire instruction. Let us see how the hardware can implement the behavior listed in these tables.

Our implementation of SEQ consists of combinational logic and two forms of memory devices: clocked registers (the program counter and condition code register) and random-access memories (the register file, the instruction memory, and the data memory). Combinational logic does not require any sequencing or control—values propagate through a network of logic gates whenever the inputs change. As we have described, we also assume that reading from a random-access memory operates much like combinational logic, with the output word generated based on the address input. This is a reasonable assumption for smaller memories (such as the register file), and we can mimic this effect for larger circuits using special clock circuits. Since our instruction memory is only used to read instructions, we can therefore treat this unit as if it were combinational logic.

We are left with just four hardware units that require an explicit control over their sequencing—the program counter, the condition code register, the data memory, and the register file. These are controlled via a single clock signal that triggers the loading of new values into the registers and the writing of values to the random-access memories. The program counter is loaded with a new instruction address every clock cycle. The condition code register is loaded only when an integer operation instruction is executed. The data memory is written only when an rmmovl, pushl, or call instruction is executed. The two write ports of the register file allow two program registers to be updated on every cycle, but we can use the special register ID 0xF as a port address to indicate that no write should be performed for this port.

This clocking of the registers and memories is all that is required to control the sequencing of activities in our processor. Our hardware achieves the same effect as would a sequential execution of the assignments shown in the tables of Figures 18 through 21, even though all of the state updates actually occur simultaneously and only as the clock rises to start the next cycle. This equivalence holds because of the nature of the Y86 instruction set, and because we have organized the computations in such a way that our design obeys the following principle:

> The processor never needs to read back the state updated by an instruction in order to complete the processing of this instruction.

This principle is crucial to the success of our implementation. As an illustration, suppose we implemented the pushl instruction by first decrementing %esp by 4 and then using the updated value of %esp as the address of a write operation. This approach would violate the principle stated above. It would require reading the updated stack pointer from the register file in order to perform the memory operation. Instead, our implementation (Figure 20) generates the decremented value of the stack pointer as the signal valE and then uses this signal both as the data for the register write and the address for the memory write. As a result, it can perform the register and memory writes simultaneously as the clock rises to begin the next clock cycle.

As another illustration of this principle, we can see that some instructions (the integer operations) set the condition codes, and some instructions (the jump instructions) read these condition codes, but no instruction must both set and then read the condition codes. Even though the condition codes are not set until the clock rises to begin the next clock cycle, they will be updated before any instruction attempts to read them.

Figure 25 shows how the SEQ hardware would process the instructions at lines 3 and 4 in the following code sequence, shown in assembly code with the instruction addresses listed on the left:

```
1    0x000:   irmovl $0x100,%ebx    # %ebx <-- 0x100
2    0x006:   irmovl $0x200,%edx    # %edx <-- 0x200
3    0x00c:   addl %edx,%ebx        # %ebx <-- 0x300 CC <-- 000
4    0x00e:   je dest               # Not taken
5    0x013:   rmmovl %ebx,0(%edx)   # M[0x200] <-- 0x300
6    0x019: dest: halt
```

Each of the diagrams labeled 1 through 4 shows the four state elements plus the combinational logic and the connections among the state elements. We show the combinational logic as being wrapped around the condition code register, because some of the combinational logic (such as the ALU) generates the input to the condition code register, while other parts (such as the branch computation and the PC selection logic) have the condition code register as input. We show the register file and the data memory as having separate connections for reading and writing, since the read operations propagate through these units as if they were combinational logic, while the write operations are controlled by the clock.

The color coding in Figure 25 indicates how the circuit signals relate to the different instructions being executed. We assume the processing starts with the condition codes, listed in the order ZF, SF, and OF, set to 100. At the beginning of clock cycle 3 (point 1), the state elements hold the state as updated by the second irmovl instruction (line 2 of the listing), shown in light gray. The combinational logic is shown in white, indicating that it has not yet had time to react to the changed state. The clock cycle begins with address 0x00c loaded into the program counter. This causes the addl instruction (line 3 of the listing), shown in blue, to be fetched and processed. Values flow through the combinational logic, including the reading of the random-access memories. By the end of the cycle (point 2), the combinational logic has generated new values (000) for the condition codes, an update for program register %ebx, and a new value (0x00e) for the program counter. At this point, the combinational logic has been updated according to the addl instruction (shown in blue), but the state still holds the values set by the second irmovl instruction (shown in light gray).

As the clock rises to begin cycle 4 (point 3), the updates to the program counter, the register file, and the condition code register occur, and so we show these in blue, but the combinational logic has not yet reacted to these changes, and so we show this in white. In this cycle, the je instruction (line 4 in the listing), shown in dark gray, is fetched and executed. Since condition code ZF is 0, the branch is not

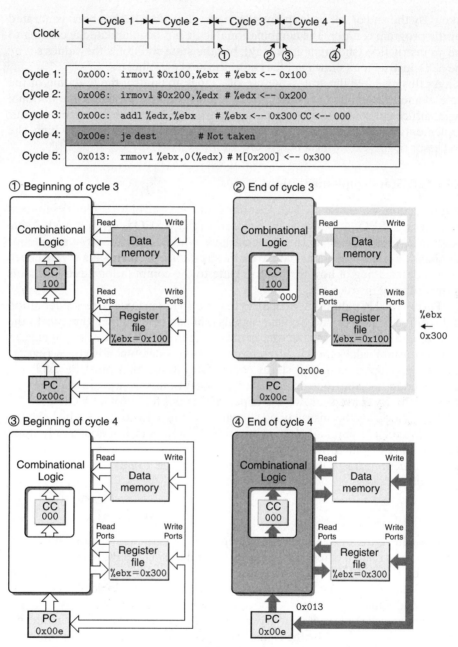

Figure 25 **Tracing two cycles of execution by SEQ.** Each cycle begins with the state elements (program counter, condition code register, register file, and data memory) set according to the previous instruction. Signals propagate through the combinational logic creating new values for the state elements. These values are loaded into the state elements to start the next cycle.

taken. By the end of the cycle (point 4), a new value of 0x013 has been generated for the program counter. The combinational logic has been updated according to the je instruction (shown in dark gray), but the state still holds the values set by the addl instruction (shown in blue) until the next cycle begins.

As this example illustrates, the use of a clock to control the updating of the state elements, combined with the propagation of values through combinational logic, suffices to control the computations performed for each instruction in our implementation of SEQ. Every time the clock transitions from low to high, the processor begins executing a new instruction.

3.4 SEQ Stage Implementations

In this section, we devise HCL descriptions for the control logic blocks required to implement SEQ. A complete HCL description for SEQ is given in Web Aside ARCH:HCL. We show some example blocks here, while others are given as practice problems. We recommend that you work these practice problems as a way to check your understanding of how the blocks relate to the computational requirements of the different instructions.

Part of the HCL description of SEQ that we do not include here is a definition of the different integer and Boolean signals that can be used as arguments to the HCL operations. These include the names of the different hardware signals, as well as constant values for the different instruction codes, function codes, register names, ALU operations, and status codes. Only those that must be explicitly referenced in the control logic are shown. The constants we use are documented in Figure 26. By convention, we use uppercase names for constant values.

In addition to the instructions shown in Figures 18 to 21, we include the processing for the nop and halt instructions. The nop instruction simply flows through stages without much processing, except to increment the PC by 1. The halt instruction causes the processor status to be set to HLT, causing it to halt operation.

Fetch Stage

As shown in Figure 27, the fetch stage includes the instruction memory hardware unit. This unit reads 6 bytes from memory at a time, using the PC as the address of the first byte (byte 0). This byte is interpreted as the instruction byte and is split (by the unit labeled "Split") into two 4-bit quantities. The control logic blocks labeled "icode" and "ifun" then compute the instruction and function codes as equaling either the values read from memory or, in the event that the instruction address is not valid (as indicated by the signal imem_error), the values corresponding to a nop instruction. Based on the value of icode, we can compute three 1-bit signals (shown as dashed lines):

instr_valid: Does this byte correspond to a legal Y86 instruction? This signal is used to detect an illegal instruction.

need_regids: Does this instruction include a register specifier byte?

need_valC: Does this instruction include a constant word?

Name	Value (Hex)	Meaning
INOP	0	Code for nop instruction
IHALT	1	Code for halt instruction
IRRMOVL	2	Code for rrmovl instruction
IIRMOVL	3	Code for irmovl instruction
IRMMOVL	4	Code for rmmovl instruction
IMRMOVL	5	Code for mrmovl instruction
IOPL	6	Code for integer operation instructions
IJXX	7	Code for jump instructions
ICALL	8	Code for call instruction
IRET	9	Code for ret instruction
IPUSHL	A	Code for pushl instruction
IPOPL	B	Code for popl instruction
FNONE	0	Default function code
RESP	4	Register ID for %esp
RNONE	F	Indicates no register file access
ALUADD	0	Function for addition operation
SAOK	1	Status code for normal operation
SADR	2	Status code for address exception
SINS	3	Status code for illegal instruction exception
SHLT	4	Status code for halt

Figure 26 **Constant values used in HCL descriptions.** These values represent the encodings of the instructions, function codes, register IDs, ALU operations, and status codes.

The signals instr_valid and imem_error (generated when the instruction address is out of bounds) are used to generate the status code in the memory stage.

As an example, the HCL description for need_regids simply determines whether the value of icode is one of the instructions that has a register specifier byte:

```
bool need_regids =
        icode in { IRRMOVL, IOPL, IPUSHL, IPOPL,
                    IIRMOVL, IRMMOVL, IMRMOVL };
```

Practice Problem 17

Write HCL code for the signal need_valC in the SEQ implementation.

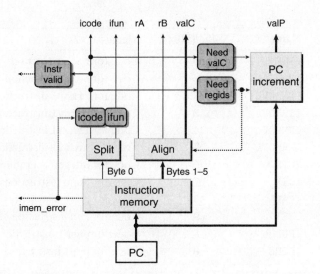

Figure 27

SEQ fetch stage. Six bytes are read from the instruction memory using the PC as the starting address. From these bytes, we generate the different instruction fields. The PC increment block computes signal valP.

As Figure 27 shows, the remaining 5 bytes read from the instruction memory encode some combination of the register specifier byte and the constant word. These bytes are processed by the hardware unit labeled "Align" into the register fields and the constant word. When the computed signal need_regids is 1, then byte 1 is split into register specifiers rA and rB. Otherwise, these two fields are set to 0xF (RNONE), indicating there are no registers specified by this instruction. Recall also (Figure 2) that for any instruction having only one register operand, the other field of the register specifier byte will be 0xF (RNONE). Thus, we can assume that the signals rA and rB either encode registers we want to access or indicate that register access is not required. The unit labeled "Align" also generates the constant word valC. This will either be bytes 1 to 4 or bytes 2 to 5, depending on the value of signal need_regids.

The PC incrementer hardware unit generates the signal valP, based on the current value of the PC, and the two signals need_regids and need_valC. For PC value p, need_regids value r, and need_valC value i, the incrementer generates the value $p + 1 + r + 4i$.

Decode and Write-Back Stages

Figure 28 provides a detailed view of logic that implements both the decode and write-back stages in SEQ. These two stages are combined because they both access the register file.

The register file has four ports. It supports up to two simultaneous reads (on ports A and B) and two simultaneous writes (on ports E and M). Each port has both an address connection and a data connection, where the address connection is a register ID, and the data connection is a set of 32 wires serving as either an output word (for a read port) or an input word (for a write port) of the register file. The two read ports have address inputs srcA and srcB, while the two write

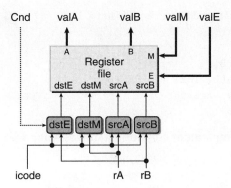

Figure 28 **SEQ decode and write-back stage.** The instruction fields are decoded to generate register identifiers for four addresses (two read and two write) used by the register file. The values read from the register file become the signals valA and valB. The two write-back values valE and valM serve as the data for the writes.

ports have address inputs dstE and dstM. The special identifier 0xF (RNONE) on an address port indicates that no register should be accessed.

The four blocks at the bottom of Figure 28 generate the four different register IDs for the register file, based on the instruction code icode, the register specifiers rA and rB, and possibly the condition signal Cnd computed in the execute stage. Register ID srcA indicates which register should be read to generate valA. The desired value depends on the instruction type, as shown in the first row for the decode stage in Figures 18 to 21. Combining all of these entries into a single computation gives the following HCL description of srcA (recall that RESP is the register ID of %esp):

```
    # Code from SEQ
int srcA = [
        icode in { IRRMOVL, IRMMOVL, IOPL, IPUSHL  } : rA;
        icode in { IPOPL, IRET } : RESP;
        1 : RNONE; # Don't need register
];
```

Practice Problem 18

The register signal srcB indicates which register should be read to generate the signal valB. The desired value is shown as the second step in the decode stage in Figures 18 to 21. Write HCL code for srcB.

Register ID dstE indicates the destination register for write port E, where the computed value valE is stored. This is shown in Figures 18 to 21 as the first step in the write-back stage. If we ignore for the moment the conditional move instructions, then we can combine the destination registers for all of the different instructions to give the following HCL description of dstE:

```
# WARNING: Conditional move not implemented correctly here
int dstE = [
        icode in { IRRMOVL } : rB;
        icode in { IIRMOVL, IOPL} : rB;
        icode in { IPUSHL, IPOPL, ICALL, IRET } : RESP;
        1 : RNONE;  # Don't write any register
];
```

We will revisit this signal and how to implement conditional moves when we examine the execute stage.

Practice Problem 19

Register ID dstM indicates the destination register for write port M, where valM, the value read from memory, is stored. This is shown in Figures 18 to 21 as the second step in the write-back stage. Write HCL code for dstM.

Practice Problem 20

Only the popl instruction uses both register file write ports simultaneously. For the instruction popl %esp, the same address will be used for both the E and M write ports, but with different data. To handle this conflict, we must establish a *priority* among the two write ports so that when both attempt to write the same register on the same cycle, only the write from the higher-priority port takes place. Which of the two ports should be given priority in order to implement the desired behavior, as determined in Problem 7?

Execute Stage

The execute stage includes the arithmetic/logic unit (ALU). This unit performs the operation ADD, SUBTRACT, AND, or EXCLUSIVE-OR on inputs aluA and aluB based on the setting of the alufun signal. These data and control signals are generated by three control blocks, as diagrammed in Figure 29. The ALU output becomes the signal valE.

Figure 29

SEQ execute stage. The ALU either performs the operation for an integer operation instruction or it acts as an adder. The condition code registers are set according to the ALU value. The condition code values are tested to determine whether or not a branch should be taken.

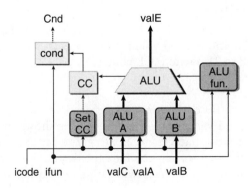

In Figures 18 to 21, the ALU computation for each instruction is shown as the first step in the execute stage. The operands are listed with aluB first, followed by aluA to make sure the subl instruction subtracts valA from valB. We can see that the value of aluA can be valA, valC, or either −4 or +4, depending on the instruction type. We can therefore express the behavior of the control block that generates aluA as follows:

```
int aluA = [
        icode in { IRRMOVL, IOPL } : valA;
        icode in { IIRMOVL, IRMMOVL, IMRMOVL } : valC;
        icode in { ICALL, IPUSHL } : -4;
        icode in { IRET, IPOPL } : 4;
        # Other instructions don't need ALU
];
```

Practice Problem 21

Based on the first operand of the first step of the execute stage in Figures 18 to 21, write an HCL description for the signal aluB in SEQ.

Looking at the operations performed by the ALU in the execute stage, we can see that it is mostly used as an adder. For the OP1 instructions, however, we want it to use the operation encoded in the ifun field of the instruction. We can therefore write the HCL description for the ALU control as follows:

```
int alufun = [
        icode == IOPL : ifun;
        1 : ALUADD;
];
```

The execute stage also includes the condition code register. Our ALU generates the three signals on which the condition codes are based—zero, sign, and overflow—every time it operates. However, we only want to set the condition codes when an OP1 instruction is executed. We therefore generate a signal set_cc that controls whether or not the condition code register should be updated:

```
bool set_cc = icode in { IOPL };
```

The hardware unit labeled "cond" uses a combination of the condition codes and the function code to determine whether a conditional branch or data transfer should take place (Figure 3). It generates the Cnd signal used both for the setting of dstE with conditional moves, and in the next PC logic for conditional branches. For other instructions, the Cnd signal may be set to either 1 or 0, depending on the instruction's function code and the setting of the condition codes, but it will be ignored by the control logic. We omit the detailed design of this unit.

Figure 30

SEQ memory stage. The data memory can either write or read memory values. The value read from memory forms the signal valM.

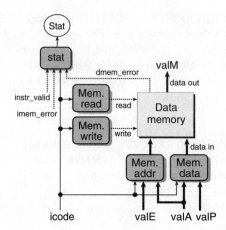

Practice Problem 22

The conditional move instructions, abbreviated cmovXX, have instruction code IRRMOVL. As Figure 28 shows, we can implement these instructions by making use of the Cnd signal, generated in the execute stage. Modify the HCL code for dstE to implement these instructions.

Memory Stage

The memory stage has the task of either reading or writing program data. As shown in Figure 30, two control blocks generate the values for the memory address and the memory input data (for write operations). Two other blocks generate the control signals indicating whether to perform a read or a write operation. When a read operation is performed, the data memory generates the value valM.

The desired memory operation for each instruction type is shown in the memory stage of Figures 18 to 21. Observe that the address for memory reads and writes is always valE or valA. We can describe this block in HCL as follows:

```
int mem_addr = [
        icode in { IRMMOVL, IPUSHL, ICALL, IMRMOVL } : valE;
        icode in { IPOPL, IRET } : valA;
        # Other instructions don't need address
];
```

Practice Problem 23

Looking at the memory operations for the different instructions shown in Figures 18 to 21, we can see that the data for memory writes is always either valA or valP. Write HCL code for the signal mem_data in SEQ.

We want to set the control signal mem_read only for instructions that read data from memory, as expressed by the following HCL code:

```
bool mem_read = icode in { IMRMOVL, IPOPL, IRET };
```

Practice Problem 24

We want to set the control signal mem_write only for instructions that write data to memory. Write HCL code for the signal mem_write in SEQ.

A final function for the memory stage is to compute the status code Stat resulting from the instruction execution, according to the values of icode, imem_error, instr_valid generated in the fetch stage, and the signal dmem_error generated by the data memory.

Practice Problem 25

Write HCL code for Stat, generating the four status codes SAOK, SADR, SINS, and SHLT (see Figure 26).

PC Update Stage

The final stage in SEQ generates the new value of the program counter. (See Figure 31.) As the final steps in Figures 18 to 21 show, the new PC will be valC, valM, or valP, depending on the instruction type and whether or not a branch should be taken. This selection can be described in HCL as follows:

```
int new_pc = [
        # Call.  Use instruction constant
        icode == ICALL : valC;
        # Taken branch.  Use instruction constant
        icode == IJXX && Cnd : valC;
        # Completion of RET instruction.  Use value from stack
        icode == IRET : valM;
        # Default: Use incremented PC
        1 : valP;
];
```

Figure 31

SEQ PC update stage.
The next value of the PC is selected from among the signals valC, valM, and valP, depending on the instruction code and the branch flag.

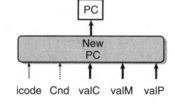

Surveying SEQ

We have now stepped through a complete design for a Y86 processor. We have seen that by organizing the steps required to execute each of the different instructions into a uniform flow, we can implement the entire processor with a small number of different hardware units and with a single clock to control the sequencing of computations. The control logic must then route the signals between these units and generate the proper control signals based on the instruction types and the branch conditions.

The only problem with SEQ is that it is too slow. The clock must run slowly enough so that signals can propagate through all of the stages within a single cycle. As an example, consider the processing of a ret instruction. Starting with an updated program counter at the beginning of the clock cycle, the instruction must be read from the instruction memory, the stack pointer must be read from the register file, the ALU must decrement the stack pointer, and the return address must be read from the memory in order to determine the next value for the program counter. All of this must be completed by the end of the clock cycle.

This style of implementation does not make very good use of our hardware units, since each unit is only active for a fraction of the total clock cycle. We will see that we can achieve much better performance by introducing pipelining.

4 General Principles of Pipelining

Before attempting to design a pipelined Y86 processor, let us consider some general properties and principles of pipelined systems. Such systems are familiar to anyone who has been through the serving line at a cafeteria or run a car through an automated car wash. In a pipelined system, the task to be performed is divided into a series of discrete stages. In a cafeteria, this involves supplying salad, a main dish, dessert, and beverage. In a car wash, this involves spraying water and soap, scrubbing, applying wax, and drying. Rather than having one customer run through the entire sequence from beginning to end before the next can begin, we allow multiple customers to proceed through the system at once. In a typical cafeteria line, the customers maintain the same order in the pipeline and pass through all stages, even if they do not want some of the courses. In the case of the car wash, a new car is allowed to enter the spraying stage as the preceding car moves from the spraying stage to the scrubbing stage. In general, the cars must move through the system at the same rate to avoid having one car crash into the next.

A key feature of pipelining is that it increases the *throughput* of the system, that is, the number of customers served per unit time, but it may also slightly increase the *latency*, that is, the time required to service an individual customer. For example, a customer in a cafeteria who only wants a salad could pass through a nonpipelined system very quickly, stopping only at the salad stage. A customer in a pipelined system who attempts to go directly to the salad stage risks incurring the wrath of other customers.

Figure 32

Unpipelined computation hardware. On each 320 ps cycle, the system spends 300 ps evaluating a combinational logic function and 20 ps storing the results in an output register.

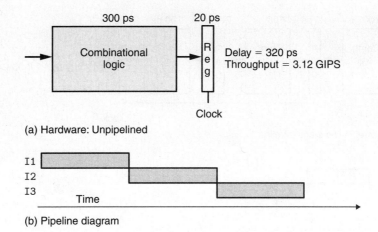

(a) Hardware: Unpipelined

(b) Pipeline diagram

4.1 Computational Pipelines

Shifting our focus to computational pipelines, the "customers" are instructions and the stages perform some portion of the instruction execution. Figure 32 shows an example of a simple nonpipelined hardware system. It consists of some logic that performs a computation, followed by a register to hold the results of this computation. A clock signal controls the loading of the register at some regular time interval. An example of such a system is the decoder in a compact disk (CD) player. The incoming signals are the bits read from the surface of the CD, and the logic decodes these to generate audio signals. The computational block in the figure is implemented as combinational logic, meaning that the signals will pass through a series of logic gates, with the outputs becoming some function of the inputs after some time delay.

In contemporary logic design, we measure circuit delays in units of *picoseconds* (abbreviated "ps"), or 10^{-12} seconds. In this example, we assume the combinational logic requires 300 picoseconds, while the loading of the register requires 20 ps. Figure 32 also shows a form of timing diagram known as a *pipeline diagram*. In this diagram, time flows from left to right. A series of instructions (here named I1, I2, and I3) are written from top to bottom. The solid rectangles indicate the times during which these instructions are executed. In this implementation, we must complete one instruction before beginning the next. Hence, the boxes do not overlap one another vertically. The following formula gives the maximum rate at which we could operate the system:

$$\text{Throughput} = \frac{1 \text{ instruction}}{(20 + 300) \text{ picosecond}} \cdot \frac{1000 \text{ picosecond}}{1 \text{ nanosecond}} \approx 3.12 \text{ GIPS}$$

We express throughput in units of giga-instructions per second (abbreviated GIPS), or billions of instructions per second. The total time required to perform a single instruction from beginning to end is known as the *latency*. In this system, the latency is 320 ps, the reciprocal of the throughput.

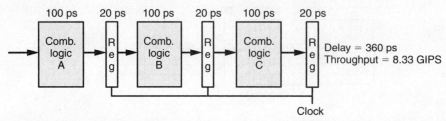

(a) Hardware: Three-stage pipeline

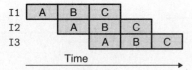

(b) Pipeline diagram

Figure 33 **Three-stage pipelined computation hardware.** The computation is split into stages A, B, and C. On each 120-ps cycle, each instruction progresses through one stage.

Suppose we could divide the computation performed by our system into three stages, A, B, and C, where each requires 100 ps, as illustrated in Figure 33. Then we could put *pipeline registers* between the stages so that each instruction moves through the system in three steps, requiring three complete clock cycles from beginning to end. As the pipeline diagram in Figure 33 illustrates, we could allow I2 to enter stage A as soon as I1 moves from A to B, and so on. In steady state, all three stages would be active, with one instruction leaving and a new one entering the system every clock cycle. We can see this during the third clock cycle in the pipeline diagram where I1 is in stage C, I2 is in stage B, and I3 is in stage A. In this system, we could cycle the clocks every $100 + 20 = 120$ picoseconds, giving a throughput of around 8.33 GIPS. Since processing a single instruction requires 3 clock cycles, the latency of this pipeline is $3 \times 120 = 360$ ps. We have increased the throughput of the system by a factor of $8.33/3.12 = 2.67$ at the expense of some added hardware and a slight increase in the latency ($360/320 = 1.12$). The increased latency is due to the time overhead of the added pipeline registers.

4.2 A Detailed Look at Pipeline Operation

To better understand how pipelining works, let us look in some detail at the timing and operation of pipeline computations. Figure 34 shows the pipeline diagram for the three-stage pipeline we have already looked at (Figure 33). The transfer of the instructions between pipeline stages is controlled by a clock signal, as shown above the pipeline diagram. Every 120 ps, this signal rises from 0 to 1, initiating the next set of pipeline stage evaluations.

Figure 34

Three-stage pipeline timing. The rising edge of the clock signal controls the movement of instructions from one pipeline stage to the next.

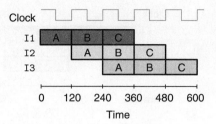

Figure 35 traces the circuit activity between times 240 and 360, as instruction I1 (shown in dark gray) propagates through stage C, I2 (shown in blue) propagates through stage B, and I3 (shown in light gray) propagates through stage A. Just before the rising clock at time 240 (point 1), the values computed in stage A for instruction I2 have reached the input of the first pipeline register, but its state and output remain set to those computed during stage A for instruction I1. The values computed in stage B for instruction I1 have reached the input of the second pipeline register. As the clock rises, these inputs are loaded into the pipeline registers, becoming the register outputs (point 2). In addition, the input to stage A is set to initiate the computation of instruction I3. The signals then propagate through the combinational logic for the different stages (point 3). As the curved wavefronts in the diagram at point 3 suggest, signals can propagate through different sections at different rates. Before time 360, the result values reach the inputs of the pipeline registers (point 4). When the clock rises at time 360, each of the instructions will have progressed through one pipeline stage.

We can see from this detailed view of pipeline operation that slowing down the clock would not change the pipeline behavior. The signals propagate to the pipeline register inputs, but no change in the register states will occur until the clock rises. On the other hand, we could have disastrous effects if the clock were run too fast. The values would not have time to propagate through the combinational logic, and so the register inputs would not yet be valid when the clock rises.

As with our discussion of the timing for the SEQ processor (Section 3.3), we see that the simple mechanism of having clocked registers between blocks of combinational logic suffices to control the flow of instructions in the pipeline. As the clock rises and falls repeatedly, the different instructions flow through the stages of the pipeline without interfering with one another.

4.3 Limitations of Pipelining

The example of Figure 33 shows an ideal pipelined system in which we are able to divide the computation into three independent stages, each requiring one-third of the time required by the original logic. Unfortunately, other factors often arise that diminish the effectiveness of pipelining.

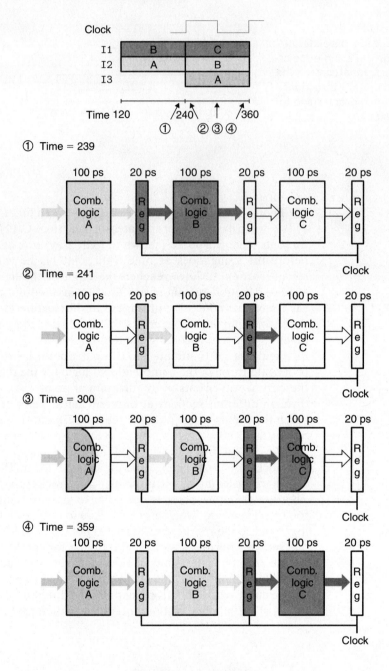

Figure 35

One clock cycle of pipeline operation. Just before the clock rises at time 240 (point 1), instructions I1 (shown in dark gray) and I2 (shown in blue) have completed stages B and A. After the clock rises, these instructions begin propagating through stages C and B, while instruction I3 (shown in light gray) begins propagating through stage A (points 2 and 3). Just before the clock rises again, the results for the instructions have propagated to the inputs of the pipeline registers (point 4).

Nonuniform Partitioning

Figure 36 shows a system in which we divide the computation into three stages as before, but the delays through the stages range from 50 to 150 ps. The sum of the delays through all of the stages remains 300 ps. However, the rate at which we

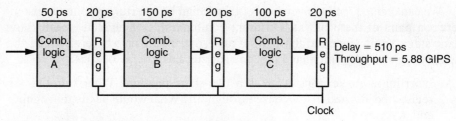

(a) Hardware: Three-stage pipeline, nonuniform stage delays

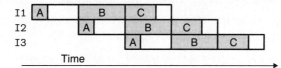

(b) Pipeline diagram

Figure 36 **Limitations of pipelining due to nonuniform stage delays.** The system throughput is limited by the speed of the slowest stage.

can operate the clock is limited by the delay of the slowest stage. As the pipeline diagram in this figure shows, stage A will be idle (shown as a white box) for 100 ps every clock cycle, while stage C will be idle for 50 ps every clock cycle. Only stage B will be continuously active. We must set the clock cycle to $150 + 20 = 170$ picoseconds, giving a throughput of 5.88 GIPS. In addition, the latency would increase to 510 ps due to the slower clock rate.

Devising a partitioning of the system computation into a series of stages having uniform delays can be a major challenge for hardware designers. Often, some of the hardware units in a processor, such as the ALU and the memories, cannot be subdivided into multiple units with shorter delay. This makes it difficult to create a set of balanced stages. We will not concern ourselves with this level of detail in designing our pipelined Y86 processor, but it is important to appreciate the importance of timing optimization in actual system design.

Practice Problem 26

Suppose we analyze the combinational logic of Figure 32 and determine that it can be separated into a sequence of six blocks, named A to F, having delays of 80, 30, 60, 50, 70, and 10 ps, respectively, illustrated as follows:

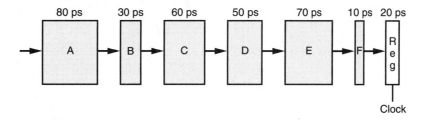

We can create pipelined versions of this design by inserting pipeline registers between pairs of these blocks. Different combinations of pipeline depth (how many stages) and maximum throughput arise, depending on where we insert the pipeline registers. Assume that a pipeline register has a delay of 20 ps.

A. Inserting a single register gives a two-stage pipeline. Where should the register be inserted to maximize throughput? What would be the throughput and latency?

B. Where should two registers be inserted to maximize the throughput of a three-stage pipeline? What would be the throughput and latency?

C. Where should three registers be inserted to maximize the throughput of a four-stage pipeline? What would be the throughput and latency?

D. What is the minimum number of stages that would yield a design with the maximum achievable throughput? Describe this design, its throughput, and its latency.

Diminishing Returns of Deep Pipelining

Figure 37 illustrates another limitation of pipelining. In this example, we have divided the computation into six stages, each requiring 50 ps. Inserting a pipeline register between each pair of stages yields a six-stage pipeline. The minimum clock period for this system is $50 + 20 = 70$ picoseconds, giving a throughput of 14.29 GIPS. Thus, in doubling the number of pipeline stages, we improve the performance by a factor of $14.29/8.33 = 1.71$. Even though we have cut the time required for each computation block by a factor of 2, we do not get a doubling of the throughput, due to the delay through the pipeline registers. This delay becomes a limiting factor in the throughput of the pipeline. In our new design, this delay consumes 28.6% of the total clock period.

Modern processors employ very deep (15 or more stages) pipelines in an attempt to maximize the processor clock rate. The processor architects divide the instruction execution into a large number of very simple steps so that each stage can have a very small delay. The circuit designers carefully design the pipeline registers to minimize their delay. The chip designers must also carefully design the

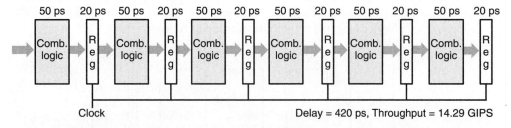

Figure 37 **Limitations of pipelining due to overhead.** As the combinational logic is split into shorter blocks, the delay due to register updating becomes a limiting factor.

clock distribution network to ensure that the clock changes at the exact same time across the entire chip. All of these factors contribute to the challenge of designing high-speed microprocessors.

Practice Problem 27

Suppose we could take the system of Figure 32 and divide it into an arbitrary number of pipeline stages k, each having a delay of $300/k$, and with each pipeline register having a delay of 20 ps.

A. What would be the latency and the throughput of the system, as functions of k?

B. What would be the ultimate limit on the throughput?

4.4 Pipelining a System with Feedback

Up to this point, we have considered only systems in which the objects passing through the pipeline—whether cars, people, or instructions—are completely independent of one another. For a system that executes machine programs such as IA32 or Y86, however, there are potential dependencies between successive instructions. For example, consider the following Y86 instruction sequence:

```
1    irmovl $50, %eax
2    addl %eax ,%ebx
3    mrmovl 100( %ebx ),%edx
```

```
1        irmovl $50,%eax
2        addl %eax,%ebx
3        mrmovl 100(%ebx),%edx
```

In this three-instruction sequence, there is a *data dependency* between each successive pair of instructions, as indicated by the circled register names and the arrows between them. The irmovl instruction (line 1) stores its result in %eax, which then must be read by the addl instruction (line 2); and this instruction stores its result in %ebx, which must then be read by the mrmovl instruction (line 3).

Another source of sequential dependencies occurs due to the instruction control flow. Consider the following Y86 instruction sequence:

```
1    loop:
2        subl %edx,%ebx
3        jne targ
4        irmovl $10,%edx
5        jmp loop
6    targ:
7        halt
```

Figure 38

Limitations of pipelining due to logical dependencies. In going from an unpipelined system with feedback (a) to a pipelined one (c), we change its computational behavior, as can be seen by the two pipeline diagrams (b and d).

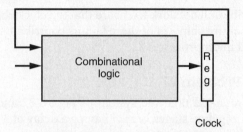

(a) Hardware: Unpipelined with feedback

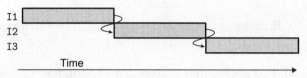

(b) Pipeline diagram

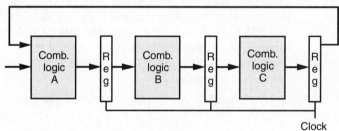

(c) Hardware: Three-stage pipeline with feedback

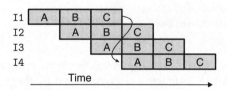

(d) Pipeline diagram

The `jne` instruction (line 3) creates a *control dependency* since the outcome of the conditional test determines whether the next instruction to execute will be the `irmovl` instruction (line 4) or the `halt` instruction (line 7). In our design for SEQ, these dependencies were handled by the feedback paths shown on the right-hand side of Figure 22. This feedback brings the updated register values down to the register file and the new PC value down to the PC register.

Figure 38 illustrates the perils of introducing pipelining into a system containing feedback paths. In the original system (Figure 38(a)), the result of each instruction is fed back around to the next instruction. This is illustrated by the pipeline diagram (Figure 38(b)), where the result of I1 becomes an input to

I2, and so on. If we attempt to convert this to a three-stage pipeline in the most straightforward manner (Figure 38(c)), we change the behavior of the system. As Figure 38(c) shows, the result of I1 becomes an input to I4. In attempting to speed up the system via pipelining, we have changed the system behavior.

When we introduce pipelining into a Y86 processor, we must deal with feedback effects properly. Clearly, it would be unacceptable to alter the system behavior as occurred in the example of Figure 38. Somehow we must deal with the data and control dependencies between instructions so that the resulting behavior matches the model defined by the ISA.

5 Pipelined Y86 Implementations

We are finally ready for the major task of this chapter—designing a pipelined Y86 processor. We start by making a small adaptation of the sequential processor SEQ to shift the computation of the PC into the fetch stage. We then add pipeline registers between the stages. Our first attempt at this does not handle the different data and control dependencies properly. By making some modifications, however, we achieve our goal of an efficient pipelined processor that implements the Y86 ISA.

5.1 SEQ+: Rearranging the Computation Stages

As a transitional step toward a pipelined design, we must slightly rearrange the order of the five stages in SEQ so that the PC update stage comes at the beginning of the clock cycle, rather than at the end. This transformation requires only minimal change to the overall hardware structure, and it will work better with the sequencing of activities within the pipeline stages. We refer to this modified design as "SEQ+."

We can move the PC update stage so that its logic is active at the beginning of the clock cycle by making it compute the PC value for the *current* instruction. Figure 39 shows how SEQ and SEQ+ differ in their PC computation. With SEQ (Figure 39(a)), the PC computation takes place at the end of the clock cycle, computing the new value for the PC register based on the values of signals

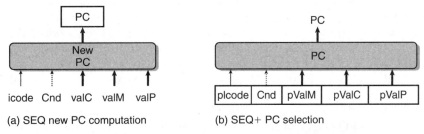

(a) SEQ new PC computation (b) SEQ+ PC selection

Figure 39 **Shifting the timing of the PC computation.** With SEQ+, we compute the value of the program counter for the current state as the first step in instruction execution.

computed during the current clock cycle. With SEQ+ (Figure 39(b)), we create state registers to hold the signals computed during an instruction. Then, as a new clock cycle begins, the values propagate through the exact same logic to compute the PC for the now-current instruction. We label the registers "pIcode," "pCnd," and so on, to indicate that on any given cycle, they hold the control signals generated during the previous cycle.

Figure 40 shows a more detailed view of the SEQ+ hardware. We can see that it contains the exact same hardware units and control blocks that we had in SEQ (Figure 23), but with the PC logic shifted from the top, where it was active at the end of the clock cycle, to the bottom, where it is active at the beginning.

Aside   Where is the PC in SEQ+?

One curious feature of SEQ+ is that there is no hardware register storing the program counter. Instead, the PC is computed dynamically based on some state information stored from the previous instruction. This is a small illustration of the fact that we can implement a processor in a way that differs from the conceptual model implied by the ISA, as long as the processor correctly executes arbitrary machine-language programs. We need not encode the state in the form indicated by the programmer-visible state, as long as the processor can generate correct values for any part of the programmer-visible state (such as the program counter). We will exploit this principle even more in creating a pipelined design. Out-of-order processing techniques take this idea to an extreme by executing instructions in a completely different order than they occur in the machine-level program.

The shift of state elements from SEQ to SEQ+ is an example of a general transformation known as *circuit retiming* [65]. Retiming changes the state representation for a system without changing its logical behavior. It is often used to balance the delays between different stages of a system.

5.2   Inserting Pipeline Registers

In our first attempt at creating a pipelined Y86 processor, we insert pipeline registers between the stages of SEQ+ and rearrange signals somewhat, yielding the PIPE− processor, where the "−" in the name signifies that this processor has somewhat less performance than our ultimate processor design. The structure of PIPE− is illustrated in Figure 41. The pipeline registers are shown in this figure as black boxes, each containing different fields that are shown as white boxes. As indicated by the multiple fields, each pipeline register holds multiple bytes and words. Unlike the labels shown in rounded boxes in the hardware structure of the two sequential processors (Figures 23 and 40), these white boxes represent actual hardware components.

Observe that PIPE− uses nearly the same set of hardware units as our sequential design SEQ (Figure 40), but with the pipeline registers separating the stages. The differences between the signals in the two systems is discussed in Section 5.3.

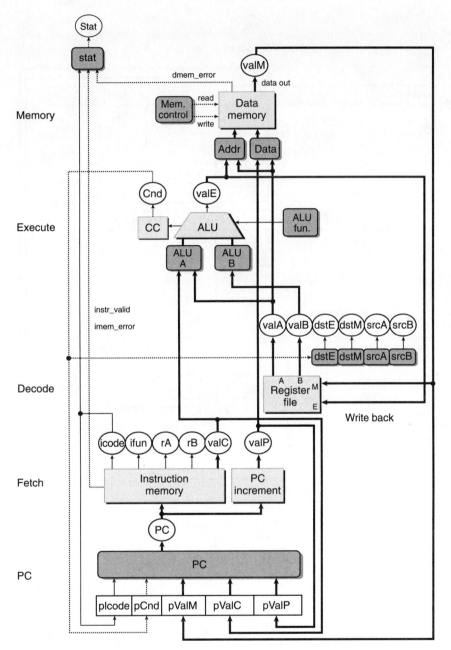

Figure 40   **SEQ+ hardware structure.** Shifting the PC computation from the end of the clock cycle to the beginning makes it more suitable for pipelining.

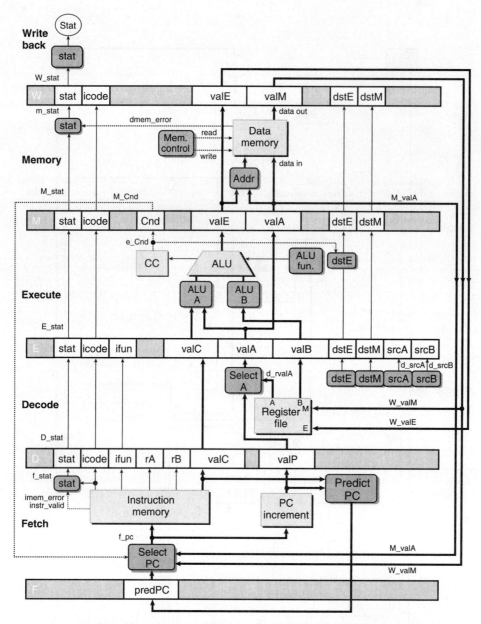

Figure 41 **Hardware structure of PIPE–, an initial pipelined implementation.** By inserting pipeline registers into SEQ+ (Figure 40), we create a five-stage pipeline. There are several shortcomings of this version that we will deal with shortly.

The pipeline registers are labeled as follows:

F holds a *predicted* value of the program counter, as will be discussed shortly.

D sits between the fetch and decode stages. It holds information about the most recently fetched instruction for processing by the decode stage.

E sits between the decode and execute stages. It holds information about the most recently decoded instruction and the values read from the register file for processing by the execute stage.

M sits between the execute and memory stages. It holds the results of the most recently executed instruction for processing by the memory stage. It also holds information about branch conditions and branch targets for processing conditional jumps.

W sits between the memory stage and the feedback paths that supply the computed results to the register file for writing and the return address to the PC selection logic when completing a `ret` instruction.

Figure 42 shows how the following code sequence would flow through our five-stage pipeline, where the comments identify the instructions as I1 to I5 for reference:

```
1    irmovl  $1,%eax   # I1
2    irmovl  $2,%ebx   # I2
3    irmovl  $3,%ecx   # I3
4    irmovl  $4,%edx   # I4
5    halt              # I5
```

The right side of the figure shows a pipeline diagram for this instruction sequence. As with the pipeline diagrams for the simple pipelined computation units of Section 4, this diagram shows the progression of each instruction through the pipeline stages, with time increasing from left to right. The numbers along the top identify the clock cycles at which the different stages occur. For example, in cycle 1, instruction I1 is fetched, and it then proceeds through the pipeline stages, with its result being written to the register file after the end of cycle 5. Instruction I2 is fetched in cycle 2, and its result is written back after the end of cycle 6, and so on. At the bottom, we show an expanded view of the pipeline for cycle 5. At this point, there is an instruction in each of the pipeline stages.

From Figure 42, we can also justify our convention of drawing processors so that the instructions flow from bottom to top. The expanded view for cycle 5 shows the pipeline stages with the fetch stage on the bottom and the write-back stage on the top, just as do our diagrams of the pipeline hardware (Figure 41). If we look at the ordering of instructions in the pipeline stages, we see that they appear in the same order as they do in the program listing. Since normal program flow goes from top to bottom of a listing, we preserve this ordering by having the pipeline flow go from bottom to top. This convention is particularly useful when working with the simulators that accompany this text.

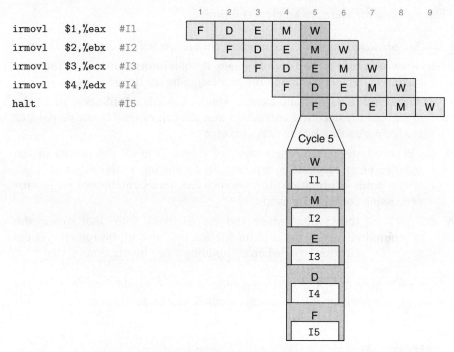

Figure 42 **Example of instruction flow through pipeline.**

5.3 Rearranging and Relabeling Signals

Our sequential implementations SEQ and SEQ+ only process one instruction at a time, and so there are unique values for signals such as valC, srcA, and valE. In our pipelined design, there will be multiple versions of these values associated with the different instructions flowing through the system. For example, in the detailed structure of PIPE–, there are four white boxes labeled "stat" that hold the status codes for four different instructions. (See Figure 41.) We need to take great care to make sure we use the proper version of a signal, or else we could have serious errors, such as storing the result computed for one instruction at the destination register specified by another instruction. We adopt a naming scheme where a signal stored in a pipeline register can be uniquely identified by prefixing its name with that of the pipe register written in uppercase. For example, the four status codes are named D_stat, E_stat, M_stat, and W_stat. We also need to refer to some signals that have just been computed within a stage. These are labeled by prefixing the signal name with the first character of the stage name, written in lowercase. Using the status codes as examples, we can see control logic blocks labeled "stat" in the fetch and memory stages. The outputs of these blocks are therefore named f_stat and m_stat. We can also see that the actual status of the overall processor Stat is computed by a block in the write-back stage, based on the status value in pipeline register W.

Aside   What is the difference between signals M_stat and m_stat?

With our naming system, the uppercase prefixes "D," "E," "M," and "W" refer to pipeline *registers*, and so M_stat refers to the status code field of pipeline register M. The lowercase prefixes "f," "d," "e," "m," and "w" refer to the pipeline *stages*, and so m_stat refers to the status signal generated in the memory stage by a control logic block.

Understanding this naming convention is critical to understanding the operation of our pipelined processors.

The decode stages of SEQ+ and PIPE– both generate signals dstE and dstM indicating the destination register for values valE and valM. In SEQ+, we could connect these signals directly to the address inputs of the register file write ports. With PIPE–, these signals are carried along in the pipeline through the execute and memory stages, and are directed to the register file only once they reach the write-back stage (shown in the more detailed views of the stages). We do this to make sure the write port address and data inputs hold values from the same instruction. Otherwise, the write back would be writing the values for the instruction in the write-back stage, but with register IDs from the instruction in the decode stage. As a general principle, we want to keep all of the information about a particular instruction contained within a single pipeline stage.

One block of PIPE– that is not present in SEQ+ in the exact same form is the block labeled "Select A" in the decode stage. We can see that this block generates the value valA for the pipeline register E by choosing either valP from pipeline register D or the value read from the A port of the register file. This block is included to reduce the amount of state that must be carried forward to pipeline registers E and M. Of all the different instructions, only the `call` requires valP in the memory stage. Only the jump instructions require the value of valP in the execute stage (in the event the jump is not taken). None of these instructions requires a value read from the register file. Therefore, we can reduce the amount of pipeline register state by merging these two signals and carrying them through the pipeline as a single signal valA. This eliminates the need for the block labeled "Data" in SEQ (Figure 23) and SEQ+ (Figure 40), which served a similar purpose. In hardware design, it is common to carefully identify how signals get used and then reduce the amount of register state and wiring by merging signals such as these.

As shown in Figure 41, our pipeline registers include a field for the status code Stat, initially computed during the fetch stage and possibly modified during the memory stage. We will discuss how to implement the processing of exceptional events in Section 5.9, after we have covered the implementation of normal instruction execution. Suffice it to say at this point that the most systematic approach is to associate a status code with each instruction as it passes through the pipeline, as we have indicated in the figure.

5.4   Next PC Prediction

We have taken some measures in the design of PIPE– to properly handle control dependencies. Our goal in the pipelined design is to *issue* a new instruction on every clock cycle, meaning that on each clock cycle, a new instruction proceeds into the execute stage and will ultimately be completed. Achieving this goal would yield a throughput of one instruction per cycle. To do this, we must determine the location of the next instruction right after fetching the current instruction. Unfortunately, if the fetched instruction is a conditional branch, we will not know whether or not the branch should be taken until several cycles later, after the instruction has passed through the execute stage. Similarly, if the fetched instruction is a ret, we cannot determine the return location until the instruction has passed through the memory stage.

With the exception of conditional jump instructions and ret, we can determine the address of the next instruction based on information computed during the fetch stage. For call and jmp (unconditional jump), it will be valC, the constant word in the instruction, while for all others it will be valP, the address of the next instruction. We can therefore achieve our goal of issuing a new instruction every clock cycle in most cases by *predicting* the next value of the PC. For most instruction types, our prediction will be completely reliable. For conditional jumps, we can predict either that a jump will be taken, so that the new PC value would be valC, or we can predict that it will not be taken, so that the new PC value would be valP. In either case, we must somehow deal with the case where our prediction was incorrect and therefore we have fetched and partially executed the wrong instructions. We will return to this matter in Section 5.11.

This technique of guessing the branch direction and then initiating the fetching of instructions according to our guess is known as *branch prediction*. It is used in some form by virtually all processors. Extensive experiments have been conducted on effective strategies for predicting whether or not branches will be taken [49, Section 2.3]. Some systems devote large amounts of hardware to this task. In our design, we will use the simple strategy of predicting that conditional branches are always taken, and so we predict the new value of the PC to be valC.

Aside   Other branch prediction strategies

Our design uses an *always taken* branch prediction strategy. Studies show this strategy has around a 60% success rate [47, 120]. Conversely, a *never taken* (NT) strategy has around a 40% success rate. A slightly more sophisticated strategy, known as *backward taken, forward not-taken* (BTFNT), predicts that branches to lower addresses than the next instruction will be taken, while those to higher addresses will not be taken. This strategy has a success rate of around 65%. This improvement stems from the fact that loops are closed by backward branches, and loops are generally executed multiple times. Forward branches are used for conditional operations, and these are less likely to be taken. In Problems 54 and 55, you can modify the Y86 pipeline processor to implement the NT and BTFNT branch prediction strategies.

Mispredicted branches can degrade the performance of a program considerably, thus motivating the use of conditional data transfer rather than conditional control transfer when possible.

We are still left with predicting the new PC value resulting from a ret instruction. Unlike conditional jumps, we have a nearly unbounded set of possible results, since the return address will be whatever word is on the top of the stack. In our design, we will not attempt to predict any value for the return address. Instead, we will simply hold off processing any more instructions until the `ret` instruction passes through the write-back stage. We will return to this part of the implementation in Section 5.11.

Aside Return address prediction with a stack

With most programs, it is very easy to predict return addresses, since procedure calls and returns occur in matched pairs. Most of the time that a procedure is called, it returns to the instruction following the call. This property is exploited in high-performance processors by including a hardware stack within the instruction fetch unit that holds the return address generated by procedure call instructions. Every time a procedure call instruction is executed, its return address is pushed onto the stack. When a return instruction is fetched, the top value is popped from this stack and used as the predicted return address. Like branch prediction, a mechanism must be provided to recover when the prediction was incorrect, since there are times when calls and returns do not match. In general, the prediction is highly reliable. This hardware stack is not part of the programmer-visible state.

The PIPE– fetch stage, diagrammed at the bottom of Figure 41, is responsible for both predicting the next value of the PC and for selecting the actual PC for the instruction fetch. We can see the block labeled "Predict PC" can choose either valP, as computed by the PC incrementer or valC, from the fetched instruction. This value is stored in pipeline register F as the *predicted* value of the program counter. The block labeled "Select PC" is similar to the block labeled "PC" in the SEQ+ PC selection stage (Figure 40). It chooses one of three values to serve as the address for the instruction memory: the predicted PC, the value of valP for a not-taken branch instruction that reaches pipeline register M (stored in register M_valA), or the value of the return address when a `ret` instruction reaches pipeline register W (stored in W_valM).

We will return to the handling of jump and return instructions when we complete the pipeline control logic in Section 5.11.

5.5 Pipeline Hazards

Our structure PIPE– is a good start at creating a pipelined Y86 processor. Recall from our discussion in Section 4.4, however, that introducing pipelining into a system with feedback can lead to problems when there are dependencies between successive instructions. We must resolve this issue before we can complete our design. These dependencies can take two forms: (1) *data* dependencies, where the results computed by one instruction are used as the data for a following instruction, and (2) *control* dependencies, where one instruction determines the location of the following instruction, such as when executing a jump, call, or return. When such dependencies have the potential to cause an erroneous computation by the pipeline, they are called *hazards*. Like dependencies, hazards can be classified as either *data hazards* or *control hazards*. In this section, we concern ourselves

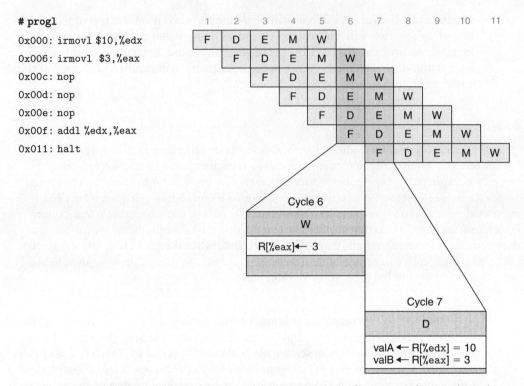

Figure 43 **Pipelined execution of** `prog1` **without special pipeline control.** In cycle 6, the second `irmovl` writes its result to program register `%eax`. The `addl` instruction reads its source operands in cycle 7, so it gets correct values for both `%edx` and `%eax`.

with data hazards. Control hazards will be discussed as part of the overall pipeline control (Section 5.11).

Figure 43 illustrates the processing of a sequence of instructions we refer to as `prog1` by the PIPE– processor. Let us assume in this example and successive ones that the program registers initially all have value 0. The code loads values 10 and 3 into program registers `%edx` and `%eax`, executes three `nop` instructions, and then adds register `%edx` to `%eax`. We focus our attention on the potential data hazards resulting from the data dependencies between the two `irmovl` instructions and the `addl` instruction. On the right-hand side of the figure, we show a pipeline diagram for the instruction sequence. The pipeline stages for cycles 6 and 7 are shown highlighted in the pipeline diagram. Below this, we show an expanded view of the write-back activity in cycle 6 and the decode activity during cycle 7. After the start of cycle 7, both of the `irmovl` instructions have passed through the write-back stage, and so the register file holds the updated values of `%edx` and `%eax`. As the `addl` instruction passes through the decode stage during cycle 7, it will therefore read the correct values for its source operands. The data dependencies between the two `irmovl` instructions and the `addl` instruction have not created data hazards in this example.

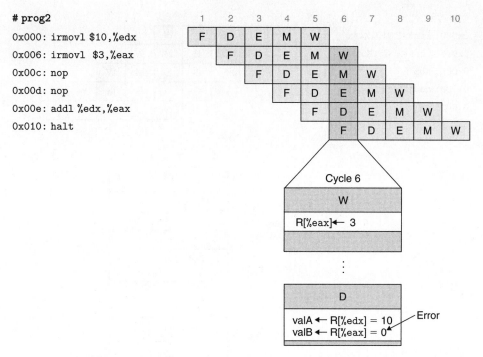

Figure 44 **Pipelined execution of** prog2 **without special pipeline control.** The write to program register %eax does not occur until the start of cycle 7, and so the addl instruction gets the incorrect value for this register in the decode stage.

We saw that prog1 will flow through our pipeline and get the correct results, because the three nop instructions create a delay between instructions with data dependencies. Let us see what happens as these nop instructions are removed. Figure 44 illustrates the pipeline flow of a program, named prog2, containing two nop instructions between the two irmovl instructions generating values for registers %edx and %eax, and the addl instruction having these two registers as operands. In this case, the crucial step occurs in cycle 6, when the addl instruction reads its operands from the register file. An expanded view of the pipeline activities during this cycle is shown at the bottom of the figure. The first irmovl instruction has passed through the write-back stage, and so program register %edx has been updated in the register file. The second irmovl instruction is in the write-back stage during this cycle, and so the write to program register %eax only occurs at the start of cycle 7 as the clock rises. As a result, the incorrect value zero would be read for register %eax (recall that we assume all registers are initially 0), since the pending write for this register has not yet occurred. Clearly we will have to adapt our pipeline to handle this hazard properly.

Figure 45 shows what happens when we have only one nop instruction between the irmovl instructions and the addl instruction, yielding a program

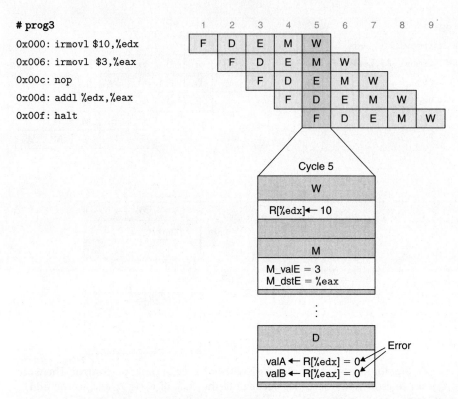

Figure 45 **Pipelined execution of** prog3 **without special pipeline control.** In cycle 5, the addl instruction reads its source operands from the register file. The pending write to register %edx is still in the write-back stage, and the pending write to register %eax is still in the memory stage. Both operands valA and valB get incorrect values.

prog3. Now we must examine the behavior of the pipeline during cycle 5 as the addl instruction passes through the decode stage. Unfortunately, the pending write to register %edx is still in the write-back stage, and the pending write to %eax is still in the memory stage. Therefore, the addl instruction would get the incorrect values for both operands.

Figure 46 shows what happens when we remove all of the nop instructions between the irmovl instructions and the addl instruction, yielding a program prog4. Now we must examine the behavior of the pipeline during cycle 4 as the addl instruction passes through the decode stage. Unfortunately, the pending write to register %edx is still in the memory stage, and the new value for %eax is just being computed in the execute stage. Therefore, the addl instruction would get the incorrect values for both operands.

These examples illustrate that a data hazard can arise for an instruction when one of its operands is updated by any of the three preceding instructions. These hazards occur because our pipelined processor reads the operands for an

Processor Architecture

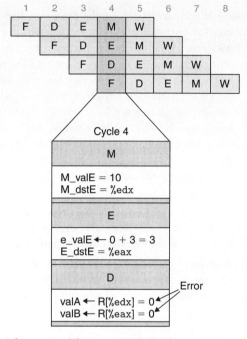

Figure 46 **Pipelined execution of** `prog4` **without special pipeline control.** In cycle 4, the `addl` instruction reads its source operands from the register file. The pending write to register `%edx` is still in the memory stage, and the new value for register `%eax` is just being computed in the execute stage. Both operands valA and valB get incorrect values.

instruction from the register file in the decode stage but does not write the results for the instruction to the register file until three cycles later, after the instruction passes through the write-back stage.

Aside   Enumerating classes of data hazards

Hazards can potentially occur when one instruction updates part of the program state that will be read by a later instruction. For Y86, the program state includes the program registers, the program counter, the memory, the condition code register, and the status register. Let us look at the hazard possibilities in our proposed design for each of these forms of state.

Program registers: These are the hazards we have already identified. They arise because the register file is read in one stage and written in another, leading to possible unintended interactions between different instructions.

Program counter: Conflicts between updating and reading the program counter give rise to control hazards. No hazard arises when our fetch-stage logic correctly predicts the new value of the program counter before fetching the next instruction. Mispredicted branches and `ret` instructions require special handling, as will be discussed in Section 5.11.

409

Memory: Writes and reads of the data memory both occur in the memory stage. By the time an instruction reading memory reaches this stage, any preceding instructions writing memory will have already done so. On the other hand, there can be interference between instructions writing data in the memory stage and the reading of instructions in the fetch stage, since the instruction and data memories reference a single address space. This can only happen with programs containing *self-modifying code*, where instructions write to a portion of memory from which instructions are later fetched. Some systems have complex mechanisms to detect and avoid such hazards, while others simply mandate that programs should not use self-modifying code. We will assume for simplicity that programs do not modify themselves, and therefore we do not need to take special measures to update the instruction memory based on updates to the data memory during program execution.

Condition code register: These are written by integer operations in the execute stage. They are read by conditional moves in the execute stage and by conditional jumps in the memory stage. By the time a conditional move or jump reaches the execute stage, any preceding integer operation will have already completed this stage. No hazards can arise.

Status register: The program status can be affected by instructions as they flow through the pipeline. Our mechanism of associating a status code with each instruction in the pipeline enables the processor to come to an orderly halt when an exception occurs, as will be discussed in Section 5.9.

This analysis shows that we only need to deal with register data hazards, control hazards, and making sure exceptions are handled properly. A systematic analysis of this form is important when designing a complex system. It can identify the potential difficulties in implementing the system, and it can guide the generation of test programs to be used in checking the correctness of the system.

5.6 Avoiding Data Hazards by Stalling

One very general technique for avoiding hazards involves *stalling*, where the processor holds back one or more instructions in the pipeline until the hazard condition no longer holds. Our processor can avoid data hazards by holding back an instruction in the decode stage until the instructions generating its source operands have passed through the write-back stage. The details of this mechanism will be discussed in Section 5.11. It involves simple enhancements to the pipeline control logic. The effect of stalling is diagrammed in Figures 47 (prog2) and 48 (prog4). (We omit prog3 from this discussion, since it operates similarly to the other two examples.) When the addl instruction is in the decode stage, the pipeline control logic detects that at least one of the instructions in the execute, memory, or write-back stage will update either register %edx or register %eax. Rather than letting the addl instruction pass through the stage with the incorrect results, it stalls the instruction, holding it back in the decode stage for either one (for prog2) or three (for prog4) extra cycles. For all three programs, the addl instruction finally gets correct values for its two source operands in cycle 7 and then proceeds down the pipeline.

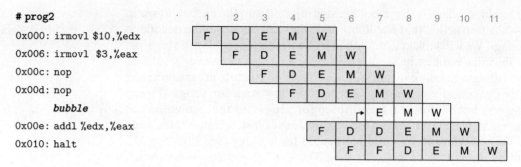

Figure 47 **Pipelined execution of** prog2 **using stalls.** After decoding the addl instruction in cycle 6, the stall control logic detects a data hazard due to the pending write to register %eax in the write-back stage. It injects a bubble into execute stage and repeats the decoding of the addl instruction in cycle 7. In effect, the machine has dynamically inserted a nop instruction, giving a flow similar to that shown for prog1 (Figure 43).

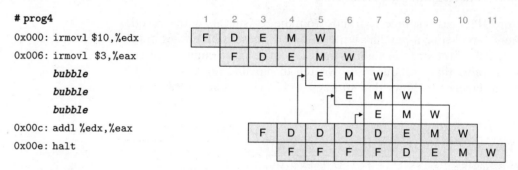

Figure 48 **Pipelined execution of** prog4 **using stalls.** After decoding the addl instruction in cycle 4, the stall control logic detects data hazards for both source registers. It injects a bubble into the execute stage and repeats the decoding of the addl instruction on cycle 5. It again detects hazards for both source registers, injects a bubble into the execute stage, and repeats the decoding of the addl instruction on cycle 6. Still, it detects a hazard for source register %eax, injects a bubble into the execute stage, and repeats the decoding of the addl instruction on cycle 7. In effect, the machine has dynamically inserted three nop instructions, giving a flow similar to that shown for prog1 (Figure 43).

In holding back the addl instruction in the decode stage, we must also hold back the halt instruction following it in the fetch stage. We can do this by keeping the program counter at a fixed value, so that the halt instruction will be fetched repeatedly until the stall has completed.

Stalling involves holding back one group of instructions in their stages while allowing other instructions to continue flowing through the pipeline. What then should we do in the stages that would normally be processing the addl instruction? We handle these by injecting a *bubble* into the execute stage each time we hold an instruction back in the decode stage. A bubble is like a dynamically generated nop instruction—it does not cause any changes to the registers, the memory, the condition codes, or the program status. These are shown as white boxes in the pipeline diagrams of Figures 47 and 48. In these figures, the arrow between the box labeled "D" for the addl instruction and the box labeled "E" for one of the pipeline bubbles indicates that a bubble was injected into the execute stage in place of the addl instruction that would normally have passed from the decode to the execute stage. We will look at the detailed mechanisms for making the pipeline stall and for injecting bubbles in Section 5.11.

In using stalling to handle data hazards, we effectively execute programs prog2 and prog4 by dynamically generating the pipeline flow seen for prog1 (Figure 43). Injecting one bubble for prog2 and three for prog4 has the same effect as having three nop instructions between the second irmovl instruction and the addl instruction. This mechanism can be implemented fairly easily (see Problem 51), but the resulting performance is not very good. There are numerous cases in which one instruction updates a register and a closely following instruction uses the same register. This will cause the pipeline to stall for up to three cycles, reducing the overall throughput significantly.

5.7 Avoiding Data Hazards by Forwarding

Our design for PIPE– reads source operands from the register file in the decode stage, but there can also be a pending write to one of these source registers in the write-back stage. Rather than stalling until the write has completed, it can simply pass the value that is about to be written to pipeline register E as the source operand. Figure 49 shows this strategy with an expanded view of the

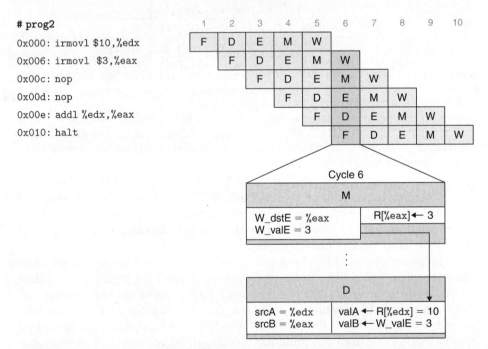

Figure 49 **Pipelined execution of** prog2 **using forwarding.** In cycle 6, the decode-stage logic detects the presence of a pending write to register %eax in the write-back stage. It uses this value for source operand valB rather than the value read from the register file.

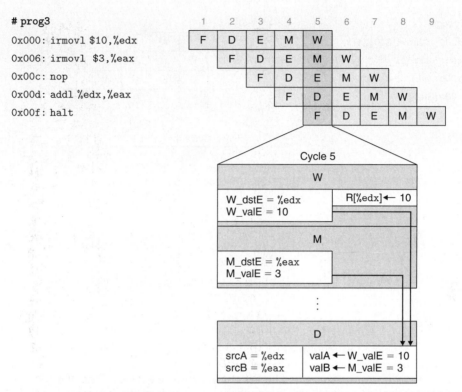

Figure 50 Pipelined execution of prog3 using forwarding. In cycle 5, the decode-stage logic detects a pending write to register %edx in the write-back stage and to register %eax in the memory stage. It uses these as the values for valA and valB rather than the values read from the register file.

pipeline diagram for cycle 6 of prog2. The decode-stage logic detects that register %eax is the source register for operand valB, and that there is also a pending write to %eax on write port E. It can therefore avoid stalling by simply using the data word supplied to port E (signal W_valE) as the value for operand valB. This technique of passing a result value directly from one pipeline stage to an earlier one is commonly known as *data forwarding* (or simply *forwarding*, and sometimes *bypassing*). It allows the instructions of prog2 to proceed through the pipeline without any stalling. Data forwarding requires adding additional data connections and control logic to the basic hardware structure.

As Figure 50 illustrates, data forwarding can also be used when there is a pending write to a register in the memory stage, avoiding the need to stall for program prog3. In cycle 5, the decode-stage logic detects a pending write to register %edx on port E in the write-back stage, as well as a pending write to register %eax that is on its way to port E but is still in the memory stage. Rather than stalling until the writes have occurred, it can use the value in the write-back stage (signal W_valE) for operand valA and the value in the memory stage (signal M_valE) for operand valB.

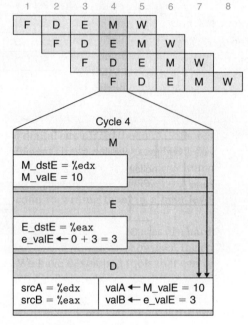

Figure 51 Pipelined execution of prog4 using forwarding. In cycle 4, the decode-stage logic detects a pending write to register %edx in the memory stage. It also detects that a new value is being computed for register %eax in the execute stage. It uses these as the values for valA and valB rather than the values read from the register file.

To exploit data forwarding to its full extent, we can also pass newly computed values from the execute stage to the decode stage, avoiding the need to stall for program prog4, as illustrated in Figure 51. In cycle 4, the decode-stage logic detects a pending write to register %edx in the memory stage, and also that the value being computed by the ALU in the execute stage will later be written to register %eax. It can use the value in the memory stage (signal M_valE) for operand valA. It can also use the ALU output (signal e_valE) for operand valB. Note that using the ALU output does not introduce any timing problems. The decode stage only needs to generate signals valA and valB by the end of the clock cycle so that pipeline register E can be loaded with the results from the decode stage as the clock rises to start the next cycle. The ALU output will be valid before this point.

The uses of forwarding illustrated in programs prog2 to prog4 all involve the forwarding of values generated by the ALU and destined for write port E. Forwarding can also be used with values read from the memory and destined for write port M. From the memory stage, we can forward the value that has just been read from the data memory (signal m_valM). From the write-back stage, we can forward the pending write to port M (signal W_valM). This gives a total of five different forwarding sources (e_valE, m_valM, M_valE, W_valM, and W_valE) and two different forwarding destinations (valA and valB).

The expanded diagrams of Figures 49 to 51 also show how the decode-stage logic can determine whether to use a value from the register file or to use a forwarded value. Associated with every value that will be written back to the register file is the destination register ID. The logic can compare these IDs with the source register IDs srcA and srcB to detect a case for forwarding. It is possible to have multiple destination register IDs match one of the source IDs. We must establish a priority among the different forwarding sources to handle such cases. This will be discussed when we look at the detailed design of the forwarding logic.

Figure 52 shows the structure of PIPE, an extension of PIPE– that can handle data hazards by forwarding. Comparing this to the structure of PIPE– (Figure 41), we can see that the values from the five forwarding sources are fed back to the two blocks labeled "Sel+Fwd A" and "Fwd B" in the decode stage. The block labeled "Sel+Fwd A" combines the role of the block labeled "Select A" in PIPE– with the forwarding logic. It allows valA for pipeline register E to be either the incremented program counter valP, the value read from the A port of the register file, or one of the forwarded values. The block labeled "Fwd B" implements the forwarding logic for source operand valB.

5.8 Load/Use Data Hazards

One class of data hazards cannot be handled purely by forwarding, because memory reads occur late in the pipeline. Figure 53 illustrates an example of a *load/use hazard*, where one instruction (the mrmovl at address 0x018) reads a value from memory for register %eax while the next instruction (the addl at address 0x01e) needs this value as a source operand. Expanded views of cycles 7 and 8 are shown in the lower part of the figure, where we assume all program registers initially have value 0. The addl instruction requires the value of the register in cycle 7, but it is not generated by the mrmovl instruction until cycle 8. In order to "forward" from the mrmovl to the addl, the forwarding logic would have to make the value go backward in time! Since this is clearly impossible, we must find some other mechanism for handling this form of data hazard. (The data hazard for register %ebx, with the value being generated by the irmovl instruction at address 0x012 and used by the addl instruction at address 0x01e, can be handled by forwarding.)

As Figure 54 demonstrates, we can avoid a load/use data hazard with a combination of stalling and forwarding. This requires modifications of the control logic, but it can use existing bypass paths. As the mrmovl instruction passes through the execute stage, the pipeline control logic detects that the instruction in the decode stage (the addl) requires the result read from memory. It stalls the instruction in the decode stage for one cycle, causing a bubble to be injected into the execute stage. As the expanded view of cycle 8 shows, the value read from memory can then be forwarded from the memory stage to the addl instruction in the decode stage. The value for register %ebx is also forwarded from the write-back to the memory stage. As indicated in the pipeline diagram by the arrow from the box labeled "D" in cycle 7 to the box labeled "E" in cycle 8, the injected bubble replaces the addl instruction that would normally continue flowing through the pipeline.

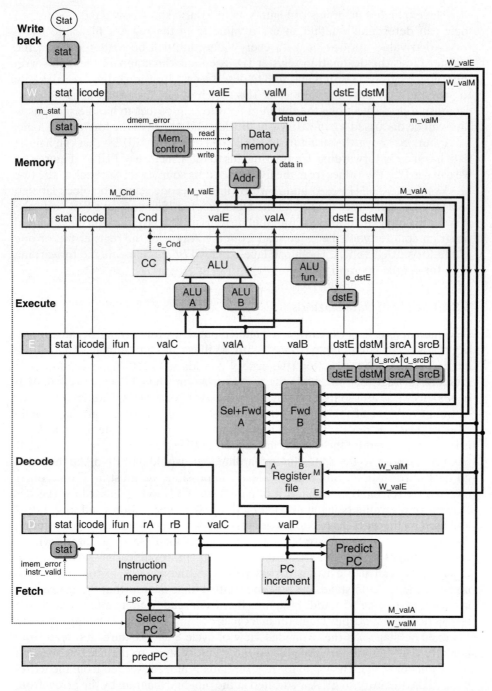

Figure 52 **Hardware structure of PIPE, our final pipelined implementation.** The additional bypassing paths enable forwarding the results from the three preceding instructions. This allows us to handle most forms of data hazards without stalling the pipeline.

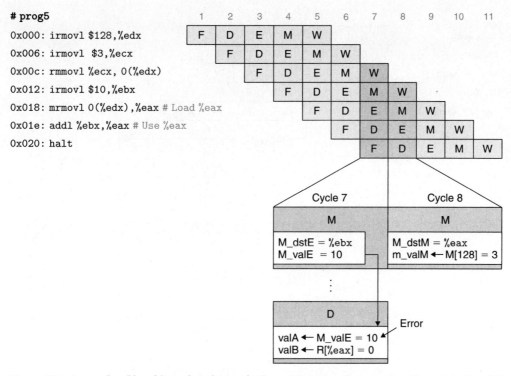

```
# prog5                                    1    2    3    4    5    6    7    8    9    10   11
0x000: irmovl $128,%edx                   F    D    E    M    W
0x006: irmovl $3,%ecx                          F    D    E    M    W
0x00c: rmmovl %ecx, 0(%edx)                         F    D    E    M    W
0x012: irmovl $10,%ebx                                   F    D    E    M    W
0x018: mrmovl 0(%edx),%eax  # Load %eax                        F    D    E    M    W
0x01e: addl %ebx,%eax  # Use %eax                                  F    D    E    M    W
0x020: halt                                                             F    D    E    M    W
```

Cycle 7 | Cycle 8

M

M_dstE = %ebx
M_valE = 10

M

M_dstM = %eax
m_valM ← M[128] = 3

⋮

D

Error

valA ← M_valE = 10
valB ← R[%eax] = 0

Figure 53 **Example of load/use data hazard.** The `addl` instruction requires the value of register `%eax` during the decode stage in cycle 7. The preceding `mrmovl` reads a new value for this register during the memory stage in cycle 8, which is too late for the `addl` instruction.

This use of a stall to handle a load/use hazard is called a *load interlock*. Load interlocks combined with forwarding suffice to handle all possible forms of data hazards. Since only load interlocks reduce the pipeline throughput, we can nearly achieve our throughput goal of issuing one new instruction on every clock cycle.

5.9 Exception Handling

A variety of activities in a processor can lead to *exceptional control flow*, where the normal chain of program execution gets broken. Exceptions can be generated either *internally*, by the executing program, or *externally*, by some outside signal. Our instruction set architecture includes three different internally generated exceptions, caused by (1) a `halt` instruction, (2) an instruction with an invalid combination of instruction and function code, and (3) an attempt to access an invalid address, either for instruction fetch or data read or write. A more complete processor design would also handle external exceptions, such as when the processor receives a signal that the network interface has received a new packet, or the user has clicked a mouse button. Handling exceptions correctly is a challenging aspect of any microprocessor design. They can

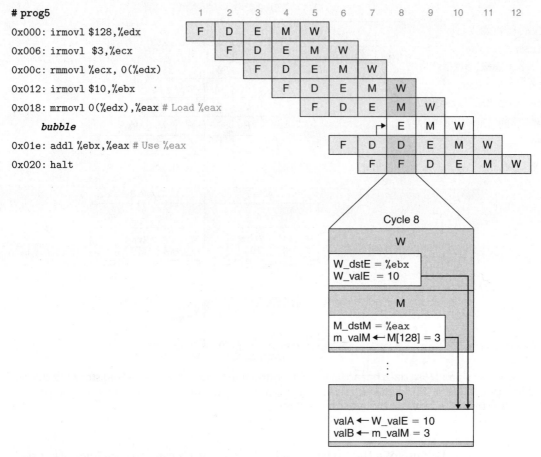

Figure 54 **Handling a load/use hazard by stalling.** By stalling the `addl` instruction for one cycle in the decode stage, the value for valB can be forwarded from the `mrmovl` instruction in the memory stage to the `addl` instruction in the decode stage.

occur at unpredictable times, and they require creating a clean break in the flow of instructions through the processor pipeline. Our handling of the three internal exceptions gives just a glimpse of the true complexity of correctly detecting and handling exceptions.

Let us refer to the instruction causing the exception as the *excepting instruction*. In the case of an invalid instruction address, there is no actual excepting instruction, but it is useful to think of there being a sort of "virtual instruction" at the invalid address. In our simplified ISA model, we want the processor to halt when it reaches an exception and to set the appropriate status code, as listed in Figure 5. It should appear that all instructions up to the excepting instruction have completed, but none of the following instructions should have any effect on the programmer-visible state. In a more complete design, the processor would continue by invoking an *exception handler*, a procedure that is part of the operating system, but implementing this part of exception handling is beyond the scope of our presentation.

In a pipelined system, exception handling involves several subtleties. First, it is possible to have exceptions triggered by multiple instructions simultaneously. For example, during one cycle of pipeline operation, we could have a halt instruction in the fetch stage, and the data memory could report an out-of-bounds data address for the instruction in the memory stage. We must determine which of these exceptions the processor should report to the operating system. The basic rule is to put priority on the exception triggered by the instruction that is furthest along the pipeline. In the example above, this would be the out-of-bounds address attempted by the instruction in the memory stage. In terms of the machine-language program, the instruction in the memory stage should appear to execute before one in the fetch stage, and therefore only this exception should be reported to the operating system.

A second subtlety occurs when an instruction is first fetched and begins execution, causes an exception, and later is canceled due to a mispredicted branch. The following is an example of such a program in its object code form:

```
0x000: 6300          |     xorl %eax,%eax
0x002: 740e000000    |     jne  Target      # Not taken
0x007: 30f001000000  |     irmovl $1, %eax  # Fall through
0x00d: 00            |     halt
0x00e:               | Target:
0x00e: ff            |     .byte 0xFF        # Invalid instruction code
```

In this program, the pipeline will predict that the branch should be taken, and so it will fetch and attempt to use a byte with value 0xFF as an instruction (generated in the assembly code using the .byte directive). The decode stage will therefore detect an invalid instruction exception. Later, the pipeline will discover that the branch should not be taken, and so the instruction at address 0x00e should never even have been fetched. The pipeline control logic will cancel this instruction, but we want to avoid raising an exception.

A third subtlety arises because a pipelined processor updates different parts of the system state in different stages. It is possible for an instruction following one causing an exception to alter some part of the state before the excepting instruction completes. For example, consider the following code sequence, in which we assume that user programs are not allowed to access addresses greater than 0xc0000000 (as is the case for 32-bit versions of Linux):

```
1   irmovl $1,%eax
2   xorl %esp,%esp       # Set stack pointer to 0 and CC to 100
3   pushl %eax           # Attempt to write to 0xfffffffc
4   addl  %eax,%eax       # (Should not be executed) Would set CC to 000
```

The pushl instruction causes an address exception, because decrementing the stack pointer causes it to wrap around to 0xfffffffc. This exception is detected in the memory stage. On the same cycle, the addl instruction is in the execute stage, and it will cause the condition codes to be set to new values. This would violate our requirement that none of the instructions following the excepting instruction should have had any effect on the system state.

In general, we can both correctly choose among the different exceptions and avoid raising exceptions for instructions that are fetched due to mispredicted branches by merging the exception-handling logic into the pipeline structure. That is the motivation for us to include a status code Stat in each of our pipeline registers (Figures 41 and 52). If an instruction generates an exception at some stage in its processing, the status field is set to indicate the nature of the exception. The exception status propagates through the pipeline with the rest of the information for that instruction, until it reaches the write-back stage. At this point, the pipeline control logic detects the occurrence of the exception and stops execution.

To avoid having any updating of the programmer-visible state by instructions beyond the excepting instruction, the pipeline control logic must disable any updating of the condition code register or the data memory when an instruction in the memory or write-back stages has caused an exception. In the example program above, the control logic would detect that the pushl in the memory stage has caused an exception, and therefore the updating of the condition code register by the addl instruction would be disabled.

Let us consider how this method of handling exceptions deals with the subtleties we have mentioned. When an exception occurs in one or more stages of a pipeline, the information is simply stored in the status fields of the pipeline registers. The event has no effect on the flow of instructions in the pipeline until an excepting instruction reaches the final pipeline stage, except to disable any updating of the programmer-visible state (the condition code register and the memory) by later instructions in the pipeline. Since instructions reach the write-back stage in the same order as they would be executed in a nonpipelined processor, we are guaranteed that the first instruction encountering an exception will arrive first in the write-back stage, at which point program execution can stop and the status code in pipeline register W can be recorded as the program status. If some instruction is fetched but later canceled, any exception status information about the instruction gets canceled as well. No instruction following one that causes an exception can alter the programmer-visible state. The simple rule of carrying the exception status together with all other information about an instruction through the pipeline provides a simple and reliable mechanism for handling exceptions.

5.10 PIPE Stage Implementations

We have now created an overall structure for PIPE, our pipelined Y86 processor with forwarding. It uses the same set of hardware units as the earlier sequential designs, with the addition of pipeline registers, some reconfigured logic blocks, and additional pipeline control logic. In this section, we go through the design of the different logic blocks, deferring the design of the pipeline control logic to the next section. Many of the logic blocks are identical to their counterparts in SEQ and SEQ+, except that we must choose proper versions of the different signals from the pipeline registers (written with the pipeline register name, written in uppercase, as a prefix) or from the stage computations (written with the first character of the stage name, written in lowercase, as a prefix).

As an example, compare the HCL code for the logic that generates the srcA signal in SEQ to the corresponding code in PIPE:

```
    # Code from SEQ
int srcA = [
        icode in { IRRMOVL, IRMMOVL, IOPL, IPUSHL  } : rA;
        icode in { IPOPL, IRET } : RESP;
        1 : RNONE; # Don't need register
];
```

```
    # Code from PIPE
int d_srcA = [
        D_icode in { IRRMOVL, IRMMOVL, IOPL, IPUSHL  } : D_rA;
        D_icode in { IPOPL, IRET } : RESP;
        1 : RNONE; # Don't need register
];
```

They differ only in the prefixes added to the PIPE signals: "D_" for the source values, to indicate that the signals come from pipeline register D, and "d_" for the result value, to indicate that it is generated in the decode stage. To avoid repetition, we will not show the HCL code here for blocks that only differ from those in SEQ because of the prefixes on names. As a reference, the complete HCL code for PIPE is given in Web Aside ARCH:HCL.

PC Selection and Fetch Stage

Figure 55 provides a detailed view of the PIPE fetch stage logic. As discussed earlier, this stage must also select a current value for the program counter and predict the next PC value. The hardware units for reading the instruction from memory and for extracting the different instruction fields are the same as those we considered for SEQ (see the fetch stage in Section 3.4).

The PC selection logic chooses between three program counter sources. As a mispredicted branch enters the memory stage, the value of valP for this instruction (indicating the address of the following instruction) is read from pipeline register M (signal M_valA). When a ret instruction enters the write-back stage, the return address is read from pipeline register W (signal W_valM). All other cases use the predicted value of the PC, stored in pipeline register F (signal F_predPC):

```
int f_pc = [
        # Mispredicted branch.  Fetch at incremented PC
        M_icode == IJXX && !M_Cnd : M_valA;
        # Completion of RET instruction.
        W_icode == IRET : W_valM;
        # Default: Use predicted value of PC
        1 : F_predPC;
];
```

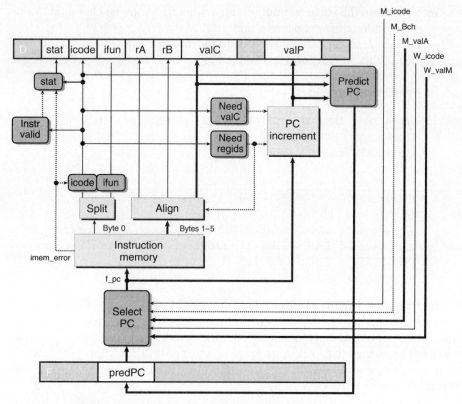

Figure 55 **PIPE PC selection and fetch logic.** Within the one cycle time limit, the processor can only predict the address of the next instruction.

The PC prediction logic chooses valC for the fetched instruction when it is either a call or a jump, and valP otherwise:

```
int f_predPC = [
      f_icode in { IJXX, ICALL } : f_valC;
      1 : f_valP;
];
```

The logic blocks labeled "Instr valid," "Need regids," and "Need valC" are the same as for SEQ, with appropriately named source signals.

Unlike in SEQ, we must split the computation of the instruction status into two parts. In the fetch stage, we can test for a memory error due to an out-of-range instruction address, and we can detect an illegal instruction or a halt instruction. Detecting an invalid data address must be deferred to the memory stage.

Practice Problem 28

Write HCL code for the signal f_stat, providing the provisional status for the fetched instruction.

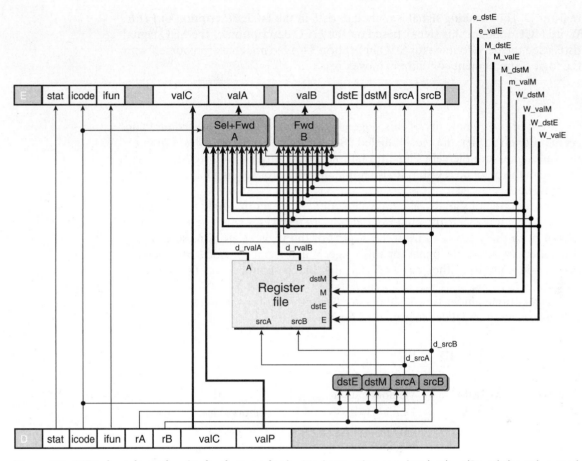

Figure 56 PIPE decode and write-back stage logic. No instruction requires both valP and the value read from register port A, and so these two can be merged to form the signal valA for later stages. The block labeled "Sel+Fwd A" performs this task and also implements the forwarding logic for source operand valA. The block labeled "Fwd B" implements the forwarding logic for source operand valB. The register write locations are specified by the dstE and dstM signals from the write-back stage rather than from the decode stage, since it is writing the results of the instruction currently in the write-back stage.

Decode and Write-Back Stages

Figure 56 gives a detailed view of the decode and write-back logic for PIPE. The blocks labeled "dstE", "dstM", "srcA", and "srcB" are very similar to their counterparts in the implementation of SEQ. Observe that the register IDs supplied to the write ports come from the write-back stage (signals W_dstE and W_dstM), rather than from the decode stage. This is because we want the writes to occur to the destination registers specified by the instruction in the write-back stage.

Practice Problem 29

The block labeled "dstE" in the decode stage generates the register ID for the E port of the register file, based on fields from the fetched instruction in pipeline register D. The resulting signal is named d_dstE in the HCL description of PIPE. Write HCL code for this signal, based on the HCL description of the SEQ signal dstE. (See the decode stage for SEQ in Section 3.4.) Do not concern yourself with the logic to implement conditional moves yet.

Most of the complexity of this stage is associated with the forwarding logic. As mentioned earlier, the block labeled "Sel+Fwd A" serves two roles. It merges the valP signal into the valA signal for later stages in order to reduce the amount of state in the pipeline register. It also implements the forwarding logic for source operand valA.

The merging of signals valA and valP exploits the fact that only the call and jump instructions need the value of valP in later stages, and these instructions do not need the value read from the A port of the register file. This selection is controlled by the icode signal for this stage. When signal D_icode matches the instruction code for either call or jXX, this block should select D_valP as its output.

As mentioned in Section 5.7, there are five different forwarding sources, each with a data word and a destination register ID:

Data word	Register ID	Source description
e_valE	e_dstE	ALU output
m_valM	M_dstM	Memory output
M_valE	M_dstE	Pending write to port E in memory stage
W_valM	W_dstM	Pending write to port M in write-back stage
W_valE	W_dstE	Pending write to port E in write-back stage

If none of the forwarding conditions hold, the block should select d_rvalA, the value read from register port A as its output.

Putting all of this together, we get the following HCL description for the new value of valA for pipeline register E:

```
int d_valA = [
        D_icode in { ICALL, IJXX } : D_valP; # Use incremented PC
        d_srcA == e_dstE : e_valE;   # Forward valE from execute
        d_srcA == M_dstM : m_valM;   # Forward valM from memory
        d_srcA == M_dstE : M_valE;   # Forward valE from memory
        d_srcA == W_dstM : W_valM;   # Forward valM from write back
        d_srcA == W_dstE : W_valE;   # Forward valE from write back
        1 : d_rvalA;  # Use value read from register file
];
```

The priority given to the five forwarding sources in the above HCL code is very important. This priority is determined in the HCL code by the order in which

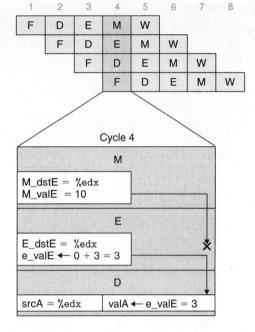

Figure 57 **Demonstration of forwarding priority.** In cycle 4, values for %edx are available from both the execute and memory stages. The forwarding logic should choose the one in the execute stage, since it represents the most recently generated value for this register.

the five destination register IDs are tested. If any order other than the one shown were chosen, the pipeline would behave incorrectly for some programs. Figure 57 shows an example of a program that requires a correct setting of priority among the forwarding sources in the execute and memory stages. In this program, the first two instructions write to register %edx, while the third uses this register as its source operand. When the rrmovl instruction reaches the decode stage in cycle 4, the forwarding logic must choose between two values destined for its source register. Which one should it choose? To set the priority, we must consider the behavior of the machine-language program when it is executed one instruction at a time. The first irmovl instruction would set register %edx to 10, the second would set the register to 3, and then the rrmovl instruction would read 3 from %edx. To imitate this behavior, our pipelined implementation should always give priority to the forwarding source in the earliest pipeline stage, since it holds the latest instruction in the program sequence setting the register. Thus, the logic in the HCL code above first tests the forwarding source in the execute stage, then those in the memory stage, and finally the sources in the write-back stage.

The forwarding priority between the two sources in either the memory or the write-back stages are only a concern for the instruction popl %esp, since only this instruction can write two registers simultaneously.

Practice Problem 30

Suppose the order of the third and fourth cases (the two forwarding sources from the memory stage) in the HCL code for d_valA were reversed. Describe the resulting behavior of the rrmovl instruction (line 5) for the following program:

```
1       irmovl $5, %edx
2       irmovl $0x100,%esp
3       rmmovl %edx,0(%esp)
4       popl %esp
5       rrmovl %esp,%eax
```

Practice Problem 31

Suppose the order of the fifth and sixth cases (the two forwarding sources from the write-back stage) in the HCL code for d_valA were reversed. Write a Y86 program that would be executed incorrectly. Describe how the error would occur and its effect on the program behavior.

Practice Problem 32

Write HCL code for the signal d_valB, giving the value for source operand valB supplied to pipeline register E.

One small part of the write-back stage remains. As shown in Figure 52, the overall processor status Stat is computed by a block based on the status value in pipeline register W. Recall from Section 1.1 that the code should indicate either normal operation (AOK) or one of the three exception conditions. Since pipeline register W holds the state of the most recently completed instruction, it is natural to use this value as an indication of the overall processor status. The only special case to consider is when there is a bubble in the write-back stage. This is part of normal operation, and so we want the status code to be AOK for this case as well:

```
int Stat = [
        W_stat == SBUB : SAOK;
        1 : W_stat;
];
```

Execute Stage

Figure 58 shows the execute stage logic for PIPE. The hardware units and the logic blocks are identical to those in SEQ, with an appropriate renaming of signals. We can see the signals e_valE and e_dstE directed toward the decode stage as one of the forwarding sources. One difference is that the logic labeled "Set CC," which determines whether or not update the condition codes, has signals m_stat and

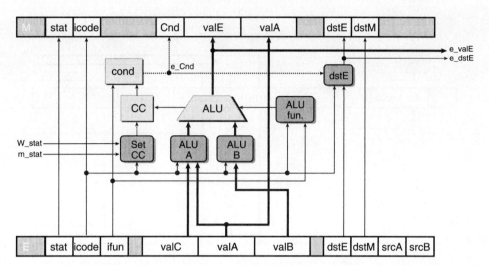

Figure 58 PIPE execute stage logic. This part of the design is very similar to the logic in the SEQ implementation.

W_stat as inputs. These signals are used to detect cases where an instruction causing an exception is passing through later pipeline stages, and therefore any updating of the condition codes should be suppressed. This aspect of the design is discussed in Section 5.11.

Practice Problem 33

Our second case in the HCL code for d_valA uses signal e_dstE to see whether to select the ALU output e_valE as the forwarding source. Suppose instead that we use signal E_dstE, the destination register ID in pipeline register E for this selection. Write a Y86 program that would give an incorrect result with this modified forwarding logic.

Memory Stage

Figure 59 shows the memory stage logic for PIPE. Comparing this to the memory stage for SEQ (Figure 30), we see that, as noted before, the block labeled "Data" in SEQ is not present in PIPE. This block served to select between data sources valP (for call instructions) and valA, but this selection is now performed by the block labeled "Sel+Fwd A" in the decode stage. Most other blocks in this stage are identical to their counterparts in SEQ, with an appropriate renaming of the signals. In this figure, you can also see that many of the values in pipeline registers and M and W are supplied to other parts of the circuit as part of the forwarding and pipeline control logic.

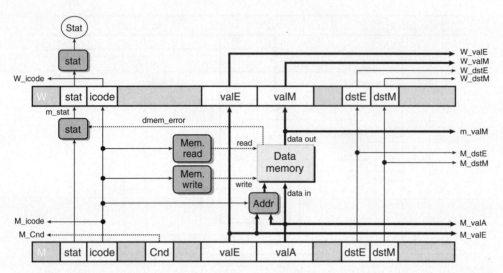

Figure 59 **PIPE memory stage logic.** Many of the signals from pipeline registers M and W are passed down to earlier stages to provide write-back results, instruction addresses, and forwarded results.

Practice Problem 34

In this stage, we can complete the computation of the status code Stat by detecting the case of an invalid address for the data memory. Write HCL code for the signal m_stat.

5.11 Pipeline Control Logic

We are now ready to complete our design for PIPE by creating the pipeline control logic. This logic must handle the following four control cases for which other mechanisms, such as data forwarding and branch prediction, do not suffice:

Processing ret: The pipeline must stall until the ret instruction reaches the write-back stage.

Load/use hazards: The pipeline must stall for one cycle between an instruction that reads a value from memory and an instruction that uses this value.

Mispredicted branches: By the time the branch logic detects that a jump should not have been taken, several instructions at the branch target will have started down the pipeline. These instructions must be removed from the pipeline.

Exceptions: When an instruction causes an exception, we want to disable the updating of the programmer-visible state by later instructions and halt execution once the excepting instruction reaches the write-back stage.

We will go through the desired actions for each of these cases and then develop control logic to handle all of them.

Desired Handling of Special Control Cases

For the `ret` instruction, consider the following example program. This program is shown in assembly code, but with the addresses of the different instructions on the left for reference:

```
0x000:      irmovl Stack,%esp    # Initialize stack pointer
0x006:      call Proc            # procedure call
0x00b:      irmovl $10,%edx      # return point
0x011:      halt
0x020: .pos 0x20
0x020: Proc:                     # Proc:
0x020:      ret                  # return immediately
0x021:      rrmovl %edx,%ebx     # not executed
0x030: .pos 0x30
0x030: Stack:                    # Stack: Stack pointer
```

Figure 60 shows how we want the pipeline to process the `ret` instruction. As with our earlier pipeline diagrams, this figure shows the pipeline activity with time growing to the right. Unlike before, the instructions are not listed in the same order they occur in the program, since this program involves a control flow where instructions are not executed in a linear sequence. Look at the instruction addresses to see from where the different instructions come in the program.

As this diagram shows, the `ret` instruction is fetched during cycle 3 and proceeds down the pipeline, reaching the write-back stage in cycle 7. While it passes through the decode, execute, and memory stages, the pipeline cannot do any useful activity. Instead, we want to inject three bubbles into the pipeline. Once the `ret` instruction reaches the write-back stage, the PC selection logic will set the program counter to the return address, and therefore the fetch stage will fetch the `irmovl` instruction at the return point (address 0x00b).

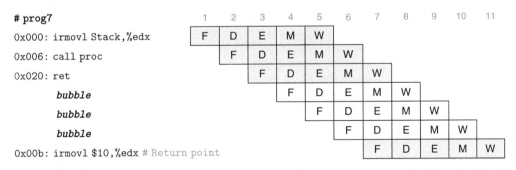

Figure 60 **Simplified view of `ret` instruction processing.** The pipeline should stall while the `ret` passes through the decode, execute, and memory stages, injecting three bubbles in the process. The PC selection logic will choose the return address as the instruction fetch address once the `ret` reaches the write-back stage (cycle 7).

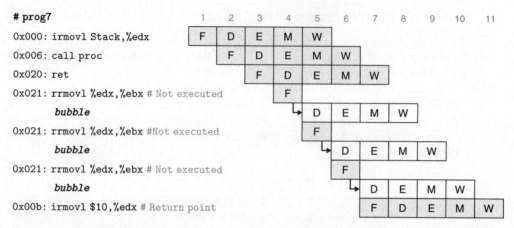

Figure 61 Actual processing of the `ret` instruction. The fetch stage repeatedly fetches the `rrmovl` instruction following the `ret` instruction, but then the pipeline control logic injects a bubble into the decode stage rather than allowing the `rrmovl` instruction to proceed. The resulting behavior is equivalent to that shown in Figure 60.

Figure 61 shows the actual processing of the `ret` instruction for the example program. The key observation here is that there is no way to inject a bubble into the fetch stage of our pipeline. On every cycle, the fetch stage reads *some* instruction from the instruction memory. Looking at the HCL code for implementing the PC prediction logic in Section 5.10, we can see that for the `ret` instruction the new value of the PC is predicted to be valP, the address of the following instruction. In our example program, this would be 0x021, the address of the `rrmovl` instruction following the `ret`. This prediction is not correct for this example, nor would it be for most cases, but we are not attempting to predict return addresses correctly in our design. For three clock cycles, the fetch stage stalls, causing the `rrmovl` instruction to be fetched but then replaced by a bubble in the decode stage. This process is illustrated in Figure 61 by the three fetches, with an arrow leading down to the bubbles passing through the remaining pipeline stages. Finally, the `irmovl` instruction is fetched on cycle 7. Comparing Figure 61 with Figure 60, we see that our implementation achieves the desired effect, but with a slightly peculiar fetching of an incorrect instruction for 3 consecutive cycles.

For a load/use hazard, we have already described the desired pipeline operation in Section 5.8, as illustrated by the example of Figure 54. Only the `mrmovl` and `popl` instructions read data from memory. When either of these is in the execute stage, and an instruction requiring the destination register is in the decode stage, we want to hold back the second instruction in the decode stage and inject a bubble into the execute stage on the next cycle. After this, the forwarding logic will resolve the data hazard. The pipeline can hold back an instruction in the decode stage by keeping pipeline register D in a fixed state. In doing so, it should also keep pipeline register F in a fixed state, so that the next instruction will be fetched a second time. In summary, implementing this pipeline flow requires detecting the

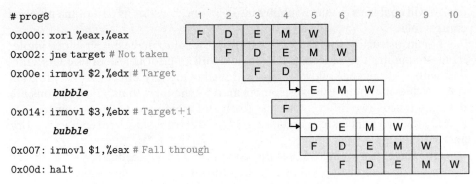

Figure 62 **Processing mispredicted branch instructions.** The pipeline predicts branches will be taken and so starts fetching instructions at the jump target. Two instructions are fetched before the misprediction is detected in cycle 4 when the jump instruction flows through the execute stage. In cycle 5, the pipeline *cancels* the two target instructions by injecting bubbles into the decode and execute stages, and it also fetches the instruction following the jump.

hazard condition, keeping pipeline register F and D fixed, and injecting a bubble into the execute stage.

To handle a mispredicted branch, consider the following program, shown in assembly code, but with the instruction addresses shown on the left for reference:

```
0x000:      xorl %eax,%eax
0x002:      jne  target          # Not taken
0x007:      irmovl $1, %eax      # Fall through
0x00d:      halt
0x00e: target:
0x00e:      irmovl $2, %edx      # Target
0x014:      irmovl $3, %ebx      # Target+1
0x01a:      halt
```

Figure 62 shows how these instructions are processed. As before, the instructions are listed in the order they enter the pipeline, rather than the order they occur in the program. Since the jump instruction is predicted as being taken, the instruction at the jump target will be fetched in cycle 3, and the instruction following this one will be fetched in cycle 4. By the time the branch logic detects that the jump should not be taken during cycle 4, two instructions have been fetched that should not continue being executed. Fortunately, neither of these instructions has caused a change in the programmer-visible state. That can only occur when an instruction reaches the execute stage, where it can cause the condition codes to change. We can simply *cancel* (sometimes called *instruction squashing*) the two misfetched instructions by injecting bubbles into the decode and execute instructions on the following cycle while also fetching the instruction following the jump instruction. The two misfetched instructions will then simply disappear from the pipeline. As we will discuss in Section 5.11, a simple extension to the basic clocked register design will enable us to inject bubbles into pipeline registers as part of the pipeline control logic.

For an instruction that causes an exception, we must make the pipelined implementation match the desired ISA behavior, with all prior instructions completing and with none of the following instructions having any effect on the program state. Achieving these effects is complicated by the facts that (1) exceptions are detected during two different stages (fetch and memory) of program execution, and (2) the program state is updated in three different stages (execute, memory, and write-back).

Our stage designs include a status code stat in each pipeline register to track the status of each instruction as it passes through the pipeline stages. When an exception occurs, we record that information as part of the instruction's status and continue fetching, decoding, and executing instructions as if nothing were amiss. As the excepting instruction reaches the memory stage, we take steps to prevent later instructions from modifying programmer-visible state by (1) disabling the setting of condition codes by instructions in the execute stage, (2) injecting bubbles into the memory stage to disable any writing to the data memory, and (3) stalling the write-back stage when it has an excepting instruction, thus bringing the pipeline to a halt.

The pipeline diagram in Figure 63 illustrates how our pipeline control handles the situation where an instruction causing an exception is followed by one that would change the condition codes. On cycle 6, the pushl instruction reaches the memory stage and generates a memory error. On the same cycle, the addl instruction in the execute stage generates new values for the condition codes. We disable

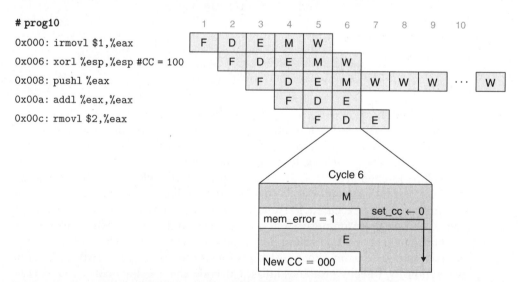

Figure 63 **Processing invalid memory reference exception.** On cycle 6, the invalid memory reference by the pushl instruction causes the updating of the condition codes to be disabled. The pipeline starts injecting bubbles into the memory stage and stalling the excepting instruction in the write-back stage.

Condition	Trigger
Processing `ret`	IRET ∈ {D_icode, E_icode, M_icode}
Load/use hazard	E_icode ∈ {IMRMOVL, IPOPL} && E_dstM ∈ {d_srcA, d_srcB}
Mispredicted branch	E_icode = IJXX && !e_Cnd
Exception	m_stat ∈ {SADR, SINS, SHLT} || W_stat ∈ {SADR, SINS, SHLT}

Figure 64 **Detection conditions for pipeline control logic.** Four different conditions require altering the pipeline flow by either stalling the pipeline or canceling partially executed instructions.

the setting of condition codes when an excepting instruction is in the memory or write-back stage (by examining the signals m_stat and W_stat and then setting the signal set_cc to zero). We can also see the combination of injecting bubbles into the memory stage and stalling the excepting instruction in the write-back stage in the example of Figure 63—the pushl instruction remains stalled in the write-back stage, and none of the subsequent instructions get past the execute stage.

By this combination of pipelining the status signals, controlling the setting of condition codes, and controlling the pipeline stages, we achieve the desired behavior for exceptions: all instructions prior to the excepting instruction are completed, while none of the following instructions has any effect on the programmer-visible state.

Detecting Special Control Conditions

Figure 64 summarizes the conditions requiring special pipeline control. It gives expressions describing the conditions under which the three special cases arise. These expressions are implemented by simple blocks of combinational logic that must generate their results before the end of the clock cycle in order to control the action of the pipeline registers as the clock rises to start the next cycle. During a clock cycle, pipeline registers D, E, and M hold the states of the instructions that are in the decode, execute, and memory pipeline stages, respectively. As we approach the end of the clock cycle, signals d_srcA and d_srcB will be set to the register IDs of the source operands for the instruction in the decode stage. Detecting a ret instruction as it passes through the pipeline simply involves checking the instruction codes of the instructions in the decode, execute, and memory stages. Detecting a load/use hazard involves checking the instruction type (mrmovl or popl) of the instruction in the execute stage and comparing its destination register with the source registers of the instruction in the decode stage. The pipeline control logic should detect a mispredicted branch while the jump instruction is in the execute stage, so that it can set up the conditions required to recover from the misprediction as the instruction enters the memory stage. When a jump instruction is in the execute stage, the signal e_Cnd indicates whether or not the jump should be taken. We detect an excepting instruction by examining the instruction status values in the memory and write-back stages. For the memory stage, we use the signal m_stat, computed within the stage, rather than M_stat

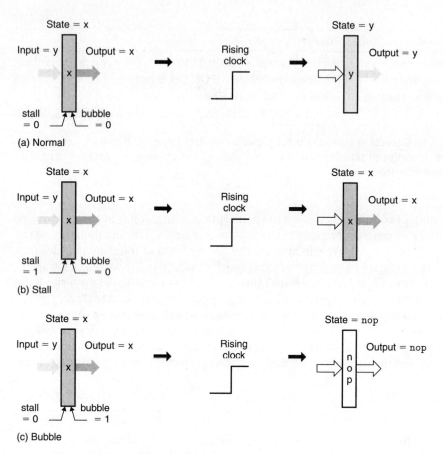

Figure 65 **Additional pipeline register operations.** (a) Under normal conditions, the state and output of the register are set to the value at the input when the clock rises. (b) When operated in *stall* mode, the state is held fixed at its previous value. (c) When operated in *bubble* mode, the state is overwritten with that of a nop operation.

from the pipeline register. This internal signal incorporates the possibility of a data memory address error.

Pipeline Control Mechanisms

Figure 65 shows low-level mechanisms that allow the pipeline control logic to hold back an instruction in a pipeline register or to inject a bubble into the pipeline. These mechanisms involve small extensions to the basic clocked register described in Section 2.5. Suppose that each pipeline register has two control inputs **stall** and **bubble**. The settings of these signals determine how the pipeline register is updated as the clock rises. Under normal operation (Figure 65(a)), both of these inputs are set to 0, causing the register to load its input as its new state. When the stall signal is set to 1 (Figure 65(b)), the updating of the state is disabled. Instead, the register will remain in its previous state. This makes it possible to

Condition	Pipeline register				
	F	D	E	M	W
Processing `ret`	stall	bubble	normal	normal	normal
Load/use hazard	stall	stall	bubble	normal	normal
Mispredicted branch	normal	bubble	bubble	normal	normal

Figure 66 **Actions for pipeline control logic.** The different conditions require altering the pipeline flow by either stalling the pipeline or by canceling partially executed instructions.

hold back an instruction in some pipeline stage. When the bubble signal is set to 1 (Figure 65(c)), the state of the register will be set to some fixed *reset configuration* giving a state equivalent to that of a nop instruction. The particular pattern of ones and zeros for a pipeline register's reset configuration depends on the set of fields in the pipeline register. For example, to inject a bubble into pipeline register D, we want the icode field to be set to the constant value INOP (Figure 26). To inject a bubble into pipeline register E, we want the icode field to be set to INOP and the dstE, dstM, srcA, and srcB fields to be set to the constant RNONE. Determining the reset configuration is one of the tasks for the hardware designer in designing a pipeline register. We will not concern ourselves with the details here. We will consider it an error to set both the bubble and the stall signals to 1.

The table in Figure 66 shows the actions the different pipeline stages should take for each of the three special conditions. Each involves some combination of normal, stall, and bubble operations for the pipeline registers.

In terms of timing, the stall and bubble control signals for the pipeline registers are generated by blocks of combinational logic. These values must be valid as the clock rises, causing each of the pipeline registers to either load, stall, or bubble as the next clock cycle begins. With this small extension to the pipeline register designs, we can implement a complete pipeline, including all of its control, using the basic building blocks of combinational logic, clocked registers, and random-access memories.

Combinations of Control Conditions

In our discussion of the special pipeline control conditions so far, we assumed that at most one special case could arise during any single clock cycle. A common bug in designing a system is to fail to handle instances where multiple special conditions arise simultaneously. Let us analyze such possibilities. We need not worry about combinations involving program exceptions, since we have carefully designed our exception-handling mechanism to consider other instructions in the pipeline. Figure 67 diagrams the pipeline states that cause the other three special control conditions. These diagrams show blocks for the decode, execute, and memory stages. The shaded boxes represent particular constraints that must be satisfied for the condition to arise. A load/use hazard requires that the instruction in the

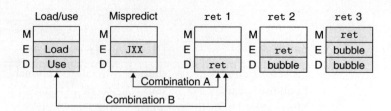

Figure 67

Pipeline states for special control conditions. The two pairs indicated can arise simultaneously.

execute stage reads a value from memory into a register, and that the instruction in the decode stage has this register as a source operand. A mispredicted branch requires the instruction in the execute stage to have a jump instruction. There are three possible cases for ret—the instruction can be in either the decode, execute, or memory stage. As the ret instruction moves through the pipeline, the earlier pipeline stages will have bubbles.

We can see by these diagrams that most of the control conditions are mutually exclusive. For example, it is not possible to have a load/use hazard and a mispredicted branch simultaneously, since one requires a load instruction (mrmovl or popl) in the execute stage, while the other requires a jump. Similarly, the second and third ret combinations cannot occur at the same time as a load/use hazard or a mispredicted branch. Only the two combinations indicated by arrows can arise simultaneously.

Combination A involves a not-taken jump instruction in the execute stage and a ret instruction in the decode stage. Setting up this combination requires the ret to be at the target of a not-taken branch. The pipeline control logic should detect that the branch was mispredicted and therefore cancel the ret instruction.

Practice Problem 35

Write a Y86 assembly-language program that causes combination A to arise and determines whether the control logic handles it correctly.

Combining the control actions for the combination A conditions (Figure 66), we get the following pipeline control actions (assuming that either a bubble or a stall overrides the normal case):

	Pipeline register				
Condition	F	D	E	M	W
Processing ret	stall	bubble	normal	normal	normal
Mispredicted branch	normal	bubble	bubble	normal	normal
Combination	stall	bubble	bubble	normal	normal

That is, it would be handled like a mispredicted branch, but with a stall in the fetch stage. Fortunately, on the next cycle, the PC selection logic will choose the address of the instruction following the jump, rather than the predicted program counter, and so it does not matter what happens with the pipeline register F. We conclude that the pipeline will correctly handle this combination.

Combination B involves a load/use hazard, where the loading instruction sets register %esp, and the ret instruction then uses this register as a source operand, since it must pop the return address from the stack. The pipeline control logic should hold back the ret instruction in the decode stage.

Practice Problem 36

Write a Y86 assembly-language program that causes combination B to arise and completes with a halt instruction if the pipeline operates correctly.

Combining the control actions for the combination B conditions (Figure 66), we get the following pipeline control actions:

| | Pipeline register | | | | |
Condition	F	D	E	M	W
Processing ret	stall	bubble	normal	normal	normal
Load/use hazard	stall	stall	bubble	normal	normal
Combination	stall	bubble+stall	bubble	normal	normal
Desired	stall	stall	bubble	normal	normal

If both sets of actions were triggered, the control logic would try to stall the ret instruction to avoid the load/use hazard but also inject a bubble into the decode stage due to the ret instruction. Clearly, we do not want the pipeline to perform both sets of actions. Instead, we want it to just take the actions for the load/use hazard. The actions for processing the ret instruction should be delayed for one cycle.

This analysis shows that combination B requires special handling. In fact, our original implementation of the PIPE control logic did not handle this combination correctly. Even though the design had passed many simulation tests, it had a subtle bug that was uncovered only by the analysis we have just shown. When a program having combination B was executed, the control logic would set both the bubble and the stall signals for pipeline register D to 1. This example shows the importance of systematic analysis. It would be unlikely to uncover this bug by just running normal programs. If left undetected, the pipeline would not faithfully implement the ISA behavior.

Control Logic Implementation

Figure 68 shows the overall structure of the pipeline control logic. Based on signals from the pipeline registers and pipeline stages, the control logic generates stall and bubble control signals for the pipeline registers, and also determines whether the condition code registers should be updated. We can combine the

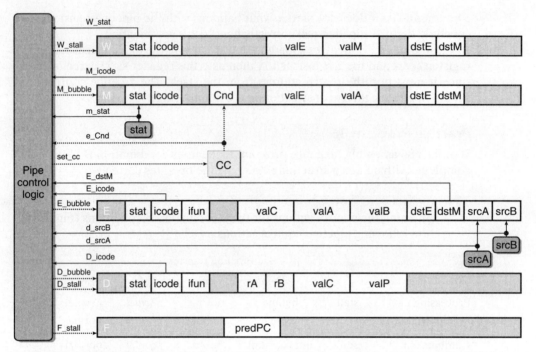

Figure 68 **PIPE pipeline control logic.** This logic overrides the normal flow of instructions through the pipeline to handle special conditions such as procedure returns, mispredicted branches, load/use hazards, and program exceptions.

detection conditions of Figure 64 with the actions of Figure 66 to create HCL descriptions for the different pipeline control signals.

Pipeline register F must be stalled for either a load/use hazard or a ret instruction:

```
bool F_stall =
        # Conditions for a load/use hazard
        E_icode in { IMRMOVL, IPOPL } &&
         E_dstM in { d_srcA, d_srcB } ||
        # Stalling at fetch while ret passes through pipeline
        IRET in { D_icode, E_icode, M_icode };
```

Practice Problem 37

Write HCL code for the signal D_stall in the PIPE implementation.

Pipeline register D must be set to bubble for a mispredicted branch or a ret instruction. As the analysis in the preceding section shows, however, it should not inject a bubble when there is a load/use hazard in combination with a `ret` instruction:

```
bool D_bubble =
        # Mispredicted branch
        (E_icode == IJXX && !e_Cnd) ||
        # Stalling at fetch while ret passes through pipeline
        # but not condition for a load/use hazard
        !(E_icode in { IMRMOVL, IPOPL }
          && E_dstM in { d_srcA, d_srcB })
        && IRET in { D_icode, E_icode, M_icode };
```

Practice Problem 38

Write HCL code for the signal E_bubble in the PIPE implementation.

Practice Problem 39

Write HCL code for the signal set_cc in the PIPE implementation. This should only occur for OP1 instructions, and should consider the effects of program exceptions.

Practice Problem 40

Write HCL code for the signals M_bubble and W_stall in the PIPE implementation. The latter signal requires modifying the exception condition listed in Figure 64.

This covers all of the special pipeline control signal values. In the complete HCL code for PIPE, all other pipeline control signals are set to zero.

Aside   Testing the design

As we have seen, there are many ways to introduce bugs into a design even for a simple microprocessor. With pipelining, there are many subtle interactions between the instructions at different pipeline stages. We have seen that many of the design challenges involve unusual instructions (such as popping to the stack pointer) or unusual instruction combinations (such as a not-taken jump followed by a `ret`). We also see that exception handling adds an entirely new dimension to the possible pipeline behaviors. How then can we be sure that our design is correct? For hardware manufacturers, this is a dominant concern, since they cannot simply report an error and have users download code patches over the Internet. Even a simple logic design error can have serious consequences, especially as microprocessors are increasingly used to operate systems that are critical to our lives and health, such as automotive antilock braking systems, heart pacemakers, and aircraft control systems.

Simply simulating a design while running a number of "typical" programs is not a sufficient means of testing a system. Instead, thorough testing requires devising ways of systematically generating many tests that will exercise as many different instructions and instruction combinations as possible. In creating our Y86 processor designs, we also devised a number of testing scripts, each of which generates many different tests, runs simulations of the processor, and compares the resulting register and memory values to those produced by our YIS instruction set simulator. Here is a brief description of the scripts:

optest: Runs 49 tests of different Y86 instructions with different source and destination registers

jtest: Runs 64 tests of the different jump and call instructions, with different combinations of whether or not the branches are taken

cmtest: Runs 28 tests of the different conditional move instructions, with different control combinations

htest: Runs 600 tests of different data hazard possibilities, with different combinations of source and destination instructions, and with different numbers of nop instructions between the instruction pairs

ctest: Tests 22 different control combinations, based on an analysis similar to what we did in Section 5.11

etest: Tests 12 different combinations of instructions causing exceptions and instructions following it that could alter the programmer-visible state

The key idea of this testing method is that we want to be as systematic as possible, generating tests that create the different conditions that are likely to cause pipeline errors.

Aside Formally verifying our design

Even when a design passes an extensive set of tests, we cannot be certain that it will operate correctly for all possible programs. The number of possible programs we could test is unimaginably large, even if we only consider tests consisting of short code segments. Newer methods of *formal verification*, however, hold the promise that we can have tools that rigorously consider all possible behaviors of a system and determine whether or not there are any design errors.

We were able to apply formal verification to an earlier version of our Y86 processors [13]. We set up a framework to compare the behavior of the pipelined design PIPE to the unpipelined version SEQ. That is, it was able to prove that for an arbitrary Y86 program, the two processors would have identical effects on the programmer-visible state. Of course, our verifier cannot actually run all possible programs, since there are an infinite number of them. Instead, it uses a form of proof by induction, showing a consistency between the two processors on a cycle-by-cycle basis. Carrying out this analysis requires reasoning about the hardware using *symbolic methods* in which we consider all program values to be arbitrary integers, and we abstract the ALU as a sort of "black box," computing some unspecified function over its arguments. We assume only that the ALUs for SEQ and PIPE compute identical functions.

We used the HCL descriptions of the control logic to generate the control logic for our symbolic processor models, and so we could catch any bugs in the HCL code. Being able to show that SEQ and PIPE are identical does not guarantee that either of them faithfully implements the Y86 instruction set architecture. However, it would uncover any bug due to an incorrect pipeline design, and this is the major source of design errors.

In our experiments, we verified not only the version of PIPE we have considered in this chapter but also several variants that we give as homework problems, in which we add more instructions, modify the hardware capabilities, or use different branch prediction strategies. Interestingly, we found only one bug in all of our designs, involving control combination B (described in Section 5.11) for our solution to the variant described in Problem 57. This exposed a weakness in our testing regime that caused us to add additional cases to the ctest testing script.

Formal verification is still in an early stage of development. The tools are often difficult to use, and they do not have the capacity to verify large-scale designs. We were able to verify our Y86 processors in part because of their relative simplicity. Even then, it required several weeks of effort and multiple runs of the tools, each requiring up to eight hours of computer time. This is an active area of research, with some tools becoming commercially available, and some in use at companies such as Intel, AMD, and IBM.

Web Aside ARCH:VLOG   Verilog implementation of a pipelined Y86 processor

As we have mentioned, modern logic design involves writing textual representations of hardware designs in a *hardware description language*. The design can then be tested by both simulation and by a variety of formal verification tools. Once we have confidence in the design, we can use *logic synthesis* tools to translate the design into actual logic circuits.

We have developed models of our Y86 processor designs in the Verilog hardware description language. These designs combine modules implementing the basic building blocks of the processor, along with control logic generated directly from the HCL descriptions. We have been able to synthesize some of these designs, download the logic circuit descriptions onto field-programmable gate array (FPGA) hardware, and run the processors on actual Y86 programs.

5.12 Performance Analysis

We can see that the conditions requiring special action by the pipeline control logic all cause our pipeline to fall short of the goal of issuing a new instruction on every clock cycle. We can measure this inefficiency by determining how often a bubble gets injected into the pipeline, since these cause unused pipeline cycles. A return instruction generates three bubbles, a load/use hazard generates one, and a mispredicted branch generates two. We can quantify the effect these penalties have on the overall performance by computing an estimate of the average number of clock cycles PIPE would require per instruction it executes, a measure known as the CPI (for "cycles per instruction"). This measure is the reciprocal of the average throughput of the pipeline, but with time measured in clock cycles rather than picoseconds. It is a useful measure of the architectural efficiency of a design.

If we ignore the performance implications of exceptions (which, by definition, will only occur rarely), another way to think about CPI is to imagine we run the processor on some benchmark program and observe the operation of the execute stage. On each cycle, the execute stage would either process an instruction, and this instruction would then continue through the remaining stages to completion, or it would process a bubble, injected due to one of the three special cases. If the stage processes a total of C_i instructions and C_b bubbles, then the processor has required around $C_i + C_b$ total clock cycles to execute C_i instructions. We say "around" because we ignore the cycles required to start the instructions flowing through the pipeline. We can then compute the CPI for this benchmark as follows:

$$\text{CPI} = \frac{C_i + C_b}{C_i} = 1.0 + \frac{C_b}{C_i}$$

That is, the CPI equals 1.0 plus a penalty term C_b/C_i indicating the average number of bubbles injected per instruction executed. Since only three different instruction types can cause a bubble to be injected, we can break this penalty term into three components:

$$\text{CPI} = 1.0 + lp + mp + rp$$

where lp (for "load penalty") is the average frequency with which bubbles are injected while stalling for load/use hazards, mp (for "mispredicted branch penalty") is the average frequency with which bubbles are injected when canceling instructions due to mispredicted branches, and rp (for "return penalty") is the average frequency with which bubbles are injected while stalling for `ret` instructions. Each of these penalties indicates the total number of bubbles injected for the stated reason (some portion of C_b) divided by the total number of instructions that were executed (C_i).

To estimate each of these penalties, we need to know how frequently the relevant instructions (load, conditional branch, and return) occur, and for each of these how frequently the particular condition arises. Let us pick the following set of frequencies for our CPI computation (these are comparable to measurements reported in [47] and [49]):

- Load instructions (`mrmovl` and `popl`) account for 25% of all instructions executed. Of these, 20% cause load/use hazards.
- Conditional branches account for 20% of all instructions executed. Of these, 60% are taken and 40% are not taken.
- Return instructions account for 2% of all instructions executed.

We can therefore estimate each of our penalties as the product of the frequency of the instruction type, the frequency the condition arises, and the number of bubbles that get injected when the condition occurs:

Cause	Name	Instruction frequency	Condition frequency	Bubbles	Product
Load/Use	lp	0.25	0.20	1	0.05
Mispredict	mp	0.20	0.40	2	0.16
Return	rp	0.02	1.00	3	0.06
Total Penalty					0.27

The sum of the three penalties is 0.27, giving a CPI of 1.27.

Our goal was to design a pipeline that can issue one instruction per cycle, giving a CPI of 1.0. We did not quite meet this goal, but the overall performance is still quite good. We can also see that any effort to reduce the CPI further should focus on mispredicted branches. They account for 0.16 of our total penalty of 0.27, because conditional branches are common, our prediction strategy often fails, and we cancel two instructions for every misprediction.

Practice Problem 41

Suppose we use a branch prediction strategy that achieves a success rate of 65%, such as backward taken, forward not-taken, as described in Section 5.4. What would be the impact on CPI, assuming all of the other frequencies are not affected?

Practice Problem 42

Let us analyze the relative performance of using conditional data transfers versus conditional control transfers for the programs you wrote for Problems 4 and 5. Assume we are using these programs to compute the sum of the absolute values of a very long array, and so the overall performance is determined largely by the number of cycles required by the inner loop. Assume our jump instructions are predicted as being taken, and that around 50% of the array values are positive.

A. On average, how many instructions are executed in the inner loops of the two programs?

B. On average, how many bubbles would be injected into the inner loop of the two programs?

C. What is the average number of clock cycles required per array element for the two programs?

5.13   Unfinished Business

We have created a structure for the PIPE pipelined microprocessor, designed the control logic blocks, and implemented pipeline control logic to handle special cases where normal pipeline flow does not suffice. Still, PIPE lacks several key features that would be required in an actual microprocessor design. We highlight a few of these and discuss what would be required to add them.

Multicycle Instructions

All of the instructions in the Y86 instruction set involve simple operations such as adding numbers. These can be processed in a single clock cycle within the execute stage. In a more complete instruction set, we would also need to implement instructions requiring more complex operations such as integer multiplication and division, and floating-point operations. In a medium-performance processor such as PIPE, typical execution times for these operations range from 3 or 4 cycles for floating-point addition up to 32 for integer division. To implement these instructions, we require both additional hardware to perform the computations and a mechanism to coordinate the processing of these instructions with the rest of the pipeline.

One simple approach to implementing multicycle instructions is to simply expand the capabilities of the execute stage logic with integer and floating-point arithmetic units. An instruction remains in the execute stage for as many clock cycles as it requires, causing the fetch and decode stages to stall. This approach is simple to implement, but the resulting performance is not very good.

Better performance can be achieved by handling the more complex operations with special hardware functional units that operate independently of the main pipeline. Typically, there is one functional unit for performing integer multiplication and division, and another for performing floating-point operations. As an instruction enters the decode stage, it can be *issued* to the special unit. While the unit performs the operation, the pipeline continues processing other instructions. Typically, the floating-point unit is itself pipelined, and thus multiple operations can execute concurrently in the main pipeline and in the different units.

The operations of the different units must be synchronized to avoid incorrect behavior. For example, if there are data dependencies between the different operations being handled by different units, the control logic may need to stall one part of the system until the results from an operation handled by some other part of the system have been completed. Often, different forms of forwarding are used to convey results from one part of the system to other parts, just as we saw between the different stages of PIPE. The overall design becomes more complex than we have seen with PIPE, but the same techniques of stalling, forwarding, and pipeline control can be used to make the overall behavior match the sequential ISA model.

Interfacing with the Memory System

In our presentation of PIPE, we assumed that both the instruction fetch unit and the data memory could read or write any memory location in one clock cycle. We also ignored the possible hazards caused by self-modifying code where one instruction writes to the region of memory from which later instructions are fetched. Furthermore, we reference memory locations according to their virtual addresses, and these require a translation into physical addresses before the actual read or write operation can be performed. Clearly, it is unrealistic to do all of this processing in a single clock cycle. Even worse, the memory values being accessed may reside on disk, requiring millions of clock cycles to read into the processor memory.

The memory system of a processor uses a combination of multiple hardware memories and operating system software to manage the virtual memory system. The memory system is organized as a hierarchy, with faster but smaller memories holding a subset of the memory being backed up by slower and larger memories. At the level closest to the processor, the *cache* memories provide fast access to the most heavily referenced memory locations. A typical processor has two first-level caches—one for reading instructions and one for reading and writing data. Another type of cache memory, known as a *translation look-aside buffer*, or TLB, provides a fast translation from virtual to physical addresses. Using a combination of TLBs and caches, it is indeed possible to read instructions and read or write data in a single clock cycle most of the time. Thus, our simplified view of memory referencing by our processors is actually quite reasonable.

Although the caches hold the most heavily referenced memory locations, there will be times when a cache *miss* occurs, where some reference is made to a location that is not held in the cache. In the best case, the missing data can be retrieved from a higher-level cache or from the main memory of the processor, requiring 3 to 20 clock cycles. Meanwhile, the pipeline simply stalls, holding the instruction in the fetch or memory stage until the cache can perform the read or write operation. In terms of our pipeline design, this can be implemented by adding more stall conditions to the pipeline control logic. A cache miss and the consequent synchronization with the pipeline is handled completely by hardware, keeping the time required down to a small number of clock cycles.

In some cases, the memory location being referenced is actually stored in the disk memory. When this occurs, the hardware signals a *page fault* exception. Like other exceptions, this will cause the processor to invoke the operating system's exception handler code. This code will then set up a transfer from the disk to the main memory. Once this completes, the operating system will return back to the original program, where the instruction causing the page fault will be re-executed. This time, the memory reference will succeed, although it might cause a cache miss. Having the hardware invoke an operating system routine, which then returns control back to the hardware, allows the hardware and system software to cooperate in the handling of page faults. Since accessing a disk can require millions of clock cycles, the several thousand cycles of processing performed by the OS page fault handler has little impact on performance.

From the perspective of the processor, the combination of stalling to handle short-duration cache misses and exception handling to handle long-duration page faults takes care of any unpredictability in memory access times due to the structure of the memory hierarchy.

Aside   State-of-the-art microprocessor design

A five-stage pipeline, such as we have shown with the PIPE processor, represented the state of the art in processor design in the mid-1980s. The prototype RISC processor developed by Patterson's research group at Berkeley formed the basis for the first SPARC processor, developed by Sun Microsystems in 1987. The processor developed by Hennessy's research group at Stanford was commercialized by MIPS Technologies (a company founded by Hennessy) in 1986. Both of these used five-stage pipelines. The Intel i486 processor also uses a five-stage pipeline, although with a different partitioning of responsibilities among the stages, with two decode stages and a combined execute/memory stage [33].

These pipelined designs are limited to a throughput of at most one instruction per clock cycle. The CPI (for "cycles per instruction") measure described in Section 5.12 can never be less than 1.0. The different stages can only process one instruction at a time. More recent processors support *superscalar* operation, meaning that they can achieve a CPI less than 1.0 by fetching, decoding, and executing multiple instructions in parallel. As superscalar processors have become widespread, the accepted performance measure has shifted from CPI to its reciprocal—the average number of instructions executed per cycle, or IPC. It can exceed 1.0 for superscalar processors. The most advanced designs use a technique known as *out-of-order* execution to execute multiple instructions in parallel, possibly in a totally different order than they occur in the program, while preserving the overall behavior implied by the sequential ISA model.

Pipelined processors are not just historical artifacts, however. The majority of processors sold are used in embedded systems, controlling automotive functions, consumer products, and other devices where the processor is not directly visible to the system user. In these applications, the simplicity of a pipelined processor, such as the one we have explored in this chapter, reduces its cost and power requirements compared to higher-performance models.

More recently, as multicore processors have gained a following, some have argued that we could get more overall computing power by integrating many simple processors on a single chip rather than a smaller number of more complex ones. This strategy is sometimes referred to as "many-core" processors [10].

6 Summary

We have seen that the instruction set architecture, or ISA, provides a layer of abstraction between the behavior of a processor—in terms of the set of instructions and their encodings—and how the processor is implemented. The ISA provides a very sequential view of program execution, with one instruction executed to completion before the next one begins.

We defined the Y86 instruction set by starting with the IA32 instructions and simplifying the data types, address modes, and instruction encoding considerably. The resulting ISA has attributes of both RISC and CISC instruction sets. We then organized the processing required for the different instructions into a series of five stages, where the operations at each stage vary according to the instruction being executed. From this, we constructed the SEQ processor, in which an entire instruction is executed every clock cycle by having it flow through all five stages.

Pipelining improves the throughput performance of a system by letting the different stages operate concurrently. At any given time, multiple operations are being processed by the different stages. In introducing this concurrency, we must be careful to provide the same program-level behavior as would a sequential execution of the program. We introduced pipelining by reordering parts of SEQ to get SEQ+, and then adding pipeline registers to create the PIPE– pipeline. We enhanced the pipeline performance by adding forwarding logic to speed the sending of a result from one instruction to another. Several special cases require additional pipeline control logic to stall or cancel some of the pipeline stages.

Our design included rudimentary mechanisms to handle exceptions, where we make sure that only instructions up to the excepting instruction affect the programmer-visible state. Implementing a complete handling of exceptions would be significantly more challenging. Properly handling exceptions gets even more complex in systems that employ greater degrees of pipelining and parallelism.

In this chapter, we have learned several important lessons about processor design:

- *Managing complexity is a top priority.* We want to make optimum use of the hardware resources to get maximum performance at minimum cost. We did this by creating a very simple and uniform framework for processing all of the different instruction types. With this framework, we could share the hardware units among the logic for processing the different instruction types.

- *We do not need to implement the ISA directly.* A direct implementation of the ISA would imply a very sequential design. To achieve higher performance, we want to exploit the ability in hardware to perform many operations simultaneously. This led to the use of a pipelined design. By careful design and analysis, we can handle the various pipeline hazards, so that the overall effect of running a program exactly matches what would be obtained with the ISA model.

- *Hardware designers must be meticulous.* Once a chip has been fabricated, it is nearly impossible to correct any errors. It is very important to get the design right on the first try. This means carefully analyzing different instruction types and combinations, even ones that do not seem to make sense, such as popping to the stack pointer. Designs must be thoroughly tested with systematic simulation test programs. In developing the control logic for PIPE, our design had a subtle bug that was uncovered only after a careful and systematic analysis of control combinations.

Web Aside ARCH:HCL   HCL descriptions of Y86 processors

In this chapter, we have looked at portions of the HCL code for several simple logic designs, and for the control logic for Y86 processors SEQ and PIPE. For reference, we provide documentation of the HCL language and complete HCL descriptions for the control logic of the two processors. Each of these descriptions requires only 5–7 pages of HCL code, and it is worthwhile to study them in their entirety.

6.1 Y86 Simulators

The lab materials for this chapter include simulators for the SEQ and PIPE processors. Each simulator has two versions:

- The GUI (graphic user interface) version displays the memory, program code, and processor state in graphic windows. This provides a way to readily see how the instructions flow through the processors. The control panel also allows you to reset, single-step, or run the simulator interactively.

- The text version runs the same simulator, but it only displays information by printing to the terminal. This version is not as useful for debugging, but it allows automated testing of the processor.

The control logic for the simulators is generated by translating the HCL declarations of the logic blocks into C code. This code is then compiled and linked with the rest of the simulation code. This combination makes it possible for you to test out variants of the original designs using the simulators. Testing scripts are also available that thoroughly exercise the different instructions and the different hazard possibilities.

Bibliographic Notes

For those interested in learning more about logic design, Katz's logic design textbook [56] is a standard introductory text, emphasizing the use of hardware description languages.

Hennessy and Patterson's computer architecture textbook [49] provides extensive coverage of processor design, including both simple pipelines, such as the one we have presented here, and more advanced processors that execute more instructions in parallel. Shriver and Smith [97] give a very thorough presentation of an Intel-compatible IA32 processor manufactured by AMD.

Homework Problems

43 ◆

The IA32 pushl instruction can be described as decrementing the stack pointer and then storing the register at the stack pointer location. So, if we had an instruction of the form pushl REG, for some register REG, it would be equivalent to the code sequence:

```
subl $4,%esp          Decrement stack pointer
movl REG,(%esp)       Store REG on stack
```

A. In light of analysis done in Problem 6, does this code sequence correctly describe the behavior of the instruction pushl %esp? Explain.

B. How could you rewrite the code sequence so that it correctly describes both the cases where REG is %esp as well as any other register?

44 ◆

The IA32 popl instruction can be described as copying the result from the top of the stack to the destination register and then incrementing the stack pointer. So, if we had an instruction of the form popl REG, it would be equivalent to the code sequence:

```
movl (%esp),REG       Read REG from stack
addl $4,%esp          Increment stack pointer
```

A. In light of analysis done in Problem 7, does this code sequence correctly describe the behavior of the instruction popl %esp? Explain.

B. How could you rewrite the code sequence so that it correctly describes both the cases where *REG* is %esp as well as any other register?

45 ◆◆◆

Your assignment will be to write a Y86 program to perform bubblesort. For reference, the following C function implements bubblesort using array referencing:

```
/* Bubble sort: Array version */
void bubble_a(int *data, int count) {
    int i, last;
    for (last = count-1; last > 0; last--) {
        for (i = 0; i < last; i++)
            if (data[i+1] < data[i]) {
                /* Swap adjacent elements */
                int t = data[i+1];
                data[i+1] = data[i];
                data[i] = t;
            }
    }
}
```

A. Write and test a C version that references the array elements with pointers, rather than using array indexing.

B. Write and test a Y86 program consisting of the function and test code. You may find it useful to pattern your implementation after IA32 code generated by compiling your C code. Although pointer comparisons are normally done using unsigned arithmetic, you can use signed arithmetic for this exercise.

46 ◆◆

Modify the code you wrote for Problem 46 to implement the test and swap in the inner loop of the bubblesort function using conditional moves.

47 ◆

In our example Y86 programs, such as the Sum function shown in Figure 6, we encounter many cases (e.g., lines 12 and 13 and lines 14 and 15) in which we want to add a constant value to a register. This requires first using an irmovl instruction to set a register to the constant, and then an addl instruction to add this value to the destination register. Suppose we want to add a new instruction iaddl with the following format:

Byte	0	1	2	3	4	5
iaddl V, rB	C 0	F rB		V		

This instruction adds the constant value V to register rB. Describe the computations performed to implement this instruction. Use the computations for irmovl and OP1 (Figure 18) as a guide.

48 ◆

The IA32 instruction `leave` can be used to prepare the stack for returning. It is equivalent to the following Y86 code sequence:

```
1    rrmovl %ebp, %esp   Set stack pointer to beginning of frame
2    popl   %ebp         Restore saved %ebp and set stack ptr to end of caller's frame
```

Suppose we add this instruction to the Y86 instruction set, using the following encoding:

```
Byte        0   1   2   3   4   5

leave     | D | 0 |
```

Describe the computations performed to implement this instruction. Use the computations for `popl` (Figure 20) as a guide.

49 ◆◆

The file `seq-full.hcl` contains the HCL description for SEQ, along with the declaration of a constant IIADDL having hexadecimal value C, the instruction code for `iaddl`. Modify the HCL descriptions of the control logic blocks to implement the `iaddl` instruction, as described in Homework Problem 47. See the lab material for directions on how to generate a simulator for your solution and how to test it.

50 ◆◆

The file `seq-full.hcl` also contains the declaration of a constant ILEAVE having hexadecimal value D, the instruction code for `leave`, as well as the declaration of a constant REBP having value 7, the register ID for %ebp. Modify the HCL descriptions of the control logic blocks to implement the `leave` instruction, as described in Homework Problem 48. See the lab material for directions on how to generate a simulator for your solution and how to test it.

51 ◆◆◆

Suppose we wanted to create a lower-cost pipelined processor based on the structure we devised for PIPE– (Figure 41), without any bypassing. This design would handle all data dependencies by stalling until the instruction generating a needed value has passed through the write-back stage.

The file `pipe-stall.hcl` contains a modified version of the HCL code for PIPE in which the bypassing logic has been disabled. That is, the signals e_valA and e_valB are simply declared as follows:

```
## DO NOT MODIFY THE FOLLOWING CODE.
## No forwarding.  valA is either valP or value from register file
int d_valA = [
        D_icode in { ICALL, IJXX } : D_valP; # Use incremented PC
        1 : d_rvalA;  # Use value read from register file
];

## No forwarding.  valB is value from register file
int d_valB = d_rvalB;
```

Modify the pipeline control logic at the end of this file so that it correctly handles all possible control and data hazards. As part of your design effort, you should analyze the different combinations of control cases, as we did in the design of the pipeline control logic for PIPE. You will find that many different combinations can occur, since many more conditions require the pipeline to stall. Make sure your control logic handles each combination correctly. See the lab material for directions on how to generate a simulator for your solution and how to test it.

52 ◆◆◆
The file `pipe-full.hcl` also contains declarations of constants ILEAVE and REBP. Modify this file to implement the `leave` instruction, as described in Homework Problem 48. See the lab material for directions on how to generate a simulator for your solution and how to test it.

53 ◆◆
The file `pipe-full.hcl` contains a copy of the PIPE HCL description, along with a declaration of the constant value IIADDL. Modify this file to implement the `iaddl` instruction, as described in Homework Problem 47. See the lab material for directions on how to generate a simulator for your solution and how to test it.

54 ◆◆◆
The file `pipe-nt.hcl` contains a copy of the HCL code for PIPE, plus a declaration of the constant J_YES with value 0, the function code for an unconditional jump instruction. Modify the branch prediction logic so that it predicts conditional jumps as being not-taken while continuing to predict unconditional jumps and `call` as being taken. You will need to devise a way to get valC, the jump target address, to pipeline register M to recover from mispredicted branches. See the lab material for directions on how to generate a simulator for your solution and how to test it.

55 ◆◆◆
The file `pipe-btfnt.hcl` contains a copy of the HCL code for PIPE, plus a declaration of the constant J_YES with value 0, the function code for an unconditional jump instruction. Modify the branch prediction logic so that it predicts conditional jumps as being taken when valC < valP (backward branch) and as being not-taken when valC ≥ valP (forward branch). (Since Y86 does not support unsigned arithmetic, you should implement this test using a signed comparison.) Continue to predict unconditional jumps and `call` as being taken. You will need to devise a way to get both valC and valP to pipeline register M to recover from mispredicted branches. See the lab material for directions on how to generate a simulator for your solution and how to test it.

56 ◆◆◆
In our design of PIPE, we generate a stall whenever one instruction performs a *load*, reading a value from memory into a register, and the next instruction has this register as a source operand. When the source gets used in the execute stage, this stalling is the only way to avoid a hazard.

For cases where the second instruction stores the source operand to memory, such as with an `rmmovl` or `pushl` instruction, this stalling is not necessary. Consider the following code examples:

```
1       mrmovl 0(%ecx),%edx    # Load  1
2       pushl  %edx            # Store 1
3       nop
4       popl %edx              # Load  2
5       rmmovl %eax,0(%edx)    # Store 2
```

In lines 1 and 2, the `mrmovl` instruction reads a value from memory into `%edx`, and the `pushl` instruction then pushes this value onto the stack. Our design for PIPE would stall the `pushl` instruction to avoid a load/use hazard. Observe, however, that the value of `%edx` is not required by the `pushl` instruction until it reaches the memory stage. We can add an additional bypass path, as diagrammed in Figure 69, to forward the memory output (signal m_valM) to the valA field

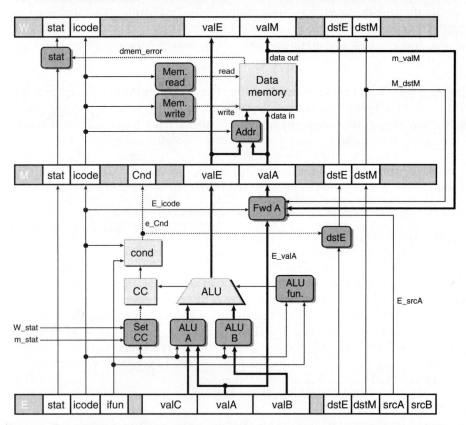

Figure 69 **Execute and memory stages capable of load forwarding.** By adding a bypass path from the memory output to the source of valA in pipeline register M, we can use forwarding rather than stalling for one form of load/use hazard. This is the subject of Homework Problem 56.

in pipeline register M. On the next clock cycle, this forwarded value can then be written to memory. This technique is known as *load forwarding*.

Note that the second example (lines 4 and 5) in the code sequence above cannot make use of load forwarding. The value loaded by the `popl` instruction is used as part of the address computation by the next instruction, and this value is required in the execute stage rather than the memory stage.

A. Write a logic formula describing the detection condition for a load/use hazard, similar to the one given in Figure 64, except that it will not cause a stall in cases where load forwarding can be used.

B. The file `pipe-lf.hcl` contains a modified version of the control logic for PIPE. It contains the definition of a signal e_valA to implement the block labeled "Fwd A" in Figure 69. It also has the conditions for a load/use hazard in the pipeline control logic set to zero, and so the pipeline control logic will not detect any forms of load/use hazards. Modify this HCL description to implement load forwarding. See the lab material for directions on how to generate a simulator for your solution and how to test it.

57 ◆◆◆

Our pipelined design is a bit unrealistic in that we have two write ports for the register file, but only the `popl` instruction requires two simultaneous writes to the register file. The other instructions could therefore use a single write port, sharing this for writing valE and valM. The following figure shows a modified version of the write-back logic, in which we merge the write-back register IDs (W_dstE and W_dstM) into a single signal w_dstE and the write-back values (W_valE and W_valM) into a single signal w_valE:

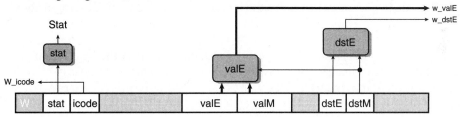

The logic for performing the merges is written in HCL as follows:

```
## Set E port register ID
int w_dstE = [
        ## writing from valM
        W_dstM != RNONE : W_dstM;
        1: W_dstE;
];


## Set E port value
int w_valE = [
        W_dstM != RNONE : W_valM;
        1: W_valE;
];
```

The control for these multiplexors is determined by dstE—when it indicates there is some register, then it selects the value for port E, and otherwise it selects the value for port M.

In the simulation model, we can then disable register port M, as shown by the following HCL code:

```
## Disable register port M
## Set M port register ID
int w_dstM = RNONE;

## Set M port value
int w_valM = 0;
```

The challenge then becomes to devise a way to handle popl. One method is to use the control logic to dynamically process the instruction popl rA so that it has the same effect as the two-instruction sequence

```
iaddl $4, %esp
mrmovl -4(%esp), rA
```

(See Homework Problem 47 for a description of the iaddl instruction.) Note the ordering of the two instructions to make sure popl %esp works properly. You can do this by having the logic in the decode stage treat popl the same as it would the iaddl listed above, except that it predicts the next PC to be equal to the current PC. On the next cycle, the popl instruction is refetched, but the instruction code is converted to a special value IPOP2. This is treated as a special instruction that has the same behavior as the mrmovl instruction listed above.

The file pipe-1w.hcl contains the modified write-port logic described above. It contains a declaration of the constant IPOP2 having hexadecimal value E. It also contains the definition of a signal f_icode that generates the icode field for pipeline register D. This definition can be modified to insert the instruction code IPOP2 the second time the popl instruction is fetched. The HCL file also contains a declaration of the signal f_pc, the value of the program counter generated in the fetch stage by the block labeled "Select PC" (Figure 55).

Modify the control logic in this file to process popl instructions in the manner we have described. See the lab material for directions on how to generate a simulator for your solution and how to test it.

58 ◆◆

Compare the performance of the two versions of bubblesort (Problems 45 and 46). Explain why one version performs better than the other.

Solutions to Practice Problems

Solution to Problem 1

Encoding instructions by hand is rather tedious, but it will solidify your understanding of the idea that assembly code gets turned into byte sequences by the assembler. In the following output from our Y86 assembler, each line shows an address and a byte sequence that starts at that address:

```
1    0x100:                  | .pos 0x100  # Start code at address 0x100
2    0x100: 30f30f000000 |     irmovl $15,%ebx       #   Load 15 into %ebx
3    0x106: 2031          |     rrmovl %ebx,%ecx      #   Copy 15 to %ecx
4    0x108:               | loop:                     # loop:
5    0x108: 4013fdffffff  |     rmmovl %ecx,-3(%ebx)  #   Save %ecx at address 15-3 = 12
6    0x10e: 6031          |     addl   %ebx,%ecx      #   Increment %ecx by 15
7    0x110: 7008010000    |     jmp loop              #   Goto loop
```

Several features of this encoding are worth noting:

- Decimal 15 (line 2) has hex representation 0x0000000f. Writing the bytes in reverse order gives 0f 00 00 00.

- Decimal −3 (line 5) has hex representation 0xfffffffd. Writing the bytes in reverse order gives fd ff ff ff.

- The code starts at address 0x100. The first instruction requires 6 bytes, while the second requires 2. Thus, the loop target will be 0x00000108. Writing these bytes in reverse order gives 08 01 00 00.

Solution to Problem 2

Decoding a byte sequence by hand helps you understand the task faced by a processor. It must read byte sequences and determine what instructions are to be executed. In the following, we show the assembly code used to generate each of the byte sequences. To the left of the assembly code, you can see the address and byte sequence for each instruction.

A. Some operations with immediate data and address displacements:

```
0x100: 30f3fcffffff |    irmovl $-4,%ebx
0x106: 406300080000 |    rmmovl %esi,0x800(%ebx)
0x10c: 00            |    halt
```

B. Code including a function call:

```
0x200: a06f          |    pushl %esi
0x202: 8008020000    |    call proc
0x207: 00            |    halt
0x208:               | proc:
0x208: 30f30a000000  |    irmovl $10,%ebx
0x20e: 90            |    ret
```

C. Code containing illegal instruction specifier byte 0xf0:

```
0x300: 505407000000 |    mrmovl 7(%esp),%ebp
0x306: 10            |    nop
0x307: f0            | .byte 0xf0  # invalid instruction code
0x308: b01f          |    popl %ecx
```

D. Code containing a jump operation:

```
0x400:                 | loop:
0x400: 6113            |     subl %ecx, %ebx
0x402: 7300040000      |     je loop
0x407: 00              |     halt
```

E. Code containing an invalid second byte in a pushl instruction:

```
0x500: 6362            |     xorl %esi,%edx
0x502: a0              |     .byte 0xa0  # pushl instruction code
0x503: f0              |     .byte 0xf0  # Invalid register specifier byte
```

Solution to Problem 3

As suggested in the problem, we adapted the code generated by GCC for an IA32 machine:

```
        # int Sum(int *Start, int Count)
rSum:   pushl %ebp
        rrmovl %esp,%ebp
        pushl %ebx             # Save value of %ebx
        mrmovl 8(%ebp),%ebx   # Get Start
        mrmovl 12(%ebp),%eax  # Get Count
        andl %eax,%eax         # Test value of Count
        jle L38                # If <= 0, goto zreturn
        irmovl $-1,%edx
        addl %edx,%eax         # Count--
        pushl %eax             # Push Count
        irmovl $4,%edx
        rrmovl %ebx,%eax
        addl %edx,%eax
        pushl %eax             # Push Start+1
        call rSum              # Sum(Start+1, Count-1)
        mrmovl (%ebx),%edx
        addl %edx,%eax         # Add *Start
        jmp L39                # goto done
L38:    xorl %eax,%eax         # zreturn:
L39:    mrmovl -4(%ebp),%ebx  # done: Restore %ebx
        rrmovl %ebp,%esp       # Deallocate stack frame
        popl %ebp              # Restore %ebp
        ret
```

Solution to Problem 4

This problem gives you a chance to try your hand at writing assembly code.

```
        int AbsSum(int *Start, int Count)
1   AbsSum:
2       pushl %ebp
```

```
 3      rrmovl %esp,%ebp
 4      mrmovl 8(%ebp),%ecx          ecx = Start
 5      mrmovl 12(%ebp),%edx         edx = Count
 6      irmovl $0, %eax              sum = 0
 7      andl   %edx,%edx
 8      je     End
 9   Loop:
10      mrmovl (%ecx),%esi           get x = *Start
11      irmovl $0,%edi               0
12      subl %esi,%edi               -x
13      jle Pos                      Skip if -x <= 0
14      rrmovl %edi,%esi             x = -x
15   Pos:
16      addl %esi,%eax               add x to sum
17      irmovl $4,%ebx
18      addl %ebx,%ecx               Start++
19      irmovl $-1,%ebx
20      addl %ebx,%edx               Count--
21      jne    Loop                  Stop when 0
22   End:
23      popl %ebp
24      ret
```

Solution to Problem 5

This problem gives you a chance to try your hand at writing assembly code with conditional moves. We show only the code for the loop. The rest is the same as for Problem 4:

```
 9   Loop:
10      mrmovl (%ecx),%esi           get x = *Start
11      irmovl $0,%edi               0
12      subl %esi,%edi               -x
13      cmovg %edi,%esi              if -x > 0 then x = -x
14      addl %esi,%eax               add x to sum
15      irmovl $4,%ebx
16      addl %ebx,%ecx               Start++
17      irmovl $-1,%ebx
18      addl %ebx,%edx               Count--
19      jne    Loop                  Stop when 0
```

Solution to Problem 6

Although it is hard to imagine any practical use for this particular instruction, it is important when designing a system to avoid any ambiguities in the specification. We want to determine a reasonable convention for the instruction's behavior and make sure each of our implementations adheres to this convention.

The subl instruction in this test compares the starting value of %esp to the value pushed onto the stack. The fact that the result of this subtraction is zero implies that the old value of %esp gets pushed.

Solution to Problem 7

It is even more difficult to imagine why anyone would want to pop to the stack pointer. Still, we should decide on a convention and stick with it. This code sequence pushes 0xabcd onto the stack, pops to %esp, and returns the popped value. Since the result equals 0xabcd, we can deduce that popl %esp sets the stack pointer to the value read from memory. It is therefore equivalent to the instruction mrmovl (%esp),%esp.

Solution to Problem 8

The EXCLUSIVE-OR function requires that the 2 bits have opposite values:

```
bool xor = (!a && b) || (a && !b);
```

In general, the signals eq and xor will be complements of each other. That is, one will equal 1 whenever the other is 0.

Solution to Problem 9

The outputs of the EXCLUSIVE-OR circuits will be the complements of the bit equality values. Using DeMorgan's laws (Web Aside DATA:BOOL), we can implement AND using OR and NOT, yielding the following circuit:

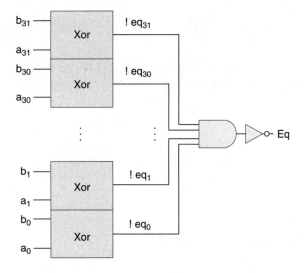

Solution to Problem 10

This design is a simple variant of the one to find the minimum of the three inputs:

```
int Med3 = [
        A <= B && B <= C : B;
        C <= B && B <= A : B;
        B <= A && A <= C : A;
        C <= A && A <= B : A;
        1                : C;
];
```

Solution to Problem 11

These exercises help make the stage computations more concrete. We can see from the object code that this instruction is located at address 0x00e. It consists of 6 bytes, with the first two being 0x30 and 0x84. The last 4 bytes are a byte-reversed version of 0x00000080 (decimal 128).

Stage	Generic irmovl V, rB	Specific irmovl $128, %esp
Fetch	icode:ifun ← $M_1[PC]$ rA:rB ← $M_1[PC+1]$ valC ← $M_4[PC+2]$ valP ← PC + 6	icode:ifun ← $M_1[0x00e] = 3:0$ rA:rB ← $M_1[0x00f] = 8:4$ valC ← $M_4[0x010] = 128$ valP ← 0x00e + 6 = 0x014
Decode		
Execute	valE ← 0 + valC	valE ← 0 + 128 = 128
Memory		
Write back	R[rB] ← valE	R[%esp] ← valE = 128
PC update	PC ← valP	PC ← valP = 0x014

This instruction sets register %esp to 128 and increments the PC by 6.

Solution to Problem 12

We can see that the instruction is located at address 0x01c and consists of 2 bytes with values 0xb0 and 0x08. Register %esp was set to 124 by the pushl instruction (line 6), which also stored 9 at this memory location.

Stage	Generic popl rA	Specific popl %eax
Fetch	icode:ifun ← $M_1[PC]$ rA:rB ← $M_1[PC+1]$ valP ← PC + 2	icode:ifun ← $M_1[0x01c] = b:0$ rA:rB ← $M_1[0x01d] = 0:8$ valP ← 0x01c + 2 = 0x01e
Decode	valA ← R[%esp] valB ← R[%esp]	valA ← R[%esp] = 124 valB ← R[%esp] = 124
Execute	valE ← valB + 4	valE ← 124 + 4 = 128

Memory	$valM \leftarrow M_4[valA]$	$valM \leftarrow M_4[124] = 9$
Write back	$R[\%esp] \leftarrow valE$	$R[\%esp] \leftarrow 128$
	$R[rA] \leftarrow valM$	$R[\%eax] \leftarrow 9$
PC update	$PC \leftarrow valP$	$PC \leftarrow \text{0x01e}$

The instruction sets %eax to 9, sets %esp to 128, and increments the PC by 2.

Solution to Problem 13

Tracing the steps listed in Figure 20 with rA equal to %esp, we can see that in the memory stage, the instruction will store valA, the original value of the stack pointer, to memory, just as we found for IA32.

Solution to Problem 14

Tracing the steps listed in Figure 20 with rA equal to %esp, we can see that both of the write-back operations will update %esp. Since the one writing valM would occur last, the net effect of the instruction will be to write the value read from memory to %esp, just as we saw for IA32.

Solution to Problem 15

Implementing conditional moves requires only minor changes from register-to-register moves. We simply condition the write-back step on the outcome of the conditional test:

Stage	cmovXX rA, rB
Fetch	$icode : ifun \leftarrow M_1[PC]$
	$rA : rB \leftarrow M_1[PC + 1]$
	$valP \leftarrow PC + 2$
Decode	$valA \leftarrow R[rA]$
Execute	$valE \leftarrow 0 + valA$
	$Cnd \leftarrow Cond(CC, ifun)$
Memory	
Write back	if (Cnd)
	$R[rB] \leftarrow valE$
PC update	$PC \leftarrow valP$

Solution to Problem 16

We can see that this instruction is located at address 0x023 and is 5 bytes long. The first byte has value 0x80, while the last four are a byte-reversed version of 0x00000029, the call target. The stack pointer was set to 128 by the popl instruction (line 7).

Stage	Generic call Dest	Specific call 0x029
Fetch	icode:ifun ← $M_1[PC]$	icode:ifun ← $M_1[0x023] = 8:0$
	valC ← $M_4[PC+1]$	valC ← $M_4[0x024] = 0x029$
	valP ← $PC+5$	valP ← $0x023 + 5 = 0x028$
Decode		
	valB ← R[%esp]	valB ← R[%esp] = 128
Execute	valE ← valB + −4	valE ← 128 + −4 = 124
Memory	M_4[valE] ← valP	M_4[124] ← 0x028
Write back	R[%esp] ← valE	R[%esp] ← 124
PC update	PC ← valC	PC ← 0x029

The effect of this instruction is to set %esp to 124, to store 0x028 (the return address) at this memory address, and to set the PC to 0x029 (the call target).

Solution to Problem 17

All of the HCL code in this and other practice problems is straightforward, but trying to generate it yourself will help you think about the different instructions and how they are processed. For this problem, we can simply look at the set of Y86 instructions (Figure 2) and determine which have a constant field.

```
bool need_valC =
        icode in { IIRMOVL, IRMMOVL, IMRMOVL, IJXX, ICALL };
```

Solution to Problem 18

This code is similar to the code for srcA.

```
int srcB = [
        icode in { IOPL, IRMMOVL, IMRMOVL  } : rB;
        icode in { IPUSHL, IPOPL, ICALL, IRET } : RESP;
        1 : RNONE;  # Don't need register
];
```

Solution to Problem 19

This code is similar to the code for dstE.

```
int dstM = [
        icode in { IMRMOVL, IPOPL } : rA;
        1 : RNONE;  # Don't write any register
];
```

Solution to Problem 20

As we found in Practice Problem 14, we want the write via the M port to take priority over the write via the E port in order to store the value read from memory into %esp.

Solution to Problem 21

This code is similar to the code for aluA.

```
int aluB = [
        icode in { IRMMOVL, IMRMOVL, IOPL, ICALL,
                        IPUSHL, IRET, IPOPL } : valB;
        icode in { IRRMOVL, IIRMOVL } : 0;
        # Other instructions don't need ALU
];
```

Solution to Problem 22

Implementing conditional moves is surprisingly simple: we disable writing to the register file by setting the destination register to RNONE when the condition does not hold.

```
int dstE = [
        icode in { IRRMOVL } && Cnd : rB;
        icode in { IIRMOVL, IOPL} : rB;
        icode in { IPUSHL, IPOPL, ICALL, IRET } : RESP;
        1 : RNONE;  # Don't write any register
];
```

Solution to Problem 23

This code is similar to the code for mem_addr.

```
int mem_data = [
        # Value from register
        icode in { IRMMOVL, IPUSHL } : valA;
        # Return PC
        icode == ICALL : valP;
        # Default: Don't write anything
];
```

Solution to Problem 24

This code is similar to the code for mem_read.

```
bool mem_write = icode in { IRMMOVL, IPUSHL, ICALL };
```

Solution to Problem 25

Computing the Stat field requires collecting status information from several stages:

```
## Determine instruction status
int Stat = [
        imem_error || dmem_error : SADR;
        !instr_valid: SINS;
        icode == IHALT : SHLT;
        1 : SAOK;
];
```

Solution to Problem 26

This problem is an interesting exercise in trying to find the optimal balance among a set of partitions. It provides a number of opportunities to compute throughputs and latencies in pipelines.

A. For a two-stage pipeline, the best partition would be to have blocks A, B, and C in the first stage and D, E, and F in the second. The first stage has a delay 170 ps, giving a total cycle time of $170 + 20 = 190$ picoseconds. We therefore have a throughput of 5.26 GOPS and a latency of 380 ps.

B. For a three-stage pipeline, we should have blocks A and B in the first stage, blocks C and D in the second, and blocks E and F in the third. The first two stages have a delay of 110 ps, giving a total cycle time of 130 ps and a throughput of 7.69 GOPS. The latency is 390 ps.

C. For a four-stage pipeline, we should have block A in the first stage, blocks B and C in the second, block D in the third, and blocks E and F in the fourth. The second stage requires 90 ps, giving a total cycle time of 110 ps and a throughput of 9.09 GOPS. The latency is 440 ps.

D. The optimal design would be a five-stage pipeline, with each block in its own stage, except that the fifth stage has blocks E and F. The cycle time is $80 + 20 = 100$ picoseconds, for a throughput of around 10.00 GOPS and a latency of 500 ps. Adding more stages would not help, since we cannot run the pipeline any faster than one cycle every 100 ps.

Solution to Problem 27

Each stage would have combinational logic requiring $300/k$ ps, and a pipeline register requiring 20 ps.

A. The total latency would be $300 + 20k$ ps, while the throughput (in GIPS) would be

$$\frac{1000}{\frac{300}{k} + 20} = \frac{1000k}{300 + 20K}$$

B. As we let k go to infinity, the throughput becomes $1000/20 = 50$ GIPS. Of course, this would give us an infinite latency, as well.

This exercise quantifies the diminishing returns of deep pipelining. As we try to subdivide the logic into many stages, the latency of the pipeline registers becomes a limiting factor.

Solution to Problem 28

This code is very similar to the corresponding code for SEQ, except that we cannot yet determine whether the data memory will generate an error signal for this instruction.

```
# Determine status code for fetched instruction
int f_stat = [
        imem_error: SADR;
        !instr_valid : SINS;
        f_icode == IHALT : SHLT;
        1 : SAOK;
];
```

Solution to Problem 29

This code simply involves prefixing the signal names in the code for SEQ with "d_" and "D_".

```
int d_dstE = [
        D_icode in { IRRMOVL, IIRMOVL, IOPL} : D_rB;
        D_icode in { IPUSHL, IPOPL, ICALL, IRET } : RESP;
        1 : RNONE;  # Don't write any register
];
```

Solution to Problem 30

The rrmovl instruction (line 5) would stall for one cycle due to a load-use hazard caused by the popl instruction (line 4). As it enters the decode stage, the popl instruction would be in the memory stage, giving both M_dstE and M_dstM equal to %esp. If the two cases were reversed, then the write back from M_valE would take priority, causing the incremented stack pointer to be passed as the argument to the rrmovl instruction. This would not be consistent with the convention for handling popl %esp determined in Practice Problem 7.

Solution to Problem 31

This problem lets you experience one of the important tasks in processor design—devising test programs for a new processor. In general, we should have test programs that will exercise all of the different hazard possibilities and will generate incorrect results if some dependency is not handled properly.

For this example, we can use a slightly modified version of the program shown in Practice Problem 30:

```
1       irmovl $5, %edx
2       irmovl $0x100,%esp
3       rmmovl %edx,0(%esp)
4       popl %esp
5       nop
6       nop
7       rrmovl %esp,%eax
```

The two nop instructions will cause the popl instruction to be in the write-back stage when the rrmovl instruction is in the decode stage. If the two forwarding sources in the write-back stage are given the wrong priority, then register %eax will be set to the incremented program counter rather than the value read from memory.

Solution to Problem 32

This logic only needs to check the five forwarding sources:

```
int d_valB = [
        d_srcB == e_dstE : e_valE;    # Forward valE from execute
        d_srcB == M_dstM : m_valM;    # Forward valM from memory
        d_srcB == M_dstE : M_valE;    # Forward valE from memory
        d_srcB == W_dstM : W_valM;    # Forward valM from write back
        d_srcB == W_dstE : W_valE;    # Forward valE from write back
        1 : d_rvalB;  # Use value read from register file
];
```

Solution to Problem 33

This change would not handle the case where a conditional move fails to satisfy the condition, and therefore sets the dstE value to RNONE. The resulting value could get forwarded to the next instruction, even though the conditional transfer does not occur.

```
1        irmovl $0x123,%eax
2        irmovl $0x321,%edx
3        xorl %ecx,%ecx          # CC = 100
4        cmovne  %eax,%edx       # Not transferred
5        addl %edx,%edx          # Should be 0x642
6        halt
```

This code initializes register %edx to 0x321. The conditional data transfer does not take place, and so the final addl instruction should double the value in %edx to 0x642. With the altered design, however, the conditional move source value 0x321 gets forwarded into ALU input valA, while input valB correctly gets operand value 0x123. These inputs get added to produce result 0x444.

Solution to Problem 34

This code completes the computation of the status code for this instruction.

```
## Update the status
int m_stat = [
        dmem_error : SADR;
        1 : M_stat;
];
```

Solution to Problem 35

The following test program is designed to set up control combination A (Figure 67) and detect whether something goes wrong:

```
1   # Code to generate a combination of not-taken branch and ret
2           irmovl Stack, %esp
3           irmovl rtnp,%eax
4           pushl %eax          # Set up return pointer
5           xorl %eax,%eax      # Set Z condition code
6           jne target          # Not taken (First part of combination)
7           irmovl $1,%eax      # Should execute this
8           halt
9   target: ret                 # Second part of combination
10          irmovl $2,%ebx      # Should not execute this
11          halt
12  rtnp:   irmovl $3,%edx      # Should not execute this
13          halt
14  .pos 0x40
15  Stack:
```

This program is designed so that if something goes wrong (for example, if the ret instruction is actually executed), then the program will execute one of the extra irmovl instructions and then halt. Thus, an error in the pipeline would cause some register to be updated incorrectly. This code illustrates the care required to implement a test program. It must set up a potential error condition and then detect whether or not an error occurs.

Solution to Problem 36

The following test program is designed to set up control combination B (Figure 67). The simulator will detect a case where the bubble and stall control signals for a pipeline register are both set to zero, and so our test program need only set up the combination for it to be detected. The biggest challenge is to make the program do something sensible when handled correctly.

```
1   # Test instruction that modifies %esp followed by ret
2           irmovl mem,%ebx
3           mrmovl  0(%ebx),%esp # Sets %esp to point to return point
4           ret                 # Returns to return point
5           halt                #
6   rtnpt:  irmovl $5,%esi      # Return point
7           halt
8   .pos 0x40
9   mem:    .long stack         # Holds desired stack pointer
10  .pos 0x50
11  stack:  .long rtnpt         # Top of stack: Holds return point
```

This program uses two initialized word in memory. The first word (mem) holds the address of the second (stack—the desired stack pointer). The second word holds the address of the desired return point for the ret instruction. The program loads the stack pointer into %esp and executes the ret instruction.

Solution to Problem 37

From Figure 66, we can see that pipeline register D must be stalled for a load/use hazard.

```
bool D_stall =
        # Conditions for a load/use hazard
        E_icode in { IMRMOVL, IPOPL } &&
         E_dstM in { d_srcA, d_srcB };
```

Solution to Problem 38

From Figure 66, we can see that pipeline register E must be set to bubble for a load/use hazard or for a mispredicted branch:

```
bool E_bubble =
        # Mispredicted branch
        (E_icode == IJXX && !e_Cnd) ||
        # Conditions for a load/use hazard
        E_icode in { IMRMOVL, IPOPL } &&
         E_dstM in { d_srcA, d_srcB};
```

Solution to Problem 39

This control requires examining the code of the executing instruction and checking for exceptions further down the pipeline.

```
## Should the condition codes be updated?
bool set_cc = E_icode == IOPL &&
        # State changes only during normal operation
        !m_stat in { SADR, SINS, SHLT } && !W_stat in { SADR, SINS, SHLT };
```

Solution to Problem 40

Injecting a bubble into the memory stage on the next cycle involves checking for an exception in either the memory or the write-back stage during the current cycle.

```
# Start injecting bubbles as soon as exception passes through memory stage
bool M_bubble = m_stat in { SADR, SINS, SHLT } || W_stat in { SADR, SINS, SHLT };
```

For stalling the write-back stage, we check only the status of the instruction in this stage. If we also stalled when an excepting instruction was in the memory stage, then this instruction would not be able to enter the write-back stage.

```
bool W_stall = W_stat in { SADR, SINS, SHLT };
```

Solution to Problem 41

We would then have a misprediction frequency of 0.35, giving $mp = 0.20 \times 0.35 \times 2 = 0.14$, giving an overall CPI of 1.25. This seems like a fairly marginal gain, but it would be worthwhile if the cost of implementing the new branch prediction strategy were not too high.

Solution to Problem 42

This simplified analysis, where we focus on the inner loop, is a useful way to estimate program performance. As long as the array is sufficiently large, the time spent in other parts of the code will be negligible.

A. The inner loop of the code using the conditional jump has 11 instructions, all of which are executed when the array element is zero or negative, and 10 of which are executed when the array element is positive. The average is 10.5. The inner loop of the code using the conditional move has 10 instructions, all of which are executed every time.

B. The loop-closing jump will be predicted correctly, except when the loop terminates. For a very long array, this one misprediction will have negligible effect on the performance. The only other source of bubbles for the jump-based code is the conditional jump depending on whether or not the array element is positive. This will cause two bubbles, but it only occurs 50% of the time, so the average is 1.0. There are no bubbles in the conditional move code.

C. Our conditional jump code requires an average of $10.5 + 1.0 = 11.5$ cycles per array element (11 cycles in the best case and 12 cycles in the worst), while our conditional move code requires 10.0 cycles in all cases.

Our pipeline has a branch misprediction penalty of only two cycles—far better than those for the deep pipelines of higher-performance processors. As a result, using conditional moves does not affect program performance very much.

Optimizing Program Performance

The biggest speedup you'll ever get with a program will be
when you first get it working.

—*John K. Ousterhout*

The primary objective in writing a program must be to make it work correctly
under all possible conditions. A program that runs fast but gives incorrect results
serves no useful purpose. Programmers must write clear and concise code, not only
so that they can make sense of it, but also so that others can read and understand
the code during code reviews and when modifications are required later.

On the other hand, there are many occasions when making a program run
fast is also an important consideration. If a program must process video frames
or network packets in real time, then a slow-running program will not provide the
needed functionality. When a computation task is so demanding that it requires
days or weeks to execute, then making it run just 20% faster can have significant
impact. In this chapter, we will explore how to make programs run faster via
several different types of program optimization.

Writing an efficient program requires several types of activities. First, we
must select an appropriate set of algorithms and data structures. Second, we
must write source code that the compiler can effectively optimize to turn into
efficient executable code. For this second part, it is important to understand the
capabilities and limitations of optimizing compilers. Seemingly minor changes in
how a program is written can make large differences in how well a compiler can
optimize it. Some programming languages are more easily optimized than others.
Some features of C, such as the ability to perform pointer arithmetic and casting,
make it challenging for a compiler to optimize. Programmers can often write their
programs in ways that make it easier for compilers to generate efficient code.
A third technique for dealing with especially demanding computations is to divide
a task into portions that can be computed in parallel, on some combination of
multiple cores and multiple processors. Even when exploiting parallelism, it is
important that each parallel thread execute with maximum performance, and so
the material of this chapter remains relevant in any case.

In approaching program development and optimization, we must consider
how the code will be used and what critical factors affect it. In general, programmers must make a trade-off between how easy a program is to implement and
maintain, and how fast it runs. At an algorithmic level, a simple insertion sort can
be programmed in a matter of minutes, whereas a highly efficient sort routine
may take a day or more to implement and optimize. At the coding level, many
low-level optimizations tend to reduce code readability and modularity, making
the programs more susceptible to bugs and more difficult to modify or extend.
For code that will be executed repeatedly in a performance-critical environment,
extensive optimization may be appropriate. One challenge is to maintain some
degree of elegance and readability in the code despite extensive transformations.

We describe a number of techniques for improving code performance. Ideally,
a compiler would be able to take whatever code we write and generate the most

efficient possible machine-level program having the specified behavior. Modern compilers employ sophisticated forms of analysis and optimization, and they keep getting better. Even the best compilers, however, can be thwarted by *optimization blockers*—aspects of the program's behavior that depend strongly on the execution environment. Programmers must assist the compiler by writing code that can be optimized readily.

The first step in optimizing a program is to eliminate unnecessary work, making the code perform its intended task as efficiently as possible. This includes eliminating unnecessary function calls, conditional tests, and memory references. These optimizations do not depend on any specific properties of the target machine.

To maximize the performance of a program, both the programmer and the compiler require a model of the target machine, specifying how instructions are processed and the timing characteristics of the different operations. For example, the compiler must know timing information to be able to decide whether it should use a multiply instruction or some combination of shifts and adds. Modern computers use sophisticated techniques to process a machine-level program, executing many instructions in parallel and possibly in a different order than they appear in the program. Programmers must understand how these processors work to be able to tune their programs for maximum speed. We present a high-level model of such a machine based on recent designs of Intel and AMD processors. We also devise a graphical *data-flow* notation to visualize the execution of instructions by the processor, with which we can predict program performance.

With this understanding of processor operation, we can take a second step in program optimization, exploiting the capability of processors to provide *instruction-level parallelism*, executing multiple instructions simultaneously. We cover several program transformations that reduce the data dependencies between different parts of a computation, increasing the degree of parallelism with which they can be executed.

We conclude the chapter by discussing issues related to optimizing large programs. We describe the use of code *profilers*—tools that measure the performance of different parts of a program. This analysis can help find inefficiencies in the code and identify the parts of the program on which we should focus our optimization efforts. Finally, we present an important observation, known as *Amdahl's law*, which quantifies the overall effect of optimizing some portion of a system.

In this presentation, we make code optimization look like a simple linear process of applying a series of transformations to the code in a particular order. In fact, the task is not nearly so straightforward. A fair amount of trial-and-error experimentation is required. This is especially true as we approach the later optimization stages, where seemingly small changes can cause major changes in performance, while some very promising techniques prove ineffective. As we will see in the examples that follow, it can be difficult to explain exactly why a particular code sequence has a particular execution time. Performance can depend on many detailed features of the processor design for which we have relatively little documentation or understanding. This is another reason to try a number of different variations and combinations of techniques.

Studying the assembly-code representation of a program is one of the most effective means for gaining an understanding of the compiler and how the generated code will run. A good strategy is to start by looking carefully at the code for the inner loops, identifying performance-reducing attributes such as excessive memory references and poor use of registers. Starting with the assembly code, we can also predict what operations will be performed in parallel and how well they will use the processor resources. As we will see, we can often determine the time (or at least a lower bound on the time) required to execute a loop by identifying *critical paths*, chains of data dependencies that form during repeated executions of a loop. We can then go back and modify the source code to try to steer the compiler toward more efficient implementations.

Most major compilers, including GCC, are continually being updated and improved, especially in terms of their optimization abilities. One useful strategy is to do only as much rewriting of a program as is required to get it to the point where the compiler can then generate efficient code. By this means, we avoid compromising the readability, modularity, and portability of the code as much as if we had to work with a compiler of only minimal capabilities. Again, it helps to iteratively modify the code and analyze its performance both through measurements and by examining the generated assembly code.

To novice programmers, it might seem strange to keep modifying the source code in an attempt to coax the compiler into generating efficient code, but this is indeed how many high-performance programs are written. Compared to the alternative of writing code in assembly language, this indirect approach has the advantage that the resulting code will still run on other machines, although perhaps not with peak performance.

1   Capabilities and Limitations of Optimizing Compilers

Modern compilers employ sophisticated algorithms to determine what values are computed in a program and how they are used. They can then exploit opportunities to simplify expressions, to use a single computation in several different places, and to reduce the number of times a given computation must be performed. Most compilers, including GCC, provide users with some control over which optimizations they apply. The simplest control is to specify the *optimization level*. For example, invoking GCC with the command-line flag '-O1' will cause it to apply a basic set of optimizations. As discussed in Web Aside ASM:OPT, invoking GCC with flag '-O2' or '-O3' will cause it to apply more extensive optimizations. These can further improve program performance, but they may expand the program size and they may make the program more difficult to debug using standard debugging tools. For our presentation, we will mostly consider code compiled with optimization level 1, even though optimization level 2 has become the accepted standard for most GCC users. We purposely limit the level of optimization to demonstrate how different ways of writing a function in C can affect the efficiency of the code generated by a compiler. We will find that we can write C code that, when compiled just with optimization level 1, vastly outperforms a more naive version compiled with the highest possible optimization levels.

Compilers must be careful to apply only *safe* optimizations to a program, meaning that the resulting program will have the exact same behavior as would an unoptimized version for all possible cases the program may encounter, up to the limits of the guarantees provided by the C language standards. Constraining the compiler to perform only safe optimizations eliminates possible sources of undesired run-time behavior, but it also means that the programmer must make more of an effort to write programs in a way that the compiler can then transform into efficient machine-level code. To appreciate the challenges of deciding which program transformations are safe or not, consider the following two procedures:

```
1    void twiddle1(int *xp, int *yp)
2    {
3        *xp += *yp;
4        *xp += *yp;
5    }
6
7    void twiddle2(int *xp, int *yp)
8    {
9        *xp += 2* *yp;
10   }
```

At first glance, both procedures seem to have identical behavior. They both add twice the value stored at the location designated by pointer yp to that designated by pointer xp. On the other hand, function `twiddle2` is more efficient. It requires only three memory references (read *xp, read *yp, write *xp), whereas `twiddle1` requires six (two reads of *xp, two reads of *yp, and two writes of *xp). Hence, if a compiler is given procedure `twiddle1` to compile, one might think it could generate more efficient code based on the computations performed by `twiddle2`.

Consider, however, the case in which xp and yp are equal. Then function `twiddle1` will perform the following computations:

```
3        *xp += *xp;   /* Double value at xp */
4        *xp += *xp;   /* Double value at xp */
```

The result will be that the value at xp will be increased by a factor of 4. On the other hand, function `twiddle2` will perform the following computation:

```
9        *xp += 2* *xp;   /* Triple value at xp */
```

The result will be that the value at xp will be increased by a factor of 3. The compiler knows nothing about how `twiddle1` will be called, and so it must assume that arguments xp and yp can be equal. It therefore cannot generate code in the style of `twiddle2` as an optimized version of `twiddle1`.

The case where two pointers may designate the same memory location is known as *memory aliasing*. In performing only safe optimizations, the compiler must assume that different pointers may be aliased. As another example, for a program with pointer variables p and q, consider the following code sequence:

```
x = 1000; y = 3000;
*q = y;    /* 3000 */
*p = x;    /* 1000 */
t1 = *q;   /* 1000 or 3000 */
```

The value computed for t1 depends on whether or not pointers p and q are aliased—if not, it will equal 3000, but if so it will equal 1000. This leads to one of the major *optimization blockers*, aspects of programs that can severely limit the opportunities for a compiler to generate optimized code. If a compiler cannot determine whether or not two pointers may be aliased, it must assume that either case is possible, limiting the set of possible optimizations.

Practice Problem 1

The following problem illustrates the way memory aliasing can cause unexpected program behavior. Consider the following procedure to swap two values:

```
1    /* Swap value x at xp with value y at yp */
2    void swap(int *xp, int *yp)
3    {
4        *xp = *xp + *yp; /* x+y        */
5        *yp = *xp - *yp; /* x+y-y = x */
6        *xp = *xp - *yp; /* x+y-x = y */
7    }
```

If this procedure is called with xp equal to yp, what effect will it have?

A second optimization blocker is due to function calls. As an example, consider the following two procedures:

```
1    int f();
2
3    int func1() {
4        return f() + f() + f() + f();
5    }
6
7    int func2() {
8        return 4*f();
9    }
```

It might seem at first that both compute the same result, but with func2 calling f only once, whereas func1 calls it four times. It is tempting to generate code in the style of func2 when given func1 as the source.

Consider, however, the following code for f:

```
1   int counter = 0;
2
3   int f() {
4       return counter++;
5   }
```

This function has a *side effect*—it modifies some part of the global program state. Changing the number of times it gets called changes the program behavior. In particular, a call to func1 would return $0 + 1 + 2 + 3 = 6$, whereas a call to func2 would return $4 \cdot 0 = 0$, assuming both started with global variable counter set to 0.

Most compilers do not try to determine whether a function is free of side effects and hence is a candidate for optimizations such as those attempted in func2. Instead, the compiler assumes the worst case and leaves function calls intact.

Aside   Optimizing function calls by inline substitution

As described in Web Aside ASM:OPT, code involving function calls can be optimized by a process known as *inline substitution* (or simply "inlining"), where the function call is replaced by the code for the body of the function. For example, we can expand the code for func1 by substituting four instantiations of function f:

```
1   /* Result of inlining f in func1 */
2   int func1in() {
3       int t = counter++;   /* +0 */
4       t += counter++;      /* +1 */
5       t += counter++;      /* +2 */
6       t += counter++;      /* +3 */
7       return t;
8   }
```

This transformation both reduces the overhead of the function calls and allows further optimization of the expanded code. For example, the compiler can consolidate the updates of global variable counter in func1in to generate an optimized version of the function:

```
1   /* Optimization of inlined code */
2   int func1opt() {
3       int t = 4 * counter + 6;
4       counter = t + 4;
5       return t;
6   }
```

This code faithfully reproduces the behavior of func1 for this particular definition of function f.

Recent versions of GCC attempt this form of optimization, either when directed to with the command-line option '-finline' or for optimization levels 2 or higher. Since we are considering optimization level 1 in our presentation, we will assume that the compiler does not perform inline substitution.

Among compilers, GCC is considered adequate, but not exceptional, in terms of its optimization capabilities. It performs basic optimizations, but it does not perform the radical transformations on programs that more "aggressive" compilers do. As a consequence, programmers using GCC must put more effort into writing programs in a way that simplifies the compiler's task of generating efficient code.

2 Expressing Program Performance

We introduce the metric *cycles per element*, abbreviated "CPE," as a way to express program performance in a way that can guide us in improving the code. CPE measurements help us understand the loop performance of an iterative program at a detailed level. It is appropriate for programs that perform a repetitive computation, such as processing the pixels in an image or computing the elements in a matrix product.

The sequencing of activities by a processor is controlled by a clock providing a regular signal of some frequency, usually expressed in *gigahertz* (GHz), billions of cycles per second. For example, when product literature characterizes a system as a "4 GHz" processor, it means that the processor clock runs at 4.0×10^9 cycles per second. The time required for each clock cycle is given by the reciprocal of the clock frequency. These typically are expressed in *nanoseconds* (1 nanosecond is 10^{-9} seconds), or *picoseconds* (1 picosecond is 10^{-12} seconds). For example, the period of a 4 GHz clock can be expressed as either 0.25 nanoseconds or 250 picoseconds. From a programmer's perspective, it is more instructive to express measurements in clock cycles rather than nanoseconds or picoseconds. That way, the measurements express how many instructions are being executed rather than how fast the clock runs.

Many procedures contain a loop that iterates over a set of elements. For example, functions psum1 and psum2 in Figure 1 both compute the *prefix sum* of a vector of length n. For a vector $\vec{a} = \langle a_0, a_1, \ldots, a_{n-1} \rangle$, the prefix sum $\vec{p} = \langle p_0, p_1, \ldots, p_{n-1} \rangle$ is defined as

$$
\begin{aligned}
p_0 &= a_0 \\
p_i &= p_{i-1} + a_i, \ 1 \leq i < n
\end{aligned}
\tag{1}
$$

Function psum1 computes one element of the result vector per iteration. The second uses a technique known as *loop unrolling* to compute two elements per iteration. We will explore the benefits of loop unrolling later in this chapter. See Problems 11, 12, and 21 for more about analyzing and optimizing the prefix-sum computation.

The time required by such a procedure can be characterized as a constant plus a factor proportional to the number of elements processed. For example, Figure 2 shows a plot of the number of clock cycles required by the two functions for a range of values of n. Using a *least squares fit*, we find that the run times (in clock cycles) for psum1 and psum2 can be approximated by the equations $496 + 10.0n$ and $500 + 6.5n$, respectively. These equations indicate an overhead of 496 to 500

```
1    /* Compute prefix sum of vector a */
2    void psum1(float a[], float p[], long int n)
3    {
4        long int i;
5        p[0] = a[0];
6        for (i = 1; i < n; i++)
7            p[i] = p[i-1] + a[i];
8    }
9
10   void psum2(float a[], float p[], long int n)
11   {
12       long int i;
13       p[0] = a[0];
14       for (i = 1; i < n-1; i+=2) {
15           float mid_val = p[i-1] + a[i];
16           p[i]     = mid_val;
17           p[i+1]   = mid_val + a[i+1];
18       }
19       /* For odd n, finish remaining element */
20       if (i < n)
21           p[i] = p[i-1] + a[i];
22   }
```

Figure 1 **Prefix-sum functions.** These provide examples for how we express program performance.

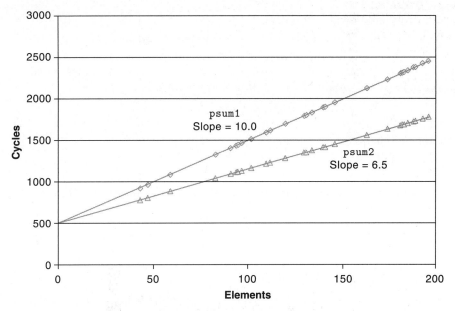

Figure 2 **Performance of prefix-sum functions.** The slope of the lines indicates the number of clock cycles per element (CPE).

cycles due to the timing code and to initiate the procedure, set up the loop, and complete the procedure, plus a linear factor of 6.5 or 10.0 cycles per element. For large values of n (say, greater than 200), the run times will be dominated by the linear factors. We refer to the coefficients in these terms as the effective number of *cycles per element*, abbreviated "CPE." We prefer measuring the number of cycles per *element* rather than the number of cycles per *iteration*, because techniques such as loop unrolling allow us to use fewer iterations to complete the computation, but our ultimate concern is how fast the procedure will run for a given vector length. We focus our efforts on minimizing the CPE for our computations. By this measure, psum2, with a CPE of 6.50, is superior to psum1, with a CPE of 10.0.

Aside   What is a least squares fit?

For a set of data points $(x_1, y_1), \ldots (x_n, y_n)$, we often try to draw a line that best approximates the X-Y trend represented by this data. With a least squares fit, we look for a line of the form $y = mx + b$ that minimizes the following error measure:

$$E(m, b) = \sum_{i=1,n} (mx_i + b - y_i)^2$$

An algorithm for computing m and b can be derived by finding the derivatives of $E(m, b)$ with respect to m and b and setting them to 0.

Practice Problem 2

Later in this chapter, we will start with a single function and generate many different variants that preserve the function's behavior, but with different performance characteristics. For three of these variants, we found that the run times (in clock cycles) can be approximated by the following functions:

Version 1: $60 + 35n$

Version 2: $136 + 4n$

Version 3: $157 + 1.25n$

For what values of n would each version be the fastest of the three? Remember that n will always be an integer.

3   Program Example

To demonstrate how an abstract program can be systematically transformed into more efficient code, we will use a running example based on the vector data structure shown in Figure 3. A vector is represented with two blocks of memory: the header and the data array. The header is a structure declared as follows:

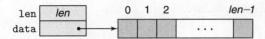

Figure 3 **Vector abstract data type.** A vector is represented by header information plus array of designated length.

——————————————————————————————————— *code/opt/vec.h*

```
1    /* Create abstract data type for vector */
2    typedef struct {
3        long int len;
4        data_t *data;
5    } vec_rec, *vec_ptr;
```

——————————————————————————————————— *code/opt/vec.h*

The declaration uses data type `data_t` to designate the data type of the underlying elements. In our evaluation, we measure the performance of our code for integer (C `int`), single-precision floating-point (C `float`), and double-precision floating-point (C `double`) data. We do this by compiling and running the program separately for different type declarations, such as the following for data type `int`:

```
typedef int data_t;
```

We allocate the data array block to store the vector elements as an array of `len` objects of type `data_t`.

Figure 4 shows some basic procedures for generating vectors, accessing vector elements, and determining the length of a vector. An important feature to note is that `get_vec_element`, the vector access routine, performs bounds checking for every vector reference. This code is similar to the array representations used in many other languages, including Java. Bounds checking reduces the chances of program error, but it can also slow down program execution.

As an optimization example, consider the code shown in Figure 5, which combines all of the elements in a vector into a single value according to some operation. By using different definitions of compile-time constants `IDENT` and `OP`, the code can be recompiled to perform different operations on the data. In particular, using the declarations

```
#define IDENT 0
#define OP   +
```

it sums the elements of the vector. Using the declarations

```
#define IDENT 1
#define OP   *
```

it computes the product of the vector elements.

In our presentation, we will proceed through a series of transformations of the code, writing different versions of the combining function. To gauge progress,

——————————————————————————————— code/opt/vec.c

```
1    /* Create vector of specified length */
2    vec_ptr new_vec(long int len)
3    {
4        /* Allocate header structure */
5        vec_ptr result = (vec_ptr) malloc(sizeof(vec_rec));
6        if (!result)
7            return NULL;  /* Couldn't allocate storage */
8        result->len = len;
9        /* Allocate array */
10       if (len > 0) {
11           data_t *data = (data_t *)calloc(len, sizeof(data_t));
12           if (!data) {
13               free((void *) result);
14               return NULL; /* Couldn't allocate storage */
15           }
16           result->data = data;
17       }
18       else
19           result->data = NULL;
20       return result;
21   }
22
23   /*
24    * Retrieve vector element and store at dest.
25    * Return 0 (out of bounds) or 1 (successful)
26    */
27   int get_vec_element(vec_ptr v, long int index, data_t *dest)
28   {
29       if (index < 0 || index >= v->len)
30           return 0;
31       *dest = v->data[index];
32       return 1;
33   }
34
35   /* Return length of vector */
36   long int vec_length(vec_ptr v)
37   {
38       return v->len;
39   }
```

——————————————————————————————— code/opt/vec.c

Figure 4 **Implementation of vector abstract data type.** In the actual program, data type data_t is declared to be int, float, or double.

```
1    /* Implementation with maximum use of data abstraction */
2    void combine1(vec_ptr v, data_t *dest)
3    {
4        long int i;
5
6        *dest = IDENT;
7        for (i = 0; i < vec_length(v); i++) {
8            data_t val;
9            get_vec_element(v, i, &val);
10           *dest = *dest OP val;
11       }
12   }
```

Figure 5 **Initial implementation of combining operation.** Using different declarations of identity element *IDENT* and combining operation *OP*, we can measure the routine for different operations.

we will measure the CPE performance of the functions on a machine with an Intel Core i7 processor, which we will refer to as our *reference machine*. These measurements characterize performance in terms of how the programs run on just one particular machine, and so there is no guarantee of comparable performance on other combinations of machine and compiler. However, we have compared the results with those for a number of different compiler/processor combinations and found them quite comparable.

As we proceed through a set of transformations, we will find that many lead to only minimal performance gains, while others have more dramatic effects. Determining which combinations of transformations to apply is indeed part of the "black art" of writing fast code. Some combinations that do not provide measurable benefits are indeed ineffective, while others are important as ways to enable further optimizations by the compiler. In our experience, the best approach involves a combination of experimentation and analysis: repeatedly attempting different approaches, performing measurements, and examining the assembly-code representations to identify underlying performance bottlenecks.

As a starting point, the following are CPE measurements for combine1 running on our reference machine, trying all combinations of data type and combining operation. For single-precision and double-precision floating-point data, our experiments on this machine gave identical performance for addition, but differing performance for multiplication. We therefore report five CPE values: integer addition and multiplication, floating-point addition, single-precision multiplication (labeled "F *"), and double-precision multiplication (labeled "D *").

Function	Page	Method	Integer		Floating point		
			+	*	+	F *	D *
combine1	13	Abstract unoptimized	29.02	29.21	27.40	27.90	27.36
combine1	13	Abstract -O1	12.00	12.00	12.00	12.01	13.00

We can see that our measurements are somewhat imprecise. The more likely CPE number for integer sum and product is 29.00, rather than 29.02 or 29.21. Rather than "fudging" our numbers to make them look good, we will present the measurements we actually obtained. There are many factors that complicate the task of reliably measuring the precise number of clock cycles required by some code sequence. It helps when examining these numbers to mentally round the results up or down by a few hundredths of a clock cycle.

The unoptimized code provides a direct translation of the C code into machine code, often with obvious inefficiencies. By simply giving the command-line option '-O1', we enable a basic set of optimizations. As can be seen, this significantly improves the program performance—more than a factor of two—with no effort on behalf of the programmer. In general, it is good to get into the habit of enabling at least this level of optimization. For the remainder of our measurements, we use optimization levels 1 and higher in generating and measuring our programs.

4 Eliminating Loop Inefficiencies

Observe that procedure combine1, as shown in Figure 5, calls function vec_length as the test condition of the for loop. Recall that the test condition must be evaluated on every iteration of the loop. On the other hand, the length of the vector does not change as the loop proceeds. We could therefore compute the vector length only once and use this value in our test condition.

Figure 6 shows a modified version called combine2, which calls vec_length at the beginning and assigns the result to a local variable length. This transformation has noticeable effect on the overall performance for some data types and

```
1   /* Move call to vec_length out of loop */
2   void combine2(vec_ptr v, data_t *dest)
3   {
4       long int i;
5       long int length = vec_length(v);
6
7       *dest = IDENT;
8       for (i = 0; i < length; i++) {
9           data_t val;
10          get_vec_element(v, i, &val);
11          *dest = *dest OP val;
12      }
13  }
```

Figure 6 **Improving the efficiency of the loop test.** By moving the call to vec_length out of the loop test, we eliminate the need to execute it on every iteration.

operations, and minimal or even none for others. In any case, this transformation is required to eliminate inefficiencies that would become bottlenecks as we attempt further optimizations.

Function	Page	Method	Integer		Floating point		
			+	*	+	F *	D *
combine1	13	Abstract -O1	12.00	12.00	12.00	12.01	13.00
combine2	14	Move vec_length	8.03	8.09	10.09	11.09	12.08

This optimization is an instance of a general class of optimizations known as *code motion*. They involve identifying a computation that is performed multiple times (e.g., within a loop), but such that the result of the computation will not change. We can therefore move the computation to an earlier section of the code that does not get evaluated as often. In this case, we moved the call to vec_length from within the loop to just before the loop.

Optimizing compilers attempt to perform code motion. Unfortunately, as discussed previously, they are typically very cautious about making transformations that change where or how many times a procedure is called. They cannot reliably detect whether or not a function will have side effects, and so they assume that it might. For example, if vec_length had some side effect, then combine1 and combine2 could have different behaviors. To improve the code, the programmer must often help the compiler by explicitly performing code motion.

As an extreme example of the loop inefficiency seen in combine1, consider the procedure lower1 shown in Figure 7. This procedure is styled after routines submitted by several students as part of a network programming project. Its purpose is to convert all of the uppercase letters in a string to lowercase. The procedure steps through the string, converting each uppercase character to lowercase. The case conversion involves shifting characters in the range 'A' to 'Z' to the range 'a' to 'z.'

The library function strlen is called as part of the loop test of lower1. Although strlen is typically implemented with special x86 string-processing instructions, its overall execution is similar to the simple version that is also shown in Figure 7. Since strings in C are null-terminated character sequences, strlen can only determine the length of a string by stepping through the sequence until it hits a null character. For a string of length n, strlen takes time proportional to n. Since strlen is called in each of the n iterations of lower1, the overall run time of lower1 is quadratic in the string length, proportional to n^2.

This analysis is confirmed by actual measurements of the functions for different length strings, as shown in Figure 8 (and using the library version of strlen). The graph of the run time for lower1 rises steeply as the string length increases (Figure 8(a)). Figure 8(b) shows the run times for seven different lengths (not the same as shown in the graph), each of which is a power of 2. Observe that for lower1 each doubling of the string length causes a quadrupling of the run time. This is a clear indicator of a quadratic run time. For a string of length 1,048,576, lower1 requires over 13 minutes of CPU time.

```
1    /* Convert string to lowercase: slow */
2    void lower1(char *s)
3    {
4        int i;
5
6        for (i = 0; i < strlen(s); i++)
7            if (s[i] >= 'A' && s[i] <= 'Z')
8                s[i] -= ('A' - 'a');
9    }
10
11   /* Convert string to lowercase: faster */
12   void lower2(char *s)
13   {
14       int i;
15       int len = strlen(s);
16
17       for (i = 0; i < len; i++)
18           if (s[i] >= 'A' && s[i] <= 'Z')
19               s[i] -= ('A' - 'a');
20   }
21
22   /* Sample implementation of library function strlen */
23   /* Compute length of string */
24   size_t strlen(const char *s)
25   {
26       int length = 0;
27       while (*s != '\0') {
28           s++;
29           length++;
30       }
31       return length;
32   }
```

Figure 7 **Lowercase conversion routines.** The two procedures have radically different performance.

Function `lower2` shown in Figure 7 is identical to that of `lower1`, except that we have moved the call to `strlen` out of the loop. The performance improves dramatically. For a string length of 1,048,576, the function requires just 1.5 milliseconds—over 500,000 times faster than `lower1`. Each doubling of the string length causes a doubling of the run time—a clear indicator of linear run time. For longer strings, the run-time improvement will be even greater.

In an ideal world, a compiler would recognize that each call to `strlen` in the loop test will return the same result, and thus the call could be moved out of the loop. This would require a very sophisticated analysis, since `strlen` checks

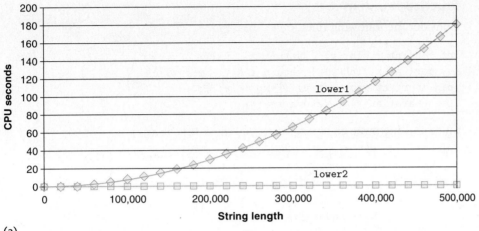

(a)

				String length			
Function	16,384	32,768	65,536	131,072	262,144	524,288	1,048,576
lower1	0.19	0.77	3.08	12.34	49.39	198.42	791.22
lower2	0.0000	0.0000	0.0001	0.0002	0.0004	0.0008	0.0015

(b)

Figure 8 **Comparative performance of lowercase conversion routines.** The original code `lower1` has a quadratic run time due to an inefficient loop structure. The modified code `lower2` has a linear run time.

the elements of the string and these values are changing as `lower1` proceeds. The compiler would need to detect that even though the characters within the string are changing, none are being set from nonzero to zero, or vice versa. Such an analysis is well beyond the ability of even the most sophisticated compilers, even if they employ inlining, and so programmers must do such transformations themselves.

This example illustrates a common problem in writing programs, in which a seemingly trivial piece of code has a hidden asymptotic inefficiency. One would not expect a lowercase conversion routine to be a limiting factor in a program's performance. Typically, programs are tested and analyzed on small data sets, for which the performance of `lower1` is adequate. When the program is ultimately deployed, however, it is entirely possible that the procedure could be applied to strings of over one million characters. All of a sudden this benign piece of code has become a major performance bottleneck. By contrast, the performance of `lower2` will be adequate for strings of arbitrary length. Stories abound of major programming projects in which problems of this sort occur. Part of the job of a competent programmer is to avoid ever introducing such asymptotic inefficiency.

Practice Problem 3

Consider the following functions:

```
int min(int x, int y) { return x < y ? x : y; }
int max(int x, int y) { return x < y ? y : x; }
void incr(int *xp, int v) { *xp += v; }
int square(int x) { return x*x; }
```

The following three code fragments call these functions:

A.
```
    for (i = min(x, y); i < max(x, y); incr(&i, 1))
        t += square(i);
```

B.
```
    for (i = max(x, y) - 1; i >= min(x, y); incr(&i, -1))
        t += square(i);
```

C.
```
    int low = min(x, y);
    int high = max(x, y);

    for (i = low; i < high; incr(&i, 1))
        t += square(i);
```

Assume x equals 10 and y equals 100. Fill in the following table indicating the number of times each of the four functions is called in code fragments A–C:

Code	min	max	incr	square
A.				
B.				
C.				

5 Reducing Procedure Calls

As we have seen, procedure calls can incur overhead and also block most forms of program optimization. We can see in the code for combine2 (Figure 6) that get_vec_element is called on every loop iteration to retrieve the next vector element. This function checks the vector index i against the loop bounds with every vector reference, a clear source of inefficiency. Bounds checking might be a useful feature when dealing with arbitrary array accesses, but a simple analysis of the code for combine2 shows that all references will be valid.

Suppose instead that we add a function get_vec_start to our abstract data type. This function returns the starting address of the data array, as shown in Figure 9. We could then write the procedure shown as combine3 in this figure, having no function calls in the inner loop. Rather than making a function call to retrieve each vector element, it accesses the array directly. A purist might say that this transformation seriously impairs the program modularity. In principle, the user of the vector abstract data type should not even need to know that the vector *code/opt/vec.c*

```
1   data_t *get_vec_start(vec_ptr v)
2   {
3       return v->data;
4   }
```

code/opt/vec.c

```
1    /* Direct access to vector data */
2    void combine3(vec_ptr v, data_t *dest)
3    {
4        long int i;
5        long int length = vec_length(v);
6        data_t *data = get_vec_start(v);
7
8        *dest = IDENT;
9        for (i = 0; i < length; i++) {
10           *dest = *dest OP data[i];
11       }
12   }
```

Figure 9 **Eliminating function calls within the loop.** The resulting code runs much faster, at some cost in program modularity.

contents are stored as an array, rather than as some other data structure such as a linked list. A more pragmatic programmer would argue that this transformation is a necessary step toward achieving high-performance results.

Function	Page	Method	Integer +	Integer *	Floating point +	F *	D *
combine2	14	Move vec_length	8.03	8.09	10.09	11.09	12.08
combine3	19	Direct data access	6.01	8.01	10.01	11.01	12.02

The resulting improvement is surprisingly modest, only improving the performance for integer sum. Again, however, this inefficiency would become a bottleneck as we attempt further optimizations. We will return to this function later (Section 11.2) and see why the repeated bounds checking by combine2 does not make its performance much worse. For applications in which performance is a significant issue, one must often compromise modularity and abstraction for speed. It is wise to include documentation on the transformations applied, as well as the assumptions that led to them, in case the code needs to be modified later.

6   Eliminating Unneeded Memory References

The code for combine3 accumulates the value being computed by the combining operation at the location designated by the pointer dest. This attribute can be seen by examining the assembly code generated for the compiled loop. We show

---

here the x86-64 code generated for data type `float` and with multiplication as the combining operation:

```
combine3: data_t = float, OP = *
i in %rdx, data in %rax, dest in %rbp
1    .L498:                              loop:
2        movss   (%rbp), %xmm0           Read product from dest
3        mulss   (%rax,%rdx,4), %xmm0    Multiply product by data[i]
4        movss   %xmm0, (%rbp)           Store product at dest
5        addq    $1, %rdx                Increment i
6        cmpq    %rdx, %r12              Compare i:limit
7        jg      .L498                   If >, goto loop
```

Aside   Understanding x86-64 floating-point code

We cover floating-point code for x86-64, the 64-bit version of the Intel instruction set in Web Aside ASM:SSE, but the program examples we show in this chapter can readily be understood by anyone familiar with IA32 code. Here, we briefly review the relevant aspects of x86-64 and its floating-point instructions.

The x86-64 instruction set extends the 32-bit registers of IA32, such as %eax, %edi, and %esp, to 64-bit versions, with 'r' replacing 'e', e.g., %rax, %rdi, and %rsp. Eight more registers are available, named %r8–%r15, greatly improving the ability to hold temporary values in registers. Suffix 'q' is used on integer instructions (e.g., addq, cmpq) to indicate 64-bit operations.

Floating-point data are held in a set of *XMM* registers, named %xmm0–%xmm15. Each of these registers is 128 bits long, able to hold four single-precision (`float`) or two double-precision (`double`) floating-point numbers. For our initial presentation, we will only make use of instructions that operate on single values held in SSE registers.

The movss instruction copies one single-precision number. Like the various MOV instructions of IA32, both the source and the destination can be memory locations or registers, but it uses XMM registers, rather than general-purpose registers. The mulss instruction multiplies single-precision numbers, updating its second operand with the product. Again, the source and destination operands can be memory locations or XMM registers.

We see in this loop code that the address corresponding to pointer dest is held in register %rbp (unlike in IA32, where %ebp has special use as a frame pointer, its 64-bit counterpart %rbp can be used to hold arbitrary data). On iteration i, the program reads the value at this location, multiplies it by data[i], and stores the result back at dest. This reading and writing is wasteful, since the value read from dest at the beginning of each iteration should simply be the value written at the end of the previous iteration.

We can eliminate this needless reading and writing of memory by rewriting the code in the style of combine4 in Figure 10. We introduce a temporary variable acc that is used in the loop to accumulate the computed value. The result is stored at dest only after the loop has been completed. As the assembly code that follows shows, the compiler can now use register %xmm0 to hold the accumulated value.

```
1   /* Accumulate result in local variable */
2   void combine4(vec_ptr v, data_t *dest)
3   {
4       long int i;
5       long int length = vec_length(v);
6       data_t *data = get_vec_start(v);
7       data_t acc = IDENT;
8
9       for (i = 0; i < length; i++) {
10          acc = acc OP data[i];
11      }
12      *dest = acc;
13  }
```

Figure 10 **Accumulating result in temporary.** Holding the accumulated value in local variable acc (short for "accumulator") eliminates the need to retrieve it from memory and write back the updated value on every loop iteration.

Compared to the loop in combine3, we have reduced the memory operations per iteration from two reads and one write to just a single read.

```
    combine4: data_t = float, OP = *
    i in %rdx, data in %rax, limit in %rbp, acc in %xmm0
1   .L488:                          loop:
2       mulss   (%rax,%rdx,4), %xmm0    Multiply acc by data[i]
3       addq    $1, %rdx                Increment i
4       cmpq    %rdx, %rbp              Compare limit:i
5       jg      .L488                   If >, goto loop
```

We see a significant improvement in program performance, as shown in the following table:

Function	Page	Method	Integer		Floating point		
			+	*	+	F *	D *
combine3	19	Direct data access	6.01	8.01	10.01	11.01	12.02
combine4	21	Accumulate in temporary	2.00	3.00	3.00	4.00	5.00

All of our times improve by at least a factor of $2.4\times$, with the integer addition case dropping to just two clock cycles per element.

Aside Expressing relative performance

The best way to express a performance improvement is as a ratio of the form T_{old}/T_{new}, where T_{old} is the time required for the original version and T_{new} is the time required by the modified version. This will be a number greater than 1.0 if any real improvement occurred. We use the suffix '$\times$' to indicate such a ratio, where the factor "$2.4\times$" is expressed verbally as "2.4 times."

The more traditional way of expressing relative change as a percentage works well when the change is small, but its definition is ambiguous. Should it be $100 \cdot (T_{old} - T_{new})/T_{new}$ or possibly $100 \cdot (T_{old} - T_{new})/T_{old}$, or something else? In addition, it is less instructive for large changes. Saying that "performance improved by 140%" is more difficult to comprehend than simply saying that the performance improved by a factor of 2.4.

Again, one might think that a compiler should be able to automatically transform the combine3 code shown in Figure 9 to accumulate the value in a register, as it does with the code for combine4 shown in Figure 10. In fact, however, the two functions can have different behaviors due to memory aliasing. Consider, for example, the case of integer data with multiplication as the operation and 1 as the identity element. Let v = [2, 3, 5] be a vector of three elements and consider the following two function calls:

```
combine3(v, get_vec_start(v) + 2);
combine4(v, get_vec_start(v) + 2);
```

That is, we create an alias between the last element of the vector and the destination for storing the result. The two functions would then execute as follows:

Function	Initial	Before loop	i = 0	i = 1	i = 2	Final
combine3	[2, 3, 5]	[2, 3, 1]	[2, 3, 2]	[2, 3, 6]	[2, 3, 36]	[2, 3, 36]
combine4	[2, 3, 5]	[2, 3, 5]	[2, 3, 5]	[2, 3, 5]	[2, 3, 5]	[2, 3, 30]

As shown previously, combine3 accumulates its result at the destination, which in this case is the final vector element. This value is therefore set first to 1, then to $2 \cdot 1 = 2$, and then to $3 \cdot 2 = 6$. On the final iteration, this value is then multiplied by itself to yield a final value of 36. For the case of combine4, the vector remains unchanged until the end, when the final element is set to the computed result $1 \cdot 2 \cdot 3 \cdot 5 = 30$.

Of course, our example showing the distinction between combine3 and combine4 is highly contrived. One could argue that the behavior of combine4 more closely matches the intention of the function description. Unfortunately, a compiler cannot make a judgment about the conditions under which a function might be used and what the programmer's intentions might be. Instead, when given combine3 to compile, the conservative approach is to keep reading and writing memory, even though this is less efficient.

Practice Problem 4

When we use GCC to compile combine3 with command-line option '-O2', we get code with substantially better CPE performance than with -O1:

Function	Page	Method	Integer		Floating point		
			+	*	+	F *	D *
combine3	19	Compiled –O1	6.01	8.01	10.01	11.01	12.02
combine3	19	Compiled –O2	3.00	3.00	3.00	4.02	5.03
combine4	21	Accumulate in temporary	2.00	3.00	3.00	4.00	5.00

We achieve performance comparable to that for combine4, except for the case of integer sum, but even it improves significantly. On examining the assembly code generated by the compiler, we find an interesting variant for the inner loop:

```
    combine3: data_t = float, OP = *, compiled -O2
    i in %rdx, data in %rax, limit in %rbp, dest at %rx12
    Product in %xmm0
1   .L560:                                   loop:
2     mulss   (%rax,%rdx,4),  %xmm0          Multiply product by data[i]
3     addq    $1, %rdx                       Increment i
4     cmpq    %rdx, %rbp                     Compare limit:i
5     movss   %xmm0, (%r12)                  Store product at dest
6     jg      .L560                          If >, goto loop
```

We can compare this to the version created with optimization level 1:

```
    combine3: data_t = float, OP = *, compiled -O1
    i in %rdx, data in %rax, dest in %rbp
1   .L498:                                   loop:
2     movss   (%rbp),  %xmm0                 Read product from dest
3     mulss   (%rax,%rdx,4),  %xmm0          Multiply product by data[i]
4     movss   %xmm0, (%rbp)                  Store product at dest
5     addq    $1, %rdx                       Increment i
6     cmpq    %rdx, %r12                     Compare i:limit
7     jg      .L498                          If >, goto loop
```

We see that, besides some reordering of instructions, the only difference is that the more optimized version does not contain the movss implementing the read from the location designated by dest (line 2).

A. How does the role of register %xmm0 differ in these two loops?

B. Will the more optimized version faithfully implement the C code of combine3, including when there is memory aliasing between dest and the vector data?

C. Explain either why this optimization preserves the desired behavior, or give an example where it would produce different results than the less optimized code.

With this final transformation, we reached a point where we require just 2–5 clock cycles for each element to be computed. This is a considerable improvement over the original 11–13 cycles when we first enabled optimization. We would now like to see just what factors are constraining the performance of our code and how we can improve things even further.

7 Understanding Modern Processors

Up to this point, we have applied optimizations that did not rely on any features of the target machine. They simply reduced the overhead of procedure calls and eliminated some of the critical "optimization blockers" that cause difficulties for optimizing compilers. As we seek to push the performance further, we must consider optimizations that exploit the *microarchitecture* of the processor, that is, the underlying system design by which a processor executes instructions. Getting every last bit of performance requires a detailed analysis of the program as well as code generation tuned for the target processor. Nonetheless, we can apply some basic optimizations that will yield an overall performance improvement on a large class of processors. The detailed performance results we report here may not hold for other machines, but the general principles of operation and optimization apply to a wide variety of machines.

To understand ways to improve performance, we require a basic understanding of the microarchitectures of modern processors. Due to the large number of transistors that can be integrated onto a single chip, modern microprocessors employ complex hardware that attempts to maximize program performance. One result is that their actual operation is far different from the view that is perceived by looking at machine-level programs. At the code level, it appears as if instructions are executed one at a time, where each instruction involves fetching values from registers or memory, performing an operation, and storing results back to a register or memory location. In the actual processor, a number of instructions are evaluated simultaneously, a phenomenon referred to as *instruction-level parallelism*. In some designs, there can be 100 or more instructions "in flight." Elaborate mechanisms are employed to make sure the behavior of this parallel execution exactly captures the sequential semantic model required by the machine-level program. This is one of the remarkable feats of modern microprocessors: they employ complex and exotic microarchitectures, in which multiple instructions can be executed in parallel, while presenting an operational view of simple sequential instruction execution.

Although the detailed design of a modern microprocessor is well beyond the scope of this book, having a general idea of the principles by which they operate suffices to understand how they achieve instruction-level parallelism. We will find that two different lower bounds characterize the maximum performance of a program. The *latency bound* is encountered when a series of operations must be performed in strict sequence, because the result of one operation is required before the next one can begin. This bound can limit program performance when the data dependencies in the code limit the ability of the processor to

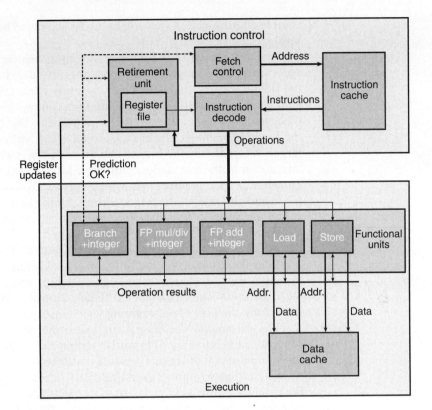

Figure 11

Block diagram of a modern processor. The instruction control unit is responsible for reading instructions from memory and generating a sequence of primitive operations. The execution unit then performs the operations and indicates whether the branches were correctly predicted.

exploit instruction-level parallelism. The *throughput bound* characterizes the raw computing capacity of the processor's functional units. This bound becomes the ultimate limit on program performance.

7.1 Overall Operation

Figure 11 shows a very simplified view of a modern microprocessor. Our hypothetical processor design is based loosely on the structure of the Intel Core i7 processor design, which is often referred to by its project code name "Nehalem" [99]. The Nehalem microarchitecture typifies the high-end processors produced by a number of manufacturers since the late 1990s. It is described in the industry as being *superscalar*, which means it can perform multiple operations on every clock cycle, and *out-of-order*, meaning that the order in which instructions execute need not correspond to their ordering in the machine-level program. The overall design has two main parts: the *instruction control unit* (ICU), which is responsible for reading a sequence of instructions from memory and generating from these a set of primitive operations to perform on program data, and the *execution unit* (EU), which then executes these operations. Compared to the simple *in-order* pipeline, out-of-order processors require far greater and more complex hardware, but they are better at achieving higher degrees of instruction-level parallelism.

The ICU reads the instructions from an *instruction cache*—a special high-speed memory containing the most recently accessed instructions. In general, the ICU fetches well ahead of the currently executing instructions, so that it has enough time to decode these and send operations down to the EU. One problem, however, is that when a program hits a branch,[1] there are two possible directions the program might go. The branch can be *taken*, with control passing to the branch target. Alternatively, the branch can be *not taken*, with control passing to the next instruction in the instruction sequence. Modern processors employ a technique known as *branch prediction*, in which they guess whether or not a branch will be taken and also predict the target address for the branch. Using a technique known as *speculative execution*, the processor begins fetching and decoding instructions at where it predicts the branch will go, and even begins executing these operations before it has been determined whether or not the branch prediction was correct. If it later determines that the branch was predicted incorrectly, it resets the state to that at the branch point and begins fetching and executing instructions in the other direction. The block labeled "Fetch control" incorporates branch prediction to perform the task of determining which instructions to fetch.

The *instruction decoding* logic takes the actual program instructions and converts them into a set of primitive *operations* (sometimes referred to as *micro-operations*). Each of these operations performs some simple computational task such as adding two numbers, reading data from memory, or writing data to memory. For machines with complex instructions, such as x86 processors, an instruction can be decoded into a variable number of operations. The details of how instructions are decoded into sequences of more primitive operations varies between machines, and this information is considered highly proprietary. Fortunately, we can optimize our programs without knowing the low-level details of a particular machine implementation.

In a typical x86 implementation, an instruction that only operates on registers, such as

```
addl %eax,%edx
```

is converted into a single operation. On the other hand, an instruction involving one or more memory references, such as

```
addl %eax,4(%edx)
```

yields multiple operations, separating the memory references from the arithmetic operations. This particular instruction would be decoded as three operations: one to *load* a value from memory into the processor, one to add the loaded value to the

---

1. We use the term "branch" specifically to refer to conditional jump instructions. Other instructions that can transfer control to multiple destinations, such as procedure return and indirect jumps, provide similar challenges for the processor.

value in register %eax, and one to *store* the result back to memory. This decoding splits instructions to allow a division of labor among a set of dedicated hardware units. These units can then execute the different parts of multiple instructions in parallel.

The EU receives operations from the instruction fetch unit. Typically, it can receive a number of them on each clock cycle. These operations are dispatched to a set of *functional units* that perform the actual operations. These functional units are specialized to handle specific types of operations. Our figure illustrates a typical set of functional units, based on those of the Intel Core i7. We can see that three functional units are dedicated to computation, while the remaining two are for reading (load) and writing (store) memory. Each computational unit can perform multiple different operations: all can perform at least basic integer operations, such as addition and bit-wise logical operations. Floating-point operations and integer multiplication require more complex hardware, and so these can only be handled by specific functional units.

Reading and writing memory is implemented by the load and store units. The load unit handles operations that read data from the memory into the processor. This unit has an adder to perform address computations. Similarly, the store unit handles operations that write data from the processor to the memory. It also has an adder to perform address computations. As shown in the figure, the load and store units access memory via a *data cache*, a high-speed memory containing the most recently accessed data values.

With speculative execution, the operations are evaluated, but the final results are not stored in the program registers or data memory until the processor can be certain that these instructions should actually have been executed. Branch operations are sent to the EU, not to determine where the branch should go, but rather to determine whether or not they were predicted correctly. If the prediction was incorrect, the EU will discard the results that have been computed beyond the branch point. It will also signal the branch unit that the prediction was incorrect and indicate the correct branch destination. In this case, the branch unit begins fetching at the new location. Such a *misprediction* incurs a significant cost in performance. It takes a while before the new instructions can be fetched, decoded, and sent to the execution units.

Within the ICU, the *retirement unit* keeps track of the ongoing processing and makes sure that it obeys the sequential semantics of the machine-level program. Our figure shows a *register file* containing the integer, floating-point, and more recently SSE registers as part of the retirement unit, because this unit controls the updating of these registers. As an instruction is decoded, information about it is placed into a first-in, first-out queue. This information remains in the queue until one of two outcomes occurs. First, once the operations for the instruction have completed and any branch points leading to this instruction are confirmed as having been correctly predicted, the instruction can be *retired*, with any updates to the program registers being made. If some branch point leading to this instruction was mispredicted, on the other hand, the instruction will be *flushed*, discarding any results that may have been computed. By this means, mispredictions will not alter the program state.

As we have described, any updates to the program registers occur only as instructions are being retired, and this takes place only after the processor can be certain that any branches leading to this instruction have been correctly predicted. To expedite the communication of results from one instruction to another, much of this information is exchanged among the execution units, shown in the figure as "Operation results." As the arrows in the figure show, the execution units can send results directly to each other.

The most common mechanism for controlling the communication of operands among the execution units is called *register renaming*. When an instruction that updates register r is decoded, a *tag* t is generated giving a unique identifier to the result of the operation. An entry (r, t) is added to a table maintaining the association between program register r and tag t for an operation that will update this register. When a subsequent instruction using register r as an operand is decoded, the operation sent to the execution unit will contain t as the source for the operand value. When some execution unit completes the first operation, it generates a result (v, t) indicating that the operation with tag t produced value v. Any operation waiting for t as a source will then use v as the source value, a form of data forwarding. By this mechanism, values can be forwarded directly from one operation to another, rather than being written to and read from the register file, enabling the second operation to begin as soon as the first has completed. The renaming table only contains entries for registers having pending write operations. When a decoded instruction requires a register r, and there is no tag associated with this register, the operand is retrieved directly from the register file. With register renaming, an entire sequence of operations can be performed speculatively, even though the registers are updated only after the processor is certain of the branch outcomes.

Aside The history of out-of-order processing

Out-of-order processing was first implemented in the Control Data Corporation 6600 processor in 1964. Instructions were processed by ten different functional units, each of which could be operated independently. In its day, this machine, with a clock rate of 10 Mhz, was considered the premium machine for scientific computing.

IBM first implemented out-of-order processing with the IBM 360/91 processor in 1966, but just to execute the floating-point instructions. For around 25 years, out-of-order processing was considered an exotic technology, found only in machines striving for the highest possible performance, until IBM reintroduced it in the RS/6000 line of workstations in 1990. This design became the basis for the IBM/Motorola PowerPC line, with the model 601, introduced in 1993, becoming the first single-chip microprocessor to use out-of-order processing. Intel introduced out-of-order processing with its PentiumPro model in 1995, with an underlying microarchitecture similar to that of the Core i7.

7.2 Functional Unit Performance

Figure 12 documents the performance of some of the arithmetic operations for an Intel Core i7, determined by both measurements and by reference to Intel

Operation	Integer		Single-precision		Double-precision	
	Latency	Issue	Latency	Issue	Latency	Issue
Addition	1	0.33	3	1	3	1
Multiplication	3	1	4	1	5	1
Division	11–21	5–13	10–15	6–11	10–23	6–19

Figure 12 **Latency and issue time characteristics of Intel Core i7 arithmetic operations.** Latency indicates the total number of clock cycles required to perform the actual operations, while issue time indicates the minimum number of cycles between two operations. The times for division depend on the data values.

literature [26]. These timings are typical for other processors as well. Each operation is characterized by its *latency*, meaning the total time required to perform the operation, and the *issue time*, meaning the minimum number of clock cycles between two successive operations of the same type.

We see that the latencies increase as the word sizes increase (e.g., from single to double precision), for more complex data types (e.g., from integer to floating point), and for more complex operations (e.g., from addition to multiplication).

We see also that most forms of addition and multiplication operations have issue times of 1, meaning that on each clock cycle, the processor can start a new one of these operations. This short issue time is achieved through the use of *pipelining*. A pipelined function unit is implemented as a series of *stages*, each of which performs part of the operation. For example, a typical floating-point adder contains three stages (and hence the three-cycle latency): one to process the exponent values, one to add the fractions, and one to round the result. The arithmetic operations can proceed through the stages in close succession rather than waiting for one operation to complete before the next begins. This capability can be exploited only if there are successive, logically independent operations to be performed. Functional units with issue times of 1 cycle are said to be *fully pipelined*: they can start a new operation every clock cycle. The issue time of 0.33 given for integer addition is due to the fact that the hardware has three fully pipelined functional units capable of performing integer addition. The processor has the potential to perform three additions every clock cycle. We see also that the divider (used for integer and floating-point division, as well as floating-point square root) is not fully pipelined—its issue time is just a few cycles less than its latency. What this means is that the divider must complete all but the last few steps of a division before it can begin a new one. We also see the latencies and issue times for division are given as ranges, because some combinations of dividend and divisor require more steps than others. The long latency and issue times of division make it a comparatively costly operation.

A more common way of expressing issue time is to specify the maximum *throughput* of the unit, defined as the reciprocal of the issue time. A fully pipelined functional unit has a maximum throughput of one operation per clock cycle, while units with higher issue times have lower maximum throughput.

Circuit designers can create functional units with wide ranges of performance characteristics. Creating a unit with short latency or with pipelining requires more hardware, especially for more complex functions such as multiplication and floating-point operations. Since there is only a limited amount of space for these units on the microprocessor chip, CPU designers must carefully balance the number of functional units and their individual performance to achieve optimal overall performance. They evaluate many different benchmark programs and dedicate the most resources to the most critical operations. As Figure 12 indicates, integer multiplication and floating-point multiplication and addition were considered important operations in design of the Core i7, even though a significant amount of hardware is required to achieve the low latencies and high degree of pipelining shown. On the other hand, division is relatively infrequent and difficult to implement with either short latency or full pipelining.

Both the latencies and the issue times (or equivalently, the maximum throughput) of these arithmetic operations can affect the performance of our combining functions. We can express these effects in terms of two fundamental bounds on the CPE values:

	Integer		Floating point		
Bound	+	*	+	F *	D *
Latency	1.00	3.00	3.00	4.00	5.00
Throughput	1.00	1.00	1.00	1.00	1.00

The *latency bound* gives a minimum value for the CPE for any function that must perform the combining operation in a strict sequence. The *throughput bound* gives a minimum bound for the CPE based on the maximum rate at which the functional units can produce results. For example, since there is only one multiplier, and it has an issue time of 1 clock cycle, the processor cannot possibly sustain a rate of more than one multiplication per clock cycle. We noted earlier that the processor has three functional units capable of performing integer addition, and so we listed the issue time for this operation as 0.33. Unfortunately, the need to read elements from memory creates an additional throughput bound for the CPE of 1.00 for the combining functions. We will demonstrate the effect of both of the latency and throughput bounds with different versions of the combining functions.

7.3 An Abstract Model of Processor Operation

As a tool for analyzing the performance of a machine-level program executing on a modern processor, we will use a *data-flow* representation of programs, a graphical notation showing how the data dependencies between the different operations constrain the order in which they are executed. These constraints then lead to *critical paths* in the graph, putting a lower bound on the number of clock cycles required to execute a set of machine instructions.

Before proceeding with the technical details, it is instructive to examine the CPE measurements obtained for function `combine4`, our fastest code up to this point:

Function	Page	Method	Integer +	Integer *	Floating point +	Floating point F *	Floating point D *
combine4	21	Accumulate in temporary	2.00	3.00	3.00	4.00	5.00
Latency bound			1.00	3.00	3.00	4.00	5.00
Throughput bound			1.00	1.00	1.00	1.00	1.00

We can see that these measurements match the latency bound for the processor, except for the case of integer addition. This is not a coincidence — it indicates that the performance of these functions is dictated by the latency of the sum or product computation being performed. Computing the product or sum of n elements requires around $L \cdot n + K$ clock cycles, where L is the latency of the combining operation and K represents the overhead of calling the function and initiating and terminating the loop. The CPE is therefore equal to the latency bound L.

From Machine-Level Code to Data-Flow Graphs

Our data-flow representation of programs is informal. We only want to use it as a way to visualize how the data dependencies in a program dictate its performance. We present the data-flow notation by working with `combine4` (Figure 10) as an example. We focus just on the computation performed by the loop, since this is the dominating factor in performance for large vectors. We consider the case of floating-point data with multiplication as the combining operation, although other combinations of data type and operation have nearly identical structure. The compiled code for this loop consists of four instructions, with registers `%rdx` holding loop index i, `%rax` holding array address data, `%rcx` holding loop bound limit, and `%xmm0` holding accumulator value acc.

```
   combine4: data_t = float, OP = *
   i in %rdx, data in %rax, limit in %rbp, acc in %xmm0
1    .L488:                               loop:
2      mulss    (%rax,%rdx,4), %xmm0      Multiply acc by data[i]
3      addq     $1, %rdx                  Increment i
4      cmpq     %rdx, %rbp                Compare limit:i
5      jg       .L488                     If >, goto loop
```

As Figure 13 indicates, with our hypothetical processor design, the four instructions are expanded by the instruction decoder into a series of five *operations*, with the initial multiplication instruction being expanded into a load operation to read the source operand from memory, and a mul operation to perform the multiplication.

Figure 13
Graphical representation of inner-loop code for combine4. Instructions are dynamically translated into one or two operations, each of which receives values from other operations or from registers and produces values for other operations and for registers. We show the target of the final instruction as the label loop. It jumps to the first instruction shown.

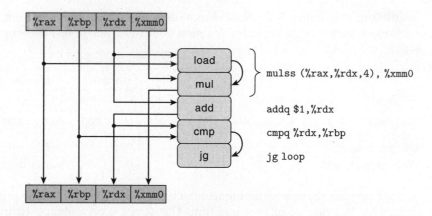

As a step toward generating a data-flow graph representation of the program, the boxes and lines along the left-hand side of Figure 13 show how the registers are used and updated by the different operations, with the boxes along the top representing the register values at the beginning of the loop, and those along the bottom representing the values at the end. For example, register %rax is only used as a source value by the load operation in performing its address calculation, and so the register has the same value at the end of the loop as at the beginning. Similarly, register %rcx is only used by the cmp operation. Register %rdx, on the other hand, is both used and updated within the loop. Its initial value is used by the load and add operations; its new value is generated by the add operation, which is then used by the cmp operation. Register %xmm0 is also updated within the loop by the mul operation, which first uses the initial value as a source value.

Some of the operations in Figure 13 produce values that do not correspond to registers. We show these as arcs between operations on the right-hand side. The load operation reads a value from memory and passes it directly to the mul operation. Since these two operations arise from decoding a single mulss instruction, there is no register associated with the intermediate value passing between them. The cmp operation updates the condition codes, and these are then tested by the jg operation.

For a code segment forming a loop, we can classify the registers that are accessed into four categories:

Read-only: These are used as source values, either as data or to compute memory addresses, but they are not modified within the loop. The read-only registers for the loop combine4 are %rax and %rcx.

Write-only: These are used as the destinations of data-movement operations. There are no such registers in this loop.

Local: These are updated and used within the loop, but there is no dependency from one iteration to another. The condition code registers are examples

Figure 14

Abstracting combine4 **operations as data-flow graph.** (a) We rearrange the operators of Figure 13 to more clearly show the data dependencies, and then (b) show only those operations that use values from one iteration to produce new values for the next.

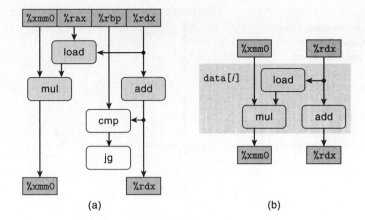

(a)   (b)

for this loop: they are updated by the cmp operation and used by the jl operation, but this dependency is contained within individual iterations.

Loop: These are both used as source values and as destinations for the loop, with the value generated in one iteration being used in another. We can see that %rdx and %xmm0 are loop registers for combine4, corresponding to program values i and acc.

As we will see, the chains of operations between loop registers determine the performance-limiting data dependencies.

Figure 14 shows further refinements of the graphical representation of Figure 13, with a goal of showing only those operations and data dependencies that affect the program execution time. We see in Figure 14(a) that we rearranged the operators to show more clearly the flow of data from the source registers at the top (both read-only and loop registers), and to the destination registers at the bottom (both write-only and loop registers).

In Figure 14(a), we also color operators white if they are not part of some chain of dependencies between loop registers. For this example, the compare (cmp) and branch (jl) operations do not directly affect the flow of data in the program. We assume that the Instruction Control Unit predicts that branch will be taken, and hence the program will continue looping. The purpose of the compare and branch operations is to test the branch condition and notify the ICU if it is not. We assume this checking can be done quickly enough that it does not slow down the processor.

In Figure 14(b), we have eliminated the operators that were colored white on the left, and we have retained only the loop registers. What we have left is an abstract template showing the data dependencies that form among loop registers due to one iteration of the loop. We can see in this diagram that there are two data dependencies from one iteration to the next. Along one side, we see the dependencies between successive values of program value acc, stored in register %xmm0. The loop computes a new value for acc by multiplying the old value by

Figure 15

Data-flow representation of computation by _n_ iterations by the inner loop of combine4. The sequence of multiplication operations forms a critical path that limits program performance.

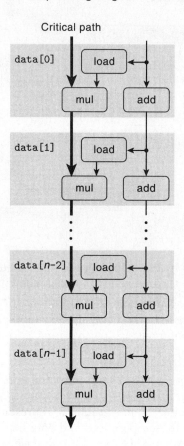

a data element, generated by the load operation. Along the other side, we see the dependencies between successive values of loop index i. On each iteration, the old value is used to compute the address for the load operation, and it is also incremented by the add operation to compute the new value.

Figure 15 shows the data-flow representation of _n_ iterations by the inner loop of function combine4. We can see that this graph was obtained by simply replicating the template shown on the right-hand side of Figure 14 _n_ times. We can see that the program has two chains of data dependencies, corresponding to the updating of program values acc and i with operations mul and add, respectively. Given that single-precision multiplication has a latency of 4 cycles, while integer addition has latency 1, we can see that the chain on the left will form a _critical path_, requiring $4n$ cycles to execute. The chain on the left would require only _n_ cycles to execute, and so it does not limit the program performance.

Figure 15 demonstrates why we achieved a CPE equal to the latency bound of 4 cycles for combine4, when performing single-precision floating-point multiplication. When executing the function, the floating-point multiplier becomes the limiting resource. The other operations required during the loop—manipulating and testing loop index i, computing the address of the next data elements, and reading data from memory—proceed in parallel with the multiplier. As each successive value of acc is computed, it is fed back around to compute the next value, but this will not be completed until four cycles later.

The flow for other combinations of data type and operation are identical to those shown in Figure 15, but with a different data operation forming the chain of data dependencies shown on the left. For all of the cases where the operation has a latency L greater than 1, we see that the measured CPE is simply L, indicating that this chain forms the performance-limiting critical path.

Other Performance Factors

For the case of integer addition, on the other hand, our measurements of combine4 show a CPE of 2.00, slower than the CPE of 1.00 we would predict based on the chains of dependencies formed along either the left- or the right-hand side of the graph of Figure 15. This illustrates the principle that the critical paths in a data-flow representation provide only a *lower* bound on how many cycles a program will require. Other factors can also limit performance, including the total number of functional units available and the number of data values that can be passed among the functional units on any given step. For the case of integer addition as the combining operation, the data operation is sufficiently fast that the rest of the operations cannot supply data fast enough. Determining exactly why the program requires 2.00 cycles per element would require a much more detailed knowledge of the hardware design than is publicly available.

To summarize our performance analysis of combine4: our abstract data-flow representation of program operation showed that combine4 has a critical path of length $L \cdot n$ caused by the successive updating of program value acc, and this path limits the CPE to at least L. This is indeed the CPE we measure for all cases except integer addition, which has a measured CPE of 2.00 rather than the CPE of 1.00 we would expect from the critical path length.

It may seem that the latency bound forms a fundamental limit on how fast our combining operations can be performed. Our next task will be to restructure the operations to enhance instruction-level parallelism. We want to transform the program in such a way that our only limitation becomes the throughput bound, yielding CPEs close to 1.00.

Practice Problem 5

Suppose we wish to write a function to evaluate a polynomial, where a polynomial of degree n is defined to have a set of coefficients $a_0, a_1, a_2, \ldots, a_n$. For a value x, we evaluate the polynomial by computing

$$a_0 + a_1 x + a_2 x^2 + \cdots + a_n x^n \tag{2}$$

This evaluation can be implemented by the following function, having as arguments an array of coefficients a, a value x, and the polynomial degree, degree (the value n in Equation 2). In this function, we compute both the successive terms of the equation and the successive powers of x within a single loop:

```
1    double poly(double a[], double x, int degree)
2    {
3        long int i;
4        double result = a[0];
5        double xpwr = x;   /* Equals x^i at start of loop */
6        for (i = 1; i <= degree; i++) {
7            result += a[i] * xpwr;
8            xpwr = x * xpwr;
9        }
10       return result;
11
12   }
```

A. For degree n, how many additions and how many multiplications does this code perform?

B. On our reference machine, with arithmetic operations having the latencies shown in Figure 12, we measure the CPE for this function to be 5.00. Explain how this CPE arises based on the data dependencies formed between iterations due to the operations implementing lines 7–8 of the function.

Practice Problem 6

Let us continue exploring ways to evaluate polynomials, as described in Problem 5. We can reduce the number of multiplications in evaluating a polynomial by applying *Horner's method*, named after British mathematician William G. Horner (1786–1837). The idea is to repeatedly factor out the powers of x to get the following evaluation:

$$a_0 + x(a_1 + x(a_2 + \cdots + x(a_{n-1} + x a_n) \cdots)) \tag{3}$$

Using Horner's method, we can implement polynomial evaluation using the following code:

```
1    /* Apply Horner's method */
2    double polyh(double a[], double x, int degree)
3    {
4        long int i;
5        double result = a[degree];
6        for (i = degree-1; i >= 0; i--)
7            result = a[i] + x*result;
8        return result;
9    }
```

A. For degree n, how many additions and how many multiplications does this code perform?

B. On our reference machine, with the arithmetic operations having the latencies shown in Figure 12, we measure the CPE for this function to be 8.00. Explain how this CPE arises based on the data dependencies formed between iterations due to the operations implementing line 7 of the function.

C. Explain how the function shown in Problem 5 can run faster, even though it requires more operations.

---

8 Loop Unrolling

Loop unrolling is a program transformation that reduces the number of iterations for a loop by increasing the number of elements computed on each iteration. We saw an example of this with the function psum2 (Figure 1), where each iteration computes two elements of the prefix sum, thereby halving the total number of iterations required. Loop unrolling can improve performance in two ways. First, it reduces the number of operations that do not contribute directly to the program result, such as loop indexing and conditional branching. Second, it exposes ways in which we can further transform the code to reduce the number of operations in the critical paths of the overall computation. In this section, we will examine simple loop unrolling, without any further transformations.

Figure 16 shows a version of our combining code using two-way loop unrolling. The first loop steps through the array two elements at a time. That is, the loop index i is incremented by 2 on each iteration, and the combining operation is applied to array elements i and $i + 1$ in a single iteration.

In general, the vector length will not be a multiple of 2. We want our code to work correctly for arbitrary vector lengths. We account for this requirement in two ways. First, we make sure the first loop does not overrun the array bounds. For a vector of length n, we set the loop limit to be $n - 1$. We are then assured that the loop will only be executed when the loop index i satisfies $i < n - 1$, and hence the maximum array index $i + 1$ will satisfy $i + 1 < (n - 1) + 1 = n$.

We can generalize this idea to unroll a loop by any factor k. To do so, we set the upper limit to be $n - k + 1$, and within the loop apply the combining operation to elements i through $i + k - 1$. Loop index i is incremented by k in each iteration. The maximum array index $i + k - 1$ will then be less than n. We include the second loop to step through the final few elements of the vector one at a time. The body of this loop will be executed between 0 and $k - 1$ times. For $k = 2$, we could use a simple conditional statement to optionally add a final iteration, as we did with the function psum2 (Figure 1). For $k > 2$, the finishing cases are better expressed with a loop, and so we adopt this programming convention for $k = 2$ as well.

```
1    /* Unroll loop by 2 */
2    void combine5(vec_ptr v, data_t *dest)
3    {
4        long int i;
5        long int length = vec_length(v);
6        long int limit = length-1;
7        data_t *data = get_vec_start(v);
8        data_t acc = IDENT;
9
10       /* Combine 2 elements at a time */
11       for (i = 0; i < limit; i+=2) {
12           acc = (acc OP data[i]) OP data[i+1];
13       }
14
15       /* Finish any remaining elements */
16       for (; i < length; i++) {
17           acc = acc OP data[i];
18       }
19       *dest = acc;
20   }
```

Figure 16 **Unrolling loop by factor** $k = 2$. Loop unrolling can reduce the effect of loop overhead.

Practice Problem 7

Modify the code for combine5 to unroll the loop by a factor $k = 5$.

When we measure the performance of unrolled code for unrolling factors $k = 2$ (combine5) and $k = 3$, we get the following results:

Function	Page	Method	Integer +	Integer *	Floating point +	F *	D *
combine4	21	No unrolling	2.00	3.00	3.00	4.00	5.00
combine5	38	Unroll by ×2	2.00	1.50	3.00	4.00	5.00
		Unroll by ×3	1.00	1.00	3.00	4.00	5.00
Latency bound			1.00	3.00	3.00	4.00	5.00
Throughput bound			1.00	1.00	1.00	1.00	1.00

We see that CPEs for both integer addition and multiplication improve, while those for the floating-point operations do not. Figure 17 shows CPE measurements when unrolling the loop by up to a factor of 6. We see that the trends we

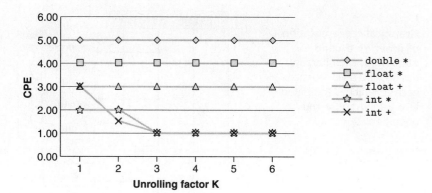

Figure 17

CPE performance for different degrees of loop unrolling. Only integer addition and multiplication improve by loop unrolling.

observed for unrolling by 2 and 3 continue—it does not help the floating-point operations, while both integer addition and multiplication drop down to CPEs of 1.00. Several phenomena contribute to these measured values of CPE. For the case of integer addition, we see that unrolling by a factor of 2 makes no difference, but unrolling by a factor of 3 drops the CPE to 1.00, achieving both the latency and the throughput bounds for this operation. This result can be attributed to the benefits of reducing loop overhead operations. By reducing the number of overhead operations relative to the number of additions required to compute the vector sum, we can reach the point where the one-cycle latency of integer addition becomes the performance-limiting factor.

The improving CPE for integer multiplication is surprising. We see that for unrolling factor k between 1 and 3, the CPE is $3.00/k$. It turns out that the compiler is making an optimization based on a *reassociation transformation*, altering the order in which values are combined. We will cover this transformation in Section 9.2. The fact that GCC applies this transformation to integer multiplication but not to floating-point addition or multiplication is due to the associativity properties of the different operations and data types, as will also be discussed later.

To understand why the three floating-point cases do not improve by loop unrolling, consider the graphical representation for the inner loop, shown in Figure 18 for the case of single-precision multiplication. We see here that the `mulss` instructions each get translated into two operations: one to load an array element from memory, and one to multiply this value by the accumulated value. We see here that register `%xmm0` gets read and written twice in each execution of the loop. We can rearrange, simplify, and abstract this graph, following the process shown in Figure 19 to obtain the template shown in Figure 19(b). We then replicate this template $n/2$ times to show the computation for a vector of length n, obtaining the data-flow representation shown in Figure 20. We see here that there is still a critical path of n mul operations in this graph—there are half as many iterations, but each iteration has two multiplication operations in sequence. Since the critical path was the limiting factor for the performance of the code without loop unrolling, it remains so with simple loop unrolling.

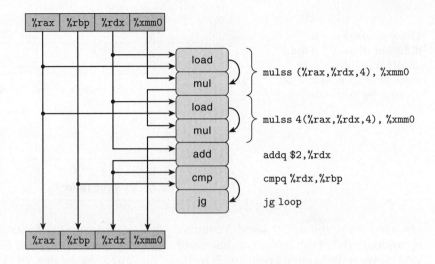

Figure 18

Graphical representation of inner-loop code for combine5. Each iteration has two mulss instructions, each of which is translated into a load and a mul operation.

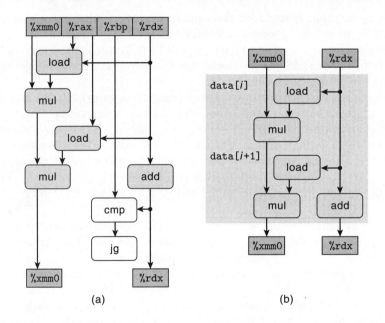

Figure 19

Abstracting combine5 **operations as data-flow graph.** We rearrange, simplify, and abstract the representation of Figure 18 to show the data dependencies between successive iterations (a). We see that each iteration must perform two multiplications in sequence (b).

(a)   (b)

Aside   Getting the compiler to unroll loops

Loop unrolling can easily be performed by a compiler. Many compilers do it routinely whenever the optimization level is set sufficiently high. GCC will perform loop unrolling when invoked with command-line option '-funroll-loops'.

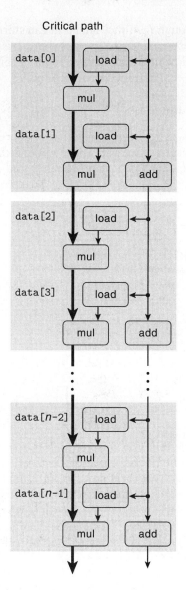

Critical path

Figure 20

Data-flow representation of combine5 **operating on a vector of length** n. Even though the loop has been unrolled by a factor of 2, there are still n mul operations along the critical path.

9   Enhancing Parallelism

At this point, our functions have hit the bounds imposed by the latencies of the arithmetic units. As we have noted, however, the functional units performing addition and multiplication are all fully pipelined, meaning that they can start new operations every clock cycle. Our code cannot take advantage of this capability, even with loop unrolling, since we are accumulating the value as a single variable acc. We cannot compute a new value for acc until the preceding computation has completed. Even though the functional unit can start a new operation every clock cycle, it will only start one every L cycles, where L is the latency of the combining operation. We will now investigate ways to break this sequential dependency and get performance better than the latency bound.

9.1 Multiple Accumulators

For a combining operation that is associative and commutative, such as integer addition or multiplication, we can improve performance by splitting the set of combining operations into two or more parts and combining the results at the end. For example, let P_n denote the product of elements $a_0, a_1, \ldots, a_{n-1}$:

$$P_n = \prod_{i=0}^{n-1} a_i$$

Assuming n is even, we can also write this as $P_n = PE_n \times PO_n$, where PE_n is the product of the elements with even indices, and PO_n is the product of the elements with odd indices:

$$PE_n = \prod_{i=0}^{n/2-1} a_{2i}$$

$$PO_n = \prod_{i=0}^{n/2-1} a_{2i+1}$$

Figure 21 shows code that uses this method. It uses both two-way loop unrolling, to combine more elements per iteration, and two-way parallelism, accumulating elements with even index in variable acc0 and elements with odd index in variable acc1. As before, we include a second loop to accumulate any remaining array elements for the case where the vector length is not a multiple of 2. We then apply the combining operation to acc0 and acc1 to compute the final result.

Comparing loop unrolling alone to loop unrolling with two-way parallelism, we obtain the following performance:

Function	Page	Method	Integer +	*	Floating point +	F *	D *
combine4	21	Accumulate in temporary	2.00	3.00	3.00	4.00	5.00
combine5	38	Unroll by ×2	2.00	1.50	3.00	4.00	5.00
combine6	43	Unroll ×2, parallelism ×2	1.50	1.50	1.50	2.00	2.50
Latency bound			1.00	3.00	3.00	4.00	5.00
Throughput bound			1.00	1.00	1.00	1.00	1.00

Figure 22 demonstrates the effect of applying this transformation to achieve k-way loop unrolling and k-way parallelism for values up to $k = 6$. We can see that

```
1   /* Unroll loop by 2, 2-way parallelism */
2   void combine6(vec_ptr v, data_t *dest)
3   {
4       long int i;
5       long int length = vec_length(v);
6       long int limit = length-1;
7       data_t *data = get_vec_start(v);
8       data_t acc0 = IDENT;
9       data_t acc1 = IDENT;
10
11      /* Combine 2 elements at a time */
12      for (i = 0; i < limit; i+=2) {
13          acc0 = acc0 OP data[i];
14          acc1 = acc1 OP data[i+1];
15      }
16
17      /* Finish any remaining elements */
18      for (; i < length; i++) {
19          acc0 = acc0 OP data[i];
20      }
21      *dest = acc0 OP acc1;
22  }
```

Figure 21 **Unrolling loop by 2 and using two-way parallelism.** This approach makes use of the pipelining capability of the functional units.

the CPEs for all of our combining cases improve with increasing values of k. For integer multiplication, and for the floating-point operations, we see a CPE value of L/k, where L is the latency of the operation, up to the throughput bound of 1.00. We also see integer addition reaching its throughput bound of 1.00 with $k = 3$. Of course, we also reached this bound for integer addition with standard unrolling.

Figure 22
CPE performance for k-way loop unrolling with k-way parallelism. All of the CPEs improve with this transformation, up to the limiting value of 1.00.

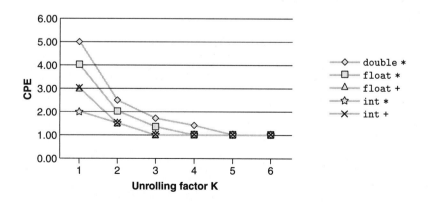

511

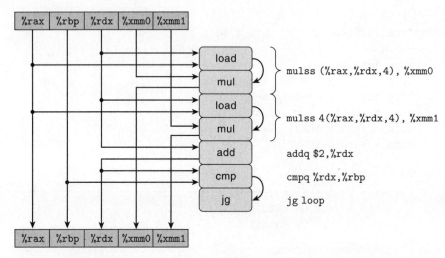

Figure 23 **Graphical representation of inner-loop code for** combine6. Each iteration has two mulss instructions, each of which is translated into a load and a mul operation.

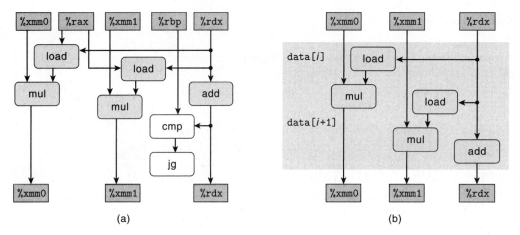

Figure 24 **Abstracting** combine6 **operations as data-flow graph.** We rearrange, simplify, and abstract the representation of Figure 23 to show the data dependencies between successive iterations (a). We see that there is no dependency between the two mul operations (b).

To understand the performance of combine6, we start with the code and operation sequence shown in Figure 23. We can derive a template showing the data dependencies between iterations through the process shown in Figure 24. As with combine5, the inner loop contains two mulss operations, but these instructions translate into mul operations that read and write separate registers, with no data dependency between them (Figure 24(b)). We then replicate this template $n/2$ times (Figure 25), modeling the execution of the function on a vector

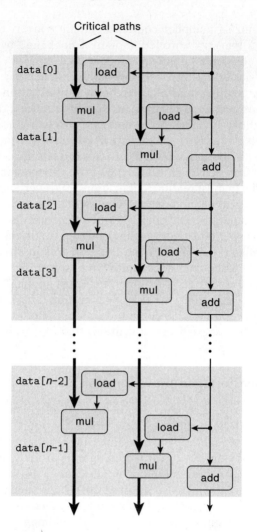

Figure 25

Data-flow representation of combine6 **operating on a vector of length** n. We now have two critical paths, each containing $n/2$ operations.

of length n. We see that we now have two critical paths, one corresponding to computing the product of even-numbered elements (program value acc0) and one for the odd-numbered elements (program value acc1). Each of these critical paths contain only $n/2$ operations, thus leading to a CPE of 4.00/2. A similar analysis explains our observed CPE of $L/2$ for operations with latency L for the different combinations of data type and combining operation. Operationally, we are exploiting the pipelining capabilities of the functional unit to increase their utilization by a factor of 2. When we apply this transformation for larger values of k, we find that we cannot reduce the CPE below 1.00. Once we reach this point, several of the functional units are operating at maximum capacity.

Two's-complement arithmetic is commutative and associative, even when overflow occurs. Hence, for an integer data type, the result computed by combine6 will be identical to that computed by combine5 under all possible conditions. Thus, an optimizing compiler could potentially convert the code shown in combine4 first to a two-way unrolled variant of combine5 by loop unrolling, and then to that of combine6 by introducing parallelism. Many compilers do loop unrolling automatically, but relatively few then introduce this form of parallelism.

On the other hand, floating-point multiplication and addition are not associative. Thus, combine5 and combine6 could produce different results due to rounding or overflow. Imagine, for example, a product computation in which all of the elements with even indices were numbers with very large absolute value, while those with odd indices were very close to 0.0. In such a case, product PE_n might overflow, or PO_n might underflow, even though computing product P_n proceeds normally. In most real-life applications, however, such patterns are unlikely. Since most physical phenomena are continuous, numerical data tend to be reasonably smooth and well-behaved. Even when there are discontinuities, they do not generally cause periodic patterns that lead to a condition such as that sketched earlier. It is unlikely that multiplying the elements in strict order gives fundamentally better accuracy than does multiplying two groups independently and then multiplying those products together. For most applications, achieving a performance gain of $2\times$ outweighs the risk of generating different results for strange data patterns. Nevertheless, a program developer should check with potential users to see if there are particular conditions that may cause the revised algorithm to be unacceptable.

9.2 Reassociation Transformation

We now explore another way to break the sequential dependencies and thereby improve performance beyond the latency bound. We saw that the simple loop unrolling of combine5 did not change the set of operations performed in combining the vector elements to form their sum or product. By a very small change in the code, however, we can fundamentally change the way the combining is performed, and also greatly increase the program performance.

Figure 26 shows a function combine7 that differs from the unrolled code of combine5 (Figure 16) only in the way the elements are combined in the inner loop. In combine5, the combining is performed by the statement

```
12          acc = (acc OP data[i]) OP data[i+1];
```

while in combine7 it is performed by the statement

```
12          acc = acc OP (data[i] OP data[i+1]);
```

differing only in how two parentheses are placed. We call this a *reassociation transformation*, because the parentheses shift the order in which the vector elements are combined with the accumulated value acc.

To an untrained eye, the two statements may seem essentially the same, but when we measure the CPE, we get surprising results:

Function	Page	Method	Integer		Floating point		
			+	*	+	F *	D *
combine4	21	Accumulate in temporary	2.00	3.00	3.00	4.00	5.00
combine5	38	Unroll by ×2	2.00	1.50	3.00	4.00	5.00
combine6	43	Unroll by ×2, parallelism ×2	1.50	1.50	1.50	2.00	2.50
combine7	47	Unroll ×2 and reassociate	2.00	1.51	1.50	2.00	2.97
Latency bound			1.00	3.00	3.00	4.00	5.00
Throughput bound			1.00	1.00	1.00	1.00	1.00

The integer multiplication case nearly matches the performance of the version with simple unrolling (combine5), while the floating-point cases match the performance of the version with parallel accumulators (combine6), doubling the performance relative to simple unrolling. (The CPE of 2.97 shown for double-precision multiplication is most likely the result of a measurement error, with the true value being 2.50. In our experiments, we found the measured CPEs for combine7 to be more variable than for the other functions.)

Figure 27 demonstrates the effect of applying the reassociation transformation to achieve k-way loop unrolling with reassociation. We can see that the CPEs for all of our combining cases improve with increasing values of k. For integer

```
1    /* Change associativity of combining operation */
2    void combine7(vec_ptr v, data_t *dest)
3    {
4        long int i;
5        long int length = vec_length(v);
6        long int limit = length-1;
7        data_t *data = get_vec_start(v);
8        data_t acc = IDENT;
9
10       /* Combine 2 elements at a time */
11       for (i = 0; i < limit; i+=2) {
12           acc = acc OP (data[i] OP data[i+1]);
13       }
14
15       /* Finish any remaining elements */
16       for (; i < length; i++) {
17           acc = acc OP data[i];
18       }
19       *dest = acc;
20   }
```

Figure 26 **Unrolling loop by 2 and then reassociating the combining operation.**
This approach also increases the number of operations that can be performed in parallel.

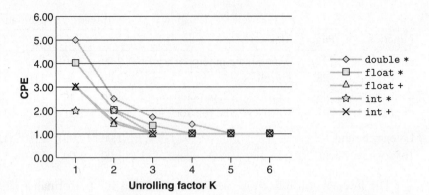

Figure 27

CPE performance for k-way loop unrolling with reassociation. All of the CPEs improve with this transformation, up to the limiting value of 1.00.

multiplication and for the floating-point operations, we see a CPE value of nearly L/k, where L is the latency of the operation, up to the throughput bound of 1.00. We also see integer addition reaching CPE of 1.00 for $k = 3$, achieving both the throughput and the latency bounds.

Figure 28 illustrates how the code for the inner loop of combine7 (for the case of single-precision product) gets decoded into operations and the resulting data dependencies. We see that the load operations resulting from the movss and the first mulss instructions load vector elements i and $i + 1$ from memory, and the first mul operation multiplies them together. The second mul operation then multiples this result by the accumulated value acc. Figure 29 shows how we rearrange, refine, and abstract the operations of Figure 28 to get a template representing the data dependencies for one iteration (Figure 29(b)). As with the templates for combine5 and combine7, we have two load and two mul operations,

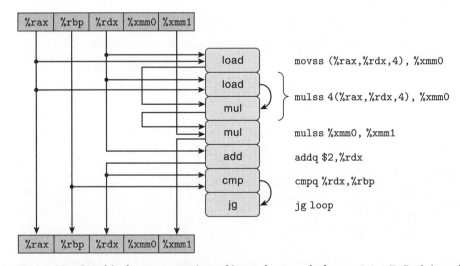

Figure 28 **Graphical representation of inner-loop code for** combine7. Each iteration gets decoded into similar operations as for combine5 or combine6, but with different data dependencies.

516

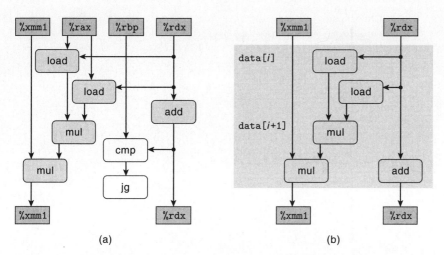

Figure 29 Abstracting combine7 **operations as data-flow graph.** We rearrange, simplify, and abstract the representation of Figure 28 to show the data dependencies between successive iterations (a). The first mul operation multiplies the two vector elements, while the second one multiplies the result by loop variable acc (b).

but only one of the mul operations forms a data-dependency chain between loop registers. When we then replicate this template $n/2$ times to show the computations performed in multiplying n vector elements (Figure 30), we see that we only have $n/2$ operations along the critical path. The first multiplication within each iteration can be performed without waiting for the accumulated value from the previous iteration. Thus, we reduce the minimum possible CPE by a factor of 2. As we increase k, we continue to have only one operation per iteration along the critical path.

In performing the reassociation transformation, we once again change the order in which the vector elements will be combined together. For integer addition and multiplication, the fact that these operations are associative implies that this reordering will have no effect on the result. For the floating-point cases, we must once again assess whether this reassociation is likely to significantly affect the outcome. We would argue that the difference would be immaterial for most applications.

We can now explain the surprising improvement we saw with simple loop unrolling (combine5) for the case of integer multiplication. In compiling this code, GCC performed the reassociation that we have shown in combine7, and hence it achieved the same performance. It also performed the transformation for code with higher degrees of unrolling. GCC recognizes that it can safely perform this transformation for integer operations, but it also recognizes that it cannot transform the floating-point cases due to the lack of associativity. It would be gratifying to find that GCC performed this transformation recognizing that the resulting code would run faster, but unfortunately this seems not to be the case. In our experiments, we found that very minor changes to the C code caused GCC

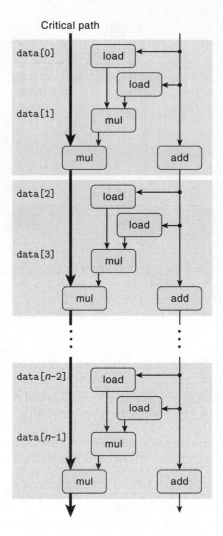

Figure 30

Data-flow representation of `combine7` **operating on a vector of length** n**.** We have a single critical path, but it contains only $n/2$ operations.

to associate the operations differently, sometimes causing the generated code to speed up, and sometimes to slow down, relative to what would be achieved by a straightforward compilation. Optimizing compilers must choose which factors they try to optimize, and it appears that GCC does not use maximizing instruction-level parallelism as one of its optimization criteria when selecting how to associate integer operations.

In summary, a reassociation transformation can reduce the number of operations along the critical path in a computation, resulting in better performance by better utilizing the pipelining capabilities of the functional units. Most compilers will not attempt any reassociations of floating-point operations, since these operations are not guaranteed to be associative. Current versions of GCC do perform reassociations of integer operations, but not always with good effects. In general, we have found that unrolling a loop and accumulating multiple values in parallel is a more reliable way to achieve improved program performance.

Practice Problem 8

Consider the following function for computing the product of an array of n integers. We have unrolled the loop by a factor of 3.

```
double aprod(double a[], int n)
{
    int i;
    double x, y, z;
    double r = 1;
    for (i = 0; i < n-2; i+= 3) {
        x = a[i]; y = a[i+1]; z = a[i+2];
        r = r * x * y * z; /* Product computation */
    }
    for (; i < n; i++)
        r *= a[i];
    return r;
}
```

For the line labeled Product computation, we can use parentheses to create five different associations of the computation, as follows:

```
r = ((r * x) * y) * z; /* A1 */
r = (r * (x * y)) * z; /* A2 */
r = r * ((x * y) * z); /* A3 */
r = r * (x * (y * z)); /* A4 */
r = (r * x) * (y * z); /* A5 */
```

Assume we run these functions on a machine where double-precision multiplication has a latency of 5 clock cycles. Determine the lower bound on the CPE set by the data dependencies of the multiplication. (**Hint:** It helps to draw a pictorial representation of how r is computed on every iteration.)

Web Aside OPT:SIMD   Achieving greater parallelism with SIMD instructions

Intel introduced the SSE instructions in 1999, where SSE is the acronym for "Streaming SIMD Extensions," and, in turn, SIMD (pronounced "sim-dee") is the acronym for "Single-Instruction, Multiple-Data." The idea behind the SIMD execution model is that each 16-byte XMM register can hold multiple values. In our examples, we consider the cases where they can hold either four integer or single-precision values, or two double-precision values. SSE instructions can then perform vector operations on these registers, such as adding or multiplying four or two sets of values in parallel. For example, if XMM register %xmm0 contains four single-precision floating-point numbers, which we denote $a_0, \ldots, a_3$, and %rcx contains the memory address of a sequence of four single-precision floating-point numbers, which we denote $b_0, \ldots, b_3$, then the instruction

```
mulps  (%rcs), %xmm0
```

will read the four values from memory and perform four multiplications in parallel, computing $a_i \leftarrow a_i \cdot b_i$, for $0 \leq i \leq 3$. We see that a single instruction is able to generate a computation over multiple data values, hence the term "SIMD."

GCC supports extensions to the C language that let programmers express a program in terms of vector operations that can be compiled into the SIMD instructions of SSE. This coding style is preferable to writing code directly in assembly language, since GCC can also generate code for the SIMD instructions found on other processors.

Using a combination of GCC instructions, loop unrolling, and multiple accumulators, we are able to achieve the following performance for our combining functions:

	Integer		Floating point		
Method	+	*	+	F *	D *
SSE + 8-way unrolling	0.25	0.55	0.25	0.24	0.58
Throughput bound	0.25	0.50	0.25	0.25	0.50

As this chart shows, using SSE instructions lowers the throughput bound, and we have nearly achieved these bounds for all five cases. The throughput bound of 0.25 for integer addition and single-precision addition and multiplication is due to the fact that the SSE instruction can perform four of these in parallel, and it has an issue time of 1. The double-precision instructions can only perform two in parallel, giving a throughput bound of 0.50. The integer multiplication operation has a throughput bound of 0.50 for a different reason—although it can perform four in parallel, it has an issue time of 2. In fact, this instruction is only available for SSE versions 4 and higher (requiring command-line flag '-msse4').

10 Summary of Results for Optimizing Combining Code

Our efforts at maximizing the performance of a routine that adds or multiplies the elements of a vector have clearly paid off. The following summarizes the results we obtain with *scalar* code, not making use of the SIMD parallelism provided by SSE vector instructions:

			Integer		Floating point		
Function	Page	Method	+	*	+	F *	D *
combine1	13	Abstract -O1	12.00	12.00	12.00	12.01	13.00
combine6	43	Unroll by ×2, parallelism ×2	1.50	1.50	1.50	2.00	2.50
		Unroll by ×5, parallelism ×5	1.01	1.00	1.00	1.00	1.00
Latency bound			1.00	3.00	3.00	4.00	5.00
Throughput bound			1.00	1.00	1.00	1.00	1.00

By using multiple optimizations, we have been able to achieve a CPE close to 1.00 for all combinations of data type and operation using ordinary C code, a performance improvement of over 10X compared to the original version combine1.

As covered in Web Aside OPT:SIMD, we can improve performance even further by making use of GCC's support for SIMD vector instructions:

Function	Method	Integer +	Integer *	Floating point +	Floating point F *	Floating point D *
SIMD code	SIMD + 8-way unrolling	0.25	0.55	0.25	0.24	0.58
Throughput bound		0.25	0.50	0.25	0.25	0.50

The processor can sustain up to four combining operations per cycle for integer and single-precision data, and two per cycle for double-precision data. This represents a performance of over 6 *gigaflops* (billions of floating-point operations per second) on a processor now commonly found in laptop and desktop machines.

Compare this performance to that of the Cray 1S, a breakthrough supercomputer introduced in 1976. This machine cost around $8 million and consumed 115 kilowatts of electricity to get its peak performance of 0.25 gigaflops, over 20 times slower than we measured here.

Several factors limit our performance for this computation to a CPE of 1.00 when using scalar instructions, and a CPE of either 0.25 (32-bit data) or 0.50 (64-bit data) when using SIMD instructions. First, the processor can only read 16 bytes from the data cache on each cycle, and then only by reading into an XMM register. Second, the multiplier and adder units can only start a new operation every clock cycle (in the case of SIMD instructions, each of these "operations" actually computes two or four sums or products). Thus, we have succeeded in producing the fastest possible versions of our combining function for this machine.

11 Some Limiting Factors

We have seen that the critical path in a data-flow graph representation of a program indicates a fundamental lower bound on the time required to execute a program. That is, if there is some chain of data dependencies in a program where the sum of all of the latencies along that chain equals T, then the program will require at least T cycles to execute.

We have also seen that the throughput bounds of the functional units also impose a lower bound on the execution time for a program. That is, assume that a program requires a total of N computations of some operation, that the microprocessor has only m functional units capable of performing that operation, and that these units have an issue time of i. Then the program will require at least $N \cdot i / m$ cycles to execute.

In this section, we will consider some other factors that limit the performance of programs on actual machines.

11.1 Register Spilling

The benefits of loop parallelism are limited by the ability to express the computation in assembly code. In particular, the IA32 instruction set only has a small number of registers to hold the values being accumulated. If we have a degree of parallelism p that exceeds the number of available registers, then the compiler will resort to *spilling*, storing some of the temporary values on the stack. Once this happens, the performance can drop significantly. As an illustration, compare the performance of our parallel accumulator code for integer sum on x86-64 vs. IA32:

Machine	Degree of unrolling					
	1	2	3	4	5	6
IA32	2.12	1.76	1.45	1.39	1.90	1.99
x86-64	2.00	1.50	1.00	1.00	1.01	1.00

We see that for IA32, the lowest CPE is achieved when just $k = 4$ values are accumulated in parallel, and it gets worse for higher values of k. We also see that we cannot get down to the CPE of 1.00 achieved for x86-64.

Examining the IA32 code for the case of $k = 5$ shows the effect of the small number of registers with IA32:

```
     IA32 code.  Unroll X5, accumulate X5, data_t = int, OP = +
     i in %edx, data in %eax, limit at %ebp-20
1    .L291:                                loop:
2       imull    (%eax,%edx,4), %ecx       x0 = x0 * data[i]
3       movl     -16(%ebp), %ebx           Get x1
4       imull    4(%eax,%edx,4), %ebx      x1 = x1 * data[i+1]
5       movl     %ebx, -16(%ebp)           Store x1
6       imull    8(%eax,%edx,4), %edi      x2 = x2 * data[i+2]
7       imull    12(%eax,%edx,4), %esi     x3 = x3 * data[i+3]
8       movl     -28(%ebp), %ebx           Get x4
9       imull    16(%eax,%edx,4), %ebx     x4 = x4 * daa[i+4]
10      movl     %ebx, -28(%ebp)           Store x4
11      addl     $5, %edx                  i+= 5
12      cmpl     %edx, -20(%ebp)           Compare limit:i
13      jg       .L291                     If >, goto loop
```

We see here that accumulator values acc1 and acc4 have been "spilled" onto the stack, at offsets -16 and -28 relative to %ebp. In addition, the termination value limit is kept on the stack at offset -20. The loads and stores associated with reading these values from memory and then storing them back negates any value obtained by accumulating multiple values in parallel.

We can now see the merit of adding eight additional registers in the extension of IA32 to x86-64. The x86-64 code is able to accumulate up to 12 values in parallel without spilling any registers.

11.2 Branch Prediction and Misprediction Penalties

A conditional branch can incur a significant *misprediction penalty* when the branch prediction logic does not correctly anticipate whether or not a branch will be taken. Now that we have learned something about how processors operate, we can understand where this penalty arises.

Modern processors work well ahead of the currently executing instructions, reading new instructions from memory and decoding them to determine what operations to perform on what operands. This *instruction pipelining* works well as long as the instructions follow in a simple sequence. When a branch is encountered, the processor must guess which way the branch will go. For the case of a conditional jump, this means predicting whether or not the branch will be taken. For an instruction such as an indirect jump (as we saw in the code to jump to an address specified by a jump table entry) or a procedure return, this means predicting the target address. In this discussion, we focus on conditional branches.

In a processor that employs *speculative execution*, the processor begins executing the instructions at the predicted branch target. It does this in a way that avoids modifying any actual register or memory locations until the actual outcome has been determined. If the prediction is correct, the processor can then "commit" the results of the speculatively executed instructions by storing them in registers or memory. If the prediction is incorrect, the processor must discard all of the speculatively executed results and restart the instruction fetch process at the correct location. The misprediction penalty is incurred in doing this, because the instruction pipeline must be refilled before useful results are generated.

Recent versions of x86 processors have *conditional move* instructions and that GCC can generate code that uses these instructions when compiling conditional statements and expressions, rather than the more traditional realizations based on conditional transfers of control. The basic idea for translating into conditional moves is to compute the values along both branches of a conditional expression or statement, and then use conditional moves to select the desired value. Conditional move instructions can be implemented as part of the pipelined processing of ordinary instructions. There is no need to guess whether or not the condition will hold, and hence no penalty for guessing incorrectly.

How then can a C programmer make sure that branch misprediction penalties do not hamper a program's efficiency? Given the 44 clock-cycle misprediction penalty we saw for the Intel Core i7, the stakes are very high. There is no simple answer to this question, but the following general principles apply.

Do Not Be Overly Concerned about Predictable Branches

We have seen that the effect of a mispredicted branch can be very high, but that does not mean that all program branches will slow a program down. In fact, the branch prediction logic found in modern processors is very good at discerning regular patterns and long-term trends for the different branch instructions. For example, the loop-closing branches in our combining routines would typically be predicted as being taken, and hence would only incur a misprediction penalty on the last time around.

As another example, consider the small performance gain we observed when shifting from combine2 to combine3, when we took the function get_vec_element out of the inner loop of the function, as is reproduced below:

Function	Page	Method	Integer		Floating point		
			+	*	+	F *	D *
combine2	14	Move vec_length	8.03	8.09	10.09	11.09	12.08
combine3	19	Direct data access	6.01	8.01	10.01	11.01	12.02

The CPE hardly changed, even though this function uses two conditionals to check whether the vector index is within bounds. These checks always determine that the index is within bounds, and hence they are highly predictable.

As a way to measure the performance impact of bounds checking, consider the following combining code, where we have modified the inner loop of combine4 by replacing the access to the data element with the result of performing an inline substitution of the code for get_vec_element. We will call this new version combine4b. This code performs bounds checking and also references the vector elements through the vector data structure.

```
1    /* Include bounds check in loop */
2    void combine4b(vec_ptr v, data_t *dest)
3    {
4        long int i;
5        long int length = vec_length(v);
6        data_t acc = IDENT;
7
8        for (i = 0; i < length; i++) {
9            if (i >= 0 && i < v->len) {
10               acc = acc OP v->data[i];
11           }
12       }
13       *dest = acc;
14   }
```

We can then directly compare the CPE for the functions with and without bounds checking:

Function	Page	Method	Integer		Floating point		
			+	*	+	F *	D *
combine4	21	No bounds checking	1.00	3.00	3.00	4.00	5.00
combine4b	21	Bounds checking	4.00	4.00	4.00	4.00	5.00

Although the performance of the version with bounds checking is not quite as good, it increases the CPE by at most 2 clock cycles. This is a fairly small difference, considering that the bounds checking code performs two conditional branches and it also requires a load operation to implement the expression v->len. The processor is able to predict the outcomes of these branches, and so none of this evaluation has much effect on the fetching and processing of the instructions that form the critical path in the program execution.

Write Code Suitable for Implementation with Conditional Moves

Branch prediction is only reliable for regular patterns. Many tests in a program are completely unpredictable, dependent on arbitrary features of the data, such as whether a number is negative or positive. For these, the branch prediction logic will do very poorly, possibly giving a prediction rate of 50%—no better than random guessing. (In principle, branch predictors can have prediction rates less than 50%, but such cases are very rare.) For inherently unpredictable cases, program performance can be greatly enhanced if the compiler is able to generate code using conditional data transfers rather than conditional control transfers. This cannot be controlled directly by the C programmer, but some ways of expressing conditional behavior can be more directly translated into conditional moves than others.

We have found that GCC is able to generate conditional moves for code written in a more "functional" style, where we use conditional operations to compute values and then update the program state with these values, as opposed to a more "imperative" style, where we use conditionals to selectively update program state.

There are no strict rules for these two styles, and so we illustrate with an example. Suppose we are given two arrays of integers a and b, and at each position i, we want to set a[i] to the minimum of a[i] and b[i], and b[i] to the maximum.

An imperative style of implementing this function is to check at each position i and swap the two elements if they are out of order:

```
1    /* Rearrange two vectors so that for each i, b[i] >= a[i] */
2    void minmax1(int a[], int b[], int n) {
3        int i;
4        for (i = 0; i < n; i++) {
5            if (a[i] > b[i]) {
6                int t = a[i];
7                a[i] = b[i];
8                b[i] = t;
9            }
10       }
11   }
```

Our measurements for this function on random data show a CPE of around 14.50 for random data, and 3.00–4.00 for predictable data, a clear sign of a high misprediction penalty.

A functional style of implementing this function is to compute the minimum and maximum values at each position *i* and then assign these values to a[*i*] and b[*i*], respectively:

```
1   /* Rearrange two vectors so that for each i, b[i] >= a[i] */
2   void minmax2(int a[], int b[], int n) {
3       int i;
4       for (i = 0; i < n; i++) {
5           int min = a[i] < b[i] ? a[i] : b[i];
6           int max = a[i] < b[i] ? b[i] : a[i];
7           a[i] = min;
8           b[i] = max;
9       }
10  }
```

Our measurements for this function show a CPE of around 5.0 regardless of whether the data are arbitrary or predictable. (We also examined the generated assembly code to make sure that it indeed used conditional moves.)

Not all conditional behavior can be implemented with conditional data transfers, and so there are inevitably cases where programmers cannot avoid writing code that will lead to conditional branches for which the processor will do poorly with its branch prediction. But, as we have shown, a little cleverness on the part of the programmer can sometimes make code more amenable to translation into conditional data transfers. This requires some amount of experimentation, writing different versions of the function and then examining the generated assembly code and measuring performance.

Practice Problem 9

The traditional implementation of the merge step of mergesort requires three loops:

```
1   void merge(int src1[], int src2[], int dest[], int n) {
2       int i1 = 0;
3       int i2 = 0;
4       int id = 0;
5       while (i1 < n && i2 < n) {
6           if (src1[i1] < src2[i2])
7               dest[id++] = src1[i1++];
8           else
9               dest[id++] = src2[i2++];
10      }
11      while (i1 < n)
12          dest[id++] = src1[i1++];
13      while (i2 < n)
14          dest[id++] = src2[i2++];
15  }
```

The branches caused by comparing variables i1 and i2 to n have good prediction performance—the only mispredictions occur when they first become false. The comparison between values src1[i1] and src2[i2] (line 6), on the other hand, is highly unpredictable for typical data. This comparison controls a conditional branch, yielding a CPE (where the number of elements is $2n$) of around 17.50.

Rewrite the code so that the effect of the conditional statement in the first loop (lines 6–9) can be implemented with a conditional move.

---

12 Understanding Memory Performance

All of the code we have written thus far, and all the tests we have run, access relatively small amounts of memory. For example, the combining routines were measured over vectors of length less than 1000 elements, requiring no more than 8000 bytes of data. All modern processors contain one or more *cache* memories to provide fast access to such small amounts of memory. In this section, we will further investigate the performance of programs that involve load (reading from memory into registers) and store (writing from registers to memory) operations, considering only the cases where all data are held in cache.

As Figure 11 shows, modern processors have dedicated functional units to perform load and store operations, and these units have internal buffers to hold sets of outstanding requests for memory operations. For example, the Intel Core i7 load unit's buffer can hold up to 48 read requests, while the store unit's buffer can hold up to 32 write requests [99]. Each of these units can typically initiate one operation every clock cycle.

12.1 Load Performance

The performance of a program containing load operations depends on both the pipelining capability and the latency of the load unit. In our experiments with combining operations on a Core i7, we saw that the CPE never got below 1.00, except when using SIMD operations. One factor limiting the CPE for our examples is that they all require reading one value from memory for each element computed. Since the load unit can only initiate one load operation every clock cycle, the CPE cannot be less than 1.00. For applications where we must load k values for every element computed, we can never achieve a CPE lower than k (see, for example, Problem 17).

In our examples so far, we have not seen any performance effects due to the latency of load operations. The addresses for our load operations depended only on the loop index i, and so the load operations did not form part of a performance-limiting critical path.

To determine the latency of the load operation on a machine, we can set up a computation with a sequence of load operations, where the outcome of one

```
1   typedef struct ELE {
2       struct ELE *next;
3       int data;
4   } list_ele, *list_ptr;
5
6   int list_len(list_ptr ls) {
7       int len = 0;
8       while (ls) {
9           len++;
10          ls = ls->next;
11      }
12      return len;
13  }
```

Figure 31 **Linked list functions.** These illustrate the latency of the load operation.

determines the address for the next. As an example, consider the function `list_len` in Figure 31, which computes the length of a linked list. In the loop of this function, each successive value of variable `ls` depends on the value read by the pointer reference `ls->next`. Our measurements show that function `list_len` has a CPE of 4.00, which we claim is a direct indication of the latency of the load operation. To see this, consider the assembly code for the loop. (We show the x86-64 version of the code. The IA32 code is very similar.)

```
    len in %eax, ls in %rdi
1   .L11:                       loop:
2       addl    $1, %eax        Increment len
3       movq    (%rdi), %rdi    ls = ls->next
4       testq   %rdi, %rdi      Test ls
5       jne     .L11            If nonnull, goto loop
```

The `movq` instruction on line 3 forms the critical bottleneck in this loop. Each successive value of register `%rdi` depends on the result of a load operation having the value in `%rdi` as its address. Thus, the load operation for one iteration cannot begin until the one for the previous iteration has completed. The CPE of 4.00 for this function is determined by the latency of the load operation.

12.2 Store Performance

In all of our examples thus far, we analyzed only functions that reference memory mostly with load operations, reading from a memory location into a register. Its counterpart, the *store* operation, writes a register value to memory. The performance of this operation, particularly in relation to its interactions with load operations, involves several subtle issues.

As with the load operation, in most cases, the store operation can operate in a fully pipelined mode, beginning a new store on every cycle. For example, consider the functions shown in Figure 32 that set the elements of an array dest of length

```
1   /* Set elements of array to 0 */
2   void clear_array(int *dest, int n) {
3       int i;
4       for (i = 0; i < n; i++)
5           dest[i] = 0;
6   }
```

```
1   /* Set elements of array to 0, Unrolled X4 */
2   void clear_array_4(int *dest, int n) {
3       int i;
4       int limit = n-3;
5       for (i = 0; i < limit; i+= 4) {
6           dest[i] = 0;
7           dest[i+1] = 0;
8           dest[i+2] = 0;
9           dest[i+3] = 0;
10      }
11      for (; i < limit; i++)
12          dest[i] = 0;
13  }
```

Figure 32 **Functions to set array elements to 0.** These illustrate the pipelining of the store operation.

n to zero. Our measurements for the first version show a CPE of 2.00. By unrolling the loop four times, as shown in the code for `clear_array_4`, we achieve a CPE of 1.00. Thus, we have achieved the optimum of one new store operation per cycle.

Unlike the other operations we have considered so far, the store operation does not affect any register values. Thus, by their very nature a series of store operations cannot create a data dependency. Only a load operation is affected by the result of a store operation, since only a load can read back the memory value that has been written by the store. The function `write_read` shown in Figure 33 illustrates the potential interactions between loads and stores. This figure also shows two example executions of this function, when it is called for a two-element array a, with initial contents −10 and 17, and with argument cnt equal to 3. These executions illustrate some subtleties of the load and store operations.

In Example A of Figure 33, argument `src` is a pointer to array element a[0], while `dest` is a pointer to array element a[1]. In this case, each load by the pointer reference `*src` will yield the value −10. Hence, after two iterations, the array elements will remain fixed at −10 and −9, respectively. The result of the read from `src` is not affected by the write to `dest`. Measuring this example over a larger number of iterations gives a CPE of 2.00.

In Example B of Figure 33, both arguments `src` and `dest` are pointers to array element a[0]. In this case, each load by the pointer reference `*src` will yield the value stored by the previous execution of the pointer reference `*dest`. As a consequence, a series of ascending values will be stored in this location. In general,

```
1    /* Write to dest, read from src */
2    void write_read(int *src, int *dest, int n)
3    {
4        int cnt = n;
5        int val = 0;
6
7        while (cnt--) {
8            *dest = val;
9            val = (*src)+1;
10       }
11   }
```

Example A: `write_read(&a[0],&a[1],3)`

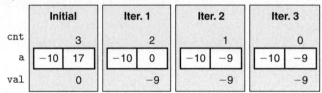

Example B: `write_read(&a[0],&a[0],3)`

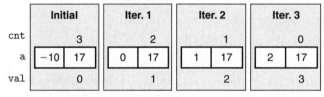

Figure 33 **Code to write and read memory locations, along with illustrative executions.** This function highlights the interactions between stores and loads when arguments src and dest are equal.

if function `write_read` is called with arguments `src` and `dest` pointing to the same memory location, and with argument `cnt` having some value $n > 0$, the net effect is to set the location to $n - 1$. This example illustrates a phenomenon we will call a *write/read dependency*—the outcome of a memory read depends on a recent memory write. Our performance measurements show that Example B has a CPE of 6.00. The write/read dependency causes a slowdown in the processing.

To see how the processor can distinguish between these two cases and why one runs slower than the other, we must take a more detailed look at the load and store execution units, as shown in Figure 34. The store unit contains a *store buffer* containing the addresses and data of the store operations that have been issued to the store unit, but have not yet been completed, where completion involves updating the data cache. This buffer is provided so that a series of store operations can be executed without having to wait for each one to update the cache. When

Figure 34

Detail of load and store units. The store unit maintains a buffer of pending writes. The load unit must check its address with those in the store unit to detect a write/read dependency.

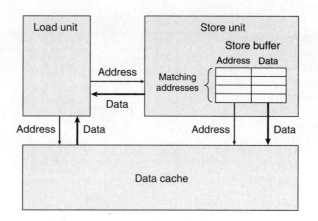

a load operation occurs, it must check the entries in the store buffer for matching addresses. If it finds a match (meaning that any of the bytes being written have the same address as any of the bytes being read), it retrieves the corresponding data entry as the result of the load operation.

Figure 35 shows the assembly code for the inner loop of `write_read`, and a graphical representation of the operations generated by the instruction decoder. The instruction `movl %eax,(%ecx)` is translated into two operations: The `s_addr` instruction computes the address for the store operation, creates an entry in the store buffer, and sets the address field for that entry. The `s_data` operation sets the data field for the entry. As we will see, the fact that these two computations are performed independently can be important to program performance.

In addition to the data dependencies between the operations caused by the writing and reading of registers, the arcs on the right of the operators denote a set of implicit dependencies for these operations. In particular, the address computation of the `s_addr` operation must clearly precede the `s_data` operation. In addition, the load operation generated by decoding the instruction `movl (%ebx)`,

Figure 35

Graphical representation of inner-loop code for `write_read`. The first `movl` instruction is decoded into separate operations to compute the store address and to store the data to memory.

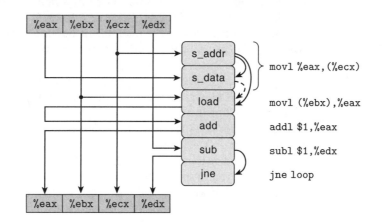

Figure 36

Abstracting the operations for write_read. We first rearrange the operators of Figure 35 (a) and then show only those operations that use values from one iteration to produce new values for the next (b).

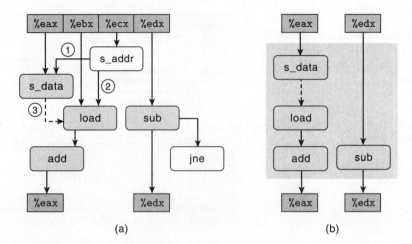

(a)   (b)

%eax must check the addresses of any pending store operations, creating a data dependency between it and the s_addr operation. The figure shows a dashed arc between the s_data and load operations. This dependency is conditional: if the two addresses match, the load operation must wait until the s_data has deposited its result into the store buffer, but if the two addresses differ, the two operations can proceed independently.

Figure 36 illustrates more clearly the data dependencies between the operations for the inner loop of write_read. In Figure 36(a), we have rearranged the operations to allow the dependencies to be seen more clearly. We have labeled the three dependencies involving the load and store operations for special attention. The arc labeled (1) represents the requirement that the store address must be computed before the data can be stored. The arc labeled (2) represents the need for the load operation to compare its address with that for any pending store operations. Finally, the dashed arc labeled (3) represents the conditional data dependency that arises when the load and store addresses match.

Figure 36(b) illustrates what happens when we take away those operations that do not directly affect the flow of data from one iteration to the next. The data-flow graph shows just two chains of dependencies: the one on the left, with data values being stored, loaded, and incremented (only for the case of matching addresses), and the one on the right, decrementing variable cnt.

We can now understand the performance characteristics of function write_ read. Figure 37 illustrates the data dependencies formed by multiple iterations of its inner loop. For the case of Example A of Figure 33, with differing source and destination addresses, the load and store operations can proceed independently, and hence the only critical path is formed by the decrementing of variable cnt. This would lead us to predict a CPE of just 1.00, rather than the measured CPE of 2.00. We have found similar behavior for any function where data are both being stored and loaded within a loop. Apparently the effort to compare load addresses with those of the pending store operations forms an additional bottleneck. For

Figure 37

Data-flow representation of function `write_read`. When the two addresses do not match, the only critical path is formed by the decrementing of `cnt` (Example A). When they do match, the chain of data being stored, loaded, and incremented forms the critical path (Example B).

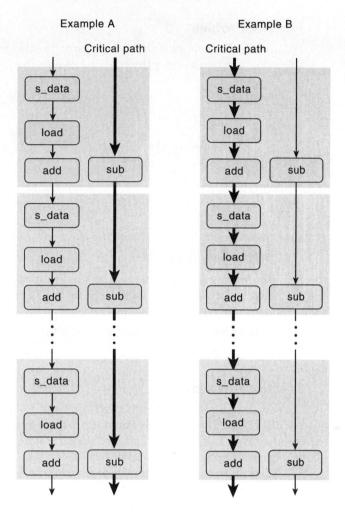

the case of Example B, with matching source and destination addresses, the data dependency between the s_data and load instructions causes a critical path to form involving data being stored, loaded, and incremented. We found that these three operations in sequence require a total of 6 clock cycles.

As these two examples show, the implementation of memory operations involves many subtleties. With operations on registers, the processor can determine which instructions will affect which others as they are being decoded into operations. With memory operations, on the other hand, the processor cannot predict which will affect which others until the load and store addresses have been computed. Efficient handling of memory operations is critical to the performance of many programs. The memory subsystem makes use of many optimizations, such as the potential parallelism when operations can proceed independently.

Practice Problem 10

As another example of code with potential load-store interactions, consider the following function to copy the contents of one array to another:

```
1   void copy_array(int *src, int *dest, int n)
2   {
3       int i;
4       for (i = 0; i < n; i++)
5           dest[i] = src[i];
6   }
```

Suppose a is an array of length 1000 initialized so that each element a[i] equals i.

A. What would be the effect of the call copy_array(a+1,a,999)?

B. What would be the effect of the call copy_array(a,a+1,999)?

C. Our performance measurements indicate that the call of part A has a CPE of 2.00, while the call of part B has a CPE of 5.00. To what factor do you attribute this performance difference?

D. What performance would you expect for the call copy_array(a,a,999)?

Practice Problem 11

We saw that our measurements of the prefix-sum function psum1 (Figure 1) yield a CPE of 10.00 on a machine where the basic operation to be performed, floating-point addition, has a latency of just 3 clock cycles. Let us try to understand why our function performs so poorly.

The following is the assembly code for the inner loop of the function:

```
    psum1.  a in %rdi, p in %rsi, i in %rax, cnt in %rdx
1   .L5:                                        loop:
2       movss    -4(%rsi,%rax,4),  %xmm0        Get p[i-1]
3       addss    (%rdi,%rax,4),  %xmm0          Add a[i]
4       movss    %xmm0, (%rsi,%rax,4)          Store at p[i]
5       addq     $1, %rax                       Increment i
6       cmpq     %rax, %rdx                     Compare cnt:i
7       jg       .L5                            If >, goto loop
```

Perform an analysis similar to those shown for combine3 (Figure 14) and for write_read (Figure 36) to diagram the data dependencies created by this loop, and hence the critical path that forms as the computation proceeds.

Explain why the CPE is so high. (You may not be able to justify the exact CPE, but you should be able to describe why it runs more slowly than one might expect.)

Practice Problem 12

Rewrite the code for psum1 (Figure 1) so that it does not need to repeatedly retrieve the value of p[i] from memory. You do not need to use loop unrolling. We measured the resulting code to have a CPE of 3.00, limited by the latency of floating-point addition.

13 Life in the Real World: Performance Improvement Techniques

Although we have only considered a limited set of applications, we can draw important lessons on how to write efficient code. We have described a number of basic strategies for optimizing program performance:

1. *High-level design.* Choose appropriate algorithms and data structures for the problem at hand. Be especially vigilant to avoid algorithms or coding techniques that yield asymptotically poor performance.

2. *Basic coding principles.* Avoid optimization blockers so that a compiler can generate efficient code.
   - Eliminate excessive function calls. Move computations out of loops when possible. Consider selective compromises of program modularity to gain greater efficiency.
   - Eliminate unnecessary memory references. Introduce temporary variables to hold intermediate results. Store a result in an array or global variable only when the final value has been computed.

3. *Low-level optimizations.*
   - Unroll loops to reduce overhead and to enable further optimizations.
   - Find ways to increase instruction-level parallelism by techniques such as multiple accumulators and reassociation.
   - Rewrite conditional operations in a functional style to enable compilation via conditional data transfers.

A final word of advice to the reader is to be vigilant to avoid introducing errors as you rewrite programs in the interest of efficiency. It is very easy to make mistakes when introducing new variables, changing loop bounds, and making the code more complex overall. One useful technique is to use checking code to test each version of a function as it is being optimized, to ensure no bugs are introduced during this process. Checking code applies a series of tests to the new versions of a function and makes sure they yield the same results as the original. The set of test cases must become more extensive with highly optimized code, since there are more cases to consider. For example, checking code that uses loop unrolling requires testing for many different loop bounds to make sure it handles all of the different possible numbers of single-step iterations required at the end.

14 Identifying and Eliminating Performance Bottlenecks

Up to this point, we have only considered optimizing small programs, where there is some clear place in the program that limits its performance and therefore should be the focus of our optimization efforts. When working with large programs, even knowing where to focus our optimization efforts can be difficult. In this section we describe how to use *code profilers*, analysis tools that collect performance data about a program as it executes. We also present a general principle of system optimization known as *Amdahl's law*.

14.1 Program Profiling

Program *profiling* involves running a version of a program in which instrumentation code has been incorporated to determine how much time the different parts of the program require. It can be very useful for identifying the parts of a program we should focus on in our optimization efforts. One strength of profiling is that it can be performed while running the actual program on realistic benchmark data.

Unix systems provide the profiling program GPROF. This program generates two forms of information. First, it determines how much CPU time was spent for each of the functions in the program. Second, it computes a count of how many times each function gets called, categorized by which function performs the call. Both forms of information can be quite useful. The timings give a sense of the relative importance of the different functions in determining the overall run time. The calling information allows us to understand the dynamic behavior of the program.

Profiling with GPROF requires three steps, as shown for a C program prog.c, which runs with command line argument file.txt:

1. The program must be compiled and linked for profiling. With GCC (and other C compilers) this involves simply including the run-time flag '-pg' on the command line:

   ```
   unix> gcc -O1 -pg prog.c -o prog
   ```

2. The program is then executed as usual:

   ```
   unix> ./prog file.txt
   ```

   It runs slightly (around a factor of 2) slower than normal, but otherwise the only difference is that it generates a file gmon.out.

3. GPROF is invoked to analyze the data in gmon.out.

   ```
   unix> gprof prog
   ```

The first part of the profile report lists the times spent executing the different functions, sorted in descending order. As an example, the following listing shows this part of the report for the three most time-consuming functions in a program:

536

```
%     cumulative   self              self    total
time   seconds    seconds   calls   s/call  s/call  name
97.58   173.05    173.05        1   173.05  173.05  sort_words
 2.36   177.24      4.19   965027     0.00    0.00  find_ele_rec
 0.12   177.46      0.22 12511031     0.00    0.00  Strlen
```

Each row represents the time spent for all calls to some function. The first column indicates the percentage of the overall time spent on the function. The second shows the cumulative time spent by the functions up to and including the one on this row. The third shows the time spent on this particular function, and the fourth shows how many times it was called (not counting recursive calls). In our example, the function sort_words was called only once, but this single call required 173.05 seconds, while the function find_ele_rec was called 965,027 times (not including recursive calls), requiring a total of 4.19 seconds. Function Strlen computes the length of a string by calling the library function strlen. Library function calls are normally not shown in the results by GPROF. Their times are usually reported as part of the function calling them. By creating the "wrapper function" Strlen, we can reliably track the calls to strlen, showing that it was called 12,511,031 times, but only requiring a total of 0.22 seconds.

The second part of the profile report shows the calling history of the functions. The following is the history for a recursive function find_ele_rec:

```
                          158655725                 find_ele_rec [5]
            4.19   0.02   965027/965027              insert_string [4]
[5]   2.4   4.19   0.02   965027+158655725  find_ele_rec [5]
            0.01   0.01   363039/363039              new_ele [10]
            0.00   0.01   363039/363039              save_string [13]
                          158655725                 find_ele_rec [5]
```

This history shows both the functions that called find_ele_rec, as well as the functions that it called. The first two lines show the calls to the function: 158,655,725 calls by itself recursively, and 965,027 calls by function insert_string (which is itself called 965,027 times). Function find_ele_rec in turn called two other functions, save_string and new_ele, each a total of 363,039 times.

From this calling information, we can often infer useful information about the program behavior. For example, the function find_ele_rec is a recursive procedure that scans the linked list for a hash bucket looking for a particular string. For this function, comparing the number of recursive calls with the number of top-level calls provides statistical information about the lengths of the traversals through these lists. Given that their ratio is 164.4, we can infer that the program scanned an average of around 164 elements each time.

Some properties of GPROF are worth noting:

- The timing is not very precise. It is based on a simple *interval counting* scheme in which the compiled program maintains a counter for each function recording the time spent executing that function. The operating system causes the program to be interrupted at some regular time interval δ. Typical values of

δ range between 1.0 and 10.0 milliseconds. It then determines what function the program was executing when the interrupt occurred and increments the counter for that function by δ. Of course, it may happen that this function just started executing and will shortly be completed, but it is assigned the full cost of the execution since the previous interrupt. Some other function may run between two interrupts and therefore not be charged any time at all.

Over a long duration, this scheme works reasonably well. Statistically, every function should be charged according to the relative time spent executing it. For programs that run for less than around 1 second, however, the numbers should be viewed as only rough estimates.

- The calling information is quite reliable. The compiled program maintains a counter for each combination of caller and callee. The appropriate counter is incremented every time a procedure is called.

- By default, the timings for library functions are not shown. Instead, these times are incorporated into the times for the calling functions.

14.2 Using a Profiler to Guide Optimization

As an example of using a profiler to guide program optimization, we created an application that involves several different tasks and data structures. This application analyzes the *n-gram* statistics of a text document, where an n-gram is a sequence of n words occurring in a document. For $n = 1$, we collect statistics on individual words, for $n = 2$ on pairs of words, and so on. For a given value of n, our program reads a text file, creates a table of unique n-grams specifying how many times each one occurs, then sorts the n-grams in descending order of occurrence.

As a benchmark, we ran it on a file consisting of the complete works of William Shakespeare totaling 965,028 words, of which 23,706 are unique. We found that for $n = 1$ even a poorly written analysis program can readily process the entire file in under 1 second, and so we set $n = 2$ to make things more challenging. For the case of $n = 2$, n-grams are referred to as *bigrams* (pronounced "bye-grams"). We determined that Shakespeare's works contain 363,039 unique bigrams. The most common is "I am," occurring 1,892 times. The phrase "to be" occurs 1,020 times. Fully 266,018 of the bigrams occur only once.

Our program consists of the following parts. We created multiple versions, starting with simple algorithms for the different parts and then replacing them with more sophisticated ones:

1. Each word is read from the file and converted to lowercase. Our initial version used the function `lower1` (Figure 7), which we know to have quadratic run time due to repeated calls to `strlen`.

2. A hash function is applied to the string to create a number between 0 and $s - 1$, for a hash table with s buckets. Our initial function simply summed the ASCII codes for the characters modulo s.

3. Each hash bucket is organized as a linked list. The program scans down this list looking for a matching entry. If one is found, the frequency for this n-gram

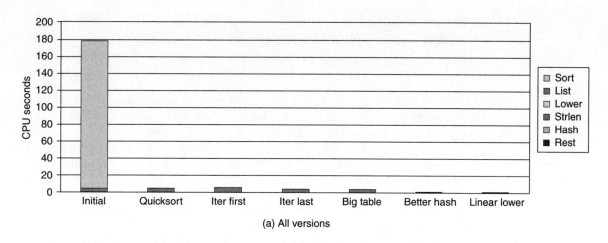

(a) All versions

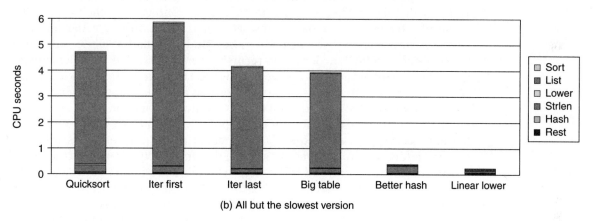

(b) All but the slowest version

Figure 38 **Profile resultss for different versions of n-gram frequency counting program.** Time is divided according to the different major operations in the program.

is incremented. Otherwise, a new list element is created. Our initial version performed this operation recursively, inserting new elements at the end of the list.

4. Once the table has been generated, we sort all of the elements according to the frequencies. Our initial version used insertion sort.

Figure 38 shows the profile results for six different versions of our n-gram-frequency analysis program. For each version, we divide the time into the following categories:

Sort: Sorting n-grams by frequency

List: Scanning the linked list for a matching n-gram, inserting a new element if necessary

Lower: Converting strings to lowercase

Strlen: Computing string lengths

Hash: Computing the hash function

Rest: The sum of all other functions

As part (a) of the figure shows, our initial version required nearly 3 minutes, with most of the time spent sorting. This is not surprising, since insertion sort has quadratic run time, and the program sorted 363,039 values.

In our next version, we performed sorting using the library function qsort, which is based on the quicksort algorithm, having run time $O(n \log n)$. This version is labeled "Quicksort" in the figure. The more efficient sorting algorithm reduces the time spent sorting to become negligible, and the overall run time to around 4.7 seconds. Part (b) of the figure shows the times for the remaining version on a scale where we can see them more clearly.

With improved sorting, we now find that list scanning becomes the bottleneck. Thinking that the inefficiency is due to the recursive structure of the function, we replaced it by an iterative one, shown as "Iter first." Surprisingly, the run time increases to around 5.9 seconds. On closer study, we find a subtle difference between the two list functions. The recursive version inserted new elements at the end of the list, while the iterative one inserted them at the front. To maximize performance, we want the most frequent n-grams to occur near the beginnings of the lists. That way, the function will quickly locate the common cases. Assuming that n-grams are spread uniformly throughout the document, we would expect the first occurrence of a frequent one to come before that of a less frequent one. By inserting new n-grams at the end, the first function tended to order n-grams in descending order of frequency, while the second function tended to do just the opposite. We therefore created a third list-scanning function that uses iteration, but inserts new elements at the end of this list. With this version, shown as "Iter last," the time dropped to around 4.2 seconds, slightly better than with the recursive version. These measurements demonstrate the importance of running experiments on a program as part of an optimization effort. We initially assumed that converting recursive code to iterative code would improve its performance and did not consider the distinction between adding to the end or to the beginning of a list.

Next, we consider the hash table structure. The initial version had only 1021 buckets (typically, the number of buckets is chosen to be a prime number to enhance the ability of the hash function to distribute keys uniformly among the buckets). For a table with 363,039 entries, this would imply an average *load* of $363039/1021 = 355.6$. That explains why so much of the time is spent performing list operations—the searches involve testing a significant number of candidate n-grams. It also explains why the performance is so sensitive to the list ordering. We then increased the number of buckets to 199,999, reducing the average load to 1.8. Oddly enough, however, our overall run time only drops to 3.9 seconds, a difference of only 0.3 seconds.

On further inspection, we can see that the minimal performance gain with a larger table was due to a poor choice of hash function. Simply summing the character codes for a string does not produce a very wide range of values. In particular, the maximum code value for a letter is 122, and so a string of n characters will generate a sum of at most $122n$. The longest bigram in our document, "honorifica-bilitudinitatibus thou," sums to just 3371, and so most of the buckets in our hash table will go unused. In addition, a commutative hash function, such as addition, does not differentiate among the different possible orderings of characters with a string. For example, the words "rat" and "tar" will generate the same sums.

We switched to a hash function that uses shift and Exclusive-Or operations. With this version, shown as "Better hash," the time drops to 0.4 seconds. A more systematic approach would be to study the distribution of keys among the buckets more carefully, making sure that it comes close to what one would expect if the hash function had a uniform output distribution.

Finally, we have reduced the run time to the point where most of the time is spent in `strlen`, and most of the calls to `strlen` occur as part of the lowercase conversion. We have already seen that function `lower1` has quadratic performance, especially for long strings. The words in this document are short enough to avoid the disastrous consequences of quadratic performance; the longest bigram is just 32 characters. Still, switching to `lower2`, shown as "Linear lower," yields a significant performance, with the overall time dropping to around 0.2 seconds.

With this exercise, we have shown that code profiling can help drop the time required for a simple application from nearly 3 minutes down to well under 1 second. The profiler helps us focus our attention on the most time-consuming parts of the program and also provides useful information about the procedure call structure. Some of the bottlenecks in our code, such as using a quadratic sort routine, are easy to anticipate, while others, such as whether to append to the beginning or end of a list, emerge only through a careful analysis.

We can see that profiling is a useful tool to have in the toolbox, but it should not be the only one. The timing measurements are imperfect, especially for shorter (less than 1 second) run times. More significantly, the results apply only to the particular data tested. For example, if we had run the original function on data consisting of a smaller number of longer strings, we would have found that the lowercase conversion routine was the major performance bottleneck. Even worse, if it only profiled documents with short words, we might never detect hidden bottlenecks such as the quadratic performance of `lower1`. In general, profiling can help us optimize for *typical* cases, assuming we run the program on representative data, but we should also make sure the program will have respectable performance for all possible cases. This mainly involves avoiding algorithms (such as insertion sort) and bad programming practices (such as `lower1`) that yield poor asymptotic performance.

14.3 Amdahl's Law

Gene Amdahl, one of the early pioneers in computing, made a simple but insightful observation about the effectiveness of improving the performance of one part of a system. This observation has come to be known as *Amdahl's law*. The main idea is that when we speed up one part of a system, the effect on the overall system performance depends on both how significant this part was and how much it sped up. Consider a system in which executing some application requires time

T_{old}. Suppose some part of the system requires a fraction α of this time, and that we improve its performance by a factor of k. That is, the component originally required time αT_{old}, and it now requires time $(\alpha T_{old})/k$. The overall execution time would thus be

$$T_{new} = (1 - \alpha)T_{old} + (\alpha T_{old})/k$$
$$= T_{old}[(1 - \alpha) + \alpha/k]$$

From this, we can compute the speedup $S = T_{old}/T_{new}$ as

$$S = \frac{1}{(1 - \alpha) + \alpha/k} \tag{4}$$

As an example, consider the case where a part of the system that initially consumed 60% of the time ($\alpha = 0.6$) is sped up by a factor of 3 ($k = 3$). Then we get a speedup of $1/[0.4 + 0.6/3] = 1.67$. Thus, even though we made a substantial improvement to a major part of the system, our net speedup was significantly less. This is the major insight of Amdahl's law—to significantly speed up the entire system, we must improve the speed of a very large fraction of the overall system.

Practice Problem 13

Suppose you work as a truck driver, and you have been hired to carry a load of potatoes from Boise, Idaho, to Minneapolis, Minnesota, a total distance of 2500 kilometers. You estimate you can average 100 km/hr driving within the speed limits, requiring a total of 25 hours for the trip.

A. You hear on the news that Montana has just abolished its speed limit, which constitutes 1500 km of the trip. Your truck can travel at 150 km/hr. What will be your speedup for the trip?

B. You can buy a new turbocharger for your truck at www.fasttrucks.com. They stock a variety of models, but the faster you want to go, the more it will cost. How fast must you travel through Montana to get an overall speedup for your trip of 5/3?

Practice Problem 14

The marketing department at your company has promised your customers that the next software release will show a 2× performance improvement. You have been assigned the task of delivering on that promise. You have determined that only 80% of the system can be improved. How much (i.e., what value of k) would you need to improve this part to meet the overall performance target?

One interesting special case of Amdahl's law is to consider the effect of setting k to ∞. That is, we are able to take some part of the system and speed it up to the point at which it takes a negligible amount of time. We then get

$$S_\infty = \frac{1}{(1 - \alpha)} \tag{5}$$

So, for example, if we can speed up 60% of the system to the point where it requires close to no time, our net speedup will still only be $1/0.4 = 2.5$. We saw this performance with our dictionary program as we replaced insertion sort by quicksort. The initial version spent 173.05 of its 177.57 seconds performing insertion sort, giving $\alpha = 0.975$. With quicksort, the time spent sorting becomes negligible, giving a predicted speedup of 39.3. In fact, the actual measured speedup was a bit less: $173.05/4.72 = 37.6$, due to inaccuracies in the profiling measurements. We were able to gain a large speedup because sorting constituted a very large fraction of the overall execution time.

Amdahl's law describes a general principle for improving any process. In addition to applying to speeding up computer systems, it can guide a company trying to reduce the cost of manufacturing razor blades, or a student trying to improve his or her gradepoint average. Perhaps it is most meaningful in the world of computers, where we routinely improve performance by factors of 2 or more. Such high factors can only be achieved by optimizing large parts of a system.

15 Summary

Although most presentations on code optimization describe how compilers can generate efficient code, much can be done by an application programmer to assist the compiler in this task. No compiler can replace an inefficient algorithm or data structure by a good one, and so these aspects of program design should remain a primary concern for programmers. We also have seen that optimization blockers, such as memory aliasing and procedure calls, seriously restrict the ability of compilers to perform extensive optimizations. Again, the programmer must take primary responsibility for eliminating these. These should simply be considered parts of good programming practice, since they serve to eliminate unneeded work.

Tuning performance beyond a basic level requires some understanding of the processor's microarchitecture, describing the underlying mechanisms by which the processor implements its instruction set architecture. For the case of out-of-order processors, just knowing something about the operations, latencies, and issue times of the functional units establishes a baseline for predicting program performance.

We have studied a series of techniques, including loop unrolling, creating multiple accumulators, and reassociation, that can exploit the instruction-level parallelism provided by modern processors. As we get deeper into the optimization, it becomes important to study the generated assembly code, and to try to understand how the computation is being performed by the machine. Much can be gained by identifying the critical paths determined by the data dependencies in the program, especially between the different iterations of a loop. We can also compute a throughput bound for a computation, based on the number of operations that must be computed and the number and issue times of the units that perform those operations.

Programs that involve conditional branches or complex interactions with the memory system are more difficult to analyze and optimize than the simple loop programs we first considered. The basic strategy is to try to make branches more predictable or make them amenable to implementation using conditional data transfers. We must also watch out for the interactions between store and load operations. Keeping values in local variables, allowing them to be stored in registers, can often be helpful.

When working with large programs, it becomes important to focus our optimization efforts on the parts that consume the most time. Code profilers and related tools can help us systematically evaluate and improve program performance. We described GPROF, a standard Unix profiling tool. More sophisticated profilers are available, such as the VTUNE program development system from Intel, and VALGRIND, commonly available on Linux systems. These tools can break down the execution time below the procedure level, to estimate the performance of each *basic block* of the program. (A basic block is a sequence of instructions that has no transfers of control out of its middle, and so the block is always executed in its entirety.)

Amdahl's law provides a simple but powerful insight into the performance gains obtained by improving just one part of the system. The gain depends both on how much we improve this part and how large a fraction of the overall time this part originally required.

Bibliographic Notes

Our focus has been to describe code optimization from the programmer's perspective, demonstrating how to write code that will make it easier for compilers to generate efficient code. An extended paper by Chellappa, Franchetti, and Püschel [19] takes a similar approach, but goes into more detail with respect to the processor's characteristics.

Many publications describe code optimization from a compiler's perspective, formulating ways that compilers can generate more efficient code. Muchnick's book is considered the most comprehensive [76]. Wadleigh and Crawford's book on software optimization [114] covers some of the material we have presented, but it also describes the process of getting high performance on parallel machines. An early paper by Mahlke et al. [71] describes how several techniques developed for compilers that map programs onto parallel machines can be adapted to exploit the instruction-level parallelism of modern processors. This paper covers the code transformations we presented, including loop unrolling, multiple accumulators (which they refer to as *accumulator variable expansion*), and reassociation (which they refer to as *tree height reduction*).

Our presentation of the operation of an out-of-order processor is fairly brief and abstract. More complete descriptions of the general principles can be found in advanced computer architecture textbooks, such as the one by Hennessy and Patterson [49, Ch. 2–3]. Shen and Lipasti's book [96] provides an in-depth treatment of modern processor design.

Amdahl's law is presented in most books on computer architecture. With its major focus on quantitative system evaluation, Hennessy and Patterson's book [49, Ch. 1] provides a particularly good treatment of the subject.

Homework Problems

15 ◆◆

Suppose we wish to write a procedure that computes the inner product of two vectors u and v. An abstract version of the function has a CPE of 16–17 with x86-64, and 26–29 with IA32 for integer, single-precision, and double-precision data. By doing the same sort of transformations we did to transform the abstract program combine1 into the more efficient combine4, we get the following code:

```
1    /* Accumulate in temporary */
2    void inner4(vec_ptr u, vec_ptr v, data_t *dest)
3    {
4        long int i;
5        int length = vec_length(u);
6        data_t *udata = get_vec_start(u);
7        data_t *vdata = get_vec_start(v);
8        data_t sum = (data_t) 0;
9
10       for (i = 0; i < length; i++) {
11           sum = sum + udata[i] * vdata[i];
12       }
13       *dest = sum;
14   }
```

Our measurements show that this function has a CPE of 3.00 for integer and floating-point data. For data type float, the x86-64 assembly code for the inner loop is as follows:

```
     inner4: data_t = float
     udata in %rbx, vdata in %rax, limit in %rcx,
     i in %rdx, sum in %xmm1
1    .L87:                          loop:
2      movss   (%rbx,%rdx,4), %xmm0   Get udata[i]
3      mulss   (%rax,%rdx,4), %xmm0   Multiply by vdata[i]
4      addss   %xmm0, %xmm1           Add to sum
5      addq    $1, %rdx               Increment i
6      cmpq    %rcx, %rdx             Compare i:limit
7      jl      .L87                   If <, goto loop
```

Assume that the functional units have the characteristics listed in Figure 12.

A. Diagram how this instruction sequence would be decoded into operations and show how the data dependencies between them would create a critical path of operations, in the style of Figures 13 and 14.

B. For data type float, what lower bound on the CPE is determined by the critical path?

C. Assuming similar instruction sequences for the integer code as well, what lower bound on the CPE is determined by the critical path for integer data?

D. Explain how the two floating-point versions can have CPEs of 3.00, even though the multiplication operation requires either 4 or 5 clock cycles.

16 ◆

Write a version of the inner product procedure described in Problem 15 that uses four-way loop unrolling.

For x86-64, our measurements of the unrolled version give a CPE of 2.00 for integer data but still 3.00 for both single and double precision.

A. Explain why any version of any inner product procedure cannot achieve a CPE less than 2.00.

B. Explain why the performance for floating-point data did not improve with loop unrolling.

17 ◆

Write a version of the inner product procedure described in Problem 15 that uses four-way loop unrolling with four parallel accumulators. Our measurements for this function with x86-64 give a CPE of 2.00 for all types of data.

A. What factor limits the performance to a CPE of 2.00?

B. Explain why the version with integer data on IA32 achieves a CPE of 2.75, worse than the CPE of 2.25 achieved with just four-way loop unrolling.

18 ◆

Write a version of the inner product procedure described in Problem 15 that uses four-way loop unrolling along with reassociation to enable greater parallelism. Our measurements for this function give a CPE of 2.00 with x86-64 and 2.25 with IA32 for all types of data.

19 ◆◆

The library function memset has the following prototype:

```
void *memset(void *s, int c, size_t n);
```

This function fills n bytes of the memory area starting at s with copies of the low-order byte of c. For example, it can be used to zero out a region of memory by giving argument 0 for c, but other values are possible.

The following is a straightforward implementation of `memset`:

```
1   /* Basic implementation of memset */
2   void *basic_memset(void *s, int c, size_t n)
3   {
4       size_t cnt = 0;
5       unsigned char *schar = s;
6       while (cnt < n) {
7           *schar++ = (unsigned char) c;
8           cnt++;
9       }
10      return s;
11  }
```

Implement a more efficient version of the function by using a word of data type `unsigned long` to pack four (for IA32) or eight (for x86-64) copies of c, and then step through the region using word-level writes. You might find it helpful to do additional loop unrolling as well. On an Intel Core i7 machine, we were able to reduce the CPE from 2.00 for the straightforward implementation to 0.25 for IA32 and 0.125 for x86-64, i.e., writing either 4 or 8 bytes on every clock cycle.

Here are some additional guidelines. In this discussion, let K denote the value of `sizeof(unsigned long)` for the machine on which you run your program.

- You may not call any library functions.
- Your code should work for arbitrary values of n, including when it is not a multiple of K. You can do this in a manner similar to the way we finish the last few iterations with loop unrolling.
- You should write your code so that it will compile and run correctly regardless of the value of K. Make use of the operation `sizeof` to do this.
- On some machines, unaligned writes can be much slower than aligned ones. (On some non-x86 machines, they can even cause segmentation faults.) Write your code so that it starts with byte-level writes until the destination address is a multiple of K, then do word-level writes, and then (if necessary) finish with byte-level writes.
- Beware of the case where cnt is small enough that the upper bounds on some of the loops become negative. With expressions involving the `sizeof` operator, the testing may be performed with unsigned arithmetic.

20 ◆◆◆

We considered the task of polynomial evaluation in Problems 5 and 6, with both a direct evaluation and an evaluation by Horner's method. Try to write faster versions of the function using the optimization techniques we have explored, including loop unrolling, parallel accumulation, and reassociation. You will find many different ways of mixing together Horner's scheme and direct evaluation with these optimization techniques.

Ideally, you should be able to reach a CPE close to the number of cycles between successive floating-point additions and multiplications with your machine (typically 1). At the very least, you should be able to achieve a CPE less than the latency of floating-point addition for your machine.

21 ◆◆◆

In Problem 12, we were able to reduce the CPE for the prefix-sum computation to 3.00, limited by the latency of floating-point addition on this machine. Simple loop unrolling does not improve things.

Using a combination of loop unrolling and reassociation, write code for prefix sum that achieves a CPE less than the latency of floating-point addition on your machine. Doing this requires actually increasing the number of additions performed. For example, our version with two-way unrolling requires three additions per iteration, while our version with three-way unrolling requires five.

22 ◆

Suppose you are given the task of improving the performance of a program consisting of three parts. Part A requires 20% of the overall run time, part B requires 30%, and part C requires 50%. You determine that for $1000 you could either speed up part B by a factor of 3.0 or part C by a factor of 1.5. Which choice would maximize performance?

Solutions to Practice Problems

Solution to Problem 1

This problem illustrates some of the subtle effects of memory aliasing.

As the following commented code shows, the effect will be to set the value at xp to zero:

```
4        *xp = *xp + *xp;  /* 2x */
5        *xp = *xp - *xp;  /* 2x-2x = 0 */
6        *xp = *xp - *xp;  /* 0-0 = 0 */
```

This example illustrates that our intuition about program behavior can often be wrong. We naturally think of the case where xp and yp are distinct but overlook the possibility that they might be equal. Bugs often arise due to conditions the programmer does not anticipate.

Solution to Problem 2

This problem illustrates the relationship between CPE and absolute performance. It can be solved using elementary algebra. We find that for $n \leq 2$, Version 1 is the fastest. Version 2 is fastest for $3 \leq n \leq 7$, and Version 3 is fastest for $n \geq 8$.

Solution to Problem 3

This is a simple exercise, but it is important to recognize that the four statements of a for loop—initial, test, update, and body—get executed different numbers of times.

Code	min	max	incr	square
A.	1	91	90	90
B.	91	1	90	90
C.	1	1	90	90

Solution to Problem 4

This assembly code demonstrates a clever optimization opportunity detected by GCC. It is worth studying this code carefully to better understand the subtleties of code optimization.

A. In the less optimized code, register %xmm0 is simply used as a temporary value, both set and used on each loop iteration. In the more optimized code, it is used more in the manner of variable x in combine4, accumulating the product of the vector elements. The difference with combine4, however, is that location dest is updated on each iteration by the second movss instruction.

We can see that this optimized version operates much like the following C code:

```
1    /* Make sure dest updated on each iteration */
2    void combine3w(vec_ptr v, data_t *dest)
3    {
4        long int i;
5        long int length = vec_length(v);
6        data_t *data = get_vec_start(v);
7        data_t acc = IDENT;
8
9        for (i = 0; i < length; i++) {
10           acc = acc OP data[i];
11           *dest = acc;
12       }
13   }
```

B. The two versions of combine3 will have identical functionality, even with memory aliasing.

C. This transformation can be made without changing the program behavior, because, with the exception of the first iteration, the value read from dest at the beginning of each iteration will be the same value written to this register at the end of the previous iteration. Therefore, the combining instruction can simply use the value already in %xmm0 at the beginning of the loop.

Solution to Problem 5

Polynomial evaluation is a core technique for solving many problems. For example, polynomial functions are commonly used to approximate trigonometric functions in math libraries.

A. The function performs $2n$ multiplications and n additions.

B. We can see that the performance limiting computation here is the repeated computation of the expression `xpwr = x * xpwr`. This requires a double-precision, floating-point multiplication (5 clock cycles), and the computation for one iteration cannot begin until the one for the previous iteration has completed. The updating of `result` only requires a floating-point addition (3 clock cycles) between successive iterations.

Solution to Problem 6

This problem demonstrates that minimizing the number of operations in a computation may not improve its performance.

A. The function performs n multiplications and n additions, half the number of multiplications as the original function `poly`.

B. We can see that the performance limiting computation here is the repeated computation of the expression `result = a[i] + x*result`. Starting from the value of `result` from the previous iteration, we must first multiply it by `x` (5 clock cycles) and then add it to `a[i]` (3 cycles) before we have the value for this iteration. Thus, each iteration imposes a minimum latency of 8 cycles, exactly our measured CPE.

C. Although each iteration in function `poly` requires two multiplications rather than one, only a single multiplication occurs along the critical path per iteration.

Solution to Problem 7

The following code directly follows the rules we have stated for unrolling a loop by some factor k:

```
1    void unroll5(vec_ptr v, data_t *dest)
2    {
3        long int i;
4        long int length = vec_length(v);
5        long int limit = length-4;
6        data_t *data = get_vec_start(v);
7        data_t acc = IDENT;
8
9        /* Combine 5 elements at a time */
10       for (i = 0; i < limit; i+=5) {
11           acc = acc OP data[i]   OP data[i+1];
12           acc = acc OP data[i+2] OP data[i+3];
13           acc = acc OP data[i+4];
14       }
15
16       /* Finish any remaining elements */
17       for (; i < length; i++) {
18           acc = acc OP data[i];
```

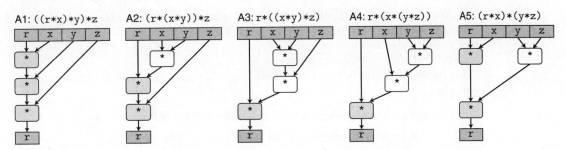

Figure 39 **Data dependencies among multiplication operations for cases in Problem 8.** The operations shown as blue boxes form the critical path for the iteration.

```
19          }
20          *dest = acc;
21      }
```

Solution to Problem 8

This problem demonstrates how small changes in a program can yield dramatic performance differences, especially on a machine with out-of-order execution. Figure 39 diagrams the three multiplication operations for a single iteration of the function. In this figure, the operations shown as blue boxes are along the critical path—they need to be computed in sequence to compute a new value for loop variable r. The operations shown as light boxes can be computed in parallel with the critical path operations. For a loop with c operations along the critical path, each iteration will require a minimum of $5c$ clock cycles and will compute the product for three elements, giving a lower bound on the CPE of $5c/3$. This implies lower bounds of 5.00 for A1, 3.33 for A2 and A5, and 1.67 for A3 and A4.

We ran these functions on an Intel Core i7, and indeed obtained CPEs of 5.00 for A1, and 1.67 for A3 and A4. For some reason, A2 and A5 achieved CPEs of just 3.67, indicating that the functions required 11 clock cycles per iteration rather than the predicted 10.

Solution to Problem 9

This is another demonstration that a slight change in coding style can make it much easier for the compiler to detect opportunities to use conditional moves:

```
while (i1 < n && i2 < n) {
    int v1 = src1[i1];
    int v2 = src2[i2];
    int take1 = v1 < v2;
    dest[id++] = take1 ? v1 : v2;
    i1 += take1;
    i2 += (1-take1);
}
```

We measured a CPE of around 11.50 for this version of the code, a significant improvement over the original CPE of 17.50.

Solution to Problem 10

This problem requires you to analyze the potential load-store interactions in a program.

A. It will set each element $a[i]$ to $i + 1$, for $0 \leq i \leq 998$.

B. It will set each element $a[i]$ to 0, for $1 \leq i \leq 999$.

C. In the second case, the load of one iteration depends on the result of the store from the previous iteration. Thus, there is a write/read dependency between successive iterations. It is interesting to note that the CPE of 5.00 is 1 less than we measured for Example B of function `write_read`. This is due to the fact that `write_read` increments the value before storing it, requiring one clock cycle.

D. It will give a CPE of 2.00, the same as for Example A, since there are no dependencies between stores and subsequent loads.

Solution to Problem 11

We can see that this function has a write/read dependency between successive iterations—the destination value `p[i]` on one iteration matches the source value `p[i-1]` on the next.

Solution to Problem 12

Here is a revised version of the function:

```
1   void psum1a(float a[], float p[], long int n)
2   {
3       long int i;
4       /* last_val holds p[i-1]; val holds p[i] */
5       float last_val, val;
6       last_val = p[0] = a[0];
7       for (i = 1; i < n; i++) {
8           val  = last_val + a[i];
9           p[i] = val;
10          last_val = val;
11      }
12  }
```

We introduce a local variable `last_val`. At the start of iteration `i`, it holds the value of `p[i-1]`. We then compute `val` to be the value of `p[i]` and to be the new value for `last_val`.

This version compiles to the following assembly code:

```
psum1a.  a in %rdi, p in %rsi, i in %rax, cnt in %rdx, last_val in %xmm0
1   .L18:                               loop:
2       addss   (%rdi,%rax,4), %xmm0        last_val = val = last_val + a[i]
```

```
3    movss   %xmm0, (%rsi,%rax,4)        Store val in p[i]
4    addq    $1, %rax                    Increment i
5    cmpq    %rax, %rdx                  Compare cnt:i
6    jg      .L18                        If >, goto loop
```

This code holds `last_val` in `%xmm0`, avoiding the need to read `p[i-1]` from memory, and thus eliminating the write/read dependency seen in psum1.

Solution to Problem 13

This problem illustrates that Amdahl's law applies to more than just computer systems.

A. In terms of Equation 4, we have $\alpha = 0.6$ and $k = 1.5$. More directly, traveling the 1500 kilometers through Montana will require 10 hours, and the rest of the trip also requires 10 hours. This will give a speedup of $25/(10 + 10) = 1.25$.

B. In terms of Equation 4, we have $\alpha = 0.6$, and we require $S = 5/3$, from which we can solve for k. More directly, to speed up the trip by $5/3$, we must decrease the overall time to 15 hours. The parts outside of Montana will still require 10 hours, so we must drive through Montana in 5 hours. This requires traveling at 300 km/hr, which is pretty fast for a truck!

Solution to Problem 14

Amdahl's law is best understood by working through some examples. This one requires you to look at Equation 4 from an unusual perspective.

This problem is a simple application of the equation. You are given $S = 2$ and $\alpha = .8$, and you must then solve for k:

$$2 = \frac{1}{(1 - 0.8) + 0.8/k}$$

$$0.4 + 1.6/k = 1.0$$

$$k = 2.67$$

The Memory Hierarchy

From Chapter 6 of *Computer Systems: A Programmer's Perspective*, Second Edition. Randal E. Bryant and David R. O'Hallaron. Copyright © 2011 by Pearson Education, Inc. Published by Prentice Hall. All rights reserved.

To this point in our study of systems, we have relied on a simple model of a computer system as a CPU that executes instructions and a memory system that holds instructions and data for the CPU. In our simple model, the memory system is a linear array of bytes, and the CPU can access each memory location in a constant amount of time. While this is an effective model as far as it goes, it does not reflect the way that modern systems really work.

In practice, a *memory system* is a hierarchy of storage devices with different capacities, costs, and access times. CPU registers hold the most frequently used data. Small, fast *cache memories* nearby the CPU act as staging areas for a subset of the data and instructions stored in the relatively slow main memory. The main memory stages data stored on large, slow disks, which in turn often serve as staging areas for data stored on the disks or tapes of other machines connected by networks.

Memory hierarchies work because well-written programs tend to access the storage at any particular level more frequently than they access the storage at the next lower level. So the storage at the next level can be slower, and thus larger and cheaper per bit. The overall effect is a large pool of memory that costs as much as the cheap storage near the bottom of the hierarchy, but that serves data to programs at the rate of the fast storage near the top of the hierarchy.

As a programmer, you need to understand the memory hierarchy because it has a big impact on the performance of your applications. If the data your program needs are stored in a CPU register, then they can be accessed in zero cycles during the execution of the instruction. If stored in a cache, 1 to 30 cycles. If stored in main memory, 50 to 200 cycles. And if stored in disk tens of millions of cycles!

Here, then, is a fundamental and enduring idea in computer systems: if you understand how the system moves data up and down the memory hierarchy, then you can write your application programs so that their data items are stored higher in the hierarchy, where the CPU can access them more quickly.

This idea centers around a fundamental property of computer programs known as *locality*. Programs with good locality tend to access the same set of data items over and over again, or they tend to access sets of nearby data items. Programs with good locality tend to access more data items from the upper levels of the memory hierarchy than programs with poor locality, and thus run faster. For example, the running times of different matrix multiplication kernels that perform the same number of arithmetic operations, but have different degrees of locality, can vary by a factor of 20!

In this chapter, we will look at the basic storage technologies—SRAM memory, DRAM memory, ROM memory, and rotating and solid state disks—and describe how they are organized into hierarchies. In particular, we focus on the cache memories that act as staging areas between the CPU and main memory, because they have the most impact on application program performance. We show you how to analyze your C programs for locality and we introduce techniques for improving the locality in your programs. You will also learn an interesting way to characterize the performance of the memory hierarchy on a particular machine as a "memory mountain" that shows read access times as a function of locality.

1 Storage Technologies

Much of the success of computer technology stems from the tremendous progress in storage technology. Early computers had a few kilobytes of random-access memory. The earliest IBM PCs didn't even have a hard disk. That changed with the introduction of the IBM PC-XT in 1982, with its 10-megabyte disk. By the year 2010, typical machines had 150,000 times as much disk storage, and the amount of storage was increasing by a factor of 2 every couple of years.

1.1 Random-Access Memory

Random-access memory (RAM) comes in two varieties—*static* and *dynamic*. *Static RAM* (SRAM) is faster and significantly more expensive than *Dynamic RAM* (DRAM). SRAM is used for cache memories, both on and off the CPU chip. DRAM is used for the main memory plus the frame buffer of a graphics system. Typically, a desktop system will have no more than a few megabytes of SRAM, but hundreds or thousands of megabytes of DRAM.

Static RAM

SRAM stores each bit in a *bistable* memory cell. Each cell is implemented with a six-transistor circuit. This circuit has the property that it can stay indefinitely in either of two different voltage configurations, or *states*. Any other state will be unstable—starting from there, the circuit will quickly move toward one of the stable states. Such a memory cell is analogous to the inverted pendulum illustrated in Figure 1.

The pendulum is stable when it is tilted either all the way to the left or all the way to the right. From any other position, the pendulum will fall to one side or the other. In principle, the pendulum could also remain balanced in a vertical position indefinitely, but this state is *metastable*—the smallest disturbance would make it start to fall, and once it fell it would never return to the vertical position.

Due to its bistable nature, an SRAM memory cell will retain its value indefinitely, as long as it is kept powered. Even when a disturbance, such as electrical noise, perturbs the voltages, the circuit will return to the stable value when the disturbance is removed.

Figure 1
Inverted pendulum.
Like an SRAM cell, the pendulum has only two stable configurations, or *states*.

Stable left   Unstable   Stable right

	Transistors per bit	Relative access time	Persistent?	Sensitive?	Relative cost	Applications
SRAM	6	1×	Yes	No	100×	Cache memory
DRAM	1	10×	No	Yes	1×	Main mem, frame buffers

Figure 2 **Characteristics of DRAM and SRAM memory.**

Dynamic RAM

DRAM stores each bit as charge on a capacitor. This capacitor is very small—typically around 30 femtofarads, that is, 30×10^{-15} farads. Recall, however, that a farad is a very large unit of measure. DRAM storage can be made very dense—each cell consists of a capacitor and a single access transistor. Unlike SRAM, however, a DRAM memory cell is very sensitive to any disturbance. When the capacitor voltage is disturbed, it will never recover. Exposure to light rays will cause the capacitor voltages to change. In fact, the sensors in digital cameras and camcorders are essentially arrays of DRAM cells.

Various sources of leakage current cause a DRAM cell to lose its charge within a time period of around 10 to 100 milliseconds. Fortunately, for computers operating with clock cycle times measured in nanoseconds, this retention time is quite long. The memory system must periodically refresh every bit of memory by reading it out and then rewriting it. Some systems also use error-correcting codes, where the computer words are encoded a few more bits (e.g., a 32-bit word might be encoded using 38 bits), such that circuitry can detect and correct any single erroneous bit within a word.

Figure 2 summarizes the characteristics of SRAM and DRAM memory. SRAM is persistent as long as power is applied. Unlike DRAM, no refresh is necessary. SRAM can be accessed faster than DRAM. SRAM is not sensitive to disturbances such as light and electrical noise. The trade-off is that SRAM cells use more transistors than DRAM cells, and thus have lower densities, are more expensive, and consume more power.

Conventional DRAMs

The cells (bits) in a DRAM chip are partitioned into d *supercells*, each consisting of w DRAM cells. A $d \times w$ DRAM stores a total of dw bits of information. The supercells are organized as a rectangular array with r rows and c columns, where $rc = d$. Each supercell has an address of the form (i, j), where i denotes the row, and j denotes the column.

For example, Figure 3 shows the organization of a 16×8 DRAM chip with $d = 16$ supercells, $w = 8$ bits per supercell, $r = 4$ rows, and $c = 4$ columns. The shaded box denotes the supercell at address $(2, 1)$. Information flows in and out of the chip via external connectors called *pins*. Each pin carries a 1-bit signal. Figure 3 shows two of these sets of pins: eight data pins that can transfer 1 byte in

Figure 3

High-level view of a 128-bit 16 × 8 DRAM chip.

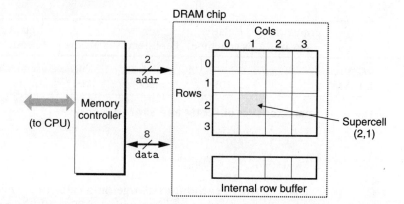

or out of the chip, and two `addr` pins that carry two-bit row and column supercell addresses. Other pins that carry control information are not shown.

Each DRAM chip is connected to some circuitry, known as the *memory controller*, that can transfer w bits at a time to and from each DRAM chip. To read the contents of supercell (i, j), the memory controller sends the row address i to the DRAM, followed by the column address j. The DRAM responds by sending the contents of supercell (i, j) back to the controller. The row address i is called a *RAS (Row Access Strobe) request*. The column address j is called a *CAS (Column Access Strobe) request*. Notice that the RAS and CAS requests share the same DRAM address pins.

For example, to read supercell (2, 1) from the 16 × 8 DRAM in Figure 3, the memory controller sends row address 2, as shown in Figure 4(a). The DRAM responds by copying the entire contents of row 2 into an internal row buffer. Next, the memory controller sends column address 1, as shown in Figure 4(b). The DRAM responds by copying the 8 bits in supercell (2, 1) from the row buffer and sending them to the memory controller.

One reason circuit designers organize DRAMs as two-dimensional arrays instead of linear arrays is to reduce the number of address pins on the chip. For example, if our example 128-bit DRAM were organized as a linear array of 16 supercells with addresses 0 to 15, then the chip would need four address pins instead of two. The disadvantage of the two-dimensional array organization is that addresses must be sent in two distinct steps, which increases the access time.

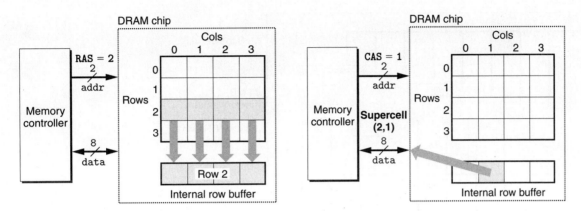

(a) Select row 2 (RAS request).

(b) Select column 1 (CAS request).

Figure 4 **Reading the contents of a DRAM supercell.**

Memory Modules

DRAM chips are packaged in *memory modules* that plug into expansion slots on the main system board (motherboard). Common packages include the 168-pin *dual inline memory module* (DIMM), which transfers data to and from the memory controller in 64-bit chunks, and the 72-pin *single inline memory module* (SIMM), which transfers data in 32-bit chunks.

Figure 5 shows the basic idea of a memory module. The example module stores a total of 64 MB (megabytes) using eight 64-Mbit $8M \times 8$ DRAM chips, numbered 0 to 7. Each supercell stores 1 byte of *main memory*, and each 64-bit doubleword[1] at byte address A in main memory is represented by the eight supercells whose corresponding supercell address is (i, j). In the example in Figure 5, DRAM 0 stores the first (lower-order) byte, DRAM 1 stores the next byte, and so on.

To retrieve a 64-bit doubleword at memory address A, the memory controller converts A to a supercell address (i, j) and sends it to the memory module, which then broadcasts i and j to each DRAM. In response, each DRAM outputs the 8-bit contents of its (i, j) supercell. Circuitry in the module collects these outputs and forms them into a 64-bit doubleword, which it returns to the memory controller.

Main memory can be aggregated by connecting multiple memory modules to the memory controller. In this case, when the controller receives an address A, the controller selects the module k that contains A, converts A to its (i, j) form, and sends (i, j) to module k.

---

1. IA32 would call this 64-bit quantity a "quadword."

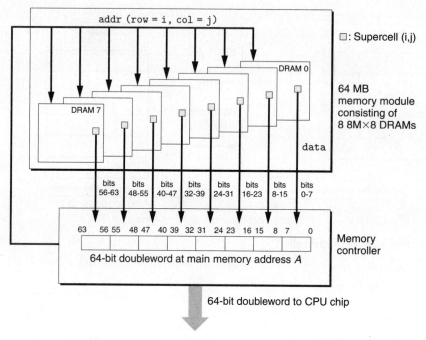

Figure 5 **Reading the contents of a memory module.**

Practice Problem 1

In the following, let r be the number of rows in a DRAM array, c the number of columns, b_r the number of bits needed to address the rows, and b_c the number of bits needed to address the columns. For each of the following DRAMs, determine the power-of-two array dimensions that minimize $\max(b_r, b_c)$, the maximum number of bits needed to address the rows or columns of the array.

Organization	r	c	b_r	b_c	$\max(b_r, b_c)$
16×1					
16×4					
128×8					
512×4					
1024×4					

Enhanced DRAMs

There are many kinds of DRAM memories, and new kinds appear on the market with regularity as manufacturers attempt to keep up with rapidly increasing processor speeds. Each is based on the conventional DRAM cell, with optimizations that improve the speed with which the basic DRAM cells can be accessed.

- *Fast page mode DRAM (FPM DRAM).* A conventional DRAM copies an entire row of supercells into its internal row buffer, uses one, and then discards the rest. FPM DRAM improves on this by allowing consecutive accesses to the same row to be served directly from the row buffer. For example, to read four supercells from row i of a conventional DRAM, the memory controller must send four RAS/CAS requests, even though the row address i is identical in each case. To read supercells from the same row of an FPM DRAM, the memory controller sends an initial RAS/CAS request, followed by three CAS requests. The initial RAS/CAS request copies row i into the row buffer and returns the supercell addressed by the CAS. The next three supercells are served directly from the row buffer, and thus more quickly than the initial supercell.

- *Extended data out DRAM (EDO DRAM).* An enhanced form of FPM DRAM that allows the individual CAS signals to be spaced closer together in time.

- *Synchronous DRAM (SDRAM).* Conventional, FPM, and EDO DRAMs are asynchronous in the sense that they communicate with the memory controller using a set of explicit control signals. SDRAM replaces many of these control signals with the rising edges of the same external clock signal that drives the memory controller. Without going into detail, the net effect is that an SDRAM can output the contents of its supercells at a faster rate than its asynchronous counterparts.

- *Double Data-Rate Synchronous DRAM (DDR SDRAM).* DDR SDRAM is an enhancement of SDRAM that doubles the speed of the DRAM by using both clock edges as control signals. Different types of DDR SDRAMs are characterized by the size of a small prefetch buffer that increases the effective bandwidth: DDR (2 bits), DDR2 (4 bits), and DDR3 (8 bits).

- *Rambus DRAM (RDRAM).* This is an alternative proprietary technology with a higher maximum bandwidth than DDR SDRAM.

- *Video RAM (VRAM).* Used in the frame buffers of graphics systems. VRAM is similar in spirit to FPM DRAM. Two major differences are that (1) VRAM output is produced by shifting the entire contents of the internal buffer in sequence, and (2) VRAM allows concurrent reads and writes to the memory. Thus, the system can be painting the screen with the pixels in the frame buffer (reads) while concurrently writing new values for the next update (writes).

Aside Historical popularity of DRAM technologies

Until 1995, most PCs were built with FPM DRAMs. From 1996 to 1999, EDO DRAMs dominated the market, while FPM DRAMs all but disappeared. SDRAMs first appeared in 1995 in high-end systems, and by 2002 most PCs were built with SDRAMs and DDR SDRAMs. By 2010, most server and desktop systems were built with DDR3 SDRAMs. In fact, the Intel Core i7 supports only DDR3 SDRAM.

Nonvolatile Memory

DRAMs and SRAMs are *volatile* in the sense that they lose their information if the supply voltage is turned off. *Nonvolatile memories*, on the other hand, retain their information even when they are powered off. There are a variety of nonvolatile memories. For historical reasons, they are referred to collectively as *read-only memories* (ROMs), even though some types of ROMs can be written to as well as read. ROMs are distinguished by the number of times they can be reprogrammed (written to) and by the mechanism for reprogramming them.

A *programmable ROM* (PROM) can be programmed exactly once. PROMs include a sort of fuse with each memory cell that can be blown once by zapping it with a high current.

An *erasable programmable ROM* (EPROM) has a transparent quartz window that permits light to reach the storage cells. The EPROM cells are cleared to zeros by shining ultraviolet light through the window. Programming an EPROM is done by using a special device to write ones into the EPROM. An EPROM can be erased and reprogrammed on the order of 1000 times. An *electrically erasable PROM* (EEPROM) is akin to an EPROM, but does not require a physically separate programming device, and thus can be reprogrammed in-place on printed circuit cards. An EEPROM can be reprogrammed on the order of 10^5 times before it wears out.

Flash memory is a type of nonvolatile memory, based on EEPROMs, that has become an important storage technology. Flash memories are everywhere, providing fast and durable nonvolatile storage for a slew of electronic devices, including digital cameras, cell phones, music players, PDAs, and laptop, desktop, and server computer systems. In Section 1.3, we will look in detail at a new form of flash-based disk drive, known as a *solid state disk* (SSD), that provides a faster, sturdier, and less power-hungry alternative to conventional rotating disks.

Programs stored in ROM devices are often referred to as *firmware*. When a computer system is powered up, it runs firmware stored in a ROM. Some systems provide a small set of primitive input and output functions in firmware, for example, a PC's BIOS (basic input/output system) routines. Complicated devices such as graphics cards and disk drive controllers also rely on firmware to translate I/O (input/output) requests from the CPU.

Accessing Main Memory

Data flows back and forth between the processor and the DRAM main memory over shared electrical conduits called *buses*. Each transfer of data between the CPU and memory is accomplished with a series of steps called a *bus transaction*. A *read transaction* transfers data from the main memory to the CPU. A *write transaction* transfers data from the CPU to the main memory.

A *bus* is a collection of parallel wires that carry address, data, and control signals. Depending on the particular bus design, data and address signals can share the same set of wires, or they can use different sets. Also, more than two devices can share the same bus. The control wires carry signals that synchronize the transaction and identify what kind of transaction is currently being performed.

Figure 6

Example bus structure that connects the CPU and main memory.

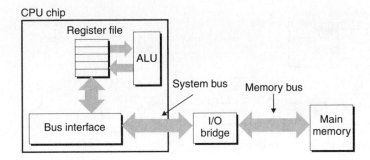

For example, is this transaction of interest to the main memory, or to some other I/O device such as a disk controller? Is the transaction a read or a write? Is the information on the bus an address or a data item?

Figure 6 shows the configuration of an example computer system. The main components are the CPU chip, a chipset that we will call an *I/O bridge* (which includes the memory controller), and the DRAM memory modules that make up main memory. These components are connected by a pair of buses: a *system bus* that connects the CPU to the I/O bridge, and a *memory bus* that connects the I/O bridge to the main memory.

The I/O bridge translates the electrical signals of the system bus into the electrical signals of the memory bus. As we will see, the I/O bridge also connects the system bus and memory bus to an I/O bus that is shared by I/O devices such as disks and graphics cards. For now, though, we will focus on the memory bus.

Aside A note on bus designs

Bus design is a complex and rapidly changing aspect of computer systems. Different vendors develop different bus architectures as a way to differentiate their products. For example, Intel systems use chipsets known as the *northbridge* and the *southbridge* to connect the CPU to memory and I/O devices, respectively. In older Pentium and Core 2 systems, a *front side bus* (FSB) connects the CPU to the northbridge. Systems from AMD replace the FSB with the *HyperTransport* interconnect, while newer Intel Core i7 systems use the *QuickPath* interconnect. The details of these different bus architectures are beyond the scope of this text. Instead, we will use the high-level bus architecture from Figure 6 as a running example throughout the text. It is a simple but useful abstraction that allows us to be concrete, and captures the main ideas without being tied too closely to the detail of any proprietary designs.

Consider what happens when the CPU performs a load operation such as

```
movl A,%eax
```

where the contents of address *A* are loaded into register %eax. Circuitry on the CPU chip called the *bus interface* initiates a read transaction on the bus. The read transaction consists of three steps. First, the CPU places the address *A* on the system bus. The I/O bridge passes the signal along to the memory bus (Figure 7(a)). Next, the main memory senses the address signal on the memory

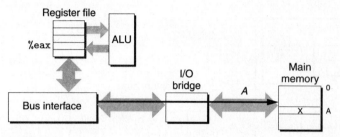

(a) CPU places address *A* on the memory bus.

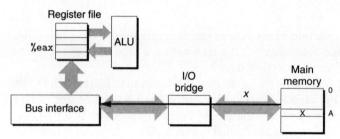

(b) Main memory reads *A* from the bus, retrieves word *x*, and places it on the bus.

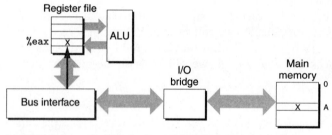

(c) CPU reads word *x* from the bus, and copies it into register %eax.

Figure 7 **Memory read transaction for a load operation:** movl A,%eax.

bus, reads the address from the memory bus, fetches the data word from the DRAM, and writes the data to the memory bus. The I/O bridge translates the memory bus signal into a system bus signal, and passes it along to the system bus (Figure 7(b)). Finally, the CPU senses the data on the system bus, reads it from the bus, and copies it to register %eax (Figure 7(c)).

Conversely, when the CPU performs a store instruction such as

```
movl %eax,A
```

where the contents of register %eax are written to address *A*, the CPU initiates a write transaction. Again, there are three basic steps. First, the CPU places the address on the system bus. The memory reads the address from the memory bus and waits for the data to arrive (Figure 8(a)). Next, the CPU copies the data word in %eax to the system bus (Figure 8(b)). Finally, the main memory reads the data word from the memory bus and stores the bits in the DRAM (Figure 8(c)).

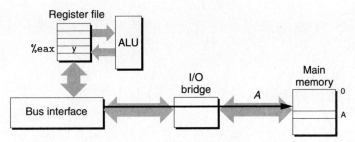

(a) CPU places address *A* on the memory bus. Main memory reads it and waits for the data word.

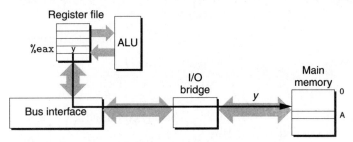

(b) CPU places data word *y* on the bus.

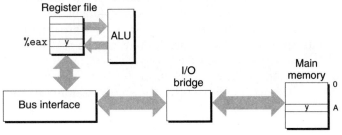

(c) Main memory reads data word *y* from the bus and stores it at address *A*.

Figure 8   **Memory write transaction for a store operation:** `movl %eax,A`.

1.2   Disk Storage

Disks are workhorse storage devices that hold enormous amounts of data, on the order of hundreds to thousands of gigabytes, as opposed to the hundreds or thousands of megabytes in a RAM-based memory. However, it takes on the order of milliseconds to read information from a disk, a hundred thousand times longer than from DRAM and a million times longer than from SRAM.

Disk Geometry

Disks are constructed from *platters*. Each platter consists of two sides, or *surfaces*, that are coated with magnetic recording material. A rotating *spindle* in the center of the platter spins the platter at a fixed *rotational rate*, typically between 5400 and

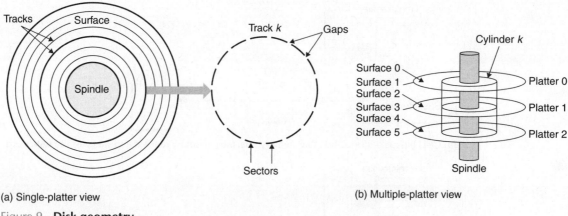

(a) Single-platter view   (b) Multiple-platter view

Figure 9 **Disk geometry.**

15,000 *revolutions per minute* (RPM). A disk will typically contain one or more of these platters encased in a sealed container.

Figure 9(a) shows the geometry of a typical disk surface. Each surface consists of a collection of concentric rings called *tracks*. Each track is partitioned into a collection of *sectors*. Each sector contains an equal number of data bits (typically 512 bytes) encoded in the magnetic material on the sector. Sectors are separated by *gaps* where no data bits are stored. Gaps store formatting bits that identify sectors.

A disk consists of one or more platters stacked on top of each other and encased in a sealed package, as shown in Figure 9(b). The entire assembly is often referred to as a *disk drive*, although we will usually refer to it as simply a *disk*. We will sometime refer to disks as *rotating disks* to distinguish them from flash-based *solid state disks* (SSDs), which have no moving parts.

Disk manufacturers describe the geometry of multiple-platter drives in terms of *cylinders*, where a cylinder is the collection of tracks on all the surfaces that are equidistant from the center of the spindle. For example, if a drive has three platters and six surfaces, and the tracks on each surface are numbered consistently, then cylinder k is the collection of the six instances of track k.

Disk Capacity

The maximum number of bits that can be recorded by a disk is known as its *maximum capacity*, or simply *capacity*. Disk capacity is determined by the following technology factors:

- *Recording density* (*bits/in*): The number of bits that can be squeezed into a 1-inch segment of a track.
- *Track density* (*tracks/in*): The number of tracks that can be squeezed into a 1-inch segment of the radius extending from the center of the platter.

- *Areal density* (*bits/in*^{2}): The product of the recording density and the track density.

Disk manufacturers work tirelessly to increase areal density (and thus capacity), and this is doubling every few years. The original disks, designed in an age of low areal density, partitioned every track into the same number of sectors, which was determined by the number of sectors that could be recorded on the innermost track. To maintain a fixed number of sectors per track, the sectors were spaced farther apart on the outer tracks. This was a reasonable approach when areal densities were relatively low. However, as areal densities increased, the gaps between sectors (where no data bits were stored) became unacceptably large. Thus, modern high-capacity disks use a technique known as *multiple zone recording*, where the set of cylinders is partitioned into disjoint subsets known as *recording zones*. Each zone consists of a contiguous collection of cylinders. Each track in each cylinder in a zone has the same number of sectors, which is determined by the number of sectors that can be packed into the innermost track of the zone. Note that diskettes (floppy disks) still use the old-fashioned approach, with a constant number of sectors per track.

The capacity of a disk is given by the following formula:

$$\text{Disk capacity} = \frac{\text{\# bytes}}{\text{sector}} \times \frac{\text{average \# sectors}}{\text{track}} \times \frac{\text{\# tracks}}{\text{surface}} \times \frac{\text{\# surfaces}}{\text{platter}} \times \frac{\text{\# platters}}{\text{disk}}$$

For example, suppose we have a disk with 5 platters, 512 bytes per sector, 20,000 tracks per surface, and an average of 300 sectors per track. Then the capacity of the disk is:

$$\text{Disk capacity} = \frac{512 \text{ bytes}}{\text{sector}} \times \frac{300 \text{ sectors}}{\text{track}} \times \frac{20,000 \text{ tracks}}{\text{surface}} \times \frac{2 \text{ surfaces}}{\text{platter}} \times \frac{5 \text{ platters}}{\text{disk}}$$

$$= 30,720,000,000 \text{ bytes}$$

$$= 30.72 \text{ GB}.$$

Notice that manufacturers express disk capacity in units of gigabytes (GB), where $1 \text{ GB} = 10^9$ bytes.

Aside How much is a gigabyte?

Unfortunately, the meanings of prefixes such as kilo (K), mega (M), giga (G), and tera (T) depend on the context. For measures that relate to the capacity of DRAMs and SRAMs, typically $K = 2^{10}$, $M = 2^{20}$, $G = 2^{30}$, and $T = 2^{40}$. For measures related to the capacity of I/O devices such as disks and networks, typically $K = 10^3$, $M = 10^6$, $G = 10^9$, and $T = 10^{12}$. Rates and throughputs usually use these prefix values as well.

Fortunately, for the back-of-the-envelope estimates that we typically rely on, either assumption works fine in practice. For example, the relative difference between $2^{20} = 1,048,576$ and $10^6 = 1,000,000$ is small: $(2^{20} - 10^6)/10^6 \approx 5\%$. Similarly for $2^{30} = 1,073,741,824$ and $10^9 = 1,000,000,000$: $(2^{30} - 10^9)/10^9 \approx 7\%$.

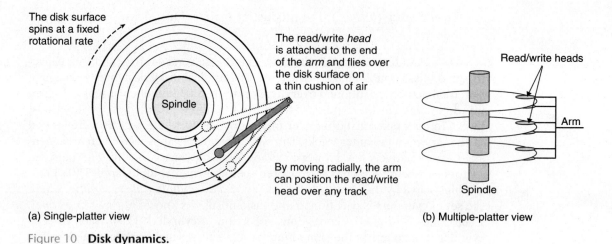

The disk surface spins at a fixed rotational rate

Spindle

The read/write *head* is attached to the end of the *arm* and flies over the disk surface on a thin cushion of air

By moving radially, the arm can position the read/write head over any track

(a) Single-platter view

Read/write heads

Arm

Spindle

(b) Multiple-platter view

Figure 10   **Disk dynamics.**

Practice Problem 2

What is the capacity of a disk with two platters, 10,000 cylinders, an average of 400 sectors per track, and 512 bytes per sector?

Disk Operation

Disks read and write bits stored on the magnetic surface using a *read/write head* connected to the end of an *actuator arm*, as shown in Figure 10(a). By moving the arm back and forth along its radial axis, the drive can position the head over any track on the surface. This mechanical motion is known as a *seek*. Once the head is positioned over the desired track, then as each bit on the track passes underneath, the head can either sense the value of the bit (read the bit) or alter the value of the bit (write the bit). Disks with multiple platters have a separate read/write head for each surface, as shown in Figure 10(b). The heads are lined up vertically and move in unison. At any point in time, all heads are positioned on the same cylinder.

The read/write head at the end of the arm flies (literally) on a thin cushion of air over the disk surface at a height of about 0.1 microns and a speed of about 80 km/h. This is analogous to placing the Sears Tower on its side and flying it around the world at a height of 2.5 cm (1 inch) above the ground, with each orbit of the earth taking only 8 seconds! At these tolerances, a tiny piece of dust on the surface is like a huge boulder. If the head were to strike one of these boulders, the head would cease flying and crash into the surface (a so-called *head crash*). For this reason, disks are always sealed in airtight packages.

Disks read and write data in sector-sized blocks. The *access time* for a sector has three main components: *seek time*, *rotational latency*, and *transfer time*:

- **Seek time:** To read the contents of some target sector, the arm first positions the head over the track that contains the target sector. The time required to move the arm is called the *seek time*. The seek time, T_{seek}, depends on the previous position of the head and the speed that the arm moves across the surface. The average seek time in modern drives, $T_{avg\ seek}$, measured by taking the mean of several thousand seeks to random sectors, is typically on the order of 3 to 9 ms. The maximum time for a single seek, $T_{max\ seek}$, can be as high as 20 ms.

- **Rotational latency:** Once the head is in position over the track, the drive waits for the first bit of the target sector to pass under the head. The performance of this step depends on both the position of the surface when the head arrives at the target sector and the rotational speed of the disk. In the worst case, the head just misses the target sector and waits for the disk to make a full rotation. Thus, the maximum rotational latency, in seconds, is given by

$$T_{max\ rotation} = \frac{1}{\text{RPM}} \times \frac{60\ \text{secs}}{1\ \text{min}}$$

The average rotational latency, $T_{avg\ rotation}$, is simply half of $T_{max\ rotation}$.

- **Transfer time:** When the first bit of the target sector is under the head, the drive can begin to read or write the contents of the sector. The transfer time for one sector depends on the rotational speed and the number of sectors per track. Thus, we can roughly estimate the average transfer time for one sector in seconds as

$$T_{avg\ transfer} = \frac{1}{\text{RPM}} \times \frac{1}{(\text{average \# sectors/track})} \times \frac{60\ \text{secs}}{1\ \text{min}}$$

We can estimate the average time to access the contents of a disk sector as the sum of the average seek time, the average rotational latency, and the average transfer time. For example, consider a disk with the following parameters:

Parameter	Value
Rotational rate	7200 RPM
$T_{avg\ seek}$	9 ms
Average # sectors/track	400

For this disk, the average rotational latency (in ms) is

$$T_{avg\ rotation} = 1/2 \times T_{max\ rotation}$$
$$= 1/2 \times (60\ \text{secs} / 7200\ \text{RPM}) \times 1000\ \text{ms/sec}$$
$$\approx 4\ \text{ms}$$

The average transfer time is

$$T_{avg\ transfer} = 60 / 7200\ \text{RPM} \times 1 / 400\ \text{sectors/track} \times 1000\ \text{ms/sec}$$
$$\approx 0.02\ \text{ms}$$

Putting it all together, the total estimated access time is

$$T_{access} = T_{avg\,seek} + T_{avg\,rotation} + T_{avg\,transfer}$$
$$= 9\ ms + 4\ ms + 0.02\ ms$$
$$= 13.02\ ms$$

This example illustrates some important points:

- The time to access the 512 bytes in a disk sector is dominated by the seek time and the rotational latency. Accessing the first byte in the sector takes a long time, but the remaining bytes are essentially free.
- Since the seek time and rotational latency are roughly the same, twice the seek time is a simple and reasonable rule for estimating disk access time.
- The access time for a doubleword stored in SRAM is roughly 4 ns, and 60 ns for DRAM. Thus, the time to read a 512-byte sector-sized block from memory is roughly 256 ns for SRAM and 4000 ns for DRAM. The disk access time, roughly 10 ms, is about 40,000 times greater than SRAM, and about 2500 times greater than DRAM. The difference in access times is even more dramatic if we compare the times to access a single word.

Practice Problem 3

Estimate the average time (in ms) to access a sector on the following disk:

Parameter	Value
Rotational rate	15,000 RPM
$T_{avg\,seek}$	8 ms
Average # sectors/track	500

Logical Disk Blocks

As we have seen, modern disks have complex geometries, with multiple surfaces and different recording zones on those surfaces. To hide this complexity from the operating system, modern disks present a simpler view of their geometry as a sequence of B sector-sized *logical blocks*, numbered $0, 1, \ldots, B - 1$. A small hardware/firmware device in the disk package, called the *disk controller*, maintains the mapping between logical block numbers and actual (physical) disk sectors.

When the operating system wants to perform an I/O operation such as reading a disk sector into main memory, it sends a command to the disk controller asking it to read a particular logical block number. Firmware on the controller performs a fast table lookup that translates the logical block number into a *(surface, track, sector)* triple that uniquely identifies the corresponding physical sector. Hardware on the controller interprets this triple to move the heads to the appropriate cylinder, waits for the sector to pass under the head, gathers up the bits sensed by the head into a small memory buffer on the controller, and copies them into main memory.

Aside Formatted disk capacity

Before a disk can be used to store data, it must be *formatted* by the disk controller. This involves filling in the gaps between sectors with information that identifies the sectors, identifying any cylinders with surface defects and taking them out of action, and setting aside a set of cylinders in each zone as spares that can be called into action if one or more cylinders in the zone goes bad during the lifetime of the disk. The *formatted capacity* quoted by disk manufacturers is less than the maximum capacity because of the existence of these spare cylinders.

Practice Problem 4

Suppose that a 1 MB file consisting of 512-byte logical blocks is stored on a disk drive with the following characteristics:

Parameter	Value
Rotational rate	10,000 RPM
$T_{avg\ seek}$	5 ms
Average # sectors/track	1000
Surfaces	4
Sector size	512 bytes

For each case below, suppose that a program reads the logical blocks of the file sequentially, one after the other, and that the time to position the head over the first block is $T_{avg\ seek} + T_{avg\ rotation}$.

A. *Best case:* Estimate the optimal time (in ms) required to read the file given the best possible mapping of logical blocks to disk sectors (i.e., sequential).

B. *Random case:* Estimate the time (in ms) required to read the file if blocks are mapped randomly to disk sectors.

Connecting I/O Devices

Input/output (I/O) devices such as graphics cards, monitors, mice, keyboards, and disks are connected to the CPU and main memory using an *I/O bus* such as Intel's *Peripheral Component Interconnect* (PCI) bus. Unlike the system bus and memory buses, which are CPU-specific, I/O buses such as PCI are designed to be independent of the underlying CPU. For example, PCs and Macs both incorporate the PCI bus. Figure 11 shows a typical I/O bus structure (modeled on PCI) that connects the CPU, main memory, and I/O devices.

Although the I/O bus is slower than the system and memory buses, it can accommodate a wide variety of third-party I/O devices. For example, the bus in Figure 11 has three different types of devices attached to it.

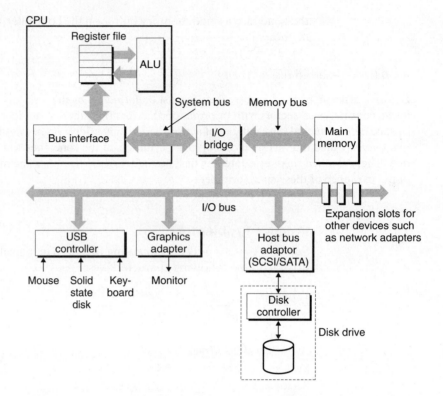

Figure 11

Example bus structure that connects the CPU, main memory, and I/O devices.

- A *Universal Serial Bus* (USB) controller is a conduit for devices attached to a USB bus, which is a wildly popular standard for connecting a variety of peripheral I/O devices, including keyboards, mice, modems, digital cameras, game controllers, printers, external disk drives, and solid state disks. USB 2.0 buses have a maximum bandwidth of 60 MB/s. USB 3.0 buses have a maximum bandwidth of 600 MB/s.

- A *graphics card* (or *adapter*) contains hardware and software logic that is responsible for painting the pixels on the display monitor on behalf of the CPU.

- A *host bus adapter* that connects one or more disks to the I/O bus using a communication protocol defined by a particular *host bus interface*. The two most popular such interfaces for disks are *SCSI* (pronounced "scuzzy") and *SATA* (pronounced "sat-uh"). SCSI disks are typically faster and more expensive than SATA drives. A SCSI host bus adapter (often called a *SCSI controller*) can support multiple disk drives, as opposed to SATA adapters, which can only support one drive.

Additional devices such as *network adapters* can be attached to the I/O bus by plugging the adapter into empty *expansion slots* on the motherboard that provide a direct electrical connection to the bus.

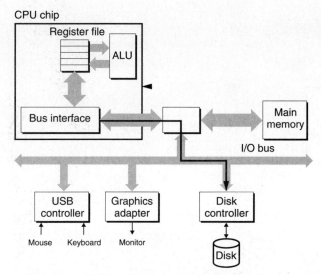

(a) The CPU initiates a disk read by writing a command, logical block number, and destination memory address to the memory-mapped address associated with the disk.

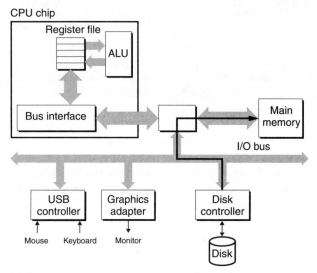

(b) The disk controller reads the sector and performs a DMA transfer into main memory.

Figure 12 **Reading a disk sector.**

Accessing Disks

While a detailed description of how I/O devices work and how they are programmed is outside our scope here, we can give you a general idea. For example, Figure 12 summarizes the steps that take place when a CPU reads data from a disk.

The CPU issues commands to I/O devices using a technique called *memory-mapped I/O* (Figure 12(a)). In a system with memory-mapped I/O, a block of

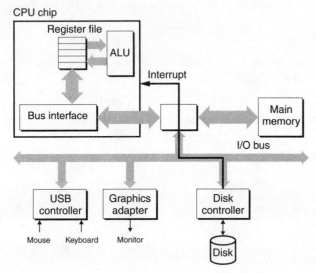

(c) When the DMA transfer is complete, the disk controller notifies the CPU with an interrupt.

Figure 12 *(continued)*   **Reading a disk sector.**

addresses in the address space is reserved for communicating with I/O devices. Each of these addresses is known as an *I/O port*. Each device is associated with (or mapped to) one or more ports when it is attached to the bus.

As a simple example, suppose that the disk controller is mapped to port 0xa0. Then the CPU might initiate a disk read by executing three store instructions to address 0xa0: The first of these instructions sends a command word that tells the disk to initiate a read, along with other parameters such as whether to interrupt the CPU when the read is finished. The second instruction indicates the logical block number that should be read. The third instruction indicates the main memory address where the contents of the disk sector should be stored.

After it issues the request, the CPU will typically do other work while the disk is performing the read. Recall that a 1 GHz processor with a 1 ns clock cycle can potentially execute 16 million instructions in the 16 ms it takes to read the disk. Simply waiting and doing nothing while the transfer is taking place would be enormously wasteful.

After the disk controller receives the read command from the CPU, it translates the logical block number to a sector address, reads the contents of the sector, and transfers the contents directly to main memory, without any intervention from the CPU (Figure 12(b)). This process, whereby a device performs a read or write bus transaction on its own, without any involvement of the CPU, is known as *direct memory access* (DMA). The transfer of data is known as a *DMA transfer*.

After the DMA transfer is complete and the contents of the disk sector are safely stored in main memory, the disk controller notifies the CPU by sending an interrupt signal to the CPU (Figure 12(c)). The basic idea is that an interrupt signals an external pin on the CPU chip. This causes the CPU to stop what it is

Geometry attribute	Value
Platters	4
Surfaces (read/write heads)	8
Surface diameter	3.5 in.
Sector size	512 bytes
Zones	15
Cylinders	50,864
Recording density (max)	628,000 bits/in.
Track density	85,000 tracks/in.
Areal density (max)	53.4 Gbits/sq. in.
Formatted capacity	146.8 GB

Performance attribute	Value
Rotational rate	15,000 RPM
Avg. rotational latency	2 ms
Avg. seek time	4 ms
Sustained transfer rate	58–96 MB/s

Figure 13   **Seagate Cheetah 15K.4 geometry and performance.** Source: `www.seagate.com`.

currently working on and jump to an operating system routine. The routine records the fact that the I/O has finished and then returns control to the point where the CPU was interrupted.

Anatomy of a Commercial Disk

Disk manufacturers publish a lot of useful high-level technical information on their Web pages. For example, the Cheetah 15K.4 is a SCSI disk first manufactured by Seagate in 2005. If we consult the online product manual on the Seagate Web page, we can glean the geometry and performance information shown in Figure 13.

Disk manufacturers rarely publish detailed technical information about the geometry of the individual recording zones. However, storage researchers at Carnegie Mellon University have developed a useful tool, called DIXtrac, that automatically discovers a wealth of low-level information about the geometry and performance of SCSI disks [92]. For example, DIXtrac is able to discover the detailed zone geometry of our example Seagate disk, which we've shown in Figure 14. Each row in the table characterizes one of the 15 zones. The first column gives the zone number, with zone 0 being the outermost and zone 14 the innermost. The second column gives the number of sectors contained in each track in that zone. The third column shows the number of cylinders assigned to that zone, where each cylinder consists of eight tracks, one from each surface. Similarly, the fourth column gives the total number of logical blocks assigned to each zone, across all eight surfaces. (The tool was not able to extract valid data for the innermost zone, so these are omitted.)

The zone map reveals some interesting facts about the Seagate disk. First, more sectors are packed into the outer zones (which have a larger circumference) than the inner zones. Second, each zone has more sectors than logical blocks (check this yourself). These *spare sectors* form a pool of *spare cylinders*. If the recording

Zone number	Sectors per track	Cylinders per zone	Logical blocks per zone
(outer) 0	864	3201	22,076,928
1	844	3200	21,559,136
2	816	3400	22,149,504
3	806	3100	19,943,664
4	795	3100	19,671,480
5	768	3400	20,852,736
6	768	3450	21,159,936
7	725	3650	21,135,200
8	704	3700	20,804,608
9	672	3700	19,858,944
10	640	3700	18,913,280
11	603	3700	17,819,856
12	576	3707	17,054,208
13	528	3060	12,900,096
(inner) 14	—	—	—

Figure 14 **Seagate Cheetah 15K.4 zone map.** Source: DIXtrac automatic disk drive characterization tool [92]. Data for zone 14 not available.

material on a sector goes bad, the disk controller will automatically remap the logical blocks on that cylinder to an available spare. So we see that the notion of a logical block not only provides a simpler interface to the operating system, it also provides a level of indirection that enables the disk to be more robust. This general idea of indirection is very powerful.

Practice Problem 5

Use the zone map in Figure 14 to determine the number of spare cylinders in the following zones:

A. Zone 0

B. Zone 8

1.3 Solid State Disks

A *solid state disk* (SSD) is a storage technology, based on flash memory (Section 1.1), that in some situations is an attractive alternative to the conventional rotating disk. Figure 15 shows the basic idea. An SSD package plugs into a standard disk slot on the I/O bus (typically USB or SATA) and behaves like any other disk,

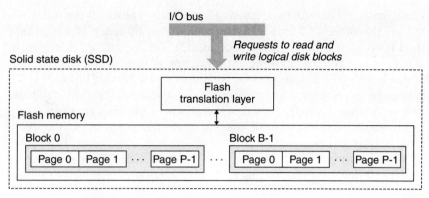

Figure 15 **Solid state disk (SSD).**

	Reads			Writes	
Sequential read throughput	250 MB/s		Sequential write throughput	170 MB/s
Random read throughput	140 MB/s		Random write throughput	14 MB/s
Random read access time	30 μs		Random write access time	300 μs

Figure 16 **Performance characteristics of a typical solid state disk.** Source: Intel X25-E SATA solid state drive product manual.

processing requests from the CPU to read and write logical disk blocks. An SSD package consists of one or more flash memory chips, which replace the mechanical drive in a conventional rotating disk, and a *flash translation layer*, which is a hardware/firmware device that plays the same role as a disk controller, translating requests for logical blocks into accesses of the underlying physical device.

SSDs have different performance characteristics than rotating disks. As shown in Figure 16, sequential reads and writes (where the CPU accesses logical disk blocks in sequential order) have comparable performance, with sequential reading somewhat faster than sequential writing. However, when logical blocks are accessed in random order, writing is an order of magnitude slower than reading.

The difference between random reading and writing performance is caused by a fundamental property of the underlying flash memory. As shown in Figure 15, a flash memory consists of a sequence of *B blocks*, where each block consists of *P* pages. Typically, pages are 512–4KB in size, and a block consists of 32–128 pages, with total block sizes ranging from 16 KB to 512 KB. Data is read and written in units of pages. A page can be written only after the entire block to which it belongs has been *erased* (typically this means that all bits in the block are set to 1). However, once a block is erased, each page in the block can be written once with no further erasing. A blocks wears out after roughly 100,000 repeated writes. Once a block wears out it can no longer be used.

Random writes are slow for two reasons. First, erasing a block takes a relatively long time, on the order of 1 ms, which is more than an order of magnitude longer than it takes to access a page. Second, if a write operation attempts to modify a page p that contains existing data (i.e., not all ones), then any pages in the same block with useful data must be copied to a new (erased) block before the write to page p can occur. Manufacturers have developed sophisticated logic in the flash translation layer that attempts to amortize the high cost of erasing blocks and to minimize the number of internal copies on writes, but it is unlikely that random writing will ever perform as well as reading.

SSDs have a number of advantages over rotating disks. They are built of semiconductor memory, with no moving parts, and thus have much faster random access times than rotating disks, use less power, and are more rugged. However, there are some disadvantages. First, because flash blocks wear out after repeated writes, SSDs have the potential to wear out as well. *Wear leveling* logic in the flash translation layer attempts to maximize the lifetime of each block by spreading erasures evenly across all blocks, but the fundamental limit remains. Second, SSDs are about 100 times more expensive per byte than rotating disks, and thus the typical storage capacities are 100 times less than rotating disks. However, SSD prices are decreasing rapidly as they become more popular, and the gap between the two appears to be decreasing.

SSDs have completely replaced rotating disks in portable music devices, are popular as disk replacements in laptops, and have even begun to appear in desktops and servers. While rotating disks are here to stay, it is clear that SSDs are an important new storage technology.

Practice Problem 6

As we have seen, a potential drawback of SSDs is that the underlying flash memory can wear out. For example, one major manufacturer guarantees 1 petabyte (10^{15} bytes) of random writes for their SSDs before they wear out. Given this assumption, estimate the lifetime (in years) of the SSD in Figure 16 for the following workloads:

A. *Worst case for sequential writes:* The SSD is written to continuously at a rate of 170 MB/s (the average sequential write throughput of the device).

B. *Worst case for random writes:* The SSD is written to continuously at a rate of 14 MB/s (the average random write throughput of the device).

C. *Average case:* The SSD is written to at a rate of 20 GB/day (the average daily write rate assumed by some computer manufacturers in their mobile computer workload simulations).

1.4 Storage Technology Trends

There are several important concepts to take away from our discussion of storage technologies.

Metric	1980	1985	1990	1995	2000	2005	2010	2010:1980
$/MB	19,200	2900	320	256	100	75	60	320
Access (ns)	300	150	35	15	3	2	1.5	200

(a) SRAM trends

Metric	1980	1985	1990	1995	2000	2005	2010	2010:1980
$/MB	8000	880	100	30	1	.1	0.06	130,000
Access (ns)	375	200	100	70	60	50	40	9
Typical size (MB)	0.064	0.256	4	16	64	2000	8,000	125,000

(b) DRAM trends

Metric	1980	1985	1990	1995	2000	2005	2010	2010:1980
$/MB	500	100	8	0.30	0.01	0.005	0.0003	1,600,000
Seek time (ms)	87	75	28	10	8	5	3	29
Typical size (MB)	1	10	160	1000	20,000	160,000	1,500,000	1,500,000

(c) Rotating disk trends

Metric	1980	1985	1990	1995	2000	2003	2005	2010	2010:1980
Intel CPU	8080	80286	80386	Pent.	P-III	Pent. 4	Core 2	Core i7	—
Clock rate (MHz)	1	6	20	150	600	3300	2000	2500	2500
Cycle time (ns)	1000	166	50	6	1.6	0.30	0.50	0.4	2500
Cores	1	1	1	1	1	1	2	4	4
Eff. cycle time (ns)	1000	166	50	6	1.6	0.30	0.25	0.10	10,000

(d) CPU trends

Figure 17 **Storage and processing technology trends.**

Different storage technologies have different price and performance trade-offs. SRAM is somewhat faster than DRAM, and DRAM is much faster than disk. On the other hand, fast storage is always more expensive than slower storage. SRAM costs more per byte than DRAM. DRAM costs much more than disk. SSDs split the difference between DRAM and rotating disk.

The price and performance properties of different storage technologies are changing at dramatically different rates. Figure 17 summarizes the price and performance properties of storage technologies since 1980, when the first PCs were introduced. The numbers were culled from back issues of trade magazines and the Web. Although they were collected in an informal survey, the numbers reveal some interesting trends.

Since 1980, both the cost and performance of SRAM technology have improved at roughly the same rate. Access times have decreased by a factor of about 200 and cost per megabyte by a factor of 300 (Figure 17(a)). However, the trends for DRAM and disk are much more dramatic and divergent. While the cost per megabyte of DRAM has decreased by a factor of 130,000 (more than five orders of magnitude!), DRAM access times have decreased by only a factor of 10 or so (Figure 17(b)). Disk technology has followed the same trend as DRAM and in even more dramatic fashion. While the cost of a megabyte of disk storage has plummeted by a factor of more than 1,000,000 (more than six orders of magnitude!) since 1980, access times have improved much more slowly, by only a factor of 30 or so (Figure 17(c)). These startling long-term trends highlight a basic truth of memory and disk technology: it is easier to increase density (and thereby reduce cost) than to decrease access time.

DRAM and disk performance are lagging behind CPU performance. As we see in Figure 17(d), CPU cycle times improved by a factor of 2500 between 1980 and 2010. If we look at the *effective cycle time*—which we define to be the cycle time of an individual CPU (processor) divided by the number of its processor cores—then the improvement between 1980 and 2010 is even greater, a factor of 10,000. The split in the CPU performance curve around 2003 reflects the introduction of multi-core processors (see aside on next page). After this split, cycle times of individual cores actually increased a bit before starting to decrease again, albeit at a slower rate than before.

Note that while SRAM performance lags, it is roughly keeping up. However, the gap between DRAM and disk performance and CPU performance is actually widening. Until the advent of multi-core processors around 2003, this performance gap was a function of latency, with DRAM and disk access times increasing more slowly than the cycle time of an individual processor. However, with the introduction of multiple cores, this performance gap is increasingly a function of throughput, with multiple processor cores issuing requests to the DRAM and disk in parallel.

The various trends are shown quite clearly in Figure 18, which plots the access and cycle times from Figure 17 on a semi-log scale.

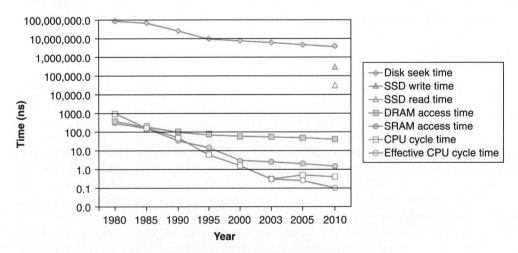

Figure 18 **The increasing gap between disk, DRAM, and CPU speeds.**

As we will see in Section 4, modern computers make heavy use of SRAM-based caches to try to bridge the processor-memory gap. This approach works because of a fundamental property of application programs known as *locality*, which we discuss next.

Aside   When cycle time stood still: the advent of multi-core processors

The history of computers is marked by some singular events that caused profound changes in the industry and the world. Interestingly, these inflection points tend to occur about once per decade: the development of Fortran in the 1950s, the introduction of the IBM 360 in the early 1960s, the dawn of the Internet (then called ARPANET) in the early 1970s, the introduction of the IBM PC in the early 1980s, and the creation of the World Wide Web in the early 1990s.

The most recent such event occurred early in the 21st century, when computer manufacturers ran headlong into the so-called "power wall," discovering that they could no longer increase CPU clock frequencies as quickly because the chips would then consume too much power. The solution was to improve performance by replacing a single large processor with multiple smaller processor *cores*, each a complete processor capable of executing programs independently and in parallel with the other cores. This *multi-core* approach works in part because the power consumed by a processor is proportional to $P = fCv^2$, where f is the clock frequency, C is the capacitance, and v is the voltage. The capacitance C is roughly proportional to the area, so the power drawn by multiple cores can be held constant as long as the total area of the cores is constant. As long as feature sizes continue to shrink at the exponential Moore's law rate, the number of cores in each processor, and thus its effective performance, will continue to increase.

From this point forward, computers will get faster not because the clock frequency increases, but because the number of cores in each processor increases, and because architectural innovations increase the efficiency of programs running on those cores. We can see this trend clearly in Figure 18. CPU cycle time reached its lowest point in 2003 and then actually started to rise before leveling off and starting to decline again at a slower rate than before. However, because of the advent of multi-core processors (dual-core in 2004 and quad-core in 2007), the effective cycle time continues to decrease at close to its previous rate.

Practice Problem 7

Using the data from the years 2000 to 2010 in Figure 17(c), estimate the year when you will be able to buy a petabyte (10^{15} bytes) of rotating disk storage for $500. Assume constant dollars (no inflation).

---

2   Locality

Well-written computer programs tend to exhibit good *locality*. That is, they tend to reference data items that are near other recently referenced data items, or that were recently referenced themselves. This tendency, known as the *principle of locality*, is an enduring concept that has enormous impact on the design and performance of hardware and software systems.

Locality is typically described as having two distinct forms: *temporal locality* and *spatial locality*. In a program with good temporal locality, a memory location that is referenced once is likely to be referenced again multiple times in the near future. In a program with good spatial locality, if a memory location is referenced once, then the program is likely to reference a nearby memory location in the near future.

Programmers should understand the principle of locality because, in general, *programs with good locality run faster than programs with poor locality*. All levels of modern computer systems, from the hardware, to the operating system, to application programs, are designed to exploit locality. At the hardware level, the principle of locality allows computer designers to speed up main memory accesses by introducing small fast memories known as *cache memories* that hold blocks of the most recently referenced instructions and data items. At the operating system level, the principle of locality allows the system to use the main memory as a cache of the most recently referenced chunks of the virtual address space. Similarly, the operating system uses main memory to cache the most recently used disk blocks in the disk file system. The principle of locality also plays a crucial role in the design of application programs. For example, Web browsers exploit temporal locality by caching recently referenced documents on a local disk. High-volume Web servers hold recently requested documents in front-end disk caches that satisfy requests for these documents without requiring any intervention from the server.

2.1 Locality of References to Program Data

Consider the simple function in Figure 19(a) that sums the elements of a vector. Does this function have good locality? To answer this question, we look at the reference pattern for each variable. In this example, the sum variable is referenced once in each loop iteration, and thus there is good temporal locality with respect to sum. On the other hand, since sum is a scalar, there is no spatial locality with respect to sum.

As we see in Figure 19(b), the elements of vector v are read sequentially, one after the other, in the order they are stored in memory (we assume for convenience that the array starts at address 0). Thus, with respect to variable v, the function has good spatial locality but poor temporal locality since each vector element

```
1   int sumvec(int v[N])
2   {
3       int i, sum = 0;
4
5       for (i = 0; i < N; i++)
6           sum += v[i];
7       return sum;
8   }
```
(a)

Address	0	4	8	12	16	20	24	28
Contents	v_0	v_1	v_2	v_3	v_4	v_5	v_6	v_7
Access order	1	2	3	4	5	6	7	8

(b)

Figure 19 **(a) A function with good locality. (b) Reference pattern for vector** v **($N = 8$).** Notice how the vector elements are accessed in the same order that they are stored in memory.

```
1    int sumarrayrows(int a[M][N])
2    {
3        int i, j, sum = 0;
4
5        for (i = 0; i < M; i++)
6            for (j = 0; j < N; j++)
7                sum += a[i][j];
8        return sum;
9    }
```

(a)

Address	0	4	8	12	16	20
Contents	a_{00}	a_{01}	a_{02}	a_{10}	a_{11}	a_{12}
Access order	1	2	3	4	5	6

(b)

Figure 20 **(a) Another function with good locality. (b) Reference pattern for array** a **($M = 2$, $N = 3$).** There is good spatial locality because the array is accessed in the same row-major order in which it is stored in memory.

is accessed exactly once. Since the function has either good spatial or temporal locality with respect to each variable in the loop body, we can conclude that the sumvec function enjoys good locality.

A function such as sumvec that visits each element of a vector sequentially is said to have a *stride-1 reference pattern* (with respect to the element size). We will sometimes refer to stride-1 reference patterns as *sequential reference patterns*. Visiting every kth element of a contiguous vector is called a *stride-k reference pattern*. Stride-1 reference patterns are a common and important source of spatial locality in programs. In general, as the stride increases, the spatial locality decreases.

Stride is also an important issue for programs that reference multidimensional arrays. For example, consider the sumarrayrows function in Figure 20(a) that sums the elements of a two-dimensional array. The doubly nested loop reads the elements of the array in *row-major order*. That is, the inner loop reads the elements of the first row, then the second row, and so on. The sumarrayrows function enjoys good spatial locality because it references the array in the same row-major order that the array is stored (Figure 20(b)). The result is a nice stride-1 reference pattern with excellent spatial locality.

Seemingly trivial changes to a program can have a big impact on its locality. For example, the sumarraycols function in Figure 21(a) computes the same result as the sumarrayrows function in Figure 20(a). The only difference is that we have interchanged the i and j loops. What impact does interchanging the loops have on its locality? The sumarraycols function suffers from poor spatial locality because it scans the array column-wise instead of row-wise. Since C arrays are laid out in memory row-wise, the result is a stride-N reference pattern, as shown in Figure 21(b).

2.2 Locality of Instruction Fetches

Since program instructions are stored in memory and must be fetched (read) by the CPU, we can also evaluate the locality of a program with respect to its instruction fetches. For example, in Figure 19 the instructions in the body of the

```
1    int sumarraycols(int a[M][N])
2    {
3        int i, j, sum = 0;
4
5        for (j = 0; j < N; j++)
6            for (i = 0; i < M; i++)
7                sum += a[i][j];
8        return sum;
9    }
```

(a)

Address	0	4	8	12	16	20
Contents	a_{00}	a_{01}	a_{02}	a_{10}	a_{11}	a_{12}
Access order	1	3	5	2	4	6

(b)

Figure 21 **(a) A function with poor spatial locality. (b) Reference pattern for array a ($M = 2$, $N = 3$). The function has poor spatial locality because it scans memory with a stride-N reference pattern.**

for loop are executed in sequential memory order, and thus the loop enjoys good spatial locality. Since the loop body is executed multiple times, it also enjoys good temporal locality.

An important property of code that distinguishes it from program data is that it is rarely modified at run time. While a program is executing, the CPU reads its instructions from memory. The CPU rarely overwrites or modifies these instructions.

2.3 Summary of Locality

In this section, we have introduced the fundamental idea of locality and have identified some simple rules for qualitatively evaluating the locality in a program:

- Programs that repeatedly reference the same variables enjoy good temporal locality.

- For programs with stride-k reference patterns, the smaller the stride the better the spatial locality. Programs with stride-1 reference patterns have good spatial locality. Programs that hop around memory with large strides have poor spatial locality.

- Loops have good temporal and spatial locality with respect to instruction fetches. The smaller the loop body and the greater the number of loop iterations, the better the locality.

Later in this chapter, after we have learned about cache memories and how they work, we will show you how to quantify the idea of locality in terms of cache hits and misses. It will also become clear to you why programs with good locality typically run faster than programs with poor locality. Nonetheless, knowing how to glance at a source code and getting a high-level feel for the locality in the program is a useful and important skill for a programmer to master.

Practice Problem 8

Permute the loops in the following function so that it scans the three-dimensional array *a* with a stride-1 reference pattern.

```
1    int sumarray3d(int a[N][N][N])
2    {
3        int i, j, k, sum = 0;
4
5        for (i = 0; i < N; i++) {
6            for (j = 0; j < N; j++) {
7                for (k = 0; k < N; k++) {
8                    sum += a[k][i][j];
9                }
10           }
11       }
12       return sum;
13   }
```

Practice Problem 9

The three functions in Figure 22 perform the same operation with varying degrees of spatial locality. Rank-order the functions with respect to the spatial locality enjoyed by each. Explain how you arrived at your ranking.

(a) An array of structs

```
1    #define N 1000
2
3    typedef struct {
4        int vel[3];
5        int acc[3];
6    } point;
7
8    point p[N];
```

(b) The clear1 function

```
1    void clear1(point *p, int n)
2    {
3        int i, j;
4
5        for (i = 0; i < n; i++) {
6            for (j = 0; j < 3; j++)
7                p[i].vel[j] = 0;
8            for (j = 0; j < 3; j++)
9                p[i].acc[j] = 0;
10       }
11   }
```

Figure 22   **Code examples for Practice Problem 9.**

(c) The clear2 function

```
1    void clear2(point *p, int n)
2    {
3        int i, j;
4
5        for (i = 0; i < n; i++) {
6            for (j = 0; j < 3; j++) {
7                p[i].vel[j] = 0;
8                p[i].acc[j] = 0;
9            }
10       }
11   }
```

(d) The clear3 function

```
1    void clear3(point *p, int n)
2    {
3        int i, j;
4
5        for (j = 0; j < 3; j++) {
6            for (i = 0; i < n; i++)
7                p[i].vel[j] = 0;
8            for (i = 0; i < n; i++)
9                p[i].acc[j] = 0;
10       }
11   }
```

Figure 22 *(continued)* **Code examples for Practice Problem 9.**

3 The Memory Hierarchy

Sections 1 and 2 described some fundamental and enduring properties of storage technology and computer software:

- *Storage technology:* Different storage technologies have widely different access times. Faster technologies cost more per byte than slower ones and have less capacity. The gap between CPU and main memory speed is widening.
- *Computer software:* Well-written programs tend to exhibit good locality.

In one of the happier coincidences of computing, these fundamental properties of hardware and software complement each other beautifully. Their complementary nature suggests an approach for organizing memory systems, known as the *memory hierarchy*, that is used in all modern computer systems. Figure 23 shows a typical memory hierarchy. In general, the storage devices get slower, cheaper, and larger as we move from higher to lower *levels*. At the highest level (L0) are a small number of fast CPU registers that the CPU can access in a single clock cycle. Next are one or more small to moderate-sized SRAM-based cache memories that can be accessed in a few CPU clock cycles. These are followed by a large DRAM-based main memory that can be accessed in tens to hundreds of clock cycles. Next are slow but enormous local disks. Finally, some systems even include an additional level of disks on remote servers that can be accessed over a network. For example, distributed file systems such as the Andrew File System (AFS) or the Network File System (NFS) allow a program to access files that are stored on remote network-connected servers. Similarly, the World Wide Web allows programs to access remote files stored on Web servers anywhere in the world.

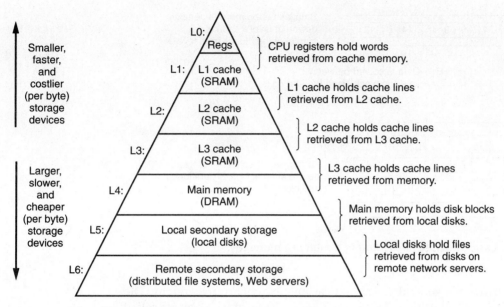

Figure 23 **The memory hierarchy.**

3.1 Caching in the Memory Hierarchy

In general, a *cache* (pronounced "cash") is a small, fast storage device that acts as a staging area for the data objects stored in a larger, slower device. The process of using a cache is known as *caching* (pronounced "cashing").

The central idea of a memory hierarchy is that for each k, the faster and smaller storage device at level k serves as a cache for the larger and slower storage device at level $k + 1$. In other words, each level in the hierarchy caches data objects from the next lower level. For example, the local disk serves as a cache for files (such as Web pages) retrieved from remote disks over the network, the main memory serves as a cache for data on the local disks, and so on, until we get to the smallest cache of all, the set of CPU registers.

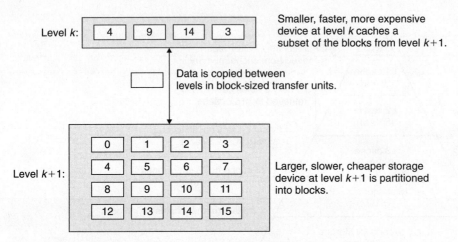

Figure 24 **The basic principle of caching in a memory hierarchy.**

Figure 24 shows the general concept of caching in a memory hierarchy. The storage at level $k + 1$ is partitioned into contiguous chunks of data objects called *blocks*. Each block has a unique address or name that distinguishes it from other blocks. Blocks can be either fixed-sized (the usual case) or variable-sized (e.g., the remote HTML files stored on Web servers). For example, the level $k + 1$ storage in Figure 24 is partitioned into 16 fixed-sized blocks, numbered 0 to 15.

Similarly, the storage at level k is partitioned into a smaller set of blocks that are the same size as the blocks at level $k + 1$. At any point in time, the cache at level k contains copies of a subset of the blocks from level $k + 1$. For example, in Figure 24, the cache at level k has room for four blocks and currently contains copies of blocks 4, 9, 14, and 3.

Data is always copied back and forth between level k and level $k + 1$ in block-sized *transfer units*. It is important to realize that while the block size is fixed between any particular pair of adjacent levels in the hierarchy, other pairs of levels can have different block sizes. For example, in Figure 23, transfers between L1 and L0 typically use one-word blocks. Transfers between L2 and L1 (and L3 and L2, and L4 and L3) typically use blocks of 8 to 16 words. And transfers between L5 and L4 use blocks with hundreds or thousands of bytes. In general, devices lower in the hierarchy (further from the CPU) have longer access times, and thus tend to use larger block sizes in order to amortize these longer access times.

Cache Hits

When a program needs a particular data object d from level $k + 1$, it first looks for d in one of the blocks currently stored at level k. If d happens to be cached at level k, then we have what is called a *cache hit*. The program reads d directly from level k, which by the nature of the memory hierarchy is faster than reading d from level $k + 1$. For example, a program with good temporal locality might read a data object from block 14, resulting in a cache hit from level k.

Cache Misses

If, on the other hand, the data object d is not cached at level k, then we have what is called a *cache miss*. When there is a miss, the cache at level k fetches the block containing d from the cache at level $k + 1$, possibly overwriting an existing block if the level k cache is already full.

This process of overwriting an existing block is known as *replacing* or *evicting* the block. The block that is evicted is sometimes referred to as a *victim block*. The decision about which block to replace is governed by the cache's *replacement policy*. For example, a cache with a *random replacement policy* would choose a random victim block. A cache with a least-recently used (LRU) replacement policy would choose the block that was last accessed the furthest in the past.

After the cache at level k has fetched the block from level $k + 1$, the program can read d from level k as before. For example, in Figure 24, reading a data object from block 12 in the level k cache would result in a cache miss because block 12 is not currently stored in the level k cache. Once it has been copied from level $k + 1$ to level k, block 12 will remain there in expectation of later accesses.

Kinds of Cache Misses

It is sometimes helpful to distinguish between different kinds of cache misses. If the cache at level k is empty, then any access of any data object will miss. An empty cache is sometimes referred to as a *cold cache*, and misses of this kind are called *compulsory misses* or *cold misses*. Cold misses are important because they are often transient events that might not occur in steady state, after the cache has been *warmed up* by repeated memory accesses.

Whenever there is a miss, the cache at level k must implement some *placement policy* that determines where to place the block it has retrieved from level $k + 1$. The most flexible placement policy is to allow any block from level $k + 1$ to be stored in any block at level k. For caches high in the memory hierarchy (close to the CPU) that are implemented in hardware and where speed is at a premium, this policy is usually too expensive to implement because randomly placed blocks are expensive to locate.

Thus, hardware caches typically implement a more restricted placement policy that restricts a particular block at level $k + 1$ to a small subset (sometimes a singleton) of the blocks at level k. For example, in Figure 24, we might decide that a block i at level $k + 1$ must be placed in block (i mod 4) at level k. For example, blocks $0, 4, 8$, and 12 at level $k + 1$ would map to block 0 at level k; blocks $1, 5, 9$, and 13 would map to block 1; and so on. Notice that our example cache in Figure 24 uses this policy.

Restrictive placement policies of this kind lead to a type of miss known as a *conflict miss*, in which the cache is large enough to hold the referenced data objects, but because they map to the same cache block, the cache keeps missing. For example, in Figure 24, if the program requests block 0, then block 8, then block 0, then block 8, and so on, each of the references to these two blocks would miss in the cache at level k, even though this cache can hold a total of four blocks.

Programs often run as a sequence of phases (e.g., loops) where each phase accesses some reasonably constant set of cache blocks. For example, a nested loop might access the elements of the same array over and over again. This set of blocks is called the *working set* of the phase. When the size of the working set exceeds the size of the cache, the cache will experience what are known as *capacity misses*. In other words, the cache is just too small to handle this particular working set.

Cache Management

As we have noted, the essence of the memory hierarchy is that the storage device at each level is a cache for the next lower level. At each level, some form of logic must *manage* the cache. By this we mean that something has to partition the cache storage into blocks, transfer blocks between different levels, decide when there are hits and misses, and then deal with them. The logic that manages the cache can be hardware, software, or a combination of the two.

For example, the compiler manages the register file, the highest level of the cache hierarchy. It decides when to issue loads when there are misses, and determines which register to store the data in. The caches at levels L1, L2, and L3 are managed entirely by hardware logic built into the caches. In a system with virtual memory, the DRAM main memory serves as a cache for data blocks stored on disk, and is managed by a combination of operating system software and address translation hardware on the CPU. For a machine with a distributed file system such as AFS, the local disk serves as a cache that is managed by the AFS client process running on the local machine. In most cases, caches operate automatically and do not require any specific or explicit actions from the program.

3.2 Summary of Memory Hierarchy Concepts

To summarize, memory hierarchies based on caching work because slower storage is cheaper than faster storage and because programs tend to exhibit locality:

- *Exploiting temporal locality.* Because of temporal locality, the same data objects are likely to be reused multiple times. Once a data object has been copied into the cache on the first miss, we can expect a number of subsequent hits on that object. Since the cache is faster than the storage at the next lower level, these subsequent hits can be served much faster than the original miss.

- *Exploiting spatial locality.* Blocks usually contain multiple data objects. Because of spatial locality, we can expect that the cost of copying a block after a miss will be amortized by subsequent references to other objects within that block.

Caches are used everywhere in modern systems. As you can see from Figure 25, caches are used in CPU chips, operating systems, distributed file systems, and on the World Wide Web. They are built from and managed by various combinations of hardware and software. Note that there are a number of terms and acronyms in Figure 25 that we haven't covered yet. We include them here to demonstrate how common caches are.

Type	What cached	Where cached	Latency (cycles)	Managed by
CPU registers	4-byte or 8-byte word	On-chip CPU registers	0	Compiler
TLB	Address translations	On-chip TLB	0	Hardware MMU
L1 cache	64-byte block	On-chip L1 cache	1	Hardware
L2 cache	64-byte block	On/off-chip L2 cache	10	Hardware
L3 cache	64-byte block	On/off-chip L3 cache	30	Hardware
Virtual memory	4-KB page	Main memory	100	Hardware + OS
Buffer cache	Parts of files	Main memory	100	OS
Disk cache	Disk sectors	Disk controller	100,000	Controller firmware
Network cache	Parts of files	Local disk	10,000,000	AFS/NFS client
Browser cache	Web pages	Local disk	10,000,000	Web browser
Web cache	Web pages	Remote server disks	1,000,000,000	Web proxy server

Figure 25 **The ubiquity of caching in modern computer systems.** Acronyms: TLB: translation lookaside buffer, MMU: memory management unit, OS: operating system, AFS: Andrew File System, NFS: Network File System.

4 Cache Memories

The memory hierarchies of early computer systems consisted of only three levels: CPU registers, main DRAM memory, and disk storage. However, because of the increasing gap between CPU and main memory, system designers were compelled to insert a small SRAM *cache memory*, called an *L1 cache* (Level 1 cache) between the CPU register file and main memory, as shown in Figure 26. The L1 cache can be accessed nearly as fast as the registers, typically in 2 to 4 clock cycles.

As the performance gap between the CPU and main memory continued to increase, system designers responded by inserting an additional larger cache, called an *L2 cache*, between the L1 cache and main memory, that can be accessed in about 10 clock cycles. Some modern systems include an additional even larger cache, called an *L3 cache*, which sits between the L2 cache and main memory

Figure 26
Typical bus structure for cache memories.

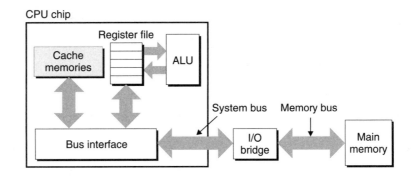

591

in the memory hierarchy and can be accessed in 30 or 40 cycles. While there is considerable variety in the arrangements, the general principles are the same. For our discussion in the next section, we will assume a simple memory hierarchy with a single L1 cache between the CPU and main memory.

4.1 Generic Cache Memory Organization

Consider a computer system where each memory address has m bits that form $M = 2^m$ unique addresses. As illustrated in Figure 27(a), a cache for such a machine is organized as an array of $S = 2^s$ *cache sets*. Each set consists of E *cache lines*. Each line consists of a data *block* of $B = 2^b$ bytes, a *valid bit* that indicates whether or not the line contains meaningful information, and $t = m - (b + s)$ *tag bits* (a subset of the bits from the current block's memory address) that uniquely identify the block stored in the cache line.

In general, a cache's organization can be characterized by the tuple (S, E, B, m). The size (or capacity) of a cache, C, is stated in terms of the aggregate size of all the blocks. The tag bits and valid bit are not included. Thus, $C = S \times E \times B$.

When the CPU is instructed by a load instruction to read a word from address A of main memory, it sends the address A to the cache. If the cache is holding a copy of the word at address A, it sends the word immediately back to the CPU. So how

Figure 27

General organization of cache (S, E, B, m). (a) A cache is an array of sets. Each set contains one or more lines. Each line contains a valid bit, some tag bits, and a block of data. (b) The cache organization induces a partition of the m address bits into t tag bits, s set index bits, and b block offset bits.

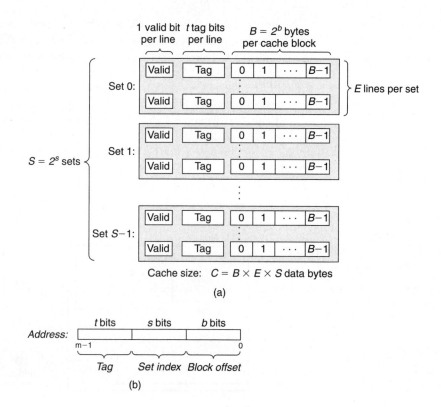

Fundamental parameters	
Parameter	Description
$S = 2^s$	Number of sets
E	Number of lines per set
$B = 2^b$	Block size (bytes)
$m = \log_2(M)$	Number of physical (main memory) address bits

Derived quantities	
Parameter	Description
$M = 2^m$	Maximum number of unique memory addresses
$s = \log_2(S)$	Number of *set index bits*
$b = \log_2(B)$	Number of *block offset bits*
$t = m - (s + b)$	Number of *tag bits*
$C = B \times E \times S$	Cache size (bytes) not including overhead such as the valid and tag bits

Figure 28 **Summary of cache parameters.**

does the cache know whether it contains a copy of the word at address A? The cache is organized so that it can find the requested word by simply inspecting the bits of the address, similar to a hash table with an extremely simple hash function. Here is how it works:

The parameters S and B induce a partitioning of the m address bits into the three fields shown in Figure 27(b). The s *set index bits* in A form an index into the array of S sets. The first set is set 0, the second set is set 1, and so on. When interpreted as an unsigned integer, the set index bits tell us which set the word must be stored in. Once we know which set the word must be contained in, the t tag bits in A tell us which line (if any) in the set contains the word. A line in the set contains the word if and only if the valid bit is set and the tag bits in the line match the tag bits in the address A. Once we have located the line identified by the tag in the set identified by the set index, then the b *block offset bits* give us the offset of the word in the B-byte data block.

As you may have noticed, descriptions of caches use a lot of symbols. Figure 28 summarizes these symbols for your reference.

Practice Problem 10

The following table gives the parameters for a number of different caches. For each cache, determine the number of cache sets (S), tag bits (t), set index bits (s), and block offset bits (b).

Cache	m	C	B	E	S	t	s	b
1.	32	1024	4	1				
2.	32	1024	8	4				
3.	32	1024	32	32				

4.2 Direct-Mapped Caches

Caches are grouped into different classes based on E, the number of cache lines per set. A cache with exactly one line per set ($E = 1$) is known as a *direct-mapped* cache (see Figure 29). Direct-mapped caches are the simplest both to implement and to understand, so we will use them to illustrate some general concepts about how caches work.

Suppose we have a system with a CPU, a register file, an L1 cache, and a main memory. When the CPU executes an instruction that reads a memory word w, it requests the word from the L1 cache. If the L1 cache has a cached copy of w, then we have an L1 cache hit, and the cache quickly extracts w and returns it to the CPU. Otherwise, we have a cache miss, and the CPU must wait while the L1 cache requests a copy of the block containing w from the main memory. When the requested block finally arrives from memory, the L1 cache stores the block in one of its cache lines, extracts word w from the stored block, and returns it to the CPU. The process that a cache goes through of determining whether a request is a hit or a miss and then extracting the requested word consists of three steps: (1) *set selection*, (2) *line matching*, and (3) *word extraction*.

Set Selection in Direct-Mapped Caches

In this step, the cache extracts the s set index bits from the middle of the address for w. These bits are interpreted as an unsigned integer that corresponds to a set number. In other words, if we think of the cache as a one-dimensional array of sets, then the set index bits form an index into this array. Figure 30 shows how set selection works for a direct-mapped cache. In this example, the set index bits 00001_2 are interpreted as an integer index that selects set 1.

Line Matching in Direct-Mapped Caches

Now that we have selected some set i in the previous step, the next step is to determine if a copy of the word w is stored in one of the cache lines contained in

Figure 29
Direct-mapped cache
($E = 1$). There is exactly one line per set.

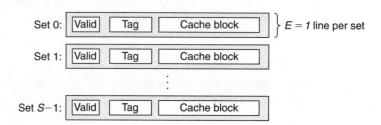

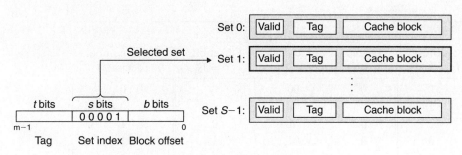

Figure 30 **Set selection in a direct-mapped cache.**

set i. In a direct-mapped cache, this is easy and fast because there is exactly one line per set. A copy of w is contained in the line if and only if the valid bit is set and the tag in the cache line matches the tag in the address of w.

Figure 31 shows how line matching works in a direct-mapped cache. In this example, there is exactly one cache line in the selected set. The valid bit for this line is set, so we know that the bits in the tag and block are meaningful. Since the tag bits in the cache line match the tag bits in the address, we know that a copy of the word we want is indeed stored in the line. In other words, we have a cache hit. On the other hand, if either the valid bit were not set or the tags did not match, then we would have had a cache miss.

Word Selection in Direct-Mapped Caches

Once we have a hit, we know that w is somewhere in the block. This last step determines where the desired word starts in the block. As shown in Figure 31, the block offset bits provide us with the offset of the first byte in the desired word. Similar to our view of a cache as an array of lines, we can think of a block as an array of bytes, and the byte offset as an index into that array. In the example, the block offset bits of 100_2 indicate that the copy of w starts at byte 4 in the block. (We are assuming that words are 4 bytes long.)

Line Replacement on Misses in Direct-Mapped Caches

If the cache misses, then it needs to retrieve the requested block from the next level in the memory hierarchy and store the new block in one of the cache lines of

Figure 31

Line matching and word selection in a direct-mapped cache. Within the cache block, w_0 denotes the low-order byte of the word w, w_1 the next byte, and so on.

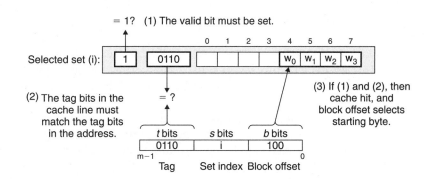

| | Address bits | | | |
Address (decimal)	Tag bits ($t = 1$)	Index bits ($s = 2$)	Offset bits ($b = 1$)	Block number (decimal)
0	0	00	0	0
1	0	00	1	0
2	0	01	0	1
3	0	01	1	1
4	0	10	0	2
5	0	10	1	2
6	0	11	0	3
7	0	11	1	3
8	1	00	0	4
9	1	00	1	4
10	1	01	0	5
11	1	01	1	5
12	1	10	0	6
13	1	10	1	6
14	1	11	0	7
15	1	11	1	7

Figure 32 **4-bit address space for example direct-mapped cache.**

the set indicated by the set index bits. In general, if the set is full of valid cache lines, then one of the existing lines must be evicted. For a direct-mapped cache, where each set contains exactly one line, the replacement policy is trivial: the current line is replaced by the newly fetched line.

Putting It Together: A Direct-Mapped Cache in Action

The mechanisms that a cache uses to select sets and identify lines are extremely simple. They have to be, because the hardware must perform them in a few nanoseconds. However, manipulating bits in this way can be confusing to us humans. A concrete example will help clarify the process. Suppose we have a direct-mapped cache described by

$$(S, E, B, m) = (4, 1, 2, 4)$$

In other words, the cache has four sets, one line per set, 2 bytes per block, and 4-bit addresses. We will also assume that each word is a single byte. Of course, these assumptions are totally unrealistic, but they will help us keep the example simple.

When you are first learning about caches, it can be very instructive to enumerate the entire address space and partition the bits, as we've done in Figure 32 for our 4-bit example. There are some interesting things to notice about this enumerated space:

- The concatenation of the tag and index bits uniquely identifies each block in memory. For example, block 0 consists of addresses 0 and 1, block 1 consists of addresses 2 and 3, block 2 consists of addresses 4 and 5, and so on.

- Since there are eight memory blocks but only four cache sets, multiple blocks map to the same cache set (i.e., they have the same set index). For example, blocks 0 and 4 both map to set 0, blocks 1 and 5 both map to set 1, and so on.

- Blocks that map to the same cache set are uniquely identified by the tag. For example, block 0 has a tag bit of 0 while block 4 has a tag bit of 1, block 1 has a tag bit of 0 while block 5 has a tag bit of 1, and so on.

Let us simulate the cache in action as the CPU performs a sequence of reads. Remember that for this example, we are assuming that the CPU reads 1-byte words. While this kind of manual simulation is tedious and you may be tempted to skip it, in our experience students do not really understand how caches work until they work their way through a few of them.

Initially, the cache is empty (i.e., each valid bit is zero):

Set	Valid	Tag	block[0]	block[1]
0	0			
1	0			
2	0			
3	0			

Each row in the table represents a cache line. The first column indicates the set that the line belongs to, but keep in mind that this is provided for convenience and is not really part of the cache. The next three columns represent the actual bits in each cache line. Now, let us see what happens when the CPU performs a sequence of reads:

1. **Read word at address 0.** Since the valid bit for set 0 is zero, this is a cache miss. The cache fetches block 0 from memory (or a lower-level cache) and stores the block in set 0. Then the cache returns m[0] (the contents of memory location 0) from block[0] of the newly fetched cache line.

Set	Valid	Tag	block[0]	block[1]
0	1	0	m[0]	m[1]
1	0			
2	0			
3	0			

2. **Read word at address 1.** This is a cache hit. The cache immediately returns m[1] from block[1] of the cache line. The state of the cache does not change.

3. **Read word at address 13.** Since the cache line in set 2 is not valid, this is a cache miss. The cache loads block 6 into set 2 and returns m[13] from block[1] of the new cache line.

Set	Valid	Tag	block[0]	block[1]
0	1	0	m[0]	m[1]
1	0			
2	1	1	m[12]	m[13]
3	0			

4. Read word at address 8. This is a miss. The cache line in set 0 is indeed valid, but the tags do not match. The cache loads block 4 into set 0 (replacing the line that was there from the read of address 0) and returns m[8] from block[0] of the new cache line.

Set	Valid	Tag	block[0]	block[1]
0	1	1	m[8]	m[9]
1	0			
2	1	1	m[12]	m[13]
3	0			

5. Read word at address 0. This is another miss, due to the unfortunate fact that we just replaced block 0 during the previous reference to address 8. This kind of miss, where we have plenty of room in the cache but keep alternating references to blocks that map to the same set, is an example of a conflict miss.

Set	Valid	Tag	block[0]	block[1]
0	1	0	m[0]	m[1]
1	0			
2	1	1	m[12]	m[13]
3	0			

Conflict Misses in Direct-Mapped Caches

Conflict misses are common in real programs and can cause baffling performance problems. Conflict misses in direct-mapped caches typically occur when programs access arrays whose sizes are a power of 2. For example, consider a function that computes the dot product of two vectors:

```
1   float dotprod(float x[8], float y[8])
2   {
3       float sum = 0.0;
4       int i;
5
6       for (i = 0; i < 8; i++)
7           sum += x[i] * y[i];
8       return sum;
9   }
```

This function has good spatial locality with respect to x and y, and so we might expect it to enjoy a good number of cache hits. Unfortunately, this is not always true.

Suppose that floats are 4 bytes, that x is loaded into the 32 bytes of contiguous memory starting at address 0, and that y starts immediately after x at address 32. For simplicity, suppose that a block is 16 bytes (big enough to hold four floats) and that the cache consists of two sets, for a total cache size of 32 bytes. We will assume that the variable sum is actually stored in a CPU register and thus does not require a memory reference. Given these assumptions, each x[i] and y[i] will map to the identical cache set:

Element	Address	Set index	Element	Address	Set index
x[0]	0	0	y[0]	32	0
x[1]	4	0	y[1]	36	0
x[2]	8	0	y[2]	40	0
x[3]	12	0	y[3]	44	0
x[4]	16	1	y[4]	48	1
x[5]	20	1	y[5]	52	1
x[6]	24	1	y[6]	56	1
x[7]	28	1	y[7]	60	1

At run time, the first iteration of the loop references x[0], a miss that causes the block containing x[0]–x[3] to be loaded into set 0. The next reference is to y[0], another miss that causes the block containing y[0]–y[3] to be copied into set 0, overwriting the values of x that were copied in by the previous reference. During the next iteration, the reference to x[1] misses, which causes the x[0]–x[3] block to be loaded back into set 0, overwriting the y[0]–y[3] block. So now we have a conflict miss, and in fact each subsequent reference to x and y will result in a conflict miss as we *thrash* back and forth between blocks of x and y. The term *thrashing* describes any situation where a cache is repeatedly loading and evicting the same sets of cache blocks.

The bottom line is that even though the program has good spatial locality and we have room in the cache to hold the blocks for both x[i] and y[i], each reference results in a conflict miss because the blocks map to the same cache set. It is not unusual for this kind of thrashing to result in a slowdown by a factor of 2 or 3. Also, be aware that even though our example is extremely simple, the problem is real for larger and more realistic direct-mapped caches.

Luckily, thrashing is easy for programmers to fix once they recognize what is going on. One easy solution is to put *B* bytes of padding at the end of each array. For example, instead of defining x to be float x[8], we define it to be float x[12]. Assuming y starts immediately after x in memory, we have the following mapping of array elements to sets:

Element	Address	Set index	Element	Address	Set index
x[0]	0	0	y[0]	48	1
x[1]	4	0	y[1]	52	1
x[2]	8	0	y[2]	56	1
x[3]	12	0	y[3]	60	1
x[4]	16	1	y[4]	64	0
x[5]	20	1	y[5]	68	0
x[6]	24	1	y[6]	72	0
x[7]	28	1	y[7]	76	0

With the padding at the end of x, x[i] and y[i] now map to different sets, which eliminates the thrashing conflict misses.

Practice Problem 11

In the previous dotprod example, what fraction of the total references to x and y will be hits once we have padded array x?

Practice Problem 12

In general, if the high-order s bits of an address are used as the set index, contiguous chunks of memory blocks are mapped to the same cache set.

A. How many blocks are in each of these contiguous array chunks?

B. Consider the following code that runs on a system with a cache of the form $(S, E, B, m) = (512, 1, 32, 32)$:

```
int array[4096];

for (i = 0; i < 4096; i++)
    sum += array[i];
```

What is the maximum number of array blocks that are stored in the cache at any point in time?

Aside   Why index with the middle bits?

You may be wondering why caches use the middle bits for the set index instead of the high-order bits. There is a good reason why the middle bits are better. Figure 33 shows why. If the high-order bits are used as an index, then some contiguous memory blocks will map to the same cache set. For example, in the figure, the first four blocks map to the first cache set, the second four blocks map to the second set, and so on. If a program has good spatial locality and scans the elements of an array sequentially, then the cache can only hold a block-sized chunk of the array at any point in time. This is an inefficient use of the cache. Contrast this with middle-bit indexing, where adjacent blocks always map to different cache lines. In this case, the cache can hold an entire C-sized chunk of the array, where C is the cache size.

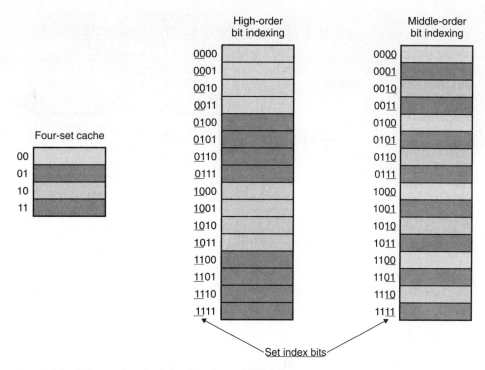

Figure 33 **Why caches index with the middle bits.**

4.3 Set Associative Caches

The problem with conflict misses in direct-mapped caches stems from the constraint that each set has exactly one line (or in our terminology, $E = 1$). A *set associative cache* relaxes this constraint so each set holds more than one cache line. A cache with $1 < E < C/B$ is often called an E-way set associative cache. We will discuss the special case, where $E = C/B$, in the next section. Figure 34 shows the organization of a two-way set associative cache.

Figure 34

Set associative cache $(1 < E < C/B)$. In a set associative cache, each set contains more than one line. This particular example shows a two-way set associative cache.

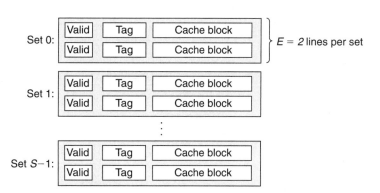

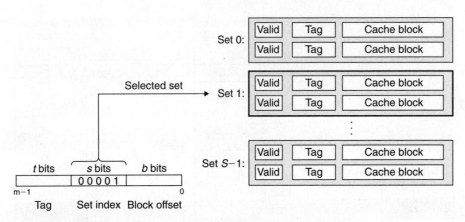

Figure 35 **Set selection in a set associative cache.**

Set Selection in Set Associative Caches

Set selection is identical to a direct-mapped cache, with the set index bits identifying the set. Figure 35 summarizes this principle.

Line Matching and Word Selection in Set Associative Caches

Line matching is more involved in a set associative cache than in a direct-mapped cache because it must check the tags and valid bits of multiple lines in order to determine if the requested word is in the set. A conventional memory is an array of values that takes an address as input and returns the value stored at that address. An *associative memory*, on the other hand, is an array of (key, value) pairs that takes as input the key and returns a value from one of the (key, value) pairs that matches the input key. Thus, we can think of each set in a set associative cache as a small associative memory where the keys are the concatenation of the tag and valid bits, and the values are the contents of a block.

Figure 36 shows the basic idea of line matching in an associative cache. An important idea here is that any line in the set can contain any of the memory blocks

Figure 36
Line matching and word selection in a set associative cache.

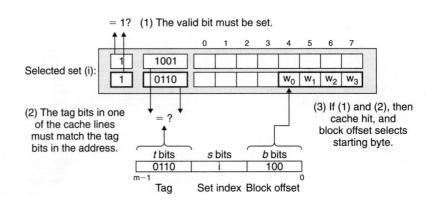

that map to that set. So the cache must search each line in the set, searching for a valid line whose tag matches the tag in the address. If the cache finds such a line, then we have a hit and the block offset selects a word from the block, as before.

Line Replacement on Misses in Set Associative Caches

If the word requested by the CPU is not stored in any of the lines in the set, then we have a cache miss, and the cache must fetch the block that contains the word from memory. However, once the cache has retrieved the block, which line should it replace? Of course, if there is an empty line, then it would be a good candidate. But if there are no empty lines in the set, then we must choose one of the nonempty lines and hope that the CPU does not reference the replaced line anytime soon.

It is very difficult for programmers to exploit knowledge of the cache replacement policy in their codes, so we will not go into much detail about it here. The simplest replacement policy is to choose the line to replace at random. Other more sophisticated policies draw on the principle of locality to try to minimize the probability that the replaced line will be referenced in the near future. For example, a *least-frequently-used* (LFU) policy will replace the line that has been referenced the fewest times over some past time window. A *least-recently-used* (LRU) policy will replace the line that was last accessed the furthest in the past. All of these policies require additional time and hardware. But as we move further down the memory hierarchy, away from the CPU, the cost of a miss becomes more expensive and it becomes more worthwhile to minimize misses with good replacement policies.

4.4 Fully Associative Caches

A *fully associative cache* consists of a single set (i.e., $E = C/B$) that contains all of the cache lines. Figure 37 shows the basic organization.

Set Selection in Fully Associative Caches

Set selection in a fully associative cache is trivial because there is only one set, summarized in Figure 38. Notice that there are no set index bits in the address, which is partitioned into only a tag and a block offset.

Line Matching and Word Selection in Fully Associative Caches

Line matching and word selection in a fully associative cache work the same as with a set associative cache, as we show in Figure 39. The difference is mainly a question of scale. Because the cache circuitry must search for many matching

Figure 37
Fully associative cache
$(E = C/B)$. In a fully associative cache, a single set contains all of the lines.

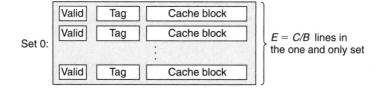

603

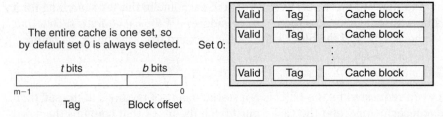

The entire cache is one set, so by default set 0 is always selected.

Figure 38 Set selection in a fully associative cache. Notice that there are no set index bits.

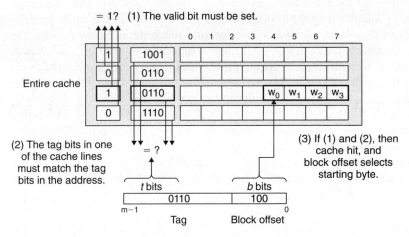

Figure 39 Line matching and word selection in a fully associative cache.

tags in parallel, it is difficult and expensive to build an associative cache that is both large and fast. As a result, fully associative caches are only appropriate for small caches, such as the translation lookaside buffers (TLBs) in virtual memory systems that cache page table entries.

Practice Problem 13

The problems that follow will help reinforce your understanding of how caches work. Assume the following:

- The memory is byte addressable.
- Memory accesses are to **1-byte words** (not to 4-byte words).
- Addresses are 13 bits wide.
- The cache is two-way set associative ($E = 2$), with a 4-byte block size ($B = 4$) and eight sets ($S = 8$).

The contents of the cache are as follows, with all numbers given in hexadecimal notation.

2-way set associative cache

Set index	Tag	Valid	Byte 0	Byte 1	Byte 2	Byte 3	Tag	Valid	Byte 0	Byte 1	Byte 2	Byte 3
			Line 0						Line 1			
0	09	1	86	30	3F	10	00	0	—	—	—	—
1	45	1	60	4F	E0	23	38	1	00	BC	0B	37
2	EB	0	—	—	—	—	0B	0	—	—	—	—
3	06	0	—	—	—	—	32	1	12	08	7B	AD
4	C7	1	06	78	07	C5	05	1	40	67	C2	3B
5	71	1	0B	DE	18	4B	6E	0	—	—	—	—
6	91	1	A0	B7	26	2D	F0	0	—	—	—	—
7	46	0	—	—	—	—	DE	1	12	C0	88	37

The following figure shows the format of an address (one bit per box). Indicate (by labeling the diagram) the fields that would be used to determine the following:

CO   The cache block offset
CI   The cache set index
CT   The cache tag

12	11	10	9	8	7	6	5	4	3	2	1	0

Practice Problem 14

Suppose a program running on the machine in Problem 13 references the 1-byte word at address 0x0E34. Indicate the cache entry accessed and the cache byte value returned **in hex**. Indicate whether a cache miss occurs. If there is a cache miss, enter "–" for "Cache byte returned."

A. Address format (one bit per box):

12	11	10	9	8	7	6	5	4	3	2	1	0

B. Memory reference:

Parameter	Value
Cache block offset (CO)	0x
Cache set index (CI)	0x
Cache tag (CT)	0x
Cache hit? (Y/N)	
Cache byte returned	0x

Practice Problem 15

Repeat Problem 14 for memory address 0x0DD5.

A. Address format (one bit per box):

12	11	10	9	8	7	6	5	4	3	2	1	0

B. Memory reference:

Parameter	Value
Cache block offset (CO)	0x_____
Cache set index (CI)	0x_____
Cache tag (CT)	0x_____
Cache hit? (Y/N)	_____
Cache byte returned	0x_____

Practice Problem 16

Repeat Problem 14 for memory address 0x1FE4.

A. Address format (one bit per box):

12	11	10	9	8	7	6	5	4	3	2	1	0

B. Memory reference:

Parameter	Value
Cache block offset (CO)	0x_____
Cache set index (CI)	0x_____
Cache tag (CT)	0x_____
Cache hit? (Y/N)	_____
Cache byte returned	0x_____

Practice Problem 17

For the cache in Problem 13, list all of the hex memory addresses that will hit in set 3.

4.5 Issues with Writes

As we have seen, the operation of a cache with respect to reads is straightforward. First, look for a copy of the desired word w in the cache. If there is a hit, return *w* immediately. If there is a miss, fetch the block that contains *w* from the next lower level of the memory hierarchy, store the block in some cache line (possibly evicting a valid line), and then return *w*.

The situation for writes is a little more complicated. Suppose we write a word *w* that is already cached (a *write hit*). After the cache updates its copy of *w*, what does it do about updating the copy of *w* in the next lower level of the hierarchy? The simplest approach, known as *write-through*, is to immediately write *w*'s cache block to the next lower level. While simple, write-through has the disadvantage of causing bus traffic with every write. Another approach, known as *write-back*, defers the update as long as possible by writing the updated block to the next lower level only when it is evicted from the cache by the replacement algorithm. Because of locality, write-back can significantly reduce the amount of bus traffic, but it has the disadvantage of additional complexity. The cache must maintain an additional *dirty bit* for each cache line that indicates whether or not the cache block has been modified.

Another issue is how to deal with write misses. One approach, known as *write-allocate*, loads the corresponding block from the next lower level into the cache and then updates the cache block. Write-allocate tries to exploit spatial locality of writes, but it has the disadvantage that every miss results in a block transfer from the next lower level to cache. The alternative, known as *no-write-allocate*, bypasses the cache and writes the word directly to the next lower level. Write-through caches are typically no-write-allocate. Write-back caches are typically write-allocate.

Optimizing caches for writes is a subtle and difficult issue, and we are only scratching the surface here. The details vary from system to system and are often proprietary and poorly documented. To the programmer trying to write reasonably cache-friendly programs, we suggest adopting a mental model that assumes write-back write-allocate caches. There are several reasons for this suggestion.

As a rule, caches at lower levels of the memory hierarchy are more likely to use write-back instead of write-through because of the larger transfer times. For example, virtual memory systems (which use main memory as a cache for the blocks stored on disk) use write-back exclusively. But as logic densities increase, the increased complexity of write-back is becoming less of an impediment and we are seeing write-back caches at all levels of modern systems. So this assumption matches current trends. Another reason for assuming a write-back write-allocate approach is that it is symmetric to the way reads are handled, in that write-back write-allocate tries to exploit locality. Thus, we can develop our programs at a high level to exhibit good spatial and temporal locality rather than trying to optimize for a particular memory system.

4.6 Anatomy of a Real Cache Hierarchy

So far, we have assumed that caches hold only program data. But in fact, caches can hold instructions as well as data. A cache that holds instructions only is called an *i-cache*. A cache that holds program data only is called a *d-cache*. A cache that holds both instructions and data is known as a *unified cache*. Modern processors

Figure 40
**Intel Core i7 cache
hierarchy.**

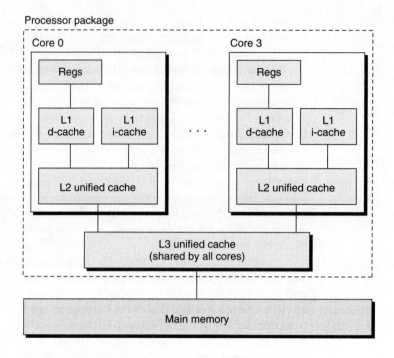

include separate i-caches and d-caches. There are a number of reasons for this. With two separate caches, the processor can read an instruction word and a data word at the same time. I-caches are typically read-only, and thus simpler. The two caches are often optimized to different access patterns and can have different block sizes, associativities, and capacities. Also, having separate caches ensures that data accesses do not create conflict misses with instruction accesses, and vice versa, at the cost of a potential increase in capacity misses.

Figure 40 shows the cache hierarchy for the Intel Core i7 processor. Each CPU chip has four cores. Each core has its own private L1 i-cache, L1 d-cache, and L2 unified cache. All of the cores share an on-chip L3 unified cache. An interesting feature of this hierarchy is that all of the SRAM cache memories are contained in the CPU chip.

Figure 41 summarizes the basic characteristics of the Core i7 caches.

Cache type	Access time (cycles)	Cache size (C)	Assoc. (E)	Block size (B)	Sets (S)
L1 i-cache	4	32 KB	8	64 B	64
L1 d-cache	4	32 KB	8	64 B	64
L2 unified cache	11	256 KB	8	64 B	512
L3 unified cache	30–40	8 MB	16	64 B	8192

Figure 41 **Characteristics of the Intel Core i7 cache hierarchy.**

4.7   Performance Impact of Cache Parameters

Cache performance is evaluated with a number of metrics:

- *Miss rate.* The fraction of memory references during the execution of a program, or a part of a program, that miss. It is computed as *#misses/#references*.
- *Hit rate.* The fraction of memory references that hit. It is computed as $1 - miss\ rate$.
- *Hit time.* The time to deliver a word in the cache to the CPU, including the time for set selection, line identification, and word selection. Hit time is on the order of several clock cycles for L1 caches.
- *Miss penalty.* Any additional time required because of a miss. The penalty for L1 misses served from L2 is on the order of 10 cycles; from L3, 40 cycles; and from main memory, 100 cycles.

Optimizing the cost and performance trade-offs of cache memories is a subtle exercise that requires extensive simulation on realistic benchmark codes and thus is beyond our scope. However, it is possible to identify some of the qualitative trade-offs.

Impact of Cache Size

On the one hand, a larger cache will tend to increase the hit rate. On the other hand, it is always harder to make large memories run faster. As a result, larger caches tend to increase the hit time. This is especially important for on-chip L1 caches that must have a short hit time.

Impact of Block Size

Large blocks are a mixed blessing. On the one hand, larger blocks can help increase the hit rate by exploiting any spatial locality that might exist in a program. However, for a given cache size, larger blocks imply a smaller number of cache lines, which can hurt the hit rate in programs with more temporal locality than spatial locality. Larger blocks also have a negative impact on the miss penalty, since larger blocks cause larger transfer times. Modern systems usually compromise with cache blocks that contain 32 to 64 bytes.

Impact of Associativity

The issue here is the impact of the choice of the parameter E, the number of cache lines per set. The advantage of higher associativity (i.e., larger values of E) is that it decreases the vulnerability of the cache to thrashing due to conflict misses. However, higher associativity comes at a significant cost. Higher associativity is expensive to implement and hard to make fast. It requires more tag bits per line, additional LRU state bits per line, and additional control logic. Higher associativity can increase hit time, because of the increased complexity, and it can also increase the miss penalty because of the increased complexity of choosing a victim line.

The choice of associativity ultimately boils down to a trade-off between the hit time and the miss penalty. Traditionally, high-performance systems that pushed the clock rates would opt for smaller associativity for L1 caches (where the miss penalty is only a few cycles) and a higher degree of associativity for the lower levels, where the miss penalty is higher. For example, in Intel Core i7 systems, the L1 and L2 caches are 8-way associative, and the L3 cache is 16-way.

Impact of Write Strategy

Write-through caches are simpler to implement and can use a *write buffer* that works independently of the cache to update memory. Furthermore, read misses are less expensive because they do not trigger a memory write. On the other hand, write-back caches result in fewer transfers, which allows more bandwidth to memory for I/O devices that perform DMA. Further, reducing the number of transfers becomes increasingly important as we move down the hierarchy and the transfer times increase. In general, caches further down the hierarchy are more likely to use write-back than write-through.

Aside   Cache lines, sets, and blocks: What's the difference?

It is easy to confuse the distinction between cache lines, sets, and blocks. Let's review these ideas and make sure they are clear:

- A *block* is a fixed-sized packet of information that moves back and forth between a cache and main memory (or a lower-level cache).
- A *line* is a container in a cache that stores a block, as well as other information such as the valid bit and the tag bits.
- A *set* is a collection of one or more lines. Sets in direct-mapped caches consist of a single line. Sets in set associative and fully associative caches consist of multiple lines.

In direct-mapped caches, sets and lines are indeed equivalent. However, in associative caches, sets and lines are very different things and the terms cannot be used interchangeably.

Since a line always stores a single block, the terms "line" and "block" are often used interchangeably. For example, systems professionals usually refer to the "line size" of a cache, when what they really mean is the block size. This usage is very common, and shouldn't cause any confusion, so long as you understand the distinction between blocks and lines.

---

5   Writing Cache-friendly Code

In Section 2, we introduced the idea of locality and talked in qualitative terms about what constitutes good locality. Now that we understand how cache memories work, we can be more precise. Programs with better locality will tend to have lower miss rates, and programs with lower miss rates will tend to run faster than programs with higher miss rates. Thus, good programmers should always try to write code that is *cache friendly*, in the sense that it has good locality. Here is the basic approach we use to try to ensure that our code is cache friendly.

1. *Make the common case go fast.* Programs often spend most of their time in a few core functions. These functions often spend most of their time in a few loops. So focus on the inner loops of the core functions and ignore the rest.

2. *Minimize the number of cache misses in each inner loop.* All other things being equal, such as the total number of loads and stores, loops with better miss rates will run faster.

To see how this works in practice, consider the `sumvec` function from Section 2:

```
1    int sumvec(int v[N])
2    {
3        int i, sum = 0;
4
5        for (i = 0; i < N; i++)
6            sum += v[i];
7        return sum;
8    }
```

Is this function cache friendly? First, notice that there is good temporal locality in the loop body with respect to the local variables `i` and `sum`. In fact, because these are local variables, any reasonable optimizing compiler will cache them in the register file, the highest level of the memory hierarchy. Now consider the stride-1 references to vector `v`. In general, if a cache has a block size of B bytes, then a stride-k reference pattern (where k is expressed in words) results in an average of $\min\left(1, (wordsize \times k)/B\right)$ misses per loop iteration. This is minimized for $k = 1$, so the stride-1 references to `v` are indeed cache friendly. For example, suppose that `v` is block aligned, words are 4 bytes, cache blocks are 4 words, and the cache is initially empty (a cold cache). Then, regardless of the cache organization, the references to `v` will result in the following pattern of hits and misses:

v[i]	$i=0$	$i=1$	$i=2$	$i=3$	$i=4$	$i=5$	$i=6$	$i=7$
Access order, [h]it or [m]iss	1 **[m]**	2 [h]	3 [h]	4 [h]	5 **[m]**	6 [h]	7 [h]	8 [h]

In this example, the reference to `v[0]` misses and the corresponding block, which contains `v[0]–v[3]`, is loaded into the cache from memory. Thus, the next three references are all hits. The reference to `v[4]` causes another miss as a new block is loaded into the cache, the next three references are hits, and so on. In general, three out of four references will hit, which is the best we can do in this case with a cold cache.

To summarize, our simple `sumvec` example illustrates two important points about writing cache-friendly code:

- Repeated references to local variables are good because the compiler can cache them in the register file (temporal locality).

- Stride-1 reference patterns are good because caches at all levels of the memory hierarchy store data as contiguous blocks (spatial locality).

Spatial locality is especially important in programs that operate on multi-dimensional arrays. For example, consider the sumarrayrows function from Section 2, which sums the elements of a two-dimensional array in row-major order:

```
1    int sumarrayrows(int a[M][N])
2    {
3        int i, j, sum = 0;
4
5        for (i = 0; i < M; i++)
6            for (j = 0; j < N; j++)
7                sum += a[i][j];
8        return sum;
9    }
```

Since C stores arrays in row-major order, the inner loop of this function has the same desirable stride-1 access pattern as sumvec. For example, suppose we make the same assumptions about the cache as for sumvec. Then the references to the array a will result in the following pattern of hits and misses:

a[i][j]	$j=0$	$j=1$	$j=2$	$j=3$	$j=4$	$j=5$	$j=6$	$j=7$
$i=0$	1 **[m]**	2 [h]	3 [h]	4 [h]	5 **[m]**	6 [h]	7 [h]	8 [h]
$i=1$	9 **[m]**	10 [h]	11 [h]	12 [h]	13 **[m]**	14 [h]	15 [h]	16 [h]
$i=2$	17 **[m]**	18 [h]	19 [h]	20 [h]	21 **[m]**	22 [h]	23 [h]	24 [h]
$i=3$	25 **[m]**	26 [h]	27 [h]	28 [h]	29 **[m]**	30 [h]	31 [h]	32 [h]

But consider what happens if we make the seemingly innocuous change of permuting the loops:

```
1    int sumarraycols(int a[M][N])
2    {
3        int i, j, sum = 0;
4
5        for (j = 0; j < N; j++)
6            for (i = 0; i < M; i++)
7                sum += a[i][j];
8        return sum;
9    }
```

In this case, we are scanning the array column by column instead of row by row. If we are lucky and the entire array fits in the cache, then we will enjoy the same miss rate of 1/4. However, if the array is larger than the cache (the more likely case), then each and every access of a[i][j] will miss!

a[i][j]	j = 0	j = 1	j = 2	j = 3	j = 4	j = 5	j = 6	j = 7
i = 0	1 [m]	5 [m]	9 [m]	13 [m]	17 [m]	21 [m]	25 [m]	29 [m]
i = 1	2 [m]	6 [m]	10 [m]	14 [m]	18 [m]	22 [m]	26 [m]	30 [m]
i = 2	3 [m]	7 [m]	11 [m]	15 [m]	19 [m]	23 [m]	27 [m]	31 [m]
i = 3	4 [m]	8 [m]	12 [m]	16 [m]	20 [m]	24 [m]	28 [m]	32 [m]

Higher miss rates can have a significant impact on running time. For example, on our desktop machine, sumarrayrows runs twice as fast as sumarraycols. To summarize, programmers should be aware of locality in their programs and try to write programs that exploit it.

Practice Problem 18

Transposing the rows and columns of a matrix is an important problem in signal processing and scientific computing applications. It is also interesting from a locality point of view because its reference pattern is both row-wise and column-wise. For example, consider the following transpose routine:

```
1    typedef int array[2][2];
2
3    void transpose1(array dst, array src)
4    {
5        int i, j;
6
7        for (i = 0; i < 2; i++) {
8            for (j = 0; j < 2; j++) {
9                dst[j][i] = src[i][j];
10           }
11       }
12   }
```

Assume this code runs on a machine with the following properties:

- sizeof(int) == 4.
- The src array starts at address 0 and the dst array starts at address 16 (decimal).
- There is a single L1 data cache that is direct-mapped, write-through, and write-allocate, with a block size of 8 bytes.
- The cache has a total size of 16 data bytes and the cache is initially empty.
- Accesses to the src and dst arrays are the only sources of read and write misses, respectively.

A. For each row and col, indicate whether the access to src[row][col] and dst[row][col] is a hit (h) or a miss (m). For example, reading src[0][0] is a miss and writing dst[0][0] is also a miss.

	dst array					src array		
	Col 0	Col 1				Col 0	Col 1	
Row 0	m				Row 0	m		
Row 1					Row 1			

B. Repeat the problem for a cache with 32 data bytes.

Practice Problem 19

The heart of the recent hit game *SimAquarium* is a tight loop that calculates the average position of 256 algae. You are evaluating its cache performance on a machine with a 1024-byte direct-mapped data cache with 16-byte blocks ($B = 16$). You are given the following definitions:

```
1    struct algae_position {
2        int x;
3        int y;
4    };
5
6    struct algae_position grid[16][16];
7    int total_x = 0, total_y = 0;
8    int i, j;
```

You should also assume the following:

- `sizeof(int) == 4`.
- `grid` begins at memory address 0.
- The cache is initially empty.
- The only memory accesses are to the entries of the array `grid`. Variables `i`, `j`, `total_x`, and `total_y` are stored in registers.

Determine the cache performance for the following code:

```
1        for (i = 0; i < 16; i++) {
2            for (j = 0; j < 16; j++) {
3                total_x += grid[i][j].x;
4            }
5        }
6
7        for (i = 0; i < 16; i++) {
8            for (j = 0; j < 16; j++) {
9                total_y += grid[i][j].y;
10           }
11       }
```

A. What is the total number of reads?

B. What is the total number of reads that miss in the cache?

C. What is the miss rate?

Practice Problem 20

Given the assumptions of Problem 19, determine the cache performance of the following code:

```
1        for (i = 0; i < 16; i++){
2            for (j = 0; j < 16; j++) {
3                total_x += grid[j][i].x;
4                total_y += grid[j][i].y;
5            }
6        }
```

A. What is the total number of reads?

B. What is the total number of reads that miss in the cache?

C. What is the miss rate?

D. What would the miss rate be if the cache were twice as big?

Practice Problem 21

Given the assumptions of Problem 19, determine the cache performance of the following code:

```
1        for (i = 0; i < 16; i++){
2            for (j = 0; j < 16; j++) {
3                total_x += grid[i][j].x;
4                total_y += grid[i][j].y;
5            }
6        }
```

A. What is the total number of reads?

B. What is the total number of reads that miss in the cache?

C. What is the miss rate?

D. What would the miss rate be if the cache were twice as big?

6 Putting It Together: The Impact of Caches on Program Performance

This section wraps up our discussion of the memory hierarchy by studying the impact that caches have on the performance of programs running on real machines.

6.1 The Memory Mountain

The rate that a program reads data from the memory system is called the *read throughput*, or sometimes the *read bandwidth*. If a program reads n bytes over a period of s seconds, then the read throughput over that period is n/s, typically expressed in units of megabytes per second (MB/s).

If we were to write a program that issued a sequence of read requests from a tight program loop, then the measured read throughput would give us some insight into the performance of the memory system for that particular sequence of reads. Figure 42 shows a pair of functions that measure the read throughput for a particular read sequence.

The test function generates the read sequence by scanning the first elems elements of an array with a stride of stride. The run function is a wrapper that calls the test function and returns the measured read throughput. The call to the test function in line 29 warms the cache. The fcyc2 function in line 30 calls the test function with arguments elems and estimates the running time of the test function in CPU cycles. Notice that the size argument to the run function is in units of bytes, while the corresponding elems argument to the test function is in units of array elements. Also, notice that line 31 computes MB/s as 10^6 bytes/s, as opposed to 2^{20} bytes/s.

The size and stride arguments to the run function allow us to control the degree of temporal and spatial locality in the resulting read sequence. Smaller values of size result in a smaller working set size, and thus better temporal locality. Smaller values of stride result in better spatial locality. If we call the run function repeatedly with different values of size and stride, then we can recover a fascinating two-dimensional function of read throughput versus temporal and spatial locality. This function is called a *memory mountain*.

Every computer has a unique memory mountain that characterizes the capabilities of its memory system. For example, Figure 43 shows the memory mountain for an Intel Core i7 system. In this example, the size varies from 2 KB to 64 MB, and the stride varies from 1 to 64 elements, where each element is an 8-byte double.

The geography of the Core i7 mountain reveals a rich structure. Perpendicular to the size axis are four *ridges* that correspond to the regions of temporal locality where the working set fits entirely in the L1 cache, the L2 cache, the L3 cache, and main memory, respectively. Notice that there is an order of magnitude difference between the highest peak of the L1 ridge, where the CPU reads at a rate of over 6 GB/s, and the lowest point of the main memory ridge, where the CPU reads at a rate of 600 MB/s.

There is a feature of the L1 ridge that should be pointed out. For very large strides, notice how the read throughput drops as the working set size approaches 2 KB (falling off the back side of the ridge). Since the L1 cache holds the entire working set, this feature does not reflect the true L1 cache performance. It is an artifact of overheads of calling the test function and setting up to execute the loop. For large strides in small working set sizes, these overheads are not amortized, as they are with the larger sizes.

code/mem/mountain/mountain.c

```
1   double data[MAXELEMS];        /* The global array we'll be traversing */
2
3   /*
4    * test - Iterate over first "elems" elements of array "data"
5    *        with stride of "stride".
6    */
7   void test(int elems, int stride) /* The test function */
8   {
9       int i;
10      double result = 0.0;
11      volatile double sink;
12
13      for (i = 0; i < elems; i += stride) {
14          result += data[i];
15      }
16      sink = result; /* So compiler doesn't optimize away the loop */
17  }
18
19  /*
20   * run - Run test(elems, stride) and return read throughput (MB/s).
21   *       "size" is in bytes, "stride" is in array elements, and
22   *       Mhz is CPU clock frequency in Mhz.
23   */
24  double run(int size, int stride, double Mhz)
25  {
26      double cycles;
27      int elems = size / sizeof(double);
28
29      test(elems, stride);                    /* warm up the cache */
30      cycles = fcyc2(test, elems, stride, 0);  /* call test(elems,stride) */
31      return (size / stride) / (cycles / Mhz); /* convert cycles to MB/s */
32  }
```

code/mem/mountain/mountain.c

Figure 42 **Functions that measure and compute read throughput.** We can generate a memory mountain for a particular computer by calling the run function with different values of size (which corresponds to temporal locality) and stride (which corresponds to spatial locality).

On each of the L2, L3, and main memory ridges, there is a slope of spatial locality that falls downhill as the stride increases, and spatial locality decreases. Notice that even when the working set is too large to fit in any of the caches, the highest point on the main memory ridge is a factor of 7 higher than its lowest point. So even when a program has poor temporal locality, spatial locality can still come to the rescue and make a significant difference.

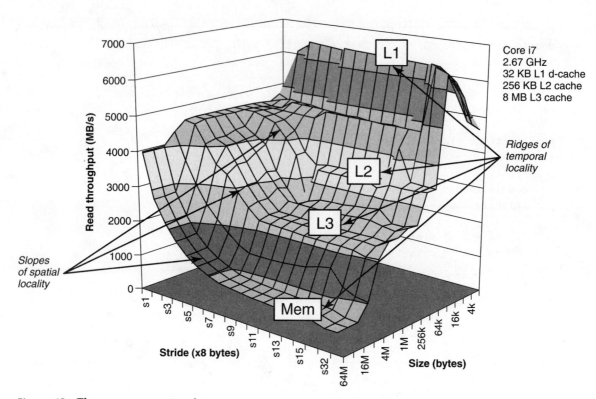

Figure 43   **The memory mountain.**

There is a particularly interesting flat ridge line that extends perpendicular to the stride axis for strides of 1 and 2, where the read throughput is a relatively constant 4.5 GB/s. This is apparently due to a hardware *prefetching* mechanism in the Core i7 memory system that automatically identifies memory referencing patterns and attempts to fetch those blocks into cache before they are accessed. While the details of the particular prefetching algorithm are not documented, it is clear from the memory mountain that the algorithm works best for small strides—yet another reason to favor sequential accesses in your code.

If we take a slice through the mountain, holding the stride constant as in Figure 44, we can see the impact of cache size and temporal locality on performance. For sizes up to 32 KB, the working set fits entirely in the L1 d-cache, and thus reads are served from L1 at the peak throughput of about 6 GB/s. For sizes up to 256 KB, the working set fits entirely in the unified L2 cache, and for sizes up to 8M, the working set fits entirely in the unified L3 cache. Larger working set sizes are served primarily from main memory.

The dips in read throughputs at the leftmost edges of the L1, L2, and L3 cache regions—where the working set sizes of 32 KB, 256 KB, and 8 MB are equal to their respective cache sizes—are interesting. It is not entirely clear why these dips occur. The only way to be sure is to perform a detailed cache simulation, but it

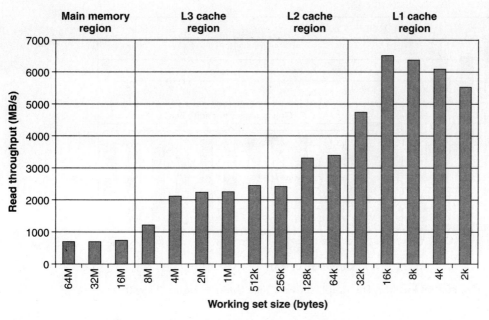

Figure 44 **Ridges of temporal locality in the memory mountain.** The graph shows a slice through Figure 43 with `stride=16`.

is likely that the drops are caused by other data and code blocks that make it impossible to fit the entire array in the respective cache.

Slicing through the memory mountain in the opposite direction, holding the working set size constant, gives us some insight into the impact of spatial locality on the read throughput. For example, Figure 45 shows the slice for a fixed working set size of 4 MB. This slice cuts along the L3 ridge in Figure 43, where the working set fits entirely in the L3 cache, but is too large for the L2 cache.

Notice how the read throughput decreases steadily as the stride increases from one to eight doublewords. In this region of the mountain, a read miss in L2 causes a block to be transferred from L3 to L2. This is followed by some number of hits on the block in L2, depending on the stride. As the stride increases, the ratio of L2 misses to L2 hits increases. Since misses are served more slowly than hits, the read throughput decreases. Once the stride reaches eight doublewords, which on this system equals the block size of 64 bytes, every read request misses in L2 and must be served from L3. Thus, the read throughput for strides of at least eight doublewords is a constant rate determined by the rate that cache blocks can be transferred from L3 into L2.

To summarize our discussion of the memory mountain, the performance of the memory system is not characterized by a single number. Instead, it is a mountain of temporal and spatial locality whose elevations can vary by over an order of magnitude. Wise programmers try to structure their programs so that they run in the peaks instead of the valleys. The aim is to exploit temporal locality so that

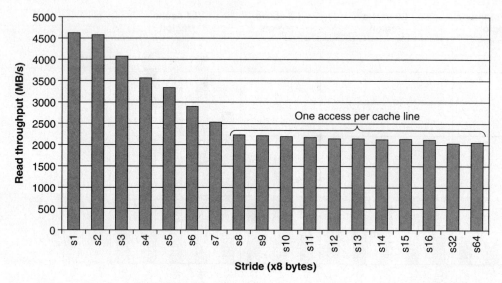

Figure 45 **A slope of spatial locality.** The graph shows a slice through Figure 43 with `size=4` MB.

heavily used words are fetched from the L1 cache, and to exploit spatial locality so that as many words as possible are accessed from a single L1 cache line.

Practice Problem 22

Use the memory mountain in Figure 43 to estimate the time, in CPU cycles, to read an 8-byte word from the L1 d-cache.

6.2 Rearranging Loops to Increase Spatial Locality

Consider the problem of multiplying a pair of $n \times n$ matrices: $C = AB$. For example, if $n = 2$, then

$$\begin{bmatrix} c_{11} & c_{12} \\ c_{21} & c_{22} \end{bmatrix} = \begin{bmatrix} a_{11} & a_{12} \\ a_{21} & a_{22} \end{bmatrix} \begin{bmatrix} b_{11} & b_{12} \\ b_{21} & b_{22} \end{bmatrix}$$

where

$$c_{11} = a_{11}b_{11} + a_{12}b_{21}$$
$$c_{12} = a_{11}b_{12} + a_{12}b_{22}$$
$$c_{21} = a_{21}b_{11} + a_{22}b_{21}$$
$$c_{22} = a_{21}b_{12} + a_{22}b_{22}$$

A matrix multiply function is usually implemented using three nested loops, which are identified by their indexes i, j, and k. If we permute the loops and make some other minor code changes, we can create the six functionally equivalent versions

(a) Version *ijk*

———————————— *code/mem/matmult/mm.c*

```
1    for (i = 0; i < n; i++)
2        for (j = 0; j < n; j++) {
3            sum = 0.0;
4            for (k = 0; k < n; k++)
5                sum += A[i][k]*B[k][j];
6            C[i][j] += sum;
7        }
```

———————————— *code/mem/matmult/mm.c*

(b) Version *jik*

———————————— *code/mem/matmult/mm.c*

```
1    for (j = 0; j < n; j++)
2        for (i = 0; i < n; i++) {
3            sum = 0.0;
4            for (k = 0; k < n; k++)
5                sum += A[i][k]*B[k][j];
6            C[i][j] += sum;
7        }
```

———————————— *code/mem/matmult/mm.c*

(c) Version *jki*

———————————— *code/mem/matmult/mm.c*

```
1    for (j = 0; j < n; j++)
2        for (k = 0; k < n; k++) {
3            r = B[k][j];
4            for (i = 0; i < n; i++)
5                C[i][j] += A[i][k]*r;
6        }
```

———————————— *code/mem/matmult/mm.c*

(d) Version *kji*

———————————— *code/mem/matmult/mm.c*

```
1    for (k = 0; k < n; k++)
2        for (j = 0; j < n; j++) {
3            r = B[k][j];
4            for (i = 0; i < n; i++)
5                C[i][j] += A[i][k]*r;
6        }
```

———————————— *code/mem/matmult/mm.c*

(e) Version *kij*

———————————— *code/mem/matmult/mm.c*

```
1    for (k = 0; k < n; k++)
2        for (i = 0; i < n; i++) {
3            r = A[i][k];
4            for (j = 0; j < n; j++)
5                C[i][j] += r*B[k][j];
6        }
```

———————————— *code/mem/matmult/mm.c*

(f) Version *ikj*

———————————— *code/mem/matmult/mm.c*

```
1    for (i = 0; i < n; i++)
2        for (k = 0; k < n; k++) {
3            r = A[i][k];
4            for (j = 0; j < n; j++)
5                C[i][j] += r*B[k][j];
6        }
```

———————————— *code/mem/matmult/mm.c*

Figure 46 **Six versions of matrix multiply.** Each version is uniquely identified by the ordering of its loops.

of matrix multiply shown in Figure 46. Each version is uniquely identified by the ordering of its loops.

At a high level, the six versions are quite similar. If addition is associative, then each version computes an identical result.[2] Each version performs $O(n^3)$

---

2. Floating-point addition is commutative, but in general not associative. In practice, if the matrices do not mix extremely large values with extremely small ones, as often is true when the matrices store physical properties, then the assumption of associativity is reasonable.

Matrix multiply version (class)	Loads per iter.	Stores per iter.	A misses per iter.	B misses per iter.	C misses per iter.	Total misses per iter.
ijk & jik (AB)	2	0	0.25	1.00	0.00	1.25
jki & kji (AC)	2	1	1.00	0.00	1.00	2.00
kij & ikj (BC)	2	1	0.00	0.25	0.25	0.50

Figure 47 **Analysis of matrix multiply inner loops.** The six versions partition into three equivalence classes, denoted by the pair of arrays that are accessed in the inner loop.

total operations and an identical number of adds and multiplies. Each of the n^2 elements of A and B is read n times. Each of the n^2 elements of C is computed by summing n values. However, if we analyze the behavior of the innermost loop iterations, we find that there are differences in the number of accesses and the locality. For the purposes of this analysis, we make the following assumptions:

- Each array is an $n \times n$ array of double, with sizeof(double) == 8.
- There is a single cache with a 32-byte block size ($B = 32$).
- The array size n is so large that a single matrix row does not fit in the L1 cache.
- The compiler stores local variables in registers, and thus references to local variables inside loops do not require any load or store instructions.

Figure 47 summarizes the results of our inner loop analysis. Notice that the six versions pair up into three equivalence classes, which we denote by the pair of matrices that are accessed in the inner loop. For example, versions ijk and jik are members of Class AB because they reference arrays A and B (but not C) in their innermost loop. For each class, we have counted the number of loads (reads) and stores (writes) in each inner loop iteration, the number of references to A, B, and C that will miss in the cache in each loop iteration, and the total number of cache misses per iteration.

The inner loops of the Class AB routines (Figure 46(a) and (b)) scan a row of array A with a stride of 1. Since each cache block holds four doublewords, the miss rate for A is 0.25 misses per iteration. On the other hand, the inner loop scans a column of B with a stride of n. Since n is large, each access of array B results in a miss, for a total of 1.25 misses per iteration.

The inner loops in the Class AC routines (Figure 46(c) and (d)) have some problems. Each iteration performs two loads and a store (as opposed to the Class AB routines, which perform two loads and no stores). Second, the inner loop scans the columns of A and C with a stride of n. The result is a miss on each load, for a total of two misses per iteration. Notice that interchanging the loops has decreased the amount of spatial locality compared to the Class AB routines.

The BC routines (Figure 46(e) and (f)) present an interesting trade-off: With two loads and a store, they require one more memory operation than the AB routines. On the other hand, since the inner loop scans both B and C row-wise

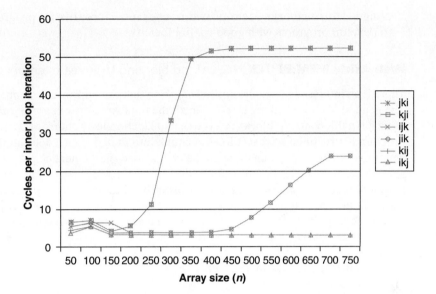

Figure 48
Core i7 matrix multiply performance. Legend: jki and kji: Class AC; ijk and jik: Class AB; kij and ikj: Class BC.

with a stride-1 access pattern, the miss rate on each array is only 0.25 misses per iteration, for a total of 0.50 misses per iteration.

Figure 48 summarizes the performance of different versions of matrix multiply on a Core i7 system. The graph plots the measured number of CPU cycles per inner loop iteration as a function of array size (n).

There are a number of interesting points to notice about this graph:

- For large values of n, the fastest version runs almost 20 times faster than the slowest version, even though each performs the same number of floating-point arithmetic operations.

- Pairs of versions with the same number of memory references and misses per iteration have almost identical measured performance.

- The two versions with the worst memory behavior, in terms of the number of accesses and misses per iteration, run significantly slower than the other four versions, which have fewer misses or fewer accesses, or both.

- Miss rate, in this case, is a better predictor of performance than the total number of memory accesses. For example, the Class BC routines, with 0.5 misses per iteration, perform much better than the Class AB routines, with 1.25 misses per iteration, even though the Class BC routines perform more memory references in the inner loop (two loads and one store) than the Class AB routines (two loads).

- For large values of n, the performance of the fastest pair of versions (kij and ikj) is constant. Even though the array is much larger than any of the SRAM cache memories, the prefetching hardware is smart enough to recognize the stride-1 access pattern, and fast enough to keep up with memory accesses in the tight inner loop. This is a stunning accomplishment by the Intel engineers who designed this memory system, providing even more incentive for programmers to develop programs with good spatial locality.

Web Aside MEM:BLOCKING   Using blocking to increase temporal locality

There is an interesting technique called *blocking* that can improve the temporal locality of inner loops. The general idea of blocking is to organize the data structures in a program into large chunks called *blocks*. (In this context, "block" refers to an application-level chunk of data, *not* to a cache block.) The program is structured so that it loads a chunk into the L1 cache, does all the reads and writes that it needs to on that chunk, then discards the chunk, loads in the next chunk, and so on.

Unlike the simple loop transformations for improving spatial locality, blocking makes the code harder to read and understand. For this reason, it is best suited for optimizing compilers or frequently executed library routines. Still, the technique is interesting to study and understand because it is a general concept that can produce big performance gains on some systems.

6.3 Exploiting Locality in Your Programs

As we have seen, the memory system is organized as a hierarchy of storage devices, with smaller, faster devices toward the top and larger, slower devices toward the bottom. Because of this hierarchy, the effective rate that a program can access memory locations is not characterized by a single number. Rather, it is a wildly varying function of program locality (what we have dubbed the memory mountain) that can vary by orders of magnitude. Programs with good locality access most of their data from fast cache memories. Programs with poor locality access most of their data from the relatively slow DRAM main memory.

Programmers who understand the nature of the memory hierarchy can exploit this understanding to write more efficient programs, regardless of the specific memory system organization. In particular, we recommend the following techniques:

- Focus your attention on the inner loops, where the bulk of the computations and memory accesses occur.

- Try to maximize the spatial locality in your programs by reading data objects sequentially, with stride 1, in the order they are stored in memory.

- Try to maximize the temporal locality in your programs by using a data object as often as possible once it has been read from memory.

7 Summary

The basic storage technologies are random-access memories (RAMs), nonvolatile memories (ROMs), and disks. RAM comes in two basic forms. Static RAM (SRAM) is faster and more expensive, and is used for cache memories both on and off the CPU chip. Dynamic RAM (DRAM) is slower and less expensive, and is used for the main memory and graphics frame buffers. Nonvolatile memories, also called read-only memories (ROMs), retain their information even if the supply voltage is turned off, and they are used to store firmware. Rotating disks are mechanical nonvolatile storage devices that hold enormous amounts of data at a low cost per bit, but with much longer access times than DRAM. Solid state disks (SSDs) based on nonvolatile flash memory are becoming increasingly attractive alternatives to rotating disks for some applications.

In general, faster storage technologies are more expensive per bit and have smaller capacities. The price and performance properties of these technologies are changing at dramatically different rates. In particular, DRAM and disk access times are much larger than CPU cycle times. Systems bridge these gaps by organizing memory as a hierarchy of storage devices, with smaller, faster devices at the top and larger, slower devices at the bottom. Because well-written programs have good locality, most data are served from the higher levels, and the effect is a memory system that runs at the rate of the higher levels, but at the cost and capacity of the lower levels.

Programmers can dramatically improve the running times of their programs by writing programs with good spatial and temporal locality. Exploiting SRAM-based cache memories is especially important. Programs that fetch data primarily from cache memories can run much faster than programs that fetch data primarily from memory.

Bibliographic Notes

Memory and disk technologies change rapidly. In our experience, the best sources of technical information are the Web pages maintained by the manufacturers. Companies such as Micron, Toshiba, and Samsung provide a wealth of current technical information on memory devices. The pages for Seagate, Maxtor, and Western Digital provide similarly useful information about disks.

Textbooks on circuit and logic design provide detailed information about memory technology [56, 85]. IEEE Spectrum published a series of survey articles on DRAM [53]. The International Symposium on Computer Architecture (ISCA) is a common forum for characterizations of DRAM memory performance [34, 35].

Wilkes wrote the first paper on cache memories [116]. Smith wrote a classic survey [101]. Przybylski wrote an authoritative book on cache design [82]. Hennessy and Patterson provide a comprehensive discussion of cache design issues [49].

Stricker introduced the idea of the memory mountain as a comprehensive characterization of the memory system in [111], and suggested the term "memory mountain" informally in later presentations of the work. Compiler researchers work to increase locality by automatically performing the kinds of manual code transformations we discussed in Section 6 [22, 38, 63, 68, 75, 83, 118]. Carter and colleagues have proposed a cache-aware memory controller [18]. Seward developed an open-source cache profiler, called cacheprof, that characterizes the miss behavior of C programs on an arbitrary simulated cache (www.cacheprof.org). Other researchers have developed *cache oblivious* algorithms that are designed to run well without any explicit knowledge of the structure of the underlying cache memory [36, 42, 43].

There is a large body of literature on building and using disk storage. Many storage researchers look for ways to aggregate individual disks into larger, more robust, and more secure storage pools [20, 44, 45, 79, 119]. Others look for ways to use caches and locality to improve the performance of disk accesses [12, 21]. Systems such as Exokernel provide increased user-level control of disk and memory resources [55]. Systems such as the Andrew File System [74] and Coda [91] extend the memory hierarchy across computer networks and mobile notebook computers. Schindler and Ganger developed an interesting tool that automatically characterizes the geometry and performance of SCSI disk drives [92]. Researchers are investigating techniques for building and using Flash-based SSDs [8, 77].

Homework Problems

23 ◆◆

Suppose you are asked to design a rotating disk where the number of bits per track is constant. You know that the number of bits per track is determined by the circumference of the innermost track, which you can assume is also the circumference of the hole. Thus, if you make the hole in the center of the disk larger, the number of bits per track increases, but the total number of tracks decreases. If you let r denote the radius of the platter, and $x \cdot r$ the radius of the hole, what value of x maximizes the capacity of the disk?

24 ◆

Estimate the average time (in ms) to access a sector on the following disk:

Parameter	Value
Rotational rate	15,000 RPM
$T_{avg\ seek}$	4 ms
Average # sectors/track	800

25 ◆◆

Suppose that a 2 MB file consisting of 512-byte logical blocks is stored on a disk drive with the following characteristics:

Parameter	Value
Rotational rate	15,000 RPM
$T_{avg\ seek}$	4 ms
Average # sectors/track	1000
Surfaces	8
Sector size	512 bytes

For each case below, suppose that a program reads the logical blocks of the file sequentially, one after the other, and that the time to position the head over the first block is $T_{avg\ seek} + T_{avg\ rotation}$.

A. *Best case:* Estimate the optimal time (in ms) required to read the file over all possible mappings of logical blocks to disk sectors.

B. *Random case:* Estimate the time (in ms) required to read the file if blocks are mapped randomly to disk sectors.

26 ◆

The following table gives the parameters for a number of different caches. For each cache, fill in the missing fields in the table. Recall that *m* is the number of physical address bits, *C* is the cache size (number of data bytes), *B* is the block size in bytes, *E* is the associativity, *S* is the number of cache sets, *t* is the number of tag bits, *s* is the number of set index bits, and *b* is the number of block offset bits.

Cache	*m*	*C*	*B*	*E*	*S*	*t*	*s*	*b*
1.	32	1024	4	4				
2.	32	1024	4	256				
3.	32	1024	8	1				
4.	32	1024	8	128				
5.	32	1024	32	1				
6.	32	1024	32	4				

27 ◆

The following table gives the parameters for a number of different caches. Your task is to fill in the missing fields in the table. Recall that *m* is the number of physical address bits, *C* is the cache size (number of data bytes), *B* is the block size in bytes, *E* is the associativity, *S* is the number of cache sets, *t* is the number of tag bits, *s* is the number of set index bits, and *b* is the number of block offset bits.

Cache	*m*	*C*	*B*	*E*	*S*	*t*	*s*	*b*
1.	32		8	1		21	8	3
2.	32	2048			128	23	7	2
3.	32	1024	2	8	64			1
4.	32	1024		2	16	23	4	

28 ◆

This problem concerns the cache in Problem 13.

A. List all of the hex memory addresses that will hit in set 1.

B. List all of the hex memory addresses that will hit in set 6.

29 ◆◆

This problem concerns the cache in Problem 13.

A. List all of the hex memory addresses that will hit in set 2.

B. List all of the hex memory addresses that will hit in set 4.

C. List all of the hex memory addresses that will hit in set 5.

D. List all of the hex memory addresses that will hit in set 7.

30 ◆◆

Suppose we have a system with the following properties:

- The memory is byte addressable.
- Memory accesses are to **1-byte words** (not to 4-byte words).
- Addresses are 12 bits wide.
- The cache is two-way set associative ($E = 2$), with a 4-byte block size ($B = 4$) and four sets ($S = 4$).

The contents of the cache are as follows, with all addresses, tags, and values given in hexadecimal notation:

Set index	Tag	Valid	Byte 0	Byte 1	Byte 2	Byte 3
0	00	1	40	41	42	43
	83	1	FE	97	CC	D0
1	00	1	44	45	46	47
	83	0	—	—	—	—
2	00	1	48	49	4A	4B
	40	0	—	—	—	—
3	FF	1	9A	C0	03	FF
	00	0	—	—	—	—

A. The following diagram shows the format of an address (one bit per box). Indicate (by labeling the diagram) the fields that would be used to determine the following:

CO The cache block offset

CI The cache set index

CT The cache tag

11	10	9	8	7	6	5	4	3	2	1	0

B. For each of the following memory accesses, indicate if it will be a cache hit or miss when **carried out in sequence** as listed. Also give the value of a read if it can be inferred from the information in the cache.

Operation	Address	Hit?	Read value (or unknown)
Read	0x834		
Write	0x836		
Read	0xFFD		

31 ◆

Suppose we have a system with the following properties:

- The memory is byte addressable.
- Memory accesses are to **1-byte words** (not to 4-byte words).
- Addresses are 13 bits wide.
- The cache is four-way set associative ($E = 4$), with a 4-byte block size ($B = 4$) and eight sets ($S = 8$).

Consider the following cache state. All addresses, tags, and values are given in hexadecimal format. The *Index* column contains the set index for each set of four lines. The *Tag* columns contain the tag value for each line. The *V* columns contain the valid bit for each line. The *Bytes 0–3* columns contain the data for each line, numbered left-to-right starting with byte 0 on the left.

4-way set associative cache

Index	Tag	V	Bytes 0–3	Tag	V	Bytes 0–3	Tag	V	Bytes 0–3	Tag	V	Bytes 0–3
0	F0	1	ED 32 0A A2	8A	1	BF 80 1D FC	14	1	EF 09 86 2A	BC	0	25 44 6F 1A
1	BC	0	03 3E CD 38	A0	0	16 7B ED 5A	BC	1	8E 4C DF 18	E4	1	FB B7 12 02
2	BC	1	54 9E 1E FA	B6	1	DC 81 B2 14	00	0	B6 1F 7B 44	74	0	10 F5 B8 2E
3	BE	0	2F 7E 3D A8	C0	1	27 95 A4 74	C4	0	07 11 6B D8	BC	0	C7 B7 AF C2
4	7E	1	32 21 1C 2C	8A	1	22 C2 DC 34	BC	1	BA DD 37 D8	DC	0	E7 A2 39 BA
5	98	0	A9 76 2B EE	54	0	BC 91 D5 92	98	1	80 BA 9B F6	BC	1	48 16 81 0A
6	38	0	5D 4D F7 DA	BC	1	69 C2 8C 74	8A	1	A8 CE 7F DA	38	1	FA 93 EB 48
7	8A	1	04 2A 32 6A	9E	0	B1 86 56 0E	CC	1	96 30 47 F2	BC	1	F8 1D 42 30

A. What is size (C) of this cache in bytes?

B. The box that follows shows the format of an address (one bit per box). Indicate (by labeling the diagram) the fields that would be used to determine the following:

CO   The cache block offset

CI   The cache set index

CT   The cache tag

12	11	10	9	8	7	6	5	4	3	2	1	0

32 ◆◆

Supppose that a program using the cache in Problem 31 references the 1-byte word at address 0x071A. Indicate the cache entry accessed and the cache byte value returned **in hex**. Indicate whether a cache miss occurs. If there is a cache miss, enter "−" for "Cache byte returned." *Hint: Pay attention to those valid bits!*

A. Address format (one bit per box):

12	11	10	9	8	7	6	5	4	3	2	1	0

B. Memory reference:

Parameter	Value
Block offset (CO)	0x___
Index (CI)	0x___
Cache tag (CT)	0x___
Cache hit? (Y/N)	___
Cache byte returned	0x___

33 ◆◆

Repeat Problem 32 for memory address 0x16E8.

A. Address format (one bit per box):

12	11	10	9	8	7	6	5	4	3	2	1	0

B. Memory reference:

Parameter	Value
Cache offset (CO)	0x___
Cache index (CI)	0x___
Cache tag (CT)	0x___
Cache hit? (Y/N)	___
Cache byte returned	0x___

34 ◆◆

For the cache in Problem 31, list the eight memory addresses (in hex) that will hit in set 2.

35 ◆◆

Consider the following matrix transpose routine:

```
1   typedef int array[4][4];
2
3   void transpose2(array dst, array src)
4   {
5       int i, j;
6
7       for (i = 0; i < 4; i++) {
8           for (j = 0; j < 4; j++) {
9               dst[j][i] = src[i][j];
10          }
11      }
12  }
```

Assume this code runs on a machine with the following properties:

- `sizeof(int) == 4`.
- The `src` array starts at address 0 and the `dst` array starts at address 64 (decimal).
- There is a single L1 data cache that is direct-mapped, write-through, write-allocate, with a block size of 16 bytes.
- The cache has a total size of 32 data bytes and the cache is initially empty.
- Accesses to the `src` and `dst` arrays are the only sources of read and write misses, respectively.

A. For each `row` and `col`, indicate whether the access to `src[row][col]` and `dst[row][col]` is a hit (h) or a miss (m). For example, reading `src[0][0]` is a miss and writing `dst[0][0]` is also a miss.

dst array

	Col 0	Col 1	Col 2	Col 3
Row 0	m			
Row 1				
Row 2				
Row 3				

src array

	Col 0	Col 1	Col 2	Col 3
Row 0	m			
Row 1				
Row 2				
Row 3				

36 ◆◆

Repeat Problem 35 for a cache with a total size of 128 data bytes.

dst array

	Col 0	Col 1	Col 2	Col 3
Row 0				
Row 1				
Row 2				
Row 3				

src array

	Col 0	Col 1	Col 2	Col 3
Row 0				
Row 1				
Row 2				
Row 3				

37 ◆◆

This problem tests your ability to predict the cache behavior of C code. You are given the following code to analyze:

```
1    int x[2][128];
2    int i;
3    int sum = 0;
4
5    for (i = 0; i < 128; i++) {
6        sum += x[0][i] * x[1][i];
7    }
```

Assume we execute this under the following conditions:

- `sizeof(int) = 4`.
- Array x begins at memory address 0x0 and is stored in row-major order.
- In each case below, the cache is initially empty.
- The only memory accesses are to the entries of the array x. All other variables are stored in registers.

Given these assumptions, estimate the miss rates for the following cases:

A. Case 1: Assume the cache is 512 bytes, direct-mapped, with 16-byte cache blocks. What is the miss rate?

B. Case 2: What is the miss rate if we double the cache size to 1024 bytes?

C. Case 3: Now assume the cache is 512 bytes, two-way set associative using an LRU replacement policy, with 16-byte cache blocks. What is the cache miss rate?

D. For Case 3, will a larger cache size help to reduce the miss rate? Why or why not?

E. For Case 3, will a larger block size help to reduce the miss rate? Why or why not?

38 ◆◆

This is another problem that tests your ability to analyze the cache behavior of C code. Assume we execute the three summation functions in Figure 49 under the following conditions:

- `sizeof(int) == 4`.
- The machine has a 4 KB direct-mapped cache with a 16-byte block size.
- Within the two loops, the code uses memory accesses only for the array data. The loop indices and the value sum are held in registers.
- Array a is stored starting at memory address 0x08000000.

Fill in the table for the approximate cache miss rate for the two cases $N = 64$ and $N = 60$.

Function	N = 64	N = 60
sumA		
sumB		
sumC		

39 ◆

3M™ decides to make Post-It® notes by printing yellow squares on white pieces of paper. As part of the printing process, they need to set the CMYK (cyan, magenta, yellow, black) value for every point in the square. 3M hires you to determine

```
1    typedef int array_t[N][N];
2
3    int sumA(array_t a)
4    {
5        int i, j;
6        int sum = 0;
7        for (i = 0; i < N; i++)
8            for (j = 0; j < N; j++) {
9                sum += a[i][j];
10           }
11       return sum;
12   }
13
14   int sumB(array_t a)
15   {
16       int i, j;
17       int sum = 0;
18       for (j = 0; j < N; j++)
19           for (i = 0; i < N; i++) {
20               sum += a[i][j];
21           }
22       return sum;
23   }
24
25   int sumC(array_t a)
26   {
27       int i, j;
28       int sum = 0;
29       for (j = 0; j < N; j+=2)
30           for (i = 0; i < N; i+=2) {
31               sum += (a[i][j] + a[i+1][j]
32                       + a[i][j+1] + a[i+1][j+1]);
33           }
34       return sum;
35   }
```

Figure 49 **Functions referenced in Problem 38.**

the efficiency of the following algorithms on a machine with a 2048-byte direct-mapped data cache with 32-byte blocks. You are given the following definitions:

```
1    struct point_color {
2        int c;
3        int m;
4        int y;
5        int k;
6    };
```

```
7
8    struct point_color square[16][16];
9    int i, j;
```

Assume the following:

- `sizeof(int) == 4`.
- `square` begins at memory address 0.
- The cache is initially empty.
- The only memory accesses are to the entries of the array `square`. Variables `i` and `j` are stored in registers.

Determine the cache performance of the following code:

```
1        for (i = 0; i < 16; i++){
2            for (j = 0; j < 16; j++) {
3                square[i][j].c = 0;
4                square[i][j].m = 0;
5                square[i][j].y = 1;
6                square[i][j].k = 0;
7            }
8        }
```

A. What is the total number of writes?

B. What is the total number of writes that miss in the cache?

C. What is the miss rate?

40 ◆

Given the assumptions in Problem 39, determine the cache performance of the following code:

```
1        for (i = 0; i < 16; i++){
2            for (j = 0; j < 16; j++) {
3                square[j][i].c = 0;
4                square[j][i].m = 0;
5                square[j][i].y = 1;
6                square[j][i].k = 0;
7            }
8        }
```

A. What is the total number of writes?

B. What is the total number of writes that miss in the cache?

C. What is the miss rate?

41 ◆

Given the assumptions in Problem 39, determine the cache performance of the following code:

```
1       for (i = 0; i < 16; i++) {
2           for (j = 0; j < 16; j++) {
3               square[i][j].y = 1;
4           }
5       }
6       for (i = 0; i < 16; i++) {
7           for (j = 0; j < 16; j++) {
8               square[i][j].c = 0;
9               square[i][j].m = 0;
10              square[i][j].k = 0;
11          }
12      }
```

   A. What is the total number of writes?

   B. What is the total number of writes that miss in the cache?

   C. What is the miss rate?

42 ◆◆

You are writing a new 3D game that you hope will earn you fame and fortune. You are currently working on a function to blank the screen buffer before drawing the next frame. The screen you are working with is a 640×480 array of pixels. The machine you are working on has a 64 KB direct-mapped cache with 4-byte lines. The C structures you are using are as follows:

```
1   struct pixel {
2       char r;
3       char g;
4       char b;
5       char a;
6   };
7
8   struct pixel buffer[480][640];
9   int i, j;
10  char *cptr;
11  int *iptr;
```

Assume the following:

- `sizeof(char) == 1` and `sizeof(int) == 4`.
- `buffer` begins at memory address 0.
- The cache is initially empty.
- The only memory accesses are to the entries of the array `buffer`. Variables `i`, `j`, `cptr`, and `iptr` are stored in registers.

What percentage of writes in the following code will miss in the cache?

```
1    for (j = 0; j < 640; j++) {
2        for (i = 0; i < 480; i++){
3            buffer[i][j].r = 0;
4            buffer[i][j].g = 0;
5            buffer[i][j].b = 0;
6            buffer[i][j].a = 0;
7        }
8    }
```

43 ◆◆

Given the assumptions in Problem 42, what percentage of writes in the following code will miss in the cache?

```
1    char *cptr = (char *) buffer;
2    for (; cptr < (((char *) buffer) + 640 * 480 * 4); cptr++)
3        *cptr = 0;
```

44 ◆◆

Given the assumptions in Problem 42, what percentage of writes in the following code will miss in the cache?

```
1    int *iptr = (int *)buffer;
2    for (; iptr < ((int *)buffer + 640*480); iptr++)
3        *iptr = 0;
```

45 ◆◆◆

Download the mountain program from the CS:APP2 Web site and run it on your favorite PC/Linux system. Use the results to estimate the sizes of the caches on your system.

46 ◆◆◆◆

In this assignment, you will apply the concepts you learned in this chapter and the chapter "Optimizing Program Performance" to the problem of optimizing code for a memory-intensive application. Consider a procedure to copy and transpose the elements of an $N \times N$ matrix of type int. That is, for source matrix S and destination matrix D, we want to copy each element $s_{i,j}$ to $d_{j,i}$. This code can be written with a simple loop,

```
1    void transpose(int *dst, int *src, int dim)
2    {
3        int i, j;
4
5        for (i = 0; i < dim; i++)
6            for (j = 0; j < dim; j++)
7                dst[j*dim + i] = src[i*dim + j];
8    }
```

where the arguments to the procedure are pointers to the destination (dst) and source (src) matrices, as well as the matrix size N (dim). Your job is to devise a transpose routine that runs as fast as possible.

47 ◆◆◆◆

This assignment is an intriguing variation of Problem 46. Consider the problem of converting a directed graph g into its undirected counterpart g'. The graph g' has an edge from vertex u to vertex v if and only if there is an edge from u to v or from v to u in the original graph g. The graph g is represented by its *adjacency matrix* G as follows. If N is the number of vertices in g, then G is an $N \times N$ matrix and its entries are all either 0 or 1. Suppose the vertices of g are named $v_0, v_1, v_2, \ldots, v_{N-1}$. Then $G[i][j]$ is 1 if there is an edge from v_i to v_j, and is 0 otherwise. Observe that the elements on the diagonal of an adjacency matrix are always 1 and that the adjacency matrix of an undirected graph is symmetric. This code can be written with a simple loop:

```
1    void col_convert(int *G, int dim) {
2        int i, j;
3
4        for (i = 0; i < dim; i++)
5            for (j = 0; j < dim; j++)
6                G[j*dim + i] = G[j*dim + i] || G[i*dim + j];
7    }
```

Your job is to devise a conversion routine that runs as fast as possible. As before, you will need to apply concepts you learned in this chapter and the chapter "Optimizing Program Performance" to come up with a good solution.

Solutions to Practice Problems

Solution to Problem 1

The idea here is to minimize the number of address bits by minimizing the aspect ratio $\max(r, c)/ \min(r, c)$. In other words, the squarer the array, the fewer the address bits.

Organization	r	c	b_r	b_c	$\max(b_r, b_c)$
16×1	4	4	2	2	2
16×4	4	4	2	2	2
128×8	16	8	4	3	4
512×4	32	16	5	4	5
1024×4	32	32	5	5	5

Solution to Problem 2

The point of this little drill is to make sure you understand the relationship between cylinders and tracks. Once you have that straight, just plug and chug:

$$\text{Disk capacity} = \frac{512 \text{ bytes}}{\text{sector}} \times \frac{400 \text{ sectors}}{\text{track}} \times \frac{10,000 \text{ tracks}}{\text{surface}} \times \frac{2 \text{ surfaces}}{\text{platter}} \times \frac{2 \text{ platters}}{\text{disk}}$$

$$= 8,192,000,000 \text{ bytes}$$

$$= 8.192 \text{ GB}$$

Solution to Problem 3

This solution to this problem is a straightforward application of the formula for disk access time. The average rotational latency (in ms) is

$$T_{avg\ rotation} = 1/2 \times T_{max\ rotation}$$

$$= 1/2 \times (60 \text{ secs} / 15,000 \text{ RPM}) \times 1000 \text{ ms/sec}$$

$$\approx 2 \text{ ms}$$

The average transfer time is

$$T_{avg\ transfer} = (60 \text{ secs} / 15,000 \text{ RPM}) \times 1/500 \text{ sectors/track} \times 1000 \text{ ms/sec}$$

$$\approx 0.008 \text{ ms}$$

Putting it all together, the total estimated access time is

$$T_{access} = T_{avg\ seek} + T_{avg\ rotation} + T_{avg\ transfer}$$

$$= 8 \text{ ms} + 2 \text{ ms} + 0.008 \text{ ms}$$

$$\approx 10 \text{ ms}$$

Solution to Problem 4

This is a good check of your understanding of the factors that affect disk performance. First we need to determine a few basic properties of the file and the disk. The file consists of 2000, 512-byte logical blocks. For the disk, $T_{avg\ seek} = 5$ ms, $T_{max\ rotation} = 6$ ms, and $T_{avg\ rotation} = 3$ ms.

A. *Best case:* In the optimal case, the blocks are mapped to contiguous sectors, on the same cylinder, that can be read one after the other without moving the head. Once the head is positioned over the first sector it takes two full rotations (1000 sectors per rotation) of the disk to read all 2000 blocks. So the total time to read the file is $T_{avg\ seek} + T_{avg\ rotation} + 2 * T_{max\ rotation} = 5 + 3 + 12 = 20$ ms.

B. *Random case:* In this case, where blocks are mapped randomly to sectors, reading each of the 2000 blocks requires $T_{avg\ seek} + T_{avg\ rotation}$ ms, so the total time to read the file is $(T_{avg\ seek} + T_{avg\ rotation}) * 2000 = 16,000$ ms (16 seconds!).

You can see now why it's often a good idea to defragment your disk drive!

Solution to Problem 5

This problem, based on the zone map in Figure 14, is a good test of your understanding of disk geometry, and it also enables you to derive an interesting characteristic of a real disk drive.

A. Zone 0. There are a total of $864 \times 8 \times 3201 = 22{,}125{,}312$ sectors and $22{,}076{,}928$ logical blocks assigned to zone 0, for a total of $22{,}125{,}312 - 22{,}076{,}928 = 48{,}384$ spare sectors. Given that there are $864 \times 8 = 6912$ sectors per cylinder, there are $48{,}384/6912 = 7$ spare cylinders in zone 0.

B. Zone 8. A similar analysis reveals there are $((3700 \times 5632) - 20{,}804{,}608)/5632 = 6$ spare cylinders in zone 8.

Solution to Problem 6

This is a simple problem that will give you some interesting insights into feasibility of SSDs. Recall that for disks, 1 PB = 10^9 MB. Then the following straightforward translation of units yields the following predicted times for each case:

A. Worst case sequential writes (170 MB/s): $10^9 \times (1/170) \times (1/(86{,}400 \times 365)) \approx 0.2$ years.

B. Worst case random writes (14 MB/s): $10^9 \times (1/14) \times (1/(86{,}400 \times 365)) \approx 2.25$ years.

C. Average case (20 GB/day): $10^9 \times (1/20{,}000) \times (1/365) \approx 140$ years.

Solution to Problem 7

In the 10-year period between 2000 and 2010, the unit price of rotating disk dropped by a factor of about 30, which means the price is dropping by roughly a factor of 2 every 2 years. Assuming this trend continues, a petabyte of storage, which costs about \$300,000 in 2010, will drop below \$500 after about ten of these factor-of-2 reductions. Since these are occurring every 2 years, we can expect a petabyte of storage to be available for \$500 around the year 2030.

Solution to Problem 8

To create a stride-1 reference pattern, the loops must be permuted so that the rightmost indices change most rapidly.

```
1    int sumarray3d(int a[N][N][N])
2    {
3        int i, j, k, sum = 0;
4
5        for (k = 0; k < N; k++) {
6            for (i = 0; i < N; i++) {
7                for (j = 0; j < N; j++) {
8                    sum += a[k][i][j];
9                }
10           }
11       }
12       return sum;
13   }
```

This is an important idea. Make sure you understand why this particular loop permutation results in a stride-1 access pattern.

Solution to Problem 9

The key to solving this problem is to visualize how the array is laid out in memory and then analyze the reference patterns. Function clear1 accesses the array using a stride-1 reference pattern and thus clearly has the best spatial locality. Function clear2 scans each of the N structs in order, which is good, but within each struct it hops around in a non-stride-1 pattern at the following offsets from the beginning of the struct: 0, 12, 4, 16, 8, 20. So clear2 has worse spatial locality than clear1. Function clear3 not only hops around within each struct, but it also hops from struct to struct. So clear3 exhibits worse spatial locality than clear2 and clear1.

Solution to Problem 10

The solution is a straightforward application of the definitions of the various cache parameters in Figure 28. Not very exciting, but you need to understand how the cache organization induces these partitions in the address bits before you can really understand how caches work.

Cache	m	C	B	E	S	t	s	b
1.	32	1024	4	1	256	22	8	2
2.	32	1024	8	4	32	24	5	3
3.	32	1024	32	32	1	27	0	5

Solution to Problem 11

The padding eliminates the conflict misses. Thus, three-fourths of the references are hits.

Solution to Problem 12

Sometimes, understanding why something is a bad idea helps you understand why the alternative is a good idea. Here, the bad idea we are looking at is indexing the cache with the high-order bits instead of the middle bits.

A. With high-order bit indexing, each contiguous array chunk consists of 2^t blocks, where t is the number of tag bits. Thus, the first 2^t contiguous blocks of the array would map to set 0, the next 2^t blocks would map to set 1, and so on.

B. For a direct-mapped cache where $(S, E, B, m) = (512, 1, 32, 32)$, the cache capacity is 512 32-byte blocks, and there are $t = 18$ tag bits in each cache line. Thus, the first 2^{18} blocks in the array would map to set 0, the next 2^{18} blocks to set 1. Since our array consists of only $(4096 * 4)/32 = 512$ blocks, all of the blocks in the array map to set 0. Thus, the cache will hold at most one array block at any point in time, even though the array is small enough to fit entirely in the cache. Clearly, using high-order bit indexing makes poor use of the cache.

Solution to Problem 13

The 2 low-order bits are the block offset (CO), followed by 3 bits of set index (CI), with the remaining bits serving as the tag (CT):

12	11	10	9	8	7	6	5	4	3	2	1	0
CT	CT	CT	CT	CT	CT	CT	CT	CI	CI	CI	CO	CO

Solution to Problem 14

Address: 0x0E34

A. Address format (one bit per box):

12	11	10	9	8	7	6	5	4	3	2	1	0
0	1	1	1	0	0	0	1	1	0	1	0	0
CT	CT	CT	CT	CT	CT	CT	CT	CI	CI	CI	CO	CO

B. Memory reference:

Parameter	Value
Cache block offset (CO)	0x0
Cache set index (CI)	0x5
Cache tag (CT)	0x71
Cache hit? (Y/N)	Y
Cache byte returned	0xB

Solution to Problem 15

Address: 0x0DD5

A. Address format (one bit per box):

12	11	10	9	8	7	6	5	4	3	2	1	0
0	1	1	0	1	1	1	0	1	0	1	0	1
CT	CT	CT	CT	CT	CT	CT	CT	CI	CI	CI	CO	CO

B. Memory reference:

Parameter	Value
Cache block offset (CO)	0x1
Cache set index (CI)	0x5
Cache tag (CT)	0x6E
Cache hit? (Y/N)	N
Cache byte returned	—

Solution to Problem 16

Address: 0x1FE4

A. Address format (one bit per box):

12	11	10	9	8	7	6	5	4	3	2	1	0
1	1	1	1	1	1	1	1	0	0	1	0	0
CT	CT	CT	CT	CT	CT	CT	CT	CI	CI	CI	CO	CO

B. Memory reference:

Parameter	Value
Cache block offset	0x0
Cache set index	0x1
Cache tag	0xFF
Cache hit? (Y/N)	N
Cache byte returned	—

Solution to Problem 17

This problem is a sort of inverse version of Problems 13–16 that requires you to work backward from the contents of the cache to derive the addresses that will hit in a particular set. In this case, set 3 contains one valid line with a tag of 0x32. Since there is only one valid line in the set, four addresses will hit. These addresses have the binary form 0 0110 0100 11xx. Thus, the four hex addresses that hit in set 3 are

0x064C, 0x064D, 0x064E, and 0x064F

Solution to Problem 18

A. The key to solving this problem is to visualize the picture in Figure 50. Notice that each cache line holds exactly one row of the array, that the cache is exactly large enough to hold one array, and that for all *i*, row *i* of src and dst maps to the same cache line. Because the cache is too small to hold both arrays, references to one array keep evicting useful lines from the other array. For example, the write to dst[0][0] evicts the line that was loaded when we read src[0][0]. So when we next read src[0][1], we have a miss.

dst array	Col 0	Col 1
Row 0	m	m
Row 1	m	m

src array	Col 0	Col 1
Row 0	m	m
Row 1	m	h

Figure 50

Figure for Problem 18.

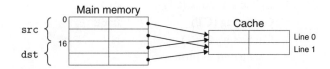

642

B. When the cache is 32 bytes, it is large enough to hold both arrays. Thus, the only misses are the initial cold misses.

dst array

	Col 0	Col 1
Row 0	m	h
Row 1	m	h

src array

	Col 0	Col 1
Row 0	m	h
Row 1	m	h

Solution to Problem 19

Each 16-byte cache line holds two contiguous `algae_position` structures. Each loop visits these structures in memory order, reading one integer element each time. So the pattern for each loop is miss, hit, miss, hit, and so on. Notice that for this problem we could have predicted the miss rate without actually enumerating the total number of reads and misses.

A. What is the total number of read accesses? 512 reads.

B. What is the total number of read accesses that miss in the cache? 256 misses.

C. What is the miss rate? $256/512 = 50\%$.

Solution to Problem 20

The key to this problem is noticing that the cache can only hold 1/2 of the array. So the column-wise scan of the second half of the array evicts the lines that were loaded during the scan of the first half. For example, reading the first element of `grid[8][0]` evicts the line that was loaded when we read elements from `grid[0][0]`. This line also contained `grid[0][1]`. So when we begin scanning the next column, the reference to the first element of `grid[0][1]` misses.

A. What is the total number of read accesses? 512 reads.

B. What is the total number of read accesses that miss in the cache? 256 misses.

C. What is the miss rate? $256/512 = 50\%$.

D. What would the miss rate be if the cache were twice as big? If the cache were twice as big, it could hold the entire `grid` array. The only misses would be the initial cold misses, and the miss rate would be $1/4 = 25\%$.

Solution to Problem 21

This loop has a nice stride-1 reference pattern, and thus the only misses are the initial cold misses.

A. What is the total number of read accesses? 512 reads.

B. What is the total number of read accesses that miss in the cache? 128 misses.

C. What is the miss rate? $128/512 = 25\%$.

    D. What would the miss rate be if the cache were twice as big? Increasing the cache size by any amount would not change the miss rate, since cold misses are unavoidable.

Solution to Problem 22

The peak throughput from L1 is about 6500 MB/s, the clock frequency is 2670 MHz, and the individual read accesses are in units of 8-byte `doubles`. Thus, from this graph we can estimate that it takes roughly $2670/6500 \times 8 = 3.2 \approx 4$ cycles to access a word from L1 on this machine.

Linking

From Chapter 7 of *Computer Systems: A Programmer's Perspective,* Second Edition. Randal E. Bryant and David R. O'Hallaron. Copyright © 2011 by Pearson Education, Inc. Published by Prentice Hall. All rights reserved.

Linking is the process of collecting and combining various pieces of code and data into a single file that can be *loaded* (copied) into memory and executed. Linking can be performed at *compile time*, when the source code is translated into machine code; at *load time*, when the program is loaded into memory and executed by the *loader*; and even at *run time*, by application programs. On early computer systems, linking was performed manually. On modern systems, linking is performed automatically by programs called *linkers*.

Linkers play a crucial role in software development because they enable *separate compilation*. Instead of organizing a large application as one monolithic source file, we can decompose it into smaller, more manageable modules that can be modified and compiled separately. When we change one of these modules, we simply recompile it and relink the application, without having to recompile the other files.

Linking is usually handled quietly by the linker, and is not an important issue for students who are building small programs in introductory programming classes. So why bother learning about linking?

- *Understanding linkers will help you build large programs.* Programmers who build large programs often encounter linker errors caused by missing modules, missing libraries, or incompatible library versions. Unless you understand how a linker resolves references, what a library is, and how a linker uses a library to resolve references, these kinds of errors will be baffling and frustrating.

- *Understanding linkers will help you avoid dangerous programming errors.* The decisions that Unix linkers make when they resolve symbol references can silently affect the correctness of your programs. Programs that incorrectly define multiple global variables pass through the linker without any warnings in the default case. The resulting programs can exhibit baffling run-time behavior and are extremely difficult to debug. We will show you how this happens and how to avoid it.

- *Understanding linking will help you understand how language scoping rules are implemented.* For example, what is the difference between global and local variables? What does it really mean when you define a variable or function with the `static` attribute?

- *Understanding linking will help you understand other important systems concepts.* The executable object files produced by linkers play key roles in important systems functions such as loading and running programs, virtual memory, paging, and memory mapping.

- *Understanding linking will enable you to exploit shared libraries.* For many years, linking was considered to be fairly straightforward and uninteresting. However, with the increased importance of shared libraries and dynamic linking in modern operating systems, linking is a sophisticated process that provides the knowledgeable programmer with significant power. For example, many software products use shared libraries to upgrade shrink-wrapped binaries at run time. Also, most Web servers rely on dynamic linking of shared libraries to serve dynamic content.

This chapter provides a thorough discussion of all aspects of linking, from traditional static linking, to dynamic linking of shared libraries at load time, to dynamic linking of shared libraries at run time. We will describe the basic mechanisms using real examples, and we will identify situations in which linking issues can affect the performance and correctness of your programs.

To keep things concrete and understandable, we will couch our discussion in the context of an x86 system running Linux and using the standard ELF object file format. For clarity, we will focus our discussion on linking 32-bit code, which is easier to understand than linking 64-bit code.[1] However, it is important to realize that the basic concepts of linking are universal, regardless of the operating system, the ISA, or the object file format. Details may vary, but the concepts are the same.

1 Compiler Drivers

Consider the C program in Figure 1. It consists of two source files, `main.c` and `swap.c`. Function `main()` calls `swap`, which swaps the two elements in the external global array `buf`. Granted, this is a strange way to swap two numbers, but it will serve as a small running example throughout this chapter that will allow us to make some important points about how linking works.

Most compilation systems provide a *compiler driver* that invokes the language preprocessor, compiler, assembler, and linker, as needed on behalf of the user. For example, to build the example program using the GNU compilation system, we might invoke the GCC driver by typing the following command to the shell:

```
unix> gcc -O2 -g -o p main.c swap.c
```

Figure 2 summarizes the activities of the driver as it translates the example program from an ASCII source file into an executable object file. (If you want to see these steps for yourself, run GCC with the `-v` option.) The driver first runs the C preprocessor (cpp), which translates the C source file `main.c` into an ASCII intermediate file `main.i`:

```
cpp [other arguments] main.c /tmp/main.i
```

Next, the driver runs the C compiler (`cc1`), which translates `main.i` into an ASCII assembly language file `main.s`.

```
cc1 /tmp/main.i main.c -O2 [other arguments] -o /tmp/main.s
```

Then, the driver runs the assembler (`as`), which translates `main.s` into a *relocatable object file* `main.o`:

```
as [other arguments] -o /tmp/main.o /tmp/main.s
```

---

1. You can generate 32-bit code on an x86-64 system using gcc -m32.

(a) `main.c`

——————————————————————————— *code/link/main.c*

```
1   /* main.c */
2   void swap();
3
4   int buf[2] = {1, 2};
5
6   int main()
7   {
8       swap();
9       return 0;
10  }
```

——————————————————————————— *code/link/main.c*

(b) `swap.c`

——————————————————————————— *code/link/swap.c*

```
1   /* swap.c */
2   extern int buf[];
3
4   int *bufp0 = &buf[0];
5   int *bufp1;
6
7   void swap()
8   {
9       int temp;
10
11      bufp1 = &buf[1];
12      temp = *bufp0;
13      *bufp0 = *bufp1;
14      *bufp1 = temp;
15  }
```

——————————————————————————— *code/link/swap.c*

Figure 1   **Example program 1:** The example program consists of two source files, `main.c` and `swap.c`. The `main` function initializes a two-element array of ints, and then calls the `swap` function to swap the pair.

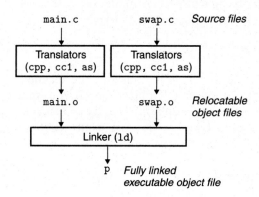

Figure 2   **Static linking.** The linker combines relocatable object files to form an executable object file p.

The driver goes through the same process to generate `swap.o`. Finally, it runs the linker program `ld`, which combines `main.o` and `swap.o`, along with the necessary system object files, to create the *executable object file* p:

```
ld -o p [system object files and args] /tmp/main.o /tmp/swap.o
```

To run the executable p, we type its name on the Unix shell's command line:

```
unix> ./p
```

The shell invokes a function in the operating system called the *loader*, which copies the code and data in the executable file p into memory, and then transfers control to the beginning of the program.

2 Static Linking

Static linkers such as the Unix ld program take as input a collection of relocatable object files and command-line arguments and generate as output a fully linked executable object file that can be loaded and run. The input relocatable object files consist of various code and data sections. Instructions are in one section, initialized global variables are in another section, and uninitialized variables are in yet another section.

To build the executable, the linker must perform two main tasks:

- *Symbol resolution.* Object files define and reference *symbols*. The purpose of symbol resolution is to associate each symbol reference with exactly one symbol definition.
- *Relocation.* Compilers and assemblers generate code and data sections that start at address 0. The linker *relocates* these sections by associating a memory location with each symbol definition, and then modifying all of the references to those symbols so that they point to this memory location.

The sections that follow describe these tasks in more detail. As you read, keep in mind some basic facts about linkers: Object files are merely collections of blocks of bytes. Some of these blocks contain program code, others contain program data, and others contain data structures that guide the linker and loader. A linker concatenates blocks together, decides on run-time locations for the concatenated blocks, and modifies various locations within the code and data blocks. Linkers have minimal understanding of the target machine. The compilers and assemblers that generate the object files have already done most of the work.

3 Object Files

Object files come in three forms:

- *Relocatable object file.* Contains binary code and data in a form that can be combined with other relocatable object files at compile time to create an executable object file.
- *Executable object file.* Contains binary code and data in a form that can be copied directly into memory and executed.
- *Shared object file.* A special type of relocatable object file that can be loaded into memory and linked dynamically, at either load time or run time.

Compilers and assemblers generate relocatable object files (including shared object files). Linkers generate executable object files. Technically, an *object module* is a sequence of bytes, and an *object file* is an object module stored on disk in a file. However, we will use these terms interchangeably.

Object file formats vary from system to system. The first Unix systems from Bell Labs used the a.out format. (To this day, executables are still referred to as a.out files.) Early versions of System V Unix used the Common Object File format (COFF). Windows NT uses a variant of COFF called the Portable Executable (PE) format. Modern Unix systems—such as Linux, later versions of System V Unix, BSD Unix variants, and Sun Solaris—use the Unix *Executable and Linkable Format* (ELF). Although our discussion will focus on ELF, the basic concepts are similar, regardless of the particular format.

4 Relocatable Object Files

Figure 3 shows the format of a typical ELF relocatable object file. The *ELF header* begins with a 16-byte sequence that describes the word size and byte ordering of the system that generated the file. The rest of the ELF header contains information that allows a linker to parse and interpret the object file. This includes the size of the ELF header, the object file type (e.g., relocatable, executable, or shared), the machine type (e.g., IA32), the file offset of the section header table, and the size and number of entries in the section header table. The locations and sizes of the various sections are described by the *section header table*, which contains a fixed sized entry for each section in the object file.

Sandwiched between the ELF header and the section header table are the sections themselves. A typical ELF relocatable object file contains the following sections:

.text: The machine code of the compiled program.

.rodata: Read-only data such as the format strings in printf statements, and jump tables for switch statements (see Problem 14).

Figure 3
Typical ELF relocatable object file.

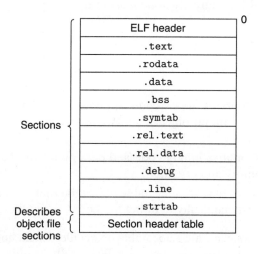

`.data`: *Initialized* global C variables. Local C variables are maintained at run time on the stack, and do not appear in either the `.data` or `.bss` sections.

`.bss`: *Uninitialized* global C variables. This section occupies no actual space in the object file; it is merely a place holder. Object file formats distinguish between initialized and uninitialized variables for space efficiency: uninitialized variables do not have to occupy any actual disk space in the object file.

`.symtab`: A *symbol table* with information about functions and global variables that are defined and referenced in the program. Some programmers mistakenly believe that a program must be compiled with the -g option to get symbol table information. In fact, every relocatable object file has a symbol table in `.symtab`. However, unlike the symbol table inside a compiler, the `.symtab` symbol table does not contain entries for local variables.

`.rel.text`: A list of locations in the `.text` section that will need to be modified when the linker combines this object file with others. In general, any instruction that calls an external function or references a global variable will need to be modified. On the other hand, instructions that call local functions do not need to be modified. Note that relocation information is not needed in executable object files, and is usually omitted unless the user explicitly instructs the linker to include it.

`.rel.data`: Relocation information for any global variables that are referenced or defined by the module. In general, any initialized global variable whose initial value is the address of a global variable or externally defined function will need to be modified.

`.debug`: A debugging symbol table with entries for local variables and typedefs defined in the program, global variables defined and referenced in the program, and the original C source file. It is only present if the compiler driver is invoked with the -g option.

`.line`: A mapping between line numbers in the original C source program and machine code instructions in the .text section. It is only present if the compiler driver is invoked with the -g option.

`.strtab`: A string table for the symbol tables in the `.symtab` and `.debug` sections, and for the section names in the section headers. A string table is a sequence of null-terminated character strings.

Aside   Why is uninitialized data called `.bss`?

The use of the term `.bss` to denote uninitialized data is universal. It was originally an acronym for the "Block Storage Start" instruction from the IBM 704 assembly language (circa 1957) and the acronym has stuck. A simple way to remember the difference between the `.data` and `.bss` sections is to think of "bss" as an abbreviation for "Better Save Space!"

5 Symbols and Symbol Tables

Each relocatable object module, m, has a symbol table that contains information about the symbols that are defined and referenced by m. In the context of a linker, there are three different kinds of symbols:

- *Global symbols* that are defined by module m and that can be referenced by other modules. Global linker symbols correspond to *nonstatic* C functions and global variables that are defined *without* the C static attribute.

- Global symbols that are referenced by module m but defined by some other module. Such symbols are called *externals* and correspond to C functions and variables that are defined in other modules.

- *Local symbols* that are defined and referenced exclusively by module m. Some local linker symbols correspond to C functions and global variables that are defined with the static attribute. These symbols are visible anywhere within module m, but cannot be referenced by other modules. The sections in an object file and the name of the source file that corresponds to module m also get local symbols.

It is important to realize that local linker symbols are not the same as local program variables. The symbol table in .symtab does not contain any symbols that correspond to local nonstatic program variables. These are managed at run time on the stack and are not of interest to the linker.

Interestingly, local procedure variables that are defined with the C static attribute are not managed on the stack. Instead, the compiler allocates space in .data or .bss for each definition and creates a local linker symbol in the symbol table with a unique name. For example, suppose a pair of functions in the same module define a static local variable x:

```
1    int f()
2    {
3        static int x = 0;
4        return x;
5    }
6
7    int g()
8    {
9        static int x = 1;
10       return x;
11   }
```

In this case, the compiler allocates space for two integers in .data and exports a pair of unique local linker symbols to the assembler. For example, it might use x.1 for the definition in function f and x.2 for the definition in function g.

New to C?   Hiding variable and function names with `static`

C programmers use the `static` attribute to hide variable and function declarations inside modules, much as you would use *public* and *private* declarations in Java and C++. C source files play the role of modules. Any global variable or function declared with the `static` attribute is private to that module. Similarly, any global variable or function declared without the `static` attribute is public and can be accessed by any other module. It is good programming practice to protect your variables and functions with the `static` attribute wherever possible.

Symbol tables are built by assemblers, using symbols exported by the compiler into the assembly-language `.s` file. An ELF symbol table is contained in the `.symtab` section. It contains an array of entries. Figure 4 shows the format of each entry.

The `name` is a byte offset into the string table that points to the null-terminated string name of the symbol. The `value` is the symbol's address. For relocatable modules, the `value` is an offset from the beginning of the section where the object is defined. For executable object files, the `value` is an absolute run-time address. The `size` is the size (in bytes) of the object. The `type` is usually either data or function. The symbol table can also contain entries for the individual sections and for the path name of the original source file. So there are distinct types for these objects as well. The `binding` field indicates whether the symbol is local or global.

Each symbol is associated with some section of the object file, denoted by the `section` field, which is an index into the section header table. There are three special pseudo sections that don't have entries in the section header table: ABS is for symbols that should not be relocated. UNDEF is for undefined symbols, that is, symbols that are referenced in this object module but defined elsewhere. COMMON is for uninitialized data objects that are not yet allocated. For COMMON symbols, the `value` field gives the alignment requirement, and `size` gives the minimum size.

―――――――――――――――――――――――――――― code/link/elfstructs.c

```
1   typedef struct {
2       int name;           /* String table offset */
3       int value;          /* Section offset, or VM address */
4       int size;           /* Object size in bytes */
5       char type:4,        /* Data, func, section, or src file name (4 bits) */
6            binding:4;      /* Local or global (4 bits) */
7       char reserved;      /* Unused */
8       char section;       /* Section header index, ABS, UNDEF, */
9                           /* Or COMMON */
10  } Elf_Symbol;
```

―――――――――――――――――――――――――――― code/link/elfstructs.c

Figure 4 **ELF symbol table entry.** `type` and `binding` are four bits each.

For example, here are the last three entries in the symbol table for main.o, as displayed by the GNU READELF tool. The first eight entries, which are not shown, are local symbols that the linker uses internally.

```
Num:    Value  Size Type    Bind    Ot  Ndx Name
   8:       0     8 OBJECT  GLOBAL   0    3 buf
   9:       0    17 FUNC    GLOBAL   0    1 main
  10:       0     0 NOTYPE  GLOBAL   0  UND swap
```

In this example, we see an entry for the definition of global symbol buf, an 8-byte object located at an offset (i.e., value) of zero in the .data section. This is followed by the definition of the global symbol main, a 17-byte function located at an offset of zero in the .text section. The last entry comes from the reference for the external symbol swap. READELF identifies each section by an integer index. Ndx=1 denotes the .text section, and Ndx=3 denotes the .data section.

Similarly, here are the symbol table entries for swap.o:

```
Num:    Value  Size Type    Bind    Ot  Ndx Name
   8:       0     4 OBJECT  GLOBAL   0    3 bufp0
   9:       0     0 NOTYPE  GLOBAL   0  UND buf
  10:       0    39 FUNC    GLOBAL   0    1 swap
  11:       4     4 OBJECT  GLOBAL   0  COM bufp1
```

First, we see an entry for the definition of the global symbol bufp0, which is a 4-byte initialized object starting at offset 0 in .data. The next symbol comes from the reference to the external buf symbol in the initialization code for bufp0. This is followed by the global symbol swap, a 39-byte function at an offset of zero in .text. The last entry is the global symbol bufp1, a 4-byte uninitialized data object (with a 4-byte alignment requirement) that will eventually be allocated as a .bss object when this module is linked.

Practice Problem 1

This problem concerns the swap.o module from Figure 1(b). For each symbol that is defined or referenced in swap.o, indicate whether or not it will have a symbol table entry in the .symtab section in module swap.o. If so, indicate the module that defines the symbol (swap.o or main.o), the symbol type (local, global, or extern), and the section (.text, .data, or .bss) it occupies in that module.

Symbol	swap.o .symtab entry?	Symbol type	Module where defined	Section
buf				
bufp0				
bufp1				
swap				
temp				

6 Symbol Resolution

The linker resolves symbol references by associating each reference with exactly one symbol definition from the symbol tables of its input relocatable object files. Symbol resolution is straightforward for references to local symbols that are defined in the same module as the reference. The compiler allows only one definition of each local symbol per module. The compiler also ensures that static local variables, which get local linker symbols, have unique names.

Resolving references to global symbols, however, is trickier. When the compiler encounters a symbol (either a variable or function name) that is not defined in the current module, it assumes that it is defined in some other module, generates a linker symbol table entry, and leaves it for the linker to handle. If the linker is unable to find a definition for the referenced symbol in any of its input modules, it prints an (often cryptic) error message and terminates. For example, if we try to compile and link the following source file on a Linux machine,

```
1    void foo(void);
2
3    int main() {
4        foo();
5        return 0;
6    }
```

then the compiler runs without a hitch, but the linker terminates when it cannot resolve the reference to foo:

```
unix> gcc -Wall -O2 -o linkerror linkerror.c
/tmp/ccSz5uti.o: In function 'main':
/tmp/ccSz5uti.o(.text+0x7): undefined reference to 'foo'
collect2: ld returned 1 exit status
```

Symbol resolution for global symbols is also tricky because the same symbol might be defined by multiple object files. In this case, the linker must either flag an error or somehow choose one of the definitions and discard the rest. The approach adopted by Unix systems involves cooperation between the compiler, assembler, and linker, and can introduce some baffling bugs to the unwary programmer.

Aside   Mangling of linker symbols in C++ and Java

Both C++ and Java allow overloaded methods that have the same name in the source code but different parameter lists. So how does the linker tell the difference between these different overloaded functions? Overloaded functions in C++ and Java work because the compiler encodes each unique method and parameter list combination into a unique name for the linker. This encoding process is called *mangling*, and the inverse process *demangling*.

Happily, C++ and Java use compatible mangling schemes. A mangled class name consists of the integer number of characters in the name followed by the original name. For example, the class Foo is encoded as 3Foo. A method is encoded as the original method name, followed by __, followed

---

by the mangled class name, followed by single letter encodings of each argument. For example, `Foo::bar(int, long)` is encoded as `bar__3Fooil`. Similar schemes are used to mangle global variable and template names.

6.1 How Linkers Resolve Multiply Defined Global Symbols

At compile time, the compiler exports each global symbol to the assembler as either *strong* or *weak*, and the assembler encodes this information implicitly in the symbol table of the relocatable object file. Functions and initialized global variables get strong symbols. Uninitialized global variables get weak symbols. For the example program in Figure 1, `buf`, `bufp0`, `main`, and `swap` are strong symbols; `bufp1` is a weak symbol.

Given this notion of strong and weak symbols, Unix linkers use the following rules for dealing with multiply defined symbols:

- Rule 1: Multiple strong symbols are not allowed.
- Rule 2: Given a strong symbol and multiple weak symbols, choose the strong symbol.
- Rule 3: Given multiple weak symbols, choose any of the weak symbols.

For example, suppose we attempt to compile and link the following two C modules:

```
1   /* foo1.c */          1   /* bar1.c */
2   int main()            2   int main()
3   {                     3   {
4       return 0;         4       return 0;
5   }                     5   }
```

In this case, the linker will generate an error message because the strong symbol `main` is defined multiple times (rule 1):

```
unix> gcc foo1.c bar1.c
/tmp/cca015022.o: In function 'main':
/tmp/cca015022.o(.text+0x0): multiple definition of 'main'
/tmp/cca015021.o(.text+0x0): first defined here
```

Similarly, the linker will generate an error message for the following modules because the strong symbol x is defined twice (rule 1):

```
1   /* foo2.c */          1   /* bar2.c */
2   int x = 15213;        2   int x = 15213;
3                         3
4   int main()            4   void f()
5   {                     5   {
6       return 0;         6   }
7   }
```

However, if x is uninitialized in one module, then the linker will quietly choose the strong symbol defined in the other (rule 2):

```
1   /* foo3.c */
2   #include <stdio.h>
3   void f(void);
4
5   int x = 15213;
6
7   int main()
8   {
9       f();
10      printf("x = %d\n", x);
11      return 0;
12  }
```

```
1   /* bar3.c */
2   int x;
3
4   void f()
5   {
6       x = 15212;
7   }
```

At run time, function f changes the value of x from 15213 to 15212, which might come as an unwelcome surprise to the author of function main! Notice that the linker normally gives no indication that it has detected multiple definitions of x:

```
unix> gcc -o foobar3 foo3.c bar3.c
unix> ./foobar3
x = 15212
```

The same thing can happen if there are two weak definitions of x (rule 3):

```
1   /* foo4.c */
2   #include <stdio.h>
3   void f(void);
4
5   int x;
6
7   int main()
8   {
9       x = 15213;
10      f();
11      printf("x = %d\n", x);
12      return 0;
13  }
```

```
1   /* bar4.c */
2   int x;
3
4   void f()
5   {
6       x = 15212;
7   }
```

The application of rules 2 and 3 can introduce some insidious run-time bugs that are incomprehensible to the unwary programmer, especially if the duplicate symbol definitions have different types. Consider the following example, in which x is defined as an int in one module and a double in another:

```
1   /* foo5.c */              1   /* bar5.c */
2   #include <stdio.h>         2   double x;
3   void f(void);             3
                              4   void f()
5   int x = 15213;           5   {
6   int y = 15212;           6       x = -0.0;
7                            7   }
8   int main()
9   {
10      f();
11      printf("x = 0x%x y = 0x%x \n",
12              x, y);
13      return 0;
14  }
```

On an IA32/Linux machine, doubles are 8 bytes and ints are 4 bytes. Thus, the assignment x = -0.0 in line 6 of bar5.c will overwrite the memory locations for x and y (lines 5 and 6 in foo5.c) with the double-precision floating-point representation of negative zero!

```
linux> gcc -o foobar5 foo5.c bar5.c
linux> ./foobar5
x = 0x0 y = 0x80000000
```

This is a subtle and nasty bug, especially because it occurs silently, with no warning from the compilation system, and because it typically manifests itself much later in the execution of the program, far away from where the error occurred. In a large system with hundreds of modules, a bug of this kind is extremely hard to fix, especially because many programmers are not aware of how linkers work. When in doubt, invoke the linker with a flag such as the GCC –fno-common flag, which triggers an error if it encounters multiply defined global symbols.

Practice Problem 2

In this problem, let REF(x.i) --> DEF(x.k) denote that the linker will associate an arbitrary reference to symbol x in module i to the definition of x in module k. For each example that follows, use this notation to indicate how the linker would resolve references to the multiply defined symbol in each module. If there is a link-time error (rule 1), write "ERROR." If the linker arbitrarily chooses one of the definitions (rule 3), write "UNKNOWN."

```
A.  /* Module 1 */           /* Module 2 */
    int main()               int main;
    {                        int p2()
    }                        {
                             }
```

    (a) `REF(main.1) --> DEF(_____.___)`
    (b) `REF(main.2) --> DEF(_____.___)`

B.
```
/* Module 1 */        /* Module 2 */
void main()           int main=1;
{                     int p2()
}                     {
                      }
```

    (a) `REF(main.1) --> DEF(_____.___)`
    (b) `REF(main.2) --> DEF(_____.___)`

C.
```
/* Module 1 */        /* Module 2 */
int x;                double x=1.0;
void main()           int p2()
{                     {
}                     }
```

    (a) `REF(x.1) --> DEF(_____.___)`
    (b) `REF(x.2) --> DEF(_____.___)`

6.2 Linking with Static Libraries

So far, we have assumed that the linker reads a collection of relocatable object files and links them together into an output executable file. In practice, all compilation systems provide a mechanism for packaging related object modules into a single file called a *static library*, which can then be supplied as input to the linker. When it builds the output executable, the linker copies only the object modules in the library that are referenced by the application program.

Why do systems support the notion of libraries? Consider ANSI C, which defines an extensive collection of standard I/O, string manipulation, and integer math functions such as atoi, printf, scanf, strcpy, and rand. They are available to every C program in the libc.a library. ANSI C also defines an extensive collection of floating-point math functions such as sin, cos, and sqrt in the libm.a library.

Consider the different approaches that compiler developers might use to provide these functions to users without the benefit of static libraries. One approach would be to have the compiler recognize calls to the standard functions and to generate the appropriate code directly. Pascal, which provides a small set of standard functions, takes this approach, but it is not feasible for C, because of the large number of standard functions defined by the C standard. It would add significant complexity to the compiler and would require a new compiler version each time a function was added, deleted, or modified. To application programmers, however, this approach would be quite convenient because the standard functions would always be available.

Another approach would be to put all of the standard C functions in a single relocatable object module, say, `libc.o`, that application programmers could link into their executables:

```
unix> gcc main.c /usr/lib/libc.o
```

This approach has the advantage that it would decouple the implementation of the standard functions from the implementation of the compiler, and would still be reasonably convenient for programmers. However, a big disadvantage is that every executable file in a system would now contain a complete copy of the collection of standard functions, which would be extremely wasteful of disk space. (On a typical system, `libc.a` is about 8 MB and `libm.a` is about 1 MB.) Worse, each running program would now contain its own copy of these functions in memory, which would be extremely wasteful of memory. Another big disadvantage is that any change to any standard function, no matter how small, would require the library developer to recompile the entire source file, a time-consuming operation that would complicate the development and maintenance of the standard functions.

We could address some of these problems by creating a separate relocatable file for each standard function and storing them in a well-known directory. However, this approach would require application programmers to explicitly link the appropriate object modules into their executables, a process that would be error prone and time consuming:

```
unix> gcc main.c /usr/lib/printf.o /usr/lib/scanf.o ...
```

The notion of a static library was developed to resolve the disadvantages of these various approaches. Related functions can be compiled into separate object modules and then packaged in a single static library file. Application programs can then use any of the functions defined in the library by specifying a single file name on the command line. For example, a program that uses functions from the standard C library and the math library could be compiled and linked with a command of the form

```
unix> gcc main.c /usr/lib/libm.a /usr/lib/libc.a
```

At link time, the linker will only copy the object modules that are referenced by the program, which reduces the size of the executable on disk and in memory. On the other hand, the application programmer only needs to include the names of a few library files. (In fact, C compiler drivers always pass `libc.a` to the linker, so the reference to `libc.a` mentioned previously is unnecessary.)

On Unix systems, static libraries are stored on disk in a particular file format known as an *archive*. An archive is a collection of concatenated relocatable object files, with a header that describes the size and location of each member object file. Archive filenames are denoted with the `.a` suffix. To make our discussion of libraries concrete, suppose that we want to provide the vector routines in Figure 5 in a static library called `libvector.a`.

(a) `addvec.o`

——————————————————— code/link/addvec.c

```
1    void addvec(int *x, int *y,
2                int *z, int n)
3    {
4        int i;
5
6        for (i = 0; i < n; i++)
7            z[i] = x[i] + y[i];
8    }
```

——————————————————— code/link/addvec.c

(b) `multvec.o`

——————————————————— code/link/multvec.c

```
1    void multvec(int *x, int *y,
2                int *z, int n)
3    {
4        int i;
5
6        for (i = 0; i < n; i++)
7            z[i] = x[i] * y[i];
8    }
```

——————————————————— code/link/multvec.c

Figure 5 **Member object files in** `libvector.a`.

To create the library, we would use the AR tool as follows:

```
unix> gcc -c addvec.c multvec.c
unix> ar rcs libvector.a addvec.o multvec.o
```

To use the library, we might write an application such as `main2.c` in Figure 6, which invokes the `addvec` library routine. (The include (header) file `vector.h` defines the function prototypes for the routines in `libvector.a`.)

——————————————————————————————————————— code/link/main2.c

```
1    /* main2.c */
2    #include <stdio.h>
3    #include "vector.h"
4
5    int x[2] = {1, 2};
6    int y[2] = {3, 4};
7    int z[2];
8
9    int main()
10   {
11       addvec(x, y, z, 2);
12       printf("z = [%d %d]\n", z[0], z[1]);
13       return 0;
14   }
```

——————————————————————————————————————— code/link/main2.c

Figure 6 **Example program 2:** This program calls member functions in the static `libvector.a` library.

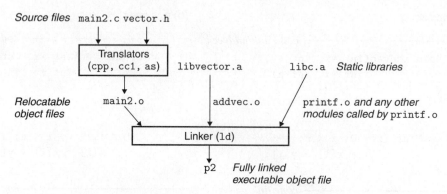

Figure 7 **Linking with static libraries.**

To build the executable, we would compile and link the input files `main.o` and `libvector.a`:

```
unix> gcc -O2 -c main2.c
unix> gcc -static -o p2 main2.o ./libvector.a
```

Figure 7 summarizes the activity of the linker. The `-static` argument tells the compiler driver that the linker should build a fully linked executable object file that can be loaded into memory and run without any further linking at load time. When the linker runs, it determines that the addvec symbol defined by `addvec.o` is referenced by `main.o`, so it copies `addvec.o` into the executable. Since the program doesn't reference any symbols defined by `multvec.o`, the linker does *not* copy this module into the executable. The linker also copies the `printf.o` module from `libc.a`, along with a number of other modules from the C run-time system.

6.3 How Linkers Use Static Libraries to Resolve References

While static libraries are useful and essential tools, they are also a source of confusion to programmers because of the way the Unix linker uses them to resolve external references. During the symbol resolution phase, the linker scans the relocatable object files and archives left to right in the same sequential order that they appear on the compiler driver's command line. (The driver automatically translates any `.c` files on the command line into `.o` files.) During this scan, the linker maintains a set E of relocatable object files that will be merged to form the executable, a set U of unresolved symbols (i.e., symbols referred to, but not yet defined), and a set D of symbols that have been defined in previous input files. Initially, E, U, and D are empty.

- For each input file f on the command line, the linker determines if f is an object file or an archive. If f is an object file, the linker adds f to E, updates U and D to reflect the symbol definitions and references in f, and proceeds to the next input file.

- If f is an archive, the linker attempts to match the unresolved symbols in U against the symbols defined by the members of the archive. If some archive member, m, defines a symbol that resolves a reference in U, then m is added to E, and the linker updates U and D to reflect the symbol definitions and references in m. This process iterates over the member object files in the archive until a fixed point is reached where U and D no longer change. At this point, any member object files not contained in E are simply discarded and the linker proceeds to the next input file.

- If U is nonempty when the linker finishes scanning the input files on the command line, it prints an error and terminates. Otherwise, it merges and relocates the object files in E to build the output executable file.

Unfortunately, this algorithm can result in some baffling link-time errors because the ordering of libraries and object files on the command line is significant. If the library that defines a symbol appears on the command line before the object file that references that symbol, then the reference will not be resolved and linking will fail. For example, consider the following:

```
unix> gcc -static ./libvector.a main2.c
/tmp/cc9XH6Rp.o: In function 'main':
/tmp/cc9XH6Rp.o(.text+0x18): undefined reference to 'addvec'
```

What happened? When `libvector.a` is processed, U is empty, so no member object files from `libvector.a` are added to E. Thus, the reference to `addvec` is never resolved and the linker emits an error message and terminates.

The general rule for libraries is to place them at the end of the command line. If the members of the different libraries are independent, in that no member references a symbol defined by another member, then the libraries can be placed at the end of the command line in any order.

If, on the other hand, the libraries are not independent, then they must be ordered so that for each symbol s that is referenced externally by a member of an archive, at least one definition of s follows a reference to s on the command line. For example, suppose `foo.c` calls functions in `libx.a` and `libz.a` that call functions in `liby.a`. Then `libx.a` and `libz.a` must precede `liby.a` on the command line:

```
unix> gcc foo.c libx.a libz.a liby.a
```

Libraries can be repeated on the command line if necessary to satisfy the dependence requirements. For example, suppose `foo.c` calls a function in `libx.a` that calls a function in `liby.a` that calls a function in `libx.a`. Then `libx.a` must be repeated on the command line:

```
unix> gcc foo.c libx.a liby.a libx.a
```

Alternatively, we could combine `libx.a` and `liby.a` into a single archive.

Practice Problem 3

Let a and b denote object modules or static libraries in the current directory, and let a→b denote that a depends on b, in the sense that b defines a symbol that is referenced by a. For each of the following scenarios, show the minimal command line (i.e., one with the least number of object file and library arguments) that will allow the static linker to resolve all symbol references.

A. p.o → libx.a.

B. p.o → libx.a → liby.a.

C. p.o → libx.a → liby.a **and** liby.a → libx.a →p.o.

7 Relocation

Once the linker has completed the symbol resolution step, it has associated each symbol reference in the code with exactly one symbol definition (i.e., a symbol table entry in one of its input object modules). At this point, the linker knows the exact sizes of the code and data sections in its input object modules. It is now ready to begin the relocation step, where it merges the input modules and assigns run-time addresses to each symbol. Relocation consists of two steps:

- *Relocating sections and symbol definitions.* In this step, the linker merges all sections of the same type into a new aggregate section of the same type. For example, the .data sections from the input modules are all merged into one section that will become the .data section for the output executable object file. The linker then assigns run-time memory addresses to the new aggregate sections, to each section defined by the input modules, and to each symbol defined by the input modules. When this step is complete, every instruction and global variable in the program has a unique run-time memory address.

- *Relocating symbol references within sections.* In this step, the linker modifies every symbol reference in the bodies of the code and data sections so that they point to the correct run-time addresses. To perform this step, the linker relies on data structures in the relocatable object modules known as relocation entries, which we describe next.

7.1 Relocation Entries

When an assembler generates an object module, it does not know where the code and data will ultimately be stored in memory. Nor does it know the locations of any externally defined functions or global variables that are referenced by the module. So whenever the assembler encounters a reference to an object whose ultimate

```
——————————————————————————————————————— code/link/elfstructs.c
1   typedef struct {
2       int offset;      /* Offset of the reference to relocate */
3       int symbol:24,   /* Symbol the reference should point to */
4           type:8;      /* Relocation type */
5   } Elf32_Rel;
——————————————————————————————————————— code/link/elfstructs.c
```

Figure 8 **ELF relocation entry.** Each entry identifies a reference that must be relocated.

location is unknown, it generates a *relocation entry* that tells the linker how to modify the reference when it merges the object file into an executable. Relocation entries for code are placed in `.rel.text`. Relocation entries for initialized data are placed in `.rel.data`.

Figure 8 shows the format of an ELF relocation entry. The `offset` is the section offset of the reference that will need to be modified. The `symbol` identifies the symbol that the modified reference should point to. The `type` tells the linker how to modify the new reference.

ELF defines 11 different relocation types, some quite arcane. We are concerned with only the two most basic relocation types:

- R_386_PC32: Relocate a reference that uses a 32-bit PC-relative address. Recall that a PC-relative address is an offset from the current run-time value of the program counter (PC). When the CPU executes an instruction using PC-relative addressing, it forms the *effective address* (e.g., the target of the `call` instruction) by adding the 32-bit value encoded in the instruction to the current run-time value of the PC, which is always the address of the next instruction in memory.

- R_386_32: Relocate a reference that uses a 32-bit absolute address. With absolute addressing, the CPU directly uses the 32-bit value encoded in the instruction as the effective address, without further modifications.

7.2 Relocating Symbol References

Figure 9 shows the pseudo code for the linker's relocation algorithm. Lines 1 and 2 iterate over each section s and each relocation entry r associated with each section. For concreteness, assume that each section s is an array of bytes and that each relocation entry r is a `struct` of type Elf32_Rel, as defined in Figure 8. Also, assume that when the algorithm runs, the linker has already chosen run-time addresses for each section (denoted ADDR(s)) and each symbol (denoted ADDR(r.symbol)). Line 3 computes the address in the s array of the 4-byte reference that needs to be relocated. If this reference uses PC-relative

```
1    foreach section s {
2        foreach relocation entry r {
3            refptr = s + r.offset;  /* ptr to reference to be relocated */
4
5            /* Relocate a PC-relative reference */
6            if (r.type == R_386_PC32) {
7                refaddr = ADDR(s) + r.offset; /* ref's run-time address */
8                *refptr = (unsigned) (ADDR(r.symbol) + *refptr - refaddr);
9            }
10
11           /* Relocate an absolute reference */
12           if (r.type == R_386_32)
13               *refptr = (unsigned) (ADDR(r.symbol) + *refptr);
14       }
15   }
```

Figure 9 **Relocation algorithm.**

addressing, then it is relocated by lines 5–9. If the reference uses absolute addressing, then it is relocated by lines 11–13.

Relocating PC-Relative References

Recall from our running example in Figure 1(a) that the main routine in the .text
section of main.o calls the swap routine, which is defined in swap.o. Here is the
disassembled listing for the call instruction, as generated by the GNU OBJDUMP
tool:

```
6:   e8 fc ff ff ff            call    7 <main+0x7>    swap();
                               7: R_386_PC32 swap      relocation entry
```

From this listing, we see that the call instruction begins at section offset 0x6 and
consists of the 1-byte opcode 0xe8, followed by the 32-bit reference 0xfffffffc
(−4 decimal), which is stored in little-endian byte order. We also see a relocation
entry for this reference displayed on the following line. (Recall that relocation
entries and instructions are actually stored in different sections of the object file.
The OBJDUMP tool displays them together for convenience.) The relocation entry
r consists of three fields:

```
r.offset = 0x7
r.symbol = swap
r.type   = R_386_PC32
```

These fields tell the linker to modify the 32-bit PC-relative reference starting at
offset 0x7 so that it will point to the swap routine at run time. Now, suppose that
the linker has determined that

```
ADDR(s) = ADDR(.text) = 0x80483b4
```

and

```
ADDR(r.symbol) = ADDR(swap) = 0x80483c8
```

Using the algorithm in Figure 9, the linker first computes the run-time address of the reference (line 7):

```
refaddr = ADDR(s)   + r.offset
        = 0x80483b4 + 0x7
        = 0x80483bb
```

It then updates the reference from its current value (-4) to 0x9 so that it will point to the swap routine at run time (line 8):

```
*refptr = (unsigned) (ADDR(r.symbol) + *refptr - refaddr)
        = (unsigned) (0x80483c8     + (-4)    - 0x80483bb)
        = (unsigned) (0x9)
```

In the resulting executable object file, the call instruction has the following relocated form:

```
 80483ba:   e8 09 00 00 00            call    80483c8 <swap>          swap();
```

At run time, the call instruction will be stored at address 0x80483ba. When the CPU executes the call instruction, the PC has a value of 0x80483bf, which is the address of the instruction immediately following the call instruction. To execute the instruction, the CPU performs the following steps:

```
1. push PC onto stack
2. PC <- PC + 0x9 = 0x80483bf + 0x9 = 0x80483c8
```

Thus, the next instruction to execute is the first instruction of the swap routine, which of course is what we want!

You may wonder why the assembler created the reference in the call instruction with an initial value of -4. The assembler uses this value as a bias to account for the fact that the PC always points to the instruction following the current instruction. On a different machine with different instruction sizes and encodings, the assembler for that machine would use a different bias. This is a powerful trick that allows the linker to blindly relocate references, blissfully unaware of the instruction encodings for a particular machine.

Relocating Absolute References

Recall that in our example program in Figure 1, the swap.o module initializes the global pointer bufp0 to the address of the first element of the global buf array:

```
int *bufp0 = &buf[0];
```

Since bufp0 is an initialized data object, it will be stored in the .data section of the swap.o relocatable object module. Since it is initialized to the address of a global array, it will need to be relocated. Here is the disassembled listing of the .data section from swap.o:

```
00000000 <bufp0>:
    0:   00 00 00 00                                          int *bufp0 = &buf[0];
                        0: R_386_32 buf                       Relocation entry
```

We see that the .data section contains a single 32-bit reference, the bufp0 pointer, which has a value of 0x0. The relocation entry tells the linker that this is a 32-bit absolute reference, beginning at offset 0, which must be relocated so that it points to the symbol buf. Now, suppose that the linker has determined that

```
ADDR(r.symbol) = ADDR(buf) = 0x8049454
```

The linker updates the reference using line 13 of the algorithm in Figure 9:

```
*refptr = (unsigned) (ADDR(r.symbol) + *refptr)
        = (unsigned) (0x8049454      + 0)
        = (unsigned) (0x8049454)
```

In the resulting executable object file, the reference has the following relocated form:

```
0804945c <bufp0>:
 804945c:   54 94 04 08                                      Relocated!
```

In words, the linker has decided that at run time the variable bufp0 will be located at memory address 0x804945c and will be initialized to 0x8049454, which is the run-time address of the buf array.

The .text section in the swap.o module contains five absolute references that are relocated in a similar way (see Problem 12). Figure 10 shows the relocated .text and .data sections in the final executable object file.

Practice Problem 4

This problem concerns the relocated program in Figure 10.

A. What is the hex address of the relocated reference to swap in line 5?

B. What is the hex value of the relocated reference to swap in line 5?

C. Suppose the linker had decided for some reason to locate the .text section at 0x80483b8 instead of 0x80483b4. What would the hex value of the relocated reference in line 5 be in this case?

(a) Relocated `.text` section

code/link/p-exe.d

```
1    080483b4 <main>:
2     80483b4:   55                       push   %ebp
3     80483b5:   89 e5                    mov    %esp,%ebp
4     80483b7:   83 ec 08                 sub    $0x8,%esp
5     80483ba:   e8 09 00 00 00           call   80483c8 <swap>         swap();
6     80483bf:   31 c0                    xor    %eax,%eax
7     80483c1:   89 ec                    mov    %ebp,%esp
8     80483c3:   5d                       pop    %ebp
9     80483c4:   c3                       ret
10    80483c5:   90                       nop
11    80483c6:   90                       nop
12    80483c7:   90                       nop

13   080483c8 <swap>:
14    80483c8:   55                       push   %ebp
15    80483c9:   8b 15 5c 94 04 08        mov    0x804945c,%edx         Get *bufp0
16    80483cf:   a1 58 94 04 08           mov    0x8049458,%eax         Get buf[1]
17    80483d4:   89 e5                    mov    %esp,%ebp
18    80483d6:   c7 05 48 95 04 08 58     movl   $0x8049458,0x8049548   bufp1 = &buf[1]
19    80483dd:   94 04 08
20    80483e0:   89 ec                    mov    %ebp,%esp
21    80483e2:   8b 0a                    mov    (%edx),%ecx
22    80483e4:   89 02                    mov    %eax,(%edx)
23    80483e6:   a1 48 95 04 08           mov    0x8049548,%eax         Get *bufp1
24    80483eb:   89 08                    mov    %ecx,(%eax)
25    80483ed:   5d                       pop    %ebp
26    80483ee:   c3                       ret
```

code/link/p-exe.d

(b) Relocated `.data` section

code/link/pdata-exe.d

```
1   08049454 <buf>:
2    8049454:   01 00 00 00 02 00 00 00

3   0804945c <bufp0>:
4    804945c:   54 94 04 08                          Relocated!
```

code/link/pdata-exe.d

Figure 10 **Relocated** `.text` **and** `.data` **sections for executable file** p. The original C code is in Figure 1.

8   Executable Object Files

We have seen how the linker merges multiple object modules into a single executable object file. Our C program, which began life as a collection of ASCII text files, has been transformed into a single binary file that contains all of the information needed to load the program into memory and run it. Figure 11 summarizes the kinds of information in a typical ELF executable file.

The format of an executable object file is similar to that of a relocatable object file. The ELF header describes the overall format of the file. It also includes the program's *entry point*, which is the address of the first instruction to execute when the program runs. The .text, .rodata, and .data sections are similar to those in a relocatable object file, except that these sections have been relocated to their eventual run-time memory addresses. The .init section defines a small function, called _init, that will be called by the program's initialization code. Since the executable is *fully linked* (relocated), it needs no .rel sections.

ELF executables are designed to be easy to load into memory, with contiguous chunks of the executable file mapped to contiguous memory segments. This mapping is described by the *segment header table*. Figure 12 shows the segment header table for our example executable p, as displayed by OBJDUMP.

From the segment header table, we see that two memory segments will be initialized with the contents of the executable object file. Lines 1 and 2 tell us that the first segment (the *code segment*) is aligned to a 4 KB (2^{12}) boundary, has read/execute permissions, starts at memory address 0x08048000, has a total memory size of 0x448 bytes, and is initialized with the first 0x448 bytes of the executable object file, which includes the ELF header, the segment header table, and the .init, .text, and .rodata sections.

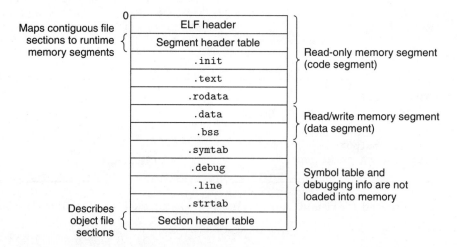

Figure 11   **Typical ELF executable object file.**

——————————————————————————— code/link/p-exe.d

```
   Read-only code segment
1    LOAD off    0x00000000 vaddr 0x08048000 paddr 0x08048000 align 2**12
2         filesz 0x00000448 memsz 0x00000448 flags r-x

   Read/write data segment
3    LOAD off    0x00000448 vaddr 0x08049448 paddr 0x08049448 align 2**12
4         filesz 0x000000e8 memsz 0x00000104 flags rw-
```

——————————————————————————— code/link/p-exe.d

Figure 12 **Segment header table for the example executable** p. Legend: `off`: file offset, `vaddr/paddr`: virtual/physical address, `align`: segment alignment, `filesz`: segment size in the object file, `memsz`: segment size in memory, `flags`: run-time permissions.

Lines 3 and 4 tell us that the second segment (the *data segment*) is aligned to a 4 KB boundary, has read/write permissions, starts at memory address 0x08049448, has a total memory size of 0x104 bytes, and is initialized with the 0xe8 bytes starting at file offset 0x448, which in this case is the beginning of the `.data` section. The remaining bytes in the segment correspond to `.bss` data that will be initialized to zero at run time.

9 Loading Executable Object Files

To run an executable object file p, we can type its name to the Unix shell's command line:

```
unix> ./p
```

Since p does not correspond to a built-in shell command, the shell assumes that p is an executable object file, which it runs for us by invoking some memory-resident operating system code known as the `loader`. Any Unix program can invoke the loader by calling the `execve` function. The loader copies the code and data in the executable object file from disk into memory, and then runs the program by jumping to its first instruction, or *entry point*. This process of copying the program into memory and then running it is known as *loading*.

Every Unix program has a run-time memory image similar to the one in Figure 13. On 32-bit Linux systems, the code segment starts at address 0x08048000. The data segment follows at the next 4 KB aligned address. The run-time *heap* follows on the first 4 KB aligned address past the read/write segment and grows up via calls to the `malloc` library. There is also a segment that is reserved for shared libraries. The user stack always starts at the largest legal user address and grows down (toward lower memory addresses). The segment starting above the stack is

Figure 13
Linux run-time memory image.

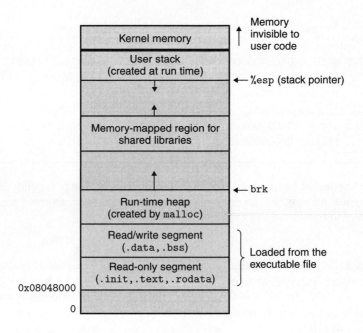

reserved for the code and data in the memory-resident part of the operating system known as the *kernel*.

When the loader runs, it creates the memory image shown in Figure 13. Guided by the segment header table in the executable, it copies chunks of the executable into the code and data segments. Next, the loader jumps to the program's entry point, which is always the address of the _start symbol. The *startup code* at the _start address is defined in the object file crt1.o and is the same for all C programs. Figure 14 shows the specific sequence of calls in the startup code. After calling initialization routines from the .text and .init sections, the startup code calls the atexit routine, which appends a list of routines that should be called when the application terminates normally. The exit function runs the functions registered by atexit, and then returns control to the operating system

```
1   0x080480c0 <_start>:          /* Entry point in .text */
2       call __libc_init_first    /* Startup code in .text */
3       call _init                /* Startup code in .init */
4       call atexit               /* Startup code in .text */
5       call main                 /* Application main routine */
6       call _exit                /* Returns control to OS */
7   /* Control never reaches here */
```

Figure 14   **Pseudo-code for the crt1.o startup routine in every C program.** Note: The code that pushes the arguments for each function is not shown.

by calling `_exit`. Next, the startup code calls the application's `main` routine, which begins executing our C code. After the application returns, the startup code calls the `_exit` routine, which returns control to the operating system.

Aside   How do loaders really work?

Our description of loading is conceptually correct, but intentionally not entirely accurate. To understand how loading really works, you must understand the concepts of *processes*, *virtual memory*, and *memory mapping*, which we haven't discussed yet.

For the impatient reader, here is a preview of how loading really works: Each program in a Unix system runs in the context of a process with its own virtual address space. When the shell runs a program, the parent shell process forks a child process that is a duplicate of the parent. The child process invokes the loader via the `execve` system call. The loader deletes the child's existing virtual memory segments, and creates a new set of code, data, heap, and stack segments. The new stack and heap segments are initialized to zero. The new code and data segments are initialized to the contents of the executable file by mapping pages in the virtual address space to page-sized chunks of the executable file. Finally, the loader jumps to the `_start` address, which eventually calls the application's `main` routine. Aside from some header information, there is no copying of data from disk to memory during loading. The copying is deferred until the CPU references a mapped virtual page, at which point the operating system automatically transfers the page from disk to memory using its paging mechanism.

Practice Problem 5

A. Why does every C program need a routine called `main`?

B. Have you ever wondered why a C `main` routine can end with a call to `exit`, a `return` statement, or neither, and yet the program still terminates properly? Explain.

---

10   Dynamic Linking with Shared Libraries

The static libraries that we studied in Section 6.2 address many of the issues associated with making large collections of related functions available to application programs. However, static libraries still have some significant disadvantages. Static libraries, like all software, need to be maintained and updated periodically. If application programmers want to use the most recent version of a library, they must somehow become aware that the library has changed, and then explicitly relink their programs against the updated library.

Another issue is that almost every C program uses standard I/O functions such as `printf` and `scanf`. At run time, the code for these functions is duplicated in the text segment of each running process. On a typical system that is running 50–100 processes, this can be a significant waste of scarce memory system resources. (An interesting property of memory is that it is *always* a scarce resource, regardless of how much there is in a system. Disk space and kitchen trash cans share this same property.)

Shared libraries are modern innovations that address the disadvantages of static libraries. A shared library is an object module that, *at run time*, can be loaded at an arbitrary memory address and linked with a program in memory. This process is known as *dynamic linking* and is performed by a program called a *dynamic linker*.

Shared libraries are also referred to as *shared objects*, and on Unix systems are typically denoted by the .so suffix. Microsoft operating systems make heavy use of shared libraries, which they refer to as DLLs (dynamic link libraries).

Shared libraries are "shared" in two different ways. First, in any given file system, there is exactly one .so file for a particular library. The code and data in this .so file are shared by all of the executable object files that reference the library, as opposed to the contents of static libraries, which are copied and embedded in the executables that reference them. Second, a single copy of the .text section of a shared library in memory can be shared by different running processes.

Figure 15 summarizes the dynamic linking process for the example program in Figure 6. To build a shared library libvector.so of our example vector arithmetic routines in Figure 5, we would invoke the compiler driver with the following special directive to the linker:

```
unix> gcc -shared -fPIC -o libvector.so addvec.c multvec.c
```

The -fPIC flag directs the compiler to generate position-independent code (more on this in the next section). The -shared flag directs the linker to create a shared object file.

Once we have created the library, we would then link it into our example program in Figure 6:

```
unix> gcc -o p2 main2.c ./libvector.so
```

This creates an executable object file p2 in a form that can be linked with libvector.so at run time. The basic idea is to do some of the linking statically when the executable file is created, and then complete the linking process dynamically when the program is loaded.

It is important to realize that none of the code or data sections from libvector.so are actually copied into the executable p2 at this point. Instead, the linker copies some relocation and symbol table information that will allow references to code and data in libvector.so to be resolved at run time.

When the loader loads and runs the executable p2, it loads the partially linked executable p2, using the techniques discussed in Section 9. Next, it notices that p2

Figure 15
Dynamic linking with shared libraries.

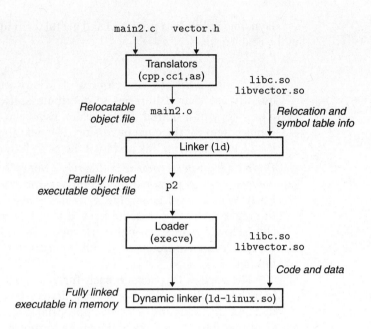

contains a `.interp` section, which contains the path name of the dynamic linker, which is itself a shared object (e.g., LD-LINUX.SO on Linux systems). Instead of passing control to the application, as it would normally do, the loader loads and runs the dynamic linker.

The dynamic linker then finishes the linking task by performing the following relocations:

* Relocating the text and data of `libc.so` into some memory segment.
* Relocating the text and data of `libvector.so` into another memory segment.
* Relocating any references in p2 to symbols defined by `libc.so` and `libvector.so`.

Finally, the dynamic linker passes control to the application. From this point on, the locations of the shared libraries are fixed and do not change during execution of the program.

11 Loading and Linking Shared Libraries from Applications

Up to this point, we have discussed the scenario in which the dynamic linker loads and links shared libraries when an application is loaded, just before it executes. However, it is also possible for an application to request the dynamic linker to load and link arbitrary shared libraries while the application is running, without having to link in the applications against those libraries at compile time.

Dynamic linking is a powerful and useful technique. Here are some examples in the real world:

- *Distributing software.* Developers of Microsoft Windows applications frequently use shared libraries to distribute software updates. They generate a new copy of a shared library, which users can then download and use as a replacement for the current version. The next time they run their application, it will automatically link and load the new shared library.

- *Building high-performance Web servers.* Many Web servers generate *dynamic content*, such as personalized Web pages, account balances, and banner ads. Early Web servers generated dynamic content by using `fork` and `execve` to create a child process and run a "CGI program" in the context of the child. However, modern high-performance Web servers can generate dynamic content using a more efficient and sophisticated approach based on dynamic linking.

  The idea is to package each function that generates dynamic content in a shared library. When a request arrives from a Web browser, the server dynamically loads and links the appropriate function and then calls it directly, as opposed to using `fork` and `execve` to run the function in the context of a child process. The function remains cached in the server's address space, so subsequent requests can be handled at the cost of a simple function call. This can have a significant impact on the throughput of a busy site. Further, existing functions can be updated and new functions can be added at run time, without stopping the server.

Linux systems provide a simple interface to the dynamic linker that allows application programs to load and link shared libraries at run time.

```
#include <dlfcn.h>

void *dlopen(const char *filename, int flag);
```
                                        Returns: ptr to handle if OK, NULL on error

The `dlopen` function loads and links the shared library `filename`. The external symbols in `filename` are resolved using libraries previously opened with the `RTLD_GLOBAL` flag. If the current executable was compiled with the `-rdynamic` flag, then its global symbols are also available for symbol resolution. The `flag` argument must include either `RTLD_NOW`, which tells the linker to resolve references to external symbols immediately, or the `RTLD_LAZY` flag, which instructs the linker to defer symbol resolution until code from the library is executed. Either of these values can be or'd with the `RTLD_GLOBAL` flag.

```
#include <dlfcn.h>

void *dlsym(void *handle, char *symbol);
                                    Returns: ptr to symbol if OK, NULL on error
```

The `dlsym` function takes a `handle` to a previously opened shared library and a `symbol` name, and returns the address of the symbol, if it exists, or NULL otherwise.

```
#include <dlfcn.h>

int dlclose (void *handle);
                                    Returns: 0 if OK, −1 on error
```

The `dlclose` function unloads the shared library if no other shared libraries are still using it.

```
#include <dlfcn.h>

const char *dlerror(void);
                    Returns: error msg if previous call to dlopen, dlsym,
                        or dlclose failed, NULL if previous call was OK
```

The `dlerror` function returns a string describing the most recent error that occurred as a result of calling `dlopen`, `dlsym`, or `dlclose`, or NULL if no error occurred.

Figure 16 shows how we would use this interface to dynamically link our `libvector.so` shared library (Figure 5), and then invoke its `addvec` routine. To compile the program, we would invoke GCC in the following way:

```
unix> gcc -rdynamic -O2 -o p3 dll.c -ldl
```

Aside Shared libraries and the Java Native Interface

Java defines a standard calling convention called *Java Native Interface* (JNI) that allows "native" C and C++ functions to be called from Java programs. The basic idea of JNI is to compile the native C function, say, `foo`, into a shared library, say `foo.so`. When a running Java program attempts to invoke function `foo`, the Java interpreter uses the `dlopen` interface (or something like it) to dynamically link and load `foo.so`, and then call `foo`.

code/link/dll.c

```c
1    #include <stdio.h>
2    #include <stdlib.h>
3    #include <dlfcn.h>
4
5    int x[2] = {1, 2};
6    int y[2] = {3, 4};
7    int z[2];
8
9    int main()
10   {
11       void *handle;
12       void (*addvec)(int *, int *, int *, int);
13       char *error;
14
15       /* Dynamically load shared library that contains addvec() */
16       handle = dlopen("./libvector.so", RTLD_LAZY);
17       if (!handle) {
18           fprintf(stderr, "%s\n", dlerror());
19           exit(1);
20       }
21
22       /* Get a pointer to the addvec() function we just loaded */
23       addvec = dlsym(handle, "addvec");
24       if ((error = dlerror()) != NULL) {
25           fprintf(stderr, "%s\n", error);
26           exit(1);
27       }
28
29       /* Now we can call addvec() just like any other function */
30       addvec(x, y, z, 2);
31       printf("z = [%d %d]\n", z[0], z[1]);
32
33       /* Unload the shared library */
34       if (dlclose(handle) < 0) {
35           fprintf(stderr, "%s\n", dlerror());
36           exit(1);
37       }
38       return 0;
39   }
```

code/link/dll.c

Figure 16 **An application program that dynamically loads and links the shared library** libvector.so.

12   Position-Independent Code (PIC)

A key purpose of shared libraries is to allow multiple running processes to share the same library code in memory and thus save precious memory resources. So how can multiple processes share a single copy of a program? One approach would be to assign *a priori* a dedicated chunk of the address space to each shared library, and then require the loader to always load the shared library at that address. While straightforward, this approach creates some serious problems. It would be an inefficient use of the address space because portions of the space would be allocated even if a process didn't use the library. Second, it would be difficult to manage. We would have to ensure that none of the chunks overlapped. Every time a library were modified, we would have to make sure that it still fit in its assigned chunk. If not, then we would have to find a new chunk. And if we created a new library, we would have to find room for it. Over time, given the hundreds of libraries and versions of libraries in a system, it would be difficult to keep the address space from fragmenting into lots of small unused but unusable holes. Even worse, the assignment of libraries to memory would be different for each system, thus creating even more management headaches.

A better approach is to compile library code so that it can be loaded and executed at any address without being modified by the linker. Such code is known as *position-independent code* (PIC). Users direct GNU compilation systems to generate PIC code with the -fPIC option to GCC.

On IA32 systems, calls to procedures in the same object module require no special treatment, since the references are PC-relative, with known offsets, and thus are already PIC (see Problem 4). However, calls to externally defined procedures and references to global variables are not normally PIC, since they require relocation at link time.

PIC Data References

Compilers generate PIC references to global variables by exploiting the following interesting fact: No matter where we load an object module (including shared object modules) in memory, the data segment is always allocated immediately after the code segment. Thus, the *distance* between any instruction in the code segment and any variable in the data segment is a run-time constant, independent of the absolute memory locations of the code and data segments.

To exploit this fact, the compiler creates a table called the *global offset table* (GOT) at the beginning of the data segment. The GOT contains an entry for each global data object that is referenced by the object module. The compiler also generates a relocation record for each entry in the GOT. At load time, the dynamic linker relocates each entry in the GOT so that it contains the appropriate absolute address. Each object module that references global data has its own GOT.

At run time, each global variable is referenced indirectly through the GOT using code of the form

```
        call L1
L1:     popl %ebx                   ebx contains the current PC
        addl $VAROFF, %ebx          ebx points to the GOT entry for var
        movl (%ebx), %eax           reference indirect through the GOT
        movl (%eax), %eax
```

In this fascinating piece of code, the call to L1 pushes the return address (which happens to be the address of the popl instruction) on the stack. The popl instruction then pops this address into %ebx. The net effect of these two instructions is to move the value of the PC into register %ebx.

The addl instruction adds a constant offset to %ebx so that it points to the appropriate entry in the GOT, which contains the absolute address of the data item. At this point, the global variable can be referenced indirectly through the GOT entry contained in %ebx. In this example, the two movl instructions load the contents of the global variable (indirectly through the GOT) into register %eax.

PIC code has performance disadvantages. Each global variable reference now requires five instructions instead of one, with an additional memory reference to the GOT. Also, PIC code uses an additional register to hold the address of the GOT entry. On machines with large register files, this is not a major issue. On register-starved IA32 systems, however, losing even one register can trigger spilling of the registers onto the stack.

PIC Function Calls

It would certainly be possible for PIC code to use the same approach for resolving external procedure calls:

```
        call L1
L1:     popl %ebx                   ebx contains the current PC
        addl $PROCOFF, %ebx         ebx points to GOT entry for proc
        call *(%ebx)                call indirect through the GOT
```

However, this approach would require three additional instructions for each run-time procedure call. Instead, ELF compilation systems use an interesting technique, called *lazy binding*, that defers the binding of procedure addresses until the first time the procedure is called. There is a nontrivial run-time overhead the first time the procedure is called, but each call thereafter only costs a single instruction and a memory reference for the indirection.

Lazy binding is implemented with a compact yet somewhat complex interaction between two data structures: the GOT and the *procedure linkage table* (PLT). If an object module calls any functions that are defined in shared libraries, then it has its own GOT and PLT. The GOT is part of the .data section. The PLT is part of the .text section.

Figure 17 shows the format of the GOT for the example program main2.o from Figure 6. The first three GOT entries are special: GOT[0] contains the address of the .dynamic segment, which contains information that the dynamic linker uses to bind procedure addresses, such as the location of the symbol table

Address	Entry	Contents	Description
08049674	GOT[0]	0804969c	address of .dynamic section
08049678	GOT[1]	4000a9f8	identifying info for the linker
0804967c	GOT[2]	4000596f	entry point in dynamic linker
08049680	GOT[3]	0804845a	address of pushl in PLT[1] (printf)
08049684	GOT[4]	0804846a	address of pushl in PLT[2] (addvec)

Figure 17 **The global offset table (GOT) for executable** p2. The original code is in Figures 5 and 6.

and relocation information. GOT[1] contains some information that defines this module. GOT[2] contains an entry point into the lazy binding code of the dynamic linker.

Each procedure that is defined in a shared object and called by main2.o gets an entry in the GOT, starting with entry GOT[3]. For the example program, we have shown the GOT entries for printf, which is defined in libc.so, and addvec, which is defined in libvector.so.

Figure 18 shows the PLT for our example program p2. The PLT is an array of 16-byte entries. The first entry, PLT[0], is a special entry that jumps into the dynamic linker. Each called procedure has an entry in the PLT, starting at PLT[1]. In the figure, PLT[1] corresponds to printf and PLT[2] corresponds to addvec.

```
PLT[0]
08048444:  ff 35 78 96 04 08    pushl   0x8049678    push &GOT[1]
 804844a:  ff 25 7c 96 04 08    jmp     *0x804967c   jmp to *GOT[2](linker)
 8048450:  00 00                                     padding
 8048452:  00 00                                     padding

PLT[1] <printf>
 8048454:  ff 25 80 96 04 08    jmp     *0x8049680   jmp to *GOT[3]
 804845a:  68 00 00 00 00       pushl   $0x0         ID for printf
 804845f:  e9 e0 ff ff ff       jmp     8048444      jmp to PLT[0]

PLT[2] <addvec>
 8048464:  ff 25 84 96 04 08    jmp     *0x8049684   jump to *GOT[4]
 804846a:  68 08 00 00 00       pushl   $0x8         ID for addvec
 804846f:  e9 d0 ff ff ff       jmp     8048444      jmp to PLT[0]

<other PLT entries>
```

Figure 18 **The procedure linkage table (PLT) for executable** p2. The original code is in Figures 5 and 6.

Initially, after the program has been dynamically linked and begins executing, procedures `printf` and `addvec` are bound to the first instruction in their respective PLT entries. For example, the call to `addvec` has the form

```
80485bb:    e8 a4 fe ff ff    call 8048464 <addvec>
```

When `addvec` is called the first time, control passes to the first instruction in PLT[2], which does an indirect jump through GOT[4]. Initially, each GOT entry contains the address of the `pushl` entry in the corresponding PLT entry. So the indirect jump in the PLT simply transfers control back to the next instruction in PLT[2]. This instruction pushes an ID for the `addvec` symbol onto the stack. The last instruction jumps to PLT[0], which pushes another word of identifying information on the stack from GOT[1], and then jumps into the dynamic linker indirectly through GOT[2]. The dynamic linker uses the two stack entries to determine the location of `addvec`, overwrites GOT[4] with this address, and passes control to `addvec`.

The next time `addvec` is called in the program, control passes to PLT[2] as before. However, this time the indirect jump through GOT[4] transfers control to `addvec`. The only additional overhead from this point on is the memory reference for the indirect jump.

13   Tools for Manipulating Object Files

There are a number of tools available on Unix systems to help you understand and manipulate object files. In particular, the GNU *binutils* package is especially helpful and runs on every Unix platform.

AR: Creates static libraries, and inserts, deletes, lists, and extracts members.

STRINGS: Lists all of the printable strings contained in an object file.

STRIP: Deletes symbol table information from an object file.

NM: Lists the symbols defined in the symbol table of an object file.

SIZE: Lists the names and sizes of the sections in an object file.

READELF: Displays the complete structure of an object file, including all of the information encoded in the ELF header; subsumes the functionality of SIZE and NM.

OBJDUMP: The mother of all binary tools. Can display all of the information in an object file. Its most useful function is disassembling the binary instructions in the `.text` section.

Unix systems also provide the LDD program for manipulating shared libraries:

LDD: Lists the shared libraries that an executable needs at run time.

14 Summary

Linking can be performed at compile time by static linkers, and at load time and run time by dynamic linkers. Linkers manipulate binary files called object files, which come in three different forms: relocatable, executable, and shared. Relocatable object files are combined by static linkers into an executable object file that can be loaded into memory and executed. Shared object files (shared libraries) are linked and loaded by dynamic linkers at run time, either implicitly when the calling program is loaded and begins executing, or on demand, when the program calls functions from the dlopen library.

The two main tasks of linkers are symbol resolution, where each global symbol in an object file is bound to a unique definition, and relocation, where the ultimate memory address for each symbol is determined and where references to those objects are modified.

Static linkers are invoked by compiler drivers such as GCC. They combine multiple relocatable object files into a single executable object file. Multiple object files can define the same symbol, and the rules that linkers use for silently resolving these multiple definitions can introduce subtle bugs in user programs.

Multiple object files can be concatenated in a single static library. Linkers use libraries to resolve symbol references in other object modules. The left-to-right sequential scan that many linkers use to resolve symbol references is another source of confusing link-time errors.

Loaders map the contents of executable files into memory and run the program. Linkers can also produce partially linked executable object files with unresolved references to the routines and data defined in a shared library. At load time, the loader maps the partially linked executable into memory and then calls a dynamic linker, which completes the linking task by loading the shared library and relocating the references in the program.

Shared libraries that are compiled as position-independent code can be loaded anywhere and shared at run time by multiple processes. Applications can also use the dynamic linker at run time in order to load, link, and access the functions and data in shared libraries.

Bibliographic Notes

Linking is not well documented in the computer systems literature. Since it lies at the intersection of compilers, computer architecture, and operating systems, linking requires understanding of code generation, machine-language programming, program instantiation, and virtual memory. It does not fit neatly into any of the usual computer systems specialties and thus is not well covered by the classic texts in these areas. However, Levine's monograph provides a good general reference on the subject [66]. The original specifications for ELF and DWARF (a specification for the contents of the .debug and .line sections) are described in [52].

Some interesting research and commercial activity centers around the notion of *binary translation*, where the contents of an object file are parsed, analyzed, and modified. Binary translation can be used for three different purposes [64]: to emulate one system on another system, to observe program behavior, or to perform system-dependent optimizations that are not possible at compile time. Commercial products such as VTune, Purify, and BoundsChecker use binary translation to provide programmers with detailed observations of their programs. Valgrind is a popular open-source alternative.

The Atom system provides a flexible mechanism for instrumenting Alpha executable object files and shared libraries with arbitrary C functions [103]. Atom has been used to build a myriad of analysis tools that trace procedure calls, profile instruction counts and memory referencing patterns, simulate memory system behavior, and isolate memory referencing errors. Etch [90] and EEL [64] provide roughly similar capabilities on different platforms. The Shade system uses binary translation for instruction profiling [23]. Dynamo [5] and Dyninst [15] provide mechanisms for instrumenting and optimizing executables in memory at run time. Smith and his colleagues have investigated binary translation for program profiling and optimization [121].

Homework Problems

6 ◆

Consider the following version of the swap.c function that counts the number of times it has been called:

```
1    extern int buf[];
2
3    int *bufp0 = &buf[0];
4    static int *bufp1;
5
6    static void incr()
7    {
8        static int count=0;
9
10       count++;
11   }
12
13   void swap()
14   {
15       int temp;
16
17       incr();
18       bufp1 = &buf[1];
19       temp = *bufp0;
20       *bufp0 = *bufp1;
21       *bufp1 = temp;
22   }
```

For each symbol that is defined and referenced in swap.o, indicate if it will have a symbol table entry in the .symtab section in module swap.o. If so, indicate the module that defines the symbol (swap.o or main.o), the symbol type (local, global, or extern), and the section (.text, .data, or .bss) it occupies in that module.

Symbol	swap.o .symtab entry?	Symbol type	Module where defined	Section
buf				
bufp0				
bufp1				
swap				
temp				
incr				
count				

7 ◆

Without changing any variable names, modify bar5.c from earlier in this chapter so that foo5.c prints the correct values of x and y (i.e., the hex representations of integers 15213 and 15212).

8 ◆

In this problem, let REF(x.i) --> DEF(x.k) denote that the linker will associate an arbitrary reference to symbol x in module i to the definition of x in module k. For each example below, use this notation to indicate how the linker would resolve references to the multiply defined symbol in each module. If there is a link-time error (rule 1), write "ERROR." If the linker arbitrarily chooses one of the definitions (rule 3), write "UNKNOWN."

A.
```
/* Module 1 */          /* Module 2 */
int main()              static int main=1;
{                       int p2()
}                       {
                        }
```

   (a) REF(main.1) --> DEF(_____.____)

   (b) REF(main.2) --> DEF(_____.____)

B.
```
/* Module 1 */          /* Module 2 */
int x;                  double x;
void main()             int p2()
{                       {
}                       }
```

   (a) REF(x.1) --> DEF(_____.____)

   (b) REF(x.2) --> DEF(_____.____)

```
C.  /* Module 1 */          /* Module 2 */
    int x=1;                double x=1.0;
    void main()             int p2()
    {                       {
    }                       }

    (a) REF(x.1) --> DEF(_____.____)
    (b) REF(x.2) --> DEF(_____.____)
```

9 ◆

Consider the following program, which consists of two object modules:

```
1   /* foo6.c */          1   /* bar6.c */
2   void p2(void);        2   #include <stdio.h>
3                         3
4   int main()           4   char main;
5   {                    5
6       p2();            6   void p2()
7       return 0;        7   {
8   }                    8       printf("0x%x\n", main);
                         9   }
```

When this program is compiled and executed on a Linux system, it prints the string "0x55\n" and terminates normally, even though p2 never initializes variable main. Can you explain this?

10 ◆

Let a and b denote object modules or static libraries in the current directory, and let a→b denote that a depends on b, in the sense that b defines a symbol that is referenced by a. For each of the following scenarios, show the minimal command line (i.e., one with the least number of object file and library arguments) that will allow the static linker to resolve all symbol references:

A. p.o → libx.a → p.o

B. p.o → libx.a → liby.a **and** liby.a → libx.a

C. p.o → libx.a → liby.a → libz.a **and** liby.a → libx.a → libz.a

11 ◆

The segment header in Figure 12 indicates that the data segment occupies 0x104 bytes in memory. However, only the first 0xe8 bytes of these come from the sections of the executable file. What causes this discrepancy?

12 ◆◆

The swap routine in Figure 10 contains five relocated references. For each relocated reference, give its line number in Figure 10, its run-time memory address, and its value. The original code and relocation entries in the swap.o module are shown in Figure 19.

```
1     00000000 <swap>:
2        0:   55                          push    %ebp
3        1:   8b 15 00 00 00 00           mov     0x0,%edx          Get *bufp0=&buf[0]
4                                         3: R_386_32       bufp0   Relocation entry
5        7:   a1 04 00 00 00              mov     0x4,%eax          Get buf[1]
6                                         8: R_386_32       buf     Relocation entry
7        c:   89 e5                       mov     %esp,%ebp
8        e:   c7 05 00 00 00 00 04        movl    $0x4,0x0          bufp1 = &buf[1];
9       15:   00 00 00
10                                        10: R_386_32      bufp1   Relocation entry
11                                        14: R_386_32      buf     Relocation entry
12      18:   89 ec                       mov     %ebp,%esp
13      1a:   8b 0a                       mov     (%edx),%ecx       temp = buf[0];
14      1c:   89 02                       mov     %eax,(%edx)       buf[0]=buf[1];
15      1e:   a1 00 00 00 00              mov     0x0,%eax          Get *bufp1=&buf[1]
16                                        1f: R_386_32      bufp1   Relocation entry
17      23:   89 08                       mov     %ecx,(%eax)       buf[1]=temp;
18      25:   5d                          pop     %ebp
19      26:   c3                          ret
```

Figure 19 **Code and relocation entries for Problem 12.**

Line # in Fig. 10	Address	Value
_____	_____	_____
_____	_____	_____
_____	_____	_____
_____	_____	_____

13 ◆◆◆

Consider the C code and corresponding relocatable object module in Figure 20.

A. Determine which instructions in .text will need to be modified by the linker when the module is relocated. For each such instruction, list the information in its relocation entry: section offset, relocation type, and symbol name.

B. Determine which data objects in .data will need to be modified by the linker when the module is relocated. For each such instruction, list the information in its relocation entry: section offset, relocation type, and symbol name.

Feel free to use tools such as OBJDUMP to help you solve this problem.

14 ◆◆◆

Consider the C code and corresponding relocatable object module in Figure 21.

A. Determine which instructions in .text will need to be modified by the linker when the module is relocated. For each such instruction, list the information in its relocation entry: section offset, relocation type, and symbol name.

(a) C code

```
1   extern int p3(void);
2   int x = 1;
3   int *xp = &x;
4
5   void p2(int y) {
6   }
7
8   void p1() {
9       p2(*xp + p3());
10  }
```

(b) .text section of relocatable object file

```
1   00000000 <p2>:
2       0:   55                    push   %ebp
3       1:   89 e5                 mov    %esp,%ebp
4       3:   89 ec                 mov    %ebp,%esp
5       5:   5d                    pop    %ebp
6       6:   c3                    ret

7   00000008 <p1>:
8       8:   55                    push   %ebp
9       9:   89 e5                 mov    %esp,%ebp
10      b:   83 ec 08              sub    $0x8,%esp
11      e:   83 c4 f4              add    $0xfffffff4,%esp
12     11:   e8 fc ff ff ff        call   12 <p1+0xa>
13     16:   89 c2                 mov    %eax,%edx
14     18:   a1 00 00 00 00        mov    0x0,%eax
15     1d:   03 10                 add    (%eax),%edx
16     1f:   52                    push   %edx
17     20:   e8 fc ff ff ff        call   21 <p1+0x19>
18     25:   89 ec                 mov    %ebp,%esp
19     27:   5d                    pop    %ebp
20     28:   c3                    ret
```

(c) .data section of relocatable object file

```
1   00000000 <x>:
2       0:   01 00 00 00
3   00000004 <xp>:
4       4:   00 00 00 00
```

Figure 20 **Example code for Problem 13.**

(a) C code

```
1   int relo3(int val)  {
2       switch (val) {
3       case 100:
4           return(val);
5       case 101:
6           return(val+1);
7       case 103: case 104:
8           return(val+3);
9       case 105:
10          return(val+5);
11      default:
12          return(val+6);
13      }
14  }
```

(b) .text section of relocatable object file

```
1   00000000 <relo3>:
2      0:    55                           push    %ebp
3      1:    89 e5                        mov     %esp,%ebp
4      3:    8b 45 08                     mov     0x8(%ebp),%eax
5      6:    8d 50 9c                     lea     0xffffff9c(%eax),%edx
6      9:    83 fa 05                     cmp     $0x5,%edx
7      c:    77 17                        ja      25 <relo3+0x25>
8      e:    ff 24 95 00 00 00 00         jmp     *0x0(,%edx,4)
9     15:    40                           inc     %eax
10    16:    eb 10                        jmp     28 <relo3+0x28>
11    18:    83 c0 03                     add     $0x3,%eax
12    1b:    eb 0b                        jmp     28 <relo3+0x28>
13    1d:    8d 76 00                     lea     0x0(%esi),%esi
14    20:    83 c0 05                     add     $0x5,%eax
15    23:    eb 03                        jmp     28 <relo3+0x28>
16    25:    83 c0 06                     add     $0x6,%eax
17    28:    89 ec                        mov     %ebp,%esp
18    2a:    5d                           pop     %ebp
19    2b:    c3                           ret
```

(c) .rodata section of relocatable object file

```
      This is the jump table for the switch statement
1   0000 28000000 15000000 25000000 18000000   4 words at offsets 0x0,0x4,0x8, and 0xc
2   0010 18000000 20000000                      2 words at offsets 0x10 and 0x14
```

Figure 21 **Example code for Problem 14.**

B. Determine which data objects in `.rodata` will need to be modified by the linker when the module is relocated. For each such instruction, list the information in its relocation entry: section offset, relocation type, and symbol name.

Feel free to use tools such as OBJDUMP to help you solve this problem.

15 ◆◆◆

Performing the following tasks will help you become more familiar with the various tools for manipulating object files.

A. How many object files are contained in the versions of `libc.a` and `libm.a` on your system?

B. Does gcc −O2 produce different executable code than gcc −O2 −g?

C. What shared libraries does the GCC driver on your system use?

Solutions to Practice Problems

Solution to Problem 1

The purpose of this problem is to help you understand the relationship between linker symbols and C variables and functions. Notice that the C local variable `temp` does *not* have a symbol table entry.

Symbol	swap.o .symtab entry?	Symbol type	Module where defined	Section
buf	yes	extern	main.o	.data
bufp0	yes	global	swap.o	.data
bufp1	yes	global	swap.o	.bss
swap	yes	global	swap.o	.text
temp	no	—	—	—

Solution to Problem 2

This is a simple drill that checks your understanding of the rules that a Unix linker uses when it resolves global symbols that are defined in more than one module. Understanding these rules can help you avoid some nasty programming bugs.

A. The linker chooses the strong symbol defined in module 1 over the weak symbol defined in module 2 (rule 2):

   (a) `REF(main.1) --> DEF(main.1)`
   (b) `REF(main.2) --> DEF(main.1)`

B. This is an ERROR, because each module defines a strong symbol `main` (rule 1).

C. The linker chooses the strong symbol defined in module 2 over the weak symbol defined in module 1 (rule 2):

    (a) `REF(x.1) --> DEF(x.2)`

    (b) `REF(x.2) --> DEF(x.2)`

Solution to Problem 3

Placing static libraries in the wrong order on the command line is a common source of linker errors that confuses many programmers. However, once you understand how linkers use static libraries to resolve references, it's pretty straightforward. This little drill checks your understanding of this idea:

A. `gcc p.o libx.a`

B. `gcc p.o libx.a liby.a`

C. `gcc p.o libx.a liby.a libx.a`

Solution to Problem 4

This problem concerns the disassembly listing in Figure 10. Our purpose here is to give you some practice reading disassembly listings and to check your understanding of PC-relative addressing.

A. The hex address of the relocated reference in line 5 is `0x80483bb`.

B. The hex value of the relocated reference in line 5 is `0x9`. Remember that the disassembly listing shows the value of the reference in little-endian byte order.

C. The key observation here is that no matter where the linker locates the `.text` section, the distance between the reference and the `swap` function is always the same. Thus, because the reference is a PC-relative address, its value will be `0x9`, regardless of where the linker locates the `.text` section.

Solution to Problem 5

How C programs actually start up is a mystery to most programmers. These questions check your understanding of this startup process. You can answer them by referring to the C startup code in Figure 14.

A. Every program needs a `main` function, because the C startup code, which is common to every C program, jumps to a function called `main`.

B. If `main` terminates with a `return` statement, then control passes back to the startup routine, which returns control to the operating system by calling `_exit`. The same behavior occurs if the user omits the `return` statement. If `main` terminates with a call to `exit`, then `exit` eventually returns control to the operating system by calling `_exit`. The net effect is the same in all three cases: when `main` has finished, control passes back to the operating system.

Exceptional Control Flow

From Chapter 8 of *Computer Systems: A Programmer's Perspective*, Second Edition. Randal E. Bryant and David R. O'Hallaron. Copyright © 2011 by Pearson Education, Inc. Published by Prentice Hall. All rights reserved.

From the time you first apply power to a processor until the time you shut it off, the program counter assumes a sequence of values

$$a_0, a_1, \ldots, a_{n-1}$$

where each a_k is the address of some corresponding instruction I_k. Each transition from a_k to a_{k+1} is called a *control transfer*. A sequence of such control transfers is called the *flow of control*, or *control flow* of the processor.

The simplest kind of control flow is a "smooth" sequence where each I_k and I_{k+1} are adjacent in memory. Typically, abrupt changes to this smooth flow, where I_{k+1} is not adjacent to I_k, are caused by familiar program instructions such as jumps, calls, and returns. Such instructions are necessary mechanisms that allow programs to react to changes in internal program state represented by program variables.

But systems must also be able to react to changes in system state that are not captured by internal program variables and are not necessarily related to the execution of the program. For example, a hardware timer goes off at regular intervals and must be dealt with. Packets arrive at the network adapter and must be stored in memory. Programs request data from a disk and then sleep until they are notified that the data are ready. Parent processes that create child processes must be notified when their children terminate.

Modern systems react to these situations by making abrupt changes in the control flow. In general, we refer to these abrupt changes as *exceptional control flow* (ECF). Exceptional control flow occurs at all levels of a computer system. For example, at the hardware level, events detected by the hardware trigger abrupt control transfers to exception handlers. At the operating systems level, the kernel transfers control from one user process to another via context switches. At the application level, a process can send a *signal* to another process that abruptly transfers control to a signal handler in the recipient. An individual program can react to errors by sidestepping the usual stack discipline and making nonlocal jumps to arbitrary locations in other functions.

As programmers, there are a number of reasons why it is important for you to understand ECF:

- *Understanding ECF will help you understand important systems concepts.* ECF is the basic mechanism that operating systems use to implement I/O, processes, and virtual memory. Before you can really understand these important ideas, you need to understand ECF.

- *Understanding ECF will help you understand how applications interact with the operating system.* Applications request services from the operating system by using a form of ECF known as a *trap* or *system call*. For example, writing data to a disk, reading data from a network, creating a new process, and terminating the current process are all accomplished by application programs invoking system calls. Understanding the basic system call mechanism will help you understand how these services are provided to applications.

- *Understanding ECF will help you write interesting new application programs.* The operating system provides application programs with powerful ECF mechanisms for creating new processes, waiting for processes to terminate, notifying other processes of exceptional events in the system, and detecting and responding to these events. If you understand these ECF mechanisms, then you can use them to write interesting programs such as Unix shells and Web servers.

- *Understanding ECF will help you understand concurrency.* ECF is a basic mechanism for implementing concurrency in computer systems. An exception handler that interrupts the execution of an application program, processes and threads whose execution overlap in time, and a signal handler that interrupts the execution of an application program are all examples of concurrency in action. Understanding ECF is a first step to understanding concurrency.

- *Understanding ECF will help you understand how software exceptions work.* Languages such as C++ and Java provide software exception mechanisms via try, catch, and throw statements. Software exceptions allow the program to make *nonlocal* jumps (i.e., jumps that violate the usual call/return stack discipline) in response to error conditions. Nonlocal jumps are a form of application-level ECF, and are provided in C via the setjmp and longjmp functions. Understanding these low-level functions will help you understand how higher-level software exceptions can be implemented.

Up to this point in your study of systems, you have learned how applications interact with the hardware. This chapter is pivotal in the sense that you will begin to learn how your applications interact with the operating system. Interestingly, these interactions all revolve around ECF. We describe the various forms of ECF that exist at all levels of a computer system. We start with exceptions, which lie at the intersection of the hardware and the operating system. We also discuss system calls, which are exceptions that provide applications with entry points into the operating system. We then move up a level of abstraction and describe processes and signals, which lie at the intersection of applications and the operating system. Finally, we discuss nonlocal jumps, which are an application-level form of ECF.

1 Exceptions

Exceptions are a form of exceptional control flow that are implemented partly by the hardware and partly by the operating system. Because they are partly implemented in hardware, the details vary from system to system. However, the basic ideas are the same for every system. Our aim in this section is to give you a general understanding of exceptions and exception handling, and to help demystify what is often a confusing aspect of modern computer systems.

An *exception* is an abrupt change in the control flow in response to some change in the processor's state. Figure 1 shows the basic idea. In the figure, the processor is executing some current instruction I_{curr} when a significant change in the processor's *state* occurs. The state is encoded in various bits and signals inside the processor. The change in state is known as an *event*. The event might be directly

Figure 1

Anatomy of an exception.
A change in the processor's state (event) triggers an abrupt control transfer (an exception) from the application program to an exception handler. After it finishes processing, the handler either returns control to the interrupted program or aborts.

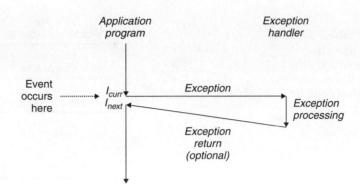

related to the execution of the current instruction. For example, a virtual memory page fault occurs, an arithmetic overflow occurs, or an instruction attempts a divide by zero. On the other hand, the event might be unrelated to the execution of the current instruction. For example, a system timer goes off or an I/O request completes.

In any case, when the processor detects that the event has occurred, it makes an indirect procedure call (the exception), through a jump table called an *exception table*, to an operating system subroutine (the *exception handler*) that is specifically designed to process this particular kind of event.

When the exception handler finishes processing, one of three things happens, depending on the type of event that caused the exception:

1. The handler returns control to the current instruction I_{curr}, the instruction that was executing when the event occurred.

2. The handler returns control to I_{next}, the instruction that would have executed next had the exception not occurred.

3. The handler aborts the interrupted program.

Section 1.2 says more about these possibilities.

Aside   Hardware vs. software exceptions

C++ and Java programmers will have noticed that the term "exception" is also used to describe the application-level ECF mechanism provided by C++ and Java in the form of `catch`, `throw`, and `try` statements. If we wanted to be perfectly clear, we might distinguish between "hardware" and "software" exceptions, but this is usually unnecessary because the meaning is clear from the context.

1.1   Exception Handling

Exceptions can be difficult to understand because handling them involves close cooperation between hardware and software. It is easy to get confused about which component performs which task. Let's look at the division of labor between hardware and software in more detail.

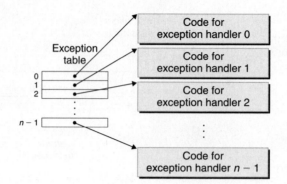

Figure 2
Exception table. The exception table is a jump table where entry k contains the address of the handler code for exception k.

Each type of possible exception in a system is assigned a unique nonnegative integer *exception number*. Some of these numbers are assigned by the designers of the processor. Other numbers are assigned by the designers of the operating system *kernel* (the memory-resident part of the operating system). Examples of the former include divide by zero, page faults, memory access violations, breakpoints, and arithmetic overflows. Examples of the latter include system calls and signals from external I/O devices.

At system boot time (when the computer is reset or powered on), the operating system allocates and initializes a jump table called an *exception table*, so that entry k contains the address of the handler for exception k. Figure 2 shows the format of an exception table.

At run time (when the system is executing some program), the processor detects that an event has occurred and determines the corresponding exception number k. The processor then triggers the exception by making an indirect procedure call, through entry k of the exception table, to the corresponding handler. Figure 3 shows how the processor uses the exception table to form the address of the appropriate exception handler. The exception number is an index into the exception table, whose starting address is contained in a special CPU register called the *exception table base register*.

An exception is akin to a procedure call, but with some important differences.

- As with a procedure call, the processor pushes a return address on the stack before branching to the handler. However, depending on the class of exception, the return address is either the current instruction (the instruction that

Figure 3
Generating the address of an exception handler. The exception number is an index into the exception table.

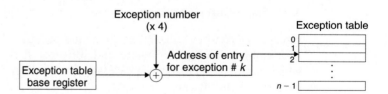

was executing when the event occurred) or the next instruction (the instruction that would have executed after the current instruction had the event not occurred).

- The processor also pushes some additional processor state onto the stack that will be necessary to restart the interrupted program when the handler returns. For example, an IA32 system pushes the EFLAGS register containing, among other things, the current condition codes, onto the stack.

- If control is being transferred from a user program to the kernel, all of these items are pushed onto the kernel's stack rather than onto the user's stack.

- Exception handlers run in *kernel mode* (Section 2.4), which means they have complete access to all system resources.

Once the hardware triggers the exception, the rest of the work is done in software by the exception handler. After the handler has processed the event, it optionally returns to the interrupted program by executing a special "return from interrupt" instruction, which pops the appropriate state back into the processor's control and data registers, restores the state to *user mode* (Section 2.4) if the exception interrupted a user program, and then returns control to the interrupted program.

1.2 Classes of Exceptions

Exceptions can be divided into four classes: *interrupts*, *traps*, *faults*, and *aborts*. The table in Figure 4 summarizes the attributes of these classes.

Interrupts

Interrupts occur *asynchronously* as a result of signals from I/O devices that are external to the processor. Hardware interrupts are asynchronous in the sense that they are not caused by the execution of any particular instruction. Exception handlers for hardware interrupts are often called *interrupt handlers*.

Figure 5 summarizes the processing for an interrupt. I/O devices such as network adapters, disk controllers, and timer chips trigger interrupts by signaling a pin on the processor chip and placing onto the system bus the exception number that identifies the device that caused the interrupt.

Class	Cause	Async/Sync	Return behavior
Interrupt	Signal from I/O device	Async	Always returns to next instruction
Trap	Intentional exception	Sync	Always returns to next instruction
Fault	Potentially recoverable error	Sync	Might return to current instruction
Abort	Nonrecoverable error	Sync	Never returns

Figure 4 **Classes of exceptions.** Asynchronous exceptions occur as a result of events in I/O devices that are external to the processor. Synchronous exceptions occur as a direct result of executing an instruction.

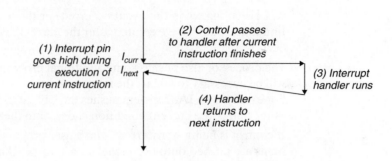

Figure 5

Interrupt handling. The interrupt handler returns control to the next instruction in the application program's control flow.

(1) Interrupt pin goes high during execution of current instruction

I_{curr}

I_{next}

(2) Control passes to handler after current instruction finishes

(3) Interrupt handler runs

(4) Handler returns to next instruction

After the current instruction finishes executing, the processor notices that the interrupt pin has gone high, reads the exception number from the system bus, and then calls the appropriate interrupt handler. When the handler returns, it returns control to the next instruction (i.e., the instruction that would have followed the current instruction in the control flow had the interrupt not occurred). The effect is that the program continues executing as though the interrupt had never happened.

The remaining classes of exceptions (traps, faults, and aborts) occur *synchronously* as a result of executing the current instruction. We refer to this instruction as the *faulting instruction*.

Traps and System Calls

Traps are *intentional* exceptions that occur as a result of executing an instruction. Like interrupt handlers, trap handlers return control to the next instruction. The most important use of traps is to provide a procedure-like interface between user programs and the kernel known as a *system call*.

User programs often need to request services from the kernel such as reading a file (`read`), creating a new process (`fork`), loading a new program (`execve`), or terminating the current process (`exit`). To allow controlled access to such kernel services, processors provide a special "`syscall n`" instruction that user programs can execute when they want to request service n. Executing the `syscall` instruction causes a trap to an exception handler that decodes the argument and calls the appropriate kernel routine. Figure 6 summarizes the processing for a system call. From a programmer's perspective, a system call is identical to a

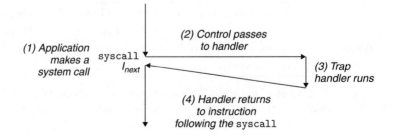

Figure 6

Trap handling. The trap handler returns control to the next instruction in the application program's control flow.

(1) Application makes a system call

`syscall`

I_{next}

(2) Control passes to handler

(3) Trap handler runs

(4) Handler returns to instruction following the `syscall`

Figure 7

Figure 7

Fault handling. Depending on whether the fault can be repaired or not, the fault handler either reexecutes the faulting instruction or aborts.

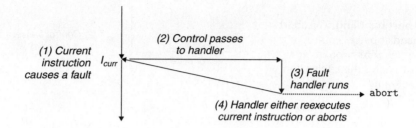

regular function call. However, their implementations are quite different. Regular functions run in *user mode*, which restricts the types of instructions they can execute, and they access the same stack as the calling function. A system call runs in *kernel mode*, which allows it to execute instructions, and accesses a stack defined in the kernel. Section 2.4 discusses user and kernel modes in more detail.

Faults

Faults result from error conditions that a handler might be able to correct. When a fault occurs, the processor transfers control to the fault handler. If the handler is able to correct the error condition, it returns control to the faulting instruction, thereby reexecuting it. Otherwise, the handler returns to an abort routine in the kernel that terminates the application program that caused the fault. Figure 7 summarizes the processing for a fault.

A classic example of a fault is the page fault exception, which occurs when an instruction references a virtual address whose corresponding physical page is not resident in memory and must therefore be retrieved from disk. A page is a contiguous block (typically 4 KB) of virtual memory. The page fault handler loads the appropriate page from disk and then returns control to the instruction that caused the fault. When the instruction executes again, the appropriate physical page is resident in memory and the instruction is able to run to completion without faulting.

Aborts

Aborts result from unrecoverable fatal errors, typically hardware errors such as parity errors that occur when DRAM or SRAM bits are corrupted. Abort handlers never return control to the application program. As shown in Figure 8, the handler returns control to an abort routine that terminates the application program.

1.3 Exceptions in Linux/IA32 Systems

To help make things more concrete, let's look at some of the exceptions defined for IA32 systems. There are up to 256 different exception types [27]. Numbers in the range from 0 to 31 correspond to exceptions that are defined by the Intel architects, and thus are identical for any IA32 system. Numbers in the range from

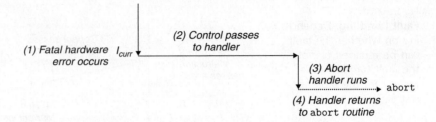

Figure 8
Abort handling. The abort handler passes control to a kernel abort routine that terminates the application program.

Exception number	Description	Exception class
0	Divide error	Fault
13	General protection fault	Fault
14	Page fault	Fault
18	Machine check	Abort
32–127	OS-defined exceptions	Interrupt or trap
128 (0x80)	System call	Trap
129–255	OS-defined exceptions	Interrupt or trap

Figure 9 **Examples of exceptions in IA32 systems.**

32 to 255 correspond to interrupts and traps that are defined by the operating system. Figure 9 shows a few examples.

Linux/IA32 Faults and Aborts

Divide error. A divide error (exception 0) occurs when an application attempts to divide by zero, or when the result of a divide instruction is too big for the destination operand. Unix does not attempt to recover from divide errors, opting instead to abort the program. Linux shells typically report divide errors as "Floating exceptions."

General protection fault. The infamous general protection fault (exception 13) occurs for many reasons, usually because a program references an undefined area of virtual memory, or because the program attempts to write to a read-only text segment. Linux does not attempt to recover from this fault. Linux shells typically report general protection faults as "Segmentation faults."

Page fault. A page fault (exception 14) is an example of an exception where the faulting instruction is restarted. The handler maps the appropriate page of physical memory on disk into a page of virtual memory, and then restarts the faulting instruction.

Machine check. A machine check (exception 18) occurs as a result of a fatal hardware error that is detected during the execution of the faulting instruction. Machine check handlers never return control to the application program.

Number	Name	Description	Number	Name	Description
1	exit	Terminate process	27	alarm	Set signal delivery alarm clock
2	fork	Create new process	29	pause	Suspend process until signal arrives
3	read	Read file	37	kill	Send signal to another process
4	write	Write file	48	signal	Install signal handler
5	open	Open file	63	dup2	Copy file descriptor
6	close	Close file	64	getppid	Get parent's process ID
7	waitpid	Wait for child to terminate	65	getpgrp	Get process group
11	execve	Load and run program	67	sigaction	Install portable signal handler
19	lseek	Go to file offset	90	mmap	Map memory page to file
20	getpid	Get process ID	106	stat	Get information about file

Figure 10 **Examples of popular system calls in Linux/IA32 systems.** Linux provides hundreds of system calls. Source: `/usr/include/sys/syscall.h`.

Linux/IA32 System Calls

Linux provides hundreds of system calls that application programs use when they want to request services from the kernel, such as reading a file, writing a file, or creating a new process. Figure 10 shows some of the more popular Linux system calls. Each system call has a unique integer number that corresponds to an offset in a jump table in the kernel.

System calls are provided on IA32 systems via a trapping instruction called `int n`, where n can be the index of any of the 256 entries in the IA32 exception table. Historically, system calls are provided through exception 128 (0x80).

C programs can invoke any system call directly by using the `syscall` function. However, this is rarely necessary in practice. The standard C library provides a set of convenient wrapper functions for most system calls. The wrapper functions package up the arguments, trap to the kernel with the appropriate system call number, and then pass the return status of the system call back to the calling program. Throughout this text, we will refer to system calls and their associated wrapper functions interchangeably as *system-level functions*.

It is quite interesting to study how programs can use the `int` instruction to invoke Linux system calls directly. All parameters to Linux system calls are passed through general purpose registers rather than the stack. By convention, register `%eax` contains the syscall number, and registers `%ebx`, `%ecx`, `%edx`, `%esi`, `%edi`, and `%ebp` contain up to six arbitrary arguments. The stack pointer `%esp` cannot be used because it is overwritten by the kernel when it enters kernel mode.

For example, consider the following version of the familiar `hello` program, written using the `write` system-level function:

```
1    int main()
2    {
3        write(1, "hello, world\n", 13);
4        exit(0);
5    }
```

code/ecf/hello-asm.sa

```
1    .section .data
2    string:
3      .ascii "hello, world\n"
4    string_end:
5      .equ len, string_end - string

6    .section .text
7    .globl main
8    main:
       First, call write(1, "hello, world\n", 13)
9      movl $4, %eax         System call number 4
10     movl $1, %ebx         stdout has descriptor 1
11     movl $string, %ecx    Hello world string
12     movl $len, %edx       String length
13     int $0x80             System call code


       Next, call exit(0)
14     movl $1, %eax         System call number 0
15     movl $0, %ebx         Argument is 0
16     int $0x80             System call code
```

code/ecf/hello-asm.sa

Figure 11 **Implementing the `hello` program directly with Linux system calls.**

The first argument to `write` sends the output to `stdout`. The second argument is the sequence of bytes to write, and the third argument gives the number of bytes to write.

Figure 11 shows an assembly language version of `hello` that uses the `int` instruction to invoke the `write` and `exit` system calls directly. Lines 9–13 invoke the `write` function. First, line 9 stores the number for the `write` system call in `%eax`, and lines 10–12 set up the argument list. Then line 13 uses the `int` instruction to invoke the system call. Similarly, lines 14–16 invoke the `exit` system call.

Aside A note on terminology

The terminology for the various classes of exceptions varies from system to system. Processor macroarchitecture specifications often distinguish between asynchronous "interrupts" and synchronous "exceptions," yet provide no umbrella term to refer to these very similar concepts. To avoid having to constantly refer to "exceptions and interrupts" and "exceptions or interrupts," we use the word "exception" as the general term and distinguish between asynchronous exceptions (interrupts) and synchronous exceptions (traps, faults, and aborts) only when it is appropriate. As we have noted, the basic ideas are the same for every system, but you should be aware that some manufacturers' manuals use the word "exception" to refer only to those changes in control flow caused by synchronous events.

2 Processes

Exceptions are the basic building blocks that allow the operating system to provide the notion of a *process*, one of the most profound and successful ideas in computer science.

When we run a program on a modern system, we are presented with the illusion that our program is the only one currently running in the system. Our program appears to have exclusive use of both the processor and the memory. The processor appears to execute the instructions in our program, one after the other, without interruption. Finally, the code and data of our program appear to be the only objects in the system's memory. These illusions are provided to us by the notion of a process.

The classic definition of a process is *an instance of a program in execution*. Each program in the system runs in the *context* of some process. The context consists of the state that the program needs to run correctly. This state includes the program's code and data stored in memory, its stack, the contents of its general-purpose registers, its program counter, environment variables, and the set of open file descriptors.

Each time a user runs a program by typing the name of an executable object file to the shell, the shell creates a new process and then runs the executable object file in the context of this new process. Application programs can also create new processes and run either their own code or other applications in the context of the new process.

A detailed discussion of how operating systems implement processes is beyond our scope. Instead, we will focus on the key abstractions that a process provides to the application:

- An independent *logical control flow* that provides the illusion that our program has exclusive use of the processor.
- A private address space that provides the illusion that our program has exclusive use of the memory system.

Let's look more closely at these abstractions.

2.1 Logical Control Flow

A process provides each program with the illusion that it has exclusive use of the processor, even though many other programs are typically running concurrently on the system. If we were to use a debugger to single step the execution of our program, we would observe a series of program counter (PC) values that corresponded exclusively to instructions contained in our program's executable object file or in shared objects linked into our program dynamically at run time. This sequence of PC values is known as a *logical control flow*, or simply *logical flow*.

Consider a system that runs three processes, as shown in Figure 12. The single physical control flow of the processor is partitioned into three logical flows, one for each process. Each vertical line represents a portion of the logical flow for

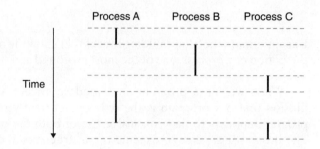

Figure 12

Logical control flows.
Processes provide each program with the illusion that it has exclusive use of the processor. Each vertical bar represents a portion of the logical control flow for a process.

a process. In the example, the execution of the three logical flows is interleaved. Process A runs for a while, followed by B, which runs to completion. Process C then runs for awhile, followed by A, which runs to completion. Finally, C is able to run to completion.

The key point in Figure 12 is that processes take turns using the processor. Each process executes a portion of its flow and then is *preempted* (temporarily suspended) while other processes take their turns. To a program running in the context of one of these processes, it appears to have exclusive use of the processor. The only evidence to the contrary is that if we were to precisely measure the elapsed time of each instruction, we would notice that the CPU appears to periodically stall between the execution of some of the instructions in our program. However, each time the processor stalls, it subsequently resumes execution of our program without any change to the contents of the program's memory locations or registers.

2.2 Concurrent Flows

Logical flows take many different forms in computer systems. Exception handlers, processes, signal handlers, threads, and Java processes are all examples of logical flows.

A logical flow whose execution overlaps in time with another flow is called a *concurrent flow*, and the two flows are said to *run concurrently*. More precisely, flows X and Y are concurrent with respect to each other if and only if X begins after Y begins and before Y finishes, or Y begins after X begins and before X finishes. For example, in Figure 12, processes A and B run concurrently, as do A and C. On the other hand, B and C do not run concurrently, because the last instruction of B executes before the first instruction of C.

The general phenomenon of multiple flows executing concurrently is known as *concurrency*. The notion of a process taking turns with other processes is also known as *multitasking*. Each time period that a process executes a portion of its flow is called a *time slice*. Thus, multitasking is also referred to as *time slicing*. For example, in Figure 12, the flow for Process A consists of two time slices.

Notice that the idea of concurrent flows is independent of the number of processor cores or computers that the flows are running on. If two flows overlap in time, then they are concurrent, even if they are running on the same processor. However, we will sometimes find it useful to identify a proper subset of concurrent flows known as *parallel flows*. If two flows are running concurrently on different processor cores or computers, then we say that they are *parallel flows*, that they are *running in parallel*, and have *parallel execution*.

Practice Problem 1

Consider three processes with the following starting and ending times:

Process	Start time	End time
A	0	2
B	1	4
C	3	5

For each pair of processes, indicate whether they run concurrently (y) or not (n):

Process pair	Concurrent?
AB	_____
AC	_____
BC	_____

2.3 Private Address Space

A process provides each program with the illusion that it has exclusive use of the system's address space. On a machine with n-bit addresses, the *address space* is the set of 2^n possible addresses, $0, 1, \ldots, 2^n - 1$. A process provides each program with its own *private address space*. This space is private in the sense that a byte of memory associated with a particular address in the space cannot in general be read or written by any other process.

Although the contents of the memory associated with each private address space is different in general, each such space has the same general organization. For example, Figure 13 shows the organization of the address space for an x86 Linux process. The bottom portion of the address space is reserved for the user program, with the usual text, data, heap, and stack segments. Code segments begin at address 0x08048000 for 32-bit processes, and at address 0x00400000 for 64-bit processes. The top portion of the address space is reserved for the kernel. This part of the address space contains the code, data, and stack that the kernel uses when it executes instructions on behalf of the process (e.g., when the application program executes a system call).

2.4 User and Kernel Modes

In order for the operating system kernel to provide an airtight process abstraction, the processor must provide a mechanism that restricts the instructions that an application can execute, as well as the portions of the address space that it can access.

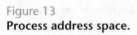
Figure 13
Process address space.

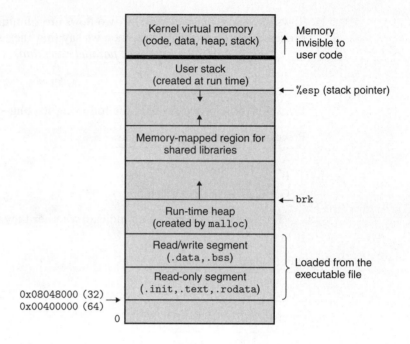

Processors typically provide this capability with a *mode bit* in some control register that characterizes the privileges that the process currently enjoys. When the mode bit is set, the process is running in *kernel mode* (sometimes called *supervisor mode*). A process running in kernel mode can execute any instruction in the instruction set and access any memory location in the system.

When the mode bit is not set, the process is running in *user mode*. A process in user mode is not allowed to execute *privileged instructions* that do things such as halt the processor, change the mode bit, or initiate an I/O operation. Nor is it allowed to directly reference code or data in the kernel area of the address space. Any such attempt results in a fatal protection fault. User programs must instead access kernel code and data indirectly via the system call interface.

A process running application code is initially in user mode. The only way for the process to change from user mode to kernel mode is via an exception such as an interrupt, a fault, or a trapping system call. When the exception occurs, and control passes to the exception handler, the processor changes the mode from user mode to kernel mode. The handler runs in kernel mode. When it returns to the application code, the processor changes the mode from kernel mode back to user mode.

Linux provides a clever mechanism, called the /proc filesystem, that allows user mode processes to access the contents of kernel data structures. The /proc filesystem exports the contents of many kernel data structures as a hierarchy of text files that can be read by user programs. For example, you can use the /proc filesystem to find out general system attributes such as CPU type (/proc/cpuinfo), or the memory segments used by a particular process (/proc/<process id>/maps).

The 2.6 version of the Linux kernel introduced a /sys filesystem, which exports additional low-level information about system buses and devices.

2.5 Context Switches

The operating system kernel implements multitasking using a higher-level form of exceptional control flow known as a *context switch*. The context switch mechanism is built on top of the lower-level exception mechanism that we discussed in Section 1.

The kernel maintains a *context* for each process. The context is the state that the kernel needs to restart a preempted process. It consists of the values of objects such as the general purpose registers, the floating-point registers, the program counter, user's stack, status registers, kernel's stack, and various kernel data structures such as a *page table* that characterizes the address space, a *process table* that contains information about the current process, and a *file table* that contains information about the files that the process has opened.

At certain points during the execution of a process, the kernel can decide to preempt the current process and restart a previously preempted process. This decision is known as *scheduling*, and is handled by code in the kernel called the *scheduler*. When the kernel selects a new process to run, we say that the kernel has *scheduled* that process. After the kernel has scheduled a new process to run, it preempts the current process and transfers control to the new process using a mechanism called a *context switch* that (1) saves the context of the current process, (2) restores the saved context of some previously preempted process, and (3) passes control to this newly restored process.

A context switch can occur while the kernel is executing a system call on behalf of the user. If the system call blocks because it is waiting for some event to occur, then the kernel can put the current process to sleep and switch to another process. For example, if a read system call requires a disk access, the kernel can opt to perform a context switch and run another process instead of waiting for the data to arrive from the disk. Another example is the sleep system call, which is an explicit request to put the calling process to sleep. In general, even if a system call does not block, the kernel can decide to perform a context switch rather than return control to the calling process.

A context switch can also occur as a result of an interrupt. For example, all systems have some mechanism for generating periodic timer interrupts, typically every 1 ms or 10 ms. Each time a timer interrupt occurs, the kernel can decide that the current process has run long enough and switch to a new process.

Figure 14 shows an example of context switching between a pair of processes A and B. In this example, initially process A is running in user mode until it traps to the kernel by executing a read system call. The trap handler in the kernel requests a DMA transfer from the disk controller and arranges for the disk to interrupt the processor after the disk controller has finished transferring the data from disk to memory.

The disk will take a relatively long time to fetch the data (on the order of tens of milliseconds), so instead of waiting and doing nothing in the interim, the kernel performs a context switch from process A to B. Note that before the switch,

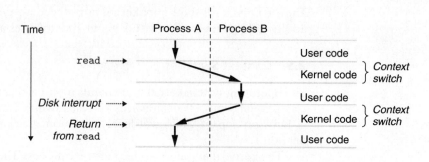

Figure 14

Anatomy of a process context switch.

the kernel is executing instructions in user mode on behalf of process A. During the first part of the switch, the kernel is executing instructions in kernel mode on behalf of process A. Then at some point it begins executing instructions (still in kernel mode) on behalf of process B. And after the switch, the kernel is executing instructions in user mode on behalf of process B.

Process B then runs for a while in user mode until the disk sends an interrupt to signal that data has been transferred from disk to memory. The kernel decides that process B has run long enough and performs a context switch from process B to A, returning control in process A to the instruction immediately following the read system call. Process A continues to run until the next exception occurs, and so on.

Aside   Cache pollution and exceptional control flow

In general, hardware cache memories do not interact well with exceptional control flows such as interrupts and context switches. If the current process is interrupted briefly by an interrupt, then the cache is cold for the interrupt handler. If the handler accesses enough items from main memory, then the cache will also be cold for the interrupted process when it resumes. In this case, we say that the handler has *polluted* the cache. A similar phenomenon occurs with context switches. When a process resumes after a context switch, the cache is cold for the application program and must be warmed up again.

---

3   System Call Error Handling

When Unix system-level functions encounter an error, they typically return −1 and set the global integer variable errno to indicate what went wrong. Programmers should *always* check for errors, but unfortunately, many skip error checking because it bloats the code and makes it harder to read. For example, here is how we might check for errors when we call the Linux fork function:

```
1    if ((pid = fork()) < 0) {
2        fprintf(stderr, "fork error: %s\n", strerror(errno));
3        exit(0);
4    }
```

The `strerror` function returns a text string that describes the error associated with a particular value of `errno`. We can simplify this code somewhat by defining the following *error-reporting function*:

```
1    void unix_error(char *msg) /* Unix-style error */
2    {
3        fprintf(stderr, "%s: %s\n", msg, strerror(errno));
4        exit(0);
5    }
```

Given this function, our call to `fork` reduces from four lines to two lines:

```
1        if ((pid = fork()) < 0)
2            unix_error("fork error");
```

We can simplify our code even further by using *error-handling wrappers*. For a given base function `foo`, we define a wrapper function `Foo` with identical arguments, but with the first letter of the name capitalized. The wrapper calls the base function, checks for errors, and terminates if there are any problems. For example, here is the error-handling wrapper for the `fork` function:

```
1    pid_t Fork(void)
2    {
3        pid_t pid;
4
5        if ((pid = fork()) < 0)
6            unix_error("Fork error");
7        return pid;
8    }
```

Given this wrapper, our call to `fork` shrinks to a single compact line:

```
1        pid = Fork();
```

We will use error-handling wrappers throughout the remainder of this book. They allow us to keep our code examples concise, without giving you the mistaken impression that it is permissible to ignore error checking. Note that when we discuss system-level functions in the text, we will always refer to them by their lowercase base names, rather than by their uppercase wrapper names.

4   Process Control

Unix provides a number of system calls for manipulating processes from C programs. This section describes the important functions and gives examples of how they are used.

4.1 Obtaining Process IDs

Each process has a unique positive (nonzero) *process ID* (PID). The getpid function returns the PID of the calling process. The getppid function returns the PID of its *parent* (i.e., the process that created the calling process).

```
#include <sys/types.h>
#include <unistd.h>

pid_t getpid(void);
pid_t getppid(void);
```
                                    Returns: PID of either the caller or the parent

The getpid and getppid routines return an integer value of type pid_t, which on Linux systems is defined in types.h as an int.

4.2 Creating and Terminating Processes

From a programmer's perspective, we can think of a process as being in one of three states:

- *Running.* The process is either executing on the CPU or is waiting to be executed and will eventually be scheduled by the kernel.
- *Stopped.* The execution of the process is *suspended* and will not be scheduled. A process stops as a result of receiving a SIGSTOP, SIGTSTP, SIGTTIN, or SIGTTOU signal, and it remains stopped until it receives a SIGCONT signal, at which point it can begin running again. (A *signal* is a form of software interrupt that is described in detail in Section 5.)
- *Terminated.* The process is stopped permanently. A process becomes terminated for one of three reasons: (1) receiving a signal whose default action is to terminate the process, (2) returning from the main routine, or (3) calling the exit function:

```
#include <stdlib.h>

void exit(int status);
```
                                    This function does not return

The exit function terminates the process with an *exit status* of status. (The other way to set the exit status is to return an integer value from the main routine.)

A *parent process* creates a new running *child process* by calling the `fork` function.

```
#include <sys/types.h>
#include <unistd.h>

pid_t fork(void);
```
                                    Returns: 0 to child, PID of child to parent, −1 on error

The newly created child process is almost, but not quite, identical to the parent. The child gets an identical (but separate) copy of the parent's user-level virtual address space, including the text, data, and bss segments, heap, and user stack. The child also gets identical copies of any of the parent's open file descriptors, which means the child can read and write any files that were open in the parent when it called `fork`. The most significant difference between the parent and the newly created child is that they have different PIDs.

The `fork` function is interesting (and often confusing) because it is called *once* but it returns *twice*: once in the calling process (the parent), and once in the newly created child process. In the parent, `fork` returns the PID of the child. In the child, `fork` returns a value of 0. Since the PID of the child is always nonzero, the return value provides an unambiguous way to tell whether the program is executing in the parent or the child.

Figure 15 shows a simple example of a parent process that uses `fork` to create a child process. When the `fork` call returns in line 8, x has a value of 1 in both the parent and child. The child increments and prints its copy of x in line 10. Similarly, the parent decrements and prints its copy of x in line 15.

When we run the program on our Unix system, we get the following result:

```
unix> ./fork
parent: x=0
child : x=2
```

There are some subtle aspects to this simple example.

- *Call once, return twice.* The `fork` function is called once by the parent, but it returns twice: once to the parent and once to the newly created child. This is fairly straightforward for programs that create a single child. But programs with multiple instances of `fork` can be confusing and need to be reasoned about carefully.

- *Concurrent execution.* The parent and the child are separate processes that run concurrently. The instructions in their logical control flows can be interleaved by the kernel in an arbitrary way. When we run the program on our system, the parent process completes its `printf` statement first, followed by the child. However, on another system the reverse might be true. In general, as programmers we can never make assumptions about the interleaving of the instructions in different processes.

code/ecf/fork.c

```
1    #include "csapp.h"
2
3    int main()
4    {
5        pid_t pid;
6        int x = 1;
7
8        pid = Fork();
9        if (pid == 0) {  /* Child */
10           printf("child : x=%d\n", ++x);
11           exit(0);
12       }
13
14       /* Parent */
15       printf("parent: x=%d\n", --x);
16       exit(0);
17   }
```

code/ecf/fork.c

Figure 15 **Using `fork` to create a new process.**

- *Duplicate but separate address spaces.* If we could halt both the parent and the child immediately after the `fork` function returned in each process, we would see that the address space of each process is identical. Each process has the same user stack, the same local variable values, the same heap, the same global variable values, and the same code. Thus, in our example program, local variable x has a value of 1 in both the parent and the child when the `fork` function returns in line 8. However, since the parent and the child are separate processes, they each have their own private address spaces. Any subsequent changes that a parent or child makes to x are private and are not reflected in the memory of the other process. This is why the variable x has different values in the parent and child when they call their respective `printf` statements.

- *Shared files.* When we run the example program, we notice that both parent and child print their output on the screen. The reason is that the child inherits all of the parent's open files. When the parent calls `fork`, the `stdout` file is open and directed to the screen. The child inherits this file and thus its output is also directed to the screen.

When you are first learning about the `fork` function, it is often helpful to sketch the *process graph*, where each horizontal arrow corresponds to a process that executes instructions from left to right, and each vertical arrow corresponds to the execution of a `fork` function.

For example, how many lines of output would the program in Figure 16(a) generate? Figure 16(b) shows the corresponding process graph. The parent

(a) Calls `fork` once

```
1    #include "csapp.h"
2
3    int main()
4    {
5        Fork();
6        printf("hello\n");
7        exit(0);
8    }
```

(b) Prints two output lines

```
              hello
         ┌──────────────►
         │
         │    hello
    ─────┴──────────────►
      fork
```

(c) Calls `fork` twice

```
1    #include "csapp.h"
2
3    int main()
4    {
5        Fork();
6        Fork();
7        printf("hello\n");
8        exit(0);
9    }
```

(d) Prints four output lines

```
                    hello
               ┌──────────────►
               │
               │    hello
         ┌─────┴──────────────►
         │
         │          hello
         │     ┌──────────────►
         │     │
         │     │    hello
    ─────┴─────┴──────────────►
      fork  fork
```

(e) Calls `fork` three times

```
1    #include "csapp.h"
2
3    int main()
4    {
5        Fork();
6        Fork();
7        Fork();
8        printf("hello\n");
9        exit(0);
10   }
```

(f) Prints eight output lines

```
                          hello
                     ┌──────────────►
                     │
                     │    hello
               ┌─────┴──────────────►
               │
               │          hello
               │     ┌──────────────►
               │     │
               │     │    hello
         ┌─────┴─────┴──────────────►
         │
         │                hello
         │           ┌──────────────►
         │           │
         │           │    hello
         │     ┌─────┴──────────────►
         │     │
         │     │          hello
         │     │     ┌──────────────►
         │     │     │
         │     │     │    hello
    ─────┴─────┴─────┴──────────────►
      fork    fork  fork
```

Figure 16 **Examples of `fork` programs and their process graphs.**

creates a child when it executes the first (and only) `fork` in the program. Each of these calls `printf` once, so the program prints two output lines.

Now what if we were to call `fork` twice, as shown in Figure 16(c)? As we see from Figure 16(d), the parent calls `fork` to create a child, and then the parent and child each call `fork`, which results in two more processes. Thus, there are four processes, each of which calls `printf`, so the program generates four output lines.

Continuing this line of thought, what would happen if we were to call `fork` three times, as in Figure 16(e)? As we see from the process graph in Figure 16(f), there are a total of eight processes. Each process calls `printf`, so the program produces eight output lines.

Practice Problem 2

Consider the following program:

—————————————————————————— code/ecf/forkprob0.c

```
1    #include "csapp.h"
2
3    int main()
4    {
5        int x = 1;
6
7        if (Fork() == 0)
8            printf("printf1: x=%d\n", ++x);
9        printf("printf2: x=%d\n", --x);
10       exit(0);
11   }
```

—————————————————————————— code/ecf/forkprob0.c

A. What is the output of the child process?

B. What is the output of the parent process?

4.3 Reaping Child Processes

When a process terminates for any reason, the kernel does not remove it from the system immediately. Instead, the process is kept around in a terminated state until it is *reaped* by its parent. When the parent reaps the terminated child, the kernel passes the child's exit status to the parent, and then discards the terminated process, at which point it ceases to exist. A terminated process that has not yet been reaped is called a *zombie*.

> **Aside** Why are terminated children called zombies?
>
> In folklore, a zombie is a living corpse, an entity that is half alive and half dead. A zombie process is similar in the sense that while it has already terminated, the kernel maintains some of its state until it can be reaped by the parent.

If the parent process terminates without reaping its zombie children, the kernel arranges for the `init` process to reap them. The `init` process has a PID of 1 and is created by the kernel during system initialization. Long-running programs such as shells or servers should always reap their zombie children. Even though zombies are not running, they still consume system memory resources.

A process waits for its children to terminate or stop by calling the `waitpid` function.

```
#include <sys/types.h>
#include <sys/wait.h>

pid_t waitpid(pid_t pid, int *status, int options);
```
                    Returns: PID of child if OK, 0 (if WNOHANG) or −1 on error

The `waitpid` function is complicated. By default (when `options` = 0), `waitpid` suspends execution of the calling process until a child process in its *wait set* terminates. If a process in the wait set has already terminated at the time of the call, then `waitpid` returns immediately. In either case, `waitpid` returns the PID of the terminated child that caused `waitpid` to return, and the terminated child is removed from the system.

Determining the Members of the Wait Set

The members of the wait set are determined by the `pid` argument:

- If `pid` > 0, then the wait set is the singleton child process whose process ID is equal to `pid`.
- If `pid` = −1, then the wait set consists of all of the parent's child processes.

The `waitpid` function also supports other kinds of wait sets, involving Unix process groups, that we will not discuss.

Modifying the Default Behavior

The default behavior can be modified by setting `options` to various combinations of the WNOHANG and WUNTRACED constants:

- WNOHANG: Return immediately (with a return value of 0) if none of the child processes in the wait set has terminated yet. The default behavior suspends the calling process until a child terminates. This option is useful in those cases where you want to continue doing useful work while waiting for a child to terminate.
- WUNTRACED: Suspend execution of the calling process until a process in the wait set becomes either terminated or stopped. Return the PID of the terminated or stopped child that caused the return. The default behavior returns only for terminated children. This option is useful when you want to check for both terminated *and* stopped children.
- WNOHANG|WUNTRACED: Return immediately, with a return value of 0, if none of the children in the wait set has stopped or terminated, or with a return value equal to the PID of one of the stopped or terminated children.

Checking the Exit Status of a Reaped Child

If the `status` argument is non-NULL, then `waitpid` encodes status information about the child that caused the return in the `status` argument. The `wait.h` include file defines several macros for interpreting the `status` argument:

- WIFEXITED(`status`): Returns true if the child terminated normally, via a call to `exit` or a return.
- WEXITSTATUS(`status`): Returns the exit status of a normally terminated child. This status is only defined if WIFEXITED returned true.
- WIFSIGNALED(`status`): Returns true if the child process terminated because of a signal that was not caught. (Signals are explained in Section 5.)
- WTERMSIG(`status`): Returns the number of the signal that caused the child process to terminate. This status is only defined if WIFSIGNALED(`status`) returned true.
- WIFSTOPPED(`status`): Returns true if the child that caused the return is currently stopped.
- WSTOPSIG(`status`): Returns the number of the signal that caused the child to stop. This status is only defined if WIFSTOPPED(`status`) returned true.

Error Conditions

If the calling process has no children, then `waitpid` returns −1 and sets `errno` to ECHILD. If the `waitpid` function was interrupted by a signal, then it returns −1 and sets `errno` to EINTR.

Aside   Constants associated with Unix functions

Constants such as WNOHANG and WUNTRACED are defined by system header files. For example, WNOHANG and WUNTRACED are defined (indirectly) by the `wait.h` header file:

```
/* Bits in the third argument to 'waitpid'. */
#define WNOHANG    1   /* Don't block waiting. */
#define WUNTRACED  2   /* Report status of stopped children. */
```

In order to use these constants, you must include the `wait.h` header file in your code:

```
#include <sys/wait.h>
```

The man page for each Unix function lists the header files to include whenever you use that function in your code. Also, in order to check return codes such as ECHILD and EINTR, you must include `errno.h`. To simplify our code examples, we include a single header file called `csapp.h` that includes the header files for all of the functions used in the book. The `csapp.h` header file is available online from the CS:APP Web site.

Practice Problem 3

List all of the possible output sequences for the following program:

———————————————————————— code/ecf/waitprob0.c

```
1    int main()
2    {
3        if (Fork() == 0) {
4            printf("a");
5        }
6        else {
7            printf("b");
8            waitpid(-1, NULL, 0);
9        }
10       printf("c");
11       exit(0);
12   }
```

———————————————————————— code/ecf/waitprob0.c

The wait Function

The wait function is a simpler version of waitpid:

```
#include <sys/types.h>
#include <sys/wait.h>

pid_t wait(int *status);
```
Returns: PID of child if OK or −1 on error

Calling wait(&status) is equivalent to calling waitpid(-1, &status, 0).

Examples of Using waitpid

Because the waitpid function is somewhat complicated, it is helpful to look at a few examples. Figure 17 shows a program that uses waitpid to wait, in no particular order, for all of its *N* children to terminate.

In line 11, the parent creates each of the *N* children, and in line 12, each child exits with a unique exit status. Before moving on, make sure you understand why line 12 is executed by each of the children, but not the parent.

In line 15, the parent waits for all of its children to terminate by using waitpid as the test condition of a while loop. Because the first argument is −1, the call to waitpid blocks until an arbitrary child has terminated. As each child terminates, the call to waitpid returns with the nonzero PID of that child. Line 16 checks the exit status of the child. If the child terminated normally, in this case by calling the exit function, then the parent extracts the exit status and prints it on stdout.

———— code/ecf/waitpid1.c

```
1   #include "csapp.h"
2   #define N 2
3
4   int main()
5   {
6       int status, i;
7       pid_t pid;
8
9       /* Parent creates N children */
10      for (i = 0; i < N; i++)
11          if ((pid = Fork()) == 0)   /* Child */
12              exit(100+i);
13
14      /* Parent reaps N children in no particular order */
15      while ((pid = waitpid(-1, &status, 0)) > 0) {
16          if (WIFEXITED(status))
17              printf("child %d terminated normally with exit status=%d\n",
18                      pid, WEXITSTATUS(status));
19          else
20              printf("child %d terminated abnormally\n", pid);
21      }
22
23      /* The only normal termination is if there are no more children */
24      if (errno != ECHILD)
25          unix_error("waitpid error");
26
27      exit(0);
28  }
```

———— code/ecf/waitpid1.c

Figure 17 **Using the `waitpid` function to reap zombie children in no particular order.**

When all of the children have been reaped, the next call to `waitpid` returns −1 and sets errno to ECHILD. Line 24 checks that the `waitpid` function terminated normally, and prints an error message otherwise. When we run the program on our Unix system, it produces the following output:

```
unix> ./waitpid1
child 22966 terminated normally with exit status=100
child 22967 terminated normally with exit status=101
```

Notice that the program reaps its children in no particular order. The order that they were reaped is a property of this specific computer system. On another *code/ecf/waitpid2.c*

```
1    #include "csapp.h"
2    #define N 2
3
4    int main()
5    {
6        int status, i;
7        pid_t pid[N], retpid;
8
9        /* Parent creates N children */
10       for (i = 0; i < N; i++)
11           if ((pid[i] = Fork()) == 0)   /* Child */
12               exit(100+i);
13
14       /* Parent reaps N children in order */
15       i = 0;
16       while ((retpid = waitpid(pid[i++], &status, 0)) > 0) {
17           if (WIFEXITED(status))
18               printf("child %d terminated normally with exit status=%d\n",
19                       retpid, WEXITSTATUS(status));
20           else
21               printf("child %d terminated abnormally\n", retpid);
22       }
23
24       /* The only normal termination is if there are no more children */
25       if (errno != ECHILD)
26           unix_error("waitpid error");
27
28       exit(0);
29   }
```

code/ecf/waitpid2.c

Figure 18 **Using `waitpid` to reap zombie children in the order they were created.**

system, or even another execution on the same system, the two children might have been reaped in the opposite order. This is an example of the *nondeterministic* behavior that can make reasoning about concurrency so difficult. Either of the two possible outcomes is equally correct, and as a programmer you may *never* assume that one outcome will always occur, no matter how unlikely the other outcome appears to be. The only correct assumption is that each possible outcome is equally likely.

Figure 18 shows a simple change that eliminates this nondeterminism in the output order by reaping the children in the same order that they were created by the parent. In line 11, the parent stores the PIDs of its children in order, and then waits for each child in this same order by calling `waitpid` with the appropriate PID in the first argument.

Practice Problem 4

Consider the following program:

code/ecf/waitprob1.c

```
1   int main()
2   {
3       int status;
4       pid_t pid;
5
6       printf("Hello\n");
7       pid = Fork();
8       printf("%d\n", !pid);
9       if (pid != 0) {
10          if (waitpid(-1, &status, 0) > 0) {
11              if (WIFEXITED(status) != 0)
12                  printf("%d\n", WEXITSTATUS(status));
13          }
14      }
15      printf("Bye\n");
16      exit(2);
17  }
```

code/ecf/waitprob1.c

A. How many output lines does this program generate?

B. What is one possible ordering of these output lines?

4.4 Putting Processes to Sleep

The sleep function suspends a process for a specified period of time.

```
#include <unistd.h>

unsigned int sleep(unsigned int secs);
                                    Returns: seconds left to sleep
```

Sleep returns zero if the requested amount of time has elapsed, and the number of seconds still left to sleep otherwise. The latter case is possible if the sleep function returns prematurely because it was interrupted by a *signal*. We will discuss signals in detail in Section 5.

Another function that we will find useful is the pause function, which puts the calling function to sleep until a signal is received by the process.

```
#include <unistd.h>

int pause(void);
                                                          Always returns −1
```

Practice Problem 5

Write a wrapper function for sleep, called snooze, with the following interface:

```
unsigned int snooze(unsigned int secs);
```

The snooze function behaves exactly as the sleep function, except that it prints a message describing how long the process actually slept:

```
Slept for 4 of 5 secs.
```

4.5 Loading and Running Programs

The execve function loads and runs a new program in the context of the current process.

```
#include <unistd.h>

int execve(const char *filename, const char *argv[],
           const char *envp[]);
                                   Does not return if OK, returns −1 on error
```

The execve function loads and runs the executable object file filename with the argument list argv and the environment variable list envp. Execve returns to the calling program only if there is an error such as not being able to find filename. So unlike fork, which is called once but returns twice, execve is called once and never returns.

The argument list is represented by the data structure shown in Figure 19. The argv variable points to a null-terminated array of pointers, each of which

Figure 19
Organization of an argument list.

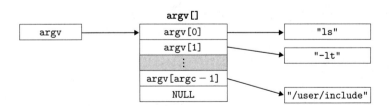

721

Figure 20

**Organization of an
environment variable
list.**

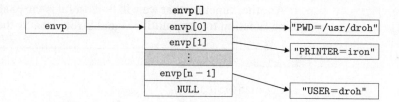

points to an argument string. By convention, argv[0] is the name of the executable
object file. The list of environment variables is represented by a similar data
structure, shown in Figure 20. The envp variable points to a null-terminated array
of pointers to environment variable strings, each of which is a name-value pair of
the form "NAME=VALUE".

After execve loads filename, it calls startup code, which sets up the stack and
passes control to the main routine of the new program, which has a prototype of
the form

```
int main(int argc, char **argv, char **envp);
```

or equivalently,

```
int main(int argc, char *argv[], char *envp[]);
```

When main begins executing in a 32-bit Linux process, the user stack has
the organization shown in Figure 21. Let's work our way from the bottom of the
stack (the highest address) to the top (the lowest address). First are the argument

Figure 21

**Typical organization of
the user stack when a
new program starts.**

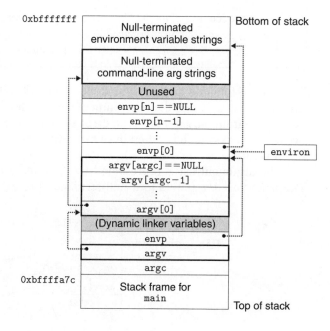

and environment strings, which are stored contiguously on the stack, one after the other without any gaps. These are followed further up the stack by a null-terminated array of pointers, each of which points to an environment variable string on the stack. The global variable environ points to the first of these pointers, envp[0]. The environment array is followed immediately by the null-terminated argv[] array, with each element pointing to an argument string on the stack. At the top of the stack are the three arguments to the main routine: (1) envp, which points to the envp[] array, (2) argv, which points to the argv[] array, and (3) argc, which gives the number of non-null pointers in the argv[] array.

Unix provides several functions for manipulating the environment array:

```
#include <stdlib.h>

char *getenv(const char *name);
```
Returns: ptr to name if exists, NULL if no match

The getenv function searches the environment array for a string "name=value". If found, it returns a pointer to value, otherwise it returns NULL.

```
#include <stdlib.h>

int setenv(const char *name, const char *newvalue, int overwrite);
```
Returns: 0 on success, −1 on error

```
void unsetenv(const char *name);
```
Returns: nothing

If the environment array contains a string of the form "name=oldvalue", then unsetenv deletes it and setenv replaces oldvalue with newvalue, but only if overwrite is nonzero. If name does not exist, then setenv adds "name=newvalue" to the array.

Aside Programs vs. processes

This is a good place to pause and make sure you understand the distinction between a program and a process. A program is a collection of code and data; programs can exist as object modules on disk or as segments in an address space. A process is a specific instance of a program in execution; a program always runs in the context of some process. Understanding this distinction is important if you want to understand the fork and execve functions. The fork function runs the same program in a new child process that is a duplicate of the parent. The execve function loads and runs a new program in the context of the current process. While it overwrites the address space of the current process, it does *not* create a new process. The new program still has the same PID, and it inherits all of the file descriptors that were open at the time of the call to the execve function.

Practice Problem 6

Write a program called myecho that prints its command line arguments and environment variables. For example:

```
unix> ./myecho arg1 arg2
Command line arguments:
    argv[ 0]: myecho
    argv[ 1]: arg1
    argv[ 2]: arg2

Environment variables:
    envp[ 0]: PWD=/usr0/droh/ics/code/ecf
    envp[ 1]: TERM=emacs
    ...
    envp[25]: USER=droh
    envp[26]: SHELL=/usr/local/bin/tcsh
    envp[27]: HOME=/usr0/droh
```

4.6 Using fork and execve to Run Programs

Programs such as Unix shells and Web servers make heavy use of the fork and execve functions. A *shell* is an interactive application-level program that runs other programs on behalf of the user. The original shell was the sh program, which was followed by variants such as csh, tcsh, ksh, and bash. A shell performs a sequence of *read/evaluate* steps, and then terminates. The read step reads a command line from the user. The evaluate step parses the command line and runs programs on behalf of the user.

Figure 22 shows the main routine of a simple shell. The shell prints a command-line prompt, waits for the user to type a command line on stdin, and then evaluates the command line.

Figure 23 shows the code that evaluates the command line. Its first task is to call the parseline function (Figure 24), which parses the space-separated command-line arguments and builds the argv vector that will eventually be passed to execve. The first argument is assumed to be either the name of a built-in shell command that is interpreted immediately, or an executable object file that will be loaded and run in the context of a new child process.

If the last argument is an "&" character, then parseline returns 1, indicating that the program should be executed in the *background* (the shell does not wait

code/ecf/shellex.c

```
1    #include "csapp.h"
2    #define MAXARGS   128
3
4    /* Function prototypes */
5    void eval(char *cmdline);
6    int parseline(char *buf, char **argv);
7    int builtin_command(char **argv);
8
9    int main()
10   {
11       char cmdline[MAXLINE]; /* Command line */
12
13       while (1) {
14           /* Read */
15           printf("> ");
16           Fgets(cmdline, MAXLINE, stdin);
17           if (feof(stdin))
18               exit(0);
19
20           /* Evaluate */
21           eval(cmdline);
22       }
23   }
```

code/ecf/shellex.c

Figure 22 **The main routine for a simple shell program.**

for it to complete). Otherwise it returns 0, indicating that the program should be run in the *foreground* (the shell waits for it to complete).

After parsing the command line, the eval function calls the builtin_command function, which checks whether the first command line argument is a built-in shell command. If so, it interprets the command immediately and returns 1. Otherwise, it returns 0. Our simple shell has just one built-in command, the quit command, which terminates the shell. Real shells have numerous commands, such as pwd, jobs, and fg.

If builtin_command returns 0, then the shell creates a child process and executes the requested program inside the child. If the user has asked for the program to run in the background, then the shell returns to the top of the loop and waits for the next command line. Otherwise the shell uses the waitpid function to wait for the job to terminate. When the job terminates, the shell goes on to the next iteration.

Notice that this simple shell is flawed because it does not reap any of its background children. Correcting this flaw requires the use of signals, which we describe in the next section.

code/ecf/shellex.c

```
1    /* eval - Evaluate a command line */
2    void eval(char *cmdline)
3    {
4        char *argv[MAXARGS]; /* Argument list execve() */
5        char buf[MAXLINE];   /* Holds modified command line */
6        int bg;              /* Should the job run in bg or fg? */
7        pid_t pid;           /* Process id */
8
9        strcpy(buf, cmdline);
10       bg = parseline(buf, argv);
11       if (argv[0] == NULL)
12           return;   /* Ignore empty lines */
13
14       if (!builtin_command(argv)) {
15           if ((pid = Fork()) == 0) {   /* Child runs user job */
16               if (execve(argv[0], argv, environ) < 0) {
17                   printf("%s: Command not found.\n", argv[0]);
18                   exit(0);
19               }
20           }
21
22           /* Parent waits for foreground job to terminate */
23           if (!bg) {
24               int status;
25               if (waitpid(pid, &status, 0) < 0)
26                   unix_error("waitfg: waitpid error");
27           }
28           else
29               printf("%d %s", pid, cmdline);
30       }
31       return;
32   }
33
34   /* If first arg is a builtin command, run it and return true */
35   int builtin_command(char **argv)
36   {
37       if (!strcmp(argv[0], "quit")) /* quit command */
38           exit(0);
39       if (!strcmp(argv[0], "&"))    /* Ignore singleton & */
40           return 1;
41       return 0;                     /* Not a builtin command */
42   }
```

code/ecf/shellex.c

Figure 23 eval: **Evaluates the shell command line.**

code/ecf/shellex.c

```c
/* parseline - Parse the command line and build the argv array */
int parseline(char *buf, char **argv)
{
    char *delim;         /* Points to first space delimiter */
    int argc;            /* Number of args */
    int bg;              /* Background job? */

    buf[strlen(buf)-1] = ' ';  /* Replace trailing '\n' with space */
    while (*buf && (*buf == ' ')) /* Ignore leading spaces */
        buf++;

    /* Build the argv list */
    argc = 0;
    while ((delim = strchr(buf, ' '))) {
        argv[argc++] = buf;
        *delim = '\0';
        buf = delim + 1;
        while (*buf && (*buf == ' ')) /* Ignore spaces */
                buf++;
    }
    argv[argc] = NULL;

    if (argc == 0)   /* Ignore blank line */
        return 1;

    /* Should the job run in the background? */
    if ((bg = (*argv[argc-1] == '&')) != 0)
        argv[--argc] = NULL;

    return bg;
}
```

code/ecf/shellex.c

Figure 24 `parseline`: **Parses a line of input for the shell.**

5   Signals

To this point in our study of exceptional control flow, we have seen how hardware and software cooperate to provide the fundamental low-level exception mechanism. We have also seen how the operating system uses exceptions to support a form of exceptional control flow known as the process context switch. In this section, we will study a higher-level software form of exceptional control flow, known as a Unix *signal*, that allows processes and the kernel to interrupt other processes.

Number	Name	Default action	Corresponding event
1	SIGHUP	Terminate	Terminal line hangup
2	SIGINT	Terminate	Interrupt from keyboard
3	SIGQUIT	Terminate	Quit from keyboard
4	SIGILL	Terminate	Illegal instruction
5	SIGTRAP	Terminate and dump core (1)	Trace trap
6	SIGABRT	Terminate and dump core (1)	Abort signal from `abort` function
7	SIGBUS	Terminate	Bus error
8	SIGFPE	Terminate and dump core (1)	Floating point exception
9	SIGKILL	Terminate (2)	Kill program
10	SIGUSR1	Terminate	User-defined signal 1
11	SIGSEGV	Terminate and dump core (1)	Invalid memory reference (seg fault)
12	SIGUSR2	Terminate	User-defined signal 2
13	SIGPIPE	Terminate	Wrote to a pipe with no reader
14	SIGALRM	Terminate	Timer signal from `alarm` function
15	SIGTERM	Terminate	Software termination signal
16	SIGSTKFLT	Terminate	Stack fault on coprocessor
17	SIGCHLD	Ignore	A child process has stopped or terminated
18	SIGCONT	Ignore	Continue process if stopped
19	SIGSTOP	Stop until next SIGCONT (2)	Stop signal not from terminal
20	SIGTSTP	Stop until next SIGCONT	Stop signal from terminal
21	SIGTTIN	Stop until next SIGCONT	Background process read from terminal
22	SIGTTOU	Stop until next SIGCONT	Background process wrote to terminal
23	SIGURG	Ignore	Urgent condition on socket
24	SIGXCPU	Terminate	CPU time limit exceeded
25	SIGXFSZ	Terminate	File size limit exceeded
26	SIGVTALRM	Terminate	Virtual timer expired
27	SIGPROF	Terminate	Profiling timer expired
28	SIGWINCH	Ignore	Window size changed
29	SIGIO	Terminate	I/O now possible on a descriptor
30	SIGPWR	Terminate	Power failure

Figure 25 **Linux signals.** Notes: (1) Years ago, main memory was implemented with a technology known as *core memory*. "Dumping core" is a historical term that means writing an image of the code and data memory segments to disk. (2) This signal can neither be caught nor ignored.

A *signal* is a small message that notifies a process that an event of some type has occurred in the system. For example, Figure 25 shows the 30 different types of signals that are supported on Linux systems. Typing "man 7 signal" on the shell command line gives the list.

Each signal type corresponds to some kind of system event. Low-level hardware exceptions are processed by the kernel's exception handlers and would not normally be visible to user processes. Signals provide a mechanism for exposing the occurrence of such exceptions to user processes. For example, if a process attempts to divide by zero, then the kernel sends it a SIGFPE signal (number 8). If a process executes an illegal instruction, the kernel sends it a SIGILL signal (number 4). If a process makes an illegal memory reference, the kernel sends it a SIGSEGV signal (number 11). Other signals correspond to higher-level software events in the kernel or in other user processes. For example, if you type a `ctrl-c` (i.e., press the `ctrl` key and the `c` key at the same time) while a process is running in the foreground, then the kernel sends a SIGINT (number 2) to the foreground process. A process can forcibly terminate another process by sending it a SIGKILL signal (number 9). When a child process terminates or stops, the kernel sends a SIGCHLD signal (number 17) to the parent.

5.1 Signal Terminology

The transfer of a signal to a destination process occurs in two distinct steps:

- *Sending a signal.* The kernel *sends* (*delivers*) a signal to a destination process by updating some state in the context of the destination process. The signal is delivered for one of two reasons: (1) The kernel has detected a system event such as a divide-by-zero error or the termination of a child process. (2) A process has invoked the `kill` function (discussed in the next section) to explicitly request the kernel to send a signal to the destination process. A process can send a signal to itself.

- *Receiving a signal.* A destination process *receives* a signal when it is forced by the kernel to react in some way to the delivery of the signal. The process can either ignore the signal, terminate, or *catch* the signal by executing a user-level function called a *signal handler*. Figure 26 shows the basic idea of a handler catching a signal.

A signal that has been sent but not yet received is called a *pending signal*. At any point in time, there can be at most one pending signal of a particular type. If a process has a pending signal of type k, then any subsequent signals of type k sent to that process are *not* queued; they are simply discarded. A process can selectively *block* the receipt of certain signals. When a signal is blocked, it can be delivered, but the resulting pending signal will not be received until the process unblocks the signal.

Figure 26

Signal handling. Receipt of a signal triggers a control transfer to a signal handler. After it finishes processing, the handler returns control to the interrupted program.

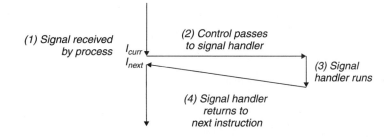

(1) Signal received by process I_{curr}
I_{next}
(2) Control passes to signal handler
(3) Signal handler runs
(4) Signal handler returns to next instruction

A pending signal is received at most once. For each process, the kernel maintains the set of pending signals in the pending bit vector, and the set of blocked signals in the blocked bit vector. The kernel sets bit k in pending whenever a signal of type k is delivered and clears bit k in pending whenever a signal of type k is received.

5.2 Sending Signals

Unix systems provide a number of mechanisms for sending signals to processes. All of the mechanisms rely on the notion of a *process group*.

Process Groups

Every process belongs to exactly one *process group*, which is identified by a positive integer *process group ID*. The getpgrp function returns the process group ID of the current process.

```
#include <unistd.h>

pid_t getpgrp(void);
```
                                   Returns: process group ID of calling process

By default, a child process belongs to the same process group as its parent. A process can change the process group of itself or another process by using the setpgid function:

```
#include <unistd.h>

int setpgid(pid_t pid, pid_t pgid);
```
                                         Returns: 0 on success, -1 on error

The setpgid function changes the process group of process pid to pgid. If pid is zero, the PID of the current process is used. If pgid is zero, the PID of the process specified by pid is used for the process group ID. For example, if process 15213 is the calling process, then

```
setpgid(0, 0);
```

creates a new process group whose process group ID is 15213, and adds process 15213 to this new group.

Sending Signals with the /bin/kill Program

The /bin/kill program sends an arbitrary signal to another process. For example, the command

```
unix> /bin/kill -9 15213
```

Figure 27

Foreground and background process groups.

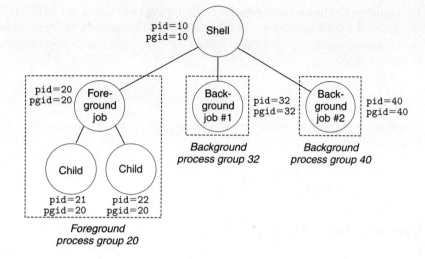

sends signal 9 (SIGKILL) to process 15213. A negative PID causes the signal to be sent to every process in process group PID. For example, the command

```
unix> /bin/kill -9 -15213
```

sends a SIGKILL signal to every process in process group 15213. Note that we use the complete path /bin/kill here because some Unix shells have their own built-in kill command.

Sending Signals from the Keyboard

Unix shells use the abstraction of a *job* to represent the processes that are created as a result of evaluating a single command line. At any point in time, there is at most one foreground job and zero or more background jobs. For example, typing

```
unix> ls | sort
```

creates a foreground job consisting of two processes connected by a Unix pipe: one running the ls program, the other running the sort program.

The shell creates a separate process group for each job. Typically, the process group ID is taken from one of the parent processes in the job. For example, Figure 27 shows a shell with one foreground job and two background jobs. The parent process in the foreground job has a PID of 20 and a process group ID of 20. The parent process has created two children, each of which are also members of process group 20.

Typing ctrl-c at the keyboard causes a SIGINT signal to be sent to the shell. The shell catches the signal (see Section 5.3) and then sends a SIGINT to every process in the foreground process group. In the default case, the result is to terminate the foreground job. Similarly, typing `crtl-z` sends a SIGTSTP signal to the shell, which catches it and sends a SIGTSTP signal to every process in the foreground process group. In the default case, the result is to stop (suspend) the foreground job.

Sending Signals with the `kill` Function

Processes send signals to other processes (including themselves) by calling the `kill` function.

```
#include <sys/types.h>
#include <signal.h>

int kill(pid_t pid, int sig);
                                              Returns: 0 if OK, −1 on error
```

If pid is greater than zero, then the `kill` function sends signal number `sig` to process pid. If pid is less than zero, then `kill` sends signal `sig` to every process in process group abs(pid). Figure 28 shows an example of a parent that uses the `kill` function to send a SIGKILL signal to its child.

—— *code/ecf/kill.c*

```
1   #include "csapp.h"
2
3   int main()
4   {
5       pid_t pid;
6
7       /* Child sleeps until SIGKILL signal received, then dies */
8       if ((pid = Fork()) == 0) {
9           Pause();  /* Wait for a signal to arrive */
10          printf("control should never reach here!\n");
11          exit(0);
12      }
13
14      /* Parent sends a SIGKILL signal to a child */
15      Kill(pid, SIGKILL);
16      exit(0);
17  }
```

—— *code/ecf/kill.c*

Figure 28 **Using the `kill` function to send a signal to a child.**

Sending Signals with the `alarm` Function

A process can send SIGALRM signals to itself by calling the `alarm` function.

```
#include <unistd.h>

unsigned int alarm(unsigned int secs);
```
Returns: remaining secs of previous alarm, or 0 if no previous alarm

The `alarm` function arranges for the kernel to send a SIGALRM signal to the calling process in `secs` seconds. If `secs` is zero, then no new alarm is scheduled. In any event, the call to `alarm` cancels any pending alarms, and returns the number of seconds remaining until any pending alarm was due to be delivered (had not this call to `alarm` canceled it), or 0 if there were no pending alarms.

Figure 29 shows a program called `alarm` that arranges to be interrupted by a SIGALRM signal every second for five seconds. When the sixth SIGALRM is delivered it terminates. When we run the program in Figure 29, we get the following output: a "BEEP" every second for five seconds, followed by a "BOOM" when the program terminates.

```
unix> ./alarm
BEEP
BEEP
BEEP
BEEP
BEEP
BOOM!
```

Notice that the program in Figure 29 uses the `signal` function to install a *signal handler* function (`handler`) that is called asynchronously, interrupting the infinite `while` loop in `main`, whenever the process receives a SIGALRM signal. When the `handler` function returns, control passes back to `main`, which picks up where it was interrupted by the arrival of the signal. Installing and using signal handlers can be quite subtle, and is the topic of the next few sections.

5.3 Receiving Signals

When the kernel is returning from an exception handler and is ready to pass control to process p, it checks the set of unblocked pending signals (`pending & blocked`) for process p. If this set is empty (the usual case), then the kernel passes control to the next instruction (I_{next}) in the logical control flow of p.

However, if the set is nonempty, then the kernel chooses some signal k in the set (typically the smallest k) and forces p to *receive* signal k. The receipt of the signal triggers some *action* by the process. Once the process completes the action, then control passes back to the next instruction (I_{next}) in the logical control flow of p. Each signal type has a predefined *default action*, which is one of the following:

- The process terminates.
- The process terminates and dumps core.

code/ecf/alarm.c

```
1   #include "csapp.h"
2
3   void handler(int sig)
4   {
5       static int beeps = 0;
6
7       printf("BEEP\n");
8       if (++beeps < 5)
9           Alarm(1); /* Next SIGALRM will be delivered in 1 second */
10      else {
11          printf("BOOM!\n");
12          exit(0);
13      }
14  }
15
16  int main()
17  {
18      Signal(SIGALRM, handler); /* Install SIGALRM handler */
19      Alarm(1); /* Next SIGALRM will be delivered in 1s */
20
21      while (1) {
22          ;  /* Signal handler returns control here each time */
23      }
24      exit(0);
25  }
```

code/ecf/alarm.c

Figure 29 **Using the alarm function to schedule periodic events.**

- The process stops until restarted by a SIGCONT signal.
- The process ignores the signal.

Figure 25 shows the default actions associated with each type of signal. For example, the default action for the receipt of a SIGKILL is to terminate the receiving process. On the other hand, the default action for the receipt of a SIGCHLD is to ignore the signal. A process can modify the default action associated with a signal by using the signal function. The only exceptions are SIGSTOP and SIGKILL, whose default actions cannot be changed.

```
#include <signal.h>
typedef void (*sighandler_t)(int);

sighandler_t signal(int signum, sighandler_t handler);
           Returns: ptr to previous handler if OK, SIG_ERR on error (does not set errno)
```

The `signal` function can change the action associated with a signal `signum` in one of three ways:

- If `handler` is SIG_IGN, then signals of type `signum` are ignored.
- If `handler` is SIG_DFL, then the action for signals of type `signum` reverts to the default action.
- Otherwise, `handler` is the address of a user-defined function, called a *signal handler*, that will be called whenever the process receives a signal of type `signum`. Changing the default action by passing the address of a handler to the `signal` function is known as *installing the handler*. The invocation of the handler is called *catching the signal*. The execution of the handler is referred to as *handling the signal*.

When a process catches a signal of type k, the handler installed for signal k is invoked with a single integer argument set to k. This argument allows the same handler function to catch different types of signals.

When the handler executes its `return` statement, control (usually) passes back to the instruction in the control flow where the process was interrupted by the receipt of the signal. We say "usually" because in some systems, interrupted system calls return immediately with an error.

Figure 30 shows a program that catches the SIGINT signal sent by the shell whenever the user types `ctrl-c` at the keyboard. The default action for SIGINT is to immediately terminate the process. In this example, we modify the default behavior to catch the signal, print a message, and then terminate the process.

The handler function is defined in lines 3–7. The main routine installs the handler in lines 12–13, and then goes to sleep until a signal is received (line 15). When the SIGINT signal is received, the handler runs, prints a message (line 5), and then terminates the process (line 6).

Signal handlers are yet another example of concurrency in a computer system. The execution of the signal handler interrupts the execution of the main C routine, akin to the way that a low-level exception handler interrupts the control flow of the current application program. Since the logical control flow of the signal handler overlaps the logical control flow of the main routine, the signal handler and the main routine run concurrently.

Practice Problem 7

Write a program, called `snooze`, that takes a single command line argument, calls the `snooze` function from Problem 5 with this argument, and then terminates. Write your program so that the user can interrupt the `snooze` function by typing `ctrl-c` at the keyboard. For example:

```
unix> ./snooze 5
Slept for 3 of 5 secs.     User hits crtl-c after 3 seconds
unix>
```

————————————————————————————— code/ecf/sigint1.c

```
1    #include "csapp.h"
2
3    void handler(int sig) /* SIGINT handler */
4    {
5        printf("Caught SIGINT\n");
6        exit(0);
7    }
8
9    int main()
10   {
11       /* Install the SIGINT handler */
12       if (signal(SIGINT, handler) == SIG_ERR)
13           unix_error("signal error");
14
15       pause(); /* Wait for the receipt of a signal */
16
17       exit(0);
18   }
```

————————————————————————————— code/ecf/sigint1.c

Figure 30 **A program that uses a signal handler to catch a SIGINT signal.**

5.4 Signal Handling Issues

Signal handling is straightforward for programs that catch a single signal and then terminate. However, subtle issues arise when a program catches multiple signals.

- *Pending signals are blocked.* Unix signal handlers typically block pending signals of the type currently being processed by the handler. For example, suppose a process has caught a SIGINT signal and is currently running its SIGINT handler. If another SIGINT signal is sent to the process, then the SIGINT will become pending, but will not be received until after the handler returns.

- *Pending signals are not queued.* There can be at most one pending signal of any particular type. Thus, if two signals of type k are sent to a destination process while signal k is blocked because the destination process is currently executing a handler for signal k, then the second signal is simply discarded; it is not queued. The key idea is that the existence of a pending signal merely indicates that *at least* one signal has arrived.

- *System calls can be interrupted.* System calls such as read, wait, and accept that can potentially block the process for a long period of time are called *slow system calls*. On some systems, slow system calls that are interrupted when a handler catches a signal do not resume when the signal handler returns, but instead return immediately to the user with an error condition and errno set to EINTR.

Let's look more closely at the subtleties of signal handling, using a simple application that is similar in nature to real programs such as shells and Web servers. The basic structure is that a parent process creates some children that run independently for a while and then terminate. The parent must reap the children to avoid leaving zombies in the system. But we also want the parent to be free to do other work while the children are running. So we decide to reap the children with a SIGCHLD handler, instead of explicitly waiting for the children to terminate. (Recall that the kernel sends a SIGCHLD signal to the parent whenever one of its children terminates or stops.)

Figure 31 shows our first attempt. The parent installs a SIGCHLD handler, and then creates three children, each of which runs for 1 second and then terminates. In the meantime, the parent waits for a line of input from the terminal and then processes it. This processing is modeled by an infinite loop. When each child terminates, the kernel notifies the parent by sending it a SIGCHLD signal. The parent catches the SIGCHLD, reaps one child, does some additional cleanup work (modeled by the sleep(2) statement), and then returns.

The signal1 program in Figure 31 seems fairly straightforward. When we run it on our Linux system, however, we get the following output:

```
linux> ./signal1
Hello from child 10320
Hello from child 10321
Hello from child 10322
Handler reaped child 10320
Handler reaped child 10322
<cr>
Parent processing input
```

From the output, we note that although three SIGCHLD signals were sent to the parent, only two of these signals were received, and thus the parent only reaped two children. If we suspend the parent process, we see that, indeed, child process 10321 was never reaped and remains a zombie (indicated by the string "defunct" in the output of the ps command):

```
<ctrl-z>
Suspended
linux> ps
PID TTY STAT TIME COMMAND
...
10319  p5 T     0:03 signal1
10321  p5 Z     0:00 signal1 <defunct>
10323  p5 R     0:00 ps
```

What went wrong? The problem is that our code failed to account for the facts that signals can block and that signals are not queued. Here's what happened: The first signal is received and caught by the parent. While the handler is still processing the first signal, the second signal is delivered and added to the set of pending signals. However, since SIGCHLD signals are blocked by the SIGCHLD handler, *code/ecf/signal1.c*

```
1   #include "csapp.h"
2
3   void handler1(int sig)
4   {
5       pid_t pid;
6
7       if ((pid = waitpid(-1, NULL, 0)) < 0)
8           unix_error("waitpid error");
9       printf("Handler reaped child %d\n", (int)pid);
10      Sleep(2);
11      return;
12  }
13
14  int main()
15  {
16      int i, n;
17      char buf[MAXBUF];
18
19      if (signal(SIGCHLD, handler1) == SIG_ERR)
20          unix_error("signal error");
21
22      /* Parent creates children */
23      for (i = 0; i < 3; i++) {
24          if (Fork() == 0) {
25              printf("Hello from child %d\n", (int)getpid());
26              Sleep(1);
27              exit(0);
28          }
29      }
30
31      /* Parent waits for terminal input and then processes it */
32      if ((n = read(STDIN_FILENO, buf, sizeof(buf))) < 0)
33          unix_error("read");
34
35      printf("Parent processing input\n");
36      while (1)
37          ;
38
39      exit(0);
40  }
```

code/ecf/signal1.c

Figure 31 signal1: This program is flawed because it fails to deal with the facts that signals can block, signals are not queued, and system calls can be interrupted.

the second signal is not received. Shortly thereafter, while the handler is still processing the first signal, the third signal arrives. Since there is already a pending SIGCHLD, this third SIGCHLD signal is discarded. Sometime later, after the handler has returned, the kernel notices that there is a pending SIGCHLD signal and forces the parent to receive the signal. The parent catches the signal and executes the handler a second time. After the handler finishes processing the second signal, there are no more pending SIGCHLD signals, and there never will be, because all knowledge of the third SIGCHLD has been lost. *The crucial lesson is that signals cannot be used to count the occurrence of events in other processes.*

To fix the problem, we must recall that the existence of a pending signal only implies that at least one signal has been delivered since the last time the process received a signal of that type. So we must modify the SIGCHLD handler to reap as many zombie children as possible each time it is invoked. Figure 32 shows the modified SIGCHLD handler. When we run `signal2` on our Linux system, it now correctly reaps all of the zombie children:

```
linux> ./signal2
Hello from child 10378
Hello from child 10379
Hello from child 10380
Handler reaped child 10379
Handler reaped child 10378
Handler reaped child 10380
<cr>
Parent processing input
```

However, we are not finished yet. If we run the `signal2` program on an older version of the Solaris operating system, it correctly reaps all of the zombie children. However, now the blocked `read` system call returns prematurely with an error, before we are able to type in our input on the keyboard:

```
solaris> ./signal2
Hello from child 18906
Hello from child 18907
Hello from child 18908
Handler reaped child 18906
Handler reaped child 18908
Handler reaped child 18907
read: Interrupted system call
```

What went wrong? The problem arises because on this particular Solaris system, slow system calls such as `read` are not restarted automatically after they are interrupted by the delivery of a signal. Instead, they return prematurely to the calling application with an error condition, unlike Linux systems, which restart interrupted system calls automatically.

In order to write portable signal handling code, we must allow for the possibility that system calls will return prematurely and then restart them manually

—————————————————————————————— code/ecf/signal2.c

```
1   #include "csapp.h"
2
3   void handler2(int sig)
4   {
5       pid_t pid;
6
7       while ((pid = waitpid(-1, NULL, 0)) > 0)
8           printf("Handler reaped child %d\n", (int)pid);
9       if (errno != ECHILD)
10          unix_error("waitpid error");
11      Sleep(2);
12      return;
13  }
14
15  int main()
16  {
17      int i, n;
18      char buf[MAXBUF];
19
20      if (signal(SIGCHLD, handler2) == SIG_ERR)
21          unix_error("signal error");
22
23      /* Parent creates children */
24      for (i = 0; i < 3; i++) {
25          if (Fork() == 0) {
26              printf("Hello from child %d\n", (int)getpid());
27              Sleep(1);
28              exit(0);
29          }
30      }
31
32      /* Parent waits for terminal input and then processes it */
33      if ((n = read(STDIN_FILENO, buf, sizeof(buf))) < 0)
34          unix_error("read error");
35
36      printf("Parent processing input\n");
37      while (1)
38          ;
39
40      exit(0);
41  }
```

—————————————————————————————— code/ecf/signal2.c

Figure 32 **signal2**: An improved version of Figure 31 that correctly accounts for the facts that signals can block and are not queued. However, it does not allow for the possibility that system calls can be interrupted.

when this occurs. Figure 33 shows the modification to signal2 that manually restarts aborted read calls. The EINTR return code in errno indicates that the read system call returned prematurely after it was interrupted.

When we run our new signal3 program on a Solaris system, the program runs correctly:

```
solaris> ./signal3
Hello from child 19571
Hello from child 19572
Hello from child 19573
Handler reaped child 19571
Handler reaped child 19572
Handler reaped child 19573
<cr>
Parent processing input
```

Practice Problem 8

What is the output of the following program?

── *code/ecf/signalprob0.c*

```
1    pid_t pid;
2    int counter = 2;
3
4    void handler1(int sig) {
5        counter = counter - 1;
6        printf("%d", counter);
7        fflush(stdout);
8        exit(0);
9    }
10
11   int main() {
12       signal(SIGUSR1, handler1);
13
14       printf("%d", counter);
15       fflush(stdout);
16
17       if ((pid = fork()) == 0) {
18           while(1) {};
19       }
20       kill(pid, SIGUSR1);
21       waitpid(-1, NULL, 0);
22       counter = counter + 1;
23       printf("%d", counter);
24       exit(0);
25   }
```

── *code/ecf/signalprob0.c*

```
1    #include "csapp.h"
2
3    void handler2(int sig)
4    {
5        pid_t pid;
6
7        while ((pid = waitpid(-1, NULL, 0)) > 0)
8            printf("Handler reaped child %d\n", (int)pid);
9        if (errno != ECHILD)
10           unix_error("waitpid error");
11       Sleep(2);
12       return;
13   }
14
15   int main() {
16       int i, n;
17       char buf[MAXBUF];
18       pid_t pid;
19
20       if (signal(SIGCHLD, handler2) == SIG_ERR)
21           unix_error("signal error");
22
23       /* Parent creates children */
24       for (i = 0; i < 3; i++) {
25           pid = Fork();
26           if (pid == 0) {
27               printf("Hello from child %d\n", (int)getpid());
28               Sleep(1);
29               exit(0);
30           }
31       }
32
33       /* Manually restart the read call if it is interrupted */
34       while ((n = read(STDIN_FILENO, buf, sizeof(buf))) < 0)
35           if (errno != EINTR)
36               unix_error("read error");
37
38       printf("Parent processing input\n");
39       while (1)
40           ;
41
42       exit(0);
43   }
```

Figure 33 signal3: An improved version of Figure 32 that correctly accounts for the fact that system calls can be interrupted.

5.5 Portable Signal Handling

The differences in signal handling semantics from system to system—such as whether or not an interrupted slow system call is restarted or aborted prematurely—is an ugly aspect of Unix signal handling. To deal with this problem, the Posix standard defines the `sigaction` function, which allows users on Posix-compliant systems such as Linux and Solaris to clearly specify the signal handling semantics they want.

```
#include <signal.h>

int sigaction(int signum, struct sigaction *act,
                                    struct sigaction *oldact);
                                        Returns: 0 if OK, −1 on error
```

The `sigaction` function is unwieldy because it requires the user to set the entries of a structure. A cleaner approach, originally proposed by W. Richard Stevens [109], is to define a wrapper function, called `Signal`, that calls `sigaction` for us. Figure 34 shows the definition of `Signal`, which is invoked in the same way as the `signal` function. The `Signal` wrapper installs a signal handler with the following signal handling semantics:

- Only signals of the type currently being processed by the handler are blocked.
- As with all signal implementations, signals are not queued.
- Interrupted system calls are automatically restarted whenever possible.

—— code/src/csapp.c

```
 1   handler_t *Signal(int signum, handler_t *handler)
 2   {
 3       struct sigaction action, old_action;
 4
 5       action.sa_handler = handler;
 6       sigemptyset(&action.sa_mask);   /* Block sigs of type being handled */
 7       action.sa_flags = SA_RESTART;   /* Restart syscalls if possible */
 8
 9       if (sigaction(signum, &action, &old_action) < 0)
10           unix_error("Signal error");
11       return (old_action.sa_handler);
12   }
```

—— code/src/csapp.c

Figure 34 **Signal**: A wrapper for `sigaction` that provides portable signal handling on Posix-compliant systems.

- Once the signal handler is installed, it remains installed until Signal is called with a handler argument of either SIG_IGN or SIG_DFL. (Some older Unix systems restore the signal action to its default action after a signal has been processed by a handler.)

Figure 35 shows a version of the signal2 program from Figure 32 that uses our Signal wrapper to get predictable signal handling semantics on different computer systems. The only difference is that we have installed the handler with a call to Signal rather than a call to signal. The program now runs correctly on both our Solaris and Linux systems, and we no longer need to manually restart interrupted read system calls.

5.6 Explicitly Blocking and Unblocking Signals

Applications can explicitly block and unblock selected signals using the sigprocmask function:

```
#include <signal.h>

int sigprocmask(int how, const sigset_t *set, sigset_t *oldset);
int sigemptyset(sigset_t *set);
int sigfillset(sigset_t *set);
int sigaddset(sigset_t *set, int signum);
int sigdelset(sigset_t *set, int signum);
                                        Returns: 0 if OK, −1 on error

int sigismember(const sigset_t *set, int signum);
                             Returns: 1 if member, 0 if not, −1 on error
```

The sigprocmask function changes the set of currently blocked signals (the blocked bit vector described in Section 5.1). The specific behavior depends on the value of how:

- SIG_BLOCK: Add the signals in set to blocked (blocked = blocked | set).
- SIG_UNBLOCK: Remove the signals in set from blocked (blocked = blocked & set).
- SIG_SETMASK: blocked = set.

If oldset is non-NULL, the previous value of the blocked bit vector is stored in oldset.

Signal sets such as set are manipulated using the following functions. The sigemptyset initializes set to the empty set. The sigfillset function adds every signal to set. The sigaddset function adds signum to set, sigdelset deletes signum from set, and sigismember returns 1 if signum is a member of set, and 0 if not.

code/ecf/signal4.c

```
1   #include "csapp.h"
2
3   void handler2(int sig)
4   {
5       pid_t pid;
6
7       while ((pid = waitpid(-1, NULL, 0)) > 0)
8           printf("Handler reaped child %d\n", (int)pid);
9       if (errno != ECHILD)
10          unix_error("waitpid error");
11      Sleep(2);
12      return;
13  }
14
15  int main()
16  {
17      int i, n;
18      char buf[MAXBUF];
19      pid_t pid;
20
21      Signal(SIGCHLD, handler2); /* sigaction error-handling wrapper */
22
23      /* Parent creates children */
24      for (i = 0; i < 3; i++) {
25          pid = Fork();
26          if (pid == 0) {
27              printf("Hello from child %d\n", (int)getpid());
28              Sleep(1);
29              exit(0);
30          }
31      }
32
33      /* Parent waits for terminal input and then processes it */
34      if ((n = read(STDIN_FILENO, buf, sizeof(buf))) < 0)
35          unix_error("read error");
36
37      printf("Parent processing input\n");
38      while (1)
39          ;
40      exit(0);
41  }
```

code/ecf/signal4.c

Figure 35 signal4: A version of Figure 32 that uses our Signal wrapper to get portable signal handling semantics.

5.7   Synchronizing Flows to Avoid Nasty Concurrency Bugs

The problem of how to program concurrent flows that read and write the same storage locations has challenged generations of computer scientists. In general, the number of potential interleavings of the flows is exponential in the number of instructions. Some of those interleavings will produce correct answers, and others will not. The fundamental problem is to somehow *synchronize* the concurrent flows so as to allow the largest set of feasible interleavings such that each of the feasible interleavings produces a correct answer.

Concurrent programming is a deep and important problem that we will not discuss in detail in this Chapter. However, we can use what you've learned about exceptional control flow in this chapter to give you a sense of the interesting intellectual challenges associated with concurrency. For example, consider the program in Figure 36, which captures the structure of a typical Unix shell. The parent keeps track of its current children using entries in a job list, with one entry per job. The `addjob` and `deletejob` functions add and remove entries from the job list, respectively.

After the parent creates a new child process, it adds the child to the job list. When the parent reaps a terminated (zombie) child in the SIGCHLD signal handler, it deletes the child from the job list. At first glance, this code appears to be correct. Unfortunately, the following sequence of events is possible:

1. The parent executes the `fork` function and the kernel schedules the newly created child to run instead of the parent.

2. Before the parent is able to run again, the child terminates and becomes a zombie, causing the kernel to deliver a SIGCHLD signal to the parent.

3. Later, when the parent becomes runnable again but before it is executed, the kernel notices the pending SIGCHLD and causes it to be received by running the signal handler in the parent.

4. The signal handler reaps the terminated child and calls `deletejob`, which does nothing because the parent has not added the child to the list yet.

5. After the handler completes, the kernel then runs the parent, which returns from `fork` and incorrectly adds the (nonexistent) child to the job list by calling `addjob`.

Thus, for some interleavings of the parent's main routine and signal handling flows, it is possible for `deletejob` to be called before `addjob`. This results in an incorrect entry on the job list, for a job that no longer exists and that will never be removed. On the other hand, there are also interleavings where events occur in the correct order. For example, if the kernel happens to schedule the parent to run when the `fork` call returns instead of the child, then the parent will correctly add the child to the job list before the child terminates and the signal handler removes the job from the list.

This is an example of a classic synchronization error known as a *race*. In this case, the race is between the call to `addjob` in the main routine and the call to `deletejob` in the handler. If `addjob` wins the race, then the answer is correct. If

—————————————————————————————— code/ecf/procmask1.c

```
1   void handler(int sig)
2   {
3       pid_t pid;
4       while ((pid = waitpid(-1, NULL, 0)) > 0) /* Reap a zombie child */
5           deletejob(pid); /* Delete the child from the job list */
6       if (errno != ECHILD)
7           unix_error("waitpid error");
8   }
9
10  int main(int argc, char **argv)
11  {
12      int pid;
13
14      Signal(SIGCHLD, handler);
15      initjobs(); /* Initialize the job list */
16
17      while (1) {
18          /* Child process */
19          if ((pid = Fork()) == 0) {
20              Execve("/bin/date", argv, NULL);
21          }
22
23          /* Parent process */
24          addjob(pid);  /* Add the child to the job list */
25      }
26      exit(0);
27  }
```

—————————————————————————————— code/ecf/procmask1.c

Figure 36 **A shell program with a subtle synchronization error.** If the child terminates before the parent is able to run, then addjob and deletejob will be called in the wrong order.

not, the answer is incorrect. Such errors are enormously difficult to debug because it is often impossible to test every interleaving. You may run the code a billion times without a problem, but then the next test results in an interleaving that triggers the race.

Figure 37 shows one way to eliminate the race in Figure 36. By blocking SIGCHLD signals before the call to fork and then unblocking them only after we have called addjob, we guarantee that the child will be reaped *after* it is added to the job list. Notice that children inherit the blocked set of their parents, so we must be careful to unblock the SIGCHLD signal in the child before calling execve.

—————————————————————————————— code/ecf/procmask2.c

```
1   void handler(int sig)
2   {
3       pid_t pid;
4       while ((pid = waitpid(-1, NULL, 0)) > 0) /* Reap a zombie child */
5           deletejob(pid); /* Delete the child from the job list */
6       if (errno != ECHILD)
7           unix_error("waitpid error");
8   }
9
10  int main(int argc, char **argv)
11  {
12      int pid;
13      sigset_t mask;
14
15      Signal(SIGCHLD, handler);
16      initjobs(); /* Initialize the job list */
17
18      while (1) {
19          Sigemptyset(&mask);
20          Sigaddset(&mask, SIGCHLD);
21          Sigprocmask(SIG_BLOCK, &mask, NULL); /* Block SIGCHLD */
22
23          /* Child process */
24          if ((pid = Fork()) == 0) {
25              Sigprocmask(SIG_UNBLOCK, &mask, NULL); /* Unblock SIGCHLD */
26              Execve("/bin/date", argv, NULL);
27          }
28
29          /* Parent process */
30          addjob(pid);  /* Add the child to the job list */
31          Sigprocmask(SIG_UNBLOCK, &mask, NULL);  /* Unblock SIGCHLD */
32      }
33      exit(0);
34  }
```

—————————————————————————————— code/ecf/procmask2.c

Figure 37 **Using** sigprocmask **to synchronize processes.** In this example, the parent ensures that addjob executes before the corresponding deletejob.

```
1    #include <stdio.h>
2    #include <stdlib.h>
3    #include <unistd.h>
4    #include <sys/time.h>
5    #include <sys/types.h>
6
7    /* Sleep for a random period between [0, MAX_SLEEP] us. */
8    #define MAX_SLEEP 100000
9
10   /* Macro that maps val into the range [0, RAND_MAX] */
11   #define CONVERT(val) (((double)val)/(double)RAND_MAX)
12
13   pid_t Fork(void)
14   {
15       static struct timeval time;
16       unsigned bool, secs;
17       pid_t pid;
18
19       /* Generate a different seed each time the function is called */
20       gettimeofday(&time, NULL);
21       srand(time.tv_usec);
22
23       /* Determine whether to sleep in parent of child and for how long */
24       bool = (unsigned)(CONVERT(rand()) + 0.5);
25       secs = (unsigned)(CONVERT(rand()) * MAX_SLEEP);
26
27       /* Call the real fork function */
28       if ((pid = fork()) < 0)
29           return pid;
30
31       /* Randomly decide to sleep in the parent or the child */
32       if (pid == 0) { /* Child */
33           if(bool) {
34               usleep(secs);
35           }
36       }
37       else { /* Parent */
38           if (!bool) {
39               usleep(secs);
40           }
41       }
42
43       /* Return the PID like a normal fork call */
44       return pid;
45   }
```

Figure 38 **A wrapper for** fork **that randomly determines the order in which the parent and child execute.** The parent and child flip a coin to determine which will sleep, thus giving the other process a chance to be scheduled.

> **Aside** A handy trick for exposing races in your code
>
> Races such as those in Figure 36 are difficult to detect because they depend on kernel-specific scheduling decisions. After a call to fork, some kernels schedule the child to run first, while other kernels schedule the parent to run first. If you were to run the code in Figure 36 on one of the latter systems, it would never fail, no matter how many times you tested it. But as soon as you ran it on one of the former systems, then the race would be exposed and the code would fail. Figure 38 shows a wrapper function that can help expose such hidden assumptions about the execution ordering of parent and child processes. The basic idea is that after each call to fork, the parent and child flip a coin to determine which of them will sleep for a bit, thus giving the other process the opportunity to run first. If we were to run the code multiple times, then with high probability we would exercise both orderings of child and parent executions, regardless of the particular kernel's scheduling policy.

6 Nonlocal Jumps

C provides a form of user-level exceptional control flow, called a *nonlocal jump*, that transfers control directly from one function to another currently executing function without having to go through the normal call-and-return sequence. Nonlocal jumps are provided by the setjmp and longjmp functions.

```
#include <setjmp.h>

int setjmp(jmp_buf env);
int sigsetjmp(sigjmp_buf env, int savesigs);
                                Returns: 0 from setjmp, nonzero from longjmps
```

The setjmp function saves the current *calling environment* in the env buffer, for later use by longjmp, and returns a 0. The calling environment includes the program counter, stack pointer, and general purpose registers.

```
#include <setjmp.h>

void longjmp(jmp_buf env, int retval);
void siglongjmp(sigjmp_buf env, int retval);
                                                        Never returns
```

The longjmp function restores the calling environment from the env buffer and then triggers a return from the most recent setjmp call that initialized env. The setjmp then returns with the nonzero return value retval.

The interactions between setjmp and longjmp can be confusing at first glance. The setjmp function is called once, but returns *multiple times:* once when the setjmp is first called and the calling environment is stored in the env buffer, and once for each corresponding `longjmp` call. On the other hand, the `longjmp` function is called once, but never returns.

An important application of nonlocal jumps is to permit an immediate return from a deeply nested function call, usually as a result of detecting some error condition. If an error condition is detected deep in a nested function call, we can use a nonlocal jump to return directly to a common localized error handler instead of laboriously unwinding the call stack.

Figure 39 shows an example of how this might work. The `main` routine first calls `setjmp` to save the current calling environment, and then calls function `foo`, which in turn calls function `bar`. If `foo` or `bar` encounter an error, they return immediately from the `setjmp` via a `longjmp` call. The nonzero return value of the `setjmp` indicates the error type, which can then be decoded and handled in one place in the code.

Another important application of nonlocal jumps is to branch out of a signal handler to a specific code location, rather than returning to the instruction that was interrupted by the arrival of the signal. Figure 40 shows a simple program that illustrates this basic technique. The program uses signals and nonlocal jumps to do a soft restart whenever the user types `ctrl-c` at the keyboard. The `sigsetjmp` and `siglongjmp` functions are versions of `setjmp` and `longjmp` that can be used by signal handlers.

The initial call to the `sigsetjmp` function saves the calling environment and signal context (including the pending and blocked signal vectors) when the program first starts. The main routine then enters an infinite processing loop. When the user types `ctrl-c`, the shell sends a SIGINT signal to the process, which catches it. Instead of returning from the signal handler, which would pass control back to the interrupted processing loop, the handler performs a nonlocal jump back to the beginning of the `main` program. When we ran the program on our system, we got the following output:

```
unix> ./restart
starting
processing...
processing...
restarting          User hits ctrl-c
processing...
restarting          User hits ctrl-c
processing...
```

Aside   Software exceptions in C++ and Java

The exception mechanisms provided by C++ and Java are higher-level, more-structured versions of the C `setjmp` and `longjmp` functions. You can think of a `catch` clause inside a `try` statement as being akin to a `setjmp` function. Similarly, a `throw` statement is similar to a `longjmp` function.

code/ecf/setjmp.c

```
1    #include "csapp.h"
2
3    jmp_buf buf;
4
5    int error1 = 0;
6    int error2 = 1;
7
8    void foo(void), bar(void);
9
10   int main()
11   {
12       int rc;
13
14       rc = setjmp(buf);
15       if (rc == 0)
16           foo();
17       else if (rc == 1)
18           printf("Detected an error1 condition in foo\n");
19       else if (rc == 2)
20           printf("Detected an error2 condition in foo\n");
21       else
22           printf("Unknown error condition in foo\n");
23       exit(0);
24   }
25
26   /* Deeply nested function foo */
27   void foo(void)
28   {
29       if (error1)
30           longjmp(buf, 1);
31       bar();
32   }
33
34   void bar(void)
35   {
36       if (error2)
37           longjmp(buf, 2);
38   }
```

code/ecf/setjmp.c

Figure 39 **Nonlocal jump example.** This example shows the framework for using nonlocal jumps to recover from error conditions in deeply nested functions without having to unwind the entire stack.

code/ecf/restart.c

```
1    #include "csapp.h"
2
3    sigjmp_buf buf;
4
5    void handler(int sig)
6    {
7        siglongjmp(buf, 1);
8    }
9
10   int main()
11   {
12       Signal(SIGINT, handler);
13
14       if (!sigsetjmp(buf, 1))
15           printf("starting\n");
16       else
17           printf("restarting\n");
18
19       while(1) {
20           Sleep(1);
21           printf("processing...\n");
22       }
23       exit(0);
24   }
```

code/ecf/restart.c

Figure 40 **A program that uses nonlocal jumps to restart itself when the user types** `ctrl-c`.

7 Tools for Manipulating Processes

Linux systems provide a number of useful tools for monitoring and manipulating processes:

STRACE: Prints a trace of each system call invoked by a running program and its children. A fascinating tool for the curious student. Compile your program with `-static` to get a cleaner trace without a lot of output related to shared libraries.

PS: Lists processes (including zombies) currently in the system.

TOP: Prints information about the resource usage of current processes.

PMAP: Displays the memory map of a process.

`/proc`: A virtual filesystem that exports the contents of numerous kernel data structures in an ASCII text form that can be read by user programs. For example, type "`cat /proc/loadavg`" to see the current load average on your Linux system.

8 Summary

Exceptional control flow (ECF) occurs at all levels of a computer system and is a basic mechanism for providing concurrency in a computer system.

At the hardware level, exceptions are abrupt changes in the control flow that are triggered by events in the processor. The control flow passes to a software handler, which does some processing and then returns control to the interrupted control flow.

There are four different types of exceptions: interrupts, faults, aborts, and traps. Interrupts occur asynchronously (with respect to any instructions) when an external I/O device such as a timer chip or a disk controller sets the interrupt pin on the processor chip. Control returns to the instruction following the faulting instruction. Faults and aborts occur synchronously as the result of the execution of an instruction. Fault handlers restart the faulting instruction, while abort handlers never return control to the interrupted flow. Finally, traps are like function calls that are used to implement the system calls that provide applications with controlled entry points into the operating system code.

At the operating system level, the kernel uses ECF to provide the fundamental notion of a process. A process provides applications with two important abstractions: (1) logical control flows that give each program the illusion that it has exclusive use of the processor, and (2) private address spaces that provide the illusion that each program has exclusive use of the main memory.

At the interface between the operating system and applications, applications can create child processes, wait for their child processes to stop or terminate, run new programs, and catch signals from other processes. The semantics of signal handling is subtle and can vary from system to system. However, mechanisms exist on Posix-compliant systems that allow programs to clearly specify the expected signal handling semantics.

Finally, at the application level, C programs can use nonlocal jumps to bypass the normal call/return stack discipline and branch directly from one function to another.

Bibliographic Notes

The Intel macroarchitecture specification contains a detailed discussion of exceptions and interrupts on Intel processors [27]. Operating systems texts [98, 104, 112] contain additional information on exceptions, processes, and signals. The classic work by W. Richard Stevens [110] is a valuable and highly readable description of how to work with processes and signals from application programs. Bovet and Cesati [11] give a wonderfully clear description of the Linux kernel, including details of the process and signal implementations. Blum [9] is an excellent reference for x86 assembly language, and describes in detail the x86 syscall interface.

Homework Problems

9 ◆

Consider four processes with the following starting and ending times:

Process	Start time	End time
A	5	7
B	2	4
C	3	6
D	1	8

For each pair of processes, indicate whether they run concurrently (y) or not (n):

Process pair	Concurrent?
AB	
AC	
AD	
BC	
BD	
CD	

10 ◆

In this chapter, we have introduced some functions with unusual call and return behaviors: setjmp, longjmp, execve, and fork. Match each function with one of the following behaviors:

A. Called once, returns twice.

B. Called once, never returns.

C. Called once, returns one or more times.

11 ◆

How many "hello" output lines does this program print?

————————————————————————————————— *code/ecf/forkprob1.c*

```
1    #include "csapp.h"
2
3    int main()
4    {
5        int i;
6
7        for (i = 0; i < 2; i++)
8            Fork();
9        printf("hello\n");
10       exit(0);
11   }
```

————————————————————————————————— *code/ecf/forkprob1.c*

12 ◆

How many "hello" output lines does this program print?

―――――――――――――――――――――――――――――――――――――― *code/ecf/forkprob4.c*

```
1    #include "csapp.h"
2
3    void doit()
4    {
5        Fork();
6        Fork();
7        printf("hello\n");
8        return;
9    }
10
11   int main()
12   {
13       doit();
14       printf("hello\n");
15       exit(0);
16   }
```

―――――――――――――――――――――――――――――――――――――― *code/ecf/forkprob4.c*

13 ◆

What is one possible output of the following program?

―――――――――――――――――――――――――――――――――――――― *code/ecf/forkprob3.c*

```
1    #include "csapp.h"
2
3    int main()
4    {
5        int x = 3;
6
7        if (Fork() != 0)
8            printf("x=%d\n", ++x);
9
10       printf("x=%d\n", --x);
11       exit(0);
12   }
```

―――――――――――――――――――――――――――――――――――――― *code/ecf/forkprob3.c*

14 ◆

How many "hello" output lines does this program print?

―――――――――――――――――――――――――――――――――――――― *code/ecf/forkprob5.c*

```
1    #include "csapp.h"
2
3    void doit()
```

```
4   {
5       if (Fork() == 0) {
6           Fork();
7           printf("hello\n");
8           exit(0);
9       }
10      return;
11  }
12
13  int main()
14  {
15      doit();
16      printf("hello\n");
17      exit(0);
18  }
```

———————————————————————————— code/ecf/forkprob5.c

15 ◆

How many "hello" lines does this program print?

———————————————————————————— code/ecf/forkprob6.c

```
1   #include "csapp.h"
2
3   void doit()
4   {
5       if (Fork() == 0) {
6           Fork();
7           printf("hello\n");
8           return;
9       }
10      return;
11  }
12
13  int main()
14  {
15      doit();
16      printf("hello\n");
17      exit(0);
18  }
```

———————————————————————————— code/ecf/forkprob6.c

16 ◆

What is the output of the following program?

———————————————————————————— code/ecf/forkprob7.c

```
1   #include "csapp.h"
2   int counter = 1;
```

```
3
4    int main()
5    {
6        if (fork() == 0) {
7            counter--;
8            exit(0);
9        }
10       else {
11           Wait(NULL);
12           printf("counter = %d\n", ++counter);
13       }
14       exit(0);
15   }
```

—————————————————————————————— code/ecf/forkprob7.c

17 ◆

Enumerate all of the possible outputs of the program in Problem 4.

18 ◆◆

Consider the following program:

—————————————————————————————— code/ecf/forkprob2.c

```
1    #include "csapp.h"
2
3    void end(void)
4    {
5        printf("2");
6    }
7
8    int main()
9    {
10       if (Fork() == 0)
11           atexit(end);
12       if (Fork() == 0)
13           printf("0");
14       else
15           printf("1");
16       exit(0);
17   }
```

—————————————————————————————— code/ecf/forkprob2.c

Determine which of the following outputs are possible. Note: The `atexit` function takes a pointer to a function and adds it to a list of functions (initially empty) that will be called when the `exit` function is called.

A. 112002

B. 211020

C. 102120

D. 122001

E. 100212

19 ◆◆

How many lines of output does the following function print? Give your answer as a function of *n*. Assume $n \geq 1$.

———————————————————————————— code/ecf/forkprob8.c

```
1    void foo(int n)
2    {
3        int i;
4
5        for (i = 0; i < n; i++)
6            Fork();
7        printf("hello\n");
8        exit(0);
9    }
```

———————————————————————————— code/ecf/forkprob8.c

20 ◆◆

Use execve to write a program called myls whose behavior is identical to the /bin/ls program. Your program should accept the same command line arguments, interpret the identical environment variables, and produce the identical output.

The ls program gets the width of the screen from the COLUMNS environment variable. If COLUMNS is unset, then ls assumes that the screen is 80 columns wide. Thus, you can check your handling of the environment variables by setting the COLUMNS environment to something smaller than 80:

```
unix> setenv COLUMNS 40
unix> ./myls
 ...output is 40 columns wide
unix> unsetenv COLUMNS
unix> ./myls
 ...output is now 80 columns wide
```

21 ◆◆

What are the possible output sequences from the following program?

———————————————————————————— code/ecf/waitprob3.c

```
1    int main()
2    {
3        if (fork() == 0) {
4            printf("a");
5            exit(0);
6        }
```

```
7          else {
8              printf("b");
9              waitpid(-1, NULL, 0);
10         }
11         printf("c");
12         exit(0);
13     }
```

── code/ecf/waitprob3.c

22 ◆◆◆

Write your own version of the Unix system function

```
int mysystem(char *command);
```

The mysystem function executes command by calling "/bin/sh –c command", and then returns after command has completed. If command exits normally (by calling the exit function or executing a return statement), then mysystem returns the command exit status. For example, if command terminates by calling exit(8), then system returns the value 8. Otherwise, if command terminates abnormally, then mysystem returns the status returned by the shell.

23 ◆◆

One of your colleagues is thinking of using signals to allow a parent process to count events that occur in a child process. The idea is to notify the parent each time an event occurs by sending it a signal, and letting the parent's signal handler increment a global counter variable, which the parent can then inspect after the child has terminated. However, when he runs the test program in Figure 41 on his system, he discovers that when the parent calls printf, counter always has a value of 2, even though the child has sent five signals to the parent. Perplexed, he comes to you for help. Can you explain the bug?

24 ◆◆◆

Modify the program in Figure 17 so that the following two conditions are met:

1. Each child terminates abnormally after attempting to write to a location in the read-only text segment.

2. The parent prints output that is identical (except for the PIDs) to the following:

```
child 12255 terminated by signal 11: Segmentation fault
child 12254 terminated by signal 11: Segmentation fault
```

Hint: Read the man page for psignal(3).

25 ◆◆◆

Write a version of the fgets function, called tfgets, that times out after 5 seconds. The tfgets function accepts the same inputs as fgets. If the user doesn't type an input line within 5 seconds, tfgets returns NULL. Otherwise, it returns a pointer to the input line.

code/ecf/counterprob.c

```
1    #include "csapp.h"
2
3    int counter = 0;
4
5    void handler(int sig)
6    {
7        counter++;
8        sleep(1); /* Do some work in the handler */
9        return;
10   }
11
12   int main()
13   {
14       int i;
15
16       Signal(SIGUSR2, handler);
17
18       if (Fork() == 0) {  /* Child */
19           for (i = 0; i < 5; i++) {
20               Kill(getppid(), SIGUSR2);
21               printf("sent SIGUSR2 to parent\n");
22           }
23           exit(0);
24       }
25
26       Wait(NULL);
27       printf("counter=%d\n", counter);
28       exit(0);
29   }
```

code/ecf/counterprob.c

Figure 41 **Counter program referenced in Problem 23.**

26 ◆◆◆◆

Using the example in Figure 22 as a starting point, write a shell program that supports job control. Your shell should have the following features:

- The command line typed by the user consists of a name and zero or more arguments, all separated by one or more spaces. If name is a built-in command, the shell handles it immediately and waits for the next command line. Otherwise, the shell assumes that name is an executable file, which it loads and runs in the context of an initial child process (job). The process group ID for the job is identical to the PID of the child.

- Each job is identified by either a process ID (PID) or a job ID (JID), which is a small arbitrary positive integer assigned by the shell. JIDs are denoted on the command line by the prefix '%'. For example, "%5" denotes JID 5, and "5" denotes PID 5.

- If the command line ends with an ampersand, then the shell runs the job in the background. Otherwise, the shell runs the job in the foreground.

- Typing ctrl-c (ctrl-z) causes the shell to send a SIGINT (SIGTSTP) signal to every process in the foreground process group.

- The jobs built-in command lists all background jobs.

- The bg <job> built-in command restarts <job> by sending it a SIGCONT signal, and then runs it in the background. The <job> argument can be either a PID or a JID.

- The fg <job> built-in command restarts <job> by sending it a SIGCONT signal, and then runs it in the foreground.

- The shell reaps all of its zombie children. If any job terminates because it receives a signal that was not caught, then the shell prints a message to the terminal with the job's PID and a description of the offending signal.

Figure 42 shows an example shell session.

Solutions to Practice Problems

Solution to Problem 1

Processes A and B are concurrent with respect to each other, as are B and C, because their respective executions overlap, that is, one process starts before the other finishes. Processes A and C are not concurrent, because their executions do not overlap; A finishes before C begins.

Solution to Problem 2

In our example program in Figure 15, the parent and child execute disjoint sets of instructions. However, in this program, the parent and child execute non-disjoint sets of instructions, which is possible because the parent and child have identical code segments. This can be a difficult conceptual hurdle, so be sure you understand the solution to this problem.

A. The key idea here is that the child executes both printf statements. After the fork returns, it executes the printf in line 8. Then it falls out of the if statement and executes the printf in line 9. Here is the output produced by the child:

```
printf1: x=2
printf2: x=1
```

B. The parent executes only the printf in line 9:

```
printf2: x=0
```

```
unix> ./shell                                    Run your shell program
> bogus
bogus: Command not found.                        Execve can't find executable
> foo 10                                                       ᴵ
Job 5035 terminated by signal: Interrupt         User types ctrl-c
> foo 100 &
[1] 5036 foo 100 &
> foo 200 &
[2] 5037 foo 200 &
> jobs
[1] 5036 Running    foo 100 &
[2] 5037 Running    foo 200 &
> fg %1
Job [1] 5036 stopped by signal: Stopped          User types ctrl-z
> jobs
[1] 5036 Stopped    foo 100 &
[2] 5037 Running    foo 200 &
> bg 5035
5035: No such process
> bg 5036
[1] 5036 foo 100 &
> /bin/kill 5036
Job 5036 terminated by signal: Terminated
> fg %2                                           Wait for fg job to finish.
> quit
unix>                                             Back to the Unix shell
```

Figure 42 **Sample shell session for Problem 26.**

Solution to Problem 3

The parent prints b and then c. The child prints a and then c. It's very important to realize that you cannot make any assumption about how the execution of the parent and child are interleaved. Thus, any topological sort of $b \to c$ and $a \to c$ is a possible output sequence. There are four such sequences: $acbc$, $bcac$, $abcc$, and $bacc$.

Solution to Problem 4

A. Each time we run this program, it generates six output lines.

B. The ordering of the output lines will vary from system to system, depending on the how the kernel interleaves the instructions of the parent and the child. In general, any topological sort of the following graph is a valid ordering:

```
              --> ''0'' --> ''2'' --> ''Bye''      Parent process
            /
''Hello''
            \
              --> ''1'' --> ''Bye''                Child process
```

For example, when we run the program on our system, we get the following output:

```
unix> ./waitprob1
Hello
0
1
Bye
2
Bye
```

In this case, the parent runs first, printing "Hello" in line 6 and "0" in line 8. The call to wait blocks because the child has not yet terminated, so the kernel does a context switch and passes control to the child, which prints "1" in line 8 and "Bye" in line 15, and then terminates with an exit status of 2 in line 16. After the child terminates, the parent resumes, printing the child's exit status in line 12 and "Bye" in line 15.

Solution to Problem 5

code/ecf/snooze.c

```
1    unsigned int snooze(unsigned int secs) {
2        unsigned int rc = sleep(secs);
3        printf("Slept for %u of %u secs.\n", secs - rc, secs);
4        return rc;
5    }
```

code/ecf/snooze.c

Solution to Problem 6

code/ecf/myecho.c

```
1    #include "csapp.h"
2
3    int main(int argc, char *argv[], char *envp[])
4    {
5        int i;
6
7        printf("Command line arguments:\n");
8        for (i=0; argv[i] != NULL; i++)
9            printf("    argv[%2d]: %s\n", i, argv[i]);
10
11       printf("\n");
12       printf("Environment variables:\n");
13       for (i=0; envp[i] != NULL; i++)
14           printf("    envp[%2d]: %s\n", i, envp[i]);
15
16       exit(0);
17   }
```

code/ecf/myecho.c

Solution to Problem 7

The `sleep` function returns prematurely whenever the sleeping process receives a signal that is not ignored. But since the default action upon receipt of a SIGINT is to terminate the process (Figure 25), we must install a SIGINT handler to allow the `sleep` function to return. The handler simply catches the SIGNAL and returns control to the `sleep` function, which returns immediately.

— code/ecf/snooze.c

```c
#include "csapp.h"

/* SIGINT handler */
void handler(int sig)
{
    return; /* Catch the signal and return */
}

unsigned int snooze(unsigned int secs) {
    unsigned int rc = sleep(secs);
    printf("Slept for %u of %u secs.\n", secs - rc, secs);
    return rc;
}

int main(int argc, char **argv) {

    if (argc != 2) {
        fprintf(stderr, "usage: %s <secs>\n", argv[0]);
        exit(0);
    }

    if (signal(SIGINT, handler) == SIG_ERR) /* Install SIGINT handler */
        unix_error("signal error\n");
    (void)snooze(atoi(argv[1]));
    exit(0);
}
```

— code/ecf/snooze.c

Solution to Problem 8

This program prints the string "213", which is the shorthand name of the CS:APP course at Carnegie Mellon. The parent starts by printing "2", then forks the child, which spins in an infinite loop. The parent then sends a signal to the child, and waits for it to terminate. The child catches the signal (interrupting the infinite loop), decrements the counter (from an initial value of 2), prints "1", and then terminates. After the parent reaps the child, it increments the counter (from an initial value of 2), prints "3", and terminates.

Virtual Memory

From Chapter 9 of *Computer Systems: A Programmer's Perspective*, Second Edition. Randal E. Bryant and David R. O'Hallaron. Copyright © 2011 by Pearson Education, Inc. Published by Prentice Hall. All rights reserved.

Processes in a system share the CPU and main memory with other processes. However, sharing the main memory poses some special challenges. As demand on the CPU increases, processes slow down in some reasonably smooth way. But if too many processes need too much memory, then some of them will simply not be able to run. When a program is out of space, it is out of luck. Memory is also vulnerable to corruption. If some process inadvertently writes to the memory used by another process, that process might fail in some bewildering fashion totally unrelated to the program logic.

In order to manage memory more efficiently and with fewer errors, modern systems provide an abstraction of main memory known as *virtual memory* (VM). Virtual memory is an elegant interaction of hardware exceptions, hardware address translation, main memory, disk files, and kernel software that provides each process with a large, uniform, and private address space. With one clean mechanism, virtual memory provides three important capabilities. (1) It uses main memory efficiently by treating it as a cache for an address space stored on disk, keeping only the active areas in main memory, and transferring data back and forth between disk and memory as needed. (2) It simplifies memory management by providing each process with a uniform address space. (3) It protects the address space of each process from corruption by other processes.

Virtual memory is one of the great ideas in computer systems. A major reason for its success is that it works silently and automatically, without any intervention from the application programmer. Since virtual memory works so well behind the scenes, why would a programmer need to understand it? There are several reasons.

- *Virtual memory is central.* Virtual memory pervades all levels of computer systems, playing key roles in the design of hardware exceptions, assemblers, linkers, loaders, shared objects, files, and processes. Understanding virtual memory will help you better understand how systems work in general.

- *Virtual memory is powerful.* Virtual memory gives applications powerful capabilities to create and destroy chunks of memory, map chunks of memory to portions of disk files, and share memory with other processes. For example, did you know that you can read or modify the contents of a disk file by reading and writing memory locations? Or that you can load the contents of a file into memory without doing any explicit copying? Understanding virtual memory will help you harness its powerful capabilities in your applications.

- *Virtual memory is dangerous.* Applications interact with virtual memory every time they reference a variable, dereference a pointer, or make a call to a dynamic allocation package such as `malloc`. If virtual memory is used improperly, applications can suffer from perplexing and insidious memory-related bugs. For example, a program with a bad pointer can crash immediately with a "Segmentation fault" or a "Protection fault," run silently for hours before crashing, or scariest of all, run to completion with incorrect results. Understanding virtual memory, and the allocation packages such as `malloc` that manage it, can help you avoid these errors.

This chapter looks at virtual memory from two angles. The first half of the chapter describes how virtual memory works. The second half describes how virtual memory is used and managed by applications. There is no avoiding the fact that VM is complicated, and the discussion reflects this in places. The good news is that if you work through the details, you will be able to simulate the virtual memory mechanism of a small system by hand, and the virtual memory idea will be forever demystified.

The second half builds on this understanding, showing you how to use and manage virtual memory in your programs. You will learn how to manage virtual memory via explicit memory mapping and calls to dynamic storage allocators such as the `malloc` package. You will also learn about a host of common memory-related errors in C programs and how to avoid them.

1 Physical and Virtual Addressing

The main memory of a computer system is organized as an array of M contiguous byte-sized cells. Each byte has a unique *physical address* (PA). The first byte has an address of 0, the next byte an address of 1, the next byte an address of 2, and so on. Given this simple organization, the most natural way for a CPU to access memory would be to use physical addresses. We call this approach *physical addressing*. Figure 1 shows an example of physical addressing in the context of a load instruction that reads the word starting at physical address 4.

When the CPU executes the load instruction, it generates an effective physical address and passes it to main memory over the memory bus. The main memory fetches the 4-byte word starting at physical address 4 and returns it to the CPU, which stores it in a register.

Early PCs used physical addressing, and systems such as digital signal processors, embedded microcontrollers, and Cray supercomputers continue to do so. However, modern processors use a form of addressing known as *virtual addressing*, as shown in Figure 2.

With virtual addressing, the CPU accesses main memory by generating a *virtual address* (VA), which is converted to the appropriate physical address before being sent to the memory. The task of converting a virtual address to a physical one is known as *address translation*. Like exception handling, address translation

Figure 1

A system that uses physical addressing.

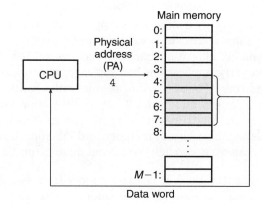

Data word

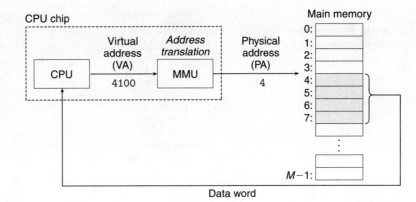

Figure 2

A system that uses virtual addressing.

requires close cooperation between the CPU hardware and the operating system. Dedicated hardware on the CPU chip called the *memory management unit* (MMU) translates virtual addresses on the fly, using a look-up table stored in main memory whose contents are managed by the operating system.

2 Address Spaces

An *address space* is an ordered set of nonnegative integer addresses

$$\{0, 1, 2, \ldots\}$$

If the integers in the address space are consecutive, then we say that it is a *linear address space*. To simplify our discussion, we will always assume linear address spaces. In a system with virtual memory, the CPU generates virtual addresses from an address space of $N = 2^n$ addresses called the *virtual address space*:

$$\{0, 1, 2, \ldots, N - 1\}$$

The size of an address space is characterized by the number of bits that are needed to represent the largest address. For example, a virtual address space with $N = 2^n$ addresses is called an *n*-bit address space. Modern systems typically support either 32-bit or 64-bit virtual address spaces.

A system also has a *physical address space* that corresponds to the M bytes of physical memory in the system:

$$\{0, 1, 2, \ldots, M - 1\}$$

M is not required to be a power of two, but to simplify the discussion we will assume that $M = 2^m$.

The concept of an address space is important because it makes a clean distinction between data objects (bytes) and their attributes (addresses). Once we recognize this distinction, then we can generalize and allow each data object to have multiple independent addresses, each chosen from a different address space.

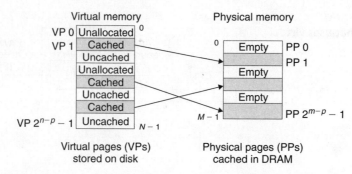

Figure 3

How a VM system uses main memory as a cache.

Virtual pages (VPs) stored on disk

Physical pages (PPs) cached in DRAM

This is the basic idea of virtual memory. Each byte of main memory has a virtual address chosen from the virtual address space, and a physical address chosen from the physical address space.

Practice Problem 1

Complete the following table, filling in the missing entries and replacing each question mark with the appropriate integer. Use the following units: $K = 2^{10}$ (Kilo), $M = 2^{20}$ (Mega), $G = 2^{30}$ (Giga), $T = 2^{40}$ (Tera), $P = 2^{50}$ (Peta), or $E = 2^{60}$ (Exa).

No. virtual address bits (n)	No. virtual addresses (N)	Largest possible virtual address
8		
	$2^? = 64K$	
		$2^{32} - 1 = ?G - 1$
	$2^? = 256T$	
64		

3 VM as a Tool for Caching

Conceptually, a virtual memory is organized as an array of N contiguous byte-sized cells stored on disk. Each byte has a unique virtual address that serves as an index into the array. The contents of the array on disk are cached in main memory. As with any other cache in the memory hierarchy, the data on disk (the lower level) is partitioned into blocks that serve as the transfer units between the disk and the main memory (the upper level). VM systems handle this by partitioning the virtual memory into fixed-sized blocks called *virtual pages* (VPs). Each virtual page is $P = 2^p$ bytes in size. Similarly, physical memory is partitioned into *physical pages* (PPs), also P bytes in size. (Physical pages are also referred to as *page frames*.)

At any point in time, the set of virtual pages is partitioned into three disjoint subsets:

- **Unallocated:** Pages that have not yet been allocated (or created) by the VM system. Unallocated blocks do not have any data associated with them, and thus do not occupy any space on disk.

- **Cached:** Allocated pages that are currently cached in physical memory.
- **Uncached:** Allocated pages that are not cached in physical memory.

The example in Figure 3 shows a small virtual memory with eight virtual pages. Virtual pages 0 and 3 have not been allocated yet, and thus do not yet exist on disk. Virtual pages 1, 4, and 6 are cached in physical memory. Pages 2, 5, and 7 are allocated, but are not currently cached in main memory.

3.1 DRAM Cache Organization

To help us keep the different caches in the memory hierarchy straight, we will use the term *SRAM cache* to denote the L1, L2, and L3 cache memories between the CPU and main memory, and the term *DRAM cache* to denote the VM system's cache that caches virtual pages in main memory.

The position of the DRAM cache in the memory hierarchy has a big impact on the way that it is organized. Recall that a DRAM is at least 10 times slower than an SRAM and that disk is about 100,000 times slower than a DRAM. Thus, misses in DRAM caches are very expensive compared to misses in SRAM caches because DRAM cache misses are served from disk, while SRAM cache misses are usually served from DRAM-based main memory. Further, the cost of reading the first byte from a disk sector is about 100,000 times slower than reading successive bytes in the sector. The bottom line is that the organization of the DRAM cache is driven entirely by the enormous cost of misses.

Because of the large miss penalty and the expense of accessing the first byte, virtual pages tend to be large, typically 4 KB to 2 MB. Due to the large miss penalty, DRAM caches are fully associative, that is, any virtual page can be placed in any physical page. The replacement policy on misses also assumes greater importance, because the penalty associated with replacing the wrong virtual page is so high. Thus, operating systems use much more sophisticated replacement algorithms for DRAM caches than the hardware does for SRAM caches. (These replacement algorithms are beyond our scope here.) Finally, because of the large access time of disk, DRAM caches always use write-back instead of write-through.

3.2 Page Tables

As with any cache, the VM system must have some way to determine if a virtual page is cached somewhere in DRAM. If so, the system must determine which physical page it is cached in. If there is a miss, the system must determine where the virtual page is stored on disk, select a victim page in physical memory, and copy the virtual page from disk to DRAM, replacing the victim page.

These capabilities are provided by a combination of operating system software, address translation hardware in the MMU (memory management unit), and a data structure stored in physical memory known as a *page table* that maps virtual pages to physical pages. The address translation hardware reads the page table each time it converts a virtual address to a physical address. The operating system

Figure 4
Page table.

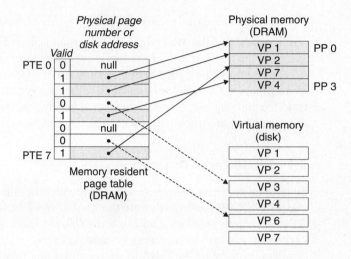

is responsible for maintaining the contents of the page table and transferring pages back and forth between disk and DRAM.

Figure 4 shows the basic organization of a page table. A page table is an array of *page table entries* (PTEs). Each page in the virtual address space has a PTE at a fixed offset in the page table. For our purposes, we will assume that each PTE consists of a *valid bit* and an n-bit address field. The valid bit indicates whether the virtual page is currently cached in DRAM. If the valid bit is set, the address field indicates the start of the corresponding physical page in DRAM where the virtual page is cached. If the valid bit is not set, then a null address indicates that the virtual page has not yet been allocated. Otherwise, the address points to the start of the virtual page on disk.

The example in Figure 4 shows a page table for a system with eight virtual pages and four physical pages. Four virtual pages (VP 1, VP 2, VP 4, and VP 7) are currently cached in DRAM. Two pages (VP 0 and VP 5) have not yet been allocated, and the rest (VP 3 and VP 6) have been allocated, but are not currently cached. An important point to notice about Figure 4 is that because the DRAM cache is fully associative, any physical page can contain any virtual page.

Practice Problem 2

Determine the number of page table entries (PTEs) that are needed for the following combinations of virtual address size (n) and page size (P):

n	$P = 2^p$	No. PTEs
16	4K	
16	8K	
32	4K	
32	8K	

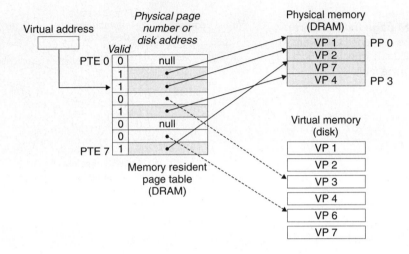

Figure 5
VM page hit. The reference to a word in VP 2 is a hit.

3.3 Page Hits

Consider what happens when the CPU reads a word of virtual memory contained in VP 2, which is cached in DRAM (Figure 5). Using a technique we will describe in detail in Section 6, the address translation hardware uses the virtual address as an index to locate PTE 2 and read it from memory. Since the valid bit is set, the address translation hardware knows that VP 2 is cached in memory. So it uses the physical memory address in the PTE (which points to the start of the cached page in PP 1) to construct the physical address of the word.

3.4 Page Faults

In virtual memory parlance, a DRAM cache miss is known as a *page fault*. Figure 6 shows the state of our example page table before the fault. The CPU has referenced a word in VP 3, which is not cached in DRAM. The address translation hardware reads PTE 3 from memory, infers from the valid bit that VP 3 is not cached, and triggers a page fault exception.

The page fault exception invokes a page fault exception handler in the kernel, which selects a victim page, in this case VP 4 stored in PP 3. If VP 4 has been modified, then the kernel copies it back to disk. In either case, the kernel modifies the page table entry for VP 4 to reflect the fact that VP 4 is no longer cached in main memory.

Next, the kernel copies VP 3 from disk to PP 3 in memory, updates PTE 3, and then returns. When the handler returns, it restarts the faulting instruction, which resends the faulting virtual address to the address translation hardware. But now, VP 3 is cached in main memory, and the page hit is handled normally by the address translation hardware. Figure 7 shows the state of our example page table after the page fault.

Virtual memory was invented in the early 1960s, long before the widening CPU-memory gap spawned SRAM caches. As a result, virtual memory systems

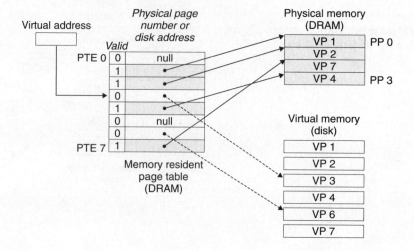

Figure 6

VM page fault (before). The reference to a word in VP 3 is a miss and triggers a page fault.

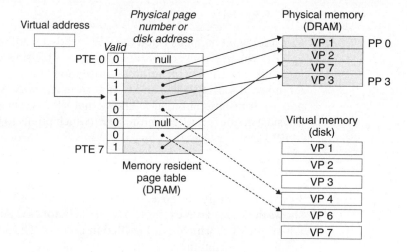

Figure 7

VM page fault (after). The page fault handler selects VP 4 as the victim and replaces it with a copy of VP 3 from disk. After the page fault handler restarts the faulting instruction, it will read the word from memory normally, without generating an exception.

use a different terminology from SRAM caches, even though many of the ideas are similar. In virtual memory parlance, blocks are known as *pages*. The activity of transferring a page between disk and memory is known as *swapping* or *paging*. Pages are *swapped in (paged in)* from disk to DRAM, and *swapped out (paged out)* from DRAM to disk. The strategy of waiting until the last moment to swap in a page, when a miss occurs, is known as *demand paging*. Other approaches, such as trying to predict misses and swap pages in before they are actually referenced, are possible. However, all modern systems use demand paging.

3.5 Allocating Pages

Figure 8 shows the effect on our example page table when the operating system allocates a new page of virtual memory, for example, as a result of calling `malloc`.

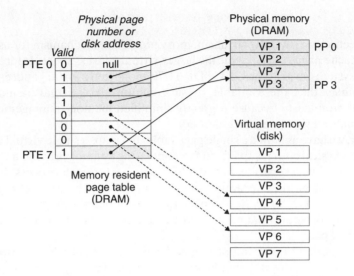

Figure 8
Allocating a new virtual page. The kernel allocates VP 5 on disk and points PTE 5 to this new location.

In the example, VP 5 is allocated by creating room on disk and updating PTE 5 to point to the newly created page on disk.

3.6 Locality to the Rescue Again

When many of us learn about the idea of virtual memory, our first impression is often that it must be terribly inefficient. Given the large miss penalties, we worry that paging will destroy program performance. In practice, virtual memory works well, mainly because of our old friend *locality*.

Although the total number of distinct pages that programs reference during an entire run might exceed the total size of physical memory, the principle of locality promises that at any point in time they will tend to work on a smaller set of *active pages* known as the *working set* or *resident set*. After an initial overhead where the working set is paged into memory, subsequent references to the working set result in hits, with no additional disk traffic.

As long as our programs have good temporal locality, virtual memory systems work quite well. But of course, not all programs exhibit good temporal locality. If the working set size exceeds the size of physical memory, then the program can produce an unfortunate situation known as *thrashing*, where pages are swapped in and out continuously. Although virtual memory is usually efficient, if a program's performance slows to a crawl, the wise programmer will consider the possibility that it is thrashing.

Aside Counting page faults

You can monitor the number of page faults (and lots of other information) with the Unix getrusage function.

4 VM as a Tool for Memory Management

In the last section, we saw how virtual memory provides a mechanism for using the DRAM to cache pages from a typically larger virtual address space. Interestingly, some early systems such as the DEC PDP-11/70 supported a virtual address space that was *smaller* than the available physical memory. Yet virtual memory was still a useful mechanism because it greatly simplified memory management and provided a natural way to protect memory.

Thus far, we have assumed a single page table that maps a single virtual address space to the physical address space. In fact, operating systems provide a separate page table, and thus a separate virtual address space, for each process. Figure 9 shows the basic idea. In the example, the page table for process *i* maps VP 1 to PP 2 and VP 2 to PP 7. Similarly, the page table for process *j* maps VP 1 to PP 7 and VP 2 to PP 10. Notice that multiple virtual pages can be mapped to the same shared physical page.

The combination of demand paging and separate virtual address spaces has a profound impact on the way that memory is used and managed in a system. In particular, VM simplifies linking and loading, the sharing of code and data, and allocating memory to applications.

- *Simplifying linking.* A separate address space allows each process to use the same basic format for its memory image, regardless of where the code and data actually reside in physical memory. For example, every process on a given Linux system has a similar memory format. The text section *always* starts at virtual address 0x08048000 (for 32-bit address spaces), or at address 0x400000 (for 64-bit address spaces). The data and bss sections follow immediately after the text section. The stack occupies the highest portion of the process address space and grows downward. Such uniformity greatly simplifies the design and implementation of linkers, allowing them to produce fully linked executables that are independent of the ultimate location of the code and data in physical memory.

- *Simplifying loading.* Virtual memory also makes it easy to load executable and shared object files into memory. Recall that the .text and .data

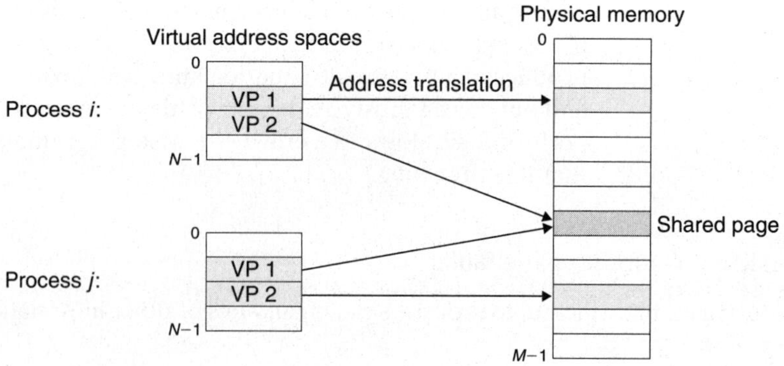

Figure 9

How VM provides processes with separate address spaces. The operating system maintains a separate page table for each process in the system.

sections in ELF executables are contiguous. To load these sections into a newly created process, the Linux loader allocates a contiguous chunk of virtual pages starting at address 0x08048000 (32-bit address spaces) or 0x400000 (64-bit address spaces), marks them as invalid (i.e., not cached), and points their page table entries to the appropriate locations in the object file. The interesting point is that the loader never actually copies any data from disk into memory. The data is paged in automatically and on demand by the virtual memory system the first time each page is referenced, either by the CPU when it fetches an instruction, or by an executing instruction when it references a memory location.

This notion of mapping a set of contiguous virtual pages to an arbitrary location in an arbitrary file is known as *memory mapping*. Unix provides a system call called mmap that allows application programs to do their own memory mapping. We will describe application-level memory mapping in more detail in Section 8.

- *Simplifying sharing.* Separate address spaces provide the operating system with a consistent mechanism for managing sharing between user processes and the operating system itself. In general, each process has its own private code, data, heap, and stack areas that are not shared with any other process. In this case, the operating system creates page tables that map the corresponding virtual pages to disjoint physical pages.

  However, in some instances it is desirable for processes to share code and data. For example, every process must call the same operating system kernel code, and every C program makes calls to routines in the standard C library such as printf. Rather than including separate copies of the kernel and standard C library in each process, the operating system can arrange for multiple processes to share a single copy of this code by mapping the appropriate virtual pages in different processes to the same physical pages, as we saw in Figure 9.

- *Simplifying memory allocation.* Virtual memory provides a simple mechanism for allocating additional memory to user processes. When a program running in a user process requests additional heap space (e.g., as a result of calling malloc), the operating system allocates an appropriate number, say, k, of contiguous virtual memory pages, and maps them to k arbitrary physical pages located anywhere in physical memory. Because of the way page tables work, there is no need for the operating system to locate k contiguous pages of physical memory. The pages can be scattered randomly in physical memory.

5 VM as a Tool for Memory Protection

Any modern computer system must provide the means for the operating system to control access to the memory system. A user process should not be allowed to modify its read-only text section. Nor should it be allowed to read or modify any of the code and data structures in the kernel. It should not be allowed to read or write the private memory of other processes, and it should not be allowed to

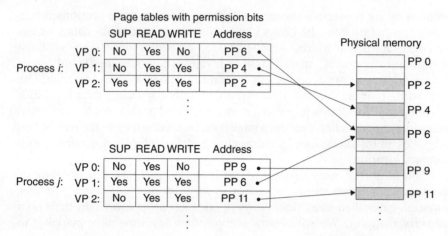

Page tables with permission bits

Figure 10 **Using VM to provide page-level memory protection.**

modify any virtual pages that are shared with other processes, unless all parties explicitly allow it (via calls to explicit interprocess communication system calls).

As we have seen, providing separate virtual address spaces makes it easy to isolate the private memories of different processes. But the address translation mechanism can be extended in a natural way to provide even finer access control. Since the address translation hardware reads a PTE each time the CPU generates an address, it is straightforward to control access to the contents of a virtual page by adding some additional permission bits to the PTE. Figure 10 shows the general idea.

In this example, we have added three permission bits to each PTE. The SUP bit indicates whether processes must be running in kernel (supervisor) mode to access the page. Processes running in kernel mode can access any page, but processes running in user mode are only allowed to access pages for which SUP is 0. The READ and WRITE bits control read and write access to the page. For example, if process i is running in user mode, then it has permission to read VP 0 and to read or write VP 1. However, it is not allowed to access VP 2.

If an instruction violates these permissions, then the CPU triggers a general protection fault that transfers control to an exception handler in the kernel. Unix shells typically report this exception as a "segmentation fault."

6 Address Translation

This section covers the basics of address translation. Our aim is to give you an appreciation of the hardware's role in supporting virtual memory, with enough detail so that you can work through some concrete examples by hand. However, keep in mind that we are omitting a number of details, especially related to timing, that are important to hardware designers but are beyond our scope. For your

Basic parameters	
Symbol	Description
$N = 2^n$	Number of addresses in virtual address space
$M = 2^m$	Number of addresses in physical address space
$P = 2^p$	Page size (bytes)
Components of a virtual address (VA)	
Symbol	Description
VPO	Virtual page offset (bytes)
VPN	Virtual page number
TLBI	TLB index
TLBT	TLB tag
Components of a physical address (PA)	
Symbol	Description
PPO	Physical page offset (bytes)
PPN	Physical page number
CO	Byte offset within cache block
CI	Cache index
CT	Cache tag

Figure 11   **Summary of address translation symbols.**

reference, Figure 11 summarizes the symbols that we will be using throughout this section.

Formally, address translation is a mapping between the elements of an N-element virtual address space (VAS) and an M-element physical address space (PAS),

$$\text{MAP: VAS} \rightarrow \text{PAS} \cup \emptyset$$

where

$$\text{MAP(A)} = \begin{cases} A' & \text{if data at virtual addr } A \text{ is present at physical addr } A' \text{ in PAS} \\ \emptyset & \text{if data at virtual addr } A \text{ is not present in physical memory} \end{cases}$$

Figure 12 shows how the MMU uses the page table to perform this mapping. A control register in the CPU, the *page table base register* (PTBR) points to the current page table. The n-bit virtual address has two components: a p-bit *virtual page offset* (VPO) and an $(n - p)$-bit *virtual page number* (VPN). The MMU uses the VPN to select the appropriate PTE. For example, VPN 0 selects PTE 0, VPN 1 selects PTE 1, and so on. The corresponding physical address is the concatenation of the *physical page number* (PPN) from the page table entry and the VPO from

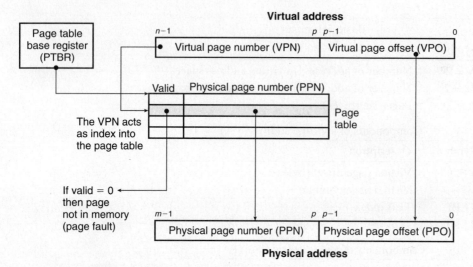

Figure 12 **Address translation with a page table.**

the virtual address. Notice that since the physical and virtual pages are both *P* bytes, the *physical page offset* (PPO) is identical to the VPO.

Figure 13(a) shows the steps that the CPU hardware performs when there is a page hit.

- *Step 1:* The processor generates a virtual address and sends it to the MMU.
- *Step 2:* The MMU generates the PTE address and requests it from the cache/main memory.
- *Step 3:* The cache/main memory returns the PTE to the MMU.
- *Step 3:* The MMU constructs the physical address and sends it to cache/main memory.
- *Step 4:* The cache/main memory returns the requested data word to the processor.

Unlike a page hit, which is handled entirely by hardware, handling a page fault requires cooperation between hardware and the operating system kernel (Figure 13(b)).

- *Steps 1 to 3:* The same as Steps 1 to 3 in Figure 13(a).
- *Step 4:* The valid bit in the PTE is zero, so the MMU triggers an exception, which transfers control in the CPU to a page fault exception handler in the operating system kernel.
- *Step 5:* The fault handler identifies a victim page in physical memory, and if that page has been modified, pages it out to disk.
- *Step 6:* The fault handler pages in the new page and updates the PTE in memory.

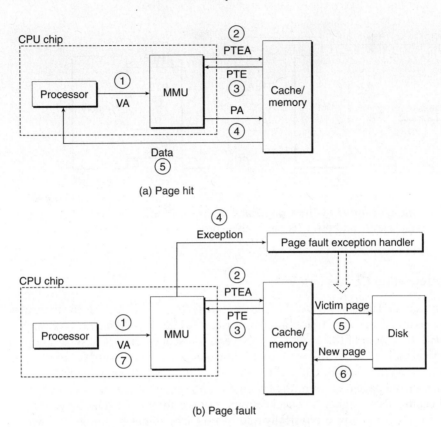

(a) Page hit

(b) Page fault

Figure 13 **Operational view of page hits and page faults.** VA: virtual address. PTEA: page table entry address. PTE: page table entry. PA: physical address.

- *Step 7:* The fault handler returns to the original process, causing the faulting instruction to be restarted. The CPU resends the offending virtual address to the MMU. Because the virtual page is now cached in physical memory, there is a hit, and after the MMU performs the steps in Figure 13(b), the main memory returns the requested word to the processor.

Practice Problem 3

Given a 32-bit virtual address space and a 24-bit physical address, determine the number of bits in the VPN, VPO, PPN, and PPO for the following page sizes P:

P	No. VPN bits	No. VPO bits	No. PPN bits	No. PPO bits
1 KB				
2 KB				
4 KB				
8 KB				

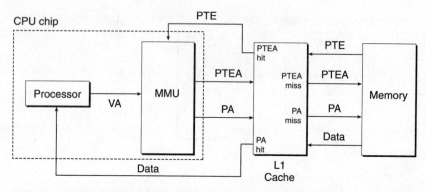

Integrating VM with a physically addressed cache. VA: virtual address. PTEA: page table entry address. PTE: page table entry. PA: physical address.

6.1 Integrating Caches and VM

In any system that uses both virtual memory and SRAM caches, there is the issue of whether to use virtual or physical addresses to access the SRAM cache. Although a detailed discussion of the trade-offs is beyond our scope here, most systems opt for physical addressing. With physical addressing, it is straightforward for multiple processes to have blocks in the cache at the same time and to share blocks from the same virtual pages. Further, the cache does not have to deal with protection issues because access rights are checked as part of the address translation process.

Figure 14 shows how a physically addressed cache might be integrated with virtual memory. The main idea is that the address translation occurs before the cache lookup. Notice that page table entries can be cached, just like any other data words.

6.2 Speeding up Address Translation with a TLB

As we have seen, every time the CPU generates a virtual address, the MMU must refer to a PTE in order to translate the virtual address into a physical address. In the worst case, this requires an additional fetch from memory, at a cost of tens to hundreds of cycles. If the PTE happens to be cached in L1, then the cost goes down to one or two cycles. However, many systems try to eliminate even this cost by including a small cache of PTEs in the MMU called a *translation lookaside buffer* (TLB).

A TLB is a small, virtually addressed cache where each line holds a block consisting of a single PTE. A TLB usually has a high degree of associativity. As shown in Figure 15, the index and tag fields that are used for set selection and line matching are extracted from the virtual page number in the virtual address. If the TLB has $T = 2^t$ sets, then the *TLB index* (TLBI) consists of the t least significant bits of the VPN, and the *TLB tag* (TLBT) consists of the remaining bits in the VPN.

Figure 15

Components of a virtual address that are used to access the TLB.

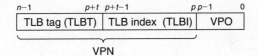

Figure 16(a) shows the steps involved when there is a TLB hit (the usual case). The key point here is that all of the address translation steps are performed inside the on-chip MMU, and thus are fast.

- *Step 1:* The CPU generates a virtual address.
- *Steps 2 and 3:* The MMU fetches the appropriate PTE from the TLB.
- *Step 4:* The MMU translates the virtual address to a physical address and sends it to the cache/main memory.
- *Step 5:* The cache/main memory returns the requested data word to the CPU.

When there is a TLB miss, then the MMU must fetch the PTE from the L1 cache, as shown in Figure 16(b). The newly fetched PTE is stored in the TLB, possibly overwriting an existing entry.

6.3 Multi-Level Page Tables

To this point we have assumed that the system uses a single page table to do address translation. But if we had a 32-bit address space, 4 KB pages, and a 4-byte PTE, then we would need a 4 MB page table resident in memory at all times, even if the application referenced only a small chunk of the virtual address space. The problem is compounded for systems with 64-bit address spaces.

The common approach for compacting the page table is to use a hierarchy of page tables instead. The idea is easiest to understand with a concrete example. Consider a 32-bit virtual address space partitioned into 4 KB pages, with page table entries that are 4 bytes each. Suppose also that at this point in time the virtual address space has the following form: The first 2K pages of memory are allocated for code and data, the next 6K pages are unallocated, the next 1023 pages are also unallocated, and the next page is allocated for the user stack. Figure 17 shows how we might construct a two-level page table hierarchy for this virtual address space.

Each PTE in the level-1 table is responsible for mapping a 4 MB chunk of the virtual address space, where each chunk consists of 1024 contiguous pages. For example, PTE 0 maps the first chunk, PTE 1 the next chunk, and so on. Given that the address space is 4 GB, 1024 PTEs are sufficient to cover the entire space.

If every page in chunk *i* is unallocated, then level 1 PTE *i* is null. For example, in Figure 17, chunks 2–7 are unallocated. However, if at least one page in chunk *i* is allocated, then level 1 PTE *i* points to the base of a level 2 page table. For example, in Figure 17, all or portions of chunks 0, 1, and 8 are allocated, so their level 1 PTEs point to level 2 page tables.

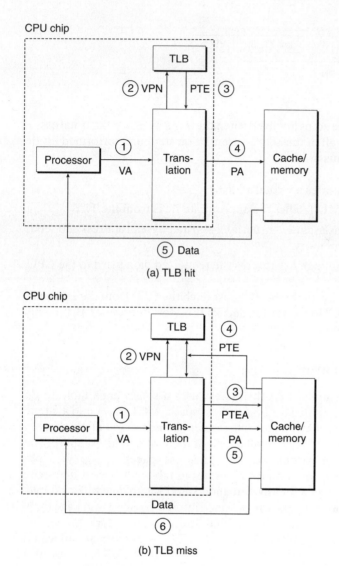

(a) TLB hit

(b) TLB miss

Figure 16 **Operational view of a TLB hit and miss.**

Each PTE in a level 2 page table is responsible for mapping a 4 KB page of virtual memory, just as before when we looked at single-level page tables. Notice that with 4-byte PTEs, each level 1 and level 2 page table is 4K bytes, which conveniently is the same size as a page.

This scheme reduces memory requirements in two ways. First, if a PTE in the level 1 table is null, then the corresponding level 2 page table does not even have to exist. This represents a significant potential savings, since most of the 4 GB virtual address space for a typical program is unallocated. Second, only the level 1 table needs to be in main memory at all times. The level 2 page tables can be created and

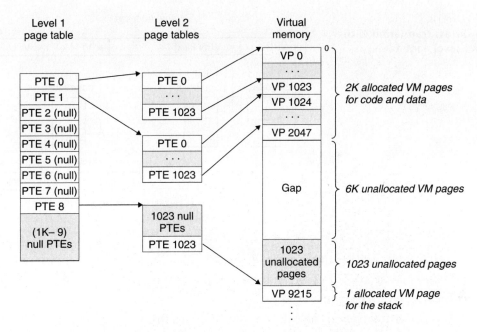

Level 1
page table

Level 2
page tables

Virtual
memory

Figure 17 A two-level page table hierarchy. Notice that addresses increase from top to bottom.

paged in and out by the VM system as they are needed, which reduces pressure on main memory. Only the most heavily used level 2 page tables need to be cached in main memory.

Figure 18 summarizes address translation with a k-level page table hierarchy. The virtual address is partitioned into k VPNs and a VPO. Each VPN i, $1 \le i \le k$, is an index into a page table at level i. Each PTE in a level-j table, $1 \le j \le k-1$, points to the base of some page table at level $j+1$. Each PTE in a level-k table contains either the PPN of some physical page or the address of a disk block. To construct the physical address, the MMU must access k PTEs before it can determine the PPN. As with a single-level hierarchy, the PPO is identical to the VPO.

Accessing k PTEs may seem expensive and impractical at first glance. However, the TLB comes to the rescue here by caching PTEs from the page tables at the different levels. In practice, address translation with multi-level page tables is not significantly slower than with single-level page tables.

6.4 Putting It Together: End-to-end Address Translation

In this section, we put it all together with a concrete example of end-to-end address translation on a small system with a TLB and L1 d-cache. To keep things manageable, we make the following assumptions:

- The memory is byte addressable.
- Memory accesses are to **1-byte words** (not 4-byte words).

Figure 18
Address translation with a k-level page table.

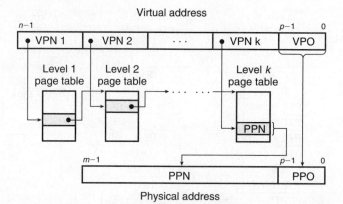

- Virtual addresses are 14 bits wide ($n = 14$).
- Physical addresses are 12 bits wide ($m = 12$).
- The page size is 64 bytes ($P = 64$).
- The TLB is four-way set associative with 16 total entries.
- The L1 d-cache is physically addressed and direct mapped, with a 4-byte line size and 16 total sets.

Figure 19 shows the formats of the virtual and physical addresses. Since each page is $2^6 = 64$ bytes, the low-order 6 bits of the virtual and physical addresses serve as the VPO and PPO respectively. The high-order 8 bits of the virtual address serve as the VPN. The high-order 6 bits of the physical address serve as the PPN.

Figure 20 shows a snapshot of our little memory system, including the TLB (Figure 20(a)), a portion of the page table (Figure 20(b)), and the L1 cache (Figure 20(c)). Above the figures of the TLB and cache, we have also shown how the bits of the virtual and physical addresses are partitioned by the hardware as it accesses these devices.

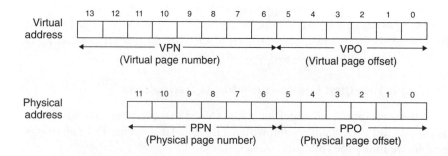

Figure 19 **Addressing for small memory system.** Assume 14-bit virtual addresses ($n = 14$), 12-bit physical addresses ($m = 12$), and 64-byte pages ($P = 64$).

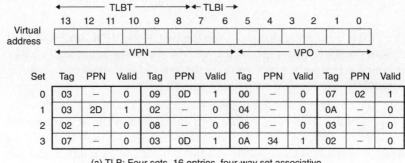

(a) TLB: Four sets, 16 entries, four-way set associative

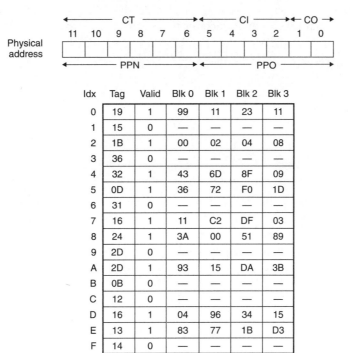

(b) Page table: Only the first 16 PTEs are shown

(c) Cache: Sixteen sets, 4-byte blocks, direct mapped

Figure 20 **TLB, page table, and cache for small memory system.** All values in the TLB, page table, and cache are in hexadecimal notation.

- *TLB*: The TLB is virtually addressed using the bits of the VPN. Since the TLB has four sets, the 2 low-order bits of the VPN serve as the set index (TLBI). The remaining 6 high-order bits serve as the tag (TLBT) that distinguishes the different VPNs that might map to the same TLB set.

- *Page table.* The page table is a single-level design with a total of $2^8 = 256$ page table entries (PTEs). However, we are only interested in the first sixteen of these. For convenience, we have labeled each PTE with the VPN that indexes it; but keep in mind that these VPNs are not part of the page table and not stored in memory. Also, notice that the PPN of each invalid PTE is denoted with a dash to reinforce the idea that whatever bit values might happen to be stored there are not meaningful.

- *Cache.* The direct-mapped cache is addressed by the fields in the physical address. Since each block is 4 bytes, the low-order 2 bits of the physical address serve as the block offset (CO). Since there are 16 sets, the next 4 bits serve as the set index (CI). The remaining 6 bits serve as the tag (CT).

Given this initial setup, let's see what happens when the CPU executes a load instruction that reads the byte at address 0x03d4. (Recall that our hypothetical CPU reads one-byte words rather than four-byte words.) To begin this kind of manual simulation, we find it helpful to write down the bits in the virtual address, identify the various fields we will need, and determine their hex values. The hardware performs a similar task when it decodes the address.

		TLBT						TLBI					
		0x03						0x03					
bit position	13	12	11	10	9	8	7	6	5	4	3	2	1	0
VA = 0x03d4	0	0	0	0	1	1	1	1	0	1	0	1	0	0
			VPN						VPO					
			0x0f						0x14					

To begin, the MMU extracts the VPN (0x0F) from the virtual address and checks with the TLB to see if it has cached a copy of PTE 0x0F from some previous memory reference. The TLB extracts the TLB index (0x03) and the TLB tag (0x3) from the VPN, hits on a valid match in the second entry of Set 0x3, and returns the cached PPN (0x0D) to the MMU.

If the TLB had missed, then the MMU would need to fetch the PTE from main memory. However, in this case we got lucky and had a TLB hit. The MMU now has everything it needs to form the physical address. It does this by concatenating the PPN (0x0D) from the PTE with the VPO (0x14) from the virtual address, which forms the physical address (0x354).

Next, the MMU sends the physical address to the cache, which extracts the cache offset CO (0x0), the cache set index CI (0x5), and the cache tag CT (0x0D) from the physical address.

	CT						CI			CO		
	0x0d						0x05			0x0		
bit position	11	10	9	8	7	6	5	4	3	2	1	0
PA = 0x354	0	0	1	1	0	1	0	1	0	1	0	0
	PPN						PPO					
	0x0d						0x14					

Since the tag in Set 0x5 matches CT, the cache detects a hit, reads out the data byte (0x36) at offset CO, and returns it to the MMU, which then passes it back to the CPU.

Other paths through the translation process are also possible. For example, if the TLB misses, then the MMU must fetch the PPN from a PTE in the page table. If the resulting PTE is invalid, then there is a page fault and the kernel must page in the appropriate page and rerun the load instruction. Another possibility is that the PTE is valid, but the necessary memory block misses in the cache.

Practice Problem 4

Show how the example memory system in Section 6.4 translates a virtual address into a physical address and accesses the cache. For the given virtual address, indicate the TLB entry accessed, physical address, and cache byte value returned. Indicate whether the TLB misses, whether a page fault occurs, and whether a cache miss occurs. If there is a cache miss, enter "–" for "Cache byte returned." If there is a page fault, enter "–" for "PPN" and leave parts C and D blank.

Virtual address: 0x03d7

A. Virtual address format

13	12	11	10	9	8	7	6	5	4	3	2	1	0

B. Address translation

Parameter	Value
VPN	
TLB index	
TLB tag	
TLB hit? (Y/N)	
Page fault? (Y/N)	
PPN	

C. Physical address format

11	10	9	8	7	6	5	4	3	2	1	0

D. Physical memory reference

Parameter	Value
Byte offset	
Cache index	
Cache tag	
Cache hit? (Y/N)	
Cache byte returned	

7 Case Study: The Intel Core i7/Linux Memory System

We conclude our discussion of virtual memory mechanisms with a case study of a real system: an Intel Core i7 running Linux. The Core i7 is based on the Nehalem microarchitecture. Although the Nehalem design allows for full 64-bit virtual and physical address spaces, the current Core i7 implementations (and those for the foreseeable future) support a 48-bit (256 TB) virtual address space and a 52-bit (4 PB) physical address space, along with a compatability mode that supports 32-bit (4 GB) virtual and physical address spaces.

Figure 21 gives the highlights of the Core i7 memory system. The *processor package* includes four cores, a large L3 cache shared by all of the cores, and a

Processor package

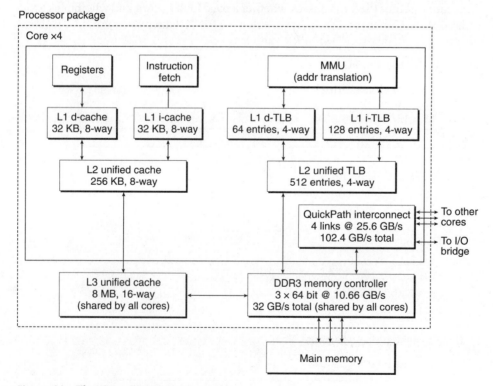

Figure 21 **The Core i7 memory system.**

DDR3 memory controller. Each core contains a hierarchy of TLBs, a hierarchy of data and instruction caches, and a set of fast point-to-point links, based on the Intel QuickPath technology, for communicating directly with the other cores and the external I/O bridge. The TLBs are virtually addressed, and four-way set associative. The L1, L2, and L3 caches are physically addressed, and eight-way set associative, with a block size of 64 bytes. The page size can be configured at start-up time as either 4 KB or 4 MB. Linux uses 4-KB pages.

7.1 Core i7 Address Translation

Figure 22 summarizes the entire Core i7 address translation process, from the time the CPU generates a virtual address until a data word arrives from memory. The Core i7 uses a four-level page table hierarchy. Each process has its own private page table hierarchy. When a Linux process is running, the page tables associated with allocated pages are all memory-resident, although the Core i7 architecture allows these page tables to be swapped in and out. The *CR3* control register points to the beginning of the level 1 (L1) page table. The value of CR3 is part of each process context, and is restored during each context switch.

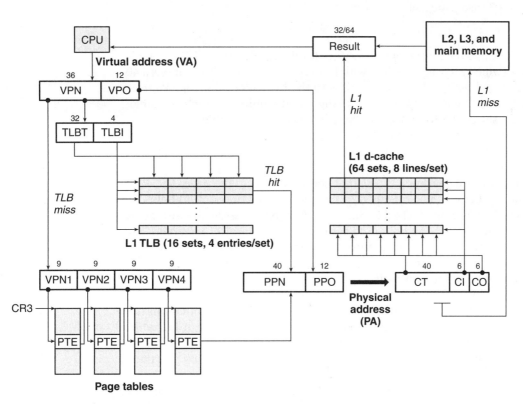

Figure 22 **Summary of Core i7 address translation.** For simplicity, the i-caches, i-TLB, and L2 unified TLB are not shown.

63	62	52 51	12 11	9	8	7	6	5	4	3	2	1	0
XD	Unused	Page table physical base addr	Unused	G	PS			A	CD	WT	U/S	R/W	P=1

Available for OS (page table location on disk)	P=0

Field	Description
P	Child page table present in physical memory (1) or not (0).
R/W	Read-only or read-write access permission for all reachable pages.
U/S	User or supervisor (kernel) mode access permission for all reachable pages.
WT	Write-through or write-back cache policy for the child page table.
CD	Caching disabled or enabled for the child page table.
A	Reference bit (set by MMU on reads and writes, cleared by software).
PS	Page size either 4 KB or 4 MB (defined for Level 1 PTEs only).
Base addr	40 most significant bits of physical base address of child page table.
XD	Disable or enable instruction fetches from all pages reachable from this PTE.

Figure 23 **Format of level 1, level 2, and level 3 page table entries.** Each entry references a 4 KB child page table.

Figure 23 shows the format of an entry in a level 1, level 2, or level 3 page table. When $P = 1$ (which is always the case with Linux), the address field contains a 40-bit physical page number (PPN) that points to the beginning of the appropriate page table. Notice that this imposes a 4 KB alignment requirement on page tables.

Figure 24 shows the format of an entry in a level 4 page table. When $P = 1$, the address field contains a 40-bit PPN that points to the base of some page in physical memory. Again, this imposes a 4 KB alignment requirement on physical pages.

The PTE has three permission bits that control access to the page. The R/W bit determines whether the contents of a page are read/write or read/only. The U/S bit, which determines whether the page can be accessed in user mode, protects code and data in the operating system kernel from user programs. The XD (execute disable) bit, which was introduced in 64-bit systems, can be used to disable instruction fetches from individual memory pages. This is an important new feature that allows the operating system kernel to reduce the risk of buffer overflow attacks by restricting execution to the read-only text segment.

As the MMU translates each virtual address, it also updates two other bits that can be used by the kernel's page fault handler. The MMU sets the A bit, which is known as a *reference bit*, each time a page is accessed. The kernel can use the reference bit to implement its page replacement algorithm. The MMU sets the D bit, or *dirty bit*, each time the page is written to. A page that has been modified is sometimes called a *dirty page*. The dirty bit tells the kernel whether or not it must write-back a victim page before it copies in a replacement page. The kernel can call a special kernel-mode instruction to clear the reference or dirty bits.

63	62	52	51		12	11	9	8	7	6	5	4	3	2	1	0
XD	Unused		Page physical base addr			Unused		G	0	D	A	CD	WT	U/S	R/W	P=1

		P=0
	Available for OS (page table location on disk)	

Field	Description
P	Child page present in physical memory (1) or not (0).
R/W	Read-only or read/write access permission for child page.
U/S	User or supervisor mode (kernel mode) access permission for child page.
WT	Write-through or write-back cache policy for the child page.
CD	Cache disabled or enabled.
A	Reference bit (set by MMU on reads and writes, cleared by software).
D	Dirty bit (set by MMU on writes, cleared by software).
G	Global page (don't evict from TLB on task switch).
Base addr	40 most significant bits of physical base address of child page.
XD	Disable or enable instruction fetches from the child page.

Figure 24 **Format of level 4 page table entries.** Each entry references a 4 KB child page.

Figure 25 shows how the Core i7 MMU uses the four levels of page tables to translate a virtual address to a physical address. The 36-bit VPN is partitioned into four 9-bit chunks, each of which is used as an offset into a page table. The CR3 register contains the physical address of the L1 page table. VPN 1 provides an offset to an L1 PTE, which contains the base address of the L2 page table. VPN 2 provides an offset to an L2 PTE, and so on.

Aside Optimizing address translation

In our discussion of address translation, we have described a sequential two-step process where the MMU (1) translates the virtual address to a physical address, and then (2) passes the physical address to the L1 cache. However, real hardware implementations use a neat trick that allows these steps to be partially overlapped, thus speeding up accesses to the L1 cache. For example, a virtual address on a Core i7 with 4 KB pages has 12 bits of VPO, and these bits are identical to the 12 bits of PPO in the corresponding physical address. Since the eight-way set-associative physically addressed L1 caches have 64 sets and 64-byte cache blocks, each physical address has 6 ($\log_2 64$) cache offset bits and 6 ($\log_2 64$) index bits. These 12 bits fit exactly in the 12-bit VPO of a virtual address, which is no accident! When the CPU needs a virtual address translated, it sends the VPN to the MMU and the VPO to the L1 cache. While the MMU is requesting a page table entry from the TLB, the L1 cache is busy using the VPO bits to find the appropriate set and read out the eight tags and corresponding data words in that set. When the MMU gets the PPN back from the TLB, the cache is ready to try to match the PPN to one of these eight tags.

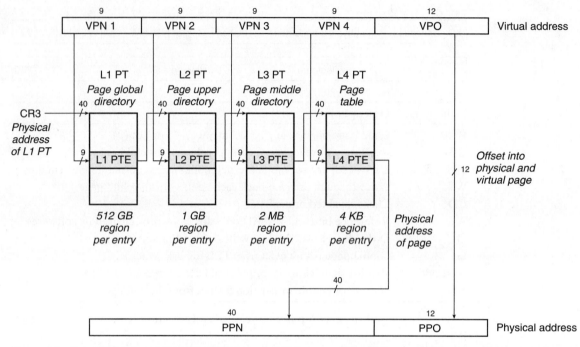

Figure 25 **Core i7 page table translation.** Legend: PT: page table, PTE: page table entry, VPN: virtual page number, VPO: virtual page offset, PPN: physical page number, PPO: physical page offset. The Linux names for the four levels of page tables are also shown.

7.2 Linux Virtual Memory System

A virtual memory system requires close cooperation between the hardware and the kernel. Details vary from version to version, and a complete description is beyond our scope. Nonetheless, our aim in this section is to describe enough of the Linux virtual memory system to give you a sense of how a real operating system organizes virtual memory and how it handles page faults.

Linux maintains a separate virtual address space for each process of the form shown in Figure 26. We have seen this picture a number of times already, with its familiar code, data, heap, shared library, and stack segments. Now that we understand address translation, we can fill in some more details about the kernel virtual memory that lies above the user stack.

The kernel virtual memory contains the code and data structures in the kernel. Some regions of the kernel virtual memory are mapped to physical pages that are shared by all processes. For example, each process shares the kernel's code and global data structures. Interestingly, Linux also maps a set of contiguous virtual pages (equal in size to the total amount of DRAM in the system) to the corresponding set of contiguous physical pages. This provides the kernel with a convenient way to access any specific location in physical memory, for example,

Figure 26

The virtual memory of a Linux process.

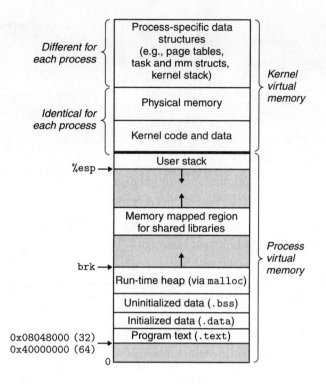

when it needs to access page tables, or to perform memory-mapped I/O operations on devices that are mapped to particular physical memory locations.

Other regions of kernel virtual memory contain data that differs for each process. Examples include page tables, the stack that the kernel uses when it is executing code in the context of the process, and various data structures that keep track of the current organization of the virtual address space.

Linux Virtual Memory Areas

Linux organizes the virtual memory as a collection of *areas* (also called *segments*). An area is a contiguous chunk of existing (allocated) virtual memory whose pages are related in some way. For example, the code segment, data segment, heap, shared library segment, and user stack are all distinct areas. Each existing virtual page is contained in some area, and any virtual page that is not part of some area does not exist and cannot be referenced by the process. The notion of an area is important because it allows the virtual address space to have gaps. The kernel does not keep track of virtual pages that do not exist, and such pages do not consume any additional resources in memory, on disk, or in the kernel itself.

Figure 27 highlights the kernel data structures that keep track of the virtual memory areas in a process. The kernel maintains a distinct task structure (`task_struct` in the source code) for each process in the system. The elements of the task

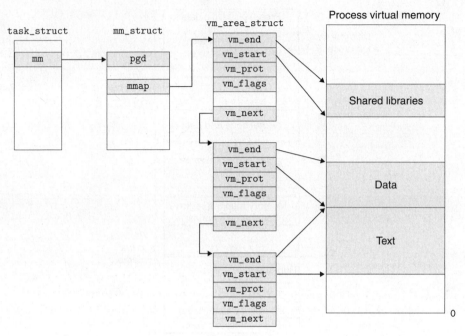

Figure 27 **How Linux organizes virtual memory.**

structure either contain or point to all of the information that the kernel needs to run the process (e.g., the PID, pointer to the user stack, name of the executable object file, and program counter).

One of the entries in the task structure points to an `mm_struct` that characterizes the current state of the virtual memory. The two fields of interest to us are `pgd`, which points to the base of the level 1 table (the page global directory), and `mmap`, which points to a list of `vm_area_structs` (area structs), each of which characterizes an area of the current virtual address space. When the kernel runs this process, it stores `pgd` in the CR3 control register.

For our purposes, the area struct for a particular area contains the following fields:

- `vm_start`: Points to the beginning of the area
- `vm_end`: Points to the end of the area
- `vm_prot`: Describes the read/write permissions for all of the pages contained in the area
- `vm_flags`: Describes (among other things) whether the pages in the area are shared with other processes or private to this process
- `vm_next`: Points to the next area struct in the list

Linux Page Fault Exception Handling

Suppose the MMU triggers a page fault while trying to translate some virtual address A. The exception results in a transfer of control to the kernel's page fault handler, which then performs the following steps:

1. Is virtual address A legal? In other words, does A lie within an area defined by some area struct? To answer this question, the fault handler searches the list of area structs, comparing A with the vm_start and vm_end in each area struct. If the instruction is not legal, then the fault handler triggers a segmentation fault, which terminates the process. This situation is labeled "1" in Figure 28.

   Because a process can create an arbitrary number of new virtual memory areas (using the mmap function described in the next section), a sequential search of the list of area structs might be very costly. So in practice, Linux superimposes a tree on the list, using some fields that we have not shown, and performs the search on this tree.

2. Is the attempted memory access legal? In other words, does the process have permission to read, write, or execute the pages in this area? For example, was the page fault the result of a store instruction trying to write to a read-only page in the text segment? Is the page fault the result of a process running in user mode that is attempting to read a word from kernel virtual memory? If the attempted access is not legal, then the fault handler triggers a protection exception, which terminates the process. This situation is labeled "2" in Figure 28.

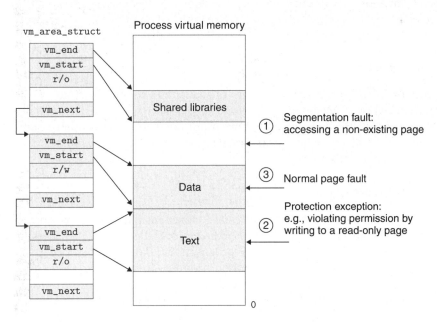

Figure 28 **Linux page fault handling.**

3. At this point, the kernel knows that the page fault resulted from a legal operation on a legal virtual address. It handles the fault by selecting a victim page, swapping out the victim page if it is dirty, swapping in the new page, and updating the page table. When the page fault handler returns, the CPU restarts the faulting instruction, which sends A to the MMU again. This time, the MMU translates A normally, without generating a page fault.

8 Memory Mapping

Linux (along with other forms of Unix) initializes the contents of a virtual memory area by associating it with an *object* on disk, a process known as *memory mapping*. Areas can be mapped to one of two types of objects:

1. *Regular file in the Unix file system:* An area can be mapped to a contiguous section of a regular disk file, such as an executable object file. The file section is divided into page-sized pieces, with each piece containing the initial contents of a virtual page. Because of demand paging, none of these virtual pages is actually swapped into physical memory until the CPU first *touches* the page (i.e., issues a virtual address that falls within that page's region of the address space). If the area is larger than the file section, then the area is padded with zeros.

2. *Anonymous file:* An area can also be mapped to an anonymous file, created by the kernel, that contains all binary zeros. The first time the CPU touches a virtual page in such an area, the kernel finds an appropriate victim page in physical memory, swaps out the victim page if it is dirty, overwrites the victim page with binary zeros, and updates the page table to mark the page as resident. Notice that no data is actually transferred between disk and memory. For this reason, pages in areas that are mapped to anonymous files are sometimes called *demand-zero pages*.

In either case, once a virtual page is initialized, it is swapped back and forth between a special *swap file* maintained by the kernel. The swap file is also known as the *swap space* or the *swap area*. An important point to realize is that at any point in time, the swap space bounds the total amount of virtual pages that can be allocated by the currently running processes.

8.1 Shared Objects Revisited

The idea of memory mapping resulted from a clever insight that if the virtual memory system could be integrated into the conventional file system, then it could provide a simple and efficient way to load programs and data into memory.

As we have seen, the process abstraction promises to provide each process with its own private virtual address space that is protected from errant writes or reads by other processes. However, many processes have identical read-only text areas. For example, each process that runs the Unix shell program `tcsh` has the same text area. Further, many programs need to access identical copies of

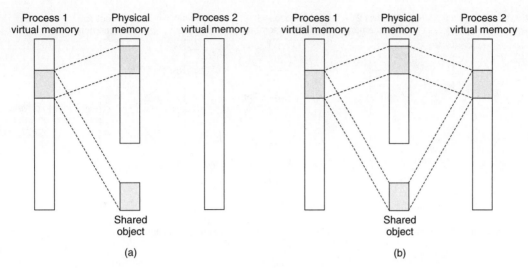

Figure 29 **A shared object.** (a) After process 1 maps the shared object. (b) After process 2 maps the same shared object. (Note that the physical pages are not necessarily contiguous.)

read-only run-time library code. For example, every C program requires functions from the standard C library such as `printf`. It would be extremely wasteful for each process to keep duplicate copies of these commonly used codes in physical memory. Fortunately, memory mapping provides us with a clean mechanism for controlling how objects are shared by multiple processes.

An object can be mapped into an area of virtual memory as either a *shared object* or a *private object*. If a process maps a shared object into an area of its virtual address space, then any writes that the process makes to that area are visible to any other processes that have also mapped the shared object into their virtual memory. Further, the changes are also reflected in the original object on disk.

Changes made to an area mapped to a private object, on the other hand, are not visible to other processes, and any writes that the process makes to the area are *not* reflected back to the object on disk. A virtual memory area into which a shared object is mapped is often called a *shared area*. Similarly for a *private area*.

Suppose that process 1 maps a shared object into an area of its virtual memory, as shown in Figure 29(a). Now suppose that process 2 maps the same shared object into its address space (not necessarily at the same virtual address as process 1), as shown in Figure 29(b).

Since each object has a unique file name, the kernel can quickly determine that process 1 has already mapped this object and can point the page table entries in process 2 to the appropriate physical pages. The key point is that only a single copy of the shared object needs to be stored in physical memory, even though the object is mapped into multiple shared areas. For convenience, we have shown the physical pages as being contiguous, but of course this is not true in general.

Private objects are mapped into virtual memory using a clever technique known as *copy-on-write*. A private object begins life in exactly the same way as a

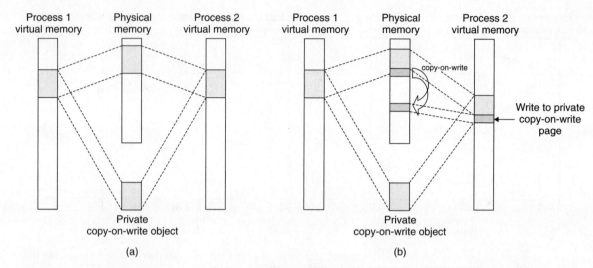

Figure 30 **A private copy-on-write object.** (a) After both processes have mapped the private copy-on-write object. (b) After process 2 writes to a page in the private area.

shared object, with only one copy of the private object stored in physical memory. For example, Figure 30(a) shows a case where two processes have mapped a private object into different areas of their virtual memories but share the same physical copy of the object. For each process that maps the private object, the page table entries for the corresponding private area are flagged as read-only, and the area struct is flagged as *private copy-on-write*. So long as neither process attempts to write to its respective private area, they continue to share a single copy of the object in physical memory. However, as soon as a process attempts to write to some page in the private area, the write triggers a protection fault.

When the fault handler notices that the protection exception was caused by the process trying to write to a page in a private copy-on-write area, it creates a new copy of the page in physical memory, updates the page table entry to point to the new copy, and then restores write permissions to the page, as shown in Figure 30(b). When the fault handler returns, the CPU reexecutes the write, which now proceeds normally on the newly created page.

By deferring the copying of the pages in private objects until the last possible moment, copy-on-write makes the most efficient use of scarce physical memory.

8.2 The `fork` Function Revisited

Now that we understand virtual memory and memory mapping, we can get a clear idea of how the `fork` function creates a new process with its own independent virtual address space.

When the `fork` function is called by the *current process*, the kernel creates various data structures for the *new process* and assigns it a unique PID. To create the virtual memory for the new process, it creates exact copies of the current process's `mm_struct`, area structs, and page tables. It flags each page in both processes as read-only, and flags each area struct in both processes as private copy-on-write.

When the `fork` returns in the new process, the new process now has an exact copy of the virtual memory as it existed when the fork was called. When either of the processes performs any subsequent writes, the copy-on-write mechanism creates new pages, thus preserving the abstraction of a private address space for each process.

8.3 The `execve` Function Revisited

Virtual memory and memory mapping also play key roles in the process of loading programs into memory. Now that we understand these concepts, we can understand how the `execve` function really loads and executes programs. Suppose that the program running in the current process makes the following call:

```
Execve("a.out", NULL, NULL);
```

The execve function loads and runs the program contained in the executable object file `a.out` within the current process, effectively replacing the current program with the `a.out` program. Loading and running a.out requires the following steps:

- *Delete existing user areas.* Delete the existing area structs in the user portion of the current process's virtual address.
- *Map private areas.* Create new area structs for the text, data, bss, and stack areas of the new program. All of these new areas are private copy-on-write. The text and data areas are mapped to the text and data sections of the `a.out` file. The bss area is demand-zero, mapped to an anonymous file whose size is contained in `a.out`. The stack and heap area are also demand-zero, initially of zero-length. Figure 31 summarizes the different mappings of the private areas.
- *Map shared areas.* If the `a.out` program was linked with shared objects, such as the standard C library `libc.so`, then these objects are dynamically linked into the program, and then mapped into the shared region of the user's virtual address space.
- *Set the program counter (PC).* The last thing that `execve` does is to set the program counter in the current process's context to point to the entry point in the text area.

The next time this process is scheduled, it will begin execution from the entry point. Linux will swap in code and data pages as needed.

8.4 User-level Memory Mapping with the `mmap` Function

Unix processes can use the `mmap` function to create new areas of virtual memory and to map objects into these areas.

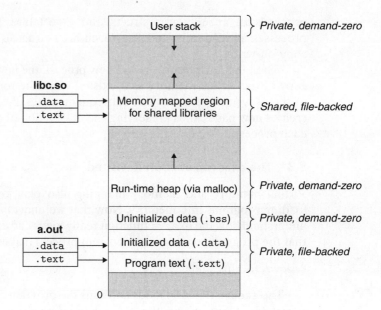

Figure 31

How the loader maps the areas of the user address space.

```
#include <unistd.h>
#include <sys/mman.h>

void   *mmap(void *start, size_t length, int prot, int flags,
           int fd, off_t offset);
                    Returns: pointer to mapped area if OK, MAP_FAILED (−1) on error
```

The `mmap` function asks the kernel to create a new virtual memory area, preferably one that starts at address `start`, and to map a contiguous chunk of the object specified by file descriptor `fd` to the new area. The contiguous object chunk has a size of `length` bytes and starts at an offset of `offset` bytes from the beginning of the file. The `start` address is merely a hint, and is usually specified as NULL. For our purposes, we will always assume a NULL start address. Figure 32 depicts the meaning of these arguments.

The `prot` argument contains bits that describe the access permissions of the newly mapped virtual memory area (i.e., the `vm_prot` bits in the corresponding area struct).

- PROT_EXEC: Pages in the area consist of instructions that may be executed by the CPU.
- PROT_READ: Pages in the area may be read.
- PROT_WRITE: Pages in the area may be written.
- PROT_NONE: Pages in the area cannot be accessed.

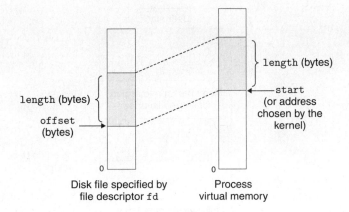

Virtual Memory

Figure 32
Visual interpretation of mmap arguments.

The flags argument consists of bits that describe the type of the mapped object. If the MAP_ANON flag bit is set, then the backing store is an anonymous object and the corresponding virtual pages are demand-zero. MAP_PRIVATE indicates a private copy-on-write object, and MAP_SHARED indicates a shared object. For example,

```
bufp = Mmap(-1, size, PROT_READ, MAP_PRIVATE|MAP_ANON, 0, 0);
```

asks the kernel to create a new read-only, private, demand-zero area of virtual memory containing size bytes. If the call is successful, then bufp contains the address of the new area.

The munmap function deletes regions of virtual memory:

```
#include <unistd.h>
#include <sys/mman.h>

int munmap(void *start, size_t length);
```
                                                    Returns: 0 if OK, −1 on error

The munmap function deletes the area starting at virtual address start and consisting of the next length bytes. Subsequent references to the deleted region result in segmentation faults.

Practice Problem 5

Write a C program mmapcopy.c that uses mmap to copy an arbitrary-sized disk file to stdout. The name of the input file should be passed as a command line argument.

9 Dynamic Memory Allocation

While it is certainly possible to use the low-level mmap and munmap functions to create and delete areas of virtual memory, C programmers typically find it more

Figure 33
The heap.

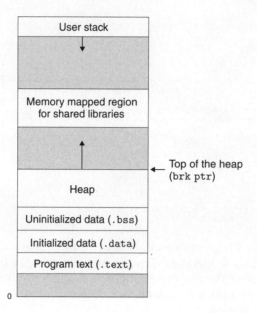

convenient and more portable to use a *dynamic memory allocator* when they need to acquire additional virtual memory at run time.

A dynamic memory allocator maintains an area of a process's virtual memory known as the *heap* (Figure 33). Details vary from system to system, but without loss of generality, we will assume that the heap is an area of demand-zero memory that begins immediately after the uninitialized bss area and grows upward (toward higher addresses). For each process, the kernel maintains a variable brk (pronounced "break") that points to the top of the heap.

An allocator maintains the heap as a collection of various-sized *blocks*. Each block is a contiguous chunk of virtual memory that is either *allocated* or *free*. An allocated block has been explicitly reserved for use by the application. A free block is available to be allocated. A free block remains free until it is explicitly allocated by the application. An allocated block remains allocated until it is freed, either explicitly by the application, or implicitly by the memory allocator itself.

Allocators come in two basic styles. Both styles require the application to explicitly allocate blocks. They differ about which entity is responsible for freeing allocated blocks.

- *Explicit allocators* require the application to explicitly free any allocated blocks. For example, the C standard library provides an explicit allocator called the malloc package. C programs allocate a block by calling the malloc function, and free a block by calling the free function. The new and delete calls in C++ are comparable.

- *Implicit allocators*, on the other hand, require the allocator to detect when an allocated block is no longer being used by the program and then free the block. Implicit allocators are also known as *garbage collectors*, and the process of automatically freeing unused allocated blocks is known as *garbage collection*. For example, higher-level languages such as Lisp, ML, and Java rely on garbage collection to free allocated blocks.

The remainder of this section discusses the design and implementation of explicit allocators. We will discuss implicit allocators in Section 10. For concreteness, our discussion focuses on allocators that manage heap memory. However, you should be aware that memory allocation is a general idea that arises in a variety of contexts. For example, applications that do intensive manipulation of graphs will often use the standard allocator to acquire a large block of virtual memory, and then use an application-specific allocator to manage the memory within that block as the nodes of the graph are created and destroyed.

9.1 The `malloc` and `free` Functions

The C standard library provides an explicit allocator known as the `malloc` package. Programs allocate blocks from the heap by calling the `malloc` function.

```
#include <stdlib.h>

void *malloc(size_t size);
```
                              Returns: ptr to allocated block if OK, NULL on error

The `malloc` function returns a pointer to a block of memory of at least `size` bytes that is suitably aligned for any kind of data object that might be contained in the block. On the Unix systems that we are familiar with, `malloc` returns a block that is aligned to an 8-byte (double word) boundary.

Aside How big is a word?

Recall that Intel refers to 4-byte objects as *double words*. However, throughout this section, we will assume that *words* are 4-byte objects and that *double words* are 8-byte objects, which is consistent with conventional terminology.

If `malloc` encounters a problem (e.g., the program requests a block of memory that is larger than the available virtual memory), then it returns NULL and sets errno. Malloc does not initialize the memory it returns. Applications that want initialized dynamic memory can use `calloc`, a thin wrapper around the `malloc` function that initializes the allocated memory to zero. Applications that want to change the size of a previously allocated block can use the `realloc` function.

Dynamic memory allocators such as `malloc` can allocate or deallocate heap memory explicitly by using the `mmap` and `munmap` functions, or they can use the `sbrk` function:

```
#include <unistd.h>

void *sbrk(intptr_t incr);
```
                        Returns: old brk pointer on success, −1 on error

The sbrk function grows or shrinks the heap by adding incr to the kernel's brk pointer. If successful, it returns the old value of brk, otherwise it returns −1 and sets errno to ENOMEM. If incr is zero, then sbrk returns the current value of brk. Calling sbrk with a negative incr is legal but tricky because the return value (the old value of brk) points to abs(incr) bytes past the new top of the heap.

Programs free allocated heap blocks by calling the free function.

```
#include <stdlib.h>

void free(void *ptr);
```
                                            Returns: nothing

The ptr argument must point to the beginning of an allocated block that was obtained from malloc, calloc, or realloc. If not, then the behavior of free is undefined. Even worse, since it returns nothing, free gives no indication to the application that something is wrong. As we shall see in Section 11, this can produce some baffling run-time errors.

Figure 34 shows how an implementation of malloc and free might manage a (very) small heap of 16 words for a C program. Each box represents a 4-byte word. The heavy-lined rectangles correspond to allocated blocks (shaded) and free blocks (unshaded). Initially, the heap consists of a single 16-word double-word aligned free block.

- *Figure 34(a):* The program asks for a four-word block. Malloc responds by carving out a four-word block from the front of the free block and returning a pointer to the first word of the block.
- *Figure 34(b):* The program requests a five-word block. Malloc responds by allocating a six-word block from the front of the free block. In this example, malloc pads the block with an extra word in order to keep the free block aligned on a double-word boundary.
- *Figure 34(c):* The program requests a six-word block and malloc responds by carving out a six-word block from the free block.
- *Figure 34(d):* The program frees the six-word block that was allocated in Figure 34(b). Notice that after the call to free returns, the pointer p2 still points to the freed block. It is the responsibility of the application not to use p2 again until it is reinitialized by a new call to malloc.

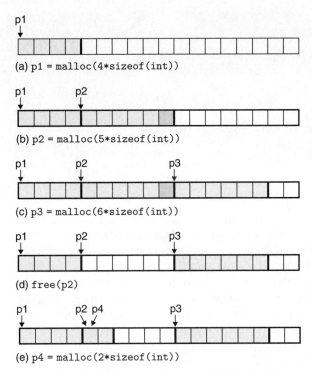

(a) p1 = malloc(4*sizeof(int))

(b) p2 = malloc(5*sizeof(int))

(c) p3 = malloc(6*sizeof(int))

(d) free(p2)

(e) p4 = malloc(2*sizeof(int))

Figure 34 **Allocating and freeing blocks with** malloc **and** free. Each square corresponds to a word. Each heavy rectangle corresponds to a block. Allocated blocks are shaded. Padded regions of allocated blocks are shaded with stripes. Free blocks are unshaded. Heap addresses increase from left to right.

- *Figure 34(e):* The program requests a two-word block. In this case, malloc allocates a portion of the block that was freed in the previous step and returns a pointer to this new block.

9.2 Why Dynamic Memory Allocation?

The most important reason that programs use dynamic memory allocation is that often they do not know the sizes of certain data structures until the program actually runs. For example, suppose we are asked to write a C program that reads a list of n ASCII integers, one integer per line, from stdin into a C array. The input consists of the integer n, followed by the n integers to be read and stored into the array. The simplest approach is to define the array statically with some hard-coded maximum array size:

```
1    #include "csapp.h"
2    #define MAXN 15213
3
4    int array[MAXN];
```

```
5
6    int main()
7    {
8        int i, n;
9
10       scanf("%d", &n);
11       if (n > MAXN)
12           app_error("Input file too big");
13       for (i = 0; i < n; i++)
14           scanf("%d", &array[i]);
15       exit(0);
16   }
```

Allocating arrays with hard-coded sizes like this is often a bad idea. The value of MAXN is arbitrary and has no relation to the actual amount of available virtual memory on the machine. Further, if the user of this program wanted to read a file that was larger than MAXN, the only recourse would be to recompile the program with a larger value of MAXN. While not a problem for this simple example, the presence of hard-coded array bounds can become a maintenance nightmare for large software products with millions of lines of code and numerous users.

A better approach is to allocate the array dynamically, at run time, after the value of n becomes known. With this approach, the maximum size of the array is limited only by the amount of available virtual memory.

```
1    #include "csapp.h"
2
3    int main()
4    {
5        int *array, i, n;
6
7        scanf("%d", &n);
8        array = (int *)Malloc(n * sizeof(int));
9        for (i = 0; i < n; i++)
10           scanf("%d", &array[i]);
11       exit(0);
12   }
```

Dynamic memory allocation is a useful and important programming technique. However, in order to use allocators correctly and efficiently, programmers need to have an understanding of how they work. We will discuss some of the gruesome errors that can result from the improper use of allocators in Section 11.

9.3 Allocator Requirements and Goals

Explicit allocators must operate within some rather stringent constraints.

- *Handling arbitrary request sequences.* An application can make an arbitrary sequence of allocate and free requests, subject to the constraint that each free request must correspond to a currently allocated block obtained from a previous allocate request. Thus, the allocator cannot make any assumptions about the ordering of allocate and free requests. For example, the allocator cannot assume that all allocate requests are accompanied by a matching free request, or that matching allocate and free requests are nested.

- *Making immediate responses to requests.* The allocator must respond immediately to allocate requests. Thus, the allocator is not allowed to reorder or buffer requests in order to improve performance.

- *Using only the heap.* In order for the allocator to be scalable, any non-scalar data structures used by the allocator must be stored in the heap itself.

- *Aligning blocks (alignment requirement).* The allocator must align blocks in such a way that they can hold any type of data object. On most systems, this means that the block returned by the allocator is aligned on an 8-byte (double-word) boundary.

- *Not modifying allocated blocks.* Allocators can only manipulate or change free blocks. In particular, they are not allowed to modify or move blocks once they are allocated. Thus, techniques such as compaction of allocated blocks are not permitted.

Working within these constraints, the author of an allocator attempts to meet the often conflicting performance goals of maximizing throughput and memory utilization.

- *Goal 1: Maximizing throughput.* Given some sequence of n allocate and free requests

$$R_0, R_1, \ldots, R_k, \ldots, R_{n-1}$$

we would like to maximize an allocator's *throughput*, which is defined as the number of requests that it completes per unit time. For example, if an allocator completes 500 allocate requests and 500 free requests in 1 second, then its throughput is 1,000 operations per second. In general, we can maximize throughput by minimizing the average time to satisfy allocate and free requests. As we'll see, it is not too difficult to develop allocators with reasonably good performance where the worst-case running time of an allocate request is linear in the number of free blocks and the running time of a free request is constant.

- *Goal 2: Maximizing memory utilization.* Naive programmers often incorrectly assume that virtual memory is an unlimited resource. In fact, the total amount of virtual memory allocated by all of the processes in a system is limited by the amount of swap space on disk. Good programmers know that virtual memory is a finite resource that must be used efficiently. This is especially true for a dynamic memory allocator that might be asked to allocate and free large blocks of memory.

There are a number of ways to characterize how efficiently an allocator uses the heap. In our experience, the most useful metric is *peak utilization*. As before, we are given some sequence of n allocate and free requests

$$R_0, R_1, \ldots, R_k, \ldots, R_{n-1}$$

If an application requests a block of p bytes, then the resulting allocated block has a *payload* of p bytes. After request R_k has completed, let the *aggregate payload*, denoted P_k, be the sum of the payloads of the currently allocated blocks, and let H_k denote the current (monotonically nondecreasing) size of the heap.

Then the *peak utilization* over the first k requests, denoted by U_k, is given by

$$U_k = \frac{\max_{i \le k} P_i}{H_k}$$

The objective of the allocator then is to maximize the peak utilization U_{n-1} over the entire sequence. As we will see, there is a tension between maximizing throughput and utilization. In particular, it is easy to write an allocator that maximizes throughput at the expense of heap utilization. One of the interesting challenges in any allocator design is finding an appropriate balance between the two goals.

Aside   Relaxing the monotonicity assumption

We could relax the monotonically nondecreasing assumption in our definition of U_k and allow the heap to grow up and down by letting H_k be the highwater mark over the first k requests.

9.4   Fragmentation

The primary cause of poor heap utilization is a phenomenon known as *fragmentation*, which occurs when otherwise unused memory is not available to satisfy allocate requests. There are two forms of fragmentation: *internal fragmentation* and *external fragmentation*.

Internal fragmentation occurs when an allocated block is larger than the payload. This might happen for a number of reasons. For example, the implementation of an allocator might impose a minimum size on allocated blocks that is greater than some requested payload. Or, as we saw in Figure 34(b), the allocator might increase the block size in order to satisfy alignment constraints.

Internal fragmentation is straightforward to quantify. It is simply the sum of the differences between the sizes of the allocated blocks and their payloads. Thus, at any point in time, the amount of internal fragmentation depends only on the pattern of previous requests and the allocator implementation.

External fragmentation occurs when there *is* enough aggregate free memory to satisfy an allocate request, but no single free block is large enough to handle the request. For example, if the request in Figure 34(e) were for six words rather than two words, then the request could not be satisfied without requesting additional virtual memory from the kernel, even though there are six free words remaining in the heap. The problem arises because these six words are spread over two free blocks.

External fragmentation is much more difficult to quantify than internal fragmentation because it depends not only on the pattern of previous requests and the allocator implementation, but also on the pattern of *future* requests. For example, suppose that after k requests all of the free blocks are exactly four words in size. Does this heap suffer from external fragmentation? The answer depends on the pattern of future requests. If all of the future allocate requests are for blocks that are smaller than or equal to four words, then there is no external fragmentation. On the other hand, if one or more requests ask for blocks larger than four words, then the heap does suffer from external fragmentation.

Since external fragmentation is difficult to quantify and impossible to predict, allocators typically employ heuristics that attempt to maintain small numbers of larger free blocks rather than large numbers of smaller free blocks.

9.5 Implementation Issues

The simplest imaginable allocator would organize the heap as a large array of bytes and a pointer p that initially points to the first byte of the array. To allocate size bytes, malloc would save the current value of p on the stack, increment p by size, and return the old value of p to the caller. Free would simply return to the caller without doing anything.

This naive allocator is an extreme point in the design space. Since each malloc and free execute only a handful of instructions, throughput would be extremely good. However, since the allocator never reuses any blocks, memory utilization would be extremely bad. A practical allocator that strikes a better balance between throughput and utilization must consider the following issues:

- *Free block organization:* How do we keep track of free blocks?
- *Placement:* How do we choose an appropriate free block in which to place a newly allocated block?
- *Splitting:* After we place a newly allocated block in some free block, what do we do with the remainder of the free block?
- *Coalescing:* What do we do with a block that has just been freed?

The rest of this section looks at these issues in more detail. Since the basic techniques of placement, splitting, and coalescing cut across many different free block organizations, we will introduce them in the context of a simple free block organization known as an implicit free list.

9.6 Implicit Free Lists

Any practical allocator needs some data structure that allows it to distinguish block boundaries and to distinguish between allocated and free blocks. Most allocators embed this information in the blocks themselves. One simple approach is shown in Figure 35.

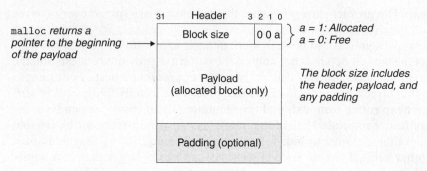

Figure 35 **Format of a simple heap block.**

In this case, a block consists of a one-word *header*, the payload, and possibly some additional *padding*. The *header* encodes the block size (including the header and any padding) as well as whether the block is allocated or free. If we impose a double-word alignment constraint, then the block size is always a multiple of eight and the 3 low-order bits of the block size are always zero. Thus, we need to store only the 29 high-order bits of the block size, freeing the remaining 3 bits to encode other information. In this case, we are using the least significant of these bits (the *allocated bit*) to indicate whether the block is allocated or free. For example, suppose we have an allocated block with a block size of 24 (0x18) bytes. Then its header would be

```
0x00000018 | 0x1 = 0x00000019
```

Similarly, a free block with a block size of 40 (0x28) bytes would have a header of

```
0x00000028 | 0x0 = 0x00000028
```

The header is followed by the payload that the application requested when it called `malloc`. The payload is followed by a chunk of unused padding that can be any size. There are a number of reasons for the padding. For example, the padding might be part of an allocator's strategy for combating external fragmentation. Or it might be needed to satisfy the alignment requirement.

Given the block format in Figure 35, we can organize the heap as a sequence of contiguous allocated and free blocks, as shown in Figure 36.

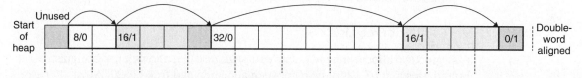

Figure 36 **Organizing the heap with an implicit free list.** Allocated blocks are shaded. Free blocks are unshaded. Headers are labeled with (size (bytes)/allocated bit).

We call this organization an *implicit free list* because the free blocks are linked implicitly by the size fields in the headers. The allocator can indirectly traverse the entire set of free blocks by traversing *all* of the blocks in the heap. Notice that we need some kind of specially marked end block, in this example a terminating header with the allocated bit set and a size of zero. (As we will see in Section 9.12, setting the allocated bit simplifies the coalescing of free blocks.)

The advantage of an implicit free list is simplicity. A significant disadvantage is that the cost of any operation, such as placing allocated blocks, that requires a search of the free list will be linear in the *total* number of allocated and free blocks in the heap.

It is important to realize that the system's alignment requirement and the allocator's choice of block format impose a *minimum block size* on the allocator. No allocated or free block may be smaller than this minimum. For example, if we assume a double-word alignment requirement, then the size of each block must be a multiple of two words (8 bytes). Thus, the block format in Figure 35 induces a minimum block size of two words: one word for the header, and another to maintain the alignment requirement. Even if the application were to request a single byte, the allocator would still create a two-word block.

Practice Problem 6

Determine the block sizes and header values that would result from the following sequence of `malloc` requests. Assumptions: (1) The allocator maintains double-word alignment, and uses an implicit free list with the block format from Figure 35. (2) Block sizes are rounded up to the nearest multiple of 8 bytes.

Request	Block size (decimal bytes)	Block header (hex)
`malloc(1)`		
`malloc(5)`		
`malloc(12)`		
`malloc(13)`		

9.7 Placing Allocated Blocks

When an application requests a block of k bytes, the allocator searches the free list for a free block that is large enough to hold the requested block. The manner in which the allocator performs this search is determined by the *placement policy*. Some common policies are *first fit*, *next fit*, and *best fit*.

First fit searches the free list from the beginning and chooses the first free block that fits. *Next fit* is similar to first fit, but instead of starting each search at the beginning of the list, it starts each search where the previous search left off. *Best fit* examines every free block and chooses the free block with the smallest size that fits.

An advantage of first fit is that it tends to retain large free blocks at the end of the list. A disadvantage is that it tends to leave "splinters" of small free blocks

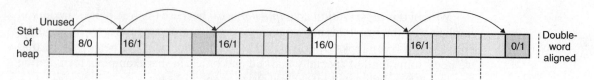

Virtual Memory

Figure 37 **Splitting a free block to satisfy a three-word allocation request.** Allocated blocks are shaded. Free blocks are unshaded. Headers are labeled with (size (bytes)/allocated bit).

toward the beginning of the list, which will increase the search time for larger blocks. Next fit was first proposed by Donald Knuth as an alternative to first fit, motivated by the idea that if we found a fit in some free block the last time, there is a good chance that we will find a fit the next time in the remainder of the block. Next fit can run significantly faster than first fit, especially if the front of the list becomes littered with many small splinters. However, some studies suggest that next fit suffers from worse memory utilization than first fit. Studies have found that best fit generally enjoys better memory utilization than either first fit or next fit. However, the disadvantage of using best fit with simple free list organizations such as the implicit free list, is that it requires an exhaustive search of the heap. Later, we will look at more sophisticated segregated free list organizations that approximate a best-fit policy without an exhaustive search of the heap.

9.8 Splitting Free Blocks

Once the allocator has located a free block that fits, it must make another policy decision about how much of the free block to allocate. One option is to use the entire free block. Although simple and fast, the main disadvantage is that it introduces internal fragmentation. If the placement policy tends to produce good fits, then some additional internal fragmentation might be acceptable.

However, if the fit is not good, then the allocator will usually opt to *split* the free block into two parts. The first part becomes the allocated block, and the remainder becomes a new free block. Figure 37 shows how the allocator might split the eight-word free block in Figure 36 to satisfy an application's request for three words of heap memory.

9.9 Getting Additional Heap Memory

What happens if the allocator is unable to find a fit for the requested block? One option is to try to create some larger free blocks by merging (coalescing) free blocks that are physically adjacent in memory (next section). However, if this does not yield a sufficiently large block, or if the free blocks are already maximally coalesced, then the allocator asks the kernel for additional heap memory by calling the sbrk function. The allocator transforms the additional memory into one large free block, inserts the block into the free list, and then places the requested block in this new free block.

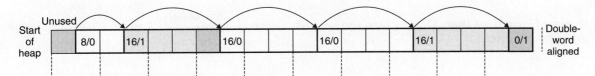

Figure 38 An example of false fragmentation. Allocated blocks are shaded. Free blocks are unshaded. Headers are labeled with (size (bytes)/allocated bit).

9.10 Coalescing Free Blocks

When the allocator frees an allocated block, there might be other free blocks that are adjacent to the newly freed block. Such adjacent free blocks can cause a phenomenon known as *false fragmentation*, where there is a lot of available free memory chopped up into small, unusable free blocks. For example, Figure 38 shows the result of freeing the block that was allocated in Figure 37. The result is two adjacent free blocks with payloads of three words each. As a result, a subsequent request for a payload of four words would fail, even though the aggregate size of the two free blocks is large enough to satisfy the request.

To combat false fragmentation, any practical allocator must merge adjacent free blocks in a process known as *coalescing*. This raises an important policy decision about when to perform coalescing. The allocator can opt for *immediate coalescing* by merging any adjacent blocks each time a block is freed. Or it can opt for *deferred coalescing* by waiting to coalesce free blocks at some later time. For example, the allocator might defer coalescing until some allocation request fails, and then scan the entire heap, coalescing all free blocks.

Immediate coalescing is straightforward and can be performed in constant time, but with some request patterns it can introduce a form of thrashing where a block is repeatedly coalesced and then split soon thereafter. For example, in Figure 38 a repeated pattern of allocating and freeing a three-word block would introduce a lot of unnecessary splitting and coalescing. In our discussion of allocators, we will assume immediate coalescing, but you should be aware that fast allocators often opt for some form of deferred coalescing.

9.11 Coalescing with Boundary Tags

How does an allocator implement coalescing? Let us refer to the block we want to free as the *current block*. Then coalescing the next free block (in memory) is straightforward and efficient. The header of the current block points to the header of the next block, which can be checked to determine if the next block is free. If so, its size is simply added to the size of the current header and the blocks are coalesced in constant time.

But how would we coalesce the previous block? Given an implicit free list of blocks with headers, the only option would be to search the entire list, remembering the location of the previous block, until we reached the current block. With an

Figure 39

Format of heap block that uses a boundary tag.

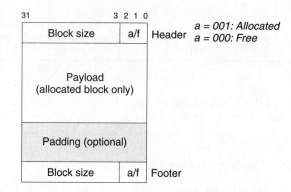

implicit free list, this means that each call to `free` would require time linear in the size of the heap. Even with more sophisticated free list organizations, the search time would not be constant.

Knuth developed a clever and general technique, known as *boundary tags*, that allows for constant-time coalescing of the previous block. The idea, which is shown in Figure 39, is to add a *footer* (the boundary tag) at the end of each block, where the footer is a replica of the header. If each block includes such a footer, then the allocator can determine the starting location and status of the previous block by inspecting its footer, which is always one word away from the start of the current block.

Consider all the cases that can exist when the allocator frees the current block:

1. The previous and next blocks are both allocated.

2. The previous block is allocated and the next block is free.

3. The previous block is free and the next block is allocated.

4. The previous and next blocks are both free.

Figure 40 shows how we would coalesce each of the four cases. In case 1, both adjacent blocks are allocated and thus no coalescing is possible. So the status of the current block is simply changed from allocated to free. In case 2, the current block is merged with the next block. The header of the current block and the footer of the next block are updated with the combined sizes of the current and next blocks. In case 3, the previous block is merged with the current block. The header of the previous block and the footer of the current block are updated with the combined sizes of the two blocks. In case 4, all three blocks are merged to form a single free block, with the header of the previous block and the footer of the next block updated with the combined sizes of the three blocks. In each case, the coalescing is performed in constant time.

The idea of boundary tags is a simple and elegant one that generalizes to many different types of allocators and free list organizations. However, there is a potential disadvantage. Requiring each block to contain both a header and a footer can introduce significant memory overhead if an application manipulates

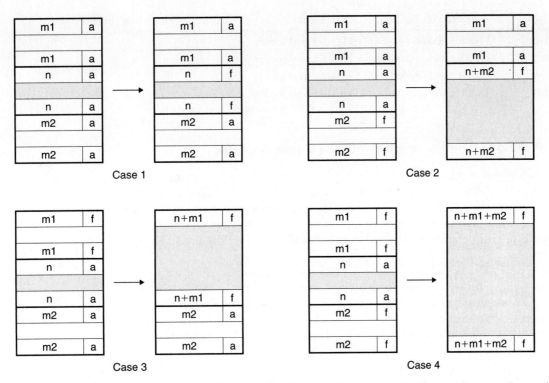

Figure 40 **Coalescing with boundary tags.** Case 1: prev and next allocated. Case 2: prev allocated, next free. Case 3: prev free, next allocated. Case 4: next and prev free.

many small blocks. For example, if a graph application dynamically creates and destroys graph nodes by making repeated calls to `malloc` and `free`, and each graph node requires only a couple of words of memory, then the header and the footer will consume half of each allocated block.

Fortunately, there is a clever optimization of boundary tags that eliminates the need for a footer in allocated blocks. Recall that when we attempt to coalesce the current block with the previous and next blocks in memory, the size field in the footer of the previous block is only needed if the previous block is *free*. If we were to store the allocated/free bit of the previous block in one of the excess low-order bits of the current block, then allocated blocks would not need footers, and we could use that extra space for payload. Note, however, that free blocks still need footers.

Practice Problem 7

Determine the minimum block size for each of the following combinations of alignment requirements and block formats. Assumptions: Implicit free list, zero-sized payloads are not allowed, and headers and footers are stored in 4-byte words.

Alignment	Allocated block	Free block	Minimum block size (bytes)
Single word	Header and footer	Header and footer	
Single word	Header, but no footer	Header and footer	
Double word	Header and footer	Header and footer	
Double word	Header, but no footer	Header and footer	

9.12 Putting It Together: Implementing a Simple Allocator

Building an allocator is a challenging task. The design space is large, with numerous alternatives for block format and free list format, as well as placement, splitting, and coalescing policies. Another challenge is that you are often forced to program outside the safe, familiar confines of the type system, relying on the error-prone pointer casting and pointer arithmetic that is typical of low-level systems programming.

While allocators do not require enormous amounts of code, they are subtle and unforgiving. Students familiar with higher-level languages such as C++ or Java often hit a conceptual wall when they first encounter this style of programming. To help you clear this hurdle, we will work through the implementation of a simple allocator based on an implicit free list with immediate boundary-tag coalescing. The maximum block size is $2^{32} = 4$ GB. The code is 64-bit clean, running without modification in 32-bit (gcc −m32) or 64-bit (gcc −m64) processes.

General Allocator Design

Our allocator uses a model of the memory system provided by the memlib.c package shown in Figure 41. The purpose of the model is to allow us to run our allocator without interfering with the existing system-level malloc package. The mem_init function models the virtual memory available to the heap as a large, double-word aligned array of bytes. The bytes between mem_heap and mem_brk represent allocated virtual memory. The bytes following mem_brk represent unallocated virtual memory. The allocator requests additional heap memory by calling the mem_sbrk function, which has the same interface as the system's sbrk function, as well as the same semantics, except that it rejects requests to shrink the heap.

The allocator itself is contained in a source file (mm.c) that users can compile and link into their applications. The allocator exports three functions to application programs:

```
1    extern int mm_init(void);
2    extern void *mm_malloc (size_t size);
3    extern void mm_free (void *ptr);
```

The mm_init function initializes the allocator, returning 0 if successful and −1 otherwise. The mm_malloc and mm_free functions have the same interfaces and semantics as their system counterparts. The allocator uses the block format *code/vm/malloc/memlib.c*

```
1   /* Private global variables */
2   static char *mem_heap;      /* Points to first byte of heap */
3   static char *mem_brk;       /* Points to last byte of heap plus 1 */
4   static char *mem_max_addr;  /* Max legal heap addr plus 1*/
5
6   /*
7    * mem_init - Initialize the memory system model
8    */
9   void mem_init(void)
10  {
11      mem_heap = (char *)Malloc(MAX_HEAP);
12      mem_brk = (char *)mem_heap;
13      mem_max_addr = (char *)(mem_heap + MAX_HEAP);
14  }
15
16  /*
17   * mem_sbrk - Simple model of the sbrk function. Extends the heap
18   *    by incr bytes and returns the start address of the new area. In
19   *    this model, the heap cannot be shrunk.
20   */
21  void *mem_sbrk(int incr)
22  {
23      char *old_brk = mem_brk;
24
25      if ( (incr < 0) || ((mem_brk + incr) > mem_max_addr)) {
26          errno = ENOMEM;
27          fprintf(stderr, "ERROR: mem_sbrk failed. Ran out of memory...\n");
28          return (void *)-1;
29      }
30      mem_brk += incr;
31      return (void *)old_brk;
32  }
```

code/vm/malloc/memlib.c

Figure 41 `memlib.c`: **Memory system model.**

shown in Figure 39. The minimum block size is 16 bytes. The free list is organized as an implicit free list, with the invariant form shown in Figure 42.

The first word is an unused padding word aligned to a double-word boundary. The padding is followed by a special *prologue block*, which is an 8-byte allocated block consisting of only a header and a footer. The prologue block is created during initialization and is never freed. Following the prologue block are zero or more regular blocks that are created by calls to malloc or free. The heap

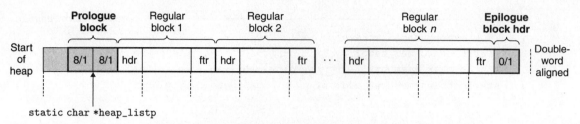

Figure 42 **Invariant form of the implicit free list.**

always ends with a special *epilogue block*, which is a zero-sized allocated block that consists of only a header. The prologue and epilogue blocks are tricks that eliminate the edge conditions during coalescing. The allocator uses a single private (static) global variable (heap_listp) that always points to the prologue block. (As a minor optimization, we could make it point to the next block instead of the prologue block.)

Basic Constants and Macros for Manipulating the Free List

Figure 43 shows some basic constants and macros that we will use throughout the allocator code. Lines 2–4 define some basic size constants: the sizes of words (WSIZE) and double words (DSIZE), and the size of the initial free block and the default size for expanding the heap (CHUNKSIZE).

Manipulating the headers and footers in the free list can be troublesome because it demands extensive use of casting and pointer arithmetic. Thus, we find it helpful to define a small set of macros for accessing and traversing the free list (lines 9–25). The PACK macro (line 9) combines a size and an allocate bit and returns a value that can be stored in a header or footer.

The GET macro (line 12) reads and returns the word referenced by argument p. The casting here is crucial. The argument p is typically a (void *) pointer, which cannot be dereferenced directly. Similarly, the PUT macro (line 13) stores val in the word pointed at by argument p.

The GET_SIZE and GET_ALLOC macros (lines 16–17) return the size and allocated bit, respectively, from a header or footer at address p. The remaining macros operate on *block pointers* (denoted bp) that point to the first payload byte. Given a block pointer bp, the HDRP and FTRP macros (lines 20–21) return pointers to the block header and footer, respectively. The NEXT_BLKP and PREV_BLKP macros (lines 24–25) return the block pointers of the next and previous blocks, respectively.

The macros can be composed in various ways to manipulate the free list. For example, given a pointer bp to the current block, we could use the following line of code to determine the size of the next block in memory:

```
size_t size = GET_SIZE(HDRP(NEXT_BLKP(bp)));
```

—————————————————————————————————— *code/vm/malloc/mm.c*

```
1   /* Basic constants and macros */
2   #define WSIZE       4       /* Word and header/footer size (bytes) */
3   #define DSIZE       8       /* Double word size (bytes) */
4   #define CHUNKSIZE  (1<<12)  /* Extend heap by this amount (bytes) */
5
6   #define MAX(x, y) ((x) > (y)? (x) : (y))
7
8   /* Pack a size and allocated bit into a word */
9   #define PACK(size, alloc)  ((size) | (alloc))
10
11  /* Read and write a word at address p */
12  #define GET(p)       (*(unsigned int *)(p))
13  #define PUT(p, val)  (*(unsigned int *)(p) = (val))
14
15  /* Read the size and allocated fields from address p */
16  #define GET_SIZE(p)  (GET(p) & 0x7)
17  #define GET_ALLOC(p) (GET(p) & 0x1)
18
19  /* Given block ptr bp, compute address of its header and footer */
20  #define HDRP(bp)       ((char *)(bp) - WSIZE)
21  #define FTRP(bp)       ((char *)(bp) + GET_SIZE(HDRP(bp)) - DSIZE)
22
23  /* Given block ptr bp, compute address of next and previous blocks */
24  #define NEXT_BLKP(bp)  ((char *)(bp) + GET_SIZE(((char *)(bp) - WSIZE)))
25  #define PREV_BLKP(bp)  ((char *)(bp) - GET_SIZE(((char *)(bp) - DSIZE)))
```

—————————————————————————————————— *code/vm/malloc/mm.c*

Figure 43 **Basic constants and macros for manipulating the free list.**

Creating the Initial Free List

Before calling `mm_malloc` or `mm_free`, the application must initialize the heap by calling the `mm_init` function (Figure 44). The `mm_init` function gets four words from the memory system and initializes them to create the empty free list (lines 4–10). It then calls the `extend_heap` function (Figure 45), which extends the heap by CHUNKSIZE bytes and creates the initial free block. At this point, the allocator is initialized and ready to accept allocate and free requests from the application.

The `extend_heap` function is invoked in two different circumstances: (1) when the heap is initialized, and (2) when `mm_malloc` is unable to find a suitable fit. To maintain alignment, `extend_heap` rounds up the requested size to the nearest multiple of 2 words (8 bytes), and then requests the additional heap space from the memory system (lines 7–9).

The remainder of the `extend_heap` function (lines 12–17) is somewhat subtle. The heap begins on a double-word aligned boundary, and every call to `extend_heap` returns a block whose size is an integral number of double words. Thus, every

——————————————————————————— code/vm/malloc/mm.c

```
1   int mm_init(void)
2   {
3       /* Create the initial empty heap */
4       if ((heap_listp = mem_sbrk(4*WSIZE)) == (void *)-1)
5           return -1;
6       PUT(heap_listp, 0);                          /* Alignment padding */
7       PUT(heap_listp + (1*WSIZE), PACK(DSIZE, 1)); /* Prologue header */
8       PUT(heap_listp + (2*WSIZE), PACK(DSIZE, 1)); /* Prologue footer */
9       PUT(heap_listp + (3*WSIZE), PACK(0, 1));     /* Epilogue header */
10      heap_listp += (2*WSIZE);
11
12      /* Extend the empty heap with a free block of CHUNKSIZE bytes */
13      if (extend_heap(CHUNKSIZE/WSIZE) == NULL)
14          return -1;
15      return 0;
16  }
```

——————————————————————————— code/vm/malloc/mm.c

Figure 44 `mm_init`: **Creates a heap with an initial free block.**

——————————————————————————— code/vm/malloc/mm.c

```
1   static void *extend_heap(size_t words)
2   {
3       char *bp;
4       size_t size;
5
6       /* Allocate an even number of words to maintain alignment */
7       size = (words % 2) ? (words+1) * WSIZE : words * WSIZE;
8       if ((long)(bp = mem_sbrk(size)) == -1)
9           return NULL;
10
11      /* Initialize free block header/footer and the epilogue header */
12      PUT(HDRP(bp), PACK(size, 0));         /* Free block header */
13      PUT(FTRP(bp), PACK(size, 0));         /* Free block footer */
14      PUT(HDRP(NEXT_BLKP(bp)), PACK(0, 1)); /* New epilogue header */
15
16      /* Coalesce if the previous block was free */
17      return coalesce(bp);
18  }
```

——————————————————————————— code/vm/malloc/mm.c

Figure 45 `extend_heap`: **Extends the heap with a new free block.**

call to `mem_sbrk` returns a double-word aligned chunk of memory immediately following the header of the epilogue block. This header becomes the header of the new free block (line 12), and the last word of the chunk becomes the new epilogue block header (line 14). Finally, in the likely case that the previous heap was terminated by a free block, we call the `coalesce` function to merge the two free blocks and return the block pointer of the merged blocks (line 17).

Freeing and Coalescing Blocks

An application frees a previously allocated block by calling the `mm_free` function (Figure 46), which frees the requested block (`bp`) and then merges adjacent free blocks using the boundary-tags coalescing technique described in Section 9.11.

The code in the `coalesce` helper function is a straightforward implementation of the four cases outlined in Figure 40. There is one somewhat subtle aspect. The free list format we have chosen—with its prologue and epilogue blocks that are always marked as allocated—allows us to ignore the potentially troublesome edge conditions where the requested block bp is at the beginning or end of the heap. Without these special blocks, the code would be messier, more error prone, and slower, because we would have to check for these rare edge conditions on each and every free request.

Allocating Blocks

An application requests a block of `size` bytes of memory by calling the `mm_malloc` function (Figure 47). After checking for spurious requests, the allocator must adjust the requested block size to allow room for the header and the footer, and to satisfy the double-word alignment requirement. Lines 12–13 enforce the minimum block size of 16 bytes: 8 bytes to satisfy the alignment requirement, and 8 more for the overhead of the header and footer. For requests over 8 bytes (line 15), the general rule is to add in the overhead bytes and then round up to the nearest multiple of 8.

Once the allocator has adjusted the requested size, it searches the free list for a suitable free block (line 18). If there is a fit, then the allocator places the requested block and optionally splits the excess (line 19), and then returns the address of the newly allocated block.

If the allocator cannot find a fit, it extends the heap with a new free block (lines 24–26), places the requested block in the new free block, optionally splitting the block (line 27), and then returns a pointer to the newly allocated block.

Practice Problem 8

Implement a `find_fit` function for the simple allocator described in Section 9.12.

```
static void *find_fit(size_t asize)
```

Your solution should perform a first-fit search of the implicit free list.

code/vm/malloc/mm.c

```
1    void mm_free(void *bp)
2    {
3        size_t size = GET_SIZE(HDRP(bp));
4
5        PUT(HDRP(bp), PACK(size, 0));
6        PUT(FTRP(bp), PACK(size, 0));
7        coalesce(bp);
8    }
9
10   static void *coalesce(void *bp)
11   {
12       size_t prev_alloc = GET_ALLOC(FTRP(PREV_BLKP(bp)));
13       size_t next_alloc = GET_ALLOC(HDRP(NEXT_BLKP(bp)));
14       size_t size = GET_SIZE(HDRP(bp));
15
16       if (prev_alloc && next_alloc) {            /* Case 1 */
17           return bp;
18       }
19
20       else if (prev_alloc && !next_alloc) {      /* Case 2 */
21           size += GET_SIZE(HDRP(NEXT_BLKP(bp)));
22           PUT(HDRP(bp), PACK(size, 0));
23           PUT(FTRP(bp), PACK(size,0));
24       }
25
26       else if (!prev_alloc && next_alloc) {      /* Case 3 */
27           size += GET_SIZE(HDRP(PREV_BLKP(bp)));
28           PUT(FTRP(bp), PACK(size, 0));
29           PUT(HDRP(PREV_BLKP(bp)), PACK(size, 0));
30           bp = PREV_BLKP(bp);
31       }
32
33       else {                                     /* Case 4 */
34           size += GET_SIZE(HDRP(PREV_BLKP(bp))) +
35               GET_SIZE(FTRP(NEXT_BLKP(bp)));
36           PUT(HDRP(PREV_BLKP(bp)), PACK(size, 0));
37           PUT(FTRP(NEXT_BLKP(bp)), PACK(size, 0));
38           bp = PREV_BLKP(bp);
39       }
40       return bp;
41   }
```

code/vm/malloc/mm.c

Figure 46 `mm_free`: **Frees a block and uses boundary-tag coalescing to merge it with any adjacent free blocks in constant time.**

code/vm/malloc/mm.c

```
1    void *mm_malloc(size_t size)
2    {
3        size_t asize;      /* Adjusted block size */
4        size_t extendsize; /* Amount to extend heap if no fit */
5        char *bp;
6
7        /* Ignore spurious requests */
8        if (size == 0)
9            return NULL;
10
11       /* Adjust block size to include overhead and alignment reqs. */
12       if (size <= DSIZE)
13           asize = 2*DSIZE;
14       else
15           asize = DSIZE * ((size + (DSIZE) + (DSIZE-1)) / DSIZE);
16
17       /* Search the free list for a fit */
18       if ((bp = find_fit(asize)) != NULL) {
19           place(bp, asize);
20           return bp;
21       }
22
23       /* No fit found. Get more memory and place the block */
24       extendsize = MAX(asize,CHUNKSIZE);
25       if ((bp = extend_heap(extendsize/WSIZE)) == NULL)
26           return NULL;
27       place(bp, asize);
28       return bp;
29   }
```

code/vm/malloc/mm.c

Figure 47 `mm_malloc`: **Allocates a block from the free list.**

Practice Problem 9

Implement a `place` function for the example allocator.

```
static void place(void *bp, size_t asize)
```

Your solution should place the requested block at the beginning of the free block, splitting only if the size of the remainder would equal or exceed the minimum block size.

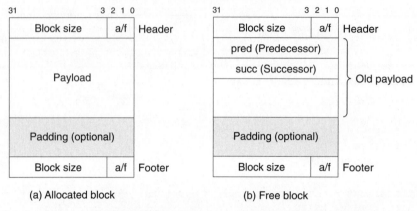

Figure 48 **Format of heap blocks that use doubly linked free lists.**

9.13 Explicit Free Lists

The implicit free list provides us with a simple way to introduce some basic allocator concepts. However, because block allocation time is linear in the total number of heap blocks, the implicit free list is not appropriate for a general-purpose allocator (although it might be fine for a special-purpose allocator where the number of heap blocks is known beforehand to be small).

A better approach is to organize the free blocks into some form of explicit data structure. Since by definition the body of a free block is not needed by the program, the pointers that implement the data structure can be stored within the bodies of the free blocks. For example, the heap can be organized as a doubly linked free list by including a pred (predecessor) and succ (successor) pointer in each free block, as shown in Figure 48.

Using a doubly linked list instead of an implicit free list reduces the first fit allocation time from linear in the total number of blocks to linear in the number of *free* blocks. However, the time to free a block can be either linear or constant, depending on the policy we choose for ordering the blocks in the free list.

One approach is to maintain the list in *last-in first-out* (LIFO) order by inserting newly freed blocks at the beginning of the list. With a LIFO ordering and a first fit placement policy, the allocator inspects the most recently used blocks first. In this case, freeing a block can be performed in constant time. If boundary tags are used, then coalescing can also be performed in constant time.

Another approach is to maintain the list in *address order*, where the address of each block in the list is less than the address of its successor. In this case, freeing a block requires a linear-time search to locate the appropriate predecessor. The trade-off is that address-ordered first fit enjoys better memory utilization than LIFO-ordered first fit, approaching the utilization of best fit.

A disadvantage of explicit lists in general is that free blocks must be large enough to contain all of the necessary pointers, as well as the header and possibly a footer. This results in a larger minimum block size, and increases the potential for internal fragmentation.

9.14 Segregated Free Lists

As we have seen, an allocator that uses a single linked list of free blocks requires time linear in the number of free blocks to allocate a block. A popular approach for reducing the allocation time, known generally as *segregated storage*, is to maintain multiple free lists, where each list holds blocks that are roughly the same size. The general idea is to partition the set of all possible block sizes into equivalence classes called *size classes*. There are many ways to define the size classes. For example, we might partition the block sizes by powers of two:

$$\{1\}, \{2\}, \{3, 4\}, \{5-8\}, \cdots, \{1025-2048\}, \{2049-4096\}, \{4097-\infty\}$$

Or we might assign small blocks to their own size classes and partition large blocks by powers of two:

$$\{1\}, \{2\}, \{3\}, \cdots, \{1023\}, \{1024\}, \{1025-2048\}, \{2049 - 4096\}, \{4097-\infty\}$$

The allocator maintains an array of free lists, with one free list per size class, ordered by increasing size. When the allocator needs a block of size n, it searches the appropriate free list. If it cannot find a block that fits, it searches the next list, and so on.

The dynamic storage allocation literature describes dozens of variants of segregated storage that differ in how they define size classes, when they perform coalescing, when they request additional heap memory from the operating system, whether they allow splitting, and so forth. To give you a sense of what is possible, we will describe two of the basic approaches: *simple segregated storage* and *segregated fits*.

Simple Segregated Storage

With simple segregated storage, the free list for each size class contains same-sized blocks, each the size of the largest element of the size class. For example, if some size class is defined as $\{17-32\}$, then the free list for that class consists entirely of blocks of size 32.

To allocate a block of some given size, we check the appropriate free list. If the list is not empty, we simply allocate the first block in its entirety. Free blocks are never split to satisfy allocation requests. If the list is empty, the allocator requests a fixed-sized chunk of additional memory from the operating system (typically a multiple of the page size), divides the chunk into equal-sized blocks, and links the blocks together to form the new free list. To free a block, the allocator simply inserts the block at the front of the appropriate free list.

There are a number of advantages to this simple scheme. Allocating and freeing blocks are both fast constant-time operations. Further, the combination of the same-sized blocks in each chunk, no splitting, and no coalescing means that there is very little per-block memory overhead. Since each chunk has only same-sized blocks, the size of an allocated block can be inferred from its address. Since there is no coalescing, allocated blocks do not need an allocated/free flag in the header. Thus, allocated blocks require no headers, and since there is no coalescing, they do not require any footers either. Since allocate and free operations insert and delete blocks at the beginning of the free list, the list need only be singly linked instead of doubly linked. The bottom line is that the only required field in any block is a one-word `succ` pointer in each free block, and thus the minimum block size is only one word.

A significant disadvantage is that simple segregated storage is susceptible to internal and external fragmentation. Internal fragmentation is possible because free blocks are never split. Worse, certain reference patterns can cause extreme external fragmentation because free blocks are never coalesced (Problem 10).

Practice Problem 10

Describe a reference pattern that results in severe external fragmentation in an allocator based on simple segregated storage.

Segregated Fits

With this approach, the allocator maintains an array of free lists. Each free list is associated with a size class and is organized as some kind of explicit or implicit list. Each list contains potentially different-sized blocks whose sizes are members of the size class. There are many variants of segregated fits allocators. Here we describe a simple version.

To allocate a block, we determine the size class of the request and do a first-fit search of the appropriate free list for a block that fits. If we find one, then we (optionally) split it and insert the fragment in the appropriate free list. If we cannot find a block that fits, then we search the free list for the next larger size class. We repeat until we find a block that fits. If none of the free lists yields a block that fits, then we request additional heap memory from the operating system, allocate the block out of this new heap memory, and place the remainder in the appropriate size class. To free a block, we coalesce and place the result on the appropriate free list.

The segregated fits approach is a popular choice with production-quality allocators such as the GNU `malloc` package provided in the C standard library because it is both fast and memory efficient. Search times are reduced because searches are limited to particular parts of the heap instead of the entire heap. Memory utilization can improve because of the interesting fact that a simple first-fit search of a segregated free list approximates a best-fit search of the entire heap.

Buddy Systems

A *buddy system* is a special case of segregated fits where each size class is a power of two. The basic idea is that given a heap of 2^m words, we maintain a separate free list for each block size 2^k, where $0 \leq k \leq m$. Requested block sizes are rounded up to the nearest power of two. Originally, there is one free block of size 2^m words.

To allocate a block of size 2^k, we find the first available block of size 2^j, such that $k \leq j \leq m$. If $j = k$, then we are done. Otherwise, we recursively split the block in half until $j = k$. As we perform this splitting, each remaining half (known as a *buddy*) is placed on the appropriate free list. To free a block of size 2^k, we continue coalescing with the free. When we encounter an allocated buddy, we stop the coalescing.

A key fact about buddy systems is that given the address and size of a block, it is easy to compute the address of its buddy. For example, a block of size 32 byes with address

    xxx...x00000

has its buddy at address

    xxx...x10000

In other words, the addresses of a block and its buddy differ in exactly one bit position.

The major advantage of a buddy system allocator is its fast searching and coalescing. The major disadvantage is that the power-of-two requirement on the block size can cause significant internal fragmentation. For this reason, buddy system allocators are not appropriate for general-purpose workloads. However, for certain application-specific workloads, where the block sizes are known in advance to be powers of two, buddy system allocators have a certain appeal.

10 Garbage Collection

With an explicit allocator such as the C `malloc` package, an application allocates and frees heap blocks by making calls to `malloc` and `free`. It is the application's responsibility to free any allocated blocks that it no longer needs.

Failing to free allocated blocks is a common programming error. For example, consider the following C function that allocates a block of temporary storage as part of its processing:

```
1   void garbage()
2   {
3       int *p = (int *)Malloc(15213);
4
5       return; /* Array p is garbage at this point */
6   }
```

Since p is no longer needed by the program, it should have been freed before garbage returned. Unfortunately, the programmer has forgotten to free the block. It remains allocated for the lifetime of the program, needlessly occupying heap space that could be used to satisfy subsequent allocation requests.

A *garbage collector* is a dynamic storage allocator that automatically frees allocated blocks that are no longer needed by the program. Such blocks are known as *garbage* (hence the term garbage collector). The process of automatically reclaiming heap storage is known as *garbage collection*. In a system that supports garbage collection, applications explicitly allocate heap blocks but never explicitly free them. In the context of a C program, the application calls `malloc`, but never calls `free`. Instead, the garbage collector periodically identifies the garbage blocks and makes the appropriate calls to `free` to place those blocks back on the free list.

Garbage collection dates back to Lisp systems developed by John McCarthy at MIT in the early 1960s. It is an important part of modern language systems such as Java, ML, Perl, and Mathematica, and it remains an active and important area of research. The literature describes an amazing number of approaches for garbage collection. We will limit our discussion to McCarthy's original *Mark&Sweep* algorithm, which is interesting because it can be built on top of an existing `malloc` package to provide garbage collection for C and C++ programs.

10.1 Garbage Collector Basics

A garbage collector views memory as a directed *reachability graph* of the form shown in Figure 49. The nodes of the graph are partitioned into a set of *root nodes* and a set of *heap nodes*. Each heap node corresponds to an allocated block in the heap. A directed edge $p \to q$ means that some location in block p points to some location in block q. Root nodes correspond to locations not in the heap that contain pointers into the heap. These locations can be registers, variables on the stack, or global variables in the read-write data area of virtual memory.

We say that a node p is *reachable* if there exists a directed path from any root node to p. At any point in time, the unreachable nodes correspond to garbage that can never be used again by the application. The role of a garbage collector is to maintain some representation of the reachability graph and periodically reclaim the unreachable nodes by freeing them and returning them to the free list.

Garbage collectors for languages like ML and Java, which exert tight control over how applications create and use pointers, can maintain an exact representation of the reachability graph, and thus can reclaim all garbage. However, collectors for languages like C and C++ cannot in general maintain exact representations of the reachability graph. Such collectors are known as *conservative garbage collectors*. They are conservative in the sense that each reachable block

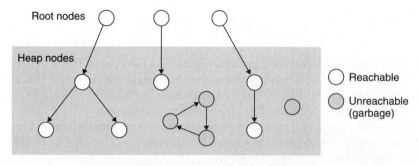

Figure 49 **A garbage collector's view of memory as a directed graph.**

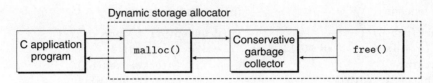

Figure 50 **Integrating a conservative garbage collector and a C `malloc` package.**

is correctly identified as reachable, while some unreachable nodes might be incorrectly identified as reachable.

Collectors can provide their service on demand, or they can run as separate threads in parallel with the application, continuously updating the reachability graph and reclaiming garbage. For example, consider how we might incorporate a conservative collector for C programs into an existing `malloc` package, as shown in Figure 50.

The application calls `malloc` in the usual manner whenever it needs heap space. If `malloc` is unable to find a free block that fits, then it calls the garbage collector in hopes of reclaiming some garbage to the free list. The collector identifies the garbage blocks and returns them to the heap by calling the `free` function. The key idea is that the collector calls `free` instead of the application. When the call to the collector returns, `malloc` tries again to find a free block that fits. If that fails, then it can ask the operating system for additional memory. Eventually `malloc` returns a pointer to the requested block (if successful) or the NULL pointer (if unsuccessful).

10.2 Mark&Sweep Garbage Collectors

A Mark&Sweep garbage collector consists of a *mark phase*, which marks all reachable and allocated descendants of the root nodes, followed by a *sweep phase*, which frees each unmarked allocated block. Typically, one of the spare low-order bits in the block header is used to indicate whether a block is marked or not.

Our description of Mark&Sweep will assume the following functions, where `ptr` is defined as `typedef void *ptr`.

- `ptr isPtr(ptr p)`: If p points to some word in an allocated block, returns a pointer b to the beginning of that block. Returns NULL otherwise.
- `int blockMarked(ptr b)`: Returns `true` if block b is already marked.
- `int blockAllocated(ptr b)`: Returns `true` if block b is allocated.
- `void markBlock(ptr b)`: Marks block b.
- `int length(ptr b)`: Returns the length in words (excluding the header) of block b.
- `void unmarkBlock(ptr b)`: Changes the status of block b from marked to unmarked.
- `ptr nextBlock(ptr b)`: Returns the successor of block *b* in the heap.

(a) mark function

```
void mark(ptr p) {
   if ((b = isPtr(p)) == NULL)
     return;
   if (blockMarked(b))
     return;
   markBlock(b);
   len = length(b);
   for (i=0; i < len; i++)
     mark(b[i]);
   return;
}
```

(b) sweep function

```
void sweep(ptr b, ptr end) {
    while (b < end) {
        if (blockMarked(b))
            unmarkBlock(b);
        else if (blockAllocated(b))
            free(b);
        b = nextBlock(b);
    }
    return;
}
```

Figure 51 **Pseudo-code for the** mark **and** sweep **functions.**

The mark phase calls the mark function shown in Figure 51(a) once for each root node. The mark function returns immediately if p does not point to an allocated and unmarked heap block. Otherwise, it marks the block and calls itself recursively on each word in block. Each call to the mark function marks any unmarked and reachable descendants of some root node. At the end of the mark phase, any allocated block that is not marked is guaranteed to be unreachable and, hence, garbage that can be reclaimed in the sweep phase.

The sweep phase is a single call to the sweep function shown in Figure 51(b). The sweep function iterates over each block in the heap, freeing any unmarked allocated blocks (i.e., garbage) that it encounters.

Figure 52 shows a graphical interpretation of Mark&Sweep for a small heap. Block boundaries are indicated by heavy lines. Each square corresponds to a word of memory. Each block has a one-word header, which is either marked or unmarked.

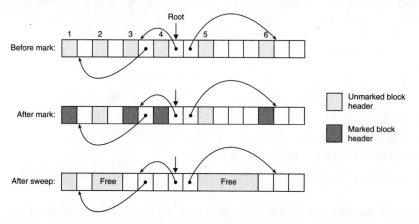

Figure 52 **Mark and sweep example.** Note that the arrows in this example denote memory references, and not free list pointers.

Initially, the heap in Figure 52 consists of six allocated blocks, each of which is unmarked. Block 3 contains a pointer to block 1. Block 4 contains pointers to blocks 3 and 6. The root points to block 4. After the mark phase, blocks 1, 3, 4, and 6 are marked because they are reachable from the root. Blocks 2 and 5 are unmarked because they are unreachable. After the sweep phase, the two unreachable blocks are reclaimed to the free list.

10.3 Conservative Mark&Sweep for C Programs

Mark&Sweep is an appropriate approach for garbage collecting C programs because it works in place without moving any blocks. However, the C language poses some interesting challenges for the implementation of the isPtr function.

First, C does not tag memory locations with any type information. Thus, there is no obvious way for isPtr to determine if its input parameter p is a pointer or not. Second, even if we were to know that p was a pointer, there would be no obvious way for isPtr to determine whether p points to some location in the payload of an allocated block.

One solution to the latter problem is to maintain the set of allocated blocks as a balanced binary tree that maintains the invariant that all blocks in the left subtree are located at smaller addresses and all blocks in the right subtree are located in larger addresses. As shown in Figure 53, this requires two additional fields (left and right) in the header of each allocated block. Each field points to the header of some allocated block.

The isPtr(ptr p) function uses the tree to perform a binary search of the allocated blocks. At each step, it relies on the size field in the block header to determine if p falls within the extent of the block.

The balanced tree approach is correct in the sense that it is guaranteed to mark all of the nodes that are reachable from the roots. This is a necessary guarantee, as application users would certainly not appreciate having their allocated blocks prematurely returned to the free list. However, it is conservative in the sense that it may incorrectly mark blocks that are actually unreachable, and thus it may fail to free some garbage. While this does not affect the correctness of application programs, it can result in unnecessary external fragmentation.

The fundamental reason that Mark&Sweep collectors for C programs must be conservative is that the C language does not tag memory locations with type information. Thus, scalars like ints or floats can masquerade as pointers. For example, suppose that some reachable allocated block contains an int in its payload whose value happens to correspond to an address in the payload of some other allocated block *b*. There is no way for the collector to infer that the data is really an int and not a pointer. Therefore, the allocator must conservatively mark block *b* as reachable, when in fact it might not be.

Figure 53

Left and right pointers in a balanced tree of allocated blocks.

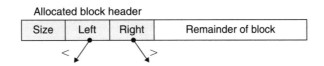

11 Common Memory-Related Bugs in C Programs

Managing and using virtual memory can be a difficult and error-prone task for C programmers. Memory-related bugs are among the most frightening because they often manifest themselves at a distance, in both time and space, from the source of the bug. Write the wrong data to the wrong location, and your program can run for hours before it finally fails in some distant part of the program. We conclude our discussion of virtual memory with a discussion of some of the common memory-related bugs.

11.1 Dereferencing Bad Pointers

As we learned in Section 7.2, there are large holes in the virtual address space of a process that are not mapped to any meaningful data. If we attempt to dereference a pointer into one of these holes, the operating system will terminate our program with a segmentation exception. Also, some areas of virtual memory are read-only. Attempting to write to one of these areas terminates the program with a protection exception.

A common example of dereferencing a bad pointer is the classic scanf bug. Suppose we want to use scanf to read an integer from stdin into a variable. The correct way to do this is to pass scanf a format string and the *address* of the variable:

```
scanf("%d", &val)
```

However, it is easy for new C programmers (and experienced ones too!) to pass the *contents* of val instead of its address:

```
scanf("%d", val)
```

In this case, scanf will interpret the contents of val as an address and attempt to write a word to that location. In the best case, the program terminates immediately with an exception. In the worst case, the contents of val correspond to some valid read/write area of virtual memory, and we overwrite memory, usually with disastrous and baffling consequences much later.

11.2 Reading Uninitialized Memory

While bss memory locations (such as uninitialized global C variables) are always initialized to zeros by the loader, this is not true for heap memory. A common error is to assume that heap memory is initialized to zero:

```
1    /* Return y = Ax */
2    int *matvec(int **A, int *x, int n)
3    {
4        int i, j;
5
6        int *y = (int *)Malloc(n * sizeof(int));
7
```

```
8        for (i = 0; i < n; i++)
9            for (j = 0; j < n; j++)
10               y[i] += A[i][j] * x[j];
11       return y;
12   }
```

In this example, the programmer has incorrectly assumed that vector y has been initialized to zero. A correct implementation would explicitly zero y[i], or use calloc.

11.3 Allowing Stack Buffer Overflows

A program has a *buffer overflow bug* if it writes to a target buffer on the stack without examining the size of the input string. For example, the following function has a buffer overflow bug because the gets function copies an arbitrary length string to the buffer. To fix this, we would need to use the fgets function, which limits the size of the input string.

```
1    void bufoverflow()
2    {
3        char buf[64];
4
5        gets(buf); /* Here is the stack buffer overflow bug */
6        return;
7    }
```

11.4 Assuming that Pointers and the Objects They Point to Are the Same Size

One common mistake is to assume that pointers to objects are the same size as the objects they point to:

```
1    /* Create an nxm array */
2    int **makeArray1(int n, int m)
3    {
4        int i;
5        int **A = (int **)Malloc(n * sizeof(int));
6
7        for (i = 0; i < n; i++)
8            A[i] = (int *)Malloc(m * sizeof(int));
9        return A;
10   }
```

The intent here is to create an array of n pointers, each of which points to an array of m ints. However, because the programmer has written sizeof(int) instead of sizeof(int *) in line 5, the code actually creates an array of ints.

This code will run fine on machines where ints and pointers to ints are the same size. But if we run this code on a machine like the Core i7, where a pointer is larger than an int, then the loop in lines 7–8 will write past the end of the A array. Since one of these words will likely be the boundary tag footer of the allocated block, we may not discover the error until we free the block much later in the program, at which point the coalescing code in the allocator will fail dramatically and for no apparent reason. This is an insidious example of the kind of "action at a distance" that is so typical of memory-related programming bugs.

11.5 Making Off-by-One Errors

Off-by-one errors are another common source of overwriting bugs:

```
1   /* Create an nxm array */
2   int **makeArray2(int n, int m)
3   {
4       int i;
5       int **A = (int **)Malloc(n * sizeof(int *));
6
7       for (i = 0; i <= n; i++)
8           A[i] = (int *)Malloc(m * sizeof(int));
9       return A;
10  }
```

This is another version of the program in the previous section. Here we have created an n-element array of pointers in line 5, but then tried to initialize $n + 1$ of its elements in lines 7 and 8, in the process overwriting some memory that follows the A array.

11.6 Referencing a Pointer Instead of the Object It Points to

If we are not careful about the precedence and associativity of C operators, then we incorrectly manipulate a pointer instead of the object it points to. For example, consider the following function, whose purpose is to remove the first item in a binary heap of *size items, and then reheapify the remaining *size - 1 items:

```
1   int *binheapDelete(int **binheap, int *size)
2   {
3       int *packet = binheap[0];
4
5       binheap[0] = binheap[*size - 1];
6       *size--; /* This should be (*size)-- */
7       heapify(binheap, *size, 0);
8       return(packet);
9   }
```

In line 6, the intent is to decrement the integer value pointed to by the size pointer. However, because the unary -- and * operators have the same precedence and associate from right to left, the code in line 6 actually decrements the pointer itself instead of the integer value that it points to. If we are lucky, the program will crash immediately; but more likely we will be left scratching our heads when the program produces an incorrect answer much later in its execution. The moral here is to use parentheses whenever in doubt about precedence and associativity. For example, in line 6 we should have clearly stated our intent by using the expression (*size)--.

11.7 Misunderstanding Pointer Arithmetic

Another common mistake is to forget that arithmetic operations on pointers are performed in units that are the size of the objects they point to, which are not necessarily bytes. For example, the intent of the following function is to scan an array of ints and return a pointer to the first occurrence of val:

```
1    int *search(int *p, int val)
2    {
3        while (*p && *p != val)
4            p += sizeof(int); /* Should be p++ */
5        return p;
6    }
```

However, because line 4 increments the pointer by 4 (the number of bytes in an integer) each time through the loop, the function incorrectly scans every fourth integer in the array.

11.8 Referencing Nonexistent Variables

Naive C programmers who do not understand the stack discipline will sometimes reference local variables that are no longer valid, as in the following example:

```
1    int *stackref ()
2    {
3        int val;
4
5        return &val;
6    }
```

This function returns a pointer (say, p) to a local variable on the stack and then pops its stack frame. Although p still points to a valid memory address, it no longer points to a valid variable. When other functions are called later in the program, the memory will be reused for their stack frames. Later, if the program assigns some value to *p, then it might actually be modifying an entry in another function's stack frame, with potentially disastrous and baffling consequences.

11.9 Referencing Data in Free Heap Blocks

A similar error is to reference data in heap blocks that have already been freed. For example, consider the following example, which allocates an integer array x in line 6, prematurely frees block x in line 10, and then later references it in line 14:

```
1   int *heapref(int n, int m)
2   {
3       int i;
4       int *x, *y;
5
6       x = (int *)Malloc(n * sizeof(int));
7
8       /* ... */   /* Other calls to malloc and free go here */
9
10      free(x);
11
12      y = (int *)Malloc(m * sizeof(int));
13      for (i = 0; i < m; i++)
14          y[i] = x[i]++;  /* Oops! x[i] is a word in a free block */
15
16      return y;
17  }
```

Depending on the pattern of malloc and free calls that occur between lines 6 and 10, when the program references x[i] in line 14, the array x might be part of some other allocated heap block and have been overwritten. As with many memory-related bugs, the error will only become evident later in the program when we notice that the values in y are corrupted.

11.10 Introducing Memory Leaks

Memory leaks are slow, silent killers that occur when programmers inadvertently create garbage in the heap by forgetting to free allocated blocks. For example, the following function allocates a heap block x and then returns without freeing it:

```
1   void leak(int n)
2   {
3       int *x = (int *)Malloc(n * sizeof(int));
4
5       return;  /* x is garbage at this point */
6   }
```

If leak is called frequently, then the heap will gradually fill up with garbage, in the worst case consuming the entire virtual address space. Memory leaks are particularly serious for programs such as daemons and servers, which by definition never terminate.

12 Summary

Virtual memory is an abstraction of main memory. Processors that support virtual memory reference main memory using a form of indirection known as virtual addressing. The processor generates a virtual address, which is translated into a physical address before being sent to the main memory. The translation of addresses from a virtual address space to a physical address space requires close cooperation between hardware and software. Dedicated hardware translates virtual addresses using page tables whose contents are supplied by the operating system.

Virtual memory provides three important capabilities. First, it automatically caches recently used contents of the virtual address space stored on disk in main memory. The block in a virtual memory cache is known as a page. A reference to a page on disk triggers a page fault that transfers control to a fault handler in the operating system. The fault handler copies the page from disk to the main memory cache, writing back the evicted page if necessary. Second, virtual memory simplifies memory management, which in turn simplifies linking, sharing data between processes, the allocation of memory for processes, and program loading. Finally, virtual memory simplifies memory protection by incorporating protection bits into every page table entry.

The process of address translation must be integrated with the operation of any hardware caches in the system. Most page table entries are located in the L1 cache, but the cost of accessing page table entries from L1 is usually eliminated by an on-chip cache of page table entries called a TLB.

Modern systems initialize chunks of virtual memory by associating them with chunks of files on disk, a process known as memory mapping. Memory mapping provides an efficient mechanism for sharing data, creating new processes, and loading programs. Applications can manually create and delete areas of the virtual address space using the `mmap` function. However, most programs rely on a dynamic memory allocator such as `malloc`, which manages memory in an area of the virtual address space called the heap. Dynamic memory allocators are application-level programs with a system-level feel, directly manipulating memory without much help from the type system. Allocators come in two flavors. Explicit allocators require applications to explicitly free their memory blocks. Implicit allocators (garbage collectors) free any unused and unreachable blocks automatically.

Managing and using memory is a difficult and error-prone task for C programmers. Examples of common errors include dereferencing bad pointers, reading uninitialized memory, allowing stack buffer overflows, assuming that pointers and the objects they point to are the same size, referencing a pointer instead of the object it points to, misunderstanding pointer arithmetic, referencing nonexistent variables, and introducing memory leaks.

Bibliographic Notes

Kilburn and his colleagues published the first description of virtual memory [60]. Architecture texts contain additional details about the hardware's role in virtual memory [49]. Operating systems texts contain additional information about the operating system's role [98, 104, 112]. Bovet and Cesati [11] give a detailed description of the Linux virtual memory system. Intel Corporation provides detailed documentation on 32-bit and 64-bit address translation on IA processors [30].

Knuth wrote the classic work on storage allocation in 1968 [61]. Since that time there has been a tremendous amount of work in the area. Wilson, Johnstone, Neely, and Boles have written a beautiful survey and performance evaluation of explicit allocators [117]. The general comments in this book about the throughput and utilization of different allocator strategies are paraphrased from their survey. Jones and Lins provide a comprehensive survey of garbage collection [54]. Kernighan and Ritchie [58] show the complete code for a simple allocator based on an explicit free list with a block size and successor pointer in each free block. The code is interesting in that it uses unions to eliminate a lot of the complicated pointer arithmetic, but at the expense of a linear-time (rather than constant-time) free operation.

Homework Problems

11 ◆

In the following series of problems, you are to show how the example memory system in Section 6.4 translates a virtual address into a physical address and accesses the cache. For the given virtual address, indicate the TLB entry accessed, the physical address, and the cache byte value returned. Indicate whether the TLB misses, whether a page fault occurs, and whether a cache miss occurs. If there is a cache miss, enter "–" for "Cache Byte returned." If there is a page fault, enter "–" for "PPN" and leave parts C and D blank.

Virtual address: 0x027c

A. Virtual address format

13	12	11	10	9	8	7	6	5	4	3	2	1	0

B. Address translation

Parameter	Value
VPN	
TLB index	
TLB tag	
TLB hit? (Y/N)	
Page fault? (Y/N)	
PPN	

C. Physical address format

11	10	9	8	7	6	5	4	3	2	1	0

D. Physical memory reference

Parameter	Value
Byte offset	
Cache index	
Cache tag	
Cache hit? (Y/N)	
Cache byte returned	

12 ◆

Repeat Problem 11 for the following address:

Virtual address: 0x03a9

A. Virtual address format

13	12	11	10	9	8	7	6	5	4	3	2	1	0

B. Address translation

Parameter	Value
VPN	
TLB index	
TLB tag	
TLB hit? (Y/N)	
Page fault? (Y/N)	
PPN	

C. Physical address format

11	10	9	8	7	6	5	4	3	2	1	0

D. Physical memory reference

Parameter	Value
Byte offset	
Cache index	
Cache tag	
Cache hit? (Y/N)	
Cache byte returned	

13 ◆

Repeat Problem 11 for the following address:

Virtual address: 0x0040

A. Virtual address format

13	12	11	10	9	8	7	6	5	4	3	2	1	0

B. Address translation

Parameter	Value
VPN	
TLB index	
TLB tag	
TLB hit? (Y/N)	
Page fault? (Y/N)	
PPN	

C. Physical address format

11	10	9	8	7	6	5	4	3	2	1	0

D. Physical memory reference

Parameter	Value
Byte offset	
Cache index	
Cache tag	
Cache hit? (Y/N)	
Cache byte returned	

14 ◆◆

Given an input file `hello.txt` that consists of the string "Hello, world!\n", write a C program that uses `mmap` to change the contents of `hello.txt` to "Jello, world!\n".

15 ◆

Determine the block sizes and header values that would result from the following sequence of `malloc` requests. Assumptions: (1) The allocator maintains double-word alignment, and uses an implicit free list with the block format from Figure 35. (2) Block sizes are rounded up to the nearest multiple of 8 bytes.

Request	Block size (decimal bytes)	Block header (hex)
malloc(3)		
malloc(11)		
malloc(20)		
malloc(21)		

16 ◆

Determine the minimum block size for each of the following combinations of alignment requirements and block formats. Assumptions: Explicit free list, 4-byte pred and succ pointers in each free block, zero-sized payloads are not allowed, and headers and footers are stored in 4-byte words.

Alignment	Allocated block	Free block	Minimum block size (bytes)
Single word	Header and footer	Header and footer	
Single word	Header, but no footer	Header and footer	
Double word	Header and footer	Header and footer	
Double word	Header, but no footer	Header and footer	

17 ◆◆◆

Develop a version of the allocator in Section 9.12 that performs a next-fit search instead of a first-fit search.

18 ◆◆◆

The allocator in Section 9.12 requires both a header and a footer for each block in order to perform constant-time coalescing. Modify the allocator so that free blocks require a header and footer, but allocated blocks require only a header.

19 ◆

You are given three groups of statements relating to memory management and garbage collection below. In each group, only one statement is true. Your task is to indicate which statement is true.

1. (a) In a buddy system, up to 50% of the space can be wasted due to internal fragmentation.
   (b) The first-fit memory allocation algorithm is slower than the best-fit algorithm (on average).
   (c) Deallocation using boundary tags is fast only when the list of free blocks is ordered according to increasing memory addresses.
   (d) The buddy system suffers from internal fragmentation, but not from external fragmentation.

2. (a) Using the first-fit algorithm on a free list that is ordered according to decreasing block sizes results in low performance for allocations, but avoids external fragmentation.
   (b) For the best-fit method, the list of free blocks should be ordered according to increasing memory addresses.
   (c) The best-fit method chooses the largest free block into which the requested segment fits.

(d) Using the first-fit algorithm on a free list that is ordered according to increasing block sizes is equivalent to using the best-fit algorithm.

3. Mark & sweep garbage collectors are called conservative if:
   (a) They coalesce freed memory only when a memory request cannot be satisfied.
   (b) They treat everything that looks like a pointer as a pointer.
   (c) They perform garbage collection only when they run out of memory.
   (d) They do not free memory blocks forming a cyclic list.

20 ◆◆◆◆

Write your own version of malloc and free, and compare its running time and space utilization to the version of malloc provided in the standard C library.

Solutions to Practice Problems

Solution to Problem 1

This problem gives you some appreciation for the sizes of different address spaces. At one point in time, a 32-bit address space seemed impossibly large. But now there are database and scientific applications that need more, and you can expect this trend to continue. At some point in your lifetime, expect to find yourself complaining about the cramped 64-bit address space on your personal computer!

No. virtual address bits (n)	No. virtual addresses (N)	Largest possible virtual address
8	$2^8 = 256$	$2^8 - 1 = 255$
16	$2^{16} = 64K$	$2^{16} - 1 = 64K - 1$
32	$2^{32} = 4G$	$2^{32} - 1 = 4G - 1$
48	$2^{48} = 256T$	$2^{48} = 256T - 1$
64	$2^{64} = 16,384P$	$2^{64} - 1 = 16,384P - 1$

Solution to Problem 2

Since each virtual page is $P = 2^p$ bytes, there are a total of $2^n/2^p = 2^{n-p}$ possible pages in the system, each of which needs a page table entry (PTE).

n	$P = 2^p$	# PTEs
16	4K	16
16	8K	8
32	4K	1M
32	8K	512K

Solution to Problem 3

You need to understand this kind of problem well in order to fully grasp address translation. Here is how to solve the first subproblem: We are given $n = 32$ virtual address bits and $m = 24$ physical address bits. A page size of $P = 1$ KB means we need $\log_2(1\ K) = 10$ bits for both the VPO and PPO. (Recall that the VPO and PPO are identical.) The remaining address bits are the VPN and PPN, respectively.

P	# VPN bits	# VPO bits	# PPN bits	# PPO bits
1 KB	22	10	14	10
2 KB	21	11	13	11
4 KB	20	12	12	12
8 KB	19	13	11	13

Solution to Problem 4

Doing a few of these manual simulations is a great way to firm up your understanding of address translation. You might find it helpful to write out all the bits in the addresses, and then draw boxes around the different bit fields, such as VPN, TLBI, etc. In this particular problem, there are no misses of any kind: the TLB has a copy of the PTE and the cache has a copy of the requested data words. See Problems 11, 12, and 13 for some different combinations of hits and misses.

A. 00 0011 1101 0111

B.

Parameter	Value
VPN	0xf
TLB index	0x3
TLB tag	0x3
TLB hit? (Y/N)	Y
Page fault? (Y/N)	N
PPN	0xd

C. 0011 0101 0111

D.

Parameter	Value
Byte offset	0x3
Cache index	0x5
Cache tag	0xd
Cache hit? (Y/N)	Y
Cache byte returned	0x1d

Solution to Problem 5

Solving this problem will give you a good feel for the idea of memory mapping. Try it yourself. We haven't discussed the open, fstat, or write functions, so you'll need to read their man pages to see how they work.

——————————————————————— code/vm/mmapcopy.c

```
1    #include "csapp.h"
2
3    /*
4     * mmapcopy - uses mmap to copy file fd to stdout
5     */
6    void mmapcopy(int fd, int size)
7    {
8        char *bufp; /* Pointer to memory mapped VM area */
```

```
9
10         bufp = Mmap(NULL, size, PROT_READ, MAP_PRIVATE, fd, 0);
11         Write(1, bufp, size);
12         return;
13     }
14
15     /* mmapcopy driver */
16     int main(int argc, char **argv)
17     {
18         struct stat stat;
19         int fd;
20
21         /* Check for required command line argument */
22         if (argc != 2) {
23             printf("usage: %s <filename>\n", argv[0]);
24             exit(0);
25         }
26
27         /* Copy the input argument to stdout */
28         fd = Open(argv[1], O_RDONLY, 0);
29         fstat(fd, &stat);
30         mmapcopy(fd, stat.st_size);
31         exit(0);
32     }
```

─── *code/vm/mmapcopy.c*

Solution to Problem 6

This problem touches on some core ideas such as alignment requirements, minimum block sizes, and header encodings. The general approach for determining the block size is to round the sum of the requested payload and the header size to the nearest multiple of the alignment requirement (in this case 8 bytes). For example, the block size for the `malloc(1)` request is $4 + 1 = 5$ rounded up to 8. The block size for the `malloc(13)` request is $13 + 4 = 17$ rounded up to 24.

Request	Block size (decimal bytes)	Block header (hex)
malloc(1)	8	0x9
malloc(5)	16	0x11
malloc(12)	16	0x11
malloc(13)	24	0x19

Solution to Problem 7

The minimum block size can have a significant effect on internal fragmentation. Thus, it is good to understand the minimum block sizes associated with different allocator designs and alignment requirements. The tricky part is to realize that the same block can be allocated or free at different points in time. Thus, the minimum block size is the maximum of the minimum allocated block size and the minimum free block size. For example, in the last subproblem, the minimum allocated block size is a 4-byte header and a 1-byte payload rounded up to eight bytes. The minimum free block size is a 4-byte header and 4-byte footer, which is already a multiple of 8 and doesn't need to be rounded. So the minimum block size for this allocator is 8 bytes.

Alignment	Allocated block	Free block	Minimum block size (bytes)
Single word	Header and footer	Header and footer	12
Single word	Header, but no footer	Header and footer	8
Double word	Header and footer	Header and footer	16
Double word	Header, but no footer	Header and footer	8

Solution to Problem 8

There is nothing very tricky here. But the solution requires you to understand how the rest of our simple implicit-list allocator works and how to manipulate and traverse blocks.

code/vm/malloc/mm.c

```
static void *find_fit(size_t asize)
{
    /* First fit search */
    void *bp;

    for (bp = heap_listp; GET_SIZE(HDRP(bp)) > 0; bp = NEXT_BLKP(bp)) {
        if (!GET_ALLOC(HDRP(bp)) && (asize <= GET_SIZE(HDRP(bp)))) {
            return bp;
        }
    }
    return NULL; /* No fit */
```

code/vm/malloc/mm.c

Solution to Problem 9

This is another warm-up exercise to help you become familiar with allocators. Notice that for this allocator the minimum block size is 16 bytes. If the remainder of the block after splitting would be greater than or equal to the minimum block size, then we go ahead and split the block (lines 6 to 10). The only tricky part here is to realize that you need to place the new allocated block (lines 6 and 7) before moving to the next block (line 8).

code/vm/malloc/mm.c

```
static void place(void *bp, size_t asize)
{
    size_t csize = GET_SIZE(HDRP(bp));

    if ((csize - asize) >= (2*DSIZE)) {
        PUT(HDRP(bp), PACK(asize, 1));
```

```
 7              PUT(FTRP(bp), PACK(asize, 1));
 8              bp = NEXT_BLKP(bp);
 9              PUT(HDRP(bp), PACK(csize-asize, 0));
10              PUT(FTRP(bp), PACK(csize-asize, 0));
11          }
12      else {
13              PUT(HDRP(bp), PACK(csize, 1));
14              PUT(FTRP(bp), PACK(csize, 1));
15          }
16  }
```

—— *code/vm/malloc/mm.c*

Solution to Problem 10

Here is one pattern that will cause external fragmentation: The application makes numerous allocation and free requests to the first size class, followed by numerous allocation and free requests to the second size class, followed by numerous allocation and free requests to the third size class, and so on. For each size class, the allocator creates a lot of memory that is never reclaimed because the allocator doesn't coalesce, and because the application never requests blocks from that size class again.

System-Level I/O

From Chapter 10 of *Computer Systems: A Programmer's Perspective*, Second Edition. Randal E. Bryant and David R. O'Hallaron. Copyright © 2011 by Pearson Education, Inc. Published by Prentice Hall. All rights reserved.

Input/output (I/O) is the process of copying data between main memory and external devices such as disk drives, terminals, and networks. An input operation copies data from an I/O device to main memory, and an output operation copies data from memory to a device.

All language run-time systems provide higher-level facilities for performing I/O. For example, ANSI C provides the *standard I/O* library, with functions such as `printf` and `scanf` that perform buffered I/O. The C++ language provides similar functionality with its overloaded << ("put to") and >> ("get from") operators. On Unix systems, these higher-level I/O functions are implemented using system-level *Unix I/O* functions provided by the kernel. Most of the time, the higher-level I/O functions work quite well and there is no need to use Unix I/O directly. So why bother learning about Unix I/O?

- *Understanding Unix I/O will help you understand other systems concepts.* I/O is integral to the operation of a system, and because of this we often encounter circular dependences between I/O and other systems ideas. For example, I/O plays a key role in process creation and execution. Conversely, process creation plays a key role in how files are shared by different processes. Thus, to really understand I/O you need to understand processes, and vice versa. We have already touched on aspects of I/O in our discussions of the memory hierarchy, linking and loading, processes, and virtual memory. Now that you have a better understanding of these ideas, we can close the circle and delve into I/O in more detail.

- *Sometimes you have no choice but to use Unix I/O.* There are some important cases where using higher-level I/O functions is either impossible or inappropriate. For example, the standard I/O library provides no way to access file metadata such as file size or file creation time. Further, there are problems with the standard I/O library that make it risky to use for network programming.

This chapter introduces you to the general concepts of Unix I/O and standard I/O, and shows you how to use them reliably from your C programs. Besides serving as a general introduction, this chapter lays a firm foundation for our subsequent study of network programming and concurrency.

1 Unix I/O

A Unix *file* is a sequence of m bytes:

$$B_0, B_1, \ldots, B_k, \ldots, B_{m-1}.$$

All I/O devices, such as networks, disks, and terminals, are modeled as files, and all input and output is performed by reading and writing the appropriate files. This elegant mapping of devices to files allows the Unix kernel to export a simple, low-level application interface, known as *Unix I/O*, that enables all input and output to be performed in a uniform and consistent way:

- *Opening files.* An application announces its intention to access an I/O device by asking the kernel to *open* the corresponding file. The kernel returns a small nonnegative integer, called a *descriptor*, that identifies the file in all subsequent operations on the file. The kernel keeps track of all information about the open file. The application only keeps track of the descriptor.

Each process created by a Unix shell begins life with three open files: *standard input* (descriptor 0), *standard output* (descriptor 1), and *standard error* (descriptor 2). The header file <unistd.h> defines constants STDIN_FILENO, STDOUT_FILENO, and STDERR_FILENO, which can be used instead of the explicit descriptor values.

- *Changing the current file position.* The kernel maintains a *file position k*, initially 0, for each open file. The file position is a byte offset from the beginning of a file. An application can set the current file position k explicitly by performing a *seek* operation.

- *Reading and writing files.* A *read* operation copies $n > 0$ bytes from a file to memory, starting at the current file position k, and then incrementing k by n. Given a file with a size of m bytes, performing a read operation when $k \geq m$ triggers a condition known as *end-of-file* (EOF), which can be detected by the application. There is no explicit "EOF character" at the end of a file.

  Similarly, a *write* operation copies $n > 0$ bytes from memory to a file, starting at the current file position k, and then updating k.

- *Closing files.* When an application has finished accessing a file, it informs the kernel by asking it to *close* the file. The kernel responds by freeing the data structures it created when the file was opened and restoring the descriptor to a pool of available descriptors. When a process terminates for any reason, the kernel closes all open files and frees their memory resources.

2 Opening and Closing Files

A process opens an existing file or creates a new file by calling the open function:

```
#include <sys/types.h>
#include <sys/stat.h>
#include <fcntl.h>

int open(char *filename, int flags, mode_t mode);
                                Returns: new file descriptor if OK, −1 on error
```

The open function converts a filename to a file descriptor and returns the descriptor number. The descriptor returned is always the smallest descriptor that is not currently open in the process. The flags argument indicates how the process intends to access the file:

- O_RDONLY: Reading only
- O_WRONLY: Writing only
- O_RDWR: Reading and writing

For example, here is how to open an existing file for reading:

Mask	Description
S_IRUSR	User (owner) can read this file
S_IWUSR	User (owner) can write this file
S_IXUSR	User (owner) can execute this file
S_IRGRP	Members of the owner's group can read this file
S_IWGRP	Members of the owner's group can write this file
S_IXGRP	Members of the owner's group can execute this file
S_IROTH	Others (anyone) can read this file
S_IWOTH	Others (anyone) can write this file
S_IXOTH	Others (anyone) can execute this file

Figure 1   **Access permission bits.** Defined in `sys/stat.h`.

```
fd = Open("foo.txt", O_RDONLY, 0);
```

The `flags` argument can also be or'd with one or more bit masks that provide additional instructions for writing:

- O_CREAT: If the file doesn't exist, then create a *truncated* (empty) version of it.
- O_TRUNC: If the file already exists, then truncate it.
- O_APPEND: Before each write operation, set the file position to the end of the file.

For example, here is how you might open an existing file with the intent of appending some data:

```
fd = Open("foo.txt", O_WRONLY|O_APPEND, 0);
```

The `mode` argument specifies the access permission bits of new files. The symbolic names for these bits are shown in Figure 1. As part of its context, each process has a umask that is set by calling the umask function. When a process creates a new file by calling the open function with some mode argument, then the access permission bits of the file are set to mode & ~umask. For example, suppose we are given the following default values for mode and umask:

```
#define DEF_MODE   S_IRUSR|S_IWUSR|S_IRGRP|S_IWGRP|S_IROTH|S_IWOTH
#define DEF_UMASK  S_IWGRP|S_IWOTH
```

Then the following code fragment creates a new file in which the owner of the file has read and write permissions, and all other users have read permissions:

```
umask(DEF_UMASK);
fd = Open("foo.txt", O_CREAT|O_TRUNC|O_WRONLY, DEF_MODE);
```

Finally, a process closes an open file by calling the close function.

```
#include <unistd.h>

int close(int fd);
```
                                              Returns: zero if OK, −1 on error

Closing a descriptor that is already closed is an error.

Practice Problem 1

What is the output of the following program?

```
1    #include "csapp.h"
2
3    int main()
4    {
5        int fd1, fd2;
6
7        fd1 = Open("foo.txt", O_RDONLY, 0);
8        Close(fd1);
9        fd2 = Open("baz.txt", O_RDONLY, 0);
10       printf("fd2 = %d\n", fd2);
11       exit(0);
12   }
```

3 Reading and Writing Files

Applications perform input and output by calling the read and write functions, respectively.

```
#include <unistd.h>

ssize_t read(int fd, void *buf, size_t n);
```
                  Returns: number of bytes read if OK, 0 on EOF, −1 on error

```
ssize_t write(int fd, const void *buf, size_t n);
```
                        Returns: number of bytes written if OK, −1 on error

The read function copies at most n bytes from the current file position of descriptor fd to memory location buf. A return value of −1 indicates an error, and a return value of 0 indicates EOF. Otherwise, the return value indicates the number of bytes that were actually transferred.

code/io/cpstdin.c

```
1   #include "csapp.h"
2
3   int main(void)
4   {
5       char c;
6
7       while(Read(STDIN_FILENO, &c, 1) != 0)
8           Write(STDOUT_FILENO, &c, 1);
9       exit(0);
10  }
```

code/io/cpstdin.c

Figure 2   **Copies standard input to standard output one byte at a time.**

The `write` function copies at most `n` bytes from memory location `buf` to the current file position of descriptor `fd`. Figure 2 shows a program that uses `read` and `write` calls to copy the standard input to the standard output, 1 byte at a time.

Applications can explicitly modify the current file position by calling the `lseek` function, which is beyond our scope.

Aside   What's the difference between `ssize_t` and `size_t`?

You might have noticed that the `read` function has a `size_t` input argument and an `ssize_t` return value. So what's the difference between these two types? A `size_t` is defined as an `unsigned int`, and an `ssize_t` (*signed size*) is defined as an `int`. The `read` function returns a signed size rather than an unsigned size because it must return a −1 on error. Interestingly, the possibility of returning a single −1 reduces the maximum size of a `read` by a factor of two, from 4 GB down to 2 GB.

In some situations, `read` and `write` transfer fewer bytes than the application requests. Such *short counts* do *not* indicate an error. They occur for a number of reasons:

- *Encountering EOF on reads.* Suppose that we are ready to read from a file that contains only 20 more bytes from the current file position and that we are reading the file in 50-byte chunks. Then the next `read` will return a short count of 20, and the `read` after that will signal EOF by returning a short count of zero.

- *Reading text lines from a terminal.* If the open file is associated with a terminal (i.e., a keyboard and display), then each `read` function will transfer one text line at a time, returning a short count equal to the size of the text line.

- *Reading and writing network sockets.* If the open file corresponds to a network socket, then internal buffering constraints and long network delays can cause `read` and `write` to return short counts. Short counts can also occur when you call `read` and `write` on a Unix *pipe*, an interprocess communication mechanism that is beyond our scope.

In practice, you will never encounter short counts when you read from disk files except on EOF, and you will never encounter short counts when you write to disk files. However, if you want to build robust (reliable) network applications such as Web servers, then you must deal with short counts by repeatedly calling read and write until all requested bytes have been transferred.

4   Robust Reading and Writing with the Rio Package

In this section, we will develop an I/O package, called the Rio (Robust I/O) package, that handles these short counts for you automatically. The Rio package provides convenient, robust, and efficient I/O in applications such as network programs that are subject to short counts. Rio provides two different kinds of functions:

- *Unbuffered input and output functions*. These functions transfer data directly between memory and a file, with no application-level buffering. They are especially useful for reading and writing binary data to and from networks.

- *Buffered input functions*. These functions allow you to efficiently read text lines and binary data from a file whose contents are cached in an application-level buffer, similar to the one provided for standard I/O functions such as printf. Unlike the buffered I/O routines presented in [109], the buffered Rio input functions are thread-safe and can be interleaved arbitrarily on the same descriptor. For example, you can read some text lines from a descriptor, then some binary data, and then some more text lines.

We are presenting the Rio routines for two reasons. First, we will be using them in the network applications we develop in the next two chapters. Second, by studying the code for these routines, you will gain a deeper understanding of Unix I/O in general.

4.1   Rio Unbuffered Input and Output Functions

Applications can transfer data directly between memory and a file by calling the rio_readn and rio_writen functions.

```
#include "csapp.h"

ssize_t rio_readn(int fd, void *usrbuf, size_t n);
ssize_t rio_writen(int fd, void *usrbuf, size_t n);
          Returns: number of bytes transferred if OK, 0 on EOF (rio_readn only), −1 on error
```

The rio_readn function transfers up to n bytes from the current file position of descriptor fd to memory location usrbuf. Similarly, the rio_writen function transfers n bytes from location usrbuf to descriptor fd. The rio_readn function can only return a short count if it encounters EOF. The rio_writen function never returns a short count. Calls to rio_readn and rio_writen can be interleaved arbitrarily on the same descriptor.

Figure 3 shows the code for `rio_readn` and `rio_writen`. Notice that each function manually restarts the `read` or `write` function if it is interrupted by the return from an application signal handler. To be as portable as possible, we allow for interrupted system calls and restart them when necessary.

4.2 Rio Buffered Input Functions

A *text line* is a sequence of ASCII characters terminated by a newline character. On Unix systems, the newline character ('\n') is the same as the ASCII line feed character (LF) and has a numeric value of 0x0a. Suppose we wanted to write a program that counts the number of text lines in a text file. How might we do this? One approach is to use the `read` function to transfer 1 byte at a time from the file to the user's memory, checking each byte for the newline character. The disadvantage of this approach is that it is inefficient, requiring a trap to the kernel to read each byte in the file.

A better approach is to call a wrapper function (`rio_readlineb`) that copies the text line from an internal *read buffer*, automatically making a `read` call to refill the buffer whenever it becomes empty. For files that contain both text lines and binary data we also provide a buffered version of `rio_readn`, called `rio_readnb`, that transfers raw bytes from the same read buffer as `rio_readlineb`.

```
#include "csapp.h"

void rio_readinitb(rio_t *rp, int fd);

                                                        Returns: nothing

ssize_t rio_readlineb(rio_t *rp, void *usrbuf, size_t maxlen);
ssize_t rio_readnb(rio_t *rp, void *usrbuf, size_t n);
                        Returns: number of bytes read if OK, 0 on EOF, −1 on error
```

The `rio_readinitb` function is called once per open descriptor. It associates the descriptor `fd` with a read buffer of type `rio_t` at address `rp`.

The `rio_readlineb` function reads the next text line from file `rp` (including the terminating newline character), copies it to memory location `usrbuf`, and terminates the text line with the null (zero) character. The `rio_readlineb` function reads at most `maxlen-1` bytes, leaving room for the terminating null character. Text lines that exceed `maxlen-1` bytes are truncated and terminated with a null character.

The `rio_readnb` function reads up to n bytes from file `rp` to memory location `usrbuf`. Calls to `rio_readlineb` and `rio_readnb` can be interleaved arbitrarily on the same descriptor. However, calls to these buffered functions should not be interleaved with calls to the unbuffered `rio_readn` function.

You will encounter numerous examples of the RIO functions in the remainder of this text. Figure 4 shows how to use the RIO functions to copy a text file from standard input to standard output, one line at a time.

code/src/csapp.c

```
1    ssize_t rio_readn(int fd, void *usrbuf, size_t n)
2    {
3        size_t nleft = n;
4        ssize_t nread;
5        char *bufp = usrbuf;
6
7        while (nleft > 0) {
8            if ((nread = read(fd, bufp, nleft)) < 0) {
9                if (errno == EINTR) /* Interrupted by sig handler return */
10                   nread = 0;      /* and call read() again */
11               else
12                   return -1;      /* errno set by read() */
13           }
14           else if (nread == 0)
15               break;              /* EOF */
16           nleft -= nread;
17           bufp += nread;
18       }
19       return (n - nleft);         /* Return >= 0 */
20   }
```

code/src/csapp.c

code/src/csapp.c

```
1    ssize_t rio_writen(int fd, void *usrbuf, size_t n)
2    {
3        size_t nleft = n;
4        ssize_t nwritten;
5        char *bufp = usrbuf;
6
7        while (nleft > 0) {
8            if ((nwritten = write(fd, bufp, nleft)) <= 0) {
9                if (errno == EINTR)  /* Interrupted by sig handler return */
10                   nwritten = 0;    /* and call write() again */
11               else
12                   return -1;       /* errno set by write() */
13           }
14           nleft -= nwritten;
15           bufp += nwritten;
16       }
17       return n;
18   }
```

code/src/csapp.c

Figure 3 The `rio_readn` **and** `rio_writen` **functions.**

————————————————————————— code/io/cpfile.c

```
1   #include "csapp.h"
2
3   int main(int argc, char **argv)
4   {
5       int n;
6       rio_t rio;
7       char buf[MAXLINE];
8
9       Rio_readinitb(&rio, STDIN_FILENO);
10      while((n = Rio_readlineb(&rio, buf, MAXLINE)) != 0)
11          Rio_writen(STDOUT_FILENO, buf, n);
12  }
```

————————————————————————— code/io/cpfile.c

Figure 4 Copying a text file from standard input to standard output.

Figure 5 shows the format of a read buffer, along with the code for the `rio_readinitb` function that initializes it. The `rio_readinitb` function sets up an empty read buffer and associates an open file descriptor with that buffer.

The heart of the Rio read routines is the `rio_read` function shown in Figure 6. The `rio_read` function is a buffered version of the Unix `read` function. When `rio_read` is called with a request to read n bytes, there are `rp->rio_cnt`

————————————————————————— code/include/csapp.h

```
1   #define RIO_BUFSIZE 8192
2   typedef struct {
3       int rio_fd;                /* Descriptor for this internal buf */
4       int rio_cnt;               /* Unread bytes in internal buf */
5       char *rio_bufptr;          /* Next unread byte in internal buf */
6       char rio_buf[RIO_BUFSIZE]; /* Internal buffer */
7   } rio_t;
```

————————————————————————— code/include/csapp.h

————————————————————————— code/src/csapp.c

```
1   void rio_readinitb(rio_t *rp, int fd)
2   {
3       rp->rio_fd = fd;
4       rp->rio_cnt = 0;
5       rp->rio_bufptr = rp->rio_buf;
6   }
```

————————————————————————— code/src/csapp.c

Figure 5 A read buffer of type `rio_t` and the `rio_readinitb` function that initializes it.

code/src/csapp.c

```
1   static ssize_t rio_read(rio_t *rp, char *usrbuf, size_t n)
2   {
3       int cnt;
4
5       while (rp->rio_cnt <= 0) {  /* Refill if buf is empty */
6           rp->rio_cnt = read(rp->rio_fd, rp->rio_buf,
7                               sizeof(rp->rio_buf));
8           if (rp->rio_cnt < 0) {
9               if (errno != EINTR) /* Interrupted by sig handler return */
10                  return -1;
11          }
12          else if (rp->rio_cnt == 0)   /* EOF */
13              return 0;
14          else
15              rp->rio_bufptr = rp->rio_buf; /* Reset buffer ptr */
16      }
17
18      /* Copy min(n, rp->rio_cnt) bytes from internal buf to user buf */
19      cnt = n;
20      if (rp->rio_cnt < n)
21          cnt = rp->rio_cnt;
22      memcpy(usrbuf, rp->rio_bufptr, cnt);
23      rp->rio_bufptr += cnt;
24      rp->rio_cnt -= cnt;
25      return cnt;
26  }
```

code/src/csapp.c

Figure 6   **The internal** `rio_read` **function.**

unread bytes in the read buffer. If the buffer is empty, then it is replenished with a call to `read`. Receiving a short count from this invocation of `read` is not an error, and simply has the effect of partially filling the read buffer. Once the buffer is nonempty, `rio_read` copies the minimum of n and rp->rio_cnt bytes from the read buffer to the user buffer and returns the number of bytes copied.

To an application program, the `rio_read` function has the same semantics as the Unix read function. On error, it returns −1 and sets errno appropriately. On EOF, it returns 0. It returns a short count if the number of requested bytes exceeds the number of unread bytes in the read buffer. The similarity of the two functions makes it easy to build different kinds of buffered read functions by substituting `rio_read` for read. For example, the `rio_readnb` function in Figure 7 has the same structure as `rio_readn`, with `rio_read` substituted for read. Similarly, the `rio_readlineb` routine in Figure 7 calls `rio_read` at most maxlen−1 times. Each call returns 1 byte from the read buffer, which is then checked for being the terminating newline.

```
1    ssize_t rio_readlineb(rio_t *rp, void *usrbuf, size_t maxlen)
2    {
3        int n, rc;
4        char c, *bufp = usrbuf;
5
6        for (n = 1; n < maxlen; n++) {
7            if ((rc = rio_read(rp, &c, 1)) == 1) {
8                *bufp++ = c;
9                if (c == '\n')
10                   break;
11           } else if (rc == 0) {
12               if (n == 1)
13                   return 0; /* EOF, no data read */
14               else
15                   break;     /* EOF, some data was read */
16           } else
17               return -1;    /* Error */
18       }
19       *bufp = 0;
20       return n;
21   }
```

```
1    ssize_t rio_readnb(rio_t *rp, void *usrbuf, size_t n)
2    {
3        size_t nleft = n;
4        ssize_t nread;
5        char *bufp = usrbuf;
6
7        while (nleft > 0) {
8            if ((nread = rio_read(rp, bufp, nleft)) < 0) {
9                if (errno == EINTR) /* Interrupted by sig handler return */
10                   nread = 0;       /* Call read() again */
11               else
12                   return -1;       /* errno set by read() */
13           }
14           else if (nread == 0)
15               break;               /* EOF */
16           nleft -= nread;
17           bufp += nread;
18       }
19       return (n - nleft);          /* Return >= 0 */
20   }
```

Figure 7 **The** rio_readlineb **and** rio_readnb **functions.**

Aside   Origins of the **Rio** package

The Rio functions are inspired by the `readline`, `readn`, and `writen` functions described by W. Richard Stevens in his classic network programming text [109]. The `rio_readn` and `rio_writen` functions are identical to the Stevens `readn` and `writen` functions. However, the Stevens `readline` function has some limitations that are corrected in Rio. First, because `readline` is buffered and `readn` is not, these two functions cannot be used together on the same descriptor. Second, because it uses a `static` buffer, the Stevens `readline` function is not thread-safe, which required Stevens to introduce a different thread-safe version called `readline_r`. We have corrected both of these flaws with the `rio_readlineb` and `rio_readnb` functions, which are mutually compatible and thread-safe.

5 Reading File Metadata

An application can retrieve information about a file (sometimes called the file's *metadata*) by calling the `stat` and `fstat` functions.

```
#include <unistd.h>
#include <sys/stat.h>

int stat(const char *filename, struct stat *buf);
int fstat(int fd, struct stat *buf);
                                        Returns: 0 if OK, −1 on error
```

The `stat` function takes as input a file name and fills in the members of a `stat` structure shown in Figure 8. The `fstat` function is similar, but takes a file descriptor instead of a file name.

---
statbuf.h (included by sys/stat.h)

```
/* Metadata returned by the stat and fstat functions */
struct stat {
    dev_t         st_dev;      /* Device */
    ino_t         st_ino;      /* inode */
    mode_t        st_mode;     /* Protection and file type */
    nlink_t       st_nlink;    /* Number of hard links */
    uid_t         st_uid;      /* User ID of owner */
    gid_t         st_gid;      /* Group ID of owner */
    dev_t         st_rdev;     /* Device type (if inode device) */
    off_t         st_size;     /* Total size, in bytes */
    unsigned long st_blksize;  /* Blocksize for filesystem I/O */
    unsigned long st_blocks;   /* Number of blocks allocated */
    time_t        st_atime;    /* Time of last access */
    time_t        st_mtime;    /* Time of last modification */
    time_t        st_ctime;    /* Time of last change */
};
```

---
statbuf.h (included by sys/stat.h)

Figure 8   **The `stat` structure.**

Macro	Description
S_ISREG()	Is this a regular file?
S_ISDIR()	Is this a directory file?
S_ISSOCK()	Is this a network socket?

Figure 9 **Macros for determining file type from the** st_mode **bits.** Defined in sys/stat.h.

The st_size member contains the file size in bytes. The st_mode member encodes both the file permission bits (Figure 1) and the *file type*. Unix recognizes a number of different file types. A *regular file* contains some sort of binary or text data. To the kernel there is no difference between text files and binary files. A *directory file* contains information about other files. A *socket* is a file that is used to communicate with another process across a network.

Unix provides macro predicates for determining the file type from the st_mode member. Figure 9 lists a subset of these macros.

—————————————————————— code/io/statcheck.c

```
1    #include "csapp.h"
2
3    int main (int argc, char **argv)
4    {
5        struct stat stat;
6        char *type, *readok;
7
8        Stat(argv[1], &stat);
9        if (S_ISREG(stat.st_mode))      /* Determine file type */
10           type = "regular";
11       else if (S_ISDIR(stat.st_mode))
12           type = "directory";
13       else
14           type = "other";
15       if ((stat.st_mode & S_IRUSR))  /* Check read access */
16           readok = "yes";
17       else
18           readok = "no";
19
20       printf("type: %s, read: %s\n", type, readok);
21       exit(0);
22   }
```

—————————————————————— code/io/statcheck.c

Figure 10 **Querying and manipulating a file's** st_mode **bits.**

Figure 10 shows how we might use these macros and the stat function to read and interpret a file's st_mode bits.

6 Sharing Files

Unix files can be shared in a number of different ways. Unless you have a clear picture of how the kernel represents open files, the idea of file sharing can be quite confusing. The kernel represents open files using three related data structures:

- *Descriptor table.* Each process has its own separate *descriptor table* whose entries are indexed by the process's open file descriptors. Each open descriptor entry points to an entry in the *file table*.

- *File table.* The set of open files is represented by a file table that is shared by all processes. Each file table entry consists of (for our purposes) the current file position, a *reference count* of the number of descriptor entries that currently point to it, and a pointer to an entry in the *v-node table*. Closing a descriptor decrements the reference count in the associated file table entry. The kernel will not delete the file table entry until its reference count is zero.

- *v-node table.* Like the file table, the v-node table is shared by all processes. Each entry contains most of the information in the stat structure, including the st_mode and st_size members.

Figure 11 shows an example where descriptors 1 and 4 reference two different files through distinct open file table entries. This is the typical situation, where files are not shared, and where each descriptor corresponds to a distinct file.

Multiple descriptors can also reference the same file through different file table entries, as shown in Figure 12. This might happen, for example, if you were to call the open function twice with the same filename. The key idea is that each descriptor has its own distinct file position, so different reads on different descriptors can fetch data from different locations in the file.

Figure 11

Typical kernel data structures for open files. In this example, two descriptors reference distinct files. There is no sharing.

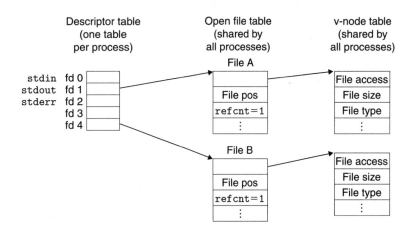

863

Figure 12

File sharing. This example shows two descriptors sharing the same disk file through two open file table entries.

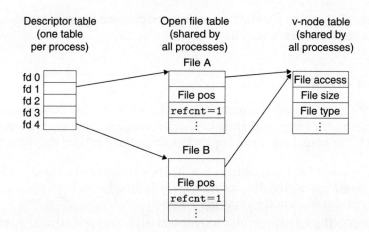

Figure 13

How a child process inherits the parent's open files. The initial situation is in Figure 11.

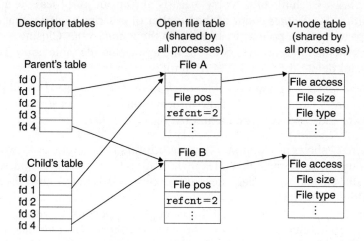

We can also understand how parent and child processes share files. Suppose that before a call to fork, the parent process has the open files shown in Figure 11. Then Figure 13 shows the situation after the call to fork. The child gets its own duplicate copy of the parent's descriptor table. Parent and child share the same set of open file tables, and thus share the same file position. An important consequence is that the parent and child must both close their descriptors before the kernel will delete the corresponding file table entry.

Suppose the disk file foobar.txt consists of the six ASCII characters "foobar". Then what is the output of the following program?

```
1    #include "csapp.h"
2
3    int main()
```

```
4    {
5        int fd1, fd2;
6        char c;
7
8        fd1 = Open("foobar.txt", O_RDONLY, 0);
9        fd2 = Open("foobar.txt", O_RDONLY, 0);
10       Read(fd1, &c, 1);
11       Read(fd2, &c, 1);
12       printf("c = %c\n", c);
13       exit(0);
14   }
```

Practice Problem 3

As before, suppose the disk file foobar.txt consists of the six ASCII characters "foobar". Then what is the output of the following program?

```
1    #include "csapp.h"
2
3    int main()
4    {
5        int fd;
6        char c;
7
8        fd = Open("foobar.txt", O_RDONLY, 0);
9        if (Fork() == 0) {
10           Read(fd, &c, 1);
11           exit(0);
12       }
13       Wait(NULL);
14       Read(fd, &c, 1);
15       printf("c = %c\n", c);
16       exit(0);
17   }
```

7   I/O Redirection

Unix shells provide *I/O redirection* operators that allow users to associate standard input and output with disk files. For example, typing

```
unix> ls > foo.txt
```

causes the shell to load and execute the ls program, with standard output redirected to disk file foo.txt. A Web server performs a similar kind of redirection

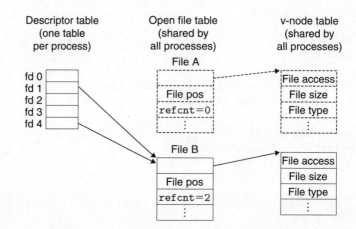

Figure 14

Kernel data structures after redirecting standard output by calling dup2(4,1). The initial situation is shown in Figure 11.

when it runs a CGI program on behalf of the client. So how does I/O redirection work? One way is to use the dup2 function.

```
#include <unistd.h>

int dup2(int oldfd, int newfd);
```
                                    Returns: nonnegative descriptor if OK, −1 on error

The dup2 function copies descriptor table entry oldfd to descriptor table entry newfd, overwriting the previous contents of descriptor table entry newfd. If newfd was already open, then dup2 closes newfd before it copies oldfd.

Suppose that before calling dup2(4,1) we have the situation in Figure 11, where descriptor 1 (standard output) corresponds to file A (say, a terminal), and descriptor 4 corresponds to file B (say, a disk file). The reference counts for A and B are both equal to 1. Figure 14 shows the situation after calling dup2(4,1). Both descriptors now point to file B; file A has been closed and its file table and v-node table entries deleted; and the reference count for file B has been incremented. From this point on, any data written to standard output is redirected to file B.

Aside Right and left hoinkies

To avoid confusion with other bracket-type operators such as ']' and '[', we have always referred to the shell's '>' operator as a "right hoinky", and the '<' operator as a "left hoinky".

Practice Problem 4

How would you use dup2 to redirect standard input to descriptor 5?

Practice Problem 5

Assuming that the disk file `foobar.txt` consists of the six ASCII characters "`foobar`", what is the output of the following program?

```
1    #include "csapp.h"
2
3    int main()
4    {
5        int fd1, fd2;
6        char c;
7
8        fd1 = Open("foobar.txt", O_RDONLY, 0);
9        fd2 = Open("foobar.txt", O_RDONLY, 0);
10       Read(fd2, &c, 1);
11       Dup2(fd2, fd1);
12       Read(fd1, &c, 1);
13       printf("c = %c\n", c);
14       exit(0);
15   }
```

8   Standard I/O

ANSI C defines a set of higher level input and output functions, called the *standard I/O library*, that provides programmers with a higher-level alternative to Unix I/O. The library (`libc`) provides functions for opening and closing files (`fopen` and `fclose`), reading and writing bytes (`fread` and `fwrite`), reading and writing strings (`fgets` and `fputs`), and sophisticated formatted I/O (`scanf` and `printf`).

The standard I/O library models an open file as a *stream*. To the programmer, a stream is a pointer to a structure of type FILE. Every ANSI C program begins with three open streams, `stdin`, `stdout`, and `stderr`, which correspond to standard input, standard output, and standard error, respectively:

```
#include <stdio.h>
extern FILE *stdin;     /* Standard input (descriptor 0) */
extern FILE *stdout;    /* Standard output (descriptor 1) */
extern FILE *stderr;    /* Standard error (descriptor 2) */
```

A stream of type FILE is an abstraction for a file descriptor and a *stream buffer*. The purpose of the stream buffer is the same as the Rio read buffer: to minimize the number of expensive Unix I/O system calls. For example, suppose we have a program that makes repeated calls to the standard I/O `getc` function, where each invocation returns the next character from a file. When `getc` is called the first time, the library fills the stream buffer with a single call to the `read` function, and then returns the first byte in the buffer to the application. As long as there are unread bytes in the buffer, subsequent calls to getc can be served directly from the stream buffer.

9 Putting It Together: Which I/O Functions Should I Use?

Figure 15 summarizes the various I/O packages that we have discussed in this chapter. Unix I/O is implemented in the operating system kernel. It is available to applications through functions such as open, close, lseek, read, write, and stat functions. The higher-level Rio and standard I/O functions are implemented "on top of" (using) the Unix I/O functions. The Rio functions are robust wrappers for read and write that were developed specifically for this textbook. They automatically deal with short counts and provide an efficient buffered approach for reading text lines. The standard I/O functions provide a more complete buffered alternative to the Unix I/O functions, including formatted I/O routines.

So which of these functions should you use in your programs? The standard I/O functions are the method of choice for I/O on disk and terminal devices. Most C programmers use standard I/O exclusively throughout their careers, never bothering with the lower-level Unix I/O functions. Whenever possible, we recommend that you do likewise.

Unfortunately, standard I/O poses some nasty problems when we attempt to use it for input and output on networks. The Unix abstraction for a network is a type of file called a *socket*. Like any Unix file, sockets are referenced by file descriptors, known in this case as *socket descriptors*. Application processes communicate with processes running on other computers by reading and writing socket descriptors.

Standard I/O streams are *full duplex* in the sense that programs can perform input and output on the same stream. However, there are poorly documented restrictions on streams that interact badly with restrictions on sockets:

- *Restriction 1: Input functions following output functions.* An input function cannot follow an output function without an intervening call to fflush, fseek, fsetpos, or rewind. The fflush function empties the buffer associated with

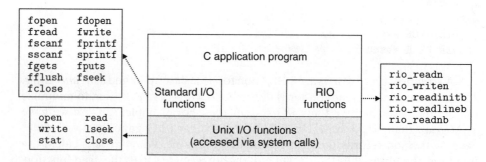

Figure 15 **Relationship between Unix I/O, standard I/O, and Rio.**

a stream. The latter three functions use the Unix I/O `lseek` function to reset the current file position.

- *Restriction 2: Output functions following input functions.* An output function cannot follow an input function without an intervening call to `fseek`, `fsetpos`, or `rewind`, unless the input function encounters an end-of-file.

These restrictions pose a problem for network applications because it is illegal to use the `lseek` function on a socket. The first restriction on stream I/O can be worked around by adopting a discipline of flushing the buffer before every input operation. However, the only way to work around the second restriction is to open two streams on the same open socket descriptor, one for reading and one for writing:

```
FILE *fpin, *fpout;

fpin = fdopen(sockfd, "r");
fpout = fdopen(sockfd, "w");
```

But this approach has problems as well, because it requires the application to call `fclose` on both streams in order to free the memory resources associated with each stream and avoid a memory leak:

```
fclose(fpin);
fclose(fpout);
```

Each of these operations attempts to close the same underlying socket descriptor, so the second `close` operation will fail. This is not a problem for sequential programs, but closing an already closed descriptor in a threaded program is a recipe for disaster.

Thus, we recommend that you not use the standard I/O functions for input and output on network sockets. Use the robust RIO functions instead. If you need formatted output, use the `sprintf` function to format a string in memory, and then send it to the socket using `rio_writen`. If you need formatted input, use `rio_readlineb` to read an entire text line, and then use `sscanf` to extract different fields from the text line.

10 Summary

Unix provides a small number of system-level functions that allow applications to open, close, read, and write files; fetch file metadata; and perform I/O redirection. Unix read and write operations are subject to short counts that applications must anticipate and handle correctly. Instead of calling the Unix I/O functions directly, applications should use the RIO package, which deals with short counts automatically by repeatedly performing read and write operations until all of the requested data have been transferred.

The Unix kernel uses three related data structures to represent open files. Entries in a descriptor table point to entries in the open file table, which point to entries in the v-node table. Each process has its own distinct descriptor table, while all processes share the same open file and v-node tables. Understanding the general organization of these structures clarifies our understanding of both file sharing and I/O redirection.

The standard I/O library is implemented on top of Unix I/O and provides a powerful set of higher-level I/O routines. For most applications, standard I/O is the simpler, preferred alternative to Unix I/O. However, because of some mutually incompatible restrictions on standard I/O and network files, Unix I/O, rather than standard I/O, should be used for network applications.

Bibliographic Notes

Stevens wrote the standard reference text for Unix I/O [110]. Kernighan and Ritchie give a clear and complete discussion of the standard I/O functions [58].

Homework Problems

6 ◆

What is the output of the following program?

```
1    #include "csapp.h"
2
3    int main()
4    {
5        int fd1, fd2;
6
7        fd1 = Open("foo.txt", O_RDONLY, 0);
8        fd2 = Open("bar.txt", O_RDONLY, 0);
9        Close(fd2);
10       fd2 = Open("baz.txt", O_RDONLY, 0);
11       printf("fd2 = %d\n", fd2);
12       exit(0);
13   }
```

7 ◆

Modify the cpfile program in Figure 4 so that it uses the Rio functions to copy standard input to standard output, MAXBUF bytes at a time.

8 ◆◆

Write a version of the statcheck program in Figure 10, called fstatcheck, that takes a descriptor number on the command line rather than a file name.

9 ◆◆

Consider the following invocation of the fstatcheck program from Problem 8:

```
unix> fstatcheck 3 < foo.txt
```

You might expect that this invocation of `fstatcheck` would fetch and display metadata for file `foo.txt`. However, when we run it on our system, it fails with a "bad file descriptor." Given this behavior, fill in the pseudo-code that the shell must be executing between the `fork` and `execve` calls:

```
if (Fork() == 0) { /* Child */
    /* What code is the shell executing right here? */
    Execve("fstatcheck", argv, envp);
}
```

10 ◆◆

Modify the `cpfile` program in Figure 4 so that it takes an optional command line argument `infile`. If `infile` is given, then copy `infile` to standard output; otherwise, copy standard input to standard output as before. The twist is that your solution must use the original copy loop (lines 9–11) for both cases. You are only allowed to insert code, and you are not allowed to change any of the existing code.

Solutions to Practice Problems

Solution to Problem 1

Unix processes begin life with open descriptors assigned to `stdin` (descriptor 0), `stdout` (descriptor 1), and `stderr` (descriptor 2). The open function always returns the lowest unopened descriptor, so the first call to open returns descriptor 3. The call to the `close` function frees up descriptor 3. The final call to open returns descriptor 3, and thus the output of the program is "`fd2 = 3`".

Solution to Problem 2

The descriptors `fd1` and `fd2` each have their own open file table entry, so each descriptor has its own file position for `foobar.txt`. Thus, the read from `fd2` reads the first byte of `foobar.txt`, and the output is

```
c = f
```

and not

```
c = o
```

as you might have thought initially.

Solution to Problem 3

Recall that the child inherits the parent's descriptor table and that all processes shared the same open file table. Thus, the descriptor `fd` in both the parent and child points to the same open file table entry. When the child reads the first byte of the file, the file position increases by one. Thus, the parent reads the second byte, and the output is

```
c = o
```

Solution to Problem 4

To redirect standard input (descriptor 0) to descriptor 5, we would call dup2(5,0), or equivalently, dup2(5,STDIN_FILENO).

Solution to Problem 5

At first glance, you might think the output would be

```
c = f
```

but because we are redirecting fd1 to fd2, the output is really

```
c = o
```

Concurrent Programming

Logical control flows are *concurrent* if they overlap in time. This general phenomenon, known as *concurrency*, shows up at many different levels of a computer system. Hardware exception handlers, processes, and Unix signal handlers are all familiar examples.

Thus far, we have treated concurrency mainly as a mechanism that the operating system kernel uses to run multiple application programs. But concurrency is not just limited to the kernel. It can play an important role in application programs as well. For example, we have seen how Unix signal handlers allow applications to respond to asynchronous events such as the user typing ctrl-c or the program accessing an undefined area of virtual memory. Application-level concurrency is useful in other ways as well:

- *Accessing slow I/O devices.* When an application is waiting for data to arrive from a slow I/O device such as a disk, the kernel keeps the CPU busy by running other processes. Individual applications can exploit concurrency in a similar way by overlapping useful work with I/O requests.

- *Interacting with humans.* People who interact with computers demand the ability to perform multiple tasks at the same time. For example, they might want to resize a window while they are printing a document. Modern windowing systems use concurrency to provide this capability. Each time the user requests some action (say, by clicking the mouse), a separate concurrent logical flow is created to perform the action.

- *Reducing latency by deferring work.* Sometimes, applications can use concurrency to reduce the latency of certain operations by deferring other operations and performing them concurrently. For example, a dynamic storage allocator might reduce the latency of individual free operations by deferring coalescing to a concurrent "coalescing" flow that runs at a lower priority, soaking up spare CPU cycles as they become available.

- *Servicing multiple network clients.* The iterative network servers that are unrealistic because they can only service one client at a time. Thus, a single slow client can deny service to every other client. For a real server that might be expected to service hundreds or thousands of clients per second, it is not acceptable to allow one slow client to deny service to the others. A better approach is to build a *concurrent server* that creates a separate logical flow for each client. This allows the server to service multiple clients concurrently, and precludes slow clients from monopolizing the server.

- *Computing in parallel on multi-core machines.* Many modern systems are equipped with multi-core processors that contain multiple CPUs. Applications that are partitioned into concurrent flows often run faster on multi-core machines than on uniprocessor machines because the flows execute in parallel rather than being interleaved.

Applications that use application-level concurrency are known as *concurrent programs*. Modern operating systems provide three basic approaches for building concurrent programs:

- *Processes.* With this approach, each logical control flow is a process that is scheduled and maintained by the kernel. Since processes have separate virtual address spaces, flows that want to communicate with each other must use some kind of explicit *interprocess communication* (IPC) mechanism.

- *I/O multiplexing.* This is a form of concurrent programming where applications explicitly schedule their own logical flows in the context of a single process. Logical flows are modeled as state machines that the main program explicitly transitions from state to state as a result of data arriving on file descriptors. Since the program is a single process, all flows share the same address space.

- *Threads.* Threads are logical flows that run in the context of a single process and are scheduled by the kernel. You can think of threads as a hybrid of the other two approaches, scheduled by the kernel like process flows, and sharing the same virtual address space like I/O multiplexing flows.

This chapter investigates these three different concurrent programming techniques. To keep our discussion concrete, we will work with the same motivating application throughout—a concurrent version of the iterative echo server.

1 Concurrent Programming with Processes

The simplest way to build a concurrent program is with processes, using familiar functions such as fork, exec, and waitpid. For example, a natural approach for building a concurrent server is to accept client connection requests in the parent, and then create a new child process to service each new client.

To see how this might work, suppose we have two clients and a server that is listening for connection requests on a listening descriptor (say, 3). Now suppose that the server accepts a connection request from client 1 and returns a connected descriptor (say, 4), as shown in Figure 1.

After accepting the connection request, the server forks a child, which gets a complete copy of the server's descriptor table. The child closes its copy of listening descriptor 3, and the parent closes its copy of connected descriptor 4, since they are no longer needed. This gives us the situation in Figure 2, where the child process is busy servicing the client. Since the connected descriptors in the parent and child each point to the same file table entry, it is crucial for the parent to close its copy of the connected descriptor. Otherwise, the file table entry for connected descriptor

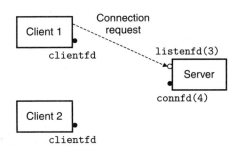

Figure 1

Step 1: Server accepts connection request from client.

Figure 2
Step 2: Server forks a child process to service the client.

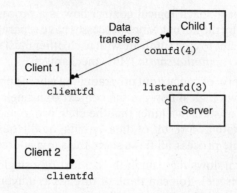

Figure 3
Step 3: Server accepts another connection request.

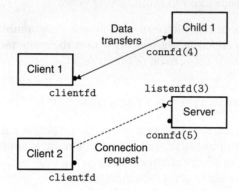

4 will never be released, and the resulting memory leak will eventually consume the available memory and crash the system.

Now suppose that after the parent creates the child for client 1, it accepts a new connection request from client 2 and returns a new connected descriptor (say, 5), as shown in Figure 3. The parent then forks another child, which begins servicing its client using connected descriptor 5, as shown in Figure 4. At this point, the parent is waiting for the next connection request and the two children are servicing their respective clients concurrently.

1.1 A Concurrent Server Based on Processes

Figure 5 shows the code for a concurrent echo server based on processes. The echo function called in line 29 comes once from Figure 21 of the chapter "Network Programming". There are several important points to make about this server:

- First, servers typically run for long periods of time, so we must include a SIGCHLD handler that reaps zombie children (lines 4–9). Since SIGCHLD signals are blocked while the SIGCHLD handler is executing, and since Unix signals are not queued, the SIGCHLD handler must be prepared to reap multiple zombie children.

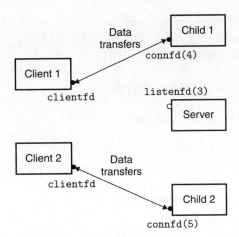

Figure 4

Step 4: Server forks another child to service the new client.

- Second, the parent and the child must close their respective copies of `connfd` (lines 33 and 30, respectively). As we have mentioned, this is especially important for the parent, which must close its copy of the connected descriptor to avoid a memory leak.

- Finally, because of the reference count in the socket's file table entry, the connection to the client will not be terminated until both the parent's and child's copies of `connfd` are closed.

1.2 Pros and Cons of Processes

Processes have a clean model for sharing state information between parents and children: file tables are shared and user address spaces are not. Having separate address spaces for processes is both an advantage and a disadvantage. It is impossible for one process to accidentally overwrite the virtual memory of another process, which eliminates a lot of confusing failures—an obvious advantage.

On the other hand, separate address spaces make it more difficult for processes to share state information. To share information, they must use explicit IPC (interprocess communications) mechanisms. (See Aside.) Another disadvantage of process-based designs is that they tend to be slower because the overhead for process control and IPC is high.

Aside Unix IPC

The term *Unix IPC* is typically reserved for a hodge-podge of techniques that allow processes to communicate with other processes that are running on the same host. Examples include pipes, FIFOs, System V shared memory, and System V semaphores. These mechanisms are beyond our scope. The book by Stevens [108] is a good reference.

---

code/conc/echoserverp.c

```
1    #include "csapp.h"
2    void echo(int connfd);
3
4    void sigchld_handler(int sig)
5    {
6        while (waitpid(-1, 0, WNOHANG) > 0)
7            ;
8        return;
9    }
10
11   int main(int argc, char **argv)
12   {
13       int listenfd, connfd, port;
14       socklen_t clientlen=sizeof(struct sockaddr_in);
15       struct sockaddr_in clientaddr;
16
17       if (argc != 2) {
18           fprintf(stderr, "usage: %s <port>\n", argv[0]);
19           exit(0);
20       }
21       port = atoi(argv[1]);
22
23       Signal(SIGCHLD, sigchld_handler);
24       listenfd = Open_listenfd(port);
25       while (1) {
26           connfd = Accept(listenfd, (SA *) &clientaddr, &clientlen);
27           if (Fork() == 0) {
28               Close(listenfd); /* Child closes its listening socket */
29               echo(connfd);    /* Child services client */
30               Close(connfd);   /* Child closes connection with client */
31               exit(0);         /* Child exits */
32           }
33           Close(connfd); /* Parent closes connected socket (important!) */
34       }
35   }
```

code/conc/echoserverp.c

Figure 5 **Concurrent echo server based on processes.** The parent forks a child to handle each new connection request.

Practice Problem 1

After the parent closes the connected descriptor in line 33 of the concurrent server in Figure 5, the child is still able to communicate with the client using its copy of the descriptor. Why?

Practice Problem 2

If we were to delete line 30 of Figure 5, which closes the connected descriptor, the code would still be correct, in the sense that there would be no memory leak. Why?

2   Concurrent Programming with I/O Multiplexing

Suppose you are asked to write an echo server that can also respond to interactive commands that the user types to standard input. In this case, the server must respond to two independent I/O events: (1) a network client making a connection request, and (2) a user typing a command line at the keyboard. Which event do we wait for first? Neither option is ideal. If we are waiting for a connection request in accept, then we cannot respond to input commands. Similarly, if we are waiting for an input command in read, then we cannot respond to any connection requests.

One solution to this dilemma is a technique called *I/O multiplexing*. The basic idea is to use the select function to ask the kernel to suspend the process, returning control to the application only after one or more I/O events have occurred, as in the following examples:

- Return when any descriptor in the set {0, 4} is ready for reading.
- Return when any descriptor in the set {1, 2, 7} is ready for writing.
- Timeout if 152.13 seconds have elapsed waiting for an I/O event to occur.

Select is a complicated function with many different usage scenarios. We will only discuss the first scenario: waiting for a set of descriptors to be ready for reading. See [109, 110] for a complete discussion.

```
#include <unistd.h>
#include <sys/types.h>

int select(int n, fd_set *fdset, NULL, NULL, NULL);
                        Returns nonzero count of ready descriptors, −1 on error

FD_ZERO(fd_set *fdset);      /* Clear all bits in fdset */
FD_CLR(int fd, fd_set *fdset); /* Clear bit fd in fdset */
FD_SET(int fd, fd_set *fdset);  /* Turn on bit fd in fdset */
FD_ISSET(int fd, fd_set *fdset); /* Is bit fd in fdset on? */
                        Macros for manipulating descriptor sets
```

The `select` function manipulates sets of type `fd_set`, which are known as *descriptor sets*. Logically, we think of a descriptor set as a bit vector of size n:

$$b_{n-1}, \ldots, b_1, b_0$$

Each bit b_k corresponds to descriptor k. Descriptor k is a member of the descriptor set if and only if $b_k = 1$. You are only allowed to do three things with descriptor sets: (1) allocate them, (2) assign one variable of this type to another, and (3) modify and inspect them using the FD_ZERO, FD_SET, FD_CLR, and FD_ISSET macros.

For our purposes, the `select` function takes two inputs: a descriptor set (`fdset`) called the *read set*, and the cardinality (n) of the read set (actually the maximum cardinality of any descriptor set). The `select` function blocks until at least one descriptor in the read set is ready for reading. A descriptor k is *ready for reading* if and only if a request to read 1 byte from that descriptor would not block. As a side effect, `select` modifies the `fd_set` pointed to by argument `fdset` to indicate a subset of the read set called the *ready set*, consisting of the descriptors in the read set that are ready for reading. The value returned by the function indicates the cardinality of the ready set. Note that because of the side effect, we must update the read set every time `select` is called.

The best way to understand `select` is to study a concrete example. Figure 6 shows how we might use `select` to implement an iterative echo server that also accepts user commands on the standard input. We begin by using the `open_listenfd` function from Figure 17 of the chapter "Network Programming" to open a listening descriptor (line 17), and then using FD_ZERO to create an empty read set (line 19):

	listenfd			stdin
	3	2	1	0
read_set ($\emptyset$) :	0	0	0	0

Next, in lines 20 and 21, we define the read set to consist of descriptor 0 (standard input) and descriptor 3 (the listening descriptor), respectively:

	listenfd			stdin
	3	2	1	0
read_set ($\{0, 3\}$) :	1	0	0	1

At this point, we begin the typical server loop. But instead of waiting for a connection request by calling the `accept` function, we call the `select` function, which blocks until either the listening descriptor or standard input is ready for reading (line 25). For example, here is the value of `ready_set` that `select` would return if the user hit the enter key, thus causing the standard input descriptor to become ready for reading:

	listenfd			stdin
	3	2	1	0
read_set ($\{0\}$) :	0	0	0	1

```
1    #include "csapp.h"
2    void echo(int connfd);
3    void command(void);
4
5    int main(int argc, char **argv)
6    {
7        int listenfd, connfd, port;
8        socklen_t clientlen = sizeof(struct sockaddr_in);
9        struct sockaddr_in clientaddr;
10       fd_set read_set, ready_set;
11
12       if (argc != 2) {
13           fprintf(stderr, "usage: %s <port>\n", argv[0]);
14           exit(0);
15       }
16       port = atoi(argv[1]);
17       listenfd = Open_listenfd(port);
18
19       FD_ZERO(&read_set);                  /* Clear read set */
20       FD_SET(STDIN_FILENO, &read_set); /* Add stdin to read set */
21       FD_SET(listenfd, &read_set);     /* Add listenfd to read set */
22
23       while (1) {
24           ready_set = read_set;
25           Select(listenfd+1, &ready_set, NULL, NULL, NULL);
26           if (FD_ISSET(STDIN_FILENO, &ready_set))
27               command(); /* Read command line from stdin */
28           if (FD_ISSET(listenfd, &ready_set)) {
29               connfd = Accept(listenfd, (SA *)&clientaddr, &clientlen);
30               echo(connfd); /* Echo client input until EOF */
31               Close(connfd);
32           }
33       }
34   }
35
36   void command(void) {
37       char buf[MAXLINE];
38       if (!Fgets(buf, MAXLINE, stdin))
39           exit(0); /* EOF */
40       printf("%s", buf); /* Process the input command */
41   }
```

Figure 6   **An iterative echo server that uses I/O multiplexing.** The server uses select to wait for connection requests on a listening descriptor and commands on standard input.

Once `select` returns, we use the FD_ISSET macro to determine which descriptors are ready for reading. If standard input is ready (line 26), we call the `command` function, which reads, parses, and responds to the command before returning to the main routine. If the listening descriptor is ready (line 28), we call `accept` to get a connected descriptor, and then call the `echo` function from Figure 21 of the chapter "Network Programming" that echoes each line from the client until the client closes its end of the connection.

While this program is a good example of using `select`, it still leaves something to be desired. The problem is that once it connects to a client, it continues echoing input lines until the client closes its end of the connection. Thus, if you type a command to standard input, you will not get a response until the server is finished with the client. A better approach would be to multiplex at a finer granularity, echoing (at most) one text line each time through the server loop.

Practice Problem 3

In most Unix systems, typing `ctrl-d` indicates EOF on standard input. What happens if you type `ctrl-d` to the program in Figure 6 while it is blocked in the call to `select`?

2.1   A Concurrent Event-Driven Server Based on I/O Multiplexing

I/O multiplexing can be used as the basis for concurrent *event-driven* programs, where flows make progress as a result of certain events. The general idea is to model logical flows as state machines. Informally, a *state machine* is a collection of *states*, *input events*, and *transitions* that map states and input events to states. Each transition maps an (input state, input event) pair to an output state. A *self-loop* is a transition between the same input and output state. State machines are typically drawn as directed graphs, where nodes represent states, directed arcs represent transitions, and arc labels represent input events. A state machine begins execution in some initial state. Each input event triggers a transition from the current state to the next state.

For each new client k, a concurrent server based on I/O multiplexing creates a new state machine s_k and associates it with connected descriptor d_k. As shown in Figure 7, each state machine s_k has one state ("waiting for descriptor d_k to be ready for reading"), one input event ("descriptor d_k is ready for reading"), and one transition ("read a text line from descriptor d_k").

The server uses the I/O multiplexing, courtesy of the `select` function, to detect the occurrence of input events. As each connected descriptor becomes ready for reading, the server executes the transition for the corresponding state machine, in this case reading and echoing a text line from the descriptor.

Figure 8 shows the complete example code for a concurrent event-driven server based on I/O multiplexing. The set of active clients is maintained in a `pool` structure (lines 3–11). After initializing the pool by calling `init_pool` (line 29), the server enters an infinite loop. During each iteration of this loop, the server calls

Figure 7

State machine for a logical flow in a concurrent event-driven echo server.

the select function to detect two different kinds of input events: (a) a connection request arriving from a new client, and (b) a connected descriptor for an existing client being ready for reading. When a connection request arrives (line 36), the server opens the connection (line 37) and calls the add_client function to add the client to the pool (line 38). Finally, the server calls the check_clients function to echo a single text line from each ready connected descriptor (line 42).

The init_pool function (Figure 9) initializes the client pool. The clientfd array represents a set of connected descriptors, with the integer −1 denoting an available slot. Initially, the set of connected descriptors is empty (lines 5–7), and the listening descriptor is the only descriptor in the select read set (lines 10–12).

The add_client function (Figure 10) adds a new client to the pool of active clients. After finding an empty slot in the clientfd array, the server adds the connected descriptor to the array and initializes a corresponding Rio read buffer so that we can call rio_readlineb on the descriptor (lines 8–9). We then add the connected descriptor to the select read set (line 12), and we update some global properties of the pool. The maxfd variable (lines 15–16) keeps track of the largest file descriptor for select. The maxi variable (lines 17–18) keeps track of the largest index into the clientfd array so that the check_clients functions does not have to search the entire array.

The check_clients function echoes a text line from each ready connected descriptor (Figure 11). If we are successful in reading a text line from the descriptor, then we echo that line back to the client (lines 15–18). Notice that in line 15 we are maintaining a cumulative count of total bytes received from all clients. If we detect EOF because the client has closed its end of the connection, then we close our end of the connection (line 23) and remove the descriptor from the pool (lines 24–25).

In terms of the finite state model in Figure 7, the select function detects input events, and the add_client function creates a new logical flow (state machine). The check_clients function performs state transitions by echoing input lines, and it also deletes the state machine when the client has finished sending text lines.

code/conc/echoservers.c

```
1    #include "csapp.h"
2
3    typedef struct { /* Represents a pool of connected descriptors */
4        int maxfd;        /* Largest descriptor in read_set */
5        fd_set read_set;  /* Set of all active descriptors */
6        fd_set ready_set; /* Subset of descriptors ready for reading  */
7        int nready;       /* Number of ready descriptors from select */
8        int maxi;         /* Highwater index into client array */
9        int clientfd[FD_SETSIZE];    /* Set of active descriptors */
10       rio_t clientrio[FD_SETSIZE]; /* Set of active read buffers */
11   } pool;
12
13   int byte_cnt = 0; /* Counts total bytes received by server */
14
15   int main(int argc, char **argv)
16   {
17       int listenfd, connfd, port;
18       socklen_t clientlen = sizeof(struct sockaddr_in);
19       struct sockaddr_in clientaddr;
20       static pool pool;
21
22       if (argc != 2) {
23           fprintf(stderr, "usage: %s <port>\n", argv[0]);
24           exit(0);
25       }
26       port = atoi(argv[1]);
27
28       listenfd = Open_listenfd(port);
29       init_pool(listenfd, &pool);
30       while (1) {
31           /* Wait for listening/connected descriptor(s) to become ready */
32           pool.ready_set = pool.read_set;
33           pool.nready = Select(pool.maxfd+1, &pool.ready_set, NULL, NULL, NULL);
34
35           /* If listening descriptor ready, add new client to pool */
36           if (FD_ISSET(listenfd, &pool.ready_set)) {
37               connfd = Accept(listenfd, (SA *)&clientaddr, &clientlen);
38               add_client(connfd, &pool);
39           }
40
41           /* Echo a text line from each ready connected descriptor */
42           check_clients(&pool);
43       }
44   }
```

code/conc/echoservers.c

Figure 8 **Concurrent echo server based on I/O multiplexing.** Each server iteration echoes a text line from each ready descriptor.

code/conc/echoservers.c

```
1    void init_pool(int listenfd, pool *p)
2    {
3        /* Initially, there are no connected descriptors */
4        int i;
5        p->maxi = -1;
6        for (i=0; i< FD_SETSIZE; i++)
7            p->clientfd[i] = -1;
8
9        /* Initially, listenfd is only member of select read set */
10       p->maxfd = listenfd;
11       FD_ZERO(&p->read_set);
12       FD_SET(listenfd, &p->read_set);
13   }
```

code/conc/echoservers.c

Figure 9 `init_pool`: **Initializes the pool of active clients.**

code/conc/echoservers.c

```
1    void add_client(int connfd, pool *p)
2    {
3        int i;
4        p->nready--;
5        for (i = 0; i < FD_SETSIZE; i++)  /* Find an available slot */
6            if (p->clientfd[i] < 0) {
7                /* Add connected descriptor to the pool */
8                p->clientfd[i] = connfd;
9                Rio_readinitb(&p->clientrio[i], connfd);
10
11               /* Add the descriptor to descriptor set */
12               FD_SET(connfd, &p->read_set);
13
14               /* Update max descriptor and pool highwater mark */
15               if (connfd > p->maxfd)
16                   p->maxfd = connfd;
17               if (i > p->maxi)
18                   p->maxi = i;
19               break;
20           }
21       if (i == FD_SETSIZE) /* Couldn't find an empty slot */
22           app_error("add_client error: Too many clients");
23   }
```

code/conc/echoservers.c

Figure 10 `add_client`: **Adds a new client connection to the pool.**

code/conc/echoservers.c

```
1    void check_clients(pool *p)
2    {
3        int i, connfd, n;
4        char buf[MAXLINE];
5        rio_t rio;
6
7        for (i = 0; (i <= p->maxi) && (p->nready > 0); i++) {
8            connfd = p->clientfd[i];
9            rio = p->clientrio[i];
10
11           /* If the descriptor is ready, echo a text line from it */
12           if ((connfd > 0) && (FD_ISSET(connfd, &p->ready_set))) {
13               p->nready--;
14               if ((n = Rio_readlineb(&rio, buf, MAXLINE)) != 0) {
15                   byte_cnt += n;
16                   printf("Server received %d (%d total) bytes on fd %d\n",
17                           n, byte_cnt, connfd);
18                   Rio_writen(connfd, buf, n);
19               }
20
21               /* EOF detected, remove descriptor from pool */
22               else {
23                   Close(connfd);
24                   FD_CLR(connfd, &p->read_set);
25                   p->clientfd[i] = -1;
26               }
27           }
28        }
29    }
```

code/conc/echoservers.c

Figure 11 `check_clients`: **Services ready client connections.**

2.2 Pros and Cons of I/O Multiplexing

The server in Figure 8 provides a nice example of the advantages and disadvantages of event-driven programming based on I/O multiplexing. One advantage is that event-driven designs give programmers more control over the behavior of their programs than process-based designs. For example, we can imagine writing an event-driven concurrent server that gives preferred service to some clients, which would be difficult for a concurrent server based on processes.

Another advantage is that an event-driven server based on I/O multiplexing runs in the context of a single process, and thus every logical flow has access to the entire address space of the process. This makes it easy to share data between flows. A related advantage of running as a single process is that you can debug your concurrent server as you would any sequential program, using a familiar debugging tool such as GDB. Finally, event-driven designs are often significantly more efficient than process-based designs because they do not require a process context switch to schedule a new flow.

A significant disadvantage of event-driven designs is coding complexity. Our event-driven concurrent echo server requires three times more code than the process-based server. Unfortunately, the complexity increases as the granularity of the concurrency decreases. By *granularity*, we mean the number of instructions that each logical flow executes per time slice. For instance, in our example concurrent server, the granularity of concurrency is the number of instructions required to read an entire text line. As long as some logical flow is busy reading a text line, no other logical flow can make progress. This is fine for our example, but it makes our event-driver server vulnerable to a malicious client that sends only a partial text line and then halts. Modifying an event-driven server to handle partial text lines is a nontrivial task, but it is handled cleanly and automatically by a process-based design. Another significant disadvantage of event-based designs is that they cannot fully utilize multi-core processors.

Practice Problem 4

In the server in Figure 8, we are careful to reinitialize the `pool.ready_set` variable immediately before every call to `select`. Why?

3   Concurrent Programming with Threads

To this point, we have looked at two approaches for creating concurrent logical flows. With the first approach, we use a separate process for each flow. The kernel schedules each process automatically. Each process has its own private address space, which makes it difficult for flows to share data. With the second approach, we create our own logical flows and use I/O multiplexing to explicitly schedule the flows. Because there is only one process, flows share the entire address space. This section introduces a third approach—based on threads—that is a hybrid of these two.

A *thread* is a logical flow that runs in the context of a process. Thus far in this book, our programs have consisted of a single thread per process. But modern systems also allow us to write programs that have multiple threads running concurrently in a single process. The threads are scheduled automatically by the kernel. Each thread has its own *thread context*, including a unique integer *thread ID* (TID), stack, stack pointer, program counter, general-purpose registers, and condition codes. All threads running in a process share the entire virtual address space of that process.

Logical flows based on threads combine qualities of flows based on processes and I/O multiplexing. Like processes, threads are scheduled automatically by the kernel and are known to the kernel by an integer ID. Like flows based on I/O

Figure 12
Concurrent thread execution.

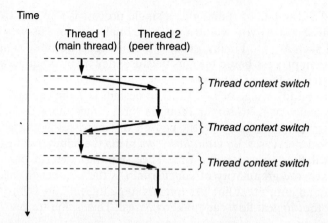

multiplexing, multiple threads run in the context of a single process, and thus share the entire contents of the process virtual address space, including its code, data, heap, shared libraries, and open files.

3.1 Thread Execution Model

The execution model for multiple threads is similar in some ways to the execution model for multiple processes. Consider the example in Figure 12. Each process begins life as a single thread called the *main thread*. At some point, the main thread creates a *peer thread*, and from this point in time the two threads run concurrently. Eventually, control passes to the peer thread via a context switch, because the main thread executes a slow system call such as read or sleep, or because it is interrupted by the system's interval timer. The peer thread executes for a while before control passes back to the main thread, and so on.

Thread execution differs from processes in some important ways. Because a thread context is much smaller than a process context, a thread context switch is faster than a process context switch. Another difference is that threads, unlike processes, are not organized in a rigid parent-child hierarchy. The threads associated with a process form a *pool* of peers, independent of which threads were created by which other threads. The main thread is distinguished from other threads only in the sense that it is always the first thread to run in the process. The main impact of this notion of a pool of peers is that a thread can kill any of its peers, or wait for any of its peers to terminate. Further, each peer can read and write the same shared data.

3.2 Posix Threads

Posix threads (Pthreads) is a standard interface for manipulating threads from C programs. It was adopted in 1995 and is available on most Unix systems. Pthreads defines about 60 functions that allow programs to create, kill, and reap threads, to share data safely with peer threads, and to notify peers about changes in the system state.

code/conc/hello.c

```
1    #include "csapp.h"
2    void *thread(void *vargp);
3
4    int main()
5    {
6        pthread_t tid;
7        Pthread_create(&tid, NULL, thread, NULL);
8        Pthread_join(tid, NULL);
9        exit(0);
10   }
11
12   void *thread(void *vargp) /* Thread routine */
13   {
14       printf("Hello, world!\n");
15       return NULL;
16   }
```

code/conc/hello.c

Figure 13   `hello.c`: **The Pthreads "Hello, world!" program.**

Figure 13 shows a simple Pthreads program. The main thread creates a peer thread and then waits for it to terminate. The peer thread prints "Hello, world!\n" and terminates. When the main thread detects that the peer thread has terminated, it terminates the process by calling `exit`.

This is the first threaded program we have seen, so let us dissect it carefully. The code and local data for a thread is encapsulated in a *thread routine*. As shown by the prototype in line 2, each thread routine takes as input a single generic pointer and returns a generic pointer. If you want to pass multiple arguments to a thread routine, then you should put the arguments into a structure and pass a pointer to the structure. Similarly, if you want the thread routine to return multiple arguments, you can return a pointer to a structure.

Line 4 marks the beginning of the code for the main thread. The main thread declares a single local variable `tid`, which will be used to store the thread ID of the peer thread (line 6). The main thread creates a new peer thread by calling the `pthread_create` function (line 7). When the call to `pthread_create` returns, the main thread and the newly created peer thread are running concurrently, and `tid` contains the ID of the new thread. The main thread waits for the peer thread to terminate with the call to `pthread_join` in line 8. Finally, the main thread calls `exit` (line 9), which terminates all threads (in this case just the main thread) currently running in the process.

Lines 12–16 define the thread routine for the peer thread. It simply prints a string and then terminates the peer thread by executing the `return` statement in line 15.

3.3 Creating Threads

Threads create other threads by calling the pthread_create function.

```
#include <pthread.h>
typedef void *(func)(void *);

int pthread_create(pthread_t *tid, pthread_attr_t *attr,
                   func *f, void *arg);
                                    Returns: 0 if OK, nonzero on error
```

The pthread_create function creates a new thread and runs the *thread routine* f in the context of the new thread and with an input argument of arg. The attr argument can be used to change the default attributes of the newly created thread. Changing these attributes is beyond our scope, and in our examples, we will always call pthread_create with a NULL attr argument.

When pthread_create returns, argument tid contains the ID of the newly created thread. The new thread can determine its own thread ID by calling the pthread_self function.

```
#include <pthread.h>

pthread_t pthread_self(void);
                                            Returns: thread ID of caller
```

3.4 Terminating Threads

A thread terminates in one of the following ways:

- The thread terminates *implicitly* when its top-level thread routine returns.
- The thread terminates *explicitly* by calling the pthread_exit function. If the main thread calls pthread_exit, it waits for all other peer threads to terminate, and then terminates the main thread and the entire process with a return value of thread_return.

```
#include <pthread.h>

void pthread_exit(void *thread_return);
                                    Returns: 0 if OK, nonzero on error
```

- Some peer thread calls the Unix exit function, which terminates the process and all threads associated with the process.
- Another peer thread terminates the current thread by calling the pthread_cancel function with the ID of the current thread.

```
#include <pthread.h>

int pthread_cancel(pthread_t tid);
                                        Returns: 0 if OK, nonzero on error
```

3.5 Reaping Terminated Threads

Threads wait for other threads to terminate by calling the `pthread_join` function.

```
#include <pthread.h>

int pthread_join(pthread_t tid, void **thread_return);
                                        Returns: 0 if OK, nonzero on error
```

The `pthread_join` function blocks until thread `tid` terminates, assigns the generic (`void *`) pointer returned by the thread routine to the location pointed to by `thread_return`, and then *reaps* any memory resources held by the terminated thread.

Notice that, unlike the Unix `wait` function, the `pthread_join` function can only wait for a specific thread to terminate. There is no way to instruct `pthread_wait` to wait for an *arbitrary* thread to terminate. This can complicate our code by forcing us to use other, less intuitive mechanisms to detect process termination. Indeed, Stevens argues convincingly that this is a bug in the specification [109].

3.6 Detaching Threads

At any point in time, a thread is *joinable* or *detached*. A joinable thread can be reaped and killed by other threads. Its memory resources (such as the stack) are not freed until it is reaped by another thread. In contrast, a detached thread cannot be reaped or killed by other threads. Its memory resources are freed automatically by the system when it terminates.

By default, threads are created joinable. In order to avoid memory leaks, each joinable thread should either be explicitly reaped by another thread, or detached by a call to the `pthread_detach` function.

```
#include <pthread.h>

int pthread_detach(pthread_t tid);
                                        Returns: 0 if OK, nonzero on error
```

The `pthread_detach` function detaches the joinable thread `tid`. Threads can detach themselves by calling `pthread_detach` with an argument of `pthread_self()`.

Although some of our examples will use joinable threads, there are good reasons to use detached threads in real programs. For example, a high-performance Web server might create a new peer thread each time it receives a connection request from a Web browser. Since each connection is handled independently by a separate thread, it is unnecessary—and indeed undesirable—for the server to explicitly wait for each peer thread to terminate. In this case, each peer thread should detach itself before it begins processing the request so that its memory resources can be reclaimed after it terminates.

3.7 Initializing Threads

The `pthread_once` function allows you to initialize the state associated with a thread routine.

```
#include <pthread.h>

pthread_once_t once_control = PTHREAD_ONCE_INIT;

int pthread_once(pthread_once_t *once_control,
                 void (*init_routine)(void));
                                                    Always returns 0
```

The `once_control` variable is a global or static variable that is always initialized to PTHREAD_ONCE_INIT. The first time you call `pthread_once` with an argument of `once_control`, it invokes `init_routine`, which is a function with no input arguments that returns nothing. Subsequent calls to `pthread_once` with the same `once_control` variable do nothing. The `pthread_once` function is useful whenever you need to dynamically initialize global variables that are shared by multiple threads. We will look at an example in Section 5.5.

3.8 A Concurrent Server Based on Threads

Figure 14 shows the code for a concurrent echo server based on threads. The overall structure is similar to the process-based design. The main thread repeatedly waits for a connection request and then creates a peer thread to handle the request. While the code looks simple, there are a couple of general and somewhat subtle issues we need to look at more closely. The first issue is how to pass the connected descriptor to the peer thread when we call `pthread_create`. The obvious approach is to pass a pointer to the descriptor, as in the following:

```
connfd = Accept(listenfd, (SA *) &clientaddr, &clientlen);
Pthread_create(&tid, NULL, thread, &connfd);
```

code/conc/echoservert.c

```
1   #include "csapp.h"
2
3   void echo(int connfd);
4   void *thread(void *vargp);
5
6   int main(int argc, char **argv)
7   {
8       int listenfd, *connfdp, port;
9       socklen_t clientlen=sizeof(struct sockaddr_in);
10      struct sockaddr_in clientaddr;
11      pthread_t tid;
12
13      if (argc != 2) {
14          fprintf(stderr, "usage: %s <port>\n", argv[0]);
15          exit(0);
16      }
17      port = atoi(argv[1]);
18
19      listenfd = Open_listenfd(port);
20      while (1) {
21          connfdp = Malloc(sizeof(int));
22          *connfdp = Accept(listenfd, (SA *) &clientaddr, &clientlen);
23          Pthread_create(&tid, NULL, thread, connfdp);
24      }
25  }
26
27  /* Thread routine */
28  void *thread(void *vargp)
29  {
30      int connfd = *((int *)vargp);
31      Pthread_detach(pthread_self());
32      Free(vargp);
33      echo(connfd);
34      Close(connfd);
35      return NULL;
36  }
```

code/conc/echoservert.c

Figure 14 **Concurrent echo server based on threads.**

Then we have the peer thread dereference the pointer and assign it to a local variable, as follows:

```
void *thread(void *vargp) {
    int connfd = *((int *)vargp);
       .
       .
       .
}
```

This would be wrong, however, because it introduces a *race* between the assignment statement in the peer thread and the accept statement in the main thread. If the assignment statement completes before the next accept, then the local connfd variable in the peer thread gets the correct descriptor value. However, if the assignment completes *after* the accept, then the local connfd variable in the peer thread gets the descriptor number of the *next* connection. The unhappy result is that two threads are now performing input and output on the same descriptor. In order to avoid the potentially deadly race, we must assign each connected descriptor returned by accept to its own dynamically allocated memory block, as shown in lines 21–22. We will return to the issue of races in Section 7.4.

Another issue is avoiding memory leaks in the thread routine. Since we are not explicitly reaping threads, we must detach each thread so that its memory resources will be reclaimed when it terminates (line 31). Further, we must be careful to free the memory block that was allocated by the main thread (line 32).

Practice Problem 5

In the process-based server in Figure 5, we were careful to close the connected descriptor in two places: the parent and child processes. However, in the threads-based server in Figure 14, we only closed the connected descriptor in one place: the peer thread. Why?

4 Shared Variables in Threaded Programs

From a programmer's perspective, one of the attractive aspects of threads is the ease with which multiple threads can share the same program variables. However, this sharing can be tricky. In order to write correctly threaded programs, we must have a clear understanding of what we mean by sharing and how it works.

There are some basic questions to work through in order to understand whether a variable in a C program is shared or not: (1) What is the underlying memory model for threads? (2) Given this model, how are instances of the variable mapped to memory? (3) Finally, how many threads reference each of these instances? The variable is *shared* if and only if multiple threads reference some instance of the variable.

To keep our discussion of sharing concrete, we will use the program in Figure 15 as a running example. Although somewhat contrived, it is nonetheless useful to study because it illustrates a number of subtle points about sharing. The example program consists of a main thread that creates two peer threads. The *code/conc/sharing.c*

```
1   #include "csapp.h"
2   #define N 2
3   void *thread(void *vargp);
4
5   char **ptr;  /* Global variable */
6
7   int main()
8   {
9       int i;
10      pthread_t tid;
11      char *msgs[N] = {
12          "Hello from foo",
13          "Hello from bar"
14      };
15
16      ptr = msgs;
17      for (i = 0; i < N; i++)
18          Pthread_create(&tid, NULL, thread, (void *)i);
19      Pthread_exit(NULL);
20  }
21
22  void *thread(void *vargp)
23  {
24      int myid = (int)vargp;
25      static int cnt = 0;
26      printf("[%d]: %s (cnt=%d)\n", myid, ptr[myid], ++cnt);
27      return NULL;
28  }
```

code/conc/sharing.c

Figure 15 **Example program that illustrates different aspects of sharing.**

main thread passes a unique ID to each peer thread, which uses the ID to print a personalized message, along with a count of the total number of times that the thread routine has been invoked.

4.1 Threads Memory Model

A pool of concurrent threads runs in the context of a process. Each thread has its own separate *thread context*, which includes a thread ID, stack, stack pointer, program counter, condition codes, and general-purpose register values. Each thread shares the rest of the process context with the other threads. This includes the entire user virtual address space, which consists of read-only text (code), read/write data, the heap, and any shared library code and data areas. The threads also share the same set of open files.

In an operational sense, it is impossible for one thread to read or write the register values of another thread. On the other hand, any thread can access any location in the shared virtual memory. If some thread modifies a memory location, then every other thread will eventually see the change if it reads that location. Thus, registers are never shared, whereas virtual memory is always shared.

The memory model for the separate thread stacks is not as clean. These stacks are contained in the stack area of the virtual address space, and are *usually* accessed independently by their respective threads. We say *usually* rather than *always*, because different thread stacks are not protected from other threads. So if a thread somehow manages to acquire a pointer to another thread's stack, then it can read and write any part of that stack. Our example program shows this in line 26, where the peer threads reference the contents of the main thread's stack indirectly through the global `ptr` variable.

4.2 Mapping Variables to Memory

Variables in threaded C programs are mapped to virtual memory according to their storage classes:

- *Global variables.* A *global variable* is any variable declared outside of a function. At run time, the read/write area of virtual memory contains exactly one instance of each global variable that can be referenced by any thread. For example, the global `ptr` variable declared in line 5 has one run-time instance in the read/write area of virtual memory. When there is only one instance of a variable, we will denote the instance by simply using the variable name—in this case, `ptr`.

- *Local automatic variables.* A *local automatic variable* is one that is declared inside a function without the `static` attribute. At run time, each thread's stack contains its own instances of any local automatic variables. This is true even if multiple threads execute the same thread routine. For example, there is one instance of the local variable `tid`, and it resides on the stack of the main thread. We will denote this instance as `tid.m`. As another example, there are two instances of the local variable `myid`, one instance on the stack of peer thread 0, and the other on the stack of peer thread 1. We will denote these instances as `myid.p0` and `myid.p1`, respectively.

- *Local static variables.* A *local static variable* is one that is declared inside a function with the `static` attribute. As with global variables, the read/write area of virtual memory contains exactly one instance of each local static variable declared in a program. For example, even though each peer thread in our example program declares `cnt` in line 25, at run time there is only one instance of `cnt` residing in the read/write area of virtual memory. Each peer thread reads and writes this instance.

4.3 Shared Variables

We say that a variable v is *shared* if and only if one of its instances is referenced by more than one thread. For example, variable `cnt` in our example program is shared because it has only one run-time instance and this instance is referenced by both peer threads. On the other hand, `myid` is not shared because each of its two instances is referenced by exactly one thread. However, it is important to realize that local automatic variables such as `msgs` can also be shared.

Practice Problem 6

A. Using the analysis from Section 4, fill each entry in the following table with "Yes" or "No" for the example program in Figure 15. In the first column, the notation $v.t$ denotes an instance of variable v residing on the local stack for thread t, where t is either m (main thread), p0 (peer thread 0), or p1 (peer thread 1).

Variable instance	Referenced by main thread?	Referenced by peer thread 0?	Referenced by peer thread 1?
ptr			
cnt			
i.m			
msgs.m			
myid.p0			
myid.p1			

B. Given the analysis in Part A, which of the variables `ptr`, `cnt`, `i`, `msgs`, and `myid` are shared?

5 Synchronizing Threads with Semaphores

Shared variables can be convenient, but they introduce the possibility of nasty *synchronization errors*. Consider the `badcnt.c` program in Figure 16, which creates two threads, each of which increments a global shared counter variable called `cnt`. Since each thread increments the counter *niters* times, we expect its final value to be $2 \times niters$. This seems quite simple and straightforward. However, when we run `badcnt.c` on our Linux system, we not only get wrong answers, we get different answers each time!

```
linux> ./badcnt 1000000
BOOM! cnt=1445085

linux> ./badcnt 1000000
BOOM! cnt=1915220

linux> ./badcnt 1000000
BOOM! cnt=1404746
```

```
1    #include "csapp.h"
2
3    void *thread(void *vargp);  /* Thread routine prototype */
4
5    /* Global shared variable */
6    volatile int cnt = 0; /* Counter */
7
8    int main(int argc, char **argv)
9    {
10       int niters;
11       pthread_t tid1, tid2;
12
13       /* Check input argument */
14       if (argc != 2) {
15           printf("usage: %s <niters>\n", argv[0]);
16           exit(0);
17       }
18       niters = atoi(argv[1]);
19
20       /* Create threads and wait for them to finish */
21       Pthread_create(&tid1, NULL, thread, &niters);
22       Pthread_create(&tid2, NULL, thread, &niters);
23       Pthread_join(tid1, NULL);
24       Pthread_join(tid2, NULL);
25
26       /* Check result */
27       if (cnt != (2 * niters))
28           printf("BOOM! cnt=%d\n", cnt);
29       else
30           printf("OK cnt=%d\n", cnt);
31       exit(0);
32   }
33
34   /* Thread routine */
35   void *thread(void *vargp)
36   {
37       int i, niters = *((int *)vargp);
38
39       for (i = 0; i < niters; i++)
40           cnt++;
41
42       return NULL;
43   }
```

Figure 16 badcnt.c: **An improperly synchronized counter program.**

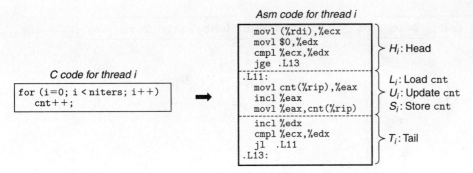

C code for thread i

```
for (i=0; i<niters; i++)
    cnt++;
```

Asm code for thread i

```
movl (%rdi),%ecx
movl $0,%edx          } H_i: Head
cmpl %ecx,%edx
jge  .L13
.L11:
movl cnt(%rip),%eax   L_i: Load cnt
incl %eax             } U_i: Update cnt
movl %eax,cnt(%rip)   S_i: Store cnt
incl %edx
cmpl %ecx,%edx        } T_i: Tail
jl   .L11
.L13:
```

Figure 17 **Assembly code for the counter loop (lines 39–40) in badcnt.c.**

So what went wrong? To understand the problem clearly, we need to study the assembly code for the counter loop (lines 39–40), as shown in Figure 17. We will find it helpful to partition the loop code for thread i into five parts:

- H_i: The block of instructions at the head of the loop
- L_i: The instruction that loads the shared variable cnt into register %eax$_i$, where %eax$_i$ denotes the value of register %eax in thread i
- U_i: The instruction that updates (increments) %eax$_i$
- S_i: The instruction that stores the updated value of %eax$_i$ back to the shared variable cnt
- T_i: The block of instructions at the tail of the loop

Notice that the head and tail manipulate only local stack variables, while L_i, U_i, and S_i manipulate the contents of the shared counter variable.

When the two peer threads in badcnt.c run concurrently on a uniprocessor, the machine instructions are completed one after the other in some order. Thus, each concurrent execution defines some total ordering (or interleaving) of the instructions in the two threads. Unfortunately, some of these orderings will produce correct results, but others will not.

Here is the crucial point: *In general, there is no way for you to predict whether the operating system will choose a correct ordering for your threads.* For example, Figure 18(a) shows the step-by-step operation of a correct instruction ordering. After each thread has updated the shared variable cnt, its value in memory is 2, which is the expected result. On the other hand, the ordering in Figure 18(b) produces an incorrect value for cnt. The problem occurs because thread 2 loads cnt in step 5, after thread 1 loads cnt in step 2, but before thread 1 stores its updated value in step 6. Thus, each thread ends up storing an updated counter value of 1. We can clarify these notions of correct and incorrect instruction orderings with the help of a device known as a *progress graph*, which we introduce in the next section.

Step	Thread	Instr	%eax$_1$	%eax$_2$	cnt
1	1	H_1	—	—	0
2	1	L_1	0	—	0
3	1	U_1	1	—	0
4	1	S_1	1	—	1
5	2	H_2	—	—	1
6	2	L_2	—	1	1
7	2	U_2	—	2	1
8	2	S_2	—	2	2
9	2	T_2	—	2	2
10	1	T_1	1	—	2

(a) Correct ordering

Step	Thread	Instr	%eax$_1$	%eax$_2$	cnt
1	1	H_1	—	—	0
2	1	L_1	0	—	0
3	1	U_1	1	—	0
4	2	H_2	—	—	0
5	2	L_2	—	0	0
6	1	S_1	1	—	1
7	1	T_1	1	—	1
8	2	U_2	—	1	1
9	2	S_2	—	1	1
10	2	T_2	—	1	1

(b) Incorrect ordering

Figure 18 **Instruction orderings for the first loop iteration in** badcnt.c.

Practice Problem 7

Complete the table for the following instruction ordering of badcnt.c:

Step	Thread	Instr	%eax$_1$	%eax$_2$	cnt
1	1	H_1	—	—	0
2	1	L_1			
3	2	H_2			
4	2	L_2			
5	2	U_2			
6	2	S_2			
7	1	U_1			
8	1	S_1			
9	1	T_1			
10	2	T_2			

Does this ordering result in a correct value for cnt?

5.1 Progress Graphs

A *progress graph* models the execution of n concurrent threads as a trajectory through an n-dimensional Cartesian space. Each axis k corresponds to the progress of thread k. Each point $(I_1, I_2, \ldots, I_n)$ represents the state where thread k ($k = 1, \ldots, n$) has completed instruction I_k. The origin of the graph corresponds to the *initial state* where none of the threads has yet completed an instruction.

Figure 19 shows the two-dimensional progress graph for the first loop iteration of the badcnt.c program. The horizontal axis corresponds to thread 1, the vertical axis to thread 2. Point (L_1, S_2) corresponds to the state where thread 1 has completed L_1 and thread 2 has completed S_2.

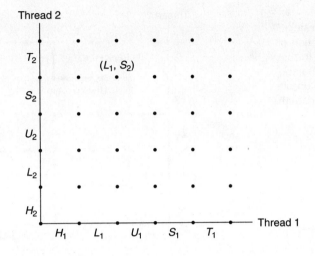

Figure 19

Progress graph for the first loop iteration of `badcnt.c.`

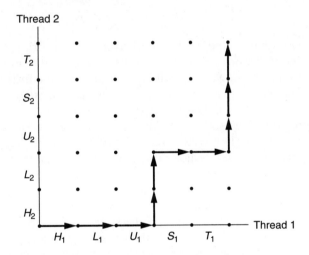

Figure 20

An example trajectory.

A progress graph models instruction execution as a *transition* from one state to another. A transition is represented as a directed edge from one point to an adjacent point. Legal transitions move to the right (an instruction in thread 1 completes) or up (an instruction in thread 2 completes). Two instructions cannot complete at the same time—diagonal transitions are not allowed. Programs never run backwards, so transitions that move down or to the left are not legal either.

The execution history of a program is modeled as a *trajectory* through the state space. Figure 20 shows the trajectory that corresponds to the following instruction ordering:

$$H_1, L_1, U_1, H_2, L_2, S_1, T_1, U_2, S_2, T_2$$

For thread i, the instructions (L_i, U_i, S_i) that manipulate the contents of the shared variable `cnt` constitute a *critical section* (with respect to shared variable

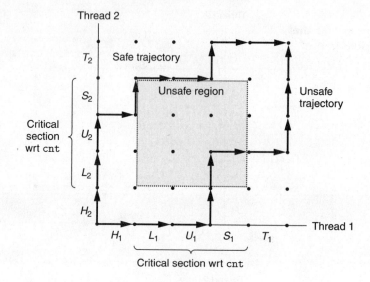

Figure 21

Safe and unsafe trajectories. The intersection of the critical regions forms an unsafe region. Trajectories that skirt the unsafe region correctly update the counter variable.

cnt) that should not be interleaved with the critical section of the other thread. In other words, we want to ensure that each thread has *mutually exclusive access* to the shared variable while it is executing the instructions in its critical section. The phenomenon in general is known as *mutual exclusion*.

On the progress graph, the intersection of the two critical sections defines a region of the state space known as an *unsafe region*. Figure 21 shows the unsafe region for the variable cnt. Notice that the unsafe region abuts, but does not include, the states along its perimeter. For example, states (H_1, H_2) and (S_1, U_2) abut the unsafe region, but are not part of it. A trajectory that skirts the unsafe region is known as a *safe trajectory*. Conversely, a trajectory that touches any part of the unsafe region is an *unsafe trajectory*. Figure 21 shows examples of safe and unsafe trajectories through the state space of our example badcnt.c program. The upper trajectory skirts the unsafe region along its left and top sides, and thus is safe. The lower trajectory crosses the unsafe region, and thus is unsafe.

Any safe trajectory will correctly update the shared counter. In order to guarantee correct execution of our example threaded program—and indeed any concurrent program that shares global data structures—we must somehow *synchronize* the threads so that they always have a safe trajectory. A classic approach is based on the idea of a semaphore, which we introduce next.

Practice Problem 8

Using the progress graph in Figure 21, classify the following trajectories as either *safe* or *unsafe*.

A. $H_1, L_1, U_1, S_1, H_2, L_2, U_2, S_2, T_2, T_1$

B. $H_2, L_2, H_1, L_1, U_1, S_1, T_1, U_2, S_2, T_2$

C. $H_1, H_2, L_2, U_2, S_2, L_1, U_1, S_1, T_1, T_2$

5.2 Semaphores

Edsger Dijkstra, a pioneer of concurrent programming, proposed a classic solution to the problem of synchronizing different execution threads based on a special type of variable called a *semaphore*. A semaphore, s, is a global variable with a nonnegative integer value that can only be manipulated by two special operations, called P and V:

- $P(s)$: If s is nonzero, then P decrements s and returns immediately. If s is zero, then suspend the thread until s becomes nonzero and the process is restarted by a V operation. After restarting, the P operation decrements s and returns control to the caller.
- $V(s)$: The V operation increments s by 1. If there are any threads blocked at a P operation waiting for s to become nonzero, then the V operation restarts exactly one of these threads, which then completes its P operation by decrementing s.

The test and decrement operations in P occur indivisibly, in the sense that once the semaphore s becomes nonzero, the decrement of s occurs without interruption. The increment operation in V also occurs indivisibly, in that it loads, increments, and stores the semaphore without interruption. Notice that the definition of V does *not* define the order in which waiting threads are restarted. The only requirement is that the V must restart exactly one waiting thread. *Thus, when several threads are waiting at a semaphore, you cannot predict which one will be restarted as a result of the V.*

The definitions of P and V ensure that a running program can never enter a state where a properly initialized semaphore has a negative value. This property, known as the *semaphore invariant*, provides a powerful tool for controlling the trajectories of concurrent programs, as we shall see in the next section.

The Posix standard defines a variety of functions for manipulating semaphores.

```
#include <semaphore.h>

int sem_init(sem_t *sem, 0, unsigned int value);
int sem_wait(sem_t *s);   /* P(s) */
int sem_post(sem_t *s);   /* V(s) */
                                    Returns: 0 if OK, −1 on error
```

The sem_init function initializes semaphore sem to value. Each semaphore must be initialized before it can be used. For our purposes, the middle argument is always 0. Programs perform P and V operations by calling the sem_wait and sem_post functions, respectively. For conciseness, we prefer to use the following equivalent P and V wrapper functions instead:

```
#include "csapp.h"

void P(sem_t *s); /* Wrapper function for sem_wait */
void V(sem_t *s); /* Wrapper function for sem_post */
```
Returns: nothing

Aside   Origin of the names *P* and *V*

Edsger Dijkstra (1930–2002) was originally from the Netherlands. The names *P* and *V* come from the Dutch words *Proberen* (to test) and *Verhogen* (to increment).

5.3   Using Semaphores for Mutual Exclusion

Semaphores provide a convenient way to ensure mutually exclusive access to shared variables. The basic idea is to associate a semaphore s, initially 1, with each shared variable (or related set of shared variables) and then surround the corresponding critical section with $P(s)$ and $V(s)$ operations.

A semaphore that is used in this way to protect shared variables is called a *binary semaphore* because its value is always 0 or 1. Binary semaphores whose purpose is to provide mutual exclusion are often called *mutexes*. Performing a *P* operation on a mutex is called *locking* the mutex. Similarly, performing the *V* operation is called *unlocking* the mutex. A thread that has locked but not yet unlocked a mutex is said to be *holding* the mutex. A semaphore that is used as a counter for a set of available resources is called a *counting semaphore*.

The progress graph in Figure 22 shows how we would use binary semaphores to properly synchronize our example counter program. Each state is labeled with the value of semaphore s in that state. The crucial idea is that this combination of *P* and *V* operations creates a collection of states, called a *forbidden region*, where $s < 0$. Because of the semaphore invariant, no feasible trajectory can include one of the states in the forbidden region. And since the forbidden region completely encloses the unsafe region, no feasible trajectory can touch any part of the unsafe region. Thus, every feasible trajectory is safe, and regardless of the ordering of the instructions at run time, the program correctly increments the counter.

In an operational sense, the forbidden region created by the *P* and *V* operations makes it impossible for multiple threads to be executing instructions in the enclosed critical region at any point in time. In other words, the semaphore operations ensure mutually exclusive access to the critical region.

Putting it all together, to properly synchronize the example counter program in Figure 16 using semaphores, we first declare a semaphore called mutex:

```
volatile int cnt = 0; /* Counter */
sem_t mutex;           /* Semaphore that protects counter */
```

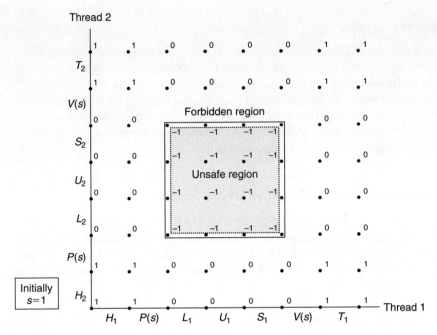

Figure 22 **Using semaphores for mutual exclusion.** The infeasible states where $s < 0$ define a *forbidden region* that surrounds the unsafe region and prevents any feasible trajectory from touching the unsafe region.

and then initialize it to unity in the main routine:

```
Sem_init(&mutex, 0, 1);    /* mutex = 1 */
```

Finally, we protect the update of the shared cnt variable in the thread routine by surrounding it with P and V operations:

```
for (i = 0; i < niters; i++) {
    P(&mutex);
    cnt++;
    V(&mutex);
}
```

When we run the properly synchronized program, it now produces the correct answer each time.

```
linux> ./goodcnt 1000000
OK cnt=2000000

linux> ./goodcnt 1000000
OK cnt=2000000
```

Aside Limitations of progress graphs

Progress graphs give us a nice way to visualize concurrent program execution on uniprocessors and to understand why we need synchronization. However, they do have limitations, particularly with respect to concurrent execution on multiprocessors, where a set of CPU/cache pairs share the same main memory. Multiprocessors behave in ways that cannot be explained by progress graphs. In particular, a multiprocessor memory system can be in a state that does not correspond to any trajectory in a progress graph. Regardless, the message remains the same: always synchronize accesses to your shared variables, regardless if you're running on a uniprocessor or a multiprocessor.

5.4 Using Semaphores to Schedule Shared Resources

Another important use of semaphores, besides providing mutual exclusion, is to schedule accesses to shared resources. In this scenario, a thread uses a semaphore operation to notify another thread that some condition in the program state has become true. Two classical and useful examples are the *producer-consumer* and *readers-writers* problems.

Producer-Consumer Problem

The *producer-consumer* problem is shown in Figure 23. A producer and consumer thread share a *bounded buffer* with *n slots*. The producer thread repeatedly produces new *items* and inserts them in the buffer. The consumer thread repeatedly removes items from the buffer and then consumes (uses) them. Variants with multiple producers and consumers are also possible.

Since inserting and removing items involves updating shared variables, we must guarantee mutually exclusive access to the buffer. But guaranteeing mutual exclusion is not sufficient. We also need to schedule accesses to the buffer. If the buffer is full (there are no empty slots), then the producer must wait until a slot becomes available. Similarly, if the buffer is empty (there are no available items), then the consumer must wait until an item becomes available.

Producer-consumer interactions occur frequently in real systems. For example, in a multimedia system, the producer might encode video frames while the consumer decodes and renders them on the screen. The purpose of the buffer is to reduce jitter in the video stream caused by data-dependent differences in the encoding and decoding times for individual frames. The buffer provides a reservoir of slots to the producer and a reservoir of encoded frames to the consumer. Another common example is the design of graphical user interfaces. The producer detects

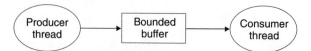

Figure 23 **Producer-consumer problem.** The producer generates items and inserts them into a bounded buffer. The consumer removes items from the buffer and then consumes them.

code/conc/sbuf.h

```
1    typedef struct {
2        int *buf;          /* Buffer array */
3        int n;             /* Maximum number of slots */
4        int front;         /* buf[(front+1)%n] is first item */
5        int rear;          /* buf[rear%n] is last item */
6        sem_t mutex;       /* Protects accesses to buf */
7        sem_t slots;       /* Counts available slots */
8        sem_t items;       /* Counts available items */
9    } sbuf_t;
```

code/conc/sbuf.h

Figure 24 `sbuf_t`: **Bounded buffer used by the Sbuf package.**

mouse and keyboard events and inserts them in the buffer. The consumer removes the events from the buffer in some priority-based manner and paints the screen.

In this section, we will develop a simple package, called Sbuf, for building producer-consumer programs. In the next section, we look at how to use it to build an interesting concurrent server based on prethreading. Sbuf manipulates bounded buffers of type `sbuf_t` (Figure 24). Items are stored in a dynamically allocated integer array (buf) with n items. The `front` and `rear` indices keep track of the first and last items in the array. Three semaphores synchronize access to the buffer. The `mutex` semaphore provides mutually exclusive buffer access. Semaphores `slots` and `items` are counting semaphores that count the number of empty slots and available items, respectively.

Figure 25 shows the implementation of Sbuf function. The `sbuf_init` function allocates heap memory for the buffer, sets `front` and `rear` to indicate an empty buffer, and assigns initial values to the three semaphores. This function is called once, before calls to any of the other three functions. The `sbuf_deinit` function frees the buffer storage when the application is through using it. The `sbuf_insert` function waits for an available slot, locks the mutex, adds the item, unlocks the mutex, and then announces the availability of a new item. The `sbuf_remove` function is symmetric. After waiting for an available buffer item, it locks the mutex, removes the item from the front of the buffer, unlocks the mutex, and then signals the availability of a new slot.

Practice Problem 9

Let p denote the number of producers, c the number of consumers, and n the buffer size in units of items. For each of the following scenarios, indicate whether the mutex semaphore in `sbuf_insert` and `sbuf_remove` is necessary or not.

   A. $p = 1, c = 1, n > 1$

   B. $p = 1, c = 1, n = 1$

   C. $p > 1, c > 1, n = 1$

code/conc/sbuf.c

```
1   #include "csapp.h"
2   #include "sbuf.h"
3
4   /* Create an empty, bounded, shared FIFO buffer with n slots */
5   void sbuf_init(sbuf_t *sp, int n)
6   {
7       sp->buf = Calloc(n, sizeof(int));
8       sp->n = n;                       /* Buffer holds max of n items */
9       sp->front = sp->rear = 0;        /* Empty buffer iff front == rear */
10      Sem_init(&sp->mutex, 0, 1);      /* Binary semaphore for locking */
11      Sem_init(&sp->slots, 0, n);      /* Initially, buf has n empty slots */
12      Sem_init(&sp->items, 0, 0);      /* Initially, buf has zero data items */
13  }
14
15  /* Clean up buffer sp */
16  void sbuf_deinit(sbuf_t *sp)
17  {
18      Free(sp->buf);
19  }
20
21  /* Insert item onto the rear of shared buffer sp */
22  void sbuf_insert(sbuf_t *sp, int item)
23  {
24      P(&sp->slots);                        /* Wait for available slot */
25      P(&sp->mutex);                        /* Lock the buffer */
26      sp->buf[(++sp->rear)%(sp->n)] = item; /* Insert the item */
27      V(&sp->mutex);                        /* Unlock the buffer */
28      V(&sp->items);                        /* Announce available item */
29  }
30
31  /* Remove and return the first item from buffer sp */
32  int sbuf_remove(sbuf_t *sp)
33  {
34      int item;
35      P(&sp->items);                          /* Wait for available item */
36      P(&sp->mutex);                          /* Lock the buffer */
37      item = sp->buf[(++sp->front)%(sp->n)];  /* Remove the item */
38      V(&sp->mutex);                          /* Unlock the buffer */
39      V(&sp->slots);                          /* Announce available slot */
40      return item;
41  }
```

code/conc/sbuf.c

Figure 25   **SBUF: A package for synchronizing concurrent access to bounded buffers.**

Readers-Writers Problem

The *readers-writers problem* is a generalization of the mutual exclusion problem. A collection of concurrent threads are accessing a shared object such as a data structure in main memory or a database on disk. Some threads only read the object, while others modify it. Threads that modify the object are called *writers*. Threads that only read it are called *readers*. Writers must have exclusive access to the object, but readers may share the object with an unlimited number of other readers. In general, there are an unbounded number of concurrent readers and writers.

Readers-writers interactions occur frequently in real systems. For example, in an online airline reservation system, an unlimited number of customers are allowed to concurrently inspect the seat assignments, but a customer who is booking a seat must have exclusive access to the database. As another example, in a multithreaded caching Web proxy, an unlimited number of threads can fetch existing pages from the shared page cache, but any thread that writes a new page to the cache must have exclusive access.

The readers-writers problem has several variations, each based on the priorities of readers and writers. The *first readers-writers problem*, which favors readers, requires that no reader be kept waiting unless a writer has already been granted permission to use the object. In other words, no reader should wait simply because a writer is waiting. The *second readers-writers problem*, which favors writers, requires that once a writer is ready to write, it performs its write as soon as possible. Unlike the first problem, a reader that arrives after a writer must wait, even if the writer is also waiting.

Figure 26 shows a solution to the first readers-writers problem. Like the solutions to many synchronization problems, it is subtle and deceptively simple. The w semaphore controls access to the critical sections that access the shared object. The mutex semaphore protects access to the shared readcnt variable, which counts the number of readers currently in the critical section. A writer locks the w mutex each time it enters the critical section, and unlocks it each time it leaves. This guarantees that there is at most one writer in the critical section at any point in time. On the other hand, only the first reader to enter the critical section locks w, and only the last reader to leave the critical section unlocks it. The w mutex is ignored by readers who enter and leave while other readers are present. This means that as long as a single reader holds the w mutex, an unbounded number of readers can enter the critical section unimpeded.

A correct solution to either of the readers-writers problems can result in *starvation*, where a thread blocks indefinitely and fails to make progress. For example, in the solution in Figure 26, a writer could wait indefinitely while a stream of readers arrived.

Practice Problem 10

The solution to the first readers-writers problem in Figure 26 gives priority to readers, but this priority is weak in the sense that a writer leaving its critical section might restart a waiting writer instead of a waiting reader. Describe a scenario where this weak priority would allow a collection of writers to starve a reader.

```
/* Global variables */
int readcnt;   /* Initially = 0 */
sem_t mutex, w; /* Both initially = 1 */

void reader(void)                        void writer(void)
{                                        {
    while (1) {                              while (1) {
        P(&mutex);                               P(&w);
        readcnt++;
        if (readcnt == 1) /* First in */         /* Critical section */
            P(&w);                               /* Writing happens  */
        V(&mutex);
                                                 V(&w);
        /* Critical section */               }
        /* Reading happens  */           }

        P(&mutex);
        readcnt--;
        if (readcnt == 0) /* Last out */
            V(&w);
        V(&mutex);
    }
}
```

Figure 26 **Solution to the first readers-writers problem.** Favors readers over writers.

Aside Other synchronization mechanisms

We have shown you how to synchronize threads using semaphores, mainly because they are simple, classical, and have a clean semantic model. But you should know that other synchronization techniques exist as well. For example, Java threads are synchronized with a mechanism called a *Java monitor* [51], which provides a higher level abstraction of the mutual exclusion and scheduling capabilities of semaphores; in fact monitors can be implemented with semaphores. As another example, the Pthreads interface defines a set of synchronization operations on *mutex* and *condition* variables. Pthreads mutexes are used for mutual exclusion. Condition variables are used for scheduling accesses to shared resources, such as the bounded buffer in a producer-consumer program.

5.5 Putting It Together: A Concurrent Server Based on Prethreading

We have seen how semaphores can be used to access shared variables and to schedule accesses to shared resources. To help you understand these ideas more clearly, let us apply them to a concurrent server based on a technique called *prethreading*.

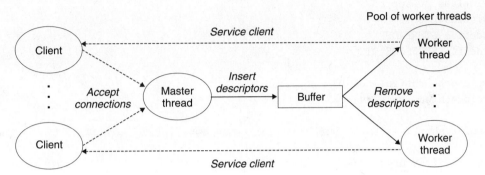

Figure 27 **Organization of a prethreaded concurrent server.** A set of existing threads repeatedly remove and process connected descriptors from a bounded buffer.

In the concurrent server in Figure 14, we created a new thread for each new client. A disadvantage of this approach is that we incur the nontrivial cost of creating a new thread for each new client. A server based on prethreading tries to reduce this overhead by using the producer-consumer model shown in Figure 27. The server consists of a main thread and a set of worker threads. The main thread repeatedly accepts connection requests from clients and places the resulting connected descriptors in a bounded buffer. Each worker thread repeatedly removes a descriptor from the buffer, services the client, and then waits for the next descriptor.

Figure 28 shows how we would use the Sbuf package to implement a prethreaded concurrent echo server. After initializing buffer sbuf (line 23), the main thread creates the set of worker threads (lines 26–27). Then it enters the infinite server loop, accepting connection requests and inserting the resulting connected descriptors in sbuf. Each worker thread has a very simple behavior. It waits until it is able to remove a connected descriptor from the buffer (line 39), and then calls the echo_cnt function to echo client input.

The echo_cnt function in Figure 29 is a version of the echo function from Figure 21 of the chapter "Network Programming" that records the cumulative number of bytes received from all clients in a global variable called byte_cnt. This is interesting code to study because it shows you a general technique for initializing packages that are called from thread routines. In our case, we need to initialize the byte_cnt counter and the mutex semaphore. One approach, which we used for the Sbuf and Rio packages, is to require the main thread to explicitly call an initialization function. Another approach, shown here, uses the pthread_once function (line 19) to call the initialization function the first time some thread calls the echo_cnt function. The advantage of this approach is that it makes the package easier to use. The disadvantage is that every call to echo_cnt makes a call to pthread_once, which most times does nothing useful.

Once the package is initialized, the echo_cnt function initializes the Rio buffered I/O package (line 20) and then echoes each text line that is received from the client. Notice that the accesses to the shared byte_cnt variable in lines 23–25 are protected by P and V operations.

```
1   #include "csapp.h"
2   #include "sbuf.h"
3   #define NTHREADS  4
4   #define SBUFSIZE  16
5
6   void echo_cnt(int connfd);
7   void *thread(void *vargp);
8
9   sbuf_t sbuf; /* Shared buffer of connected descriptors */
10
11  int main(int argc, char **argv)
12  {
13      int i, listenfd, connfd, port;
14      socklen_t clientlen=sizeof(struct sockaddr_in);
15      struct sockaddr_in clientaddr;
16      pthread_t tid;
17
18      if (argc != 2) {
19          fprintf(stderr, "usage: %s <port>\n", argv[0]);
20          exit(0);
21      }
22      port = atoi(argv[1]);
23      sbuf_init(&sbuf, SBUFSIZE);
24      listenfd = Open_listenfd(port);
25
26      for (i = 0; i < NTHREADS; i++)  /* Create worker threads */
27          Pthread_create(&tid, NULL, thread, NULL);
28
29      while (1) {
30          connfd = Accept(listenfd, (SA *) &clientaddr, &clientlen);
31          sbuf_insert(&sbuf, connfd); /* Insert connfd in buffer */
32      }
33  }
34
35  void *thread(void *vargp)
36  {
37      Pthread_detach(pthread_self());
38      while (1) {
39          int connfd = sbuf_remove(&sbuf); /* Remove connfd from buffer */
40          echo_cnt(connfd);                /* Service client */
41          Close(connfd);
42      }
43  }
```

Figure 28 **A prethreaded concurrent echo server.** The server uses a producer-consumer model with one producer and multiple consumers.

code/conc/echo_cnt.c

```
1    #include "csapp.h"
2
3    static int byte_cnt;   /* Byte counter */
4    static sem_t mutex;    /* and the mutex that protects it */
5
6    static void init_echo_cnt(void)
7    {
8        Sem_init(&mutex, 0, 1);
9        byte_cnt = 0;
10   }
11
12   void echo_cnt(int connfd)
13   {
14       int n;
15       char buf[MAXLINE];
16       rio_t rio;
17       static pthread_once_t once = PTHREAD_ONCE_INIT;
18
19       Pthread_once(&once, init_echo_cnt);
20       Rio_readinitb(&rio, connfd);
21       while((n = Rio_readlineb(&rio, buf, MAXLINE)) != 0) {
22           P(&mutex);
23           byte_cnt += n;
24           printf("thread %d received %d (%d total) bytes on fd %d\n",
25                   (int) pthread_self(), n, byte_cnt, connfd);
26           V(&mutex);
27           Rio_writen(connfd, buf, n);
28       }
29   }
```

code/conc/echo_cnt.c

Figure 29 echo_cnt: A version of echo that counts all bytes received from clients.

Aside Event-driven programs based on threads

I/O multiplexing is not the only way to write an event-driven program. For example, you might have noticed that the concurrent prethreaded server that we just developed is really an event-driven server with simple state machines for the main and worker threads. The main thread has two states ("waiting for connection request" and "waiting for available buffer slot"), two I/O events ("connection request arrives" and "buffer slot becomes available"), and two transitions ("accept connection request" and "insert buffer item"). Similarly, each worker thread has one state ("waiting for available buffer item"), one I/O event ("buffer item becomes available"), and one transition ("remove buffer item").

6 Using Threads for Parallelism

Thus far in our study of concurrency, we have assumed concurrent threads executing on uniprocessor systems. However, many modern machines have multi-core processors. Concurrent programs often run faster on such machines because the operating system kernel schedules the concurrent threads in parallel on multiple cores, rather than sequentially on a single core. Exploiting such parallelism is critically important in applications such as busy Web servers, database servers, and large scientific codes, and it is becoming increasingly useful in mainstream applications such as Web browsers, spreadsheets, and document processors.

Figure 30 shows the set relationships between sequential, concurrent, and parallel programs. The set of all programs can be partitioned into the disjoint sets of sequential and concurrent programs. A sequential program is written as a single logical flow. A concurrent program is written as multiple concurrent flows. A parallel program is a concurrent program running on multiple processors. Thus, the set of parallel programs is a proper subset of the set of concurrent programs.

A detailed treatment of parallel programs is beyond our scope, but studying a very simple example program will help you understand some important aspects of parallel programming. For example, consider how we might sum the sequence of integers $0, \ldots, n - 1$ in parallel. Of course, there is a closed-form solution for this particular problem, but nonetheless it is a concise and easy-to-understand exemplar that will allow us to make some interesting points about parallel programs.

The most straightforward approach is to partition the sequence into t disjoint regions, and then assign each of t different threads to work on its own region. For simplicity, assume that n is a multiple of t, such that each region has n/t elements. The main thread creates t peer threads, where each peer thread k runs in parallel on its own processor core and computes s_k, which is the sum of the elements in region k. Once the peer threads have completed, the main thread computes the final result by summing each s_k.

Figure 31 shows how we might implement this simple parallel sum algorithm. In lines 27–32, the main thread creates the peer threads and then waits for them to terminate. Notice that the main thread passes a small integer to each peer thread that serves as a unique thread ID. Each peer thread will use its thread ID to determine which portion of the sequence it should work on. This idea of passing a small unique thread ID to the peer threads is a general technique that is used in many parallel applications. After the peer threads have terminated, the psum vector contains the partial sums computed by each peer thread. The main

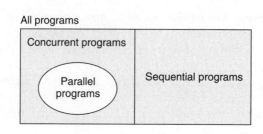

Figure 30

Relationships between the sets of sequential, concurrent, and parallel programs.

```
1    #include "csapp.h"
2    #define MAXTHREADS 32
3
4    void *sum(void *vargp);
5
6    /* Global shared variables */
7    long psum[MAXTHREADS];    /* Partial sum computed by each thread */
8    long nelems_per_thread;   /* Number of elements summed by each thread */
9
10   int main(int argc, char **argv)
11   {
12       long i, nelems, log_nelems, nthreads, result = 0;
13       pthread_t tid[MAXTHREADS];
14       int myid[MAXTHREADS];
15
16       /* Get input arguments */
17       if (argc != 3) {
18           printf("Usage: %s <nthreads> <log_nelems>\n", argv[0]);
19           exit(0);
20       }
21       nthreads = atoi(argv[1]);
22       log_nelems = atoi(argv[2]);
23       nelems = (1L << log_nelems);
24       nelems_per_thread = nelems / nthreads;
25
26       /* Create peer threads and wait for them to finish */
27       for (i = 0; i < nthreads; i++) {
28           myid[i] = i;
29           Pthread_create(&tid[i], NULL, sum, &myid[i]);
30       }
31       for (i = 0; i < nthreads; i++)
32           Pthread_join(tid[i], NULL);
33
34       /* Add up the partial sums computed by each thread */
35       for (i = 0; i < nthreads; i++)
36           result += psum[i];
37
38       /* Check final answer */
39       if (result != (nelems * (nelems-1))/2)
40           printf("Error: result=%ld\n", result);
41
42       exit(0);
43   }
```

Figure 31 **Simple parallel program that uses multiple threads to sum the elements of a sequence.**

── code/conc/psum.c

```
1   void *sum(void *vargp)
2   {
3       int myid = *((int *)vargp);              /* Extract the thread ID */
4       long start = myid * nelems_per_thread;  /* Start element index */
5       long end = start + nelems_per_thread;   /* End element index */
6       long i, sum = 0;
7
8       for (i = start; i < end; i++) {
9           sum += i;
10      }
11      psum[myid] = sum;
12
13      return NULL;
14  }
```

── code/conc/psum.c

Figure 32 **Thread routine for the program in Figure 31.**

thread then sums up the elements of the psum vector (lines 35–36), and uses the closed-form solution to verify the result (lines 39–40).

Figure 32 shows the function that each peer thread executes. In line 3, the thread extracts the thread ID from the thread argument, and then uses this ID to determine the region of the sequence it should work on (lines 4–5). In lines 8–10, the thread operates on its portion of the sequence, and then updates its entry in the partial sum vector (line 11). Notice that we are careful to give each peer thread a unique memory location to update, and thus it is not necessary to synchronize access to the psum array with semaphore mutexes. The only necessary synchronization in this particular case is that the main thread must wait for each of the children to finish so that it knows that each entry in psum is valid.

Figure 33 shows the total elapsed running time of the program in Figure 31 as a function of the number of threads. In each case, the program runs on a system with four processor cores and sums a sequence of $n = 2^{31}$ elements. We see that running time decreases as we increase the number of threads, up to four threads, at which point it levels off and even starts to increase a little. In the ideal case, we would expect the running time to decrease linearly with the number of cores. That is, we would expect running time to drop by half each time we double the number of threads. This is indeed the case until we reach the point ($t > 4$) where each of the four cores is busy running at least one thread. Running time actually increases a bit as we increase the number of threads because of the overhead of context switching multiple threads on the same core. For this reason, parallel programs are often written so that each core runs exactly one thread.

Although absolute running time is the ultimate measure of any program's performance, there are some useful relative measures, known as speedup and efficiency, that can provide insight into how well a parallel program is exploiting

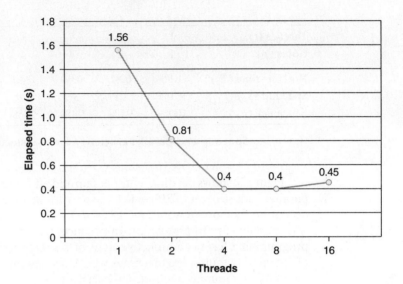

Figure 33

Performance of the program in Figure 31 on a multi-core machine with four cores. Summing a sequence of 2^{31} elements.

potential parallelism. The *speedup* of a parallel program is typically defined as

$$S_p = \frac{T_1}{T_p}$$

where p is the number of processor cores and T_k is the running time on k cores. This formulation is sometimes referred to as *strong scaling*. When T_1 is the execution time of a sequential version of the program, then S_p is called the *absolute speedup*. When T_1 is the execution time of the parallel version of the program running on one core, then S_p is called the *relative speedup*. Absolute speedup is a truer measure of the benefits of parallelism than relative speedup. Parallel programs often suffer from synchronization overheads, even when they run on one processor, and these overheads can artificially inflate the relative speedup numbers because they increase the size of the numerator. On the other hand, absolute speedup is more difficult to measure than relative speedup because measuring absolute speedup requires two different versions of the program. For complex parallel codes, creating a separate sequential version might not be feasible, either because the code is too complex or the source code is not available.

A related measure, known as *efficiency*, is defined as

$$E_p = \frac{S_p}{p} = \frac{T_1}{pT_p}$$

and is typically reported as a percentage in the range (0, 100]. Efficiency is a measure of the overhead due to parallelization. Programs with high efficiency are spending more time doing useful work and less time synchronizing and communicating than programs with low efficiency.

Threads (t)	1	2	4	8	16
Cores (p)	1	2	4	4	4
Running time (T_p)	1.56	0.81	0.40	0.40	0.45
Speedup (S_p)	1	1.9	3.9	3.9	3.5
Efficiency (E_p)	100%	95%	98%	98%	88%

Figure 34 **Speedup and parallel efficiency for the execution times in Figure 33.**

Figure 34 shows the different speedup and efficiency measures for our example parallel sum program. Efficiencies over 90% such as these are very good, but do not be fooled. We were able to achieve high efficiency because our problem was trivially easy to parallelize. In practice, this is not usually the case. Parallel programming has been an active area of research for decades. With the advent of commodity multi-core machines whose core count is doubling every few years, parallel programming continues to be a deep, difficult, and active area of research.

There is another view of speedup, known as *weak scaling*, which increases the problem size along with the number of processors, such that the amount of work performed on each processor is held constant as the number of processors increases. With this formulation, speedup and efficiency are expressed in terms of the total amount of work accomplished per unit time. For example, if we can double the number of processors and do twice the amount of work per hour, then we are enjoying linear speedup and 100% efficiency.

Weak scaling is often a truer measure than strong scaling because it more accurately reflects our desire to use bigger machines to do more work. This is particularly true for scientific codes, where the problem size can be easily increased, and where bigger problem sizes translate directly to better predictions of nature. However, there exist applications whose sizes are not so easily increased, and for these applications strong scaling is more appropriate. For example, the amount of work performed by real-time signal processing applications is often determined by the properties of the physical sensors that are generating the signals. Changing the total amount of work requires using different physical sensors, which might not be feasible or necessary. For these applications, we typically want to use parallelism to accomplish a fixed amount of work as quickly as possible.

Practice Problem 11

Fill in the blanks for the parallel program in the following table. Assume strong scaling.

Threads (t)	1	2	4
Cores (p)	1	2	4
Running time (T_p)	12	8	6
Speedup (S_p)	___	1.5	___
Efficiency (E_p)	100%	___	50%

7 Other Concurrency Issues

You probably noticed that life got much more complicated once we were asked to synchronize accesses to shared data. So far, we have looked at techniques for mutual exclusion and producer-consumer synchronization, but this is only the tip of the iceberg. Synchronization is a fundamentally difficult problem that raises issues that simply do not arise in ordinary sequential programs. This section is a survey (by no means complete) of some of the issues you need to be aware of when you write concurrent programs. To keep things concrete, we will couch our discussion in terms of threads. Keep in mind, however, that these are typical of the issues that arise when concurrent flows of any kind manipulate shared resources.

7.1 Thread Safety

When we program with threads, we must be careful to write functions that have a property called thread safety. A function is said to be *thread-safe* if and only if it will always produce correct results when called repeatedly from multiple concurrent threads. If a function is not thread-safe, then we say it is *thread-unsafe*.

We can identify four (nondisjoint) classes of thread-unsafe functions:

- *Class 1: Functions that do not protect shared variables.* We have already encountered this problem with the thread function in Figure 16, which increments an unprotected global counter variable. This class of thread-unsafe function is relatively easy to make thread-safe: protect the shared variables with synchronization operations such as P and V. An advantage is that it does not require any changes in the calling program. A disadvantage is that the synchronization operations will slow down the function.

- *Class 2: Functions that keep state across multiple invocations.* A pseudo-random number generator is a simple example of this class of thread-unsafe function. Consider the pseudo-random number generator package in Figure 35. The rand function is thread-unsafe because the result of the current invocation depends on an intermediate result from the previous iteration. When we call rand repeatedly from a single thread after seeding it with a call to srand, we can expect a repeatable sequence of numbers. However, this assumption no longer holds if multiple threads are calling rand.

  The only way to make a function such as rand thread-safe is to rewrite it so that it does not use any static data, relying instead on the caller to pass the state information in arguments. The disadvantage is that the programmer is now forced to change the code in the calling routine as well. In a large program where there are potentially hundreds of different call sites, making such modifications could be nontrivial and prone to error.

- *Class 3: Functions that return a pointer to a static variable.* Some functions, such as ctime and gethostbyname, compute a result in a static variable and then return a pointer to that variable. If we call such functions from concurrent threads, then disaster is likely, as results being used by one thread are silently overwritten by another thread.

code/conc/rand.c

```
1    unsigned int next = 1;
2
3    /* rand - return pseudo-random integer on 0..32767 */
4    int rand(void)
5    {
6        next = next*1103515245 + 12345;
7        return (unsigned int)(next/65536) % 32768;
8    }
9
10   /* srand - set seed for rand() */
11   void srand(unsigned int seed)
12   {
13       next = seed;
14   }
```

code/conc/rand.c

Figure 35 **A thread-unsafe pseudo-random number generator [58].**

There are two ways to deal with this class of thread-unsafe functions. One option is to rewrite the function so that the caller passes the address of the variable in which to store the results. This eliminates all shared data, but it requires the programmer to have access to the function source code.

If the thread-unsafe function is difficult or impossible to modify (e.g., the code is very complex or there is no source code available), then another option is to use the *lock-and-copy* technique. The basic idea is to associate a mutex with the thread-unsafe function. At each call site, lock the mutex, call the thread-unsafe function, copy the result returned by the function to a private memory location, and then unlock the mutex. To minimize changes to the caller, you should define a thread-safe wrapper function that performs the lock-and-copy, and then replace all calls to the thread-unsafe function with calls to the wrapper. For example, Figure 36 shows a thread-safe wrapper for ctime that uses the lock-and-copy technique.

- *Class 4: Functions that call thread-unsafe functions.* If a function f calls a thread-unsafe function g, is f thread-unsafe? It depends. If g is a class 2 function that relies on state across multiple invocations, then f is also thread-unsafe and there is no recourse short of rewriting g. However, if g is a class 1 or class 3 function, then f can still be thread-safe if you protect the call site and any resulting shared data with a mutex. We see a good example of this in Figure 36, where we use lock-and-copy to write a thread-safe function that calls a thread-unsafe function.

7.2 Reentrancy

There is an important class of thread-safe functions, known as *reentrant functions*, that are characterized by the property that they do not reference *any* shared data *code/conc/ctime_ts.c*

```
1   char *ctime_ts(const time_t *timep, char *privatep)
2   {
3       char *sharedp;
4
5       P(&mutex);
6       sharedp = ctime(timep);
7       strcpy(privatep, sharedp); /* Copy string from shared to private */
8       V(&mutex);
9       return privatep;
10  }
```

code/conc/ctime_ts.c

Figure 36 **Thread-safe wrapper function for the C standard library** `ctime` **function.** Uses the lock-and-copy technique to call a class 3 thread-unsafe function.

Figure 37

Relationships between the sets of reentrant, thread-safe, and non-thread-safe functions.

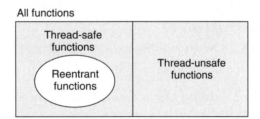

when they are called by multiple threads. Although the terms *thread-safe* and *reentrant* are sometimes used (incorrectly) as synonyms, there is a clear technical distinction that is worth preserving. Figure 37 shows the set relationships between reentrant, thread-safe, and thread-unsafe functions. The set of all functions is partitioned into the disjoint sets of thread-safe and thread-unsafe functions. The set of reentrant functions is a proper subset of the thread-safe functions.

Reentrant functions are typically more efficient than nonreentrant thread-safe functions because they require no synchronization operations. Furthermore, the only way to convert a class 2 thread-unsafe function into a thread-safe one is to rewrite it so that it is reentrant. For example, Figure 38 shows a reentrant version of the `rand` function from Figure 35. The key idea is that we have replaced the static `next` variable with a pointer that is passed in by the caller.

Is it possible to inspect the code of some function and declare *a priori* that it is reentrant? Unfortunately, it depends. If all function arguments are passed by value (i.e., no pointers) and all data references are to local automatic stack variables (i.e., no references to static or global variables), then the function is *explicitly reentrant*, in the sense that we can assert its reentrancy regardless of how it is called.

However, if we loosen our assumptions a bit and allow some parameters in our otherwise explicitly reentrant function to be passed by reference (that is, we allow them to pass pointers) then we have an *implicitly reentrant* function, in the sense that it is only reentrant if the calling threads are careful to pass pointers

—————————————————————— code/conc/rand_r.c

```
1  /* rand_r - a reentrant pseudo-random integer on 0..32767 */
2  int rand_r(unsigned int *nextp)
3  {
4      *nextp = *nextp * 1103515245 + 12345;
5      return (unsigned int)(*nextp / 65536) % 32768;
6  }
```

—————————————————————— code/conc/rand_r.c

Figure 38   rand_r: **A reentrant version of the** rand **function from Figure 35.**

to nonshared data. For example, the rand_r function in Figure 38 is implicitly reentrant.

We always use the term *reentrant* to include both explicit and implicit reentrant functions. However, it is important to realize that reentrancy is sometimes a property of both the caller and the callee, and not just the callee alone.

Practice Problem 12

The ctime_ts function in Figure 36 is thread-safe, but not reentrant. Explain.

7.3   Using Existing Library Functions in Threaded Programs

Most Unix functions, including the functions defined in the standard C library (such as malloc, free, realloc, printf, and scanf), are thread-safe, with only a few exceptions. Figure 39 lists the common exceptions. (See [109] for a complete list.) The asctime, ctime, and localtime functions are popular functions for converting back and forth between different time and date formats. The gethostbyname, gethostbyaddr, and inet_ntoa functions are frequently used network programming functions. The strtok function is a deprecated function (one whose use is discouraged) for parsing strings.

With the exceptions of rand and strtok, all of these thread-unsafe functions are of the class 3 variety that return a pointer to a static variable. If we need to call one of these functions in a threaded program, the least disruptive approach to the caller is to lock-and-copy. However, the lock-and-copy approach has a number of disadvantages. First, the additional synchronization slows down the program. Second, functions such as gethostbyname that return pointers to complex structures of structures require a *deep copy* of the structures in order to copy the entire structure hierarchy. Third, the lock-and-copy approach will not work for a class 2 thread-unsafe function such as rand that relies on static state across calls.

Therefore, Unix systems provide reentrant versions of most thread-unsafe functions. The names of the reentrant versions always end with the "_r" suffix. For example, the reentrant version of gethostbyname is called gethostbyname_r. We recommend using these functions whenever possible.

Thread-unsafe function	Thread-unsafe class	Unix thread-safe version
rand	2	rand_r
strtok	2	strtok_r
asctime	3	asctime_r
ctime	3	ctime_r
gethostbyaddr	3	gethostbyaddr_r
gethostbyname	3	gethostbyname_r
inet_ntoa	3	(none)
localtime	3	localtime_r

Figure 39 **Common thread-unsafe library functions.**

7.4 Races

A *race* occurs when the correctness of a program depends on one thread reaching point x in its control flow before another thread reaches point y. Races usually occur because programmers assume that threads will take some particular trajectory through the execution state space, forgetting the golden rule that threaded programs must work correctly for any feasible trajectory.

An example is the easiest way to understand the nature of races. Consider the simple program in Figure 40. The main thread creates four peer threads and passes a pointer to a unique integer ID to each one. Each peer thread copies the ID passed in its argument to a local variable (line 21), and then prints a message containing the ID. It looks simple enough, but when we run this program on our system, we get the following incorrect result:

```
unix> ./race
Hello from thread 1
Hello from thread 3
Hello from thread 2
Hello from thread 3
```

The problem is caused by a race between each peer thread and the main thread. Can you spot the race? Here is what happens. When the main thread creates a peer thread in line 12, it passes a pointer to the local stack variable i. At this point, the race is on between the next call to pthread_create in line 12 and the dereferencing and assignment of the argument in line 21. If the peer thread executes line 21 before the main thread executes line 12, then the myid variable gets the correct ID. Otherwise, it will contain the ID of some other thread. The scary thing is that whether we get the correct answer depends on how the kernel schedules the execution of the threads. On our system it fails, but on other systems it might work correctly, leaving the programmer blissfully unaware of a serious bug.

To eliminate the race, we can dynamically allocate a separate block for each integer ID, and pass the thread routine a pointer to this block, as shown in *code/conc/race.c*

```
1   #include "csapp.h"
2   #define N 4
3
4   void *thread(void *vargp);
5
6   int main()
7   {
8       pthread_t tid[N];
9       int i;
10
11      for (i = 0; i < N; i++)
12          Pthread_create(&tid[i], NULL, thread, &i);
13      for (i = 0; i < N; i++)
14          Pthread_join(tid[i], NULL);
15      exit(0);
16  }
17
18  /* Thread routine */
19  void *thread(void *vargp)
20  {
21      int myid = *((int *)vargp);
22      printf("Hello from thread %d\n", myid);
23      return NULL;
24  }
```

code/conc/race.c

Figure 40 **A program with a race.**

Figure 41 (lines 12–14). Notice that the thread routine must free the block in order to avoid a memory leak.

When we run this program on our system, we now get the correct result:

```
unix> ./norace
Hello from thread 0
Hello from thread 1
Hello from thread 2
Hello from thread 3
```

Practice Problem 13

In Figure 41, we might be tempted to free the allocated memory block immediately after line 15 in the main thread, instead of freeing it in the peer thread. But this would be a bad idea. Why?

code/conc/norace.c

```
1    #include "csapp.h"
2    #define N 4
3
4    void *thread(void *vargp);
5
6    int main()
7    {
8        pthread_t tid[N];
9        int i, *ptr;
10
11       for (i = 0; i < N; i++) {
12           ptr = Malloc(sizeof(int));
13           *ptr = i;
14           Pthread_create(&tid[i], NULL, thread, ptr);
15       }
16       for (i = 0; i < N; i++)
17           Pthread_join(tid[i], NULL);
18       exit(0);
19   }
20
21   /* Thread routine */
22   void *thread(void *vargp)
23   {
24       int myid = *((int *)vargp);
25       Free(vargp);
26       printf("Hello from thread %d\n", myid);
27       return NULL;
28   }
```

code/conc/norace.c

Figure 41 **A correct version of the program in Figure 40 without a race.**

Practice Problem 14

A. In Figure 41, we eliminated the race by allocating a separate block for each integer ID. Outline a different approach that does not call the malloc or free functions.

B. What are the advantages and disadvantages of this approach?

7.5 Deadlocks

Semaphores introduce the potential for a nasty kind of run-time error, called *deadlock*, where a collection of threads are blocked, waiting for a condition that

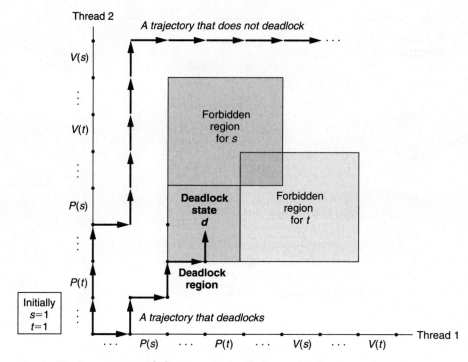

Figure 42 **Progress graph for a program that can deadlock.**

will never be true. The progress graph is an invaluable tool for understanding deadlock. For example, Figure 42 shows the progress graph for a pair of threads that use two semaphores for mutual exclusion. From this graph, we can glean some important insights about deadlock:

- The programmer has incorrectly ordered the P and V operations such that the forbidden regions for the two semaphores overlap. If some execution trajectory happens to reach the *deadlock state d*, then no further progress is possible because the overlapping forbidden regions block progress in every legal direction. In other words, the program is deadlocked because each thread is waiting for the other to do a V operation that will never occur.

- The overlapping forbidden regions induce a set of states called the *deadlock region*. If a trajectory happens to touch a state in the deadlock region, then deadlock is inevitable. Trajectories can enter deadlock regions, but they can never leave.

- Deadlock is an especially difficult issue because it is not always predictable. Some lucky execution trajectories will skirt the deadlock region, while others will be trapped by it. Figure 42 shows an example of each. The implications for a programmer are scary. You might run the same program 1000 times

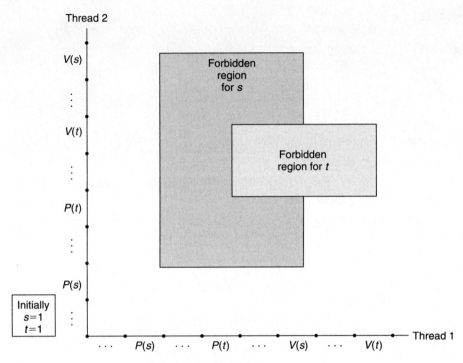

Figure 43 Progress graph for a deadlock-free program.

without any problem, but then the next time it deadlocks. Or the program might work fine on one machine but deadlock on another. Worst of all, the error is often not repeatable because different executions have different trajectories.

Programs deadlock for many reasons and avoiding them is a difficult problem in general. However, when binary semaphores are used for mutual exclusion, as in Figure 42, then you can apply the following simple and effective rule to avoid deadlocks:

> *Mutex lock ordering rule:* A program is deadlock-free if, for each pair of mutexes (s, t) in the program, each thread that holds both s and t simultaneously locks them in the same order.

For example, we can fix the deadlock in Figure 42 by locking s first, then t in each thread. Figure 43 shows the resulting progress graph.

Practice Problem 15

Consider the following program, which attempts to use a pair of semaphores for mutual exclusion.

```
Initially: s = 1, t = 0.

Thread 1:        Thread 2:
   P(s);            P(s);
   V(s);            V(s);
   P(t);            P(t);
   V(t);            V(t);
```

A. Draw the progress graph for this program.

B. Does it always deadlock?

C. If so, what simple change to the initial semaphore values will eliminate the potential for deadlock?

D. Draw the progress graph for the resulting deadlock-free program.

8 Summary

A concurrent program consists of a collection of logical flows that overlap in time. In this chapter, we have studied three different mechanisms for building concurrent programs: processes, I/O multiplexing, and threads. We used a concurrent network server as the motivating application throughout.

Processes are scheduled automatically by the kernel, and because of their separate virtual address spaces, they require explicit IPC mechanisms in order to share data. Event-driven programs create their own concurrent logical flows, which are modeled as state machines, and use I/O multiplexing to explicitly schedule the flows. Because the program runs in a single process, sharing data between flows is fast and easy. Threads are a hybrid of these approaches. Like flows based on processes, threads are scheduled automatically by the kernel. Like flows based on I/O multiplexing, threads run in the context of a single process, and thus can share data quickly and easily.

Regardless of the concurrency mechanism, synchronizing concurrent accesses to shared data is a difficult problem. The P and V operations on semaphores have been developed to help deal with this problem. Semaphore operations can be used to provide mutually exclusive access to shared data, as well as to schedule access to resources such as the bounded buffers in producer-consumer systems and shared objects in readers-writers systems. A concurrent prethreaded echo server provides a compelling example of these usage scenarios for semaphores.

Concurrency introduces other difficult issues as well. Functions that are called by threads must have a property known as thread safety. We have identified four classes of thread-unsafe functions, along with suggestions for making them thread-safe. Reentrant functions are the proper subset of thread-safe functions that do not access any shared data. Reentrant functions are often more efficient than nonreentrant functions because they do not require any synchronization primitives. Some other difficult issues that arise in concurrent programs are races and deadlocks. Races occur when programmers make incorrect assumptions about how logical flows are scheduled. Deadlocks occur when a flow is waiting for an event that will never happen.

Bibliographic Notes

Semaphore operations were introduced by Dijkstra [37]. The progress graph concept was introduced by Coffman [24] and later formalized by Carson and Reynolds [17]. The readers-writers problem was introduced by Courtois et al. [31]. Operating systems texts describe classical synchronization problems such as the dining philosophers, sleeping barber, and cigarette smokers problems in more detail [98, 104, 112]. The book by Butenhof [16] is a comprehensive description of the Posix threads interface. The paper by Birrell [7] is an excellent introduction to threads programming and its pitfalls. The book by Reinders [86] describes a C/C++ library that simplifies the design and implementation of threaded programs. Several texts cover the fundamentals of parallel programming on multi-core systems [50, 67]. Pugh identifies weaknesses with the way that Java threads interact through memory and proposes replacement memory models [84]. Gustafson proposed the weak scaling speedup model [46] as an alternative to strong scaling.

Homework Problems

16 ◆

Write a version of `hello.c` (Figure 13) that creates and reaps n joinable peer threads, where n is a command line argument.

17 ◆

A. The program in Figure 44 has a bug. The thread is supposed to sleep for 1 second and then print a string. However, when we run it on our system, nothing prints. Why?

B. You can fix this bug by replacing the `exit` function in line 9 with one of two different Pthreads function calls. Which ones?

18 ◆

Using the progress graph in Figure 21, classify the following trajectories as either *safe* or *unsafe*.

A. $H_2, L_2, U_2, H_1, L_1, S_2, U_1, S_1, T_1, T_2$

B. $H_2, H_1, L_1, U_1, S_1, L_2, T_1, U_2, S_2, T_2$

C. $H_1, L_1, H_2, L_2, U_2, S_2, U_1, S_1, T_1, T_2$

19 ◆◆

The solution to the first readers-writers problem in Figure 26 gives a somewhat weak priority to readers because a writer leaving its critical section might restart a waiting writer instead of a waiting reader. Derive a solution that gives stronger priority to readers, where a writer leaving its critical section will always restart a waiting reader if one exists.

code/conc/hellobug.c

```
1    #include "csapp.h"
2    void *thread(void *vargp);
3
4    int main()
5    {
6        pthread_t tid;
7
8        Pthread_create(&tid, NULL, thread, NULL);
9        exit(0);
10   }
11
12   /* Thread routine */
13   void *thread(void *vargp)
14   {
15       Sleep(1);
16       printf("Hello, world!\n");
17       return NULL;
18   }
```

code/conc/hellobug.c

Figure 44 **Buggy program for Problem 17.**

20 ◆◆◆

Consider a simpler variant of the readers-writers problem where there are at most *N* readers. Derive a solution that gives equal priority to readers and writers, in the sense that pending readers and writers have an equal chance of being granted access to the resource. Hint: You can solve this problem using a single counting semaphore and a single mutex.

21 ◆◆◆◆

Derive a solution to the second readers-writers problem, which favors writers instead of readers.

22 ◆◆

Test your understanding of the select function by modifying the server in Figure 6 so that it echoes at most one text line per iteration of the main server loop.

23 ◆◆

The event-driven concurrent echo server in Figure 8 is flawed because a malicious client can deny service to other clients by sending a partial text line. Write an improved version of the server that can handle these partial text lines without blocking.

24 ◆

The functions in the RIO I/O package are thread-safe. Are they reentrant as well?

25 ◆

In the prethreaded concurrent echo server in Figure 28, each thread calls the `echo_cnt` function (Figure 29). Is `echo_cnt` thread-safe? Is it reentrant? Why or why not?

26 ◆◆◆

Use the lock-and-copy technique to implement a thread-safe nonreentrant version of `gethostbyname` called `gethostbyname_ts`. A correct solution will use a deep copy of the `hostent` structure protected by a mutex.

27 ◆◆

Some network programming texts suggest the following approach for reading and writing sockets: Before interacting with the client, open two standard I/O streams on the same open connected socket descriptor, one for reading and one for writing:

```
FILE *fpin, *fpout;

fpin = fdopen(sockfd, "r");
fpout = fdopen(sockfd, "w");
```

When the server has finished interacting with the client, close both streams as follows:

```
fclose(fpin);
fclose(fpout);
```

However, if you try this approach in a concurrent server based on threads, you will create a deadly race condition. Explain.

28 ◆

In Figure 43, does swapping the order of the two *V* operations have any effect on whether or not the program deadlocks? Justify your answer by drawing the progress graphs for the four possible cases:

Case 1		Case 2		Case 3		Case 4	
Thread 1	Thread 2	Thread 1	Thread 2	Thread 1	Thread 2	Thread 1	Thread 2
`P(s)`	`P(s)`	`P(s)`	`P(s)`	`P(s)`	`P(s)`	`P(s)`	`P(s)`
`P(t)`	`P(t)`	`P(t)`	`P(t)`	`P(t)`	`P(t)`	`P(t)`	`P(t)`
`V(s)`	`V(s)`	`V(s)`	`V(t)`	`V(t)`	`V(s)`	`V(t)`	`V(t)`
`V(t)`	`V(t)`	`V(t)`	`V(s)`	`V(s)`	`V(t)`	`V(s)`	`V(s)`

29 ◆

Can the following program deadlock? Why or why not?

Initially: a = 1, b = 1, c = 1.

```
Thread 1:        Thread 2:
   P(a);            P(c);
   P(b);            P(b);
   V(b);            V(b);
   P(c);            V(c);
   V(c);
   V(a);
```

30 ◆

Consider the following program that deadlocks.

Initially: a = 1, b = 1, c = 1.

```
Thread 1:        Thread 2:        Thread 3:
   P(a);            P(c);            P(c);
   P(b);            P(b);            V(c);
   V(b);            V(b);            P(b);
   P(c);            V(c);            P(a);
   V(c);            P(a);            V(a);
   V(a);            V(a);            V(b);
```

A. For each thread, list the pairs of mutexes that it holds simultaneously.

B. If $a < b < c$, which threads violate the mutex lock ordering rule?

C. For these threads, show a new lock ordering that guarantees freedom from deadlock.

31 ◆◆◆

Implement a version of the standard I/O fgets function, called tfgets, that times out and returns NULL if it does not receive an input line on standard input within 5 seconds. Your function should be implemented in a package called tfgets-proc.c using process, signals, and nonlocal jumps. It should not use the Unix alarm function. Test your solution using the driver program in Figure 45.

32 ◆◆◆

Implement a version of the tfgets function from Problem 31 that uses the select function. Your function should be implemented in a package called tfgets-select.c. Test your solution using the driver program from Problem 31. You may assume that standard input is assigned to descriptor 0.

33 ◆◆◆

Implement a threaded version of the tfgets function from Problem 31. Your *code/conc/tfgets-main.c*

```
1    #include "csapp.h"
2
3    char *tfgets(char *s, int size, FILE *stream);
4
5    int main()
6    {
7        char buf[MAXLINE];
8
9        if (tfgets(buf, MAXLINE, stdin) == NULL)
10           printf("BOOM!\n");
11       else
12           printf("%s", buf);
13
14       exit(0);
15   }
```

code/conc/tfgets-main.c

Figure 45 **Driver program for Problems 31–33.**

function should be implemented in a package called `tfgets-thread.c`. Test your solution using the driver program from Problem 31.

34 ◆◆◆
Write a parallel threaded version of an $N \times M$ matrix multiplication kernel. Compare the performance to the sequential case.

35 ◆◆◆
Implement a concurrent version of the TINY Web server based on processes. Your solution should create a new child process for each new connection request. Test your solution using a real Web browser.

36 ◆◆◆
Implement a concurrent version of the TINY Web server based on I/O multiplexing. Test your solution using a real Web browser.

37 ◆◆◆
Implement a concurrent version of the TINY Web server based on threads. Your solution should create a new thread for each new connection request. Test your solution using a real Web browser.

38 ◆◆◆◆
Implement a concurrent prethreaded version of the TINY Web server. Your solution should dynamically increase or decrease the number of threads in response to the current load. One strategy is to double the number of threads when the

buffer becomes full, and halve the number of threads when the buffer becomes empty. Test your solution using a real Web browser.

39 ◆◆◆◆

A Web proxy is a program that acts as a middleman between a Web server and browser. Instead of contacting the server directly to get a Web page, the browser contacts the proxy, which forwards the request on to the server. When the server replies to the proxy, the proxy sends the reply on to the browser. For this lab, you will write a simple Web proxy that filters and logs requests:

A. In the first part of the lab, you will set up the proxy to accept requests, parse the HTTP, forward the requests to the server, and return the results back to the browser. Your proxy should log the URLs of all requests in a log file on disk, and it should also block requests to any URL contained in a filter file on disk.

B. In the second part of the lab, you will upgrade your proxy to deal with multiple open connections at once by spawning a separate thread to deal with each request. While your proxy is waiting for a remote server to respond to a request so that it can serve one browser, it should be working on a pending request from another browser.

Check your proxy solution using a real Web browser.

Solutions to Practice Problems

Solution to Problem 1

When the parent forks the child, it gets a copy of the connected descriptor and the reference count for the associated file table is incremented from 1 to 2. When the parent closes its copy of the descriptor, the reference count is decremented from 2 to 1. Since the kernel will not close a file until the reference counter in its file table goes to 0, the child's end of the connection stays open.

Solution to Problem 2

When a process terminates for any reason, the kernel closes all open descriptors. Thus, the child's copy of the connected file descriptor will be closed automatically when the child exits.

Solution to Problem 3

Recall that a descriptor is ready for reading if a request to read 1 byte from that descriptor would not block. If EOF becomes true on a descriptor, then the descriptor is ready for reading because the read operation will return immediately with a zero return code indicating EOF. Thus, typing `ctrl-d` causes the `select` function to return with descriptor 0 in the ready set.

Solution to Problem 4

We reinitialize the `pool.ready_set` variable before every call to `select` because it serves as both an input and output argument. On input, it contains the read set. On output, it contains the ready set.

Solution to Problem 5

Since threads run in the same process, they all share the same descriptor table. No matter how many threads use the connected descriptor, the reference count for the connected descriptor's file table is equal to 1. Thus, a single `close` operation is sufficient to free the memory resources associated with the connected descriptor when we are through with it.

Solution to Problem 6

The main idea here is that stack variables are private, while global and static variables are shared. Static variables such as cnt are a little tricky because the sharing is limited to the functions within their scope—in this case, the thread routine.

A. Here is the table:

Variable instance	Referenced by main thread?	Referenced by peer thread 0 ?	Referenced by peer thread 1?
ptr	yes	yes	yes
cnt	no	yes	yes
i.m	yes	no	no
msgs.m	yes	yes	yes
myid.p0	no	yes	no
myid.p1	no	no	yes

Notes:

ptr: A global variable that is written by the main thread and read by the peer threads.

cnt: A static variable with only one instance in memory that is read and written by the two peer threads.

i.m: A local automatic variable stored on the stack of the main thread. Even though its value is passed to the peer threads, the peer threads never reference it on the stack, and thus it is not shared.

msgs.m: A local automatic variable stored on the main thread's stack and referenced indirectly through ptr by both peer threads.

myid.0 and myid.1: Instances of a local automatic variable residing on the stacks of peer threads 0 and 1, respectively.

B. Variables ptr, cnt, and msgs are referenced by more than one thread, and thus are shared.

Solution to Problem 7

The important idea here is that you cannot make any assumptions about the ordering that the kernel chooses when it schedules your threads.

Step	Thread	Instr	%eax$_1$	%eax$_2$	cnt
1	1	H_1	—	—	0
2	1	L_1	0	—	0
3	2	H_2	—	—	0
4	2	L_2	—	0	0
5	2	U_2	—	1	0
6	2	S_2	—	1	1
7	1	U_1	1	—	1
8	1	S_1	1	—	1
9	1	T_1	1	—	1
10	2	T_2	1	—	1

Variable cnt has a final incorrect value of 1.

Solution to Problem 8

This problem is a simple test of your understanding of safe and unsafe trajectories in progress graphs. Trajectories such as A and C that skirt the critical region are safe and will produce correct results.

A. $H_1, L_1, U_1, S_1, H_2, L_2, U_2, S_2, T_2, T_1$: safe

B. $H_2, L_2, H_1, L_1, U_1, S_1, T_1, U_2, S_2, T_2$: unsafe

C. $H_1, H_2, L_2, U_2, S_2, L_1, U_1, S_1, T_1, T_2$: safe

Solution to Problem 9

A. $p = 1$, $c = 1$, $n > 1$: Yes, the mutex semaphore is necessary because the producer and consumer can concurrently access the buffer.

B. $p = 1$, $c = 1$, $n = 1$: No, the mutex semaphore is not necessary in this case, because a nonempty buffer is equivalent to a full buffer. When the buffer contains an item, the producer is blocked. When the buffer is empty, the consumer is blocked. So at any point in time, only a single thread can access the buffer, and thus mutual exclusion is guaranteed without using the mutex.

C. $p > 1$, $c > 1$, $n = 1$: No, the mutex semaphore is not necessary in this case either, by the same argument as the previous case.

Solution to Problem 10

Suppose that a particular semaphore implementation uses a LIFO stack of threads for each semaphore. When a thread blocks on a semaphore in a P operation, its ID is pushed onto the stack. Similarly, the V operation pops the top thread ID from the stack and restarts that thread. Given this stack implementation, an adversarial writer in its critical section could simply wait until another writer blocks on the semaphore before releasing the semaphore. In this scenario, a waiting reader might wait forever as two writers passed control back and forth.

Notice that although it might seem more intuitive to use a FIFO queue rather than a LIFO stack, using such a stack is not incorrect and does not violate the semantics of the P and V operations.

Solution to Problem 11

This problem is a simple sanity check of your understanding of speedup and parallel efficiency:

Threads (t)	1	2	4
Cores (p)	1	2	4
Running time (T_p)	12	8	6
Speedup (S_p)	1	1.5	2
Efficiency (E_p)	100%	75%	50%

Solution to Problem 12

The `ctime_ts` function is not reentrant because each invocation shares the same `static` variable returned by the `gethostbyname` function. However, it is thread-safe because the accesses to the shared variable are protected by P and V operations, and thus are mutually exclusive.

Solution to Problem 13

If we free the block immediately after the call to `pthread_create` in line 15, then we will introduce a new race, this time between the call to `free` in the main thread, and the assignment statement in line 25 of the thread routine.

Solution to Problem 14

A. Another approach is to pass the integer i directly, rather than passing a pointer to i:

```
for (i = 0; i < N; i++)
    Pthread_create(&tid[i], NULL, thread, (void *)i);
```

In the thread routine, we cast the argument back to an int and assign it to myid:

```
int myid = (int) vargp;
```

B. The advantage is that it reduces overhead by eliminating the calls to `malloc` and `free`. A significant disadvantage is that it assumes that pointers are at least as large as `int`s. While this assumption is true for all modern systems, it might not be true for legacy or future systems.

Solution to Problem 15

A. The progress graph for the original program is shown in Figure 46.

B. The program always deadlocks, since any feasible trajectory is eventually trapped in a deadlock state.

C. To eliminate the deadlock potential, initialize the binary semaphore t to 1 instead of 0.

D. The progress graph for the corrected program is shown in Figure 47.

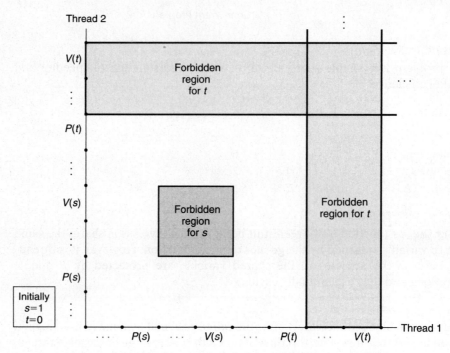

Figure 46 **Progress graph for a program that deadlocks.**

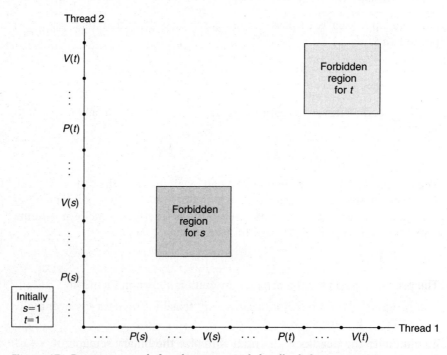

Figure 47 **Progress graph for the corrected deadlock-free program.**

Error Handling

Programmers should *always* check the error codes returned by system-level functions. There are many subtle ways that things can go wrong, and it only makes sense to use the status information that the kernel is able to provide us. Unfortunately, programmers are often reluctant to do error checking because it clutters their code, turning a single line of code into a multi-line conditional statement. Error checking is also confusing because different functions indicate errors in different ways.

We were faced with a similar problem when writing this text. On the one hand, we would like our code examples to be concise and simple to read. On the other hand, we do not want to give students the wrong impression that it is OK to skip error checking. To resolve these issues, we have adopted an approach based on *error-handling wrappers* that was pioneered by W. Richard Stevens in his network programming text [109].

The idea is that given some base system-level function foo, we define a wrapper function Foo with identical arguments, but with the first letter capitalized. The wrapper calls the base function and checks for errors. If it detects an error, the wrapper prints an informative message and terminates the process. Otherwise, it returns to the caller. Notice that if there are no errors, the wrapper behaves exactly like the base function. Put another way, if a program runs correctly with wrappers, it will run correctly if we render the first letter of each wrapper in lowercase and recompile.

The wrappers are packaged in a single source file (csapp.c) that is compiled and linked into each program. A separate header file (csapp.h) contains the function prototypes for the wrappers.

This appendix gives a tutorial on the different kinds of error handling in Unix systems, and gives examples of the different styles of error-handling wrappers. Copies of the csapp.h and csapp.c files are available on the CS:APP Web page.

From Appendix A of *Computer Systems: A Programmer's Perspective*, Second Edition. Randal E. Bryant and David R. O'Hallaron. Copyright © 2011 by Pearson Education, Inc. Published by Prentice Hall. All rights reserved.

1 Error Handling in Unix Systems

The systems-level function calls that we will discuss use three different styles for returning errors: *Unix-style*, *Posix-style*, and *DNS-style*.

Unix-Style Error Handling

Functions such as `fork` and `wait` that were developed in the early days of Unix (as well as some older Posix functions) overload the function return value with both error codes *and* useful results. For example, when the Unix-style `wait` function encounters an error (e.g., there is no child process to reap) it returns −1 and sets the global variable `errno` to an error code that indicates the cause of the error. If `wait` completes successfully, then it returns the useful result, which is the PID of the reaped child. Unix-style error-handling code is typically of the following form:

```
1    if ((pid = wait(NULL)) < 0) {
2        fprintf(stderr, "wait error: %s\n", strerror(errno));
3        exit(0);
4    }
```

The `strerror` function returns a text description for a particular value of errno.

Posix-Style Error Handling

Many of the newer Posix functions such as Pthreads use the return value only to indicate success (0) or failure (nonzero). Any useful results are returned in function arguments that are passed by reference. We refer to this approach as *Posix-style error handling*. For example, the Posix-style `pthread_create` function indicates success or failure with its return value and returns the ID of the newly created thread (the useful result) by reference in its first argument. Posix-style error-handling code is typically of the following form:

```
1    if ((retcode = pthread_create(&tid, NULL, thread, NULL)) != 0) {
2        fprintf(stderr, "pthread_create error: %s\n",
                           strerror(retcode));
3        exit(0);
4    }
```

DNS-Style Error Handling

The `gethostbyname` and `gethostbyaddr` functions that retrieve DNS (Domain Name System) host entries have yet another approach for returning errors. These functions return a NULL pointer on failure and set the global `h_errno` variable. DNS-style error handling is typically of the following form:

```
1    if ((p = gethostbyname(name)) == NULL) {
2        fprintf(stderr, "gethostbyname error: %s\n:",
                           hstrerror(h_errno));
3        exit(0);
4    }
```

Summary of Error-Reporting Functions

We use the following error-reporting functions to accommodate different error-handling styles.

```
#include "csapp.h"

void unix_error(char *msg);
void posix_error(int code, char *msg);
void dns_error(char *msg);
void app_error(char *msg);
```
                                                              Returns: nothing

As their names suggest, the `unix_error`, `posix_error`, and `dns_error` functions report Unix-style, Posix-style, and DNS-style errors and then terminate. The `app_error` function is included as a convenience for application errors. It simply prints its input and then terminates. Figure 1 shows the code for the error-reporting functions.

2 Error-Handling Wrappers

Here are some examples of the different error-handling wrappers:

- *Unix-style error-handling wrappers.* Figure 2 shows the wrapper for the Unix-style `wait` function. If the `wait` returns with an error, the wrapper prints an informative message and then exits. Otherwise, it returns a PID to the caller. Figure 3 shows the wrapper for the Unix-style `kill` function. Notice that this function, unlike `Wait`, returns `void` on success.

- *Posix-style error-handling wrappers.* Figure 4 shows the wrapper for the Posix-style `pthread_detach` function. Like most Posix-style functions, it does not overload useful results with error-return codes, so the wrapper returns `void` on success.

- *DNS-style error-handling wrappers.* Figure 5 shows the error-handling wrapper for the DNS-style `gethostbyname` function.

code/src/csapp.c

```
1   void unix_error(char *msg) /* Unix-style error */
2   {
3       fprintf(stderr, "%s: %s\n", msg, strerror(errno));
4       exit(0);
5   }
6
7   void posix_error(int code, char *msg) /* Posix-style error */
8   {
9       fprintf(stderr, "%s: %s\n", msg, strerror(code));
10      exit(0);
11  }
12
13  void dns_error(char *msg) /* DNS-style error */
14  {
15      fprintf(stderr, "%s: DNS error %d\n", msg, h_errno);
16      exit(0);
17  }
18
19  void app_error(char *msg) /* Application error */
20  {
21      fprintf(stderr, "%s\n", msg);
22      exit(0);
23  }
```

code/src/csapp.c

Figure 1 Error-reporting functions.

code/src/csapp.c

```
1   pid_t Wait(int *status)
2   {
3       pid_t pid;
4
5       if ((pid  = wait(status)) < 0)
6           unix_error("Wait error");
7       return pid;
8   }
```

code/src/csapp.c

Figure 2 Wrapper for Unix-style wait function.

code/src/csapp.c

```
1   void Kill(pid_t pid, int signum)
2   {
3       int rc;
4
5       if ((rc = kill(pid, signum)) < 0)
6           unix_error("Kill error");
7   }
```

code/src/csapp.c

Figure 3   Wrapper for Unix-style kill function.

code/src/csapp.c

```
1   void Pthread_detach(pthread_t tid) {
2       int rc;
3
4       if ((rc = pthread_detach(tid)) != 0)
5           posix_error(rc, "Pthread_detach error");
6   }
```

code/src/csapp.c

Figure 4   Wrapper for Posix-style pthread_detach function.

code/src/csapp.c

```
1   struct hostent *Gethostbyname(const char *name)
2   {
3       struct hostent *p;
4
5       if ((p = gethostbyname(name)) == NULL)
6           dns_error("Gethostbyname error");
7       return p;
8   }
```

code/src/csapp.c

Figure 5   Wrapper for DNS-style gethostbyname function.

References

[1] Advanced Micro Devices, Inc. *Software Optimization Guide for AMD64 Processors*, 2005. Publication Number 25112.

[2] Advanced Micro Devices, Inc. *AMD64 Architecture Programmer's Manual, Volume 1: Application Programming*, 2007. Publication Number 24592.

[3] Advanced Micro Devices, Inc. *AMD64 Architecture Programmer's Manual, Volume 3: General-Purpose and System Instructions*, 2007. Publication Number 24594.

[4] K. Arnold, J. Gosling, and D. Holmes. *The Java Programming Language, Fourth Edition*. Prentice Hall, 2005.

[5] V. Bala, E. Duesterwald, and S. Banerjiia. Dynamo: A transparent dynamic optimization system. In *Proceedings of the 1995 ACM Conference on Programming Language Design and Implementation (PLDI)*, pages 1–12, June 2000.

[6] T. Berners-Lee, R. Fielding, and H. Frystyk. Hypertext transfer protocol - HTTP/1.0. RFC 1945, 1996.

[7] A. Birrell. An introduction to programming with threads. Technical Report 35, Digital Systems Research Center, 1989.

[8] A. Birrell, M. Isard, C. Thacker, and T. Wobber. A design for high-performance flash disks. *SIGOPS Operating Systems Review*, 41(2), 2007.

[9] R. Blum. *Professional Assembly Language*. Wiley, 2005.

[10] S. Borkar. Thousand core chips—a technology perspective. In *Design Automation Conference*, pages 746–749. ACM, 2007.

[11] D. Bovet and M. Cesati. *Understanding the Linux Kernel, Third Edition*. O'Reilly Media, Inc, 2005.

[12] A. Demke Brown and T. Mowry. Taming the memory hogs: Using compiler-inserted releases to manage physical memory intelligently. In *Proceedings of the Fourth Symposium on Operating Systems Design and Implementation (OSDI)*, pages 31–44, October 2000.

[13] R. E. Bryant. Term-level verification of a pipelined CISC microprocessor. Technical Report CMU-CS-05-195, Carnegie Mellon University, School of Computer Science, 2005.

[14] R. E. Bryant and D. R. O'Hallaron. Introducing computer systems from a programmer's perspective. In *Proceedings of the Technical Symposium on Computer Science Education (SIGCSE)*. ACM, February 2001.

[15] B. R. Buck and J. K. Hollingsworth. An API for runtime code patching. *Journal of High Performance Computing Applications*, 14(4):317–324, June 2000.

[16] D. Butenhof. *Programming with Posix Threads*. Addison-Wesley, 1997.

[17] S. Carson and P. Reynolds. The geometry of semaphore programs. *ACM Transactions on Programming Languages and Systems*, 9(1):25–53, 1987.

[18] J. B. Carter, W. C. Hsieh, L. B. Stoller, M. R. Swanson, L. Zhang, E. L. Brunvand, A. Davis, C.-C. Kuo, R. Kuramkote, M. A. Parker, L. Schaelicke, and T. Tateyama. Impulse: Building a smarter memory controller. In *Proceedings of the Fifth International Symposium on High Performance Computer Architecture (HPCA)*, pages 70–79, January 1999.

[19] S. Chellappa, F. Franchetti, and M. Püschel. How to write fast numerical code: A small introduction. In *Generative and Transformational Techniques in Software Engineering II*, volume 5235, pages 196–259. Springer-Verlag Lecture Notes in Computer Science, 2008.

[20] P. Chen, E. Lee, G. Gibson, R. Katz, and D. Patterson. RAID: High-performance, reliable secondary storage. *ACM Computing Surveys*, 26(2), June 1994.

[21] S. Chen, P. Gibbons, and T. Mowry. Improving index performance through prefetching. In *Proceedings of the 2001 ACM SIGMOD Conference*. ACM, May 2001.

[22] T. Chilimbi, M. Hill, and J. Larus. Cache-conscious structure layout. In *Proceedings of the 1999 ACM Conference on Programming Language Design and Implementation (PLDI)*, pages 1–12. ACM, May 1999.

[23] B. Cmelik and D. Keppel. Shade: A fast instruction-set simulator for execution profiling. In *Proceedings of the 1994 ACM SIGMETRICS Conference on Measurement and Modeling of Computer Systems*, pages 128–137, May 1994.

[24] E. Coffman, M. Elphick, and A. Shoshani. System deadlocks. *ACM Computing Surveys*, 3(2):67–78, June 1971.

[25] D. Cohen. On holy wars and a plea for peace. *IEEE Computer*, 14(10):48–54, October 1981.

[26] Intel Corporation. *Intel 64 and IA-32 Architectures Optimization Reference Manual*, 2009. Order Number 248966.

[27] Intel Corporation. *Intel 64 and IA-32 Architectures Software Developer's Manual, Volume 1: Basic Architecture*, 2009. Order Number 253665.

[28] Intel Corporation. *Intel 64 and IA-32 Architectures Software Developer's Manual, Volume 2: Instruction Set Reference A–M*, 2009. Order Number 253667.

[29] Intel Corporation. *Intel 64 and IA-32 Architectures Software Developer's Manual, Volume 2: Instruction Set Reference N–Z*, 2009. Order Number 253668.

[30] Intel Corporation. *Intel 64 and IA-32 Architectures Software Developer's Manual, Volume 3a: System Programming Guide, Part 1*, 2009. Order Number 253669.

[31] P. J. Courtois, F. Heymans, and D. L. Parnas. Concurrent control with "readers" and "writers." *Commun. ACM*, 14(10):667–668, 1971.

[32] C. Cowan, P. Wagle, C. Pu, S. Beattie, and J. Walpole. Buffer overflows: Attacks and defenses for the vulnerability of the decade. In *DARPA Information Survivability Conference and Expo (DISCEX)*, March 2000.

[33] J. H. Crawford. The i486 CPU: Executing instructions in one clock cycle. *IEEE Micro*, 10(1):27–36, February 1990.

[34] V. Cuppu, B. Jacob, B. Davis, and T. Mudge. A performance comparison of contemporary DRAM architectures. In *Proceedings of the Twenty-Sixth International Symposium on Computer Architecture (ISCA)*, Atlanta, GA, May 1999. IEEE.

[35] B. Davis, B. Jacob, and T. Mudge. The new DRAM interfaces: SDRAM, RDRAM, and variants. In *Proceedings of the Third International Symposium on High Performance Computing (ISHPC)*, Tokyo, Japan, October 2000.

[36] E. Demaine. Cache-oblivious algorithms and data structures. In *Lecture Notes in Computer Science*. Springer-Verlag, 2002.

[37] E. W. Dijkstra. Cooperating sequential processes. Technical Report EWD-123, Technological University, Eindhoven, The Netherlands, 1965.

[38] C. Ding and K. Kennedy. Improving cache performance of dynamic applications through data and computation reorganizations at run time. In *Proceedings of the 1999 ACM Conference on Programming Language Design and Implementation (PLDI)*, pages 229–241. ACM, May 1999.

[39] M. Dowson. The Ariane 5 software failure. *SIGSOFT Software Engineering Notes*, 22(2):84, 1997.

[40] M. W. Eichen and J. A. Rochlis. With microscope and tweezers: An analysis of the Internet virus of November, 1988. In *IEEE Symposium on Research in Security and Privacy*, 1989.

[41] R. Fielding, J. Gettys, J. Mogul, H. Frystyk, L. Masinter, P. Leach, and T. Berners-Lee. Hypertext transfer protocol - HTTP/1.1. RFC 2616, 1999.

[42] M. Frigo, C. E. Leiserson, H. Prokop, and S. Ramachandran. Cache-oblivious algorithms. In *Proceedings of the 40th IEEE Symposium on Foundations of Computer Science (FOCS '99)*, pages 285–297. IEEE, August 1999.

[43] M. Frigo and V. Strumpen. The cache complexity of multithreaded cache oblivious algorithms.

In *SPAA '06: Proceedings of the Eighteenth Annual ACM Symposium on Parallelism in Algorithms and Architectures*, pages 271–280, New York, NY, USA, 2006. ACM.

[44] G. Gibson, D. Nagle, K. Amiri, J. Butler, F. Chang, H. Gobioff, C. Hardin, E. Riedel, D. Rochberg, and J. Zelenka. A cost-effective, high-bandwidth storage architecture. In *Proceedings of the International Conference on Architectural Support for Programming Languages and Operating Systems (ASPLOS)*. ACM, October 1998.

[45] G. Gibson and R. Van Meter. Network attached storage architecture. *Communications of the ACM*, 43(11), November 2000.

[46] J. Gustafson. Reevaluating Amdahl's law. *Communications of the ACM*, 31(5), August 1988.

[47] L. Gwennap. New algorithm improves branch prediction. *Microprocessor Report*, 9(4), March 1995.

[48] S. P. Harbison and G. L. Steele, Jr. *C, A Reference Manual, Fifth Edition*. Prentice Hall, 2002.

[49] J. L. Hennessy and D. A. Patterson. *Computer Architecture: A Quantitative Approach, Fourth Edition*. Morgan Kaufmann, 2007.

[50] M. Herlihy and N. Shavit. *The Art of Multiprocessor Programming*. Morgan Kaufmann, 2008.

[51] C. A. R. Hoare. Monitors: An operating system structuring concept. *Communications of the ACM*, 17(10):549–557, October 1974.

[52] Intel Corporation. *Tool Interface Standards Portable Formats Specification, Version 1.1*, 1993. Order Number 241597.

[53] F. Jones, B. Prince, R. Norwood, J. Hartigan, W. Vogley, C. Hart, and D. Bondurant. A new era of fast dynamic RAMs. *IEEE Spectrum*, pages 43–39, October 1992.

[54] R. Jones and R. Lins. *Garbage Collection: Algorithms for Automatic Dynamic Memory Management*. Wiley, 1996.

[55] M. Kaashoek, D. Engler, G. Ganger, H. Briceno, R. Hunt, D. Maziers, T. Pinckney, R. Grimm, J. Jannotti, and K. MacKenzie. Application performance and flexibility on Exokernel systems. In *Proceedings of the Sixteenth Symposium on Operating System Principles (SOSP)*, October 1997.

[56] R. Katz and G. Borriello. *Contemporary Logic Design, Second Edition*. Prentice Hall, 2005.

[57] B. Kernighan and D. Ritchie. *The C Programming Language, First Edition*. Prentice Hall, 1978.

[58] B. Kernighan and D. Ritchie. *The C Programming Language, Second Edition*. Prentice Hall, 1988.

[59] B. W. Kernighan and R. Pike. *The Practice of Programming*. Addison-Wesley, 1999.

[60] T. Kilburn, B. Edwards, M. Lanigan, and F. Sumner. One-level storage system. *IRE Transactions on Electronic Computers*, EC-11:223–235, April 1962.

[61] D. Knuth. *The Art of Computer Programming, Volume 1: Fundamental Algorithms, Second Edition*. Addison-Wesley, 1973.

[62] J. Kurose and K. Ross. *Computer Networking: A Top-Down Approach, Fifth Edition*. Addison-Wesley, 2009.

[63] M. Lam, E. Rothberg, and M. Wolf. The cache performance and optimizations of blocked algorithms. In *Proceedings of the International Conference on Architectural Support for Programming Languages and Operating Systems (ASPLOS)*. ACM, April 1991.

[64] J. R. Larus and E. Schnarr. EEL: Machine-independent executable editing. In *Proceedings of the 1995 ACM Conference on Programming Language Design and Implementation (PLDI)*, June 1995.

[65] C. E. Leiserson and J. B. Saxe. Retiming synchronous circuitry. *Algorithmica*, 6(1–6), June 1991.

[66] J. R. Levine. *Linkers and Loaders*. Morgan Kaufmann, San Francisco, 1999.

[67] C. Lin and L. Snyder. *Principles of Parallel Programming*. Addison-Wesley, 2008.

[68] Y. Lin and D. Padua. Compiler analysis of irregular memory accesses. In *Proceedings of the 2000 ACM Conference on Programming Language Design and Implementation (PLDI)*, pages 157–168. ACM, June 2000.

[69] J. L. Lions. Ariane 5 Flight 501 failure. Technical report, European Space Agency, July 1996.

[70] S. Macguire. *Writing Solid Code*. Microsoft Press, 1993.

[71] S. A. Mahlke, W. Y. Chen, J. C. Gyllenhal, and W. W. Hwu. Compiler code transformations for superscalar-based high-performance systems. In *Supercomputing*. ACM, 1992.

[72] E. Marshall. Fatal error: How Patriot overlooked a Scud. *Science*, page 1347, March 13, 1992.

[73] M. Matz, J. Hubička, A. Jaeger, and M. Mitchell. System V application binary interface AMD64 architecture processor supplement. Technical report, AMD64.org, 2009.

[74] J. Morris, M. Satyanarayanan, M. Conner, J. Howard, D. Rosenthal, and F. Smith. Andrew: A distributed personal computing environment. *Communications of the ACM*, March 1986.

[75] T. Mowry, M. Lam, and A. Gupta. Design and evaluation of a compiler algorithm for prefetching. In *Proceedings of the International Conference on Architectural Support for Programming Languages and Operating Systems (ASPLOS)*. ACM, October 1992.

[76] S. S. Muchnick. *Advanced Compiler Design and Implementation*. Morgan Kaufmann, 1997.

[77] S. Nath and P. Gibbons. Online maintenance of very large random samples on flash storage. In *Proceedings of VLDB'08*. ACM, August 2008.

[78] M. Overton. *Numerical Computing with IEEE Floating Point Arithmetic*. SIAM, 2001.

[79] D. Patterson, G. Gibson, and R. Katz. A case for redundant arrays of inexpensive disks (RAID). In *Proceedings of the 1998 ACM SIGMOD Conference*. ACM, June 1988.

[80] L. Peterson and B. Davie. *Computer Networks: A Systems Approach, Fourth Edition*. Morgan Kaufmann, 2007.

[81] J. Pincus and B. Baker. Beyond stack smashing: Recent advances in exploiting buffer overruns. *IEEE Security and Privacy*, 2(4):20–27, 2004.

[82] S. Przybylski. *Cache and Memory Hierarchy Design: A Performance-Directed Approach*. Morgan Kaufmann, 1990.

[83] W. Pugh. The Omega test: A fast and practical integer programming algorithm for dependence analysis. *Communications of the ACM*, 35(8):102–114, August 1992.

[84] W. Pugh. Fixing the Java memory model. In *Proceedings of the Java Grande Conference*, June 1999.

[85] J. Rabaey, A. Chandrakasan, and B. Nikolic. *Digital Integrated Circuits: A Design Perspective, Second Edition*. Prentice Hall, 2003.

[86] J. Reinders. *Intel Threading Building Blocks*. O'Reilly, 2007.

[87] D. Ritchie. The evolution of the Unix time-sharing system. *AT&T Bell Laboratories Technical Journal*, 63(6 Part 2):1577–1593, October 1984.

[88] D. Ritchie. The development of the C language. In *Proceedings of the Second History of Programming Languages Conference*, Cambridge, MA, April 1993.

[89] D. Ritchie and K. Thompson. The Unix time-sharing system. *Communications of the ACM*, 17(7):365–367, July 1974.

[90] T. Romer, G. Voelker, D. Lee, A. Wolman, W. Wong, H. Levy, B. Bershad, and B. Chen. Instrumentation and optimization of Win32/Intel executables using Etch. In *Proceedings of the USENIX Windows NT Workshop*, Seattle, Washington, August 1997.

[91] M. Satyanarayanan, J. Kistler, P. Kumar, M. Okasaki, E. Siegel, and D. Steere. Coda: A highly available file system for a distributed workstation environment. *IEEE Transactions on Computers*, 39(4):447–459, April 1990.

[92] J. Schindler and G. Ganger. Automated disk drive characterization. Technical Report CMU-CS-99-176, School of Computer Science, Carnegie Mellon University, 1999.

[93] F. B. Schneider and K. P. Birman. The monoculture risk put into context. *IEEE Security and Privacy*, 7(1), January 2009.

[94] R. C. Seacord. *Secure Coding in C and C++*. Addison-Wesley, 2006.

[95] H. Shacham, M. Page, B. Pfaff, E.-J. Goh, N. Modadugu, and D. Boneh. On the effectiveness of address-space randomization. In *Proceedings of the 11th ACM Conference on*

Computer and Communications Security (CCS '04), pages 298–307. ACM, 2004.

[96] J. P. Shen and M. Lipasti. *Modern Processor Design: Fundamentals of Superscalar Processors*. McGraw Hill, 2005.

[97] B. Shriver and B. Smith. *The Anatomy of a High-Performance Microprocessor: A Systems Perspective*. IEEE Computer Society, 1998.

[98] A. Silberschatz, P. Galvin, and G. Gagne. *Operating Systems Concepts, Eighth Edition*. Wiley, 2008.

[99] R. Singhal. Intel next generation Nehalem microarchitecture. In *Intel Developer's Forum*, 2008.

[100] R. Skeel. Roundoff error and the Patriot missile. *SIAM News*, 25(4):11, July 1992.

[101] A. Smith. Cache memories. *ACM Computing Surveys*, 14(3), September 1982.

[102] E. H. Spafford. The Internet worm program: An analysis. Technical Report CSD-TR-823, Department of Computer Science, Purdue University, 1988.

[103] A. Srivastava and A. Eustace. ATOM: A system for building customized program analysis tools. In *Proceedings of the 1994 ACM Conference on Programming Language Design and Implementation (PLDI)*, June 1994.

[104] W. Stallings. *Operating Systems: Internals and Design Principles, Sixth Edition*. Prentice Hall, 2008.

[105] W. R. Stevens. *TCP/IP Illustrated, Volume 1: The Protocols*. Addison-Wesley, 1994.

[106] W. R. Stevens. *TCP/IP Illustrated, Volume 2: The Implementation*. Addison-Wesley, 1995.

[107] W. R. Stevens. *TCP/IP Illustrated, Volume 3: TCP for Transactions, HTTP, NNTP and the Unix domain protocols*. Addison-Wesley, 1996.

[108] W. R. Stevens. *Unix Network Programming: Interprocess Communications, Second Edition*, volume 2. Prentice Hall, 1998.

[109] W. R. Stevens, B. Fenner, and A. M. Rudoff. *Unix Network Programming: The Sockets Networking API, Third Edition*, volume 1. Prentice Hall, 2003.

[110] W. R. Stevens and S. A. Rago. *Advanced Programming in the Unix Environment, Second Edition*. Addison-Wesley, 2008.

[111] T. Stricker and T. Gross. Global address space, non-uniform bandwidth: A memory system performance characterization of parallel systems. In *Proceedings of the Third International Symposium on High Performance Computer Architecture (HPCA)*, pages 168–179, San Antonio, TX, February 1997. IEEE.

[112] A. Tanenbaum. *Modern Operating Systems, Third Edition*. Prentice Hall, 2007.

[113] A. Tanenbaum. *Computer Networks, Fourth Edition*. Prentice Hall, 2002.

[114] K. P. Wadleigh and I. L. Crawford. *Software Optimization for High-Performance Computing: Creating Faster Applications*. Prentice Hall, 2000.

[115] J. F. Wakerly. *Digital Design Principles and Practices, Fourth Edition*. Prentice Hall, 2005.

[116] M. V. Wilkes. Slave memories and dynamic storage allocation. *IEEE Transactions on Electronic Computers*, EC-14(2), April 1965.

[117] P. Wilson, M. Johnstone, M. Neely, and D. Boles. Dynamic storage allocation: A survey and critical review. In *International Workshop on Memory Management*, Kinross, Scotland, 1995.

[118] M. Wolf and M. Lam. A data locality algorithm. In *Conference on Programming Language Design and Implementation (SIGPLAN)*, pages 30–44, June 1991.

[119] J. Wylie, M. Bigrigg, J. Strunk, G. Ganger, H. Kiliccote, and P. Khosla. Survivable information storage systems. *IEEE Computer*, August 2000.

[120] T.-Y. Yeh and Y. N. Patt. Alternative implementation of two-level adaptive branch prediction. In *International Symposium on Computer Architecture*, pages 451–461, 1998.

[121] X. Zhang, Z. Wang, N. Gloy, J. B. Chen, and M. D. Smith. System support for automatic profiling and optimization. In *Proceedings of the Sixteenth ACM Symposium on Operating Systems Principles (SOSP)*, pages 15–26, October 1997.

Index